Handbook of
ATMOSPHERIC ELECTRODYNAMICS

Volume II

Edited by Hans Volland
Radioastronomical Institute
University of Bonn
Bonn, Germany

CRC Press
Taylor & Francis Group
Boca Raton London New York

CRC Press is an imprint of the
Taylor & Francis Group, an **informa** business

First published 1995 by CRC Press
Taylor & Francis Group
6000 Broken Sound Parkway NW, Suite 300
Boca Raton, FL 33487-2742

Reissued 2018 by CRC Press

© 1995 by Taylor & Francis
CRC Press is an imprint of Taylor & Francis Group, an Informa business

No claim to original U.S. Government works

A Library of Congress record exists under LC control number: 94039127

Publisher's Note
The publisher has gone to great lengths to ensure the quality of this reprint but points out that some imperfections in the original copies may be apparent.

Disclaimer
The publisher has made every effort to trace copyright holders and welcomes correspondence from those they have been unable to contact.

ISBN 13: 978-1-138-10569-0 (hbk)
ISBN 13: 978-1-138-55902-8 (pbk)
ISBN 13: 978-0-203-71329-7 (ebk)

Visit the Taylor & Francis Web site at http://www.taylorandfrancis.com and the
CRC Press Web site at http://www.crcpress.com

PREFACE

Atmospheric Electrodynamics is a term coined to emphasize the importance of unifying two often separately treated subjects of research: geoelectricity, which deals with low frequency electric fields and currents within the lower atmosphere (in particular thunderstorms and related phenomena), and low frequency electric and magnetic fields of upper atmospheric origin. For physicists of the 19th and early 20th centuries, geoelectricity and geomagnetic variations of external origin were generally considered to be related subjects, which we presently refer to as low frequency electromagnetic fields, excited by various sources throughout the atmosphere. This traditional connection is still evident from the choice of names of scientific journals. For instance, there still exists the Japanese *Journal of Geomagnetism and Geoelectricity* and the former name of the present American *Journal of Geophysical Research* was *Terrestrial Magnetism and Atmospheric Electricity.*

Whereas geomagnetism became the root of modern magnetospheric physics, culminating in the space age exploration of the earth's environment, geoelectricity evolved as a stepchild of meteorology. The reason for this is clear. The quasistatic atmospheric electric field, generated by thunderstorm activity and observed on the ground, is intimately associated with the local weather and all of its frustrating unpredictability. However, the variable external geomagnetic field, which can also be measured on the ground, is a useful indicator of ionospheric and magnetospheric electric current systems.

Only in the past three decades have ionospheric and magnetospheric physicists rediscovered the importance of electric fields of upper atmospheric origin. Following the development of new instruments and their carriers (balloons, rockets, satellites), electric fields and currents of lower and upper atmospheric origin are now measured throughout the atmosphere from the ground to the magnetosphere and beyond. These recent technological advances basically closed the gap between geoelectricity and geomagnetism that existed for more than half a century.

This handbook is the extension of the two-volume *CRC Handbook of Atmospherics,* published in 1982, which covered only the first subject: geoelectricity, with particular emphasis on lightning and sferics phenomena. The present handbook updates the 1982 edition and also includes the second subject: low-frequency electric and magnetic fields and currents in the ionosphere and magnetosphere. Twenty-eight experts in their fields review a broad range of research in this area. Hopefully it will help to enhance the mutual understanding between lower and upper atmospheric physicists.

Ken Bullough, author or Chapter II/10, "Power Line Harmonic Radiation: Sources and Environmental Effects", died suddenly of a heart attack on March 9, 1994. Ken is recognized as a leading authority in his field. He was a kind and good-natured person. He will be greatly missed by all who knew him.

Hans Volland
Bonn, Germany

THE EDITOR

Hans Volland studied Geophysics and Meteorology at the Humboldt University in Berlin from 1948 until 1952. He was scientist at the Heinrich-Hertz-Institut in Berlin (East) from 1952 to 1958 and in the Heinrich-Hertz-Institut Berlin (West) from 1958 to 1964 working in the fields of geomagnetism, ionospheric physics, and electromagnetic wave propagation. In 1964 he became a professor at the Radioastronomical Institute, University of Bonn, where his main research subjects were solar radioastronomy, solar-terrestrial physics, and atmospheric physics. He has been retired since 1990. He is author of numerous scientific articles, three books, and editor of the *CRC Handbook of Atmospherics*.

CONTRIBUTORS

Wolfgang Baumjohann
Max-Planck Institute of extraterrestrial
   Physics
Postfach 1603
85740 Garching, Germany

Ken Bullough*
Department of Physics
University of Sheffield
P.O. Box 597
Sheffield S10 2UN, England

Arthur A. Few, Jr.
Space Physics & Astronomy
P.O. Box 1892
Rice University
Houston, Texas 77251

Karl-Heinz Glassmeier
Institute of Geophysics
TU Braunschweig
Mendelsohnstr. 3
38106 Braunschweig, Germany

Masashi Hayakawa
The University of Electro-Communications
1-5-1 Chofugaoka
Chofu-shi, Tokyo 182, Japan

Conrad L. Longmire
Mission Research Corporation
735 State Street
P.O. Drawer 719
Santa Barbara, CA 93102-0719

Michel Parrot
LPCE/CNRS
3A av de la Recherche Scientifique
45071 Orléans Cedex 2, France

Gerd W. Prölss
Institute of Astrophysics and Extraterrestrial
   Physics
University of Bonn
Auf dem Huegel 71
53121 Bonn, Germany

Reinhold Reiter
Consulting Bureau Reiter
Fritz-Mueller Str. 54
82467 Garmisch-Partenkirchen
Germany

Arthur D. Richmond
NCAR-HAO
P.O. Box 3000
Boulder, CO 80307-3000

Vikas S. Sonwalkar
STAR Laboratory, Durand 319
Stanford University
Stanford, CA 94305-4055

Gerd-Hannes Voigt
Fachbereich 06
Fachhochschule Aachen
Hohenstaufenallee 6
52064 Aachen, Germany

Hans Volland
Radioastronomical Institute
University of Bonn
Auf dem Huegel 71
53121 Bonn, Germany

J. Wiesinger
Electrotechnical Faculty
University of the German
   Armed Forces
85577 Neubiberg, Germany

W. Zischank
Electrotechnical Faculty
University of the German
   Armed Forces
85577 Neubiberg, Germany

---

* Deceased

CONTENTS

Chapter 1

Acoustic Radiations from Lightning

Arthur A. Few, Jr.

CONTENTS

1.   ACOUSTIC SOURCES IN THUNDERSTORMS

Severe electrical storms produce a variety of acoustic emissions. In this chapter we will describe these processes to the extent that they are understood. The acoustic emissions can be broadly divided into two categories: (1) those that are related to electrical processes (i.e., they correlate with lightning) and (2) those that either do not depend upon cloud electricity or no correlations with electrical changes have been observed. This second group will be discussed in the last section of this chapter.

Of the acoustic emissions correlated with electrical processes, we consider two types: thunder, which is produced by the explosively heated lightning discharge channel, and electrostatic emissions, which are produced by electrostatic fields throughout the charged regions of the cloud. This electrostatically produced acoustic signal is frequently called infrasonic thunder. Our initial discussion will be directed to the acoustic emissions from hot lightning discharge channels. Thunder is probably the most common of all loud natural sounds, while these other acoustic emissions are not ordinarily observed without special devices and careful preparations.

2.   THUNDER — THE RADIATION FROM HOT CHANNELS

In this section we will confine our discussion to the acoustic waves generated by the hot channels in the lightning process. Spectrographic studies of lightning return strokes[45] show that this electrical discharge process heats the channel gases to a temperature in the 24,000 K range. We believe that there may be differences in the breakdown and propagation of lightning inside and

0-8493-2520-X/95/$0.00+$.50
© 1995 by CRC Press

outside of clouds, but once a discharge is formed we see no reason to believe the thermodynamics and hydrodynamics are basically any different, other than differences in energy input. For this reason we do not treat the thunder generation theory differently for the two types of discharges.

The timing of the electrical and hydrodynamic processes is fundamental to the theory of thunder generation. At 30,000 K the expansion speed of the shock wave is roughly 3×10^3 m/s and decreases rapidly as the shock wave expands; in comparison the various measured speeds for lightning breakdown processes range from 10^4 to 10^8 m/s.[59,64] The electrical breakdown process in a given discharge event (stroke, etc.) is finished and completed before the hydrodynamic responses are fully organized. There are other electrical processes occurring over larger periods, for example, continuing currents, but the energy input to the hot channel is strongly weighted toward the early breakdown processes when channel resistance is higher. The energy input by the currents following the hydrodynamic expansion are thought to be small.[30]

2.1. SHOCK WAVE FORMATION AND EXPANSION

The starting point for developing a theory for a shock wave expansion to form thunder is the hot ($\sim$24,000 K) high pressure ($>10^6$ Pa) channel left by the electric discharge. Hill's[30] computer simulation indicates that approximately 95% of the total channel energy is deposited within the first 20 μs with the peak power dissipation occurring at 2 μs; during the 20 μs period of energy input, the shock wave can only move approximately 5 cm. This simulation may actually be slower than real lightning, because Hill used a slower current rise time than indicated by modern measurements.[65]

The time-resolved spectra of return strokes[45] show the effective temperature dropping from $\sim$30,000 to $\sim$10,000 K in a period of 40 μs, and the pressure of the luminous channel dropping to atmospheric in this same time frame. During this period the shock wave can expand roughly 0.1 m.

Even though channel luminosities and currents can continue for periods exceeding 100 μs, the processes that are important to the generation of thunder occur very quickly ($<$10 μs) and in a very confined volume (radius $<$5 cm). The strong shock wave propagates outward beyond the luminous channel, which returns to atmosphere pressure within 40 μs. The channel remnant cools slowly by conduction, convection, and radiation and becomes nonconducting at temperatures between 2000 and 4000 K, perhaps 100 ms later.[63]

Turning our attention now to the shock wave itself, we can divide its history into three periods: strong shock; weak shock; and acoustic. The division between strong and weak can be related physically to the energy input to the channel; the weak shock transition to acoustic is much more arbitrary. Calculations and measurements here show that the radiated energy is on the order of 1% of the total channel energy (e.g., Uman[59]); hence, most of the available energy is in the form of internal heat energy behind the shock wave.

As the strong shock wave expands, it must do thermodynamic work (PdV) on the surrounding fluid. The maximum distance through which the strong shock wave can expand will be the distance at which all of the internal thermal energy has been expended in doing the work of expansion. Few[13] proposed that this distance, which he called the "relaxation radius", would be the appropriate scaling parameter for comparing different sources and different geometries. Brode,[7] Lin,[38] and Sakurai[54] had used scaling values similar to these in their work with strong shock wave expansion. If we ignore the initial volume before shock wave expansion compared to the final volume, we get Few's[13] expressions for the spherical, R_s, and cylindrical, R_c, relaxation radii,

$$R_s = (3E_t/4\pi P_0)^{1/3} \tag{1}$$

$$R_c = (E_\ell/\pi P_0)^{1/2} \tag{2}$$

Table 1.2.1 **Altitude and Energy Effects on Relaxation Radii**

Eℓ(J/m)	10^4	$2 \cdot 10^4$	$5 \cdot 10^4$	10^5	$2 \cdot 10^5$	$5 \cdot 10^5$	10^6
R_c(m) $P_0 = 100$ kPa $\sim$ surface	0.18	0.25	0.40	0.56	.80	1.26	1.78
R_c(m) $P_0 = 60$ kPa Height $\sim$4 km	0.23	0.32	0.52	0.72	1.03	1.63	2.30
R_c(m) $P_0 = 30$ kPa Height $\sim$9 km	0.33	0.46	0.73	1.02	1.46	2.30	3.25

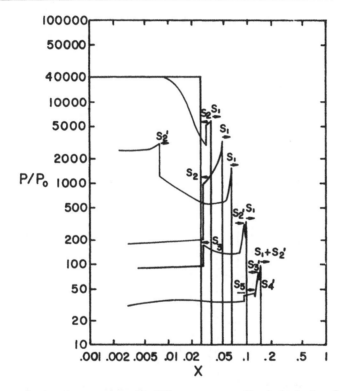

Figure 1.2.1 Strong shock region pressure profiles.[12] These pressure profiles depict the time development of the spherical shock wave from a high temperature, high pressure sphere.[7] The nondimensional coordinate X is the radius divided by R$_s$ as defined in Equation 1.

where E_t is the total energy for the spherical shock wave, E_ℓ is the energy per unit length for the cylindrical shock wave, and P_0 is the environmental atmospheric pressure. Table 1.2.1 gives R_c over a range of values that have been suggested in the literature for E_ℓ.

Figure 1.2.1 from Few[12] displays the early strong shock solution for spherical geometry from Brode.[7] Although this particular region of Brode's solution is inappropriate for direct comparison with the lightning shock wave, it does provide a good presentation of some of the internal processes of the strong shock behavior. We see here that due to the very high internal temperatures there are numerous internal waves (numbers S_2, S_3, etc.) that travel back and forth between the shock front, S_1, and the origin. The shock front itself advances somewhat slower than these internal waves because it is moving into cold stationary air. Figure 1.2.2 from Hill[30] shows qualitatively similar internal waves for the cylindrical case and much smaller initial energy input. The function

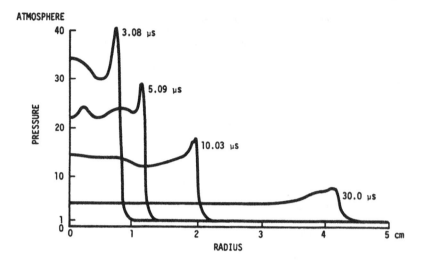

Figure 1.2.2 Strong shock wave development for a simulated lightning return stroke using cylindrical geometry.[30] The computed electrical energy input into this simulated channel was 1.5×10^4 J/m. Early development of the shock wave shows several internal waves prior to final shaping of the shock wave.

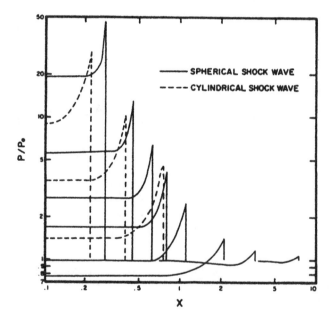

Figure 1.2.3 The expansion of spherical and cylindrical shock waves from the strong shock region into the weak shock region.[12] The radii of both spherical and cylindrical geometries have been nondimensionalized using the relaxation radii defined in Equations 1 and 2. The spherical shock wave is that of Brode,[7] and the cylindrical shock wave is from a similarity solution by Sakurai.[53]

of these internal waves is to organize the fluid behind the shock, so that the resulting distribution of pressure and temperature allows the steady progression of the strong shock; in fact, both the spherical and cylindrical solutions do approach the same qualitative shape predicted by their analogous similarity solutions, which are based upon the assumption that the flow behind the shock is analytically uniform and predictable. Brode's numerical simulations are almost identical to the similarity solutions of Taylor[55] by the time the shock wave reaches $X = 0.2$.

Figure 1.2.3 from Few[12] shows the propagation of the strong shock into the transition region $(X \sim 1)$ and beyond into the weak shock region. Again, Brode's data is used for the spherical

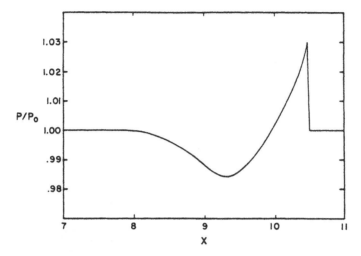

Figure 1.2.4 The weak shock wave formed from the spherical strong shock wave.[12] This is the final pressure profile computed by Brode.[7] For an energy input of 10^5J/m ($R_c = 0.56$ m for $P_0 = 10^5$ Pa) this weak shock wave would be approximately 6 m from the lightning channel.

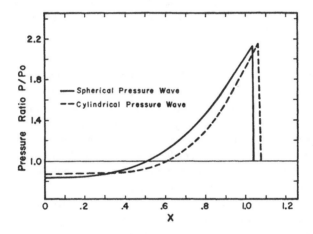

Figure 1.2.5 Comparison of spherical and cylindrical shock wave shapes near $X = 1$.[12] These profiles are for the point-source, ideal-gas solutions of Brode[6] and Plooster.[48] In the transition region of strong shock to weak shock, these wave shapes are nearly identical.

shock wave, and a similarity solution by Sakurai[53] for the cylindrical strong shock is shown for comparison. Throughout this region of strong shock expansion ($0.2 < X < 1$) the shape of the pressure profile is essentially that of the similarity solution. Note that as the shock front passes the relaxation radius ($X = 1$) and the central pressure passes through ambient pressure as postulated in the definition of the relaxation radius. The momentum gained by the gas during the expansion carries it beyond $X = 1$ and forces the central pressure to momentarily go below atmospheric. At this point the now weak shock pulse decouples from the hot channel remnant and propagates outward as an "N" wave. Figure 1.2.4 shows on a linear coordinate system the final output from Brode's numerical solution, the weak shock pulse or "N" wave at a radius of $X = 10.5$.

Another of the interesting consequences of the strong shock behavior and the relaxation radius scaling suggested by Few is the comparable forms of the spherical and cylindrical shock waves in the $X = 1$ region. Figure 1.2.5 from Few[13] shows Plooster's[48] cylindrical shock wave near $X = 1$ with Brode's[6] spherical shock wave. The effects of channel tortuosity will be discussed

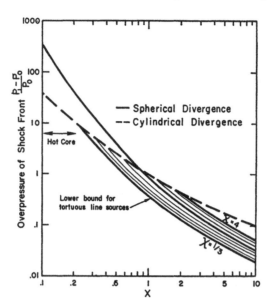

Figure 1.2.6 Line source shock wave expansion.[13] The overpressure of the shock front is given for spherical[7] and cylindrical[48] shock waves. Line sources must initially follow cylindrical behavior, but upon expanding to distances of the same size as line irregularities, they change to spherical expansion following curves similar to the curves depicted in this figure.

in greater detail later; for now we note that due to tortuosity we cannot expect the shock wave to continue to perform as a cylindrical wave once it has propagated beyond a distance equal to the effective straight section of the channel that generated it. If the transition from cylindrical to spherical occurs near $X = 1$ as suggested by Few,[13] then the spherical weak shock solutions of Brode provide a good means of estimating the wave shapes of lightning-caused pulses.

Figure 1.2.6 from Few[13] presents a graphical summary of the various transitions that are thought to take place. The initial strong shock will behave cylindrically following the dashed line based upon Plooster;[48] this must be the case for the line source regardless of the tortuosity, because the high speed internal waves depicted in Figures 1.2.1 and 1.2.2 will hydrodynamically adjust the shape of the channel during this phase. The transition from strong shock to weak shock occurs near $X = 1$ and the transitions from cylindrical divergence to spherical divergence will occur somewhere beyond $X = 0.3$ and probably beyond $X = 1$, depending upon the particular geometry of the channel at this point. The family of lines labeled χ in Figure 1.2.6 represents transitions occurring at different points. χ is the effective length, L, of the cylindrical source divided by R_c; it is approximately equal to the value of X at which the transition to spherical divergence takes place.

$$\chi = L/R_c \qquad (3)$$

2.2. COMPARISONS WITH NUMERICAL SIMULATIONS AND EXPERIMENTS

Hill[30–33] and Plooster[49,50] have questioned the application of the theory of the preceding section to return strokes as the generators of thunder. Although this difference has been interpreted as a controversy by some scientists,[59] the difference is not with the mechanism of shock formation and the behavior of the expanding shock waves. The main point of difference concerns the energy deposition by lightning return strokes. In the numerical solutions of Plooster[49,50] and Hill,[30] the energy inputs to the cylindrical problem were computed as a function of time for specified current wave shapes and channel resistance obtained from the predictions of the numerical model. These model results predicted that the energy input to the lightning channel was an order of magnitude or more below the values one obtains from electrostatic estimates or from other indirect measurements of lightning energy. It has been their contention that the channel resistance is too low for the currents in their model to produce the required energy per unit length. By their reasoning the energy in the lightning stroke must be expended by some process other than the return stroke. They are not very specific about where and how this is to be accomplished.

The potential between the negative charge of the cloud and the ground can be shown from simple models[59] and direct measurement[22] to be of the order of 10^8V. A lightning stroke process (leader included) that lowers -5C of charge must expend on the order of 5×10^8 J in the process. If the length of the channel is 5 km, then the average energy per unit length must be approximately 10^5J/m ($R_c \sim 0.5$ m). Optical, spectroscopic, and electromagnetic radiations all indicate that the return stroke itself is the most energetic part of the event.

The acoustic processes are much too slow to be useful in discriminating between energy expended by a leader or a return stroke. In fact, by the theory of the preceding section the thunder would be the same, irrespective of the deposition process so long as it was sufficiently fast and concentrated to produce a strong shock. If, however, most of the lightning stroke energy is deposited in a large volume (radius of several meters) so that a strong shock is never formed, then a new theory for the production of thunder from weak shock initial conditions would need to be formulated.

Hill[33] has made some rough calculations for the sound that would be produced by this type of diffuse mechanism and finds that the pressure amplitude will be a factor of 80 below the return stroke wave even using the 10^4J/m energy for the return stroke. This estimate was based upon the electrostatic theory of Colgate and McKee.[9] Hill's value of E_ℓ is 1.5×10^4J/m and Plooster's (for conditions similar to those of Hill) is 2.4×10^3J/m; both are far below the total stroke energy (10^5J/m) estimated from electrostatic computations and indirect measurements of lightning.[59] Still, it is just as significant that the two numerical simulations differ from each other by nearly an order of magnitude in the energy measurement. Hill[33] records these two theoretical measurements side by side without comment regarding the source of the differences. Not only are the energy computations different, but to the extent that their solutions can be compared because of variations in formats used for presenting the results; all of their detailed calculations differ by factors of approximately 3. Hill's[33] claim that these theoretical results are not highly pulse shape and model dependent does not appear to stand when intercomparisons are made with his and Plooster's results for similar conditions.

Close examinations of the fine structure of return stroke wave forms with modern high speed electronic equipment[65] reveal two parts to the return stroke current wave form. An initial slow current increase is followed by a very fast current increase to the peak value. The first part is thought to be associated with the upward propagating streamers from the ground below the stepped leader tip. The second and faster part is associated with the actual return stroke propagation. Prior estimates of current rise times probably averaged these two parts. The measured rise time for the true return strokes was found to be typically 0.2 µs approaching in some cases the 0.15 µs system limit. This recent data obtained with fast response time equipment yields current rise times for natural cloud-to-ground lightning in the 35 to 50 kA/µs range. These values are considered as representative of normal strokes; extraordinary strokes have been measured with current rise times in the 100 to 200 kA/µs range. By way of comparison, Hill's[30] current rise time was 2.5 kA/µs. Hill[33] argues that the energy input to the channel is not a function of the current rise time, yet when he used a current rise time of 35 kA/µs in one of his computer runs he obtained $E_\ell = 2.3 \times 10^5$J/m. He attributed the high energy to the large peak current (300 kA at 8.5 µs) not to rapid current rise time. This interpretation seems to be inconsistent with his earlier findings that showed the energy disposition rate peaking very early in the discharge process (2 µs power peak vs. 8.5 µs current peak).

The numerical experiments of Plooster and Hill have served a very valuable purpose in forcing us to reexamine the details of the physics of the discharge process and to look at the return strokes

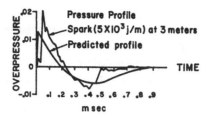

Figure 1.2.7   Comparison of theory with a pressure wave from a long spark.[13] The measured pressure wave from a long spark[61] is compared with the predicted pressure from a section of a mesotortuous channel having the same energy per unit length. χ (Equation 3) is assumed to be 4/3.

in finer detail. There is still significant research to be done before this question of the return stroke energy can be satisfactorily answered.[17a,d]

Natural thunder does not provide a good source for testing the details of the shock wave theory of its origin because

1.   The thunder signal is a superposition of multiple pulses from multiple leaders and strokes.
2.   The thunder signal also convolves the pulses from the numerous tortuous channel segments and branches.
3.   The strike point is not controllable, making systematic investigation difficult.
4.   Each event is unique and the range in parameters is wide, making the experiment nonrepeatable.
5.   One rarely has the quantitative information on the lightning parameters needed for predicting the thunder signal.

We can find some support for the theory by making enough simplifying assumptions regarding the nature of the thunder signal to test it with real lightning; or, we can employ laboratory simulations for a more controlled test.

The laboratory simulations approach has been successfully performed in a series of experiments conducted at Westinghouse Research Laboratories. In these tests a 6.4×10^6V impulse generator was used to produce 4-m spark discharges in air.[61] Circuit instrumentation allowed the measurement of the spark gap voltage and current from instrumentation allowed the measurement of the spark gap voltage and current from which the power deposition can be computed, and calibrated microphones were used to measure the shock wave from the spark as a function of distance. The results of the research[61] were compared with the theory of Few[13] and with other possible interpretations. In general the data were found to be consistent with the theory developed by Few.

Figure 1.2.7 from Few[13] compares a measured spark pressure pulse with the profile that is predicted from the theory. In this figure $\chi = 4/3$ was used to obtain the theoretical pressure profile; that the peak measured pressure exceeded the predictions was interpreted by Few[13] as indicating that $\chi = 4$ would be the appropriate value for the measured wave.

Figures 1.2.8 and 1.2.9 from Uman et al.[61] summarize the extensive series of spark measurements. Figure 1.2.8 is in the same format as Figure 1.2.6. The center line passing through the nce scattered points and labeled L = 0.5 m corresponds (using the measured energy input of 5×10^3J/m giving $R_c = 0.126$ m) to $\chi = 4$ in Figure 1.2.6. The two boundary lines L = 6.25 cm and L = 4.0 m would correspond to χ values 0.5 and 32. The lower bound is very close to the one third value lower limit indicated in Figure 1.2.6. The upper bound of Figure 1.2.8 ($\chi \sim 32$) is too large to be depicted in Figure 1.2.6 where $\chi = 4$ is the last line shown.

The data points of Figure 1.2.8 corresponding to the larger χ or L values could represent situations where the shock wave expansion was following the cylindrical behavior over a long distance, hence large χ. However, if the expansions were truly cylindrical to that extent, then the length of the pulse would be much longer as required by the cylindrical wave predictions. The

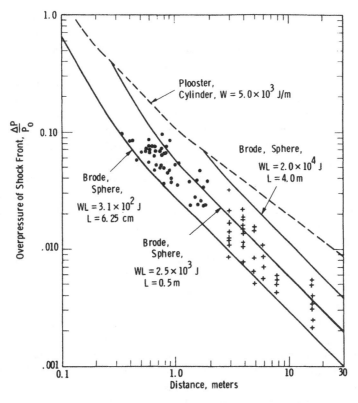

Figure 1.2.8 Shock front overpressure as a function of distance from the spark.[61] The dots represent data obtained with a piezoelectric microphone; the crosses represent data obtained with a capacitor microphone. The total electrical energy per unit length computed from measurement of the spark voltage and current is 5×10^3 J/m. Also shown are theoretical values for cylindrical and spherical shock waves.

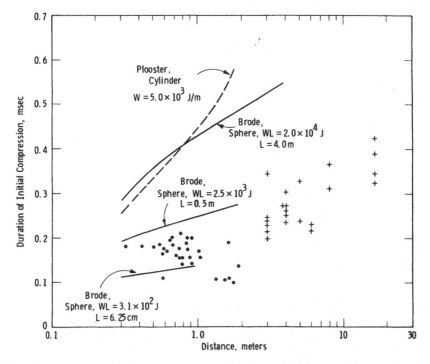

Figure 1.2.9 Duration of positive part of the shock wave from the long spark.[61] For the same data from Figure 1.2.8, we see here the length of the positive pressure pulse for the 5×10^3 J/m sparks at various distances.

data of Figure 1.2.9 indicate that this cannot be the case. The lengths of the positive pressure pulses shown in Figure 1.2.9 are clearly not in the cylindrical regime; if anything, they tend to be even shorter than predicted by the spherical expansion (see also Figure 1.2.7).

It is obvious from both the spark photographs and wave forms in Uman et al.[61] that the spark is tortuous and produces multiple pulses. They found that the waveforms more distant from the spark where pulse transit times were most similar showed evidence of an in-phase superposition of pulses; at closer range the pulses exhibited greater relative phase shifts and more multiplicity aspects. The in-phase superposition of spherical waves would produce exactly the distribution shown in Figures 1.2.8 and 1.2.9. The pressure amplitude would be increased relative to a single pulse, but the wave length would not be substantially affected.

The spark waveforms measured by Uman et al.[61] were systematically shorter than predicted by the theory. As shown in Figure 1.2.7, the tail of the wave was compressed, and the data of Figure 1.2.9 indicate that the positive pulse was similarly shortened. This shortening could be due simply to an inadequacy in the numerical shock wave model; we think, instead, that the difference results from the energy input being instantaneous in the one case and of longer duration for the spark case. If energy, even in small quantities, continues to be input into the low density channel core after the shock front has moved outward, then the core will be kept at temperatures much higher than predicted by the theories having an instantaneous energy input followed by expansion. Owing to the elevated sound speed associated with the higher core temperature, the part of the wave following the shock front will form and propagate outward faster than predicted by theory. We expect, therefore, that the elevated core temperature associated with sparks and lightning can reasonably produce the shortened waveforms measured by Uman et al.[61]

Plooster[49] and Hill[33] give an alternate interpretation to the spark data described above. Again the difference centers on how much of the total energy was deposited in the hot channel. Plooster suggests that there may have been an error in the measurement of the spark energy, but the error would need to be one order of magnitude. If the order-of-magnitude-lower spark energy is used in Plooster's computer simulation, one gets a relatively good correspondence between theory and measurements as shown in Figure 1.2.10. Hill contends that one tenth of the electrical energy appears in the hot spark channel and produces the observed wave forms as true cylindrical waves. The other nine tenths of the spark energy is by Hill's suggestion diffusely deposited around the spark channel and produces no observed acoustic emission; had one been produced it would have been detected in the experiment. We do not accept either of these interpretations of the experiment because

1. The spark channel is known to be tortuous and does not behave as a mathematical cylindrical source.
2. It is very unlikely that 4.5×10^3J/m can be rapidly and yet silently deposited.
3. We do not believe that an order of magnitude error was made in the energy measurement.

Upon close examination of Figure 1.2.10 we suggest that the fit of the data with the cylindrical source at one tenth of the measured input energy is not very good. Most of the data points lie above the cylindrical line for radii less than 3 m, while most of the data points lie below the cylindrical line at radii longer than 3 m. It is clear that the pressure is decreasing more rapidly than predicted by the cylindrical theory. The same data are plotted in Figure 1.2.8 where we see that the spherically divergent curves closely parallel the data points, and it is not necessary to dismiss nine tenths of the energy to obtain a reasonable fit to the data.

Other laboratory experiments with sparks have confirmed the majority of the results and interpretations of Uman et al.[61] The work of Page and McKelvie[47] confirms the spherical nature of the expanding shock wave in the weak shock zone. They also found a variation in the apparent efficiency of conversion of electrical energy to shock wave energy; the efficiency is found to vary as the inverse square root of the electrical energy. These efficiencies are, however, inferred from

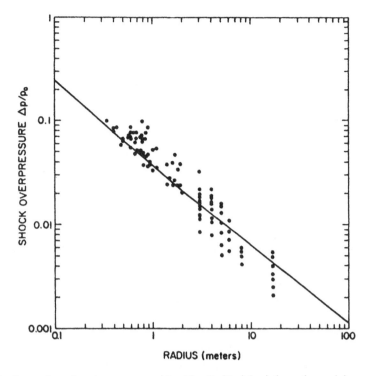

Figure 1.2.10 Comparison of spark overpressure data with cylindrical shock theory for much lower energy.[49] The data from Figure 1.2.8 are compared here with a cylindrical shock wave expansion with an order of magnitude lower energy input.

the peak pressure of the weak shock wave not directly measured; they attribute the losses to radiation.

For the reasons described previously, thunder observations are extremely difficult to quantify with respect to theory of generation by shock waves. If the generating mechanism produces strong shock, then rough estimates of the lightning stroke energy can be obtained, but the attribution of the energy to the return stroke or some other process is not possible.

A measurement of the power spectrum of thunder provides one method of energy measurement[13] if the conditions and assumptions employed in the theory are met. When the conditions are not met, the power spectrum measurement is still useful for ascribing limits to the lightning energy. These are the conditions and assumptions incorporated in Few's thunder power spectrum theory:

1. Lightning channels are highly tortuous; therefore, cylindrical expansion is applicable only in the strong shock zone and spherical expansion is appropriate beyond. Specifically Few assumed $\chi = 4/3$; see Figure 1.2.6. There is some evidence to support this assumption,[11,29] but there will probably be straight channel sections that are longer than this. Longer straight sections will have the effect of decreasing the energy estimated by the power spectrum peak. In other words, the power spectrum peak will overestimate the channel energy.

2. Spherical wave propagation beyond $X = 10$ can be treated with linear acoustic theory. This assumption also raises the energy estimates to higher values because it is well known that "N" waves lengthen as they propagate. If this lengthening had been included by Few, the resulting power spectrum estimate would have been range dependent. We show in the section on propagation that this effect is significant. Also, the Few theory did not include attenuation of the acoustic signal; this can also influence the spectrum.

3. There is uniform energy disposition and random tortuous features along the channel. Few proposed that, if the thunder signal was the linear superposition of basically identical pressure pulses whose relative phases were random, the power spectrum of the resulting composite signal would be essentially identical in shape to that of a single pulse. If the signal contains mixed sources the power spectrum of the sum will be weighted toward the more energetic, hence longer, pulses. Also special geometric situations can produce special effects in the power spectrum that are beyond the scope of Few's proposal.

4. The lightning flash producing the thunder can be treated as if it had only a single return stroke. Multiple pulses produced by multiple strokes from the same geometric source produce thunder signatures containing repetitive sequences of pulses which can alter the resulting power spectrum. The multiple pulses are each responsible for a spectral shape which is summed in the total thunder spectrum; and, there will be cross-product terms among the pulses with spectral components corresponding to the time lapses between strokes. Few[13] noted that the first return stroke in multiple stroke lightning usually was more energetic and would dominate the resulting power spectrum. He also gave a qualitative description of the power spectrum of multistroke thunder; it appears that a quantitative theory for the multiple stroke thunder has not been developed but is certainly feasible.

The foregoing assumptions all affect the thunder spectrum in the same sense; the peak of the theoretical spectrum will occur at higher frequencies than the peak of the real thunder spectrum. The lightning channel energy that one estimates from the peak will therefore be an overestimate of the actual lightning channel energy. Few[13] provided a series of measured thunder spectra from a single thunder record; these had peaks around 40 Hz, which according to the theory corresponded to $E_\ell \approx 2 \times 10^6 J/m$. It was noted that this energy estimate was an order of magnitude higher than indicated by other energy estimates. Holmes et al.[34] have provided the most complete published thunder spectra to date; these spectra show a lot of variation. Most of the spectra are consistent with the qualitative expectations of thunder produced by multiple stroke lightning, but a few of them exhibit very low frequency ($\sim$1 Hz) components that are dominant during portions of the record and appear to be totally inconsistent with the thunder generation theory from the hot explosive channel. Dessler,[10] Bohannon et al.,[5] and Balachandran[1] have suggested that these lower frequency components might be electrostatic in origin; Holmes et al.[34] also considered that this was a possible explanation. Most of the researchers agree that the thunder spectra published prior to the 1968 period were unreliable primarily due to inadequate instrumentation (see Hill[32] for a discussion of these instruments).

2.3. TORTUOSITY AND THE THUNDER SIGNATURE

With respect to the effects of lightning channel tortuosity on the thunder signal there is almost unanimous agreement among researchers. Lightning channels are undeniably tortuous and are tortuous apparently on all scales (Evans and Walker;[11] Hill;[29] Few et al.[18]). For convenience in discussing channel tortuosity Few[13] employed the terms micro-, meso-, and macro-tortuosity relative to the relaxation radius of the lightning shock wave.

For a lightning channel having an internal energy of $10^5 J/m$ (see Table 1.2.1), $R_c \approx 1/2$ m. The micro-tortuous features smaller than R_c although optically resolvable, are probably not important to the shock wave as measured at a distance, because the high speed internal waves illustrated in Figures 1.2.1 and 1.2.2 are capable of rearranging the distribution of internal energy along the channel while the shock remains in the strong shock regime. At the meso-tortuous scale ($\sim R_c$) the outward propagating shock wave decouples from the irregular line source because the acoustic waves from the extended line source can no longer catch up with the shock wave. Somewhere in this meso-tortuous range the divergence of the shock waves makes the transition

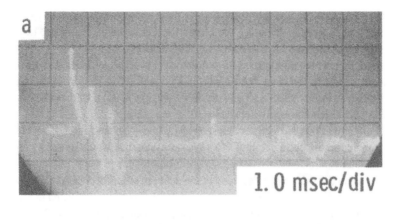

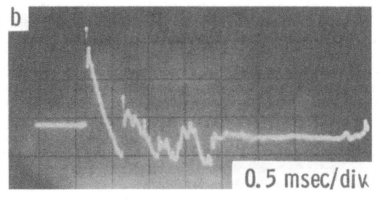

Figure 1.2.11 Typical weak shock waves from the long spark measured in the transition region between cylindrical and spherical divergence.[61] Multiple pulses arriving from different parts of the tortuous spark source are seen at this range.

from cylindrical to spherical. Figure 1.2.11 from Uman et al.[61] shows that measured pressure waves from tortuous sparks exhibit multiple pulse characteristics rather than a single pulse that is characteristic of a simple cylindrical wave.

Whereas the meso-tortuous channel segments are important in the formation and shaping of the individual pulses being emitted by the channel, the macro-tortuous segments are fundamental to the overall organization of the pulses and the amplitude modulation of the resulting thunder signature. Few[15] computed, based upon the experimental results of Wright and Medendorp,[69] that 80% of the acoustic energy from a short spark was confined to within ±30° of the plane perpendicular to the short line source. A macro-tortuous segment of a lightning channel will direct the acoustic radiations from its constituent meso-tortuous, pulse-emitting segments into a limited annular zone. An observer located in this zone (near the perpendicular plane bisecting the macro-tortuous segment) will perceive the group of pulses as a loud clap of thunder; whereas another observer outside the zone will perceive this same source as a lower amplitude rumbling thunder. This relationship between claps, rumbles, and channel macro-tortuosity has been confirmed by experiment[14] and in computer simulations.[52,62]

Hill[32] describes experiments performed by Brook[70] with line explosive charges; in these experiments it was found that a tortuous line explosion was necessary to achieve thunder-like signals.

Another feature found in the spark experiments of Wright and Medendorp[69] was the change in the shape of the pulse when measured out of the perpendicular plane. For the short spark they found that the positive and negative parts of the N wave became separated at large angles and that the shape of pulses became more rounded. If these characteristics are also valid for long continuous lightning channels, then the "rumbling" thunder would be composed of sources that were somewhat different from the sources comprising thunder claps.

Loud claps of thunder are produced as mentioned above near the perpendicular plane of macro-tortuous channel segments; there are three contributory effects[15,17] to the formation of the thunder claps. The directed acoustic radiation pattern described previously is one of the contributing factors and this effect is distributed roughly between ±30° of the plane.

A second effect that occurs only very close to the plane is juxtaposition of several pulses in phase, which increases the pulse amplitude to a greater extent than would a random arrival of the same pulses. Uman et al.[61] found that when the spark pulses such as those in Figure 1.2.11 were observed at a greater distance, the multiplicity of the pressure pulses decreased. At these distances the travel time differences from the midpoint and ends of the spark are small and in-phase arrivals are expected. Plooster[50a] has pointed out that nonlinear propagation effects will cause these superimposed waves in this region to fuse together and reform a single wave, in contrast to linear acoustic theory which would essentially have them pass through one another maintaining their initial direction of propagation. The extent to which this nonlinear process works depends on channel tortuosity and the amplitudes of the pulses; its final effect will be a reduction in the number of individual pulses that are detected at a distance. Whether by linear or nonlinear processes, there will be in-phase superposition of pulses in the perpendicular plane bisecting the macro-tortuous channel segments, producing in these special positions a large amplitude pulse. There are other special geometric arrangements of this type that can occur which produce particular phase relationships between the pulses, such as focusing on the inside of curved channels and out-of-phase superposition at special angles. None of these special effects have been quantitatively incorporated in the theoretical description of thunder signatures.

The third effect contributing to thunder clap formation is simply the bunching in time of the pulses. In a given period of time more pulses will be received from a nearly perpendicular macro-tortuous segment of channel than from an equally long segment that is perceived at a greater angle owing to the overall difference in the travel times of the composite pulses.

In this section we have examined the complex nature of the formation of individual pulses from hot lightning channels and how a tortuous line source arranges and directs the pulses to form a thunder signature. The resulting thunder signature depends upon:

1. The number and energy of each rapid channel heating event (leaders and strokes)
2. The tortuous and branched configuration of the individual lightning channel
3. The relative position of the observer with respect to the lightning channel

In the following section, we will examine the alterations to this signal that are produced while propagating through the air.

3. PROPAGATION EFFECTS

Once generated, the acoustic pulses from the lightning channel must propagate for long distances through the atmosphere, which is a nonhomogeneous, anisotropic, turbulent medium. Some of the propagation effects can be estimated by modeling the propagation using appropriate simplifying assumptions; however, other effects are too unpredictable to be reasonably modeled and must be considered in individual situations.

Three of the largest propagation effects, finite amplitude propagation, attenuation by air, and thermal refraction, can be treated with appropriate models to account for average atmosphere effects. Reflections from the flat ground can also be easily treated. If one has measured the horizontal wind structure between the source and the receiver then the refractive effects of wind shear and improved transient times may also be calculated. Beyond these effects, elements such as vertical winds, nonsteady storm-related horizontal winds, turbulence, aerosol effects, reflections from irregular terrain, etc. produce complications, which must be either ignored or examined on a case-by-case basis.

3.1. FINITE AMPLITUDE PROPAGATION

As large amplitude acoustic waves propagate through air, theory predicts that the shape of the wave must evolve with time. A single pulse will evolve to the shape of an N wave (see, for example, the spark wave in Figure 1.2.7); further propagation of the wave produces a lengthening of this N wave.[66] The best theoretical treatment of this process for application to the thunder problem is the one developed by Otterman.[46] In this formulation he looks at the lengthening of a Brode-type pulse such as Figure 1.2.4 from an initial state at an initial altitude H_0 down to the surface; his treatment differs from many others that do not include the change of ambient pressure with altitude. The result for the length of the positive pressure pulse at the ground, L_g, is given by:

$$2/3 \, (L_g^{3/2} - L_0^{3/2}) = \frac{\gamma + 1}{4\gamma} R_0 L_0^{1/2} \Pi_0 \{[-E_i(-x_0)] - [-E_i(-x_g)]\} \tag{4}$$

where L_0 is the initial length of the positive pulse at a distance R_0 from the source.

$$\pi_0 = \frac{\delta P_0}{P_0} \tag{5}$$

gives the over pressure, δP_0, at ambient pressure P_0, for the initial conditions. E_i is the exponential integral

$$-E_i(-x) = \int_x^\infty \frac{e^{-t}}{t} \, dt \tag{6}$$

By another notation the exponential integral can also be expressed[25]

$$E_1(x) = -E_i(-x) \tag{7}$$

In Equation 4, x_0 and x_g have been normalized using the atmospheric scale height, H_g, as follows,

$$x_0 = R_0 \cos\theta / 2H_g \tag{8}$$

$$x_g = H_0 / 2H_g \tag{9}$$

These expressions with $\cos\theta = 1$ would apply to a pulse propagating vertically downward,[46] but have been modified here for downward propagation at an angle θ with respect to the vertical by redefining x_0. Values for $2H_g$ are typically 16 km, $R_0 \sim 5$ m, and H_0 is in the range 0 to 10 km. Hence, x_g is less than one and x_0 is much less than one ($x_0 < 10^{-3}$). For values of the variables less than one we can employ accurate polynomial expressions for the exponential integrals;[25] we can show that when $x_0 < x_g$

$$E_1(x_0) - E_1(x_g) \approx \ln \frac{x_g}{x_0} - (a_1 x_g + a_2 x_g^2 + a_3 x_g^3 + a_4 x_g^4 + a_5 x_g^5) \tag{10}$$

where $a_1 = 0.99999$, $a_2 = -0.24991$, $a_3 = 0.05520$, $a_4 = -0.00976$, and $a_5 = 0.00108$.

If $x_g < 0.75$, corresponding to $H_0 < 12$ km, the second order term in x_g is less than 20% of the first order term. Hence, a good approximation for lightning

$$E_1(x_0) - E_1(x_g) \approx \ell n \frac{x_g}{x_0} - x_g$$

Using Equations 8 and 9,

$$E_1(x_0) - E_1(x_g) = \ell n \left(\frac{H_0}{R_0 \cos\theta}\right) - \frac{H_0}{2 H_g} \tag{11}$$

When substituting Equation 11 into Equation 4 we get

$$\frac{2}{3} (L_g^{3/2} - L_0^{3/2}) = \frac{\gamma + 1}{4\gamma} R_0 L_0^{1/2} \Pi_0 \left[\ell n \left(\frac{H_0}{R_0 \cos\theta}\right) - \frac{H_0}{2H_g} \right] \tag{12}$$

Equation 12 provides the finite amplitude stretching that should be applied to the waves predicted by strong shock theory. Uman et al.[61] demonstrated that pulse stretching occurred beyond Brode's final pressure profile shown in Figure 1.2.4; we see this clearly in Figure 1.2.9. Few[13] used linear propagation beyond the profile of Figure 1.2.4 to estimate the power spectrum of thunder but commented that nonlinear effects may be important. The need for application of nonlinear or finite amplitude theory to the thunder signal has been voiced in a number of papers in addition to these mentioned above (e.g., Few[17]; Holmes et al.;[34] and Hill[32]).

If we use the Brode pressure pulse shown in Figure 1.2.4 as the initial conditions for the finite amplitude propagation effect given by Equation 12, we find the following values for input to Equation 12:

$$R_0 = 10.45 R_c$$
$$L_0 = 0.53 R_c \tag{13}$$
$$\Pi_0 = 0.03$$

In addition, if we use for the other constants in Equation 12, $\gamma = 1.4$ and $H_g = 8 \times 10^3$ m, we obtain the following equation:

$$L_g = R_c \left\{ 0.386 + 0.147 \left[\ell n \left(\frac{H_0}{10.46 R_c \cos\theta}\right) - \frac{H_0}{16 \times 10^3 \text{ m}} \right] \right\}^{2/3} \tag{14}$$

Equation 14 has been used to generate the values of Table 1.3.1. The relaxation radii, R_c, cover the entire range of values for R_c in Table 1.2.1. Three values for θ are represented, as are three heights for the source. We see in general that the finite amplitude propagation causes a doubling in the length of the positive pulse within the first kilometer, but that beyond this range the wavelength remains approximately constant. This theory developed by Otterman[46] did not include attenuation of the signal; because attenuation reduces wave energy, which in turn reduces the wave stretching, this theory should be viewed as a maximum estimator of the pulse length.

The finite-amplitude propagation effect does, however, help resolve the overestimate of lightning channel energy made by acoustic power spectra measurements. Few[13] noted that the thunder spectrum method yielded a value for E_ℓ that was an order of magnitude greater than an optical measurement by Krider et al.[36] By assuming a doubling in wavelength by the finite amplitude propagation the energy estimate is reduced by a factor of 4, bringing the two measurements into a range of natural variations and measurement precision. These two estimates are still an order

Table 1.3.1 **Wave Lengthening Owing to Finite Amplitude Propagation**

											(Angle degrees)
R_c(m)	0.20	0.40	0.60	0.80	1.00	1.50	2.00	2.50	3.00	3.50	θ
L_0(m)	0.11	0.21	0.32	0.42	0.53	0.80	1.06	1.33	1.59	1.86	θ
L_g(m),H_0 = 1 km	0.24	0.45	0.65	0.84	1.03	1.49	1.93	2.35	2.77	3.17	0
L_g(m),H_0 = 4 km	0.26	0.49	0.71	0.93	1.14	1.66	2.16	2.65	3.12	3.59	0
L_g(m),H_0 = 8 km	0.26	0.51	0.74	0.96	1.18	1.72	2.24	2.75	3.25	3.74	0
L_g(m),H_0 = 1 km	0.24	0.46	0.67	0.87	1.06	1.54	1.99	2.44	2.87	3.30	45
L_g(m),H_0 = 4 km	0.26	0.50	0.73	0.96	1.18	1.71	2.22	2.73	3.22	3.71	45
L_g(m),H_0 = 8 km	0.27	0.52	0.76	0.99	1.22	1.77	2.31	2.83	3.35	3.85	45
L_g(m),H_0 = 1 km	0.25	0.47	0.69	0.89	1.10	1.59	2.06	2.52	2.98	3.42	60
L_g(m),H_0 = 4 km	0.27	0.51	0.75	0.98	1.21	1.76	2.29	2.81	3.32	3.82	60
L_g(m),H_0 = 8 km	0.28	0.53	0.77	1.01	1.25	1.81	2.37	2.91	3.44	3.97	60

of magnitude greater than the value proposed by Hill.[33] Hill's proposed value (10^4J/m) corresponds (at an altitude of 4 km) to an $R_c \sim 0.23$ m, which when arriving at the ground will be $L_g \sim 0.3$ m from Equation 14 using θ = 45°. If L_g represents half of the wave length of the dominant frequency, f_m, then $f_m \sim 570$ Hz. These high pitched sounds are not the dominant sound in thunder; measured spectra[13,34] indicate the spectral content of f > 500 Hz is more than an order of magnitude below the content for f < 100 Hz. Hill's energy values seem to be totally out of line.

3.2. ATTENUATION

In the previous section the pressure amplitude of the pulse decreased because of spherical divergence of the wave, and because as the wave lengthened, the total energy was spread over a longer distance. These contributors to wave amplitude decay are sometimes included in discussions of attenuation, but we have treated them separately. Otterman[46] assumed in his derivations that the total energy of the wave was conserved; hence, energy losses from the wave cause a more rapid amplitude decay and decrease the lengthening predicted by the theory. The results of the section on finite amplitude propagation are, therefore, maximum effects.

There are three processes on the molecular scale that attenuate the signal by actual energy dissipation; the wave energy is transferred to heat. Viscosity and heat conduction represent the molecular diffusion of wave momentum and wave internal energy from the condensation to the rarefaction parts of the wave. The so-called molecular attenuation results from the transfer of part of the wave energy from the translational motion of molecules to internal molecular vibrational energy of the O_2 molecules during the condensation part of the wave and back out during the rarefaction part of the wave. The phase lag of the energy transfer relative to the wave causes some of the energy being retrieved from the O_2 to appear at an inappropriate phase; thus it goes into heat rather than the wave. These three processes can be treated theoretically within a common framework.[35] The amplitude of a plane wave, δP, as a function of the distance, x, from the coordinate origin (x = 0) is given by:

$$\delta P = \delta P_0 e^{-\alpha x} \tag{15}$$

where δP_0 is the wave amplitude at the origin. The coefficient of attenuation, α, can be shown in the low frequency regime to be

$$\alpha = \frac{\omega^2 \tau}{2c} \tag{16}$$

In Equation 16, ω is the angular frequency and τ is the relaxation time (or e-folding time) for the molecular process being considered; c is the speed of sound. The low frequency condition above assumes that $\omega\tau \ll 1$. The expressions for τ depend upon the particular molecular processes under consideration; it is important to note, however, that α is proportional to ω^2 for the assumed conditions; hence, attenuation alters the spectral shape of the propagating signals.

When molecular viscosity is considered[35] we obtain,

$$\tau_v = \frac{4\eta}{3\rho_0 c^2} \tag{17}$$

in which η is the coefficient of viscosity and ρ_0 is the ambient density. For air at 20°C and 1 atm, $\tau_v = 1.7 \times 10^{-10}$ s.

When heat conduction is considered[35] we obtain,

$$\tau_c = \frac{(\gamma - 1)\, K}{\rho_0 c^2 c_p} \tag{18}$$

where c_p is the specific heat at constant pressure, γ is the ratio of specific heats, and K is the thermal conductivity. For air at 15°C and 1 atm, $\tau_c = 0.71 \times 10^{-10}$ s.

The third contributor is molecular attenuation; Landau and Lifshitz[37] include this process along with chemical reactions and phase changes in a term which they call second viscosity. The actual process by which the energy transfer to the vibrational states of O_2 occurs is a complex one. In water-free air the process is extremely inefficient and the e-folding time for equilibrium vibrational excitation of O_2 is on the order of several seconds. This transfer time is drastically shortened by the presence of water molecules in air owing to the much larger interaction cross section between water and O_2. Unfortunately, the resulting attenuation is strongly dependent upon relative humidity.

We can obtain rough quantitative estimates of molecular attenuation for comparison with other processes based upon the discussions in Kinsler and Frey[35] and Harris.[28] The coefficient of attenuation, α, has the form of Equation 16 in the low frequency regime; when examined over wider values of ω, α is observed to maximize at $\omega = \omega_m$, where ω_m may be in the megacycle range for high humidity. ω_m is found to vary quadratically with the relative humidity; however, the ratio $(\alpha/\omega)_m$ at the maximum in α is found to be nearly constant with respect to changes in relative humidity. Harris[28] has made a power series fit to experimental data providing an estimator of ω_m as a function of humidity, h

$$\omega_m = 4.96\, h^2 + 9.24\, h - 0.94 \tag{19}$$

where the humidity is in units of gram per cubic meter and ω_m is in kilohertz. The data of Harris and that quoted in Kinsler and Frey for the constant, $(\alpha/\omega)_m \approx 5 \times 10^{-7}$ s/m, agree within 6%. In the low frequency regime the τ_m appropriate for use in Equation 16 can be approximated (from Kinsler and Frey)

$$\tau_m = \frac{4c}{\omega_m} (\alpha/\omega)_m \tag{20}$$

Using Equations 19 and 20 we can estimate τ_m, but we must be aware of certain limitations: (1) the constant $(\alpha/\omega)_m$ varies with temperature; we are using 20°C and (2) the values used for h in the experiments were low compared to thunderstorm conditions requiring an extrapolation of Equation 19 beyond the region in which it has been verified.

We will look at two cases — high humidity, $h_h = 15$ gm/m^3, and low humidity, $h_\ell = 5$ gm/m^3, — both at 20°C and 1000 mb corresponding, respectively, to 85 and 28% relative humidity. We obtain by this procedure

$$\tau_m(h_h) \approx 5.5 \times 10^{-10} \text{ s}$$
$$\tau_m(h_\ell) \approx 4.1 \times 10^{-9} \text{ s}$$

These results show that the low humidity case produces an order-of-magnitude greater attenuation than does the high humidity case; this occurs because the attenuation peak is shifted to lower frequencies with lower humidities. With respect to thunder measurements these results indicate that we might expect differences in the thunder measured in coastal areas with high surface humidities compared to that measured in dry or mountainous areas where lower humidities are common.

We note also that, even in the high humidity case, the attenuation due to the molecular absorption is larger than the sum, which is called classical attenuation, of viscous and conduction attenuations.

For thunder at frequencies below 100 Hz we can show that the total attenuation is insignificant. If we let $\tau = 10^{-9}$s, $\omega = 2\pi \cdot 100$ Hz, and $x = 10$ km, we find using Equations 16 and 15 that $\delta P/\delta P_0 = 0.994$. However, for the many small branches having much lower energy than the main channel, the frequencies will be much higher and attenuation is important. For the conditions mentioned with a frequency of 1 kHz, we find that $\delta P/\delta P_0 \sim 0.5$. Because of lower initial acoustic energies, spherical divergence, and attenuation, it is unlikely that acoustics emitted by the smaller branches and channels can be easily detected.

The most important effect of attenuation on thunder involves its interaction with the finite amplitude propagation effect. The "N" wave is continuously attempting to maintain a sharp pressure jump or shock at the front and rear at the wave; attenuation is most effective for just this condition and is continuously eroding these shocks to form more rounded pressure waves. Thus, in the weak shock zone, attenuation can effectively reduce the energy of even the long pulses by continuously dissipating the shock. When the pulses have propagated beyond the weak shock region, attenuation will reshape the shocks and produce a rounded pulse. At distances beyond a few kilometers of a lightning channel, one never hears the high frequency components that are heard when close to the source.

3.3. SCATTERING AND AEROSOL EFFECTS

The scattering of acoustic waves from the cloud particles is very similar to the scattering of radar waves from the particles; both are very strongly dependent upon wavelength. In the case of sound waves Rayleigh[51] shows that the intensity of the scattered waves from plane acoustic waves of wavelength, λ, incident upon a hard stationary sphere of radius a is proportional to $(\pi a^2) (a/\lambda)^4$, which is the same relationship of these dimensions that appears in the radar cross section expression. For thunder wavelengths (~ 1 m) and cloud particles ($\sim 10^{-3}$) the ratio $(a/\lambda)^4$ is 10^{-12}. The cloud is, therefore, transparent to low frequency thunder just as it is to meter wavelength electromagnetic waves, although insignificant fractions of the energy do get scattered.

There are, however, the high frequency components in the shocks of the pulses. These components will be scattered more effectively by the cloud particles. The scattered part is distributed broadly from the scatterer, hence it removes energy from the incident beam. The effect of cloud particles acting as scatterers is similar to attenuation; the higher frequencies are preferentially attenuated ($\propto \omega^2$).

There are, however, turbulent eddies in the same size range as low frequency thunder wavelengths, and these features, owing to small thermal changes and flow shears, produce a distortion of wave fronts and scattering type effects. For the part of the turbulent spectrum having wavelengths smaller than the acoustic wavelengths of interest, the turbulence can be treated statistically by scattering theory. Larger scale turbulence must be described with geometric acoustics.

Brown and Clifford[8] have recently reviewed the turbulent attenuation problem and have developed a theory that accounts for much of the great disparity that exists in the observational data. Among the points discussed in their paper are

1. The very strong dependence of the attenuation on atmospheric conditions with the attenuation changing by several orders of magnitude; the greater the turbulence, the greater is the attenuation.
2. Turbulence close to the sound source is many times more effective at scattering than the same turbulence close to the receiver.
3. Turbulent attenuation is not a strong function of frequency as is the case for other attenuation sources; they found α varied as $f^{1/3}$ for turbulent scattering vs. f^2 for attenuation due to classical, molecular, and small particle scattering.

Based upon the previously noted properties of turbulent scattering and the turbulence in the medium in which thunder is produced, we think that for the low frequency thunder, attenuation by turbulent scattering will be orders-of-magnitude greater than any of the ω^2-dependent mechanisms. Turbulence is a viable attenuator of thunder, and it will in some cases be significant and important. Because each situation is complex and individual this process is impossible to quantify for a general case.

In the first part of this section we treated the cloud particles as sources of acoustic scattering; there are other and probably more important ways in which these aerosol components interact with the acoustic waves. The surface area within a volume provided by the cloud particles provides preferred sites for enhanced viscosity and heat conduction; hence the presence of particles increases the "classical" attenuation coefficient.

There is a totally different process producing attenuation that results from wave-associated phase changes. The changes in thermodynamic parameters associated with the passage of the acoustic wave through the medium can produce changes in the equilibrium vapor pressures over the surfaces of cloud particles and shifts in the local vapor to liquid or solid conversion rates. For example, during the compressional part of the wave, the air temperature is increased and the relative humidity is decreased relative to equilibrium; the droplets partially evaporate in response and withdraw some energy from the wave to accomplish it. The opposite situation occurs during the expansion part of the wave. Because the phase change energy is ideally 180° out of phase with the acoustic wave energy, this process produces attenuation. Landau and Lifshitz[37] include this effect in their "second viscosity" term. This attenuation process differs from the other microscopic processes in that it can be very effective at the lower frequencies. The magnitude of this effect plus the enhanced viscous and heat conduction exceeds that of particle-free air by a factor of 10 or greater, depending upon the type, size, and concentration of the cloud particles.[35]

Finally, there is a mass loading effect with respect to the cloud particles that we must consider. The amplitude of the fluid displacement, ζ, produced by an acoustic wave of pressure amplitude δp and angular frequency ω is[35]

$$\zeta = \frac{\delta p}{\rho_0 c \omega} \tag{21}$$

Using 50 Pa as a representative value of δP for thunder inside a cloud we find for a 100 Hz frequency that $\zeta = 100$ μm. The part of the cloud particle population whose diameter is much smaller than this, say 10 μm, should, due to viscous drag, come into dynamic equilibrium with the wave flow. These cloud particles, which participate in the wave motion, add their mass to the effective mass of the air; this affects both the speed of sound and the impedance of the medium. For higher frequency waves, fewer cloud particles participate so the effect is reduced; whereas lower frequency waves include greater percentages of the population and are more strongly affected. Clouds are therefore dispersive with respect to low frequency waves. Also, the cloud boundary acts as a partial reflector of the low frequency acoustic signals because of the impedance change at the boundary. Assuming a total water content of the order 5 g/m^3, we estimate the order of magnitude of the effect on sound speed and impedance is 10^{-3}, which is not large, but it may be detectable.

We have seen in these discussions that the cloud aerosols interact with the acoustic waves in three different ways depending upon their size relative to the amplitude of air motion of the sound. The smallest fraction "ride with the wave"; altering the wave propagation parameters. The largest particles are stationary and act as scatterers of the acoustic waves. The particles on the middle range provide a transition scale for the above effects but are primarily responsible for enhanced viscous attenuation.

In summary, we see that there are several processes that can effectively attenuate higher frequency components of thunder; this is in support of the conclusions of the previous subsection. We have, in addition, found three processes that affect the low frequency components. Low frequencies can be attenuated by turbulent scattering, and in the cloud, by coupling wave energy to phase changes. We have also found that low frequencies interact with the cloud population dynamically; as a result cloud boundaries may act as partial reflectors and in-cloud propagation may be dispersive.

3.4. REFRACTION

There is a very wide range of refractive effects in the environment of thunderstorms. In the preceding section we found that turbulence on the scale of the acoustic wavelength and smaller could be treated with scattering theory. Turbulence larger than acoustic wavelengths, up to and including storm scale motions, should be describable by geometric acoustics, or ray theory, To actually do this is impractical because it requires detailed information (down to the turbulent scale) of temperature and velocity of the air everywhere along the path between the source and the observer. Because the thunder sources are widely distributed, we would require complete knowledge of the storm environment down to the meter scale in order to accurately trace the path of an individual acoustic ray. We can be relieved of these requirements if we relax our expectations regarding the accuracy of our ray path somewhat and require only the information of its average or net position.

The basic assumption regarding turbulence that makes this simplification possible is that turbulent eddies are closed circulation elements, which produce compensating effects on acoustic rays traversing them. The three fluid properties that cause an acoustic ray to change its direction of propagation are the components of thermal gradient, velocity gradient, and velocity that is perpendicular to the direction of propagation. The individual eddies participate in vertical motions of air parcels and contribute to the production of the overall thermal structure of the environment, which will be approximately adiabatic. We do not expect that the thermal perturbations due to turbulence beyond this will be systematic; hence, an acoustic ray propagating through turbulence should not deviate markedly due to thermal gradients associated with the turbulence from the path predicted by the overall thermal structure of the environment. Similarly, velocity and velocity gradients should produce a zero net effect on the acoustic ray propagating through the turbulence.

This argument of compensating effects is not valid for circulation elements (large eddies) whose dimensions are equal to or greater than the path length of the ray because the ray traverses only part of the closed circulation. We can obtain a worst-case estimate of these effects by examining the horizontal ray propagation from a source at the center of an updraft of 30 m/s through 2 km to the cloud boundary where the vertical velocity is assumed to be zero. The ray will be "advected" by 90 m upward during this transit, while the direction of propagation of the ray will be rotated through 5° downward. Due to this rotation the "apparent" source by straight ray path would be 180 m above the real source. These two effects have been estimated independently, when in fact they are coupled and are to some extent compensatory; when we merely add them, the result is an overestimate of the apparent source shift, which in this example is 270 m. If this worst case is the total error in propagation to the receiver at 5 km then this error represents 5% of the range; over the length in which it occurs, 2 km, it represents 13% error.

Now we turn our attention to the large scale refraction effects that can be incorporated in an atmospheric model that employs horizontal stratification. The two strongest refractive effects of the atmosphere — the vertical thermal gradient and boundary layer wind shears — fall into this category along with other winds and wind shears of less importance.

The nearly adiabatic thermal structure of the atmosphere during thunderstorm conditions has been recognized for a long time as a strong influence on thunder propagation.[23] This thermal gradient is very effective because it is spatially persistent and unavoidable. Even though the temperature in updrafts and downdrafts inside and outside the cloud may differ, sometimes significantly, the thermal gradients in all parts of the system will be near the adiabatic limit (or pseudoadiabatic in some cases) because of the vertical motion. Hence, the acoustic rays propagate in this strong thermal gradient throughout its existence.

We can employ a simplified version of ray theory to illustrate some of the consequences of this thermal structure. If we assume no wind, a constant lapse rate Γ ($\Gamma = -\partial T/\partial z$, and $\Delta T/T_0 \ll 1$, where ΔT is the change in temperature and T_0 is the maximum temperature along the path, then the ray path may be described as a segment of a parabola

$$\ell^2 = \frac{4T_0}{\Gamma} h \qquad (22)$$

In Equation 22 the maximum ray path temperature, T_0, also corresponds to the vertex of the parabola where the ray slope passes through zero and starts climbing. Respectively, h and ℓ are the height above the vertex and the horizontal displacement from the vertex. In order to apply Equation 22 to all rays, it is necessary to ignore (mathematically) the presence of the ground, because the vertices of rays reflecting from the ground are mathematically below ground, and we must, in other cases, visualize rays extending backward beyond the source, in order to locate their mathematical vertices.

If we now set T_0 equal to the surface temperature, then we define a special acoustic ray that is tangent to the surface when it reaches the surface; this is depicted in Figure 1.3.1. This same ray is applicable to any source, such as S_1, S_2, or S_3, that lies on this ray path. For the conditions assumed in this approximation it is not possible for rays from a point source to cross one another (except those that reflect from the surface). The other acoustic rays emanating from S_2 must pass over the point on the ground where the tangent ray makes contact, this is also true for rays reflecting from the surface inside the tangent point. The shaded zone in Figure 1.3.1 corresponds to a shadow zone which receives no sound from any point source on the tangent ray beyond the tangent point. Point sources below the tangent ray, such as source S_4 in Figure 1.3.1, have their tangent ray shifted to the left in this representation and similarly cannot be detected in the shadow zone. However, sources above the tangent ray, S_5, for example, can be detected in some parts of the shadow zone.

For each observation point on the ground one can define about the vertical a parabaloid of revolution generated by the tangent ray through the observation point; the observer can only detect sounds originating above this parabolic surface. For this reason we usually hear only the thunder from the higher parts of the lightning channel unless we are very close to the point of a ground strike. For evening storms, which can be seen at long distances, it is common to observe copious lightning activity but hear no thunder at all; thermal refraction is the probable cause of

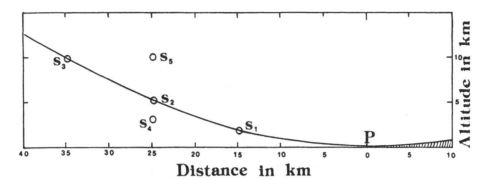

Figure 1.3.1 Parabolic acoustic ray from sources S_1, S_2, or S_3, tangent to the surface at P. This ray was generated utilizing Equation 22 with $T_0 = 30°C$ and $\Gamma = 9.8$ K/km. Observers on the surface to the right of P cannot detect sound from sources S_1, S_2, S_3, or S_4; an observer at P can only detect sound originating on or above the parabolic ray shown.

this phenomenon. For $T_0 = 30°C$, $\Gamma = 9.8$ K/km, and h = 5 km we find that $\ell \approx 25$ km; as noted by Fleagle[23] thunder is seldom heard beyond 25 km.

Winds and wind shears also produce curved ray paths but are more difficult to describe because these effects are vectors; whereas, the temperature was a scalar effect. If you are downwind of a source and the wind has positive vertical shear ($\partial u/\partial z > 0$) the rays will be curved downward by the shear; on the upwind side the rays are curved upward. Wind shears are very strong close to the surface and can effectively bend the acoustic rays that propagate nearly parallel to the surface. The combined effects of temperature gradients, winds, and wind shears can best be handled with a ray-tracing program on a computer. With such a program one can accurately trace ray paths through a multilevel atmosphere with many variations in the parameters; it is usually necessary in these programs to assume horizontal stratification of the atmosphere. The accuracy of the ray tracing by these techniques can be very high, usually exceeding the accuracy with which temperature and wind profiles can be determined.

4. MEASUREMENTS AND APPLICATIONS

A number of the experimental and theoretical research papers dealing with thunder have been discussed in earlier sections of this chapter and will not be repeated in this section. This is particularly true for the shock waves and power spectra discussion in Section 2. In this section we describe research results, techniques, and papers that have not been adequately covered previously.

4.1. PROPAGATION EFFECTS EVALUATION

The reader should have, at this point, an appreciation for the difficulty in quantitatively dealing with the propagation effects on both the spectral distribution of thunder and the amplitude of the signal. If, however, we are willing to forfeit the information content in the higher frequency (>100 Hz) portion of the thunder signal, which is most strongly effected by propagation, we can recover some of the original acoustic properties from the low frequency thunder signal. Bass and Losey[2] have attempted to restore the high frequency end of the spectrum by estimating the molecular and classical attenuation occurring along the path of the thunder to the ground. Their technique undoubtedly restores lost frequency content in the signal but introduces errors involving the assumed conditions along the path. Furthermore, their treatment does not address turbulent, aerosol, or finite amplitude effects; hence, it is incomplete and represents only a partial restoration of the original spectrum.

If we assume that the peak in the original power spectrum of thunder is roughly below 100 Hz, then we can safely conclude that the ω^2-attenuation effects deplete the higher frequencies without shifting the position of the peak. Because most spectral peaks of thunder tend to be around or below 50 Hz, this assumption appears to be safe even with finite amplitude stretching effects considered. We must further assume that the spectra are not substantially altered by turbulent scattering and cloud aerosols. To the extent that these assumptions are valid, the finite amplitude stretching can be removed from the thunder signal and its peak frequency at the source can be estimated. This technique enables us to obtain an estimate of the energy per unit length of the stroke corrected for first order propagation effects. Holmes et al.[34] found that the spectral peak overestimated the channel energy using Few's[13] method; if corrected for stretching, these measurements are in closer agreement, with the exception of those events containing other lower frequency acoustic sources.

There are a number of experiments that could and should be done to evaluate the propagation effects. Using thunder as the acoustic source, several widely separated arrays of microphones could compare signals from the same source at several distances. If carefully executed this experiment could quantify some of the propagation effects. Another approach would be to employ a combination of active and passive experiments such as point source explosions inside clouds from either balloons or rockets. This experiment provides an additional controllable factor that can yield more precise data; it also involves greater cost and hazard.

There has not been much new research of the amplitude frequency information in thunder since the Holmes et al.[34] paper; this is probably a result of the realization that any improvements to theory and interpretation will require addressing the propagation effects. In the past decade there have been a number of triggered lightning experiments performed in which natural lightning is initiated by launching a small rocket trailing a wire toward a thundercloud producing large electric fields at the ground. In these experiments the bottom portion of the lightning channel is more or less controlled by the rocket-wire trajectory. Using triggered cloud-to-ground lightning as a source for thunder acoustics, research offers a good situation in which arrays of microphones could be strategically placed to obtain the optimum data for researching the propagation effects on the thunder signal and the energy content of lightning.

4.2. ACOUSTIC RECONSTRUCTION OF LIGHTNING CHANNELS

In the section on refraction, we mentioned the utility of ray-tracing computer programs that could accurately calculate the curved path of an acoustic ray from its source to a receiver; the accuracy is limited to the precision with which we are able to define the atmosphere. An obvious application of thunder measurements is to invert this process; one measures thunder then traces it backward from the point of observation along the appropriate ray to its position at the time of the flash. Few[12,14] showed that by performing this reverse ray propagation for many sources in a thunder record, it was possible to reconstruct in three dimensions the lightning channel producing the thunder signal. The sources in this case were defined by dividing the thunder record in short ($\sim$0.5 s) intervals and associating the acoustics in a given time interval with a source on the channel.

Within each time interval the direction of propagation of the acoustic rays are found by cross-correlating the signals recorded by an array of microphones. The position of the peak in the cross-correlation function gives the difference in time of arrival of the wave fronts at the microphones; from this and the geometry of the array, one calculates the direction of propagation. At least three noncolinear microphones are required. Close spacing of the microphones produces higher correlations and shorter intervals; thus more sources; however, the pointing accuracy of a small array is less than that of a large array. Based upon experiences with several array shapes and sizes, a 50-m square has been adopted as the optimum by the Rice University Group (see Few[15]).

We now derive the basic equations that are used by ray-tracing programs; a good general reference is Landau and Lifshitz.[37] Few[12] gives a description of the ray-tracing theory applied to thunder; however, there is an error in his equations. MacGorman[39] provides a succinct statement of both the correlation analysis and the correct ray-tracing equations. In a coordinate system moving with the local flow an acoustic wave has a speed, c, and propagates in a direction, n̂. If the local flow has a velocity $\vec{v}$ then the velocity of propagation of the ray element (the group velocity) is:

$$\vec{v}_p = c\hat{n} + \vec{v} \tag{23}$$

Denoting with α, β, and γ the angles between $\hat{n}$ and the x-, y-, and z-axes we can write the laws of refraction for the wave phase at a boundary in the fluid. Selecting the z-axis perpendicular to a boundary across which c and $\vec{v}$ change c_1 to and $\vec{v}_1$ we find,

$$\frac{c_1 + v_{1x}\cos\alpha_1 + v_{1y}\cos\beta_1}{\cos\alpha_1} = \frac{c + v_x\cos\alpha + v_y\cos\beta}{\cos\alpha} \tag{24}$$

In Equation 24 the flow velocity components along the x- and y-axes are denoted by v_x and v_y. Equation 24 is "Snell's law" applied along the x-axis; there is a similar equation for the y-axis.

$$\frac{c_1 + v_{1x}\cos\alpha_1 + v_{1y}\cos\beta_1}{\cos\beta_1} = \frac{c + v_x\cos\alpha + v_y\cos\beta}{\cos\beta} \tag{25}$$

In Equations 24 and 25 the numerators are the phase speeds of the waves and the terms of Equations 24 and 25 are the trace speeds along the x- and y-axes. From Equations 24 and 25 we also obtain the relation

$$\frac{\cos\alpha_1}{\cos\beta_1} = \frac{\cos\alpha}{\cos\beta} \tag{26}$$

Equations 23 through 26 provide a complete set of equations that allow us to integrate Equation 23; this integral is the ray path. Given $\vec{v}\,(\vec{r}, t)$ and $c(\vec{r}, t)$, Equations 24 and 25 provide $\alpha\,(\vec{r}, t)$ and $\beta\,(\vec{r}, t)$ everywhere for given initial conditions. Noting that $\cos^2\alpha + \cos^2\beta + \cos^2\gamma = 1$, $\alpha\,(\vec{r}, t)$ and $\beta\,(\vec{r}, t)$ provide $\hat{n}\,(\vec{r}, t)$; hence, Equation 23 can be integrated. In most applications one assumes constant atmospheric conditions during the ray propagation and horizontal stratification of the atmosphere. Thus, $\vec{v}(\vec{r}, t) = \vec{v}(z) = v_x(z)\,\hat{i} + v_y(z)\hat{j}$, where $\hat{i}$ and $\hat{j}$ are the unit vectors along the x- and y- axes, and $c\,(\vec{r}, t) = c(z)$. It follows then that $\alpha(z)$, $\beta(z)$, and $\hat{n}(z)$; furthermore, time can be eliminated, if desired, to get $x(z)$ and $y(z)$, where x and y are the coordinate of the ray path as a function of z.

The reconstruction of lightning channels by ray tracing was described by Few,[12,14] Teer,[56,57] Nakano,[44] and MacGorman.[39,40] A discussion of the accuracy and problems of the technique is given in Few and Teer[19] and also in Teer[56,57] and MacGorman.[39,40]

In the work of Teer[56] and Few and Teer[19] acoustically reconstructed channels are found to agree closely with photographs of the channels below the clouds. The point is dramatically made in these comparisons that the visual part of the lightning channel is merely the "tip of the iceberg".

Nakano[44] reconstructed, with only a few points per channel, 14 events from a single storm. Teer[56,57] and Teer and Few[58] reconstructed all of the events during an active period of a thunderstorm cell. MacGorman[39,40] has similarly done a whole storm analysis by acoustic channel reconstruction. Reconstructed lightning channels by ray tracing have been used to support other electrical observations of thunderstorms at Langmuir Laboratory by Weber[64] and Winn et al.[68]

A second technique for reconstructing lightning channels has been developed that is called thunder ranging. This ray-tracing technique has many advantages but suffers from one major disadvantage; the analysis process is very time consuming both in manpower and computer time and it requires special computer hardware and software. As a result there is a long time lag between recording the thunder data in the field and obtaining plots of the lightning channels.

Thunder ranging was developed to provide a quick coarse view of the channels within minutes after the lightning, if necessary. Thunder ranging requires thunder data from at least three non-colinear microphones separated by distances on the order of kilometers. Our experience with cross-correlation analysis of thunder signals has shown us that the signals become spacially incoherent at separations greater than 100 m owing to differences in perspective and propagation path. However, the envelope of the thunder signals and the gross features such as claps remain coherent for distances on the order of kilometers. As discussed earlier these gross features are produced by the large-scale tortuous sections of the lightning channel. Thunder ranging works as follows:

1. The investigator identifies features in the signals (such as claps) that are common to the three thunder signatures on an oscillograph.
2. The time lags between the flash and the arrival of each thunder feature at each measurement point are determined.
3. The ranges to the lightning channel segments producing each thunder feature are computed.
4. The three ranges from the three separated observation points for each thunder feature define three spheres, which should have a unique point in space that is common to all of them.
5. The set of points give the locations of the channel segments producing the thunder features.

Few[16] described thunder ranging and application of the technique to a right triangular arrangement of sensors. MacGorman[39] performed a very detailed analysis of the errors involved in thunder ranging and compared the locations of acoustic sources determined by ray tracing with thunder ranging. Bohannon[3] generalized a thunder ranging theory to allow for arbitrary locations of the sensors and included an error analysis of individual points; this version of thunder ranging has been programmed for interactive use with a computer and was used to reconstruct the lightning channel in Uman et al.[60] The basic criticism we have of the thunder ranging is that the selection of thunder features is the subjective judgment of the researcher; for many features the selection is unambiguous for other features that are close together — they may appear separated at one location and merged at another. The program developed by Bohannon[3] allows us to include these uncertainties in the estimation of errors associated with such points. Most of the recent thunder research has used a combination of ranging and ray tracing.[22a,42a]

The whole-storm studies in which an extended series of channels are reconstructed have proven to be the most valuable use of thunder data to date. They define the volume of the cloud actually producing lightning, the evolution of the lightning producing volume with time, and the relationship of individual channels with other cloud observables such as radar reflectivity and environmental winds.[16,20,21,39–42,44,57,58]

5. ELECTROSTATIC ACOUSTIC EMISSIONS

The concept of electrostatically produced acoustic waves from thunderclouds goes all the way back to the writings of Benjamin Franklin in the 18th century.[24] Wilson[67] provided a rough quantitative estimate of the magnitude of the electrostatically produced pressure wave. McGehee[43] and Dessler[10] developed quantitative models for this phenomenon; McGehee for spherical symmetry and Dessler for spherical, cylindrical, and disk symmetries. The theory developed by Dessler is of particular importance because it made specific predictions regarding the directivity and shape of the wave. The predictions were subsequently verified in part by Bohannon,[3] Bohannon et al.,[5] and Balachandran.[1] Bohannon[4] has extended Dessler's theory to include cloud changes other than the concentrated negative thundercloud charge and concludes that the positive pressure excursion preceding the main negative phase of the wave can be produced by a positive charge below the main negative charge.

We should comment that the experimental search for electrostatic pressure waves has been difficult. The wave is low frequency ($\sim$1 Hz), small amplitude ($\sim$1 Pa), and buried in large background pressure variations produced by wind, turbulence, and thunder. Prior to Dessler's prediction of beaming, one wondered why the signal was not more frequently seen in thunder measurements. Holmes et al.[34] had measured a low frequency component in a few of their thunder power spectra but found these components completely missing in others. Dessler showed that that signal would be beamed for cylindrical and disk geometry; the disk case would require that the detectors be placed directly underneath the charged volume for observation. This relationship has been observed by Bohannon[3] and by Balachandran.[1]

Colgate and McKee[9] theoretically describe an electrostatic pressure pulse using this same mechanism but applied to a volume of charged air surrounding a stepped leader. This particular signature has not been experimentally verified because it has the regular thunder signal, which is 300 times more energetic, superimposed upon it.

The brief theoretical description that follows is similar to the derivation by Dessler.[10] In the presence of an electric field, $\vec{E}$, charged particles (and ions) will drift with the net average velocity such that the imposed electrical force is balanced by an opposite drag force. Dessler shows that following an instantaneous change in $\vec{E}$ a new equilibrium velocity is established within 1 ms for cloud particles. The moving particle (or ion) must exert a force on the atmosphere equal to the drag force that it experiences; hence, the electrical force is ultimately imposed on the air by way of drag forces. If ρ represents the net charge concentration then the force/unit volume is $\rho\vec{E}$, and if we let P_E be the electrical pressure experienced by the volume of air then

$$-\vec{\nabla} P_E = \rho\vec{E} \tag{27}$$

If we let $\rho(\vec{r})$ specify all of the charge considered in the system then $\vec{E}(\vec{r})$ is related to $\rho(\vec{r})$ by

$$\vec{\nabla} \cdot \vec{E} = \rho/\varepsilon_0 \tag{28}$$

Equations 27 and 28 can be combined to yield

$$-\vec{\nabla} P_E = \varepsilon_0 (\vec{\nabla} \cdot \vec{E})\vec{E} \tag{29}$$

For symmetric charge distributions in which ρ and therefore $\vec{E}$ are functions of a single variable, r

$$-\frac{\partial}{\partial r} (P_E) = \frac{\partial}{\partial r} \left(\frac{\varepsilon_0 E^2}{2} \right) + \frac{2n}{r} \left(\frac{\varepsilon_0 E^2}{2} \right) \tag{30}$$

where $n = 0$, 1, or 2, respectively, for plane, cylindrical, and spherical geometries. For the planar case ($n = 0$) we see in Equation 30 that $P_E = -\varepsilon_0 E^2/2$, a familiar expression for electrostatic pressure. Furthermore, if the symmetric charge distribution $\rho(r)$, is assumed to be constant, $\rho(r) = \rho_0$, then Equation 30 becomes

$$-\frac{\partial}{\partial r} (P_E) = (n + 1) \frac{\partial}{\partial r} \left(\frac{\varepsilon_0 E^2}{2} \right)$$

from which we identify

$$P_E = -(n + 1) \frac{\varepsilon_0 E^2}{2} \tag{31}$$

The simple geometries and assumptions incorporated in Equation 31 define the class of theoretical models that have been solved by Dessler[10] and McGehee.[43] For more complex problems associated with realistic cloud models we need to return to Equation 27 to obtain solutions.

The experimental search for electrostatically generated pressure waves has been difficult. The wave is low frequency ($\sim$1 Hz), which requires special microphones, and is buried in large background pressure variations produced by wind, turbulence, and thunder. Prior to Dessler's[10] prediction of beaming, one wondered why the signal was not more frequently seen in thunder measurements. Holmes et al.[34] measured a few low frequency components in their thunder power spectra, but the infrasonic components were missing in many others. Dessler showed that the signal would be beamed for cylindrical and disk geometry; the disk case, which is most likely, requires that the acoustic sensors be placed directly underneath the charged volume participating in the lightning flash in order to be detected. This relationship has been observed by Bohannon et al.[5] and by Balachandran[1,1a]

The electrostatically produced pressure wave predicted by the theory discussed above is a negative pulse. The measured acoustic signals thought to be the verification of the theory invariably exhibit a positive pulse followed by a larger negative pulse.[1,1a,3–5,17b–d] The negative pulse appears to fit the theory, but the theory is deficient in that the positive precursor is not described. Few[17b,c] modified the basic electrostatic theory by including consideration of the diabatic heating of the air in the discharged volume that must occur when positive streamers propagate through the volume. This heating creates a positive pressure change in the same volume of the negative electrostatic pressure. Few further showed that at some distance near the edge of the discharged volume the positive pressure that is produced by the streamer heating must exceed the negative pressure due to electrostatic forces. This is the proper geometric relationship needed to produce an infrasonic wave with a leading positive pressure pulse followed by a larger negative pressure pulse. This modified theory appears to have resolved the questions of the source and characteristics of the infrasonic component of thunder.

6. NONELECTRICAL SOURCES OF INFRASOUND FROM THUNDERSTORMS

In this final section we extensively utilize the two reviews of this topic by T. M. Georges[26,27] "Infrasound from Convective Storms: Examining the Evidence" and "Infrasound from Convective Storms, Part II: A Critique of Source Candidates." The researcher interested in pursuing this topic should read these reviews and the sources referenced therein.

This type of severe storm infrasound has been observed by two different techniques, but as yet it cannot be conclusively proven that the two observations are of the same source of phenomenon. The direct observation is made at the surface with microbarograph arrays; these waves have a typical period between 12 to 60 s. The indirect observations are made of the Doppler motions of electrons in the F layer (200 to 300 km) produced by nearly vertically propagating waves with wave periods in the 120 to 300 s range. The difference in the periods of these two observations can be interpreted as indicating different sources for the waves; however, the evidence shows such a strong association of the waves with similar severe storm conditions that it is argued that either both phenomena are produced by the same source or that a common severe storm process drives the sources that produce both phenomena.

Some of the pertinent common observations of the source of these waves are

1. The waves can always be traced to a severe storm or storm system with radar tops above 40,000 ft. The surface infrasonic waves can be detected at distances of 1500 km under proper conditions; whereas the more vertical ionospheric waves are detected only within 300 km of the source.
2. Not all severe storms produce these waves. There is a subset having special, as yet undefined, characteristics that are responsible for the wave production.

3. The wave-producing storms are almost exclusively observed in the central part of the U.S., even though peak thunderstorm activity is in Florida and along the Gulf Coast.
4. The typical duration for these wave events is 2 h. There is no consistent identifiable characteristic in either the wave signature or the spectra of either type of observation.

Conditions 2, 3, and 4 tend to rule out an electrostatic source, which is responsible for the infrasonic pulses observed in the 0.5- to 10-s range. The height of the storms and the observed correlation with tropopause penetration are indicative of gravity wave generation; the gravity wave period, however, is 15 min (900 s) which is too long to be responsible for the observed waves. Georges,[27] after considering numerous candidate sources, concludes that the most likely source is the mesocyclone rotation that is observed in tornadic storms in the central U.S.

There is ample evidence that tornadic storms form a large vortex in the upper part of the cloud (the mesocyclone) about which smaller vortices are formed and rotate. The large destructive tornadoes that develop downward from this region also carry with them the multiple corotating vortices. This type of system, even before developing a tornado, can produce infrasound, which has characteristics most nearly matching the severe storm infrasound observations.

ACKNOWLEDGMENT

The author's research into the acoustic radiations from lightning has been supported under various grants and contracts from the Meteorology Program, Division of Atmospheric Sciences, National Science Foundation, the Atmospheric Sciences Program of the Office of Naval Research, and the Charles Conly Research Fund of Rice University; their support is gratefully acknowledged.

REFERENCES

1. Balachandran, N. K., Infrasonic signals from thunder, *J. Geophys. Res.*, 84, 1735, 1979.
1a. Balachandran, N. K., Acoustic and electric signals from lightning, *J. Geophys. Res.*, 88, 3879, 1983.
2. Bass, H. E. and Losey, R. E., Effect of atmospheric absorption on the acoustic power spectrum of thunder, *J. Acoust. Soc. Am.*, 57, 822, 1975.
3. Bohannon, J. L., Infrasonic Pulses from Thunderstorms, M.S. thesis, Rice University, Houston, Texas, 1978.
4. Bohannon, J. L., Infrasonic Thunder: Explained, Ph.D. thesis, Rice University, Houston, Texas, 1980.
5. Bohannon, J. L., Few, A. A., and Dessler, A. J., Detection of infrasonic pulses from thunderclouds, *Geophys. Res. Lett.*, 4, 49, 1977.
6. Brode, H. L., Numerical solutions of spherical blast waves, *J. Appl. Phys.*, 26, 766, 1955.
7. Brode, H. L., The Blast Wave in Air Resulting from a High Temperature, High Pressure Sphere of Air, RM-1825-AEC, Rand Corp., Santa Monica, CA, 1956.
8. Brown, E. H. and Clifford, S. F., On the attenuation of sound by turbulence, *J. Acoust. Soc. Am.*, 60, 788, 1976.
9. Colgate, S. A. and McKee, C., Electrostatic sound in clouds and lightning, *J. Geophys. Res.*, 74, 5379, 1969.
10. Dessler, A. J., Infrasonic thunder, *J. Geophys. Res.*, 78, 1889, 1973.
11. Evans, W. H. and Walker, R. L., High speed photographs of lightning at close range, *J. Geophys. Res.*, 68, 4455, 1963.
12. Few, A. A., Thunder, Ph.D. thesis, Rice University, Houston, Texas, 1968.
13. Few, A. A., Power spectrum of thunder, *J. Geophys. Res.*, 74, 6926, 1969.
14. Few, A. A., Lightning channel reconstruction from thunder measurements, *J. Geophys. Res.*, 75, 7517, 1970.
15. Few, A. A., Thunder signatures, *EOS Trans. A.G.U.*, 55, 508, 1974.
16. Few, A. A., Lightning sources in severe thunderstorms, Preprint Volume, *Conference on Cloud Physics*, American Meteorological Society, Boston, MA, 1974, 387.
17. Few, A. A., Thunder, *Sci. Am.*, 233 (1), 80, 1975.
17a. Few, A. A., Acoustic radiations from lightning, in *Handbook of Atmospherics, Vol. II*, Volland, H., Ed., CRC Press, Boca Raton, FL, 1982, 257.
17b. Few, A. A., Lightning-associated infrasonic acoustic sources, in *Preprints: VII International Conference of Atmospheric Electricity*, American Meteorological Society, Boston, MA, 1984, 484.

17c. Few, A. A., The production of lightning-associated infrasonic acoustic sources in thunderclouds, *J. Geophys. Res.*, 90, 6175, 1985.

17d. Few, A. A., Acoustic radiations from lightning, in *The Earth's Electrical Environment*, National Academy Press, Washington, D.C., 1986, 46.

18. Few, A. A., Garrett, H. B., Uman, M. A., and Salanave, L. E., Comments on letter by W. W. Troutman, "Numerical calculation of the pressure pulse from a lightning stroke," *J. Geophys. Res.*, 75, 4192, 1970.

19. Few, A. A. and Teer, T. L., The accuracy of acoustic reconstructions of lightning channels, *J. Geophys. Res.*, 79, 5007, 1974.

20. Few, A. A., Teer, T. L., and MacGorman, D. R., Advances in a decade of thunder research, in *Electrical Processes in Atmospheres*, Dolezelak, H. and Reiter, R., Eds., Steinkopff Verlag, Darmstadt, 1977, 628.

21. Few, A. A., MacGorman, D. R., and Bohannon, J. L., Thundercloud Charge Distributions, Inferences from the Intracloud Structure of Lightning Channels, Conf. Cloud Phys. Atmos. Electr., American Meteorological Society, Boston, MA, 1978, 591.

22. Few, A. A., Weber, M. E., and Christian, H. J., Vector Electric Field Measurements inside Thunderstorms — Vertical Profiles with Electric Field Sensors, Preprint Volume — Conf. Cloud Phys. Atmos. Electr., American Meteorological Society, Boston, MA, 1978.

22a. Few, A. A., and other TRIP 1979 participants, TRIP 1979 case study of August 7 thunderstorm over Langmuir Laboratory, in *Proc. Atmos. Electr.*, Ruhnke, L. and Latham, J., Eds., A Deepak Publishing, Hampton, VA, 1983, 301.

23. Fleagle, R. G., The audibility of thunder, *J. Acoust. Soc. Am.*, 21, 411, 1949.

24. Franklin, B., Experiments and observations on electricity, in *Benjamin Franklin's Experiments*, Cohen, I. B., Ed., Harvard University Press, Cambridge, MA, 1941.

25. Gautschi, W. and Cahill, W. F., Exponential integral and related functions, in Handbook of Mathematical Functions, Abramowitz, M. and Stegun, I. A., Eds., National Bureau of Standards, Appl. Math. Series 55, U.S. Government Printing Office, Washington, D.C., 1964.

26. Georges, T. M., Infrasound from convective storms: examining the evidence, *Rev. Geophys. Space Phys.*, 11, 571, 1973.

27. Georges, T. M., Infrasound from Convective Storms. II. A Critique of Some Candidates, NOAA Technical Report ERL 380-WPL 49, U.S. Department of Commerce, 1976.

28. Harris, C. M., Absorption of Sound in Air Versus Humidity and Temperature, NASA-CR-647, Columbia University, NY, 1967.

29. Hill, R. D., Analysis of irregular paths of lightning channels, *J. Geophys. Res.*, 73, 1897, 1968.

30. Hill, R. D., Channel heating in return-stroke lightning, *J. Geophys. Res.*, 76, 637, 1971.

31. Hill, R. D., Energy dissipation in lightning, *J. Geophys. Res.*, 82, 4967, 1977.

32. Hill, R. D., Thunder, in *Lightning*, Golde, R. H., Ed., Academic Press, New York, 1977, 385.

33. Hill, R. D., A survey of lightning energy estimates, *Rev. Geophys. Space Phys.*, 17, 155, 1979.

34. Holmes, C. R., Brook, M., Krehbiel, P., and McCrory, R. A., On the power spectrum and mechanism of thunder, *J. Geophys. Res.*, 76, 2106, 1971.

35. Kinsler, L. E. and Frey, A. R., *Fundamentals of Acoustics*, 2nd ed., John Wiley & Sons, New York, 1962.

36. Krider, E. P., Dawson, G. A., and Uman, M. A., Peak power and energy dissipation in a single-stroke lightning flash, *J. Geophys. Res.*, 73, 3335, 1968.

37. Landau, L. D. and Lifshitz, E. M., *Fluid Mechanics*, Pergamon Press, London, 1959.

38. Lin, S. C., Cylindrical shock waves produced by an instantaneous energy release, *J. Appl. Phys.*, 25, 54, 1954.

39. MacGorman, D. L., Lightning Location in a Colorado Thunderstorm, M.S. thesis, Rice University, Houston, TX, 1977.

40. MacGorman, D. L., Lightning Location in a Storm with Strong Wind Shear, Ph.D. thesis, Rice University, Houston, TX, 1978.

41. MacGorman, D. R. and Few, A. A., Correlations between radar reflectivity contours and lightning channels for a Colorado storm on 25 July 1972, *Conf. Cloud Phys. Atmos. Electr.*, American Meteorological Society, Boston, MA, 1978, 597.

42. MacGorman, D. R., Few, A. A., and Teer, T. L., Layered lightning activity, *J. Geophys. Res.*, in press.

42a. MacGorman, D. R., Taylor, W. L., and Few., A. A., Lightning location from acoustic and VHF techniques relative to storm structure from 10-cm radar, in *Proc. Atmos. Electr.*, Ruhnke, L. and Latham, J., A Deepak Publishing, Hampton, VA, 1983, 301.

43. McGehee, R. M., The influence of thunderstorm space charges on pressure, *J. Geophys. Res.*, 69, 1033, 1964.

44. Nakano, Minoru, Lightning channel determined by thunder, *Proc. Res. Inst. of Atmospherica, Nagoya University*, 20, 1, 1973.

45. Orville, R. E., A high-speed time-resolved spectroscopic study of the lightning return stroke, I, II, III, *J. Atmos. Sci.*, 25, 827, 1968.

46. Otterman, J., Finite-amplitude propagation effect on shock-wave travel times from explosions at high altitudes, *J. Acoust. Soc. Am.*, 31, 470, 1959.

47. Page, N. W. and McKelvie, P. I., Shock Waves Generated by Spark Discharge, 6th Australasian Hydraulics and Fluid Mechanics Conf., Adelaide, Australia, December 1977, 221.

48. Plooster, M. N., Shock Waves from Line Sources, NCAR-TN-37, National Center for Atmospheric Research, Boulder, CO, 1968.

49. Plooster, M. N., Numerical simulation of spark discharges in air, *Phys. Fluids,* 14, 2111, 1971.

50. Plooster, M. N., Numerical model of the return stroke of the lightning discharge, *Phys. Fluids,* 14, 2124, 1971.

50a. Plooster, M. N., personal communication, 1970.

51. Rayleigh, J. W. S., *The Theory of Sound,* Vol. 2, Dover Publications, New York, 1945.

52. Ribner, H. S., Lam, F., Leung, K. A., Kurtz, D., and Ellis, N. D., Computer Model of the Lightning-Thunder Process with Audible Demonstration, AIAA paper 75–548, AIAA 2nd Aero-Acoustics Conference, American Institute of Aeronautics and Astronautics, New York, 1975.

53. Sakurai, A., On the propagation and structure of the blast wave. II, *J. Phys. Soc. Jpn.,* 9, 256, 1954.

54. Sakurai, A., On exact solution of the blast wave problem, *J. Phys. Soc. Jpn.,* 10, 827, 1955.

55. Taylor, G. I., The formation of a blast wave by a very intense explosion. II. The atomic explosion of 1945, *Proc. R. Soc. London A,* 201, 159, 1950.

56. Teer, T. L., Acoustic Profiling — A Technique for Lightning Channel Reconstruction, M.S. thesis, Rice University, Houston, TX, 1972.

57. Teer, T. L., Lightning Channel Structure inside an Arizona Thunderstorm, Ph.D. thesis, Rice University, Houston, TX, 1973.

58. Teer, T. L. and Few, A. A., Horizontal lightning, *J. Geophys. Res.,* 79, 3436, 1974.

59. Uman, M. A., *The Lightning Discharge,* Academic Press, Orlando, FL, 1987.

60. Uman, M. A., Beasley, W. H., Tiller, J. A., Lin, Y. T., Krider, E. P., Weidmann, C. D., Krehbiel, P. R., Brook, M., Few, A. A., Bohannon, J. L., Lennon, C. L., Poehler, H. A., Jafferis, W., Gulick, J. R., and Nicholson, J. R., An unusual lightning flash at Kennedy Space Center, *Science,* 201, 9, 1978.

61. Uman, M. A., Cookson, A. H., and Moreland, J. B., Shock wave from a four-meter spark, *J. Appl. Phys.,* 41, 3148, 1970.

62. Uman, M. A., McLain, D. K., and Myers, F., Sound from Line Sources with Applications to Thunder, Res. Rep. 68-9E4-HIVOL-R1, Westinghouse Research Laboratories, Pittsburgh, PA, 1968.

63. Uman, M. A. and Voshall, R. E., Time interval between lightning strokes and the initiation of dart leaders, *J. Geophys. Res.,* 73, 497, 1968.

64. Weber, M. E., Thundercloud Electric Field Soundings with Instrumented Free Balloons, Ph.D. thesis, Rice University, Houston, TX, 1980.

65. Weidman, C. D. and Krider, E. P., The fine structure of lightning return stroke wave forms, *J. Geophys. Res.,* 83, 6239, 1978.

66. Whitham, G. B., On the propagation of weak shock waves, *J. Fluid Mech.,* 1, 290, 1956.

67. Wilson, C. T. R., Investigations on lightning discharges and on the electric field of thunderstorms, *Phil. Trans. R. Soc. London Ser. A,* 221, 73, 1920.

68. Winn, W. P., Moore, C. B., Holmes, C. R., and Byerly, L. G., Thunderstorm on July 16, 1975, over Langmuir Laboratory: a case study, *J. Geophys. Res.,* 83, 3079, 1978.

69. Wright, W. M. and Medendorp, N. W., Acoustic radiation from a finite line source with N-wave excitation, *J. Acoust. Soc. Am.,* 43, 966, 1967.

70. Brook, M. et al., personal communication, 1969.

Chapter 2

Lightning Protection

J. Wiesinger and W. Zischank

CONTENTS

1. INTRODUCTION

1.1. HISTORY

Up to the 17th century the lightning phenomenon as well as lightning protection measures were dominated by mythological imaginations. The physicist and engineer Otto von Guericke (1602–1686), who built an electrostatic machine by a hand-driven sphere of sulfur in 1670 in Magdeburg, Germany, declared for the first time the analogy of electrostatic discharge in the laboratory and lightning discharge.

Benjamin Franklin (1706–1790), the famous statesman, author, and scientist, proposed in a letter of July 29, 1750, to Peter Collinson at the Royal Society London an experiment by which the hypothesis of the electrical nature of lightning should be proved: an insulated standing person should be charged under a thundercloud by help of a metal rod and then discharged by a spark. A modified experiment was carried out near Paris, France, by Francois Dalibard on May 10, 1752: an approximately 12-m high iron rod, insulated against earth by bottles of glass, was installed; during a thunderstorm an assistant of Dalibard, Coffier, standing on earth, produced sparks of some centimeters in length to the base of the rod. By this successful experiment, showing evidently the analogy of a spark from an object electrostatically charged in the laboratory and a spark from an object charged by a thunderstorm, the electrical nature of lightning was accepted in the scientific world as proved.

On the basis of this physical background, for the first time specific measures against lightning strikes were applicable to protect persons, livestock, and houses. In 1760 at the house of the businessman West in Philadelphia, very likely the first Franklin rod for the interception of lightning strikes was installed. In Germany, the first lightning protection system was built at the St. Jacobi church in Hamburg.

As an example and a substitute for a series of instructions dealing with the erection of lightning protection systems in the second part of the 18th century, a proposal from 1778, very likely of the philosopher and experimental physicist G. Ch. Lichtenberg, is shown (Ettinger, 1778). The proposed lightning protection system for a house already contains air terminals in the form of rods and wires, down conductors, and earth terminals made of iron or copper (Figure 2.1.1). These are the main components for today's lighting protection system for a common structure.

Figure 2.1.1   Proposal of a lightning protection system for houses, dated 1778.

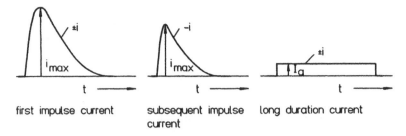

first impulse current   subsequent impulse   long duration current
current

Figure 2.2.1 Components of lightning currents of flashes to earth.

It is amazing to see how in a very short time by the method of trial and error a proper and effective dimensioning of a protection system for houses has been reached.

1.2. RECENT ACTIVITY

The recent lighting protection activity with worldwide validity is dominated by Technical Committee 81 of the International Electrotechnical Commission (IEC) where standards mainly for the protection of structures against lightning are elaborated. In 1992, the general principles of the protection for common structures by lighting protection systems (lps) were published (IEC 1024–1, 1992) and some subsequent standards were prepared. In the following sections these international and draft standards are preferably considered. A list of symbols and abbreviations used throughout this chapter can be found in the appendix.

2. PARAMETERS OF LIGHTNING CURRENTS AND LIGHTNING ELECTROMAGNETIC FIELDS

2.1. CHARACTERISTICS OF LIGHTNING CURRENTS

For protection purposes of ground-based structures three typical components of the lightning currents in a flash are to be considered, which may not occur simultaneously (Figure 2.2.1): (1) a first impulse current of a positive or negative flash; (2) a subsequent impulse current of a negative flash; and (3) a long duration (continuing) current of a positive or negative flash. These currents are to be considered as impressed currents into the object struck.

The first and subsequent impulse current can be described (IEC 81, 1992) by:

$$i = (i_{max}/\eta) \cdot \{(t/\tau_1)^{10}/[1 + (t/\tau_1)^{10}]\}\exp(-t/\tau_2) \text{ in A for } t \geq 0 \qquad (2.1)$$

The long duration current can be described by:

$$i = I_a \text{ in A for } 0 \leq t \leq T_l \qquad (2.2)$$

The parameters of Equations 2.1 and 2.2 are given in Table 2.2.1. In addition the front time T_1 and the time to the half value T_2 of the impulse currents are shown (for definitions see Figure 2.2.2).

In IEC 1024-1, 1992, for the protection of a structure, protection levels (pl) I to IV are defined, where pl I covers about 99% of all flashes to earth and pl III covers about 95%.

In Table 2.2.2 the maximum values according to the individual pl of the first and subsequent impulse currents, i_{max}, and of the long duration current, I_a, are given. In addition the main parameters of the three lightning current components relevant to the individual pl are shown (Hasse et al., 1992):

Table 2.2.1   **Parameters of Equations 2.1 and 2.2**

	First impulse current	Subsequent impulse current	Long duration current
η (1)	0.93	0.993	—
τ_1 (s)	$19 \cdot 10^{-6}$	$0.454 \cdot 10^{-6}$	—
τ_2 (s)	$485 \cdot 10^{-6}$	$143 \cdot 10^{-6}$	—
T_d (s)	—	—	0.5
T_1 (s)	$10 \cdot 10^{-6}$	$0.25 \cdot 10^{-6}$	—
T_2 (s)	$350 \cdot 10^{-6}$	$100 \cdot 10^{-6}$	—

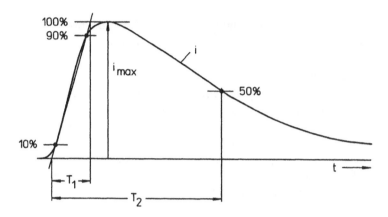

Figure 2.2.2   Definitions of T_1 and T_2.

Table 2.2.2   **Parameters of Impulse and Long Duration Currents**

Protection level	First impulse current			Subsequent impulse current			Long duration current		
	I	II	III–IV	I	II	III–IV	I	II	III–IV
i_{max} (A)	$200 \cdot 10^3$	$150 \cdot 10^3$	$100 \cdot 10^3$	$50 \cdot 10^3$	$37.5 \cdot 10^3$	$25 \cdot 10^3$	—	—	—
I_a (A)	—	—	—	—	—	—	400	300	200
Q_i (C)	100	75	50	—	—	—	—	—	—
W/R (J/Ω)	$10 \cdot 10^6$	$5.6 \cdot 10^6$	$2.5 \cdot 10^6$	—	—	—	—	—	—
i_{max}/T_1 (A/s)	—	—	—	$200 \cdot 10^9$	$150 \cdot 10^9$	$100 \cdot 10^9$	—	—	—

- Q_i charge $\int i \cdot dt$ of the first impulse current (representing nearly the total charge of the impulse components)
- Q_l charge $I_a \cdot T_l$ of the long duration current
- W/R specific energy $\int i^2 \cdot dt$ of the first impulse current (representing nearly the total specific energy of the impulse components)
- i_{max}/T_1 average front steepness of the subsequent impulse current

2.1.1. Effect of the Maximum Value of the Current

The maximum value of the first impulse current, i_{max}, is responsible for the maximum value of the voltage between a struck object, having an earthing resistance R_e and distant earth:

$$u_{e/max} = R_e \cdot i_{max} \tag{2.3}$$

Table 2.2.3 **Parameters of Equations 2.5 and 2.6**

	Aluminum	Iron, steel	Copper
γ (kg/m³)	2700	7700	8920
c_w [J/(kg · k)]	908	469	385
c_s (J/kg)	$397 \cdot 10^3$	$272 \cdot 10^3$	$209 \cdot 10^3$
θ_s (°C)	658	1530	1080
ρ_o (Ωm) at 20°C	$29 \cdot 10^{-9}$	$120 \cdot 10^{-9}$	$17.8 \cdot 10^{-9}$
α (1/K)	$4 \cdot 10^{-3}$	$6.5 \cdot 10^{-3}$	$3.92 \cdot 10^{-3}$

The value of R_e depends on the configuration of the earthing system and is proportional to the soil resistivity ρ_e. The DC value of R_e may be diminished during a lightning strike by arc discharges in the earth starting from the earthing system; on the other hand, R_e of an earthing system, built up by long horizontal wires, may be increased due to a reduced effective length in case of transient lightning currents. In practice, values of R_e for structures to be protected are in the range of 0.1 to 100 Ω corresponding to values of $u_{e/max}$ up to more than 10 MV.

2.1.2. Effect of the Charge of the Current

The charge Q of the first impulse current and the long duration current, respectively, is responsible for melting effects to metals at the point of strike. The energy arising at the transition from the flash arc to the metal, where the anode, or the cathode voltage drop $U_{A,K}$ exists, is given by:

$$W = U_{A,K} \cdot Q \text{ in J} \tag{2.4}$$

A part of this energy, $k \cdot W$, is transferred to the temperature rise up to the melting temperature and the melting energy. Thus, the melted volume of the metal can be calculated:

$$V = k \cdot U_{A,K} \cdot Q/\{\gamma[c_w(\theta_m - \theta_o) + c_m]\} \text{ in m}^3 \tag{2.5}$$

The parameters of Equation 2.5 are shown in Table 2.2.3. Experiments have shown that $U_{A,K}$ can be taken as 30 V and the factor k as 0.7 (Kern, 1990).

Furthermore, the charge of the long duration current, Q_l, has proven to be responsible for the puncturing of metal sheets and tank walls and can therefore be taken as a relevant charge for calculations and experiments dealing with melting effects. (The charge of the first impulse current, Q_i, usually causes melting on metal surfaces large in diameter, but not deep, thus, not forming holes.)

The thickness s of metal sheets or walls, necessary to avoid puncturing even in the case of protection level I is: 4 mm for aluminum and 3 mm for iron, steel, and copper (Kern, 1990).

2.1.3. Effect of the Specific Energy of the Current

The specific energy of the first impulse current is responsible for heating effects to metal conductors (i.e., down conductors of an lps) and for electrodynamic forces to those conductors. Theory and experiments have shown that the final temperature of a conductor at the end of an impulse current can be calculated by assuming an adiabatic energy transfer, considering the temperature-dependent resistivity and neglecting skin effects (Steinbigler, 1977):

$$\theta_f = \theta_o + (1/\alpha) \cdot \{\exp[(W/R) \cdot \alpha \cdot \rho/(q^2 \cdot \gamma \cdot c_w)] - 1\} \text{ in °C} \tag{2.6}$$

The parameters of Equation 2.6 are given in Table 2.2.3.

Table 2.2.4 **Final Temperatures in °C of Conductors**

Protection level	Aluminum			Iron, steel			Copper		
	I	**II**	**III–IV**	**I**	**II**	**III–IV**	**I**	**II**	**III–IV**
W/R (J/Ω)	$2.5 \cdot 10^6$	$5.6 \cdot 10^6$	$10 \cdot 10^6$	$2.5 \cdot 10^6$	$5.6 \cdot 10^6$	$10 \cdot 10^6$	$2.5 \cdot 10^6$	$5.6 \cdot 10^6$	$10 \cdot 10^6$
10 mm²	564	a	a	a	a	a	169	542	a
16 mm²	146	454	a	1120	a	a	56	143	309
25 mm²	52	132	283	211	913	a	22	51	98
50 mm²	12	28	52	37	96	211	5	12	22
100 mm²	3	7	12	9	20	37	1	3	5

[a] Melting and evaporating, respectively.

Table 2.2.4 shows the final temperatures, θ_f, of conductors with standardized cross sections, if a lightning current according to pl I to III to IV flows through them. It is evident that the total lightning current even for pl I can be conducted by a wire of 16 mm² copper, 25 mm² aluminum, and 50 mm² iron, or steel. These are the minimum dimensions given by IEC (IEC 1024-1, 1992) for down conductors of an lps.

The electrodynamic impulse to two parallel conductors with the distance, s, related to the length is given by:

$$(\smallint F \cdot dt)' = 10^{-6} \cdot (W/R)/(20 \cdot d) \text{ in Ns/m} \qquad (2.7)$$

where W/R is the specific energy of the current through both conductors.

Normally, the electrodynamic forces do not dominate the sizing of elements in an lps.

2.1.4. Effect of the Current Steepness

The average current steepness during the rise of a subsequent impulse current, i_{max}/T_1, is responsible for magnetic induction effects, i.e., for induced voltages and resulting currents in any kind of loops in the neighborhood of conductors of the lps. The coupling mechanism between those conductors and the loops considered (e.g., built by electric installations inside the structure to be protected) is described for lightning protection purposes by the mutual inductance M (Hasse et al., 1992).

For the induced voltage in a loop near a conductor with the partial lightning current i_c:

$$u_1 = M \cdot di_c/dt \text{ in V}$$

or

$$U_1 = M \cdot i_{c/max}/T_1 \text{ in V for } 0 \le t \le T_1 \qquad (2.8)$$

For the resulting current i_1 in a short circuit loop with the self-inductance L and neglected resistivity, corresponding to a current transformer:

$$i_1 = i_c \cdot M/L \text{ in A}$$

or

$$i_{1/max} = i_{c/max} \cdot M/L \text{ in A} \qquad (2.9)$$

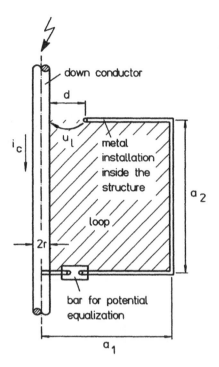

Figure 2.2.3 Arrangement of a loop bonded to a down conductor.

For a typical arrangement, where a metal installation inside a structure approaches a down conductor as given in Figure 2.2.3:

$$M = 0.2 \cdot 10^{-6} \cdot a_2 \cdot \ln(a_1/r) \text{ in H}$$

where $a \gg r$.

As a consequence the minimum distance d (between the installation and the down conductor of the lps with the partial lightning current i_c) necessary to avoid dangerous flashover can be calculated. Experiments have shown that the electric breakdown field strength E_b is about $3.0 \cdot 10^6$ V/m in air and about $1.5 \cdot 10^6$ V/m in solid materials like walls of stone or concrete, respectively. The values of E_b are valid for the time T_1 of 0.25 μs (see Table 2.2.1) for which U_1 exists.

For the minimum distance:

$$d = U_l/E_b = (M \cdot i_{c/max}/T_1)E_b \text{ in m} \tag{2.11}$$

Equation 2.11, based on a simplified quasistationary presumption, became the basis for the so-called safety distances in order to avoid dangerous sparking, which was demanded in the IEC standard (IEC 1024-1, 1992).

For a further typical arrangement where an installation loop is separated from a down conductor, as shown in Figure 2.2.4:

$$M = 0.2 \cdot 10^{-6} \cdot a_2 \cdot \ln[(a_1 + a_3)/a_3] \text{ in H} \tag{2.12}$$

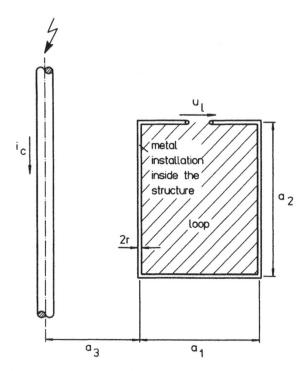

Figure 2.2.4 Arrangement of a loop isolated from a down conductor.

In order to calculate the maximum current in the loop, $i_{l/max}$, according to Equation 2.9, the self-inductance L must be known as:

$$L = 0.4 \cdot 10^{-6}\{2\sqrt{a_1^2 + a_2^2} - 2(a_1 + a_2) + a_1 \cdot \ln[2a_2/(r + r\sqrt{1 + (a_2/a_1)^2})] $$
$$+ a_2 \cdot \ln[2a_1/(r + r\sqrt{1 + (a_1/a_2)^2})]\} \text{ in H} \quad (2.13)$$

where $a \gg r$.

2.2 CHARACTERISTICS OF LIGHTNING ELECTROMAGNETIC FIELDS

The lightning electromagnetic field is characterized in IEC (IEC 81, 1992) as LEMP (lightning electromagnetic impulse). Most important for induction effects is the magnetic field, lightning magnetic impulse (LMP), caused by the lightning channel and the conductors of the lps, in which the lightning current or a part of it flows. In the near field region, up to distances in the range of 100 m, relevant for protection tasks, the magnetic field H(t) is proportional to the lightning current.

Figure 2.2.5 shows the amplitude density spectrum of the lightning current (IEC 81, 1992) from which the proportional spectrum of the LMP can be deduced.

As a simple example, for H(t) in a distance s from an infinitely long, straight conductor, in which the lightning current i flows:

$$H(t) = i/(2\pi s) \text{ in A/m} \quad (2.14)$$

The electric field, lightning electric impulse (LEP) is less important for protection tasks, except for influences to exposed electric antennas. The maximum values in the near field region reach up to the range of 1 MV/m, while the maximum steepness is up to the range of 1 (MV/m)/μs.

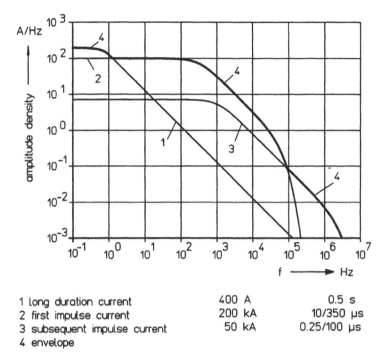

1 long duration current   400 A   0.5 s
2 first impulse current   200 kA   10/350 µs
3 subsequent impulse current   50 kA   0.25/100 µs
4 envelope

Figure 2.2.5   Amplitude density spectra of lightning currents.

3. "ROLLING SPHERE" METHOD

The method of the "rolling sphere" is applied to identify volumes protected against direct lightning strikes, especially the volumes protected by an air termination system of an lps.

In accordance with observations, it is postulated that the downward leader of a cloud-to-ground flash approaches the Earth or earthed objects to a certain critical distance, where an upward leader starts from the earth and connects the downward leader with the earth (Figure 2.3.1). This connection mechanism is described as "final jump". After connection the charge, stored in the downward leader around the thermoionized core in a cylindrical volume with a diameter in the range of 100 m, is carried to Earth causing an impulse current to the struck object. This discharging process propagates along the downward leader with a nearly constant velocity v of about 1/2 to 1/3 of the velocity of light and is described as "return stroke" (Uman, 1988).

It is assumed that the current i at the point of strike is proportional to $Q' \cdot v$, where Q' is the length-related charge of the downward leader. From this:

$$i_{max} \sim Q' \tag{3.1}$$

Furthermore, it is assumed that the electric field strength E_r at the point of strike at that moment when the critical distance r is reached and the upward leader starts, is proportional to Q'/r.

From this simplified assumption it follows that r is proportional to Q' and therefore to i_{max} (Equation 3.1)

$$r = k \cdot i_{max} \text{ in m} \tag{3.2}$$

Whereas Equation 3.2 results from very simplified assumptions, the Working Group 33 of the International Conference on Large High Voltage Electric Systems (CIGRE) has developed an

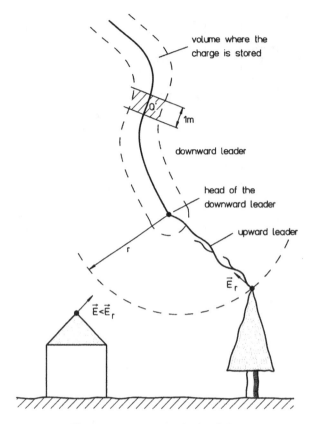

Figure 2.3.1 Connecting leader phase.

extended formula based on many years of worldwide observations on overhead power lines (Darveniza et al., 1975)

$$r = k_1 \cdot i_{max} + k_2 + k_3 \cdot \exp(k_4 \cdot i_{max}) \text{ in m} \tag{3.3}$$

The constants were found as: $k_1 = +2 \cdot 10^{-3}$ m/A; $k_2 = +30$ m; $k_3 = -30$ m; and $k_4 = -0.15 \cdot 10^{-3}$ 1/A.

From the above shown hypothesis it follows that

1. The head of the downward leader cannot come closer to an earthed object than the critical distance r.
2. If the critical distance r is reached for an earthed object, its upward leader will contact the downward leader thus fixing the striking point.
3. Other earthed objects in the vicinity not reaching the critical distance r are not struck due to reduced field strength (see Figure 2.3.1) and the missing or late-starting upward leader.

The downward leader is considered to approach the Earth "blindly" and its head can take any possible position around the earthed objects. Therefore the head of the leader can be thought to build the center of a sphere with a radius r.

If this imaged sphere is rolled around in any possible position on the Earth and its objects, every point which is touched by the sphere is identified as a possible point of strike, whereas volumes not being touched are excluded from strikes. In practice this evaluation can be done for a modeled structure by a real, adequately sized sphere (see Figure 2.3.2).

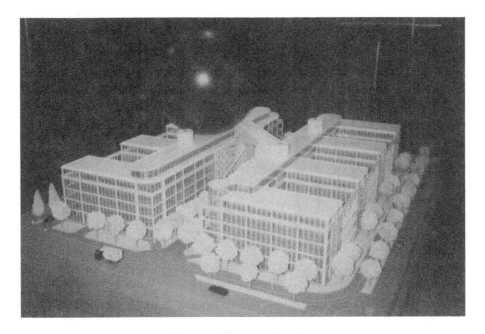

Figure 2.3.2 The rolling sphere method applied to a model of an office building.

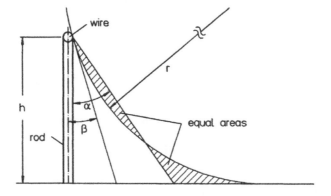

Figure 2.3.3 Protection angles derived from the rolling sphere method.

In order to find the worst case conditions for the evaluation of the protected volume the smallest possible values of i_{max} corresponding to the smallest possible values of r have to be looked at.

In IEC (IEC 1024-1, 1992) the values of r for several pls have been fixed where for pl I more than 99% of the cloud-to-ground flashes show a critical distance greater than the fixed value of r; the corresponding percentages are 95% for pl II, 90% for pl III, and 80% for pl IV

- r = 20 m for pl I
- r = 30 m for pl II
- r = 45 m for pl III
- r = 60 m for pl IV

For simple air terminations like rods or single horizontal wires from the universal rolling sphere method, so-called protection angles can be derived. According to Figure 2.3.3 the volume protected against direct lightning strikes can be described by a very conservative tangential angle β or by an averaged angle α (according to IEC 1024-1, 1992).

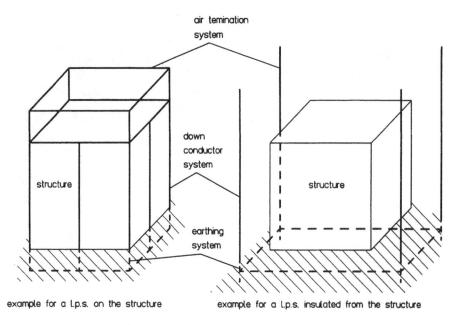

Figure 2.4.1 Lightning protection system (lps).

For a vertical rod or a horizontal wire with the height, h, where h is smaller than the radius, r, of the rolling sphere,

$$\beta = \arcsin(1 - h/r) \text{ in grd} \tag{3.4}$$

$$\alpha = \arctan\{\sqrt{2rh - h^2}/h + r \cdot \sqrt{2rh - h^2}/h^2 - (r/h)^2 \cdot \arccos[(r-h)/h]\} \text{ in grd} \tag{3.5}$$

4. PROTECTION OF COMMON STRUCTURES

An lps for a structure to protect a building and persons or livestock inside, according to Figure 2.4.1 consists of:

1. An air termination system of metal components which can be directly struck by lightning
2. A down conductor system of metal components which can conduct the lightning current from the air termination system to the earthing system
3. An earthing system of metal components which can conduct the lightning current into the Earth without dangerous sparking to neighboring buried services

The lps can be installed directly on the structure or be insulated from it.

4.1. AIR TERMINATION SYSTEM

The air termination system has to fix the points of strikes and conduct the lightning currents to the down conductor system. It can optimally be positioned by the universal rolling sphere method where the rolling sphere may not touch any part of the structure to be protected. In order to avoid unacceptable melting at the point of strike and unacceptable heating while conducting the current, minimum dimensions are fixed by the IEC (IEC 1024-1, 1992) independently from the individual pl, as shown in Table 2.4.1.

An air termination system can be built up by rods, wires, or their combination (see Figure 2.4.2). Also, natural metal components of the structure, e.g., metal roofs, masts, or attics, can be used as the single or supplementary air termination components.

Table 2.4.1 **Minimum Dimensions of the Metal Elements of the Air Termination System and the Down Conductor System**

		Aluminum	Iron/steel	Copper
Air termination system	Cross section in mm² of conductors	70	50	30
	Thickness in mm of sheets or pipes if melting through is to be excluded	7	4	5
Down conductor system	Cross section in mm² of conductors	25	50	16

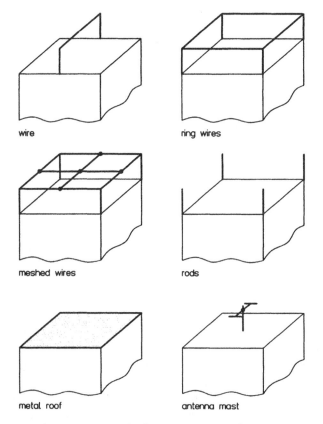

Figure 2.4.2 Examples for air termination configurations.

An air termination system can be optionally installed insulated from the structure (see Figure 2.4.1); normally its earthing system is interconnected with the earthing system of the structure, if at all.

4.2. DOWN CONDUCTOR SYSTEM
The down conductor system is the link between the air termination system and the earthing system. If has to conduct the lightning currents without unacceptable heating of the conductors. For minimum dimensions, independent of the individual pl, see Table 2.4.1. The maximum distances between the preferably vertical down conductors are 10 m for pl I, 15 m for pl II, 20 m for pl III, and 25 m for pl IV (IEC 1024-1, 1992).

The down conductors must be interconnected by horizontal ring conductors every 20 m of height of the structure. Also, natural metal components of the structure, e.g., metal facade sheets

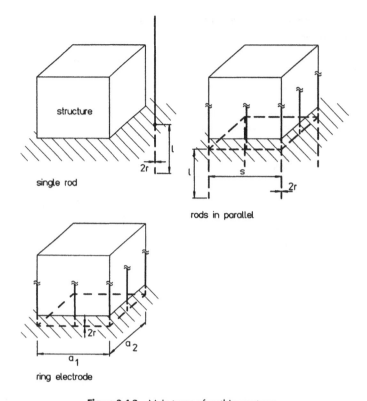

Figure 2.4.3 Main types of earthing systems.

or definitely interconnected metal reinforcement in concrete, can be used as the single or supplementary down conductor component.

4.3. EARTHING SYSTEM

The earthing system has to conduct the lightning current into the soil by avoiding undesirable arc discharges to neighboring buried installations. The earthing resistance R_e depends on the configuration of the earthing system and is proportional to the soil resistivity ρ_e which varies from approximately 10 to 1000 Ωm and should therefore be measured in individual cases.

The main types of earthing systems are (Figure 2.4.3) (1) a single vertical rod or rods connected in parallel and (2) a ring Earth electrode which may be laid into the concrete foundation.

For a vertically buried rod electrode with its length l and its radius r (Hasse et al., 1992)

$$R_e \approx [\rho_e/2\pi l)] \cdot \ln(l/r) \text{ in } \Omega \tag{4.1}$$

For n parallel rod electrodes with the distance s between neighboring rods (Hasse et al., 1992)

$$R_e \approx (1/n)[\rho_e/(2\pi l)] \cdot \ln(l/r) \text{ in } \Omega \tag{4.2}$$

for $s \geq n \cdot l$.

For a ring electrode in a rectangular configuration with the lengths a_1 and a_2 and the radius r (Hasse et al., 1992)

$$R_e \approx \rho_E \cdot \{\ln[2(a_1 + a_2)/r]\}/[2\pi(a_1 + a_2)] \text{ in } \Omega \tag{4.3}$$

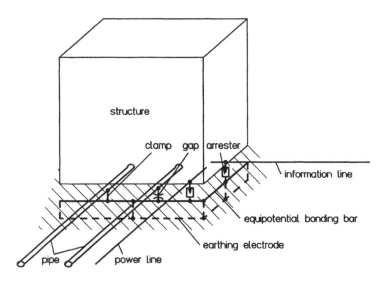

Figure 2.4.4 Bonding of external services near ground level.

The knowledge of R_e is important in determining the safety distance d between the earthing system and neighboring buried metal services, e.g., pipes or cables, insulated from the earthing system

$$d \geq i_{max} \cdot R_e/E_b \text{ in m} \tag{4.4}$$

where E_b is the average breakdown field strength of the soil being approximately $500 \cdot 10^3$ V/m.

4.4. POTENTIAL EQUALIZATION FOR LIGHTNING PROTECTION REASONS

If a structure is struck by lightning a potential difference to the far surroundings arises in the order of 100 kV to 1 MV. This voltage would arise between the structure and the entering external metal services earthed far away. It cannot be controlled by insulation measures, therefore, potential equalization for these services must be performed to the lps at the point of entrance (Figure 2.4.4). The services are bonded to the lps near ground level by means of an equipotential bonding bar. Thus partial lightning currents are accepted to flow along these services to the distant earth. Bonding is done by clamps or gaps for all kinds of pipes and by suitable arresters for electrical lines and cables (IEC 1024-1, 1992).

Between extended metal installations inside the structure and the lps, dangerous voltages would arise mainly by magnetical induction effects. Therefore these installations are also bonded where at least every 20 m in a structure's height additional bonding is performed. Therefore in addition to the equipotential area in the basement every 20 m in height additional bonding areas are installed. By this measure the induced voltages are sufficiently suppressed. Nevertheless, safety distances between the internal installation and the lps are to be guaranteed, as described in Section 4.5, in order to avoid dangerous sparking by the remaining induced voltages (IEC 1024-1, 1992).

4.5. SAFETY DISTANCES OF INSTALLATIONS TO THE LIGHTNING PROTECTION SYSTEM

When services enter a structure with an lps they are bonded at the point of entry; potential equalization is performed to the lps. Also, all extended internal installations are bonded. Thus induction loops are built up, as shown in principle in Figure 2.4.5, where induced voltages arise between the lps and the installation at the "proximity". Here, a safety distance must be assured

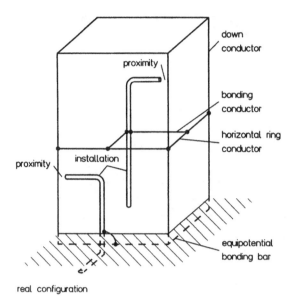

Figure 2.4.5 Proximities between the lps and the internal installations.

in order to avoid dangerous sparking by the induced voltage. For simplification in IEC (IEC 1024–1, 1992) any possible configuration of a loop was reduced to one typical configuration, as shown in Figure 2.4.5.

From calculations and experiments for this reference configuration a simplified formula was derived for the necessary safety distance d where only the neighboring down conductor of the lps with its partial lightning current, i_c is considered to cause the magnetic field in the induction loop.

$$d = k_i \cdot k_c \cdot k_m \cdot l \text{ in m} \tag{4.5}$$

k_i is proportional to i_{max}/T_1 of the subsequent impulse current and therefore dependent from the pl to be applied. Further, k_i contains the mutual inductance M between down conductor and installation and the breakdown field strength of air in the submicrosecond range.

$k_i = 0.1$ for pl I; $k_i = 0.075$ for pl II; $k_i = 0.05$ for pl III and IV.

$k_c = i_c/i$; for the configuration of Figure 2.4.5 $k_c = 0.44$.

k_m describes the ratio of the breakdown field strength of air to the material in the proximity: $k_m = 1$ for air; $k_m = 2$ for any solid material, e.g., a wall of bricks. l is defined in Figure 2.4.5.

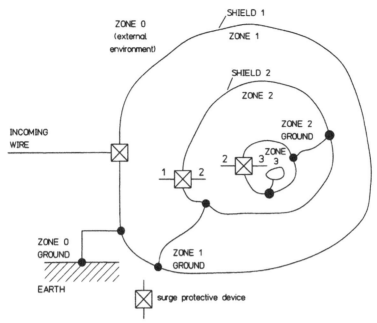

Figure 2.5.1 General "zone" concept. (Adapted from Vance, E.F., 1980, *IEEE Trans. Electromagn.*, 22, 319–328.)

Similar considerations are valid for the erection of an insulated lps. Here, the necessary distances between the lps and the structure to be protected are also determined by the safety distances according to the described induction effects.

5. PROTECTION OF STRUCTURES WITH ELECTRIC AND ELECTRONIC EQUIPMENT

The rapidly increasing installation of electronic equipment and its widely meshed networks in the public and industrial field has led to the demand for a new and advanced quality of lightning protection from not only the immediate damages of electronic systems by lightning effects — in Germany this damage is supposed to be more than half a billion dollars in 1991 — but mainly the unacceptable breakdown of electronic installations makes it necessary to develop adequate protection.

According to the EMC demands the electronic systems must function or at least survive in the environment determined by the electromagnetic lightning impact. Consequently, management methods and procedures common in the EMC world must be applied to these adapted lightning protection measures apart from the measures taken to avoid only endangering of structures and persons inside (Hasse et al., 1993).

5.1. PHILOSOPHY OF "LIGHTNING PROTECTION ZONES"

In the seventies a general "zone" concept was developed by C.E. Baum, F.M. Tesche, E.F. Vance et al. in order to protect any kind of structure with any kind of electronic equipment from electromagnetic impacts, including atmospheric discharges (see Figure 2.5.1; Vance, 1980). This general concept has proven to be so effective that it has been adapted, in a somewhat modified version, as general principle in the IEC work of TC 81; see Figure 2.5.2 [IEC 81(Sec.) 44, 1992].

A structure with its electronic equipment to be protected against lightning disturbances is divided into so-called lightning protection zones (LPZ); see Figure 2.5.3. The volume LPZ O_A

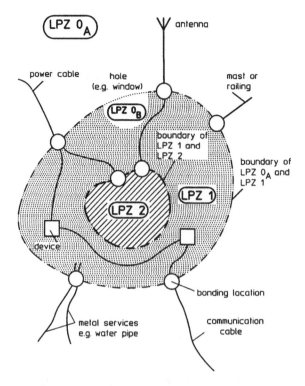

Figure 2.5.2 Lightning protection zone (LPZ) concept.

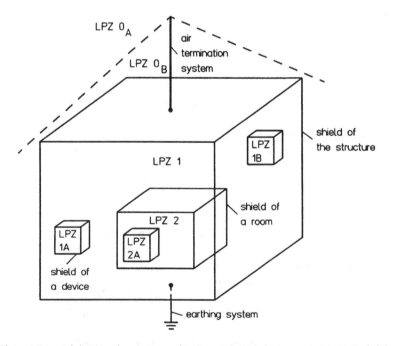

Figure 2.5.3 Subdivision of a structure to be protected into lightning protection zones (LPZ).

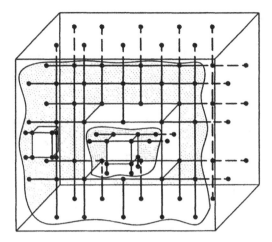

Figure 2.5.4 Meshed equipotential system.

is subject to direct lightning strikes and unattenuated LEMP. The volume LPZ O_B is characterized by not being directly struck, nevertheless, the original LEMP exists there. LPZ O_B can be identified by the rolling sphere method (see Section 3). The volume LPZ 1 is surrounded by a closed electromagnetic shield typically represented by the natural components of the structure such as metal roofs, facades, or reinforcement in the concrete ceilings, walls, and floors. Although this shield may not be perfect and have holes like windows, inside LPZ 1 the LEMP is essentially reduced compared with LPZ O_A and O_B. A further reduction of the LEMP by shielding is reached in LPZ 2 and possibly the local zones, e.g., LPZ 1A, 1B, and 2A.

The planner is free to define for each individual situation an adequate number of LPZs and to fix their boundaries, taking into account the structural facilities and the LEMP immunity of the installed electronic equipment.

At any time if a metal service, including power and telecommunication lines, penetrates the shield of an LPZ it must be bonded at this boundary by adequate components such as clamps or surge protective devices in order to reduce the conducted disturbances along the services coming from an LPZ with a higher interference level. Such disturbances may be partial lightning currents or induced overvoltages.

It would be ideal if all services could enter the individual LPZ at the same point, but normally — especially in very extended industrial sites — this ideal condition cannot be reached and entering at different locations must be accepted in practice.

5.2. EQUIPOTENTIAL SYSTEM INSIDE THE STRUCTURE

While in case of ideal electromagnetic shields a starpoint configuration of the equipotential system inside the structure to be protected could be installed (see Figure 2.5.1), for the actual situation where shields are used built up mainly by natural components with more or less extended holes (as windows, doors, and openings for air condition), a meshed configuration of the equipotential system has proven to be a very effective alternative. This meshed system, which also integrates as many natural components as possible interconnects as often as possible all shields of the LPZs and all metal components inside the LPZs (Figure 2.5.4). Thus the remaining partial lightning currents in the individual conductors of the equipotential system are very effectively reduced by the great amount of parallel paths, and the resulting multiple short-circuited loops acting as so-called reduction loops support the shielding effect of the LPZs' shields. The three-dimensional equipotential system guarantees, at any location of the structure, a low impedance reference point

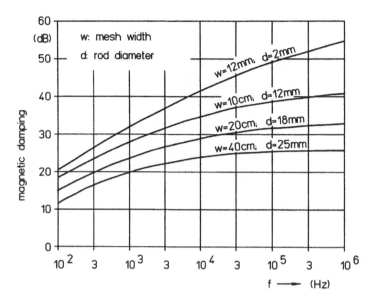

Figure 2.5.5　Damping characteristic for the magnetic field of grids built by the reinforcement.

with connection to Earth for any kind of local equipotential measure, especially for the bonding elements of the services at the boundaries of the LPZs.

5.3.　PROTECTION AGAINST LIGHTNING ELECTROMAGNETIC FIELDS

The most relevant component of the LEMP is the near magnetic field caused by the partial lightning currents in all kinds of metal conductors such as

1. Conductors of the air termination system and the down conductor system of an isolated lps (see Section 4.1)
2. The shield of LPZ 1, acting as an lps, where natural components (e.g., metal roofs, facades, and reinforcement in the basement) are integrated into the shield
3. Conductors of the equipotential system

Because the remaining magnetic fields inside the LPZs are mainly caused by immediately coupling through the holes in the shields of the LPZs, in a first approximation the near magnetic field in the LPZs are supposed to have the same shape as the lightning current (see Figure 2.2.1) resp. an amplitude density spectrum being proportional to the spectrum of the lightning current (see Figure 2.2.5).

In practice the magnetic field is damped at every boundary of an LPZ by 20 to 60 dB. An idea of the frequency-dependent magnetic damping behavior of reinforcement is given by Figure 2.5.5 (VG 96 907, 1986).

5.4.　PROTECTION AGAINST CONDUCTED TRANSIENTS

The induced transients to the electronic plants inside LPZ 1 and the subsequent LPZs are adequately reduced by building up spatial shields and a meshed equipotential system with its reduction loops.

Beside that, the conducted transients along the metal services, especially along the lines and cables of the power and telecommunication systems, are also to be adequately diminished at every entrance to LPZ 1 and the subsequent LPZs. Therefore, every service is to be bonded at the location where it penetrates the boundary of a LPZ by clamps or adapted surge protective devices (see Figure 2.5.6).

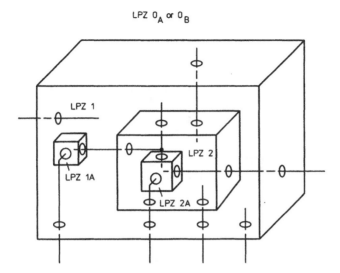

LPZ 0_A or 0_B

LPZ 1

LPZ 1A

LPZ 2

LPZ 2A

Figure 2.5.6 Bonding locations of services at the boundaries of the LPZs.

The main task is to keep essential parts of the lightning current on the services in LPZ 0_A apart from LPZ 1 by adequate bonding at the boundary of LPZ 0_A and LPZ 1 (see also Section 4.4 and Figure 2.4.4).

As a first approximation (and, as calculations and experiments have shown, as a "worst case" assumption) it can be assumed in accordance to IEC [IEC 81(Sec.)44, 1992] that in case of a lightning strike to a structure, 50% of the total lightning current will enter the Earth via the earthing system and 50% will enter the Earth via the services, including power and telecommunication lines being connected to the far Earth. These 50% are equally subdivided to the several services and again equally subdivided to the individual conductors of a line.

The currents flowing through the bonding elements at the boundary of LPZ 0_A and LPZ 1 have the same shape as the original lightning currents. The arresters at this place, which have to feed partial lightning currents into the lines connected to the far Earth are called lightning current arresters.

These bonding elements are to be connected as close as possible to the shield of LPZ 1, if necessary even by help of a metal plate (see example in Figure 2.5.7).

Inside LPZ 1 and the succeeding LPZs induced voltages and the resulting currents dominate as transients. They can be approximated and adequately simulated by a hybrid source which produces a double exponential open circuit voltage 1.2/50 μs (front time T_1: 1.2 μs; time to the half value T_2: 50 μs) and a damped oscillating short circuit current 8/20 μs (T_1: 8 μs; T_2 of the first half circle: 20 μs, see Figure 2.2.2).

The surge protective devices at the boundaries of LPZ 1 and higher are called overvoltage arresters. Typically, overvoltage arresters consist of two stages (see Figure 2.5.8). The primary element, for instance, consists of a gas discharge tube or a varistor, the secondary element of a transzorb or a diode, where the decoupling element is an inductor or a resistor. Still, overvoltage arresters can also be designed as insulating transformers, filters, or optical coupling links.

Lightning current arresters need a heavy duty element such as a gap or an adequately dimensioned varistor to carry the partial lightning currents.

At any transition from one LPZ to another, controlled by a surge protective device, the transient energy along a line is typically reduced by 20 to 40 dB, according to the magnetic damping characteristic of the LPZ's shield. The surge protective devices, successively installed along a line at the LPZ boundaries, must be carefully coordinated by their input and output characteristics.

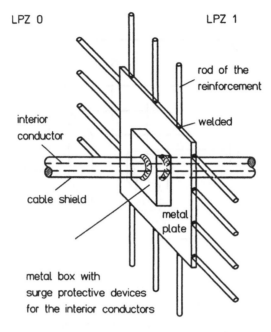

Figure 2.5.7   Bonding by the help of a metal plate.

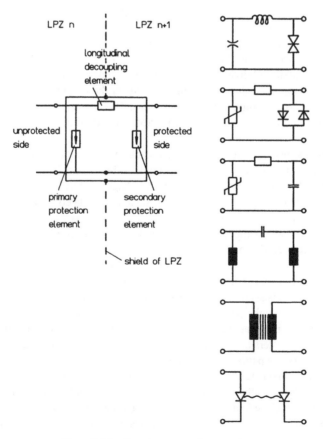

Figure 2.5.8   Two stage overvoltage arresters.

6. PROTECTION OF AIRBORNE OBJECTS

Hazardous lightning strikes to aircraft and aerospace rockets were reported several times. For example:

1. In 1987, a Boeing 747 aircraft immediately after takeoff was struck by four flashes within a few minutes near London airport. Consequently, beside severe mechanical damages, avionic systems like autopilot, radar, and radio transmission were damaged. A nearby flying jet received a mayday call and guided the aircraft back to London Gatwick for emergency landing (Sad, 1987).
2. In 1969, Apollo 12 was struck by lightning during takeoff. Consequently, some instrumentation sensors were damaged. Temporary upsets of communication systems, disturbances on instrumentation, illumination of warning lights and alarm, disconnection of three full cells from the busses, and loss of attitude reference by the inertial platform were reported (Uman, 1988).
3. In 1987, the Atlas Centaur 67 went out of control after a lightning strike. There was a memory upset in the vehicle guidance system which caused the vehicle to commence an unplanned yaw rotation. The Safety Officer then ordered the missile to be destroyed.

Aircraft are additionally endangered indirectly if a tower on an airport is struck and the air traffic control breaks down. Such critical situations were reported, for example, in 1987 from the Barcelona airport (Spain) and twice in 1992 from the Frankfurt airport (Germany).

The periodical *International Aerospace and Ground Conference on Lightning and Static Electricity* deals with hazard analyses and protection measures including test procedures for aircraft and aerospace vehicles.

If aircraft or rockets are flying in or near a thunderstorm cell where a sufficiently high electric field exists, the field strength at the extremities is enhanced to a value where lightning discharges can be triggered by starting leaders. This trigger effect caused by the airborne object itself is the reason for the relative high number of lightning strikes to aircraft and rockets (Cliffors et al., 1982b; Mazur, 1989).

Passengers and crew are protected against immediate lightning strike impact by the aluminum skin around aircraft, but there is an indirect endangerment by possible interference to electronic equipment caused by the residual electromagnetic field inside due to direct or nearby lightning strikes. These interferences become more serious because of the increasing use of carbon composite material for components of fuselage instead of aluminum.

Hardening of electronic equipment in aircraft and rockets is based on electromagnetic shielding, adequate routing of the wiring, an adapted mass concept, and the use of overvoltage protection devices at the entrance of electronic devices.

Besides the lightning protection measures on the airborne objects themselves, preventive measures are applied:

1. Thunderstorm electrification is carefully evaluated by multipoint field measurements on ground and computer analysis during the countdown period of launches, whereby the start possibly has to be postponed (Jacobson et al., 1976).
2. Aircraft pilots are instructed to pass by a thunderstorm cell and to avoid starting or landing during an active thunderstorm near the airport.

Lightning testing techniques and facilities for aircraft and rockets are quite sophisticated (Cliffors et al., 1982a). Components or even total aircraft are tested with simulated lightning currents covering parameters similar to those given in Table 2.2.2. The laboratory facilities are similar to those described in Section 7. Recently, the traditional tests are supplemented by burst-tests. These multiple impulse currents, rising in the nanosecond to kiloampere range and decaying in the submicrosecond range, are caused by negative leader discharges starting from the extremities of an aircraft and thus are an additional source of interference for electronic systems (Boulay, 1991).

7. TEST FACILITIES

7.1. GENERAL ASPECTS

Lightning current generators will be designed to reliably duplicate the parameters of natural lightning currents (Section 7.2) in the laboratory. There are basically two types of tests: (1) Direct lightning effects where the current is injected into the object under test and (2) Indirect lightning effects where the object under test is exposed to the electromagnetic field of a nearby stroke.

Direct lightning effects, like degrading or physical damage, are mainly related to the parameters i_{max}, Q_i, Q_l, and W/R of the first impulse current and the long duration current. In some cases the current steepness of first and subsequent impulse currents have to be considered too, e.g., when evaluating the limiting voltage of a surge protective device.

Indirect effects are governed by the magnetic field coupling. Relevant parameters are the current steepness of a subsequent impulse current and the peak value of a first impulse current.

The following sections separately contain the description of test generators for the simulation of first and subsequent impulse and long duration currents. To approach natural lightning phenomena as closely as possible may require the combination of different generator types to one test facility for specific cases.

7.2. SIMULATION OF THE FIRST IMPULSE CURRENT

Design criteria for a first impulse current simulator are i_{max}, Q_i, and W/R. Typical values are given in Table 2.2.2. The front time T_1 usually is of minor interest as it does not affect degrading or physical damage. Nevertheless, T_1 should be typical for a positive return stroke, i.e., in the order of about 10 μs.

7.2.1. RLC Circuits

Surge current generators usually consist of a set of large, high voltage capacitors in parallel (Figure 2.7.1). The basic configuration is shown in Figure 2.7.2. The generator is characterized by its maximum charging voltage U_c and the stored energy $W = (1/2) C_s \cdot U_c^2$.

The capacitor bank C_s is slowly charged by a DC source to a high voltage (e.g., 100 kV) and then rapidly discharged into the load via a starting switch, S1. The internal resistance R_i and inductance L_i of the generator, external wave forming elements R_{ex} and L_{ex}, and the load characteristics R_l and L_l can be summed up as

$$R = R_i + R_e + R_l \tag{7.1}$$

$$L = L_i + L_e + L_l \tag{7.2}$$

The current i of this RLC circuit is given by the differential equation

$$(d^2i/dt^2) + (R/L)(di/dt) + i/(LC_s) = 0 \tag{7.3}$$

The resulting current waveform depends on the ratio of the circuit elements R, L, and C_s. Examples for the three different waveforms are shown in Figure 2.7.3.

To obtain maximum current, RLC surge generators should be operated in an underdamped mode, but the resulting oscillatory waveform is contrary to the unidirectional currents of natural lightning. To get a unidirectional waveform critical (or overcritical) damping is needed. Critical damping is obtained by increasing the circuit resistance. This in turn means reducing the peak

Figure 2.7.1 Lightning current test generator at the University of the Federal Armed Forces, Munich.

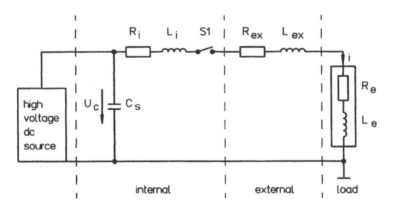

Figure 2.7.2 Basic configuration of an RLC circuit.

current and wasting most of the energy initially stored in the capacitor bank by heating the generator's damping resistors.

As an example, the simulation of a first impulse current, according to pl III (100 kA, 50 C, and 2.5 MJ/Ω, see Table 2.2.2) with a critically damped RLC generator would necessitate an extraordinarily large capacitor bank of, for instance, $C_s = 200$ μF operated at $U_c = 250$ kV.

7.2.2. Crowbar Circuits
A very effective way to obtain a unidirectional current with a tolerable size of capacitor bank is the use of a crowbar switch in an RLC circuit. The basic circuit diagram is shown in Figure 2.7.4

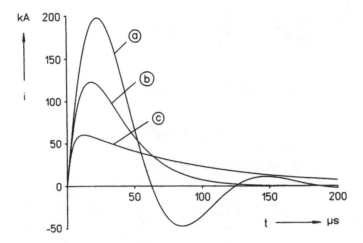

Figure 2.7.3 Resulting waveforms of an RLC circuit. (a) Undercritically damped: $0 < R < 2\sqrt{L/C_s}$. (b) Critically damped: $R = 2\sqrt{L/C_s}$. (c) Overcritically damped: $R > 2\sqrt{L/C_s}$.

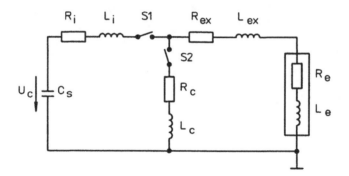

Figure 2.7.4 Test generator with a crowbar switch.

An external inductance, L_{ex}, significantly higher than the internal inductance, L_i, is connected to the generator. The discharge is initiated by a starting gap S1 at $t = 0$, while the crowbar switch S2 is open at first. To obtain maximum peak currents the generator is operated in a strong underdamped mode with low resistance. Ideally, the maximum value of the current of an undamped RLC circuit is $\exp(1) = 2.7$ times greater than that of a critically damped circuit.

At the instant of the crest value of the current, t_{cr}, S2 is closed. Most of the energy initially stored in the capacitors is then transferred to the external inductance L_{ex} and to the inductance of the load, L_l. By shorting out the capacitor with the crowbar switch at t_{cr} the current turns from an oscillatory to an exponentially decaying waveform, having a decay time constant

$$\tau = (L_{ex} + L_l + L_c)/(R_{ex} + R_l + R_c) \tag{7.4}$$

where L_c is the self-inductance and R_c is the resistance of the crowbar branch.

With τ being high, a long duration of the current is achieved; with R_{ex} and R_c being low, a major part of the energy is transferred into the object under test.

A crucial element of this generator type is the crowbar switch. At $t = 0$, when open, it has to withstand nearly the whole charging voltage, U_c. At the time of shorting, $t = t_{cr}$, when the energy is transferred to the inductances L_{ex} and L_l, the voltage across the capacitor C_s, and thus across the crowbar switch, is near zero. Closing a crowbar switch near or at zero voltage calls for advanced technologies. Further, a crowbar switch must have the abilities to handle currents in the 200-kA range with decay times of about 100 µs.

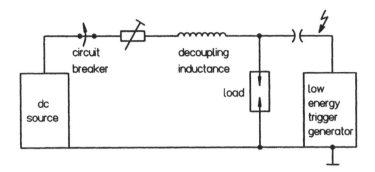

Figure 2.7.5 Long duration current generator.

Mechanical switches are usually too slow to operate at a microsecond time scale. During the last decade a number of crowbar techniques have been developed and successfully applied:

1. Three electrode spark gaps in ambient air, under pressurized gases (e.g., SF_6), or vacuum, triggered by a high voltage impulse (Zischank, 1984 and 1987; Pettinga et al., 1991)
2. Two electrode spark gaps triggered by a UV laser beam (Landry et al., 1984)
3. High power diodes (Hourtane, 1989)

A capacitive impulse current generator using a crowbar switch transfers a major part of the energy stored in the capacitor bank into the object, therefore the costs of the capacitor bank are considerably reduced compared to a critically damped circuit. The inductance of the object under test is not critical as the "external inductance" (typically 10 μH) can be partly or completely replaced by the self-inductance of the tested object. If some internal damping R_i (typically 100 mΩ) is also used the stress to the capacitor is significantly reduced; the high, long lasting current (after t_{cr}) flows only through the crowbar switch and the object under test.

7.3. SIMULATION OF THE LONG DURATION CURRENT

Long duration currents are characterized by averaged currents of some 100 A lasting up to some 100 ms resulting in a charge transfer Q_l of some 100 C (Table 2.2.2). Simulation of long duration currents can be done by critically damped capacitor discharges or, preferably, by using a DC source. Possible DC sources are storage batteries connected in series, AC transformers with rectifier and smoothing capacitor, or rotating machines. The basic circuit diagram is shown in Figure 2.7.5. The current is adjusted by a variable resistor. The duration is determined by interrupting the current with either a circuit breaker or a power semiconductor (e.g., thyristor).

Often the injection of a current into an object being tested is to be done via a spark, simulating the lightning channel at the point of strike. To avoid undesirable influences of the spark electrode to the object being tested, a gap spacing of at least 50 mm should be kept (Cliffors et al., 1982a). The voltage of the DC source, therefore, has to be high enough to prevent premature extinction of the arc. Preferable values are in the range of 1000 to 2000 V. To fire the injection spark gap a high voltage, low energy trigger generator can be used. It is also possible to initiate the discharge by bridging the spark gap with a thin starter wire, evaporizing during the very beginning of current flow.

7.4. SIMULATION OF SUBSEQUENT IMPULSE CURRENTS

The dominating current parameter of subsequent impulse currents is the current steepness with values in the range of 100 to 200 kA/μs (Table 2.2.2). Peak current, charge, and specific energy are much lower than that of a first impulse current and therefore, consequently, must not be considered here.

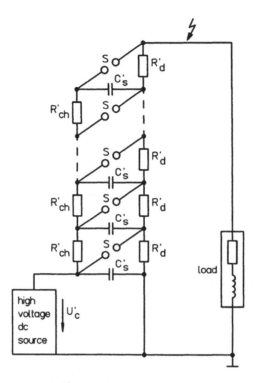

Figure 2.7.6 Marx generator.

The maximum current steepness of an RLC circuit is equal to the ratio of the charging voltage U_c to the total inductance L:

$$(di/dt)_{max} = U_c/L \qquad (7.5)$$

The crucial design criterion therefore is the inductance of an object to be tested, which determines, for a given value of di/dt, the charging voltage U_c and thus the generator size needed. Given a certain load inductance, attempts to simply increase the charging voltage yields little benefit. Increasing the charging voltage requires more insulation spacing, which in turn increases the circuit internal inductance. Therefore, specific measures to keep the internal inductance low or to steepen the current front have to be applied.

Small objects, like surge protective devices, having inductances of just a few 100 nH, usually can be tested with generators having charging voltages of a few 100 kV.

Large test objects, like parts of an aircraft, as well as indirect effects testing where an object is exposed to the electromagnetic field of a lightning current in a down conductor of at least some meters in length, usually require generators in the megavolt range. Load inductances of up to 10 µH have to be taken into account.

7.4.1. Marx Generators

As shown above (Equation 7.5), for testing of large objects at high current steepnesses voltages in the megavolt range are necessary. Single capacitors usually are rated to voltages up to about 200 kV. Megavolt output can be produced by slowly charging a group of capacitors (C_s') in parallel via the charging resistors R_{ch}' and then rapidly discharging them in series via the spark gaps S into the load (Figure 2.7.6). R_d' represents the internal discharging resistors of the generator. These generators are called Marx generators, after their inventor.

Marx generators operated under ambient conditions have a high internal inductance of about 4 µH/MV (Modrusan et al., 1985); therefore, they are not very useful in producing fast rising

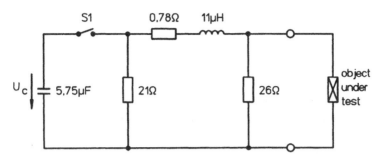

Figure 2.7.7 Example of a combination wave generator.

currents. To significantly lower the internal inductance, the generators are embedded in a (metal) tank filled with a high insulating medium like oil (White, 1984) or SF_6.

7.4.2. Peaking Circuits

To avoid sophisticated insulation techniques, peaking circuits are added to conventional surge current generators to steepen the current rise. Capacitive peaking circuits can be built either by lumped capacitances (Fisher et al., 1989) or by distributed capacitances in the form of a large grid, e.g., 50 m × 50 m wide, 10 m above ground (Perala et al., 1984).

Another possible way to steepen the current rise is by using exploding wires together with an external inductance L_{ex} of about 10 µH (Salge et al., 1970; Zischank, 1992). The rapid change of current during the abrupt explosion of the wire results in a very high voltage across the inductance, much higher than the charging voltage (e.g., 10 times). This voltage is used to drive a very fast rising current into the object under test.

7.5. SIMULATION OF CONDUCTED TRANSIENTS INSIDE INTERIOR LIGHTNING PROTECTION ZONES

For testing equipment or protective elements located inside LPZ 1 or higher (see Section 5.4) against conducted transients, two waveforms are conventionally used: (1) 1.2/50 µs impulse voltage and (2) 8/20 µs impulse current. Although not being representative of all of the possible waveforms inside a real installation, for many years these waveforms have proven to be a realistic stress.

When testing complex devices, often of unknown internal design, it is uncertain whether a voltage or a current wave should be applied. In recognition of this problem, a hybrid or combination wave has been proposed (Richman, 1983; Wiesinger, 1983). A circuit design for a combination wave generator is given in Figure 2.7.7 (Wiesinger, 1983).

A combination wave is characterized by specifying both an open circuit voltage and a short circuit current. The open circuit voltage has a waveshape of 1.2/50 µs and the short circuit current a waveshape 8/20 µs. The ratio of the peak open circuit voltage u_{max} to the peak short circuit current i_{max} is called fictive impedance, Z_f. A value of $Z_f = 2 \ \Omega$ is thought to be representative for conducted transients in LPZ 1 or higher.

The 1.2/50 µs voltage waveform applies to high impedance test samples; the 8/20 µs current waveform applies to low impedance test samples. For any intermediate impedance the device being tested will form its specific voltage/current waveshapes and peak values in interaction with the combination wave generator, as it would in a real installation in interaction with a surge.

8. CONCLUSIONS

The primary task of today's lightning protection is no longer to avoid hazards to persons or damage to buildings, but to ensure the infrastructure of electric power, telecommunications, and data processing on which our modern society is totally dependent. Along with the task of preventing power and electronic systems from damage, it is more important to guarantee their availability. It became evident that catastrophic situations can arise if vital systems fail; as an example the blackout of the control system in an airport tower can result in unacceptable situations in air traffic.

From this new dimension of lightning protection needs the necessity arises of applying rules and management methods of EMC to lightning protection techniques. The philosophy of protection zones is the link between EMC and lightning protection.

The basic data of lightning interference, like transient lightning currents and electromagnetic fields, are to be evaluated by continued research of the natural lightning phenomenon. These data have to be transferred into adequate laboratory test techniques in order to ensure effective protection measures.

APPENDIX

LIST OF SYMBOLS

a	length (m)
C_s	surge capacitance (F)
c_s	specific melting energy (J/kg)
c_w	specific heat capacity [J/(kg · K)]
d	distance (m)
E(t)	electric field (V/m)
E_b	break down field strength (V/m)
E_r	field strength at the point of strike (V/m)
H(t)	magnetic field (A/m)
h	height (m)
I_a	average current (A)
i	current (A)
i_c	partial lightning current (A)
i_l	short circuit currents in a loop (A)
k	constant (1), proportional factor
L	self-inductance (H)
L_c	inductance of the crowbar switch (H)
L_{ex}	external inductance (H)
L_i	internal inductance (H)
L_l	load inductance (H)
l	length (m)
M	mutual inductance (H)
n	number (1)
Q	charge of a current (C)
Q_i	charge of the first impulse current (C)
q	cross section of a conductor (m²)
Q_l	charge of the long duration current (C)
R	resistance (Ω)
R_c	resistance of the crowbar switch (Ω)
R_{ch}	charging resistor (Ω)
R_d	discharging resistor (Ω)
R_e	earthing resistance (Ω)
R_{ex}	external resistance (Ω)
R_i	internal resistance (Ω)

R_l	load resistance (Ω)
r	radius of a conductor (m), radius of the rolling sphere (m)
s	thickness (m)
T_l	time duration of a long duration current (s)
T_1	front time (s)
T_2	time to the half value (s)
t	time (s)
t_{cr}	time for the crest value (s)
$U_{A,K}$	anode, respectively, cathode voltage drop (V)
U_c	charging voltage (V)
u	voltage (V)
u_e	voltage at the earthing resistance (V)
u_l	induced voltage in a loop (v)
v	velocity (m/s)
W	energy (J)
W/R	specific energy of the first impulse current ($A^2s = J/\Omega$)
Z_f	Fictive impedance (Ω)
α	temperature coefficient of the resistivity (1/K)
α,β	protective angles (grd)
γ	mass density (kg/m^3)
η	constant (1)
θ_f	final temperature (°C)
θ_o	ambient temperature (°C)
θ_s	melting temperature (°C)
ρ_e	soil resistivity (Ωm)
ρ_o	resistivity at ambient temperature (Ωm)
τ_1	time, characterizing the front time of an impulse current (s)
τ_2	time characterizing the decay time of an impulse current (s)

LIST OF ABBREVIATIONS

DC	direct current
AC	alternating current
EMC	electromagnetic compatibility
LEMP	lightning electromagnetic impulse
LEP	lightning electric impulse
LMP	lightning magnetic impulse
lps	lightning protection system
LPZ	lightning protection zone
pl	protection level, according to IEC 1024-1(1992)

LIST OF INDICES

max	maximum value

REFERENCES

Boulay, J.-L. (1991): Current wave-form observed during lightning strikes on aircraft. *Int. Aerospace Ground Conf. Lightning Static Electricity*, Cocoa Beach, 4-1.

Cliffors, D.W., Crouch, K.E., Schulte, E.H. (1982a): Lightning simulation and testing. *IEEE Trans. Electromagn. Compat.*, EMC-24 No. 2, 209.

Cliffors, D.W., Kasemir, H.W. (1982b): Triggered lightning. *IEEE Trans. Electromagn. Compat.*, EMC-24 No. 2, 112.

Craven, J.D., Knaur, J.A., Moore, T.W., Shumpert, Th.H. (1991): A simulated lightning effects test facility for testing live and inert missiles and components. Int. Aerospace Ground Conf. Lightning Static Electricity, Cocoa Beach, 108-1.

Darveniza, M., Popolansky, F., Whitehead, E.R. (1975): Lightning protection of UHV transmission lines. *Electra*, 41, 36–69.

Ettinger C.W. (1778): *Verhaltungs=Regeln by nahen Donnerwettern.* 3rd ed., Gotha, Germany.

Fisher, F.A., Plumer, J.A., Perala, R.A. (1989): *Aircraft Lightning Protection Handbook.* Federal Aviation Administration, Technical Center, DOT/FAA/CT-89/22.

Hasse, P., Wiesinger, J. (1992): *Handbuch für Blitzschutz und Erdung.* 4th ed., Pflaum Verlag, München; VDE-Verlag Berlin/Offenbach.

Hasse, P., Wiesinger, J. (1993): *EMV Blitz-Schutzzonen-Konzept.* Pflaum-Verlag, München; VDE-Verlag Berlin/Offenbach.

Hourtane, J.-L. (1989): DICOM: Current generator delivering the A, B, C, D waveform for direct lightning effects simulation on aircraft. Int. Aerospace Ground Conf. Lightning Static Electricity, Bath, England, 5A.5.

Jacobson, E.A., Krider, E.P. (1976): Electrostatic field changes produced by Florida lightning. *J. Atmos. Sci.* 33, 103.

IEC 81(Secr.)44 (1992): Protection against LEMP. Part 1. General principles. (Committee draft).

IEC 1024-1 (1992): Protection of structures against lightning. Part 1. General principles. (Standard).

Kern, A. (1990): Theoretische und experimentelle Untersuchung der Erwärmung von Metallblechen bei Blitzeinschlag. Doctoral thesis, University of the German Armed Forces, Munich, Germany.

Landry, M.J., Brigham, W.P. (1984): UV laser triggering and crowbars used in the Sandia lightning simulator. Int. Aerospace Ground Conf. Lightning Static Electricity, Orlando, FL, 46.

Mazur, V. (1989): Triggered lightning strikes to aircraft and natural intracloud discharges. *J. Geophys. Res.*, Vol. 94, No. D3, 3311.

Modrusan, M., Walther, P. (1985): Aircraft testing with simulated lightning currents of high amplitude and high rate of rise time. Int. Aerospace Ground Conf. Lightning Static Electricity, Paris, 3A-5.

Perala, R.A., Rudolph, T.H., McKenna, P.M., Robb, J.D. (1984): The use of a distributed peaking capacitor and a marx generator for increasing current rise rates and the electric field for lightning simulation. Int. Aerospace Ground Conf. Lightning Static Electricity, Orlando, FL, 45.

Pettinga, J.A.J., Damstra, G.C. (1991): A lightning current pulse source of 160 kV/120 kJ. 7th Int. Symp. High Voltage Engineering, Dresden, Germany, 51.06.

Richman, P. (1983): Single output, voltage and current test waveform, *IEEE Symp. Electromagn. Compat.*, 47.

Sad: Springer-Auslandsdienst (1987): Jumbo wurde viermal von Blitz getroffen. 12. Nov. 1987.

Salge, J., Pauls, N., Neumann, K.-K. (1970): Drahtexplosionsexperimente in Kondensator-Entladekreisen mit großer Induktivität. *Z. Angew. Phys.*, 29, 339.

Steinbigler, H. (1977): Die Stoßstromerwärmung unmagnetischer und ferromagnetischer Leiter in Blitzschutzanlagen. Postdoctoral thesis, Technical University Munich, Germany.

Uman, M.A. (1988): Natural and artificially-initiated lightning and lightning test standards. *Proc. IEEE*, Vol. 76, No. 12, 1548.

Vance, E.F. (1980): Electromagnetic interference control. *IEEE Trans. Electromagn.*, 22, 319–328.

VG 96 907 Part 2 (1986): Nuclear electromagnetic pulse (NEMP) and lightning protection. Beuth Verlag GmbH, Köln, 1.

White, R.A. (1984): Lightning simulator circuit parameters and performance for severe threat, high-action-integral testing. Int. Aerospace Ground Conf. Lightning and Static Electricity, Orlando, FL, 40-1.

Wiesinger, J. (1983): Hybrid-Generator für die Isolationskoordination. etz, 104, 1102.

Zischank, W. (1984): Simulation von Blitzströmen bei direkten Einschlägen. etz, 105, 12.

Zischank, W. (1987): A surge current generator with a double-crowbar sparkgap for the simulation of direct lightning stroke effects. 5th Int. Symp. High Voltage Engineering, Braunschweig, 61.07.

Zischank, W. (1992): Simulation of fast rate-of-rise lightning currents using exploding wires. 21st Int. Conf. Lightning Protection, Berlin, 5.01.

Chapter 3

Longwave Sferics Propagation within the Atmospheric Waveguide

Hans Volland

CONTENTS

1. INTRODUCTION

Transient electric currents within lighting channels during return strokes (R strokes) and intracloud strokes (K strokes) are the main sources for the generation of impulse-type electromagnetic radiation known as atmospherics or sferics (definitions of the various types of lightning strokes are given in Chapter I/4*). While this impulsive radiation dominates at frequencies less than about 100 kHz (loosely called long waves), a continuous noise component becomes increasingly important at higher frequencies. This chapter deals with the radiation properties of longwave electromagnetic impulses generated by lightning within the waveguide between the Earth's surface

---

* These numbers refer to volume number and chapter number, respectively, throughout this volume.

and the ionospheric D layer and lower E region (atmospheric waveguide). Propagation effects of radio noise from lightning at higher frequencies will be treated in Chapter I/15.

2. SOURCE PROPERTIES

2.1. LIGHTNING STROKE AS RESONANT WAVEGUIDE

2.1.1. Basic Stroke Parameters

In a typical cloud-to-ground lightning return stroke (R stroke) negative electric charge of the order $Q = 1$ C stored within the lightning channel is lowered to the ground within a typical impulse time interval of $\tau = 100$ μs. This corresponds to an average electric current flowing within the channel of the order $J = Q/\tau = 10$ kA. Maximum spectral energy is, therefore, generated within a frequency band near $f = 1/\tau = 10$ kHz (e.g., Serhan et al., 1980) or at a wavelength of $\lambda = c/f = 30$ km (c is the speed of light). In typical intracloud K strokes positive electric charge of the order $Q = 10$ mC stored in the upper part of the channel and an equal amount of negative charge in its lower part neutralize within a typical time interval of $\tau = 25$ μs. The corresponding values for average electric current, frequency, and wavelength are $J = 400$ A, $f = 40$ kHz, and $\lambda = 7.5$ km. The energy of K strokes is, in general, two orders of magnitude weaker than the energy of R strokes.

As we will see later, the typical length of lightning channels can be estimated to be of the order $\ell = \lambda/4 \simeq 8$ km for R strokes and $\ell = \lambda/2 \simeq 4$ km for K strokes. Often, a continuing current component flows between successive R strokes (Uman, 1987). Its "pulse" time typically varies between about 10 and 200 ms, and its electric current may be of the order $J = 100$ A, corresponding to numbers of $Q = 1$ to 20 C, $f = 5$ to 100 Hz, and $\lambda = 3$ to 60 Mm.

The visible part of an R stroke channel has a typical length of about 5 km. Another part of comparable length may be hidden within the cloud and may have a significant horizontal branch. Evidently, the dominant wavelength of the electromagnetic wave of the R stroke is larger than the length of its channel. The physics of electromagnetic wave propagation within the channel must thus be derived from full wave theory, because the ray concept breaks down in this case.

2.1.2. Electric Channel Current

The channel of an R stroke can be considered as a thin isolated wire of length ℓ and diameter d, in which negative electric charge (electrons) has been stored. This electric charge is lowered to the ground when the wire approaches the electrically well-conducting earth. In terms of electric circuit theory, one can adopt a simple transmission line model with a capacitor, where the charge is stored, a resistance of the channel, and an inductance simulating the electric properties of the channel.

If the contact with the ground starts at time $t = 0$, an infinite number of individual modes of wavenumbers m will be excited just as in the case of a piano string or any other resonant system. For the mth mode, the downward flow of the electrons at $t > 0$ corresponds to an upward directed electric current varying in time t and height z according to (Volland, 1984)

$$J_m = Q_m F_m(t) \cos(K_m z/\ell) \quad (t \geq 0) \tag{2.1}$$

where

$$F_m = -\frac{\alpha_m \beta_m}{(\beta_m - \alpha_m)}(e^{-\alpha_m t} - e^{-\beta_m t}) = -\frac{(\gamma_m^2 + \delta_m^2)}{\delta_m}e^{-\gamma_m t} \sin \delta_m t \tag{2.2}$$

Here, Q_m is the total electric charge of the mth mode stored within the channel at time $t = 0$, $\alpha_m = \gamma_m - i\delta_m$, and $\beta_m = \gamma_m + i\delta_m$ are eigenvalues, which may be real or complex numbers depending on the properties of the channel, and $K_m = (2m - 1)\pi/2$ with m a positive integer. The first expression in Equation 2.2 — the famous Bruce-Golde formula (Golde, 1977) — is valid when α_m and β_m become real (or δ_m imaginary). In the second equation, γ_m and δ_m are real.

In reality, only the modes of lowest order ($m \le 3$) are well defined. The fundamental mode with $m = 1$ has a wavelength of $\lambda_1 = 4\ell$ so that it does not "feel" the contorted configuration of the actual lightning channel. The channel behaves like a quarter wave antenna with maximum current at the bottom and zero current at the top. The electric charge density stored within the channel has a maximum at the top of the channel and zero at the bottom (Figure 3.2.1a). The higher order modes have wavelengths of $4\ell/3$, $4\ell/5$, etc. For these modes, charge is transported to the ground only in the lower part of the channel while the rest is internally redistributed. The first mode is therefore the most efficient one for lowering charge to the ground and heating the channel. It also has the longest lifetime and thus essentially determines the general waveform of the R stroke. Higher order modes decay more rapidly and are not excited as strongly. They increasingly contribute to the high frequency radio noise with increasing mode number. We thus consider only the first mode of an R stroke in detail suppressing the subscript "1" in the equations to follow.

K strokes centered at a height H above ground have a channel length of 2ℓ. Current and charge configuration for the two first modes are shown in Figure 3.2.1b. Here, positive charge is stored in the upper half of the channel and negative charge in the lower half, so that the current flow is downward.

2.1.3. Total Electric Energy

The total dissipated electric energy of the first mode of an R stroke is given by

$$W_{\text{dis}} = \frac{1}{\ell} \int_0^\infty \int_0^\ell RJ^2 \, dzdt = \frac{RQ^2\alpha\beta}{4(\alpha + \beta)} \tag{2.3}$$

with R the total resistance of the channel. The same amount is radiated from the channel (Volland, 1981) so that the total electric energy of the first mode is

$$W_{\text{tot}} = 2W_{\text{dis}} \tag{2.4}$$

Mode one has the greatest efficiency for heating the channel up to temperatures beyond 10,000 K where luminous events can be observed. This wave mode is a standing wave, excited instantaneously along the channel after breakdown. This would apparently disagree with the observational fact that a luminous wave front propagates upward with a velocity of a fraction of the speed of light (see Chapter I/4). One must bear in mind, however, that luminosity and electric current are two distinct phenomena. The temperature is proportional to the square of the current integrated over the time. Because the current amplitude is largest at the bottom of the channel, the point where the critical temperature of luminosity is reached apparently "travels" upward from the bottom of the channel (Volland, 1984).

A more advanced model which considers the channel to be a cylindrical wire of finite thickness, finite electric conductivity, and finite length is a resonant waveguide that supports propagation of transverse magnetic waves (Volland, 1981, 1982). It is possible in this case to relate the channel parameters length ℓ and diameter d to the observed waveform parameters α and β. Given an observed sferic waveform, which is proportional to the time derivative of J in Equation 2.1, one can determine the channel parameters ℓ and d. If the electric conductivity σ of the channel is known, then the total resistance of the channel is given by

$$R = \frac{4\ell}{\sigma\pi d^2} \tag{2.5}$$

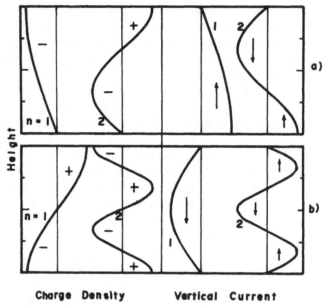

Charge Density   Vertical Current

Figure 3.2.1 Height distribution of charge density (left) and vertical electric current (right) for the two modes of order $m = 1$ and 2. (a) Return stroke model. (b) K stroke model. (From Volland, H., *Atmospheric Electrodynamics*, Springer-Verlag, Heidelberg, 1984. With permission.)

Table 3.2.1   **Parameters of Lightning Strokes of Figures 3.2.2**

Stroke	$G - R_1$	$G - R_2$	$G - R_3$	$C - K$
α (μs^{-1})	0.02	0.1	—	—
β (μs^{-1})	0.2	0.5	—	—
λ (μs^{-1})	—	—	0.012	0.1
δ (μs^{-1})	—	—	0.024	0.2
I_o (kA)	30.0	4.0	11.0	-2.0
d (cm)	1.6	1.0	4.5	1.6
ℓ (km)	7.9	2.2	18.0	4.4
R (kΩ)	3.9	2.8	1.1	2.2
Q (C)	-1.35	-0.032	-0.73	0.016
M (C km)	-6.8	-0.045	-8.4	0.045
W_{tot} (MJ)	65.1	0.12	9.1	0.070
W_{rad}/W_{tot}	0.01	0.02	0.04	0.08
f_{max} (kHz)	10.1	35.6	4.3	35.6

Note: α, β, γ, δ are from Equation 2.2, I_o from Equation 2.8. d and ℓ are diameter and length of the lightning channel. R is its resistance (Equation 2.5). Q is the electric charge stored within the channel, M the dipole moment of the channel (Equation 2.6). W_{tot} is the total electric energy of the lightning stroke (Equation 2.4), W_{rad} is the radiated energy (Equation 2.9), and f_{max} is the maximum spectral energy (Equation 2.13).

A typical value for the channel conductivity is $\sigma = 10^4$ S/m (Uman, 1987), which is used to estimate the channel parameters in Table 3.2.1. Of course, one should be cautious not to over-interpret such a model, which presents only a crude simulation of the basic macroscopic process, leaving the details unexplained. More about lightning theories may be found in Chapter I/4.

2.2. LIGHTNING CHANNEL AS HERTZ DIPOLE

The impulsive electric currents flowing in lightning channels with lengths of several kilometers behave like huge antennas generating electromagnetic LF-VLF-ELF radiation (LF = "low frequency'': 30 to 300 kHz; VLF = ''very low frequency''; 3 to 30 kHz; ELF = ''extremely low frequency'': 3 Hz to 3 kHz). The continuing currents generate mainly Schumann resonances (7.5 Hz and harmonics; see Chapter I/11). These electromagnetic pulses called sferics propagate within the dispersive and lossy atmospheric waveguide.

2.2.1. Electric and Magnetic Radiation Fields

The electromagnetic radiation of R and K strokes at distances r much larger than the channel length ℓ can be described in a first approximation by the radiation component of a vertical electric Hertz dipole. If it is located at a height H above the perfectly conducting flat ground, it generates transverse magnetic waves with the vertical electric field strength E_z at the ground ($z = 0$) given by (e.g., Sommerfeld, 1952)

$$E_z = -\frac{M\mu}{2\pi r}\frac{dF}{dt}\sin^2\theta \quad (\ell \ll r) \tag{2.6}$$

and its azimuthal magnetic field strength B_ϕ

$$B_\phi = \frac{M\mu}{2\pi c r}\frac{dF}{dt}\sin\theta \quad (\ell \ll r) \tag{2.7}$$

with $M = Q\bar{\ell}$ the dipole moment, $\bar{\ell} = \ell/K = 2\ell/\pi$ an effective antenna length, $\mu = 4\pi \cdot 10^{-7}$ H/m the permeability of free space, F from Equation 2.1, θ the angle between dipole axis and radius vector to the observer, $\sin\theta = \rho/r$, r the distance between dipole and observer, and ρ the ground distance. For a dipole on the ground ($\sin\theta = 1$), it is according to Equation 2.2

$$E_z = -cB_\phi = \frac{\bar{\ell}\mu I_0}{2\pi\rho}(\alpha e^{-\alpha t} - \beta e^{-\beta t}) = \frac{\bar{\ell}\mu I_0}{\pi\rho}e^{-\gamma t}(\gamma\sin\delta t - \delta\cos\delta t) \tag{2.8}$$

with the amplitude factors $I_0 = -Q\alpha\beta/(\beta - \alpha)$ and $I_0 = -Q(\gamma^2 + \delta^2)/2\delta$, respectively ($I_0$ in A).

The first equation in Equation 2.8 yields an aperiodic waveform (type 1), the second equation a damped oscillator (type 2), often referred to as a bipolar waveform in the literature (Weidman and Krider, 1979). Indeed, both waveforms have been observed, although the bipolar waveform seems to dominate. The type 2 waveforms are generated if the channel resistance is typically $R < 2.5$ kΩ, depending on the parameter γ. Type 1 waveforms exist for larger R.

For typical K strokes where positive electric charge is lowered within a channel at height H without connection to the earth, $Q > 0$, and Equation 2.8 is still approximately valid provided $\rho \gg H$ and that ℓ is replaced by 2ℓ.

2.2.2. Electromagnetic Radiation Energy

The total radiated energy of the first mode in the far field of an R stroke into the half space above ground is (Sommerfeld, 1952)

$$W_{\text{rad}} = \frac{2\pi c}{\mu}\int_0^\infty\int_0^{\pi/2} B_\phi^2 r^2\sin\theta d\theta dt = \frac{\mu M^2\alpha^2\beta^2}{6\pi c(\alpha + \beta)} \tag{2.9}$$

and the ratio between radiated energy and total energy of the first mode (Equation 2.4) is then

$$\frac{W_{\text{rad}}}{W_{\text{tot}}} = \frac{\mu\bar{\ell}^2\alpha\beta}{3\pi c R} \tag{2.10}$$

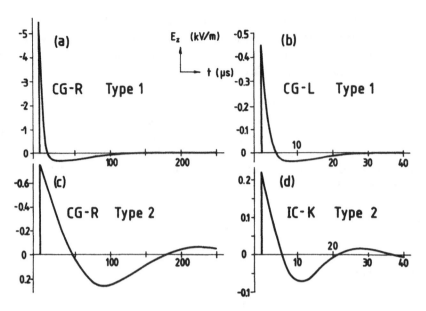

Figure 3.2.2 Electric radiation fields of radiation of the vertical electric dipoles from Table 3.2.1 derived from Equation 2.8. The electric field is normalized to a distance of 1 km. (a) Return stroke of type 1 ($G - R_1$; aperiodic wave) with $\alpha = 0.02$ μs^{-1} and $\beta = 0.2$ μs^{-1} (Dennis and Pierce, 1964). (b) Return stroke of type 1 ($G - R_2$) with $\alpha = 0.1$ μs^{-1} and $\beta = 0.5$ μs^{-1} (Krider et al., 1977). (c) Return stroke of type 2 ($G - R_3$; damped oscillator) with $\gamma = 0.012$ μs^{-1} and $\delta = 0.024$ μs^{-1} (Lin et al., 1979). (d) K stroke of type 2 ($C - K$) with $\gamma = 0.1$ μs^{-1} and $\delta = 0.2$ μs^{-1}. (Weidman and Krider, 1979.)

Figure 3.2.2 shows four examples of electric radiation waveforms calculated from Equation 2.8 and based on the numerical values in Table 3.2.1. They simulate the radiation fields of the type 1 Bruce-Golde current ($G - R_1$), an R stroke of type 1 ($G - R_2$), an R stroke of type 2 ($G - R_3$), and a K stroke of type 2 ($C - K$). They are remarkably similar to observed waveforms. Table 3.2.1 also contains parameters of these four strokes derived from the observed waveforms and amplitudes. The ratio between radiated energy and total energy of the first mode is smaller than 10%.

The amplitudes in Figure 3.2.2 increase abruptly to a maximum at the beginning of the stroke. However, the maximum amplitudes of real waveforms are rounded and are delayed by a few microseconds because of the finite channel lengths and the different arrival times of the signal from the bottom and the top of the channel (see Figure 3.3.5). Furthermore, the finite electric conductivity of the ground will cause the higher frequencies of the waveforms to be attenuated more strongly than the lower frequencies. This has essentially the same effect of rounding the spikes in Figure 3.2.2 (see Section 3.2.1).

2.2.3. Spectral Function

Maxwell's equations are linear. Therefore, they maintain their validity for each spectral component derived from any temporal variation of the electric field via a Fourier integral transformation. From system theory it is well known (e.g., Stein and Jones, 1967) that an original signal $E_o(t)$ at the input of a transmission line is modified after propagating through that transmission line into an output signal $E(t)$ given by

$$E(t) = \frac{1}{2\pi} \int_{-\infty}^{\infty} \hat{E}_o(\omega)T(\omega)e^{-i\omega t}\, d\omega \qquad (2.11)$$

with $\omega = 2\pi f$ the angular frequency, $\hat{E}_o(\omega)$ the Fourier transform of the input signal, and $T(\omega)$ the transfer function of the transmission line.

In the next section we will consider the transfer function of the atmospheric waveguide. We define this transfer function to be $T = 1$ over a flat, perfectly conducting ground. The transfer function of the atmospheric waveguide in Equation 2.11 thus describes the deviation from ideal propagation conditions.

The Fourier transform of the vertical electric field in Equation 2.8 at the distance ρ from the origin over a flat, perfectly conducting ground is given by

$$\hat{E}_o(\omega) = \int_{\rho/c}^{\infty} E_z(t_r)e^{i\omega t}dt = \frac{i\alpha\beta\mu\omega M}{2\pi\rho(\alpha - i\omega)(\beta - i\omega)} e^{i\omega\rho/c} \tag{2.12}$$

where $t_r = t - \rho/c$ is the retarded time.

The maximum spectral amplitude occurs at the frequency

$$f_{max} = \frac{1}{2\pi} \sqrt{\alpha\beta} \tag{2.13}$$

3. TRANSFER FUNCTION OF ATMOSPHERIC WAVEGUIDE

3.1. Overview

The ground with its finite electric conductivity as well as the ionospheric D layer modify the sferic waveform at greater distances. We describe the properties of the waveguide between earth and ionosphere by a transfer function $T(\omega, \rho)$ (see Equation 2.11) that depends on the spectral angular frequency ω and the distance ρ between lightning stroke and receiver. Furthermore, the transmission may depend on local time, season, and geographic and geomagnetic latitude. This transfer function can be approximated applying the two concepts of ray and mode theory (e.g., Wait, 1970, 1981; Galej, 1972; Harth, 1982).

Lightning strokes generate essentially transverse magnetic waves in the far field (TM waves: horizontal magnetic field orthogonal to the direction of propagation). In the following, the transfer function T is derived for the vertical electric field of TM waves. Transverse electric fields (TE waves: horizontal electric field component orthogonal to the direction of propagation) are only of minor importance for longwave sferics.

3.1.1. Ray Approximation

Ray theory describes the propagation of a pulse as the sum of a directly arriving ground wave and sky waves reflected at the base of the ionosphere near 70 to 90 km height, depending on time of day and season (Figure 3.3.1). The first hop sky wave arrives at the time

$$\Delta t = (r_1 - \rho)/c \tag{3.1}$$

later than the ground wave and has nearly the same pulse form (r_1 is the ray path of the first hop sky wave). Pulses with pulse lengths smaller than Δt can then be separated into a ground pulse and a first hop sky pulse at distances between about 200 and 500 km (McDonald et al., 1979).

Ray theory in its simplest form assumes an ideal ground wave and a first hop wave. The transfer function of the vertical electric field of an electric vertical dipole then becomes

$$T = 1 + 2R_i \sin^3 \theta e^{ik(r_1 - \rho)} \tag{3.2}$$

with θ the angle of incidence (see Figure 3.3.1), R_i the ionospheric reflection factor, and $k = \omega/c$ the wavenumber. The ground wave dominates at smaller distances. Beyond a critical distance (to be defined below), the first hop sky wave begins to dominate.

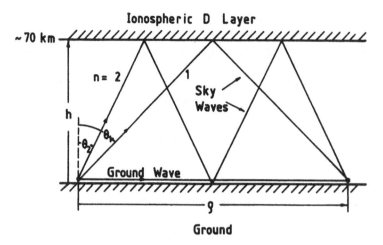

Figure 3.3.1   Geometry of ray propagation within the atmospheric waveguide between earth and ionospheric D region. Ground wave and two sky waves are displayed. (From Volland, H., *Atmospheric Electrodynamics,* Springer-Verlag, Heidelberg, 1984. With permission.)

3.1.2.   Wave Mode Approximation

Because the multihop components become increasingly important at greater distances, the wave concept is more convenient in this case. The transfer function here is composed of a series of wave modes of wavenumbers n. The transfer function of mode theory in its simplest form assumes an interference between the two first modes:

$$T = K_1 e^{(S_1 - 1)ik\rho} + K_2 e^{(S_2 - 1)ik\rho} \tag{3.3}$$

with K_n amplitude factors and S_n propagation factors. Within the ELF/VLF range the attenuation of the modes increases with increasing mode number (see Section 3.3.2). The propagation factor must be determined from an eigenvalue equation which is (in the case of a flat earth)

$$R_i R_e e^{2ikhC_n - 2in\pi} = 1 \tag{3.4}$$

with C_n an eigenvalue (related to $\cos\theta$ in Figure 3.3.1), R_e the reflection factor of the ground, R_i the reflection factor of the ionosphere, h the virtual reflection height, n the mode number, and

$$S_n = \sqrt{1 - C_n^2} \approx 1 - C_n^2/2 \tag{3.5}$$

where the last approximation is valid for $|C_n^2| \ll 1$.

For VLF waves, a first approximation is to set $R_e = 1$ (ideal electric wall) and $R_i = -1$ (ideal magnetic wall), in which case the eigenvalue in Equation 3.4 becomes

$$C_n = \overline{C}_{n2} = \frac{(2n - 1)\pi}{2kh} \quad (n = 1, 2, \dots) \tag{3.6}$$

The vertical mode structure of electric and magnetic field components is shown in Figure 3.3.2 (lower panel). The horizontal electric field disappears at the electric wall (ground), and the horizontal magnetic field disappears at the magnetic wall (ionospheric D layer). The wave modes become evanescent for $C_n \geq 1$, i.e., S_n in Equation 3.5 becomes imaginary. For the first and second mode this occurs at frequencies $f_1 < 1$ kHz and $f_2 < 3$ kHz, respectively. Evanescent modes are strongly attenuated and cannot contribute to electromagnetic energy transfer. The

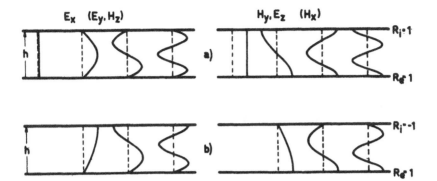

Figure 3.3.2 Vertical mode structure of electric (E) and magnetic (H) field components within an ideal plane waveguide with an electric wall at the bottom ($R_e = 1$) and a magnetic wall at the top ($R_i = -1$) simulating the atmospheric waveguide for VLF propagation (lower panel), and waveguide with two electric walls simulating ELF propagation (upper panel). The first four modes are displayed. x component in the direction of propagation, y orthogonal, and z vertical. (From Volland, H., *Atmospheric Electrodynamics*, Springer-Verlag. Heidelberg, 1984. With permission.)

frequency where a wave mode becomes evanescent is called the cut-off frequency (e.g., Sukhorukov et al., 1992).

In the ELF range the ionosphere changes to an electric wall ($R_i = 1$) in a first approximation (see Figure 3.3.2; upper panel), and the eigenvalues in Equation 3.4 now become

$$C_n = \overline{C}_{n1} = \frac{n\pi}{kh} \quad (n = 0, 1, \ldots) \tag{3.7}$$

Evidently, all modes with $n > 0$ are evanescent, and only the zeroth mode remains a propagation mode for ELF waves.

The ground changes to a magnetic wall ($R_e = -1$) for LF waves, the cut-off frequency depending on the electric ground conductivity. The eigenvalues in Equation 3.4 now become

$$C_n = \overline{C}_{n3} = \frac{(n-1)\pi}{kh} \quad (n = 2, 3, \ldots) \tag{3.8}$$

with the second mode predominating at about $40 < f < 80$ kHz, replaced by the third mode at about $80 < f < 110$ kHz, etc. (Volland, 1968).

3.1.3. Range Estimates for Ray and Mode Theory

The concepts of ray and mode theory are equivalent, representing each different approximations of a vector potential developed into series which are valid within two different ranges of convergence (Volland, 1968, 1984). This can be illustrated in the case of VLF waves.

The interference minima of the two rays in Equation 3.2 occur at

$$k(r_1 - \rho_m) - \pi = (2m - 1)\pi \quad (m = 1, 2, \ldots) \tag{3.9}$$

(with $r_1 = \sqrt{\rho^2 + 4h^2} \simeq \rho + 2h^2/\rho$) or at distances $\rho_m \simeq kh^2/(m\pi)$. On the other hand, according to Equations 3.5 and 3.6, the interference minima of the two modes in Equation 3.3 occur at

$$(C_2^2 - C_1^2)k\rho_m/2 = (2m' - 1)\pi \quad (m' = 1, 2, \ldots)$$

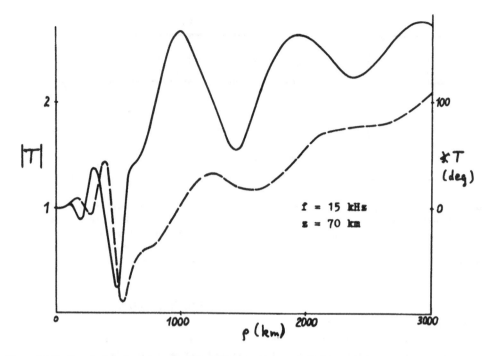

Figure 3.3.3 Transfer function in magnitude (solid line) and phase (dashed line) vs. distance ρ. The transmitter is a vertical dipole located on the ground and radiating at 15 kHz. The virtual reflection height is at 70 km simulating daytime conditions. The minimum near 500 km is the last interference minimum between ground wave and first hop sky wave. It is also the first interference minimum between first and second mode. (From Volland, H., *Atmospheric Electrodynamics,* Springer-Verlag, Heidelberg, 1984. With permission.)

or at distances $\rho_{m'} \simeq (2m' - 1)kh^2/\pi$. Evidently, the last interference minimum of ray theory ($m = 1$) at the critical distance of

$$\rho_c = \rho_1 = kh^2/\pi \qquad (3.11)$$

is also the first interference minimum of mode theory ($m' = 1$). For VLF waves ($3 < f < 30$ kHz) and a virtual height of $h = 70$ km (daytime conditions at mid-latitudes), this critical distance is in the range $100 < \rho_c < 1000$ km.

Figure 3.3.3 shows the transfer function of the atmospheric waveguide vs. distance for a frequency $f = 15$ kHz and an ionospheric reflection height of $h = 70$ km. The minima at $\rho \leq 500$ km are due to destructive interference between ground wave and first sky wave in Equation 3.2. The equally spaced interference minima at distances 500, 1500, and 2500 km are due to destructive interference between the two modes in Equation 3.3. At distances greater than about 3000 km, only the first mode remains of importance. Ray theory is thus a good approximation for $\rho < \rho_c$ while mode theory is appropriate at greater distances. ρ_c depends on frequency and local time. Ray theory has no meaning in the ELF range, even at small distances.

3.1.4. Estimate of Schumann Resonances

For Schumann resonances, the zeroth order mode is predominant, and the atmospheric waveguide behaves like a resonance cavity if the horizontal wavelength is an integral multiple m of the Earth's circumference of $2\pi a = 40$ Mm ($a = 6371$ km is the earth's radius). This yields resonance frequencies of

$$f_m \simeq \frac{mc}{2\pi a} = 7.5m \quad (m = 1, 2, \ldots)$$

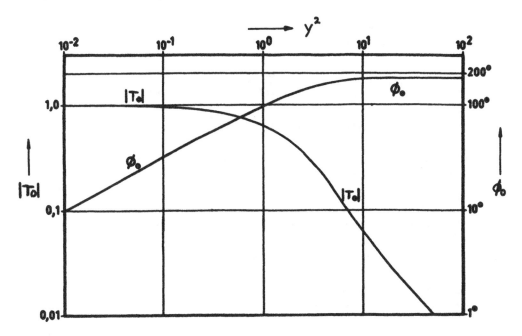

Figure 3.3.4   Approximate transfer function T_0 of ground wave in magnitude and phase vs. numerical distance y^2 (see also Equation 3.13).

The first Schumann resonance is at 7.5 Hz in reasonable agreement with the observations. The theory of Schumann resonances will be considered in some detail in Chapter I/11.

3.2.   RAY THEORY
3.2.1.   Ground Wave
In the VLF range, the ground wave dominates at distances smaller than about 1000 km depending on frequency (see Equation 3.11). Its transfer function for the vertical electric field of TM waves has been calculated by Sommerfeld (1952) (see also Wait, 1970; Richter, 1990; Mclean and Wu, 1993). Its amplitude and phase have been tabulated in various CCIR reports (CCIR, 1986a,b).

For a quick estimate, the following approximation is appropriate (Volland, 1982):

$$T_o(\omega, \rho) = \frac{X^2}{2(C - i\omega)(D - i\omega)} = \frac{1}{(1 - iy/A)(1 - 2iAy)} \tag{3.13}$$

with $X = \sqrt{2c\sigma_e/\varepsilon_0\rho}$, σ_e the electric conductivity of the ground [which is of the order $10^{-4} - 10^{-2}$ S/m for the solid ground and about 4 S/m for sea water (e.g., CCIR, 1986c)], $\varepsilon_0 = 8.854 \cdot 10^{-12}$ F/m the permeability of free space, $C = AX$, $D = X/2A$, $A = 0.49 + i \, 0.85$ an empirical complex number, and $y = \omega/X$. The quantity $y^2 = \omega^2/X^2 = 5.82 \cdot 10^{-19} \, \rho f^2/\sigma_e$ is called the numerical distance. For large ω and/or ρ, y becomes large, and $T_o \rightarrow -1/2y^2$ which is proportional to $1/\rho f^2$. In Figure 3.3.4, T_o is plotted in magnitude and phase vs. the numerical distance. For numerical distances larger than 0.5, the effect of the curved earth must be taken into account (e.g., Wait, 1970). For VLF waves at intermediate distances ($\rho < 1000$ km), y becomes small ($y^2 < 0.17$ and $|T_o| > 0.9$ if $\sigma_e = 3 \cdot 10^{-3}$ S/m), and the effect of the finite ground conductivity is of minor importance for sferics except for regions of low ground conductivity (Ferguson, 1992). However, it may be of some significance for LF/VLF navigation systems like LORAN-C or OMEGA (e.g., Cooray and Orville, 1989; Swanson, 1982).

The vertical electric radiation field of the electric dipole in Equation 2.8 is now modified according to Equations 2.11, 2.12, and 3.13 as

$$E_z(\rho, t) = \frac{\bar{\varrho}\mu I_o X^2}{4\pi\rho} \, Real \left\{ \frac{\alpha e^{-\alpha t}}{(C - \alpha)(D - \alpha)} - \frac{\beta e^{-\beta t}}{(C - \beta)(D - \beta)} \right. $$
$$\left. + \frac{(\beta - \alpha)}{(D - C)} \left[\frac{C e^{-Ct}}{(\alpha - C)(\beta - C)} - \frac{D e^{-Dt}}{(\alpha - D)(\beta - D)} \right] \right\} \quad (3.14)$$

Figure 3.3.5 shows the effect of the finite electric conductivity on the waveform of the K stroke in Figure 3.2.2. The maximum amplitude has decreased in amplitude and is delayed by a few microseconds (the delay depending on the numerical value of σ_e). Also shown in Figure 3.3.5 is the effect of the finite channel length centered above ground at $H = 5$ km (dotted curve). Radiation from the lower part of the channel reaching down to a height of 2.8 km arrives about 1 μs earlier at the receiver than the signal from the point at height H. The result of Figure 3.3.5 shows that the finite ground conductivity affects mainly the higher frequencies of the waveform at $f \geq 30$ kHz (see also Figure 3.3.9).

On the ground, a horizontal electric field component of the TM mode in ρ direction exists which is related to E_z as (e.g., Sommerfeld, 1952; Wait, 1970)

$$E_\rho = E_z/n_e \quad (3.15)$$

where

$$n_e = \sqrt{\varepsilon' + \frac{i\sigma_e}{\varepsilon_o \omega}} \simeq (1 + i)/b_e \quad (3.16)$$

(with $b_e = \sqrt{2\omega\varepsilon_o/\sigma_e} = \frac{1}{3000} \sqrt{f/\sigma_e}$; σ_e in S/m; f in kHz) is the refractive index of the ground. ε' is the relative dielectric constant which is of the order 1 to 10 for the crust and 80 for sea water (e.g., CCIR, 1986c). For long waves, ε' can be neglected because the imaginary term in the square root of Equation 3.16 becomes much larger then ε', and $|n_e| > 40$.

E_z and E_ρ move on a polarization ellipse. Its measurement yields an estimate for the ground conductivity σ_e (e.g., Labson et al., 1985). VLF waves penetrate almost vertically into the ground. The skin depth for these waves, which is the e-folding depth of the amplitude, is given by

$$z_m = c \sqrt{\frac{2\varepsilon_o}{\sigma_e \omega}} = \frac{503}{\sqrt{\sigma_e f}} \quad (3.17)$$

(z_m in m; σ_e in S/m; f in Hz) and thus larger than about 30 m for long waves decreasing proportional to the square root of the reciprocal frequency. In a vertically layered ground with layers of different values of the electric conductivity, one can still use formulae 3.15 to 3.17 if an effective conductivity is introduced (e.g., Grosskopf, 1970; Wait, 1970). Such effective conductivity depends on frequency. If the lower layer is more conductive than the upper layer, the effective conductivity will decrease with increasing frequency because of the decreasing skin depth.

Propagation of the ground wave over a mixed path (e.g., sea-continent or grounds with different electric conductivities) leads to the so-called "recovery" effect. If the propagation path is partly over a region with lower electric conductivity (e.g., a continent), partly over a region with higher conductivity (e.g., an ocean), and if the transmitter is located on the continent, the field strength increases suddenly when the signal passes the coastline and then follows the behavior of a signal as if it had propagated exclusively over the ocean. If, on the other hand, the transmitter is located

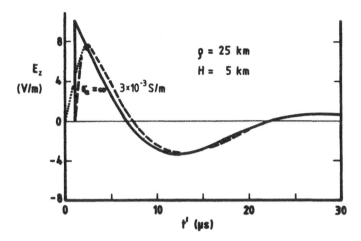

Figure 3.3.5 Waveform of radiation component of vertical electric field of the type 2 intracloud K stroke $(C - K)$ from Table 3.2.1 vs. time. The distance from the lightning channel is assumed to be 25 km. Solid line: dipole radiation over a perfectly conducting ground (Equation 2.8). Dashed line: dipole radiation over an imperfectly conducting ground ($\sigma_e = 0.003$ S/m; see Equation 3.14). Dotted line: radiation from channel of length $2\ell = 4.4$ km located at height $H = 5$ km above ground. (Adopted from Volland, H., *Handbook of Atmospherics,* Vol. I, CRC Press, Boca Raton, FL, 1982, 179. With permission.)

on the ocean and the receiver on the continent, the field strength decreases drastically as compared to the propagation over the ocean. For more details of propagation over multisection terrain, see Furutsu (1982) and Mclean and Wu (1993).

Oblique propagation over a coastline can lead to serious bearing errors. Other bearing errors of VLF waves depend on the environment of the antenna (large buildings), on mountain areas, etc. (Wait, 1992).

The vertical electric radiation field (far field) of the ground wave of a horizontal electric dipole (TE wave) located at the ground is smaller by the factor $\cos\phi/n_e$ (with n_e from Equation 3.16) compared to the vertical electric field of the TM wave of the vertical dipole (Sommerfeld, 1952; Wait, 1970); ϕ is the angle between ray direction and dipole axis. Evidently, strongest radiation is in the direction of the dipole axis. Its amplitude becomes very small compared to the TM wave in the longwave range ($|E_{TE}/E_{TM}| < 1/40$). This is the reason why the horizontal branches of lightning strokes do not contribute very much to the longwave radio noise, apart perhaps from excitation of whistler signals (see Chapter II/7).

3.2.2. Sky Waves

Taking multihop sky waves into account in the ray approach, the transfer function of the vertical electric field of a vertical electric dipole in the far field region ($\rho \ll \lambda$) becomes (e.g., Volland, 1968)

$$T = T_0(\omega, \rho) + \sum_{n=1}^{\infty} R_i^n R_e^{n-1} \frac{(1 + R_e)^2}{2} \sin^3\theta_n e^{ik(r_n - \rho)} \quad (\rho \gg \lambda) \qquad (3.18)$$

where T_0 is the transfer function of the ground wave from Equation 3.13, ρ the distance between dipole and receiver (both assumed to be located on the ground), θ_n the angle of incidence of the nth hop sky wave ($\tan\theta_n = \rho/2nh$), r_n its phase path ($r_n^2 = \rho^2 + 4n^2h^2$), h the virtual reflection height (see Figure 3.3.1), and R_e and R_i the reflection factors of the ground and ionosphere, respectively.

For distances larger than about 500 km, the curved earth must be taken into account. The angles of incidence at the ground and at the ionosphere differ by $\sin\theta_{ni}/\sin\theta_{ne} = a/(a + h)$, and r_n depends on the distance ρ as

$$r_n^2 = 4n^2[(a + h)^2 + a^2 - 2a(a + h)\cos(\rho/2na)] \simeq 4n^2h^2 + \rho^2(1 + h/a)$$

Clearly, ray approximation breaks down for distances beyond the grazing angle $\theta_{ne} = 90°$.

R_e can be described by the Fresnel reflection factor for TM waves:

$$R_e = \frac{\eta_e \cos\theta - 1}{\eta_e \cos\theta + 1} \tag{3.19}$$

with

$$\eta_e = \frac{n_e^2}{\sqrt{n_e^2 - \sin^2\theta_n}} \simeq n_e$$

a wave impedance (proportional to the ratio between the horizontal components of the magnetic and the electric field strength on the ground), and n_e the refractive index from Equation 3.16. Since n_e is complex, the magnitude of R_e has a minimum at a quasi-Brewster angle at $\theta_{Be} \simeq \arccos(1/|n_e|)$. For $\theta < \theta_{Be}$ one can approximate Equation 3.19 by

$$R_e \simeq e^{-(1-i)b_e/\cos\theta} \quad (\theta < \theta_{Be}) \tag{3.20}$$

(with b_e from Equation 3.16). This is a good approximation over the whole ELF/VLF range. Evidently, $R_e \to 1$ for small b_e (electric wall; see Section 3.1.2). For $\theta > \theta_{Be}$, we use the approximation

$$R_e \simeq e^{-2(1+i)\cos\theta/b_e + i\pi} \quad (\theta > \theta_{Be}) \tag{3.21}$$

This expression will be needed in the mode theory of LF waves (see Section 3.3.2). Here, $R_e \to -1$ for $\theta \to 90°$ (magnetic wall).

The plasma of the ionospheric D region where longwaves are reflected is maintained by solar X and EUV radiation at wavelengths smaller than 122 nm (e.g., Hargreaves, 1979). For long-waves, the ionospheric D layer can be simulated in a first approximation by an equivalent sharply bounded homogeneous and isotropic half space at the virtual height h. Its reflection factor has then the same form as Equation 3.19 if one replaces n_e by an effective refraction index of the ionospheric D layer given by

$$n_i^2 = 1 - \frac{\omega_p^2}{\omega^2 + \nu^2} + \frac{i\omega_p^2\nu}{(\omega^2 + \nu^2)\omega} = \varepsilon_i + \frac{i\sigma_i}{\varepsilon_0\omega} \tag{3.22}$$

with

$$\varepsilon_i = 1 - \frac{\omega_p^2}{\omega^2 + \nu^2} \simeq 1$$

an effective dielectric "constant",

$$\sigma_i = \frac{\omega_p^2\nu\varepsilon_0}{\omega^2 + \nu^2} \simeq \frac{\omega_p^2\varepsilon_0}{\nu}$$

Table 3.3.1   Ionospheric Parametric Numbers of the Formulas in Equation 3.27 for Average Day and Night Conditions

	VLF				ELF			
	g_o	Ψ	h_o	μ	c_o	ν	h_o	κ
Day	0.65	0.29	0.44	0.99	0.46	0.68	58	0.065
Night	0.46	0.17	0.26	0.76	0.32	0.69	91	0.065

Note: The ELF parameters are valid in the frequency interval 10 Hz to 2 kHz and are adapted to the data of Bannister (1993).

an effective ionospheric conductivity, $\omega_p = \sqrt{Ne^2/m\varepsilon_o} = 56.4\sqrt{N}$ (with ω_p in s^{-1} and N in m^{-3}) the plasma frequency of electrons, N the electron number density, m their mass, e the electric elementary charge, and ν the collision rate between electrons and neutrals ($\nu > \omega$ for long waves) (e.g., Volland, 1968; Wait, 1970; Harth, 1982).

The virtual reflection height of VLF waves reacts to the diurnal and seasonal change of the ionospheric plasma in a rather regular manner during undisturbed conditions. The virtual reflection height of waves at the frequency $f = 16$ kHz over short distances at north hemispheric mid-latitudes varies like

$$h(\chi) = h_o + H/\cos \chi \quad (\chi < 85°)$$

with $h_o = 69$ km, χ the solar zenith angle, and the scale height H varies from 4 km in January to 7 km in June (Straker, 1955). There exists a slight dependency on the 11-year solar cycle with EUV flux increasing from solar minimum to solar maximum. Furthermore, h depends on geomagnetic activity. Over larger distances, the variation of the reflection height between day and night is somewhat smaller (e.g., Thomson, 1993).

The virtual reflection height of ELF waves during sunlit hours varies between about 45 and 60 km increasing with frequency (Bannister, 1993); (see also h in Equation 3.27 and in Table 3.3.1).

3.2.3.   Ionospheric Reflection Factors

VLF waves are reflected in the height range between about 70 and 90 km. At these altitudes, N increases and ν decreases almost exponentially with altitude depending on local time, season, and geographic location. The ionospheric reflection factor must therefore be determined numerically, and it may vary appreciably depending on the actual conditions of the ionosphere (e.g., Budden, 1985).

For a quick estimate of its behavior, it is often sufficient to assume that plane TM waves are reflected on a flat, vertically layered isotropic plasma with exponential profiles of its parameters N and ν (Wait and Spies, 1964; Harth, 1982):

$$N = N_o e^{p(z-h_p)}; \quad \nu = \nu_o e^{-q(z-h_q)} \quad (z < 100 \text{ km}) \tag{3.23}$$

with $N_o = 3 \cdot 10^8$ m^{-3}; $p = 0.15$ km^{-1}; $h_p = 70$ km ($p = 0.35$ km^{-1}; $h_p = 85$ km) during day (night), and $\nu_o = 5 \cdot 10^6$ s^{-1}; $q = 0.15$ km^{-1}; $h_q = 70$ km, so that the effective conductivity in Equation 3.22 is almost independent of frequency and increases with height as

$$\sigma_i \simeq \frac{\omega_p^2 \varepsilon_o}{\nu} = \sigma_o e^{r(z-h_q)} \tag{3.24}$$

with $\sigma_o = 1.7 \cdot 10^{-6}$ S/m; $r = 0.3$ km^{-1} for daytime conditions, and $\sigma_o = 1.5 \cdot 10^{-9}$ S/m; $r = 0.5$ km^{-1} for nighttime conditions.

The height where $\sigma_i = 2.2 \cdot 10^{-6}$ S/m has been found to be a convenient measure of the virtual reflection height of the ionosphere for VLF waves (Wait and Spies, 1964). Using Equation 3.24, this occurs at $h = 71$ and 85 km for day and night time conditions, respectively.

Figure 3.3.6a,b shows the magnitudes of the reflection factors of TM waves at different frequencies for day and night time conditions, respectively, calculated with model parameters of the form of Equation 3.24 valid for mid-latitudes. The minimum of $|R_i|$ at the quasi-Brewster angle θ_{Bi} depends on frequency. For a rough estimate, one can approximate the ionospheric reflection factor by the expressions

$$R_i \simeq e^{-(b_i - ic_i)/\cos\theta} \quad (\theta < \theta_{Bi}) \tag{3.25}$$

$$R_i \simeq e^{-2g_i \cos\theta + i\pi} \quad (\theta > \theta_{Bi}) \tag{3.26}$$

with θ_{Bi} the quasi-Brewster angle, and $b_i - ic_i \simeq 1/2n_i$. The form from Equation 3.25 is valid in the ELF range and that of Equation 3.26 in the VLF and LF range.

Empirical determinations of the average frequency dependence of the parameters in Equations 3.25 and 3.26 and of the virtual reflection height h of ELF waves are as follows:

$$g_i \simeq g_o f^{\psi}; \quad b_i \simeq b_o f^{\mu}; \quad c_i \simeq c_o f^{\nu}; \quad h \simeq h_o f^{\kappa} \tag{3.27}$$

(f in kHz). The numerical values of the coefficients are listed in Table 3.3.1 for day and night conditions.

In order to simulate the VLF reflection characteristics of the ionospheric D layer sufficiently well for a given propagation path, frequency, and actual ionospheric conditions, it is necessary to determine from a comparison between observations and theory an effective ionospheric homogeneous layer with effective values of ε_i and σ_i at a virtual height h (e.g., Thomson, 1993).

The geomagnetic field makes the ionospheric plasma anisotropic. Coupling occurs between ordinary and extraordinary waves. Both are elliptically polarized, their polarization angles rotating in opposite directions. A pure TM wave from below, after reflection at the lower boundary of the ionosphere, can be transformed into a superposition of a TM wave and a TE wave. This effect can be described by a reflection matrix of the form (e.g., Budden, 1985)

$$R_i = \begin{pmatrix} R_{11} & R_{12} \\ R_{21} & R_{22} \end{pmatrix}_i$$

where the symbol "1" means a TM wave (with horizontal polarization of the magnetic field), and "2" means a TE-wave (with horizontal polarization of the electric field). $(R_{11})_i$ is therefore the ratio between the reflected TM wave and the incident TM wave and is identical with R_i in Equation 3.18. $(R_{21})_i$ is a conversion factor describing the ratio between the reflected TE wave and the incident TM wave before reflection, etc. For VLF and ELF waves, only these two matrix components are of importance.

These components have been determined numerically (e.g., Johler and Harper, 1962; Volland, 1968; Wait, 1970; Harth, 1982) and have been verified by observations. They depend on the direction of the geomagnetic field relative to the propagation path. Reflection is weaker for waves propagating from east to west than for waves propagating from west to east. Moreover, reflection is stronger during night rather than during day and in winter rather than in summer (e.g., Bickel et al., 1970; Pappert and Hitney, 1988).

For short distances (<300 km), the ratio R_{21}/R_{11} may become of the order one (larger during the night than during the day) with a 90° phase shift (leading to reflected circularly polarized

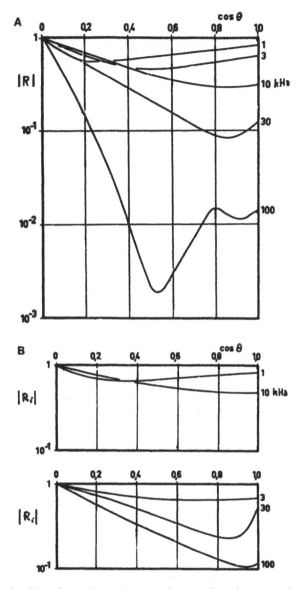

Figure 3.3.6 Magnitudes of the reflection factors R_i vs. cosθ for ionospheric daytime conditions (Figure 3.3.6a) and nighttime conditions (Figure 3.3.6b) at northern mid-latitudes calculated for different frequencies.

waves). Over larger distances, the polarization remains nearly linear. The excitation of TE waves after ionospheric reflection of TM waves at shorter distances may thus cause serious bearing errors, since direction finding methods are based on the measurement of the horizontal magnetic component of the electromagnetic wave (e.g., Watt, 1967).

Equation 3.18 can still be used if one replaces R_e with a reflection matrix for the ground of the form

$$R_e = \begin{pmatrix} R_{11} & 0 \\ 0 & R_{22} \end{pmatrix}_e$$

with $(R_{11})_e$ corresponding to R_e in Equation 3.19, and $(R_{22})_e$ corresponding to a Fresnel reflection factor for TE waves.

3.3. FULL WAVE THEORY

3.3.1. Basic Equation

Full wave theory is appropriate for the description of the propagation of VLF waves at distances $\rho > \rho_c$ (see Equation 3.11) and for ELF waves. The transfer function of the vertical electric field of TM waves in terms of full wave theory in the far field ($\rho \gg \lambda$) is (see e.g., Wait, 1970, 1981; Galej, 1972)

$$T = \frac{1}{h} \sqrt{\frac{i\rho\lambda\Theta}{\sin\Theta}} \sum_n K_n e^{-A_n\rho + iB_n k\rho + i(1+B_n)hk\rho/2a} \quad (\rho \gg \lambda) \qquad (3.28)$$

(valid for $\Theta < \pi/2$) with $\lambda = c/f$ the wavelength, $\Theta = \rho/a$ the polar distance between transmitter and receiver (both assumed to be located on the ground), ($a = 6371$ km is the Earth's radius), K_n amplitude factors, A_n attenuation factors, and B_n phase factors. The geometric factor $\Theta/\sin\Theta$ in Equation 3.28 accounts for the focusing effect of the sphere. The last exponential factor in Equation 3.28 is responsible for the enlarged phase path within the spherical waveguide. In the near field ($\rho < \lambda$), the exponential functions in Equation 3.28 must be replaced by Hankel functions in the case of VLF waves or by hypergeometric functions in the case of ELF waves where the wavelengths became already comparable with the dimension of the spherical waveguide (e.g., Booker, 1980; Bannister, 1986; Burrows, 1978; Greifinger and Greifinger, 1978, 1979).

At distances greater than about 10 Mm, the transfer function T in Equation 3.28 should be replaced by more complicated formulae, taking into account the spherical configuration of the atmospheric waveguide. This is necessary in particular for the lower ELF band. Near the antipode ($\Theta = \pi$), the field strength remains limited there and slightly increases (Wait, 1970). Moreover, the low attenuation rate of ELF waves allows bidirectional propagation over both the direct and the antipodal path, resulting in a spatial interference pattern (e.g., Bannister, 1974).

More sophisticated calculations of the transfer function with transmitter and receiver above the ground and/or for a horizontal electric dipole can be found in the books of Wait (1970) and Galej (1972) (see also Field et al., 1986a). Computer programs exist to calculate the parameters of the transfer function 3.28 for various conditions (Morfitt, 1978; Morfitt and Shellman, 1976; Shellman, 1986; Ferguson and Snyder, 1980).

The $\sqrt{\rho}$ dependence of T in Equation 3.28 is remarkable. The electric field strength of the pulse in Equations 2.11 and 2.12 thus decreases as $1/\sqrt{\rho}$ in the flat waveguide due to the ducting of electromagnetic energy within this bounded waveguide. In the case of the spherical waveguide and at larger distances (where $\sin\Theta \approx 1$), T is proportional to ρ. This almost compensates the $1/\rho$ dependence of the source function in Equation 2.12 so that only the attenuation factor A_n in Equation 3.28 is responsible for the field strength decrease with distance.

The amplitude factors K_n in Equation 3.28 must be determined from the exact solution of the vector potential (e.g., Wait, 1970). In Figure 3.3.7, the magnitude of K_n is plotted vs. frequency for daytime conditions at mid-latitudes using model parameters of the ionosphere of the form in Equation 3.24. The phase factors B_n and the attenuation factors A_n in Equation 3.28 must be derived from the eigenvalue Equation 3.4. It is convenient to express the factor A_n in units of dB/Mm (1 Mm = 1000 km) using the conversion (e.g., Davies, 1990)

$$A_n^* = 8.686 \cdot 10^6 A_n \qquad (3.29)$$

(with A_n^* in dB/Mm and A_n in m^{-1}).

Figure 3.3.8 shows the frequency dependence of the attenuation factors A_n^*. Again, daytime conditions at mid-latitudes are assumed. It follows from Figure 3.3.8 that the zeroth order mode dominates at frequencies $f < 2$ kHz, while the first mode dominates between about 2 and 40 kHz

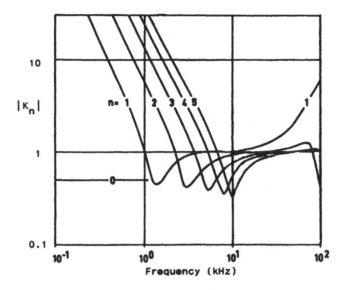

Figure 3.3.7 Magnitude of amplitude factors K_n vs. frequency for ionospheric daytime conditions at northern mid-latitudes for different wavemodes n.

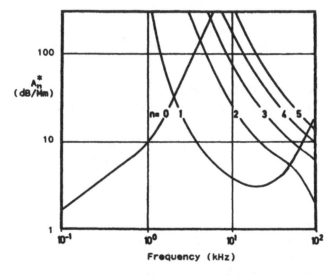

Figure 3.3.8 Attenuation factors A_n^* vs. frequency for ionospheric daytime conditions at northern mid-latitudes for different wavemodes n.

with minimum attenuation near 20 kHz. The higher order modes become more important at $f > 40$ kHz. The atmospheric waveguide therefore has two windows, one in the ELF range and the other one in the VLF range. This can be seen in Figure 3.3.9, which shows the magnitude of the spectral field strength $|\hat{E}_o T_n|$ vs. frequency at different distances for the sferic $G - R_3$ of the return stroke from Table 3.2.1. The transfer function used in Equation 2.11 is that of mode 0 for $f < 2$ kHz and of mode 1 for $f > 2$ kHz. Since $T \cong 1$ for $f < 30$ kHz and $\rho < 200$ km, the upper curve in Figure 3.3.9 is the magnitude of $\hat{E}_o$ in Equation 2.12. The difference between the solid and the dashed line indicates the attenuation of the ground wave at frequencies >30 kHz due to the finite electric conductivity of the ground. The bandpass effect shifts the maximum of the spectral amplitude from 4 kHz at distances <200 km to 12 kHz at $\rho \cong 10$ Mm.

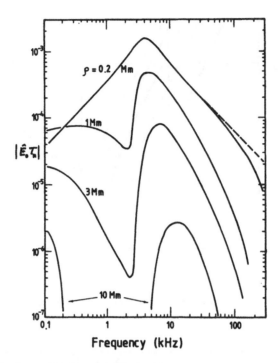

Figure 3.3.9 Magnitude of spectral field strength $|\hat{E}_o T_n|$ of type 2 return stroke $G - R_3$ from Table 3.2.1 vs. frequency at different distances. Dashed curve at 200 km distance yields for a perfectly conducting ground.

3.3.2. Approximations of Mode Parameters

The amplitude factors in Equation 3.28 can be written as (e.g., Volland, 1968)

$$K_n = \frac{S_n^{3/2}\,(1 + R_e + 1/R_e)}{1 + \dfrac{i}{2hkR_iR_e}\left[\dfrac{\partial(R_iR_e)}{\partial C}\right]_{C=C_n}} \tag{3.30}$$

(see below for C_n and S_n). The dominant amplitude factors are $K_0 \simeq 0.5$ in the ELF range, $K_1 \simeq 1$ in the VLF range, and $K_2 \simeq 1$ in the lower LF range (see Figure 3.3.7).

For an approximation of the phase and attenuation factors in Equation 3.28, we use the simplified expressions in Equations 3.20, 3.21, 3.25, and 3.26, replace $\cos\theta$ by C, and apply the parametric values of Equation 3.27 and Table 3.3.1. C is now an eigenvalue in Equation 3.4 which may be complex and larger than one (see Section 3.1.2).

For ELF waves (strictly speaking, for $|C| > C_{Bi}, C_{Be}$) we introduce Equation 3.20 and Equation 3.25 into Equation 3.4 and obtain

$$C_0 = \sqrt{\frac{-(c_i + b_e) - i(b_i + b_e)}{2kh}} \tag{3.31}$$

$$C_n = \frac{n\pi}{2kh}\left\{1 + \sqrt{1 - 2[c_i + b_e + i(b_i + b_e)]kh/(n^2\pi^2)}\right\} \tag{3.32}$$

$$\simeq \overline{C}_{n1} - \frac{c_i + b_e}{2n\pi} - \frac{i(b_i + b_e)}{2n\pi} \quad (n \geq 1)$$

with $\overline{C}_{n1}$ from Equation 3.7. The last approximation is valid as long as $b_i/(kh\overline{C}_{n1}^2) \ll 1$. Combining Equations 3.5 and 3.28 to obtain

$$B_n + iA_n/k = [\sqrt{1 - C_n^2} - 1] \tag{3.33}$$

it follows that

$$B_0 \simeq \frac{c_i + b_e}{4hk}; \quad A_0 \simeq \frac{b_i + b_e}{4h} \tag{3.34}$$

and

$$B_n \simeq \frac{b_i + b_e}{2n\pi} - 1; \, A_n \simeq k\overline{C}_{n1} \quad (n \geq 1) \tag{3.35}$$

The approximation, Equation 3.34 is valid for $b_i/(kh)$, $c_i/(kh) < 1$. In general, $b_e \ll b_i$, c_i so that the influence of the ground conductivity is negligible for ELF waves.

For VLF waves (strictly speaking, for $|C| < C_{Bi}$; $|C| > C_{Be}$), we introduce the expressions 3.20 and 3.26 into 3.4 and obtain

$$\begin{aligned}
C_n &= \frac{(2n - 1)\pi}{4(kh + ig_i)} \left\{ 1 + \sqrt{1 - \frac{2(1 + i)b_e(kh + ig_i)}{(n - 1/2)^2\pi^2}} \right\} \\
&\simeq \overline{C}_{n2} - \frac{b_e}{(2n - 1)\pi} - i \left\{ \frac{\overline{C}_{n2}g_i}{kh} + \frac{b_e}{(2n - 1)\pi} \right\} \quad (n \geq 1)
\end{aligned} \tag{3.36}$$

with $\overline{C}_{n2}$ from Equation 3.6. The last approximation is valid as long as $g_i/(kh) \ll 1$. Furthermore,

$$B_n \simeq -\frac{(\overline{C}_{n2})^2}{2} + \frac{b_e}{2kh}; \quad A_n \simeq \frac{(\overline{C}_{n2})^2 g_i}{h} + \frac{b_e}{2h} \tag{3.37}$$

The contributions from the ground conductivity and the ionosphere to A_n in Equation 3.37 are of the same order of magnitude for VLF waves. The attenuation rate A_1 of the first mode in Equation 3.37 has a minimum at the frequency (see Figure 3.3.8)

$$f_{min} \simeq \left[6.75 \cdot 10^7 \frac{(2 - \psi)g_0\sqrt{\sigma_e}}{h^2} \right]^{\frac{1}{2.5 - \psi}} \tag{3.38}$$

(f_{min} in kHz; h in km). This yields $f_{min} = 21$ kHz (13 kHz) during the day (night) for the parametric values $\sigma_e = 0.003$ S/m and $h = 70$ (90) km. The ground conductivity dominates for $f > f_{min}$, while for $f < f_{min}$ the ionosphere contributes mainly to the attenuation. f_{min} depends critically on the numerical value of the ground conductivity. For VLF propagation over the oceans ($\sigma_e \simeq 4$ S/m), b_e can be neglected. However, for propagation across the arctic or antarctic polar ice caps, where σ_e may be as low as 10^{-5} S/m, the attenuation rate can increase by a factor of ten or more. The ionospheric component of A_n, proportional to n^2, produces greater attenuation for higher order modes. A_n is also proportional to $1/h^3$, resulting in smaller attenuation during the night than during sunlit hours. Typical values for A_1^* are 2 to 4 dB/Mm during the day and 0.5 to 2 dB/Mm during the night (e.g., Galej, 1972; Davies, 1990). Optimum propagation conditions exist for west-to-east propagation at mid-latitudes over the ocean at night. Tweek-atmospherics, which are narrow band signals near 2 to 4 kHz lasting for several tens of milliseconds, may be generated near cutoff of the first mode during night time conditions (e.g., Yano et al., 1989; Sukhorukov et al., 1992).

Finally, for LF waves it is generally true that $|C| < C_{Bi}, C_{Be}$. From Equations 3.21, 3.26, and 3.4 it now yields

$$C_n \simeq \frac{kh\overline{C}_{n3}}{kh + ig_i - (1 - i)/b_e} \tag{3.39}$$

with $\overline{C}_{n3}$ from Equation 3.8, and with $g_i \ll 1/b_e$ follows

$$B_n \simeq -\frac{\overline{C}_{n3}}{2}\left(1 + \frac{2}{khb_e}\right); \quad A_n \simeq \frac{\overline{C}_{n3}}{hb_e} \tag{3.40}$$

Although the attenuation factors of the LF wave modes decrease with frequency, the amplitude factors K_n in Equation 3.28 also decrease so that the higher order modes assume dominance with increasing frequency (depending on distance) and give rise to a complicated interference pattern (Volland, 1968). A simple empirical formula from Austin (1914) is quite appropriate for estimating the attenuation rate of LF waves at distances greater than about 1500 km and at frequencies between about 50 and 150 kHz:

$$|T| = \frac{1.2}{h}\sqrt{\frac{\rho\lambda\Theta}{\sin\Theta}}e^{-A\rho} \tag{3.41}$$

with $A = A_o \cdot 10^{-5}f^{0.6}$ (f in kHz) and $A_o = 4.9$ (2.9) km^{-1} for day (night) conditions.

The effect of the geomagnetic field and the curvature of the earth must be included in more sophisticated models (e.g., Wait, 1970; Galej, 1972; Harth, 1982; Pappert and Hitley, 1988). The geomagnetic field makes the ionosphere anisotropic so that the factors A_n depend on the direction of the propagation path with respect to the geomagnetic field. This is most pronounced under nighttime conditions at mid-latitudes and for long paths across the geomagnetic equator (e.g., Araki, 1973; Hildebrand, 1993).

Figure 3.3.10 shows calculated attenuation factors A_1^* of the first mode vs. an anisotropic factor Ω for a daytime model (upper curves) and a nighttime model (lower curves) at northern mid-latitudes for different frequencies. Solid lines are calculated for a flat model, dashed lines for a spherical model. The anisotropic factor is related to the ratio between the electron gyrofrequency ω_H and the collision rate ν (see Equation 3.23) at a height near the reflection point:

$$\frac{\omega_H}{\nu} = \frac{|eB_i|}{m\nu} = 1.76 \cdot 10^{11} |B_i|/\nu \tag{3.42}$$

with e and m electric charge and mass of the electron, respectively, and B_i (in Tesla) the geomagnetic field (assumed to be homogeneous within the area of wave propagation). The number $\Omega = 1$ corresponds to east-to-west propagation at northern mid-latitudes ($B_i \simeq -50\,\mu T$; dip angle $\simeq 67°$). The number $\Omega = -1$ corresponds to west-to-east propagation, and the number $\Omega = 0$ to north-to-south or south-to-north propagation. Clearly, attenuation is weaker during the night than during the day, and weaker for west-to-east propagation than for east-to-west propagation. This has been verified by observations (e.g., Bickel et al., 1970; Grandt, 1992).

For propagation paths crossing the day-night terminator, it is often convenient to simulate such mixed paths by using the arithmetic mean of the phase and attenuation factors:

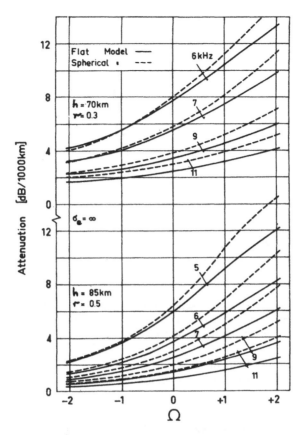

Figure 3.3.10 Attenuation factor A_1^* vs. anisotropic factor Ω calculated for a daytime model (upper panel) and a nighttime model (lower panel) for northern mid-latitudes at different VLF frequencies. Solid curves are based on a flat model of the atmospheric waveguide. Dashed curves are based on a spherical model. (Adopted from Harth, W., *Handbook of Atmospherics,* Vol. II, CRC Press, Boca Raton, FL, 1982, 133. With permission.)

$$\bar{B}_{np} \simeq (B_n)_1 \rho_1 + (B_n)_2 \rho_2; \quad \bar{A}_n \rho \simeq (A_n)_1 \rho_1 + (A_n)_2 \rho_2 \tag{3.43}$$

and geometric means of the amplitude factors K_n in Equation 3.28. ρ_1 and ρ_2 are the distances in darkness and in daylight, respectively, and the $(A_n)_i$ and $(B_n)_i$ are the appropriate values for the attenuation and phase factors in the areas of darkness and daylight.

Mode conversion, e.g., between mode one and mode two, can occur preferentially near the terminator between day and night (e.g., Reder and Westerlund, 1970; Pappert and Ferguson, 1986; Wait, 1991; Kossey and Lewis, 1992; Hildebrand, 1993). If both modes (or ground wave and first sky wave) are comparable in magnitude and out of phase by about π, cycle slippage can occur, which means a gain or loss of 2π (e.g., Frisius, 1970; Kikuchi and Ohtani, 1984; Reder and Westerlund, 1970). Azimuthal inhomogeneity of the ground and/or the D region may lead to diffraction of the signals. For instance, VLF waves pathing the polar ice cap are diffracted around the edge of the ice cap (e.g., Barr, 1987).

No exact theory of mode propagation in the real atmospheric waveguide presently exists, which accounts for horizontal inhomogeneities (e.g., land-ocean coverage and/or day-night paths) and the real geomagnetic dipole field. However, path segmentation is possile in computer programs that are based on reasonable model representations of the actual atmospheric waveguide (Ferguson and Snyder, 1980).

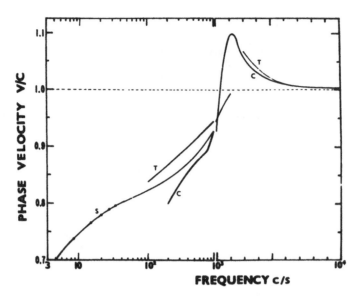

Figure 3.3.11 Phase velocity vs. frequency observed during daytime conditions. Results are derived from Schumann resonances (S), from single-station observations (T), and two-station observations (C). (From Chapman et al., *Radio Sci.*, 1, 1273, 1966. With permission. Copyright by American Geophysical Union.)

3.3.3. Phase and Group Velocities

The phase velocity υ_p of the wave modes in Equation 3.28 can be derived from the phase of the product $\hat{E}_oT_n$ which is

$$\Phi_n = -\omega t + (1 + B_n)(1 + h/2a)k\rho \qquad (3.44)$$

From the condition $\partial\Phi/\partial t = 0$ follows $\upsilon_p = \partial\rho/\partial t$ and

$$\frac{c}{(\upsilon_p)_n} = \left(1 + \frac{h}{2a}\right)(1 + B_n) \qquad (3.45)$$

$(1 + h/2a)_\rho$ is the mean phase path in the spherical waveguide. Their group velocity υ_g is determined from the condition $\partial\Phi/\partial\omega = 0$ as $\upsilon_g = \rho/t$ and thus,

$$\frac{c}{(\upsilon_g)_n} = \left(1 + \frac{h}{2a}\right)\left(1 + \frac{\partial(\omega B_n)}{\partial\omega}\right) \qquad (3.46)$$

It follows from Equation 3.37 that B_1 is negative in the lower VLF range so that for frequencies smaller than about 10 kHz the phase velocity of the first mode is larger than the speed of light. In the ELF range, B_0 in Equation 3.34 is positive so that its phase velocity is smaller than c. This is verified by the observations (Figure 3.3.11). According to Equations 3.34, 3.37, and 3.46, the group velocities of modes 0 and 1, respectively, are

$$\frac{c}{(\upsilon_g)_0} = \left(1 + \frac{h}{2a}\right)\left[1 + \frac{(\nu - \kappa)c_i\lambda}{8h\pi}\right] \quad \text{(ELF)} \qquad (3.47)$$

$$\frac{c}{(\upsilon_g)_1} = \left(1 + \frac{h}{2a)}\right)\left[1 + \frac{\lambda^2}{32h^2} + \frac{b_e\lambda}{8h\pi}\right] \quad \text{(VLF)} \qquad (3.48)$$

($\lambda = 2\pi/k$ is the wavelength) which are, as expected, smaller than c. The influence of the electric conductivity of the ground on the group velocity is small in the case of ELF waves, but of importance for VLF waves.

The frequency dependence of υ_g of VLF waves in Equation 3.48 can be used to measure the distance of sferics. If one records the spectral phases of a sferic at three adjacent frequencies (e.g., at 5, 7, and 9 kHz) and orders their difference in the following manner one obtains (provided the first mode predominates)

$$GDD = \Phi(9) + \Phi(5) - 2\Phi(7) \propto \frac{\rho}{h^2 f^3} \tag{3.49}$$

This quantity (the group delay time difference "*GDD*") is an approximation of the second derivative of Φ in Equation 3.44 with respect to the angular frequency and is proportional to the distance ρ (e.g., Grandt, 1992). The group velocity at 15 kHz is about 0.992 c, while that at 1 kHz is about 0.940 c. In slow tail sferic wave forms at distances larger than 5000 km, the ELF part thus arrives more than 1 ms later than the VLF part (e.g., Taylor and Sao, 1970; Sao and Jindoh, 1974; Sukhorukov, 1992).

A weak dependence of phase and group velocity on the geomagnetic field exists with the tendency of larger phase velocities for propagation from west to east than vice versa, depending on local time and geographic latitude (Galej, 1972).

3.3.4. ELF/VLF Disturbances

The D region is disturbed either directly by incoming enhanced solar radiation (solar flare effects; enhanced EUV flux from active regions on the sun) and solar protons (polar cap absorption), or indirectly by the variable solar wind interaction with the magnetosphere (storm after effects; auroral disturbances) (e.g., Hargreaves, 1979). Sporadic E events at heights above 90 km, which are caused by fluctuating atmospheric winds, can disturb ELF propagation because these waves can penetrate to those heights even during the day (Barr, 1977; Pappert, 1985).

A solar flare typically lasting for 30 to 60 min illuminates the dayside hemisphere and can enhance the electron density of the D layer by a factor of ten or more. During this time, the reflection height for VLF waves can shift down by 5 km and more. This will affect the interference pattern between ground wave and sky wave or between two modes (sudden phase anomaly; SPA), and will increase the attenuation factors of the wave modes. Depending on the distance between transmitter and receiver and on frequency, there may be an enhancement (sudden enhancement of signal strength; SES) or a decrease of the field strength (sudden decrease of signal strength; SDS), (e.g., Davies, 1990). Solar flares are much more frequent during solar activity maximum than during solar activity minimum.

If one observes the integrated sferics rate with a narrowband receiver at a given frequency during sunlit hours, local thunderstorm centers at distances <2000 km will contribute mainly to the sferics noise. During a solar flare, one observes a sudden enhancement of atmospherics (SEA) at frequencies larger than about 10 to 15 kHz and smaller than about 1 kHz, while there is a sudden decrease of atmospherics (SDA) in the frequency range $1 < f < 10$ kHz (Sao et al., 1970).

For a qualitative explanation of this behavior, we write the ratio between the magnitudes of the transfer function of the dominant mode during disturbed (symbol "D") and normal conditions (symbol "N"). According to Equation 3.28 this is

$$\frac{(|T_i|)_D}{(|T_i|)_N} = \frac{h_N}{h_D} e^{-(A_D - A_N)\rho} \tag{3.50}$$

In the range of minimum attenuation in the upper VLF range and the lower ELF range, the influence of the height change outweighs that of the attenuation factors. The amplitude ratio thus increases. In the frequency range of high attenuation in between, the effect of the change of the attenuation factors dominates, resulting in a decrease of the amplitude ratio.

Nuclear explosions can create disturbances in the height distribution of the ionospheric D region. The effects of VLF propagation are therefore similar to those during solar flares (Pierce, 1965; see also Chapter II/6).

Solar flares produce outbursts of highly energetic solar cosmic rays (mainly protons). These protons reach the upper atmosphere of the earth several hours later than the flare associated X and EUV radiation. Their propagation is modulated by the interplanetary magnetic field and, in particular, by the geomagnetic field. If their energy exceeds 30 MeV, they can reach the ionospheric D region within the polar caps and can ionize the neutral air. The so-enhanced electron density is responsible for excessive HF radio absorption (polar cap absorption; PCA), and also influences VLF propagation.

Large solar proton events, occurring on the average of once a month (mainly during solar maximum activity), can cover the entire polar cap above 60° magnetic latitude and can last up to 12 days. The effects of these events include a decrease in the virtual reflection height of VLF waves by as much as 10 km (e.g., Westerlund et al., 1969; Burgess, 1970; Sauer et al., 1987), as well as focusing and diffracting of VLF/ELF waves (e.g., Field et al., 1985, 1986b).

Excessive electron precipitation into the auroral regions during magnetospheric storms is responsible for an enhanced plasma density within the upper part of the D layer. This event is localized in space and time. VLF propagation through these regions is affected mainly during the night. Auroral effects are small during sunlit hours (Reder and Westerlund, 1970).

Energetic electrons, which may have been built up within the inner part of the magnetosphere, can be released during a geomagnetic storm. These electrons precipitate into the D region during the storm and continue to precipitate for several days after the geomagnetic data have returned to normal. This is called a storm after effect and is generally confined to latitudes poleward of about 50°. The reflection height of VLF waves is reduced during such events, particularly during the night (Belrose, 1964).

Lightning pulses can directly penetrate into the ionosphere (mainly during the night) and can locally heat the electrons producing secondary ionization at 90 to 95 km altitude (e.g., Inan et al., 1991). This effect may account for short (<50 ms) perturbations of VLF signals. In addition, lightning-induced whistlers can generate electron precipitation from the magnetosphere, also leading to a localized enhancement of the D layer ionization (see also Chapters II/7 and II/13) with the result of VLF propagation disturbances (≈0.4 dB in amplitude). Such events, sometimes called "Trimpi", involve rapidly (<2s) followed by a slower recovery (10 to 100 s) (e.g., Dowden and Adams, 1990).

REFERENCES

Araki, T. (1973). Anomalous diurnal changes of transequatorial VLF radio waves, *J. Atm. Terr. Phys.* 35, 693.

Austin, L.W. (1914). Quantitative experiments in radio telegraphic transmitters, *Bull. Bur. Standards (U.S.A.)* 11, 69.

Bannister, P.R. (1974). Far-field extremely low frequency (ELF) propagation measurements, 1970–1972, *IEEE Trans. Commun.* COM-22, 468.

Banister, P.R. (1986). Simplified formulas for ELF propagation at shorter distances, *Radio Sci.* 21, 529.

Bannister, P.R. (1993). ELF propagation highlights, in Belrose, J.S. (Ed.): *ELF/VLF/LF Radio Propagation and System Aspects*, AGARD Conference Proc. No. 529, Advisory Group on Aerospace Research and Development, NATO, 2–1.

Barr, R. (1977). The effect of sporadic-E on the nocturnal propagation of ELF radio waves, *J. Atm. Terr. Phys.* 39, 1379.

Barr, R. (1987). The diffraction of VLF radio waves by the Antarctic ice cap, *J. Atm. Terr. Phys.* 49, 1.

Belrose, J.S. (1964). Present knowledge of the lowest ionosphere, in Blackband, W.T. (Ed.): *Propagation of Radio Waves at Frequencies below 300 kc/s,* AGARDograph 74, Macmillan, New York, 3.

Bickel, J.E., Ferguson, J.A., and Stanley, G.V. (1970). Experimental observations of magnetic field effects on VLF propagation at night, *Radio Sci.* 5, 19.

Booker, H.G. (1980). A simplified theory of ELF propagation in the earth-ionosphere transmission line, *J. Atm. Terr. Phys.* 42, 929.

Budden, K.G. (1985). *The Theory of Wave Propagation,* Cambridge University Press, Cambridge, U.K.

Burgess, B. (1970). VLF phase delay variability and the design of long range navigation aids, in Davies, K. (Ed.): *Phase and Frequency Instabilities,* AGARD Conference Proc. No. 33, Advisory Group on Aerospace Research and Development, NATO, 19.

Burrows, M.L. (1978). *ELF Communication Antennas,* Peter Peregrinus, Stevenage, England.

CCIR (1986a). *Ground Wave Propagation Curves for Frequencies between 10 kHz and 30 MHz,* CCIR Report 368-5, Geneva.

CCIR (1986b). *The Phase of the Ground Wave,* CCIR-Report 716-2, Geneva.

CCIR (1986c). *World Atlas of Ground Conductivities,* CCIR-Report 717-2, Geneva.

Chapman, F.W., Jones, D.L., Todd, S.D.W., and Challinor, R.A. (1966). Observations on the propagation constant of the earth-ionosphere waveguide in the frequency band 8 c/s to 16 kc/s, *Radio Sci.* 1, 1273.

Cooray, V., and Orville, R.E. (1989). LORAN-C timing errors caused by propagation over finitely conducting ground, *Radio Sci.* 24, 179.

Davies, K. (1990). *Ionospheric Radio,* Peter Peregrinus, Exeter, England.

Dennis, A.S., and Pierce, E.T. (1964). The return stroke of the lightning flash to earth as a source of VLF atmospherics, *Radio Sci.* 68D, 777.

Dowden, R.L., and Adams, C.D.D. (1990). Lightning-induced perturbations on VLF subionospheric transmissions, *J. Atm. Terr. Phys.* 52, 357.

Ferguson, J.A. (1992). The effect of variability of ground conductivity on the calculation of atmospheric noise using thunderstorm-based models, *Radio Sci.* 27, 63.

Ferguson, J.A., and Snyder, F.P. (1980). *Approximate VLF/LF Waveguide Mode Conversion Model Computer Applications: FASTMC and BUMP,* Technical Document 400, Naval Ocean Systems Center, San Diego.

Field, E.C., Warber, C.R., and Joiner, R.G. (1985). Effects of the ionosphere on ELF signals during polar cap absorption events: comparison of theory and experiment, in Soicher, H. (Ed.): *Propagation Effects on Military Systems in the High Latitude Region,* AGARD Conference Proc. No. 382, Advisory Group on Aerospace Research and Development, NATO, p. 8.3-1.

Field, E.C., Warber, C.R., Kossey, P.A., Lewis, E.A., and Harrison, R.P. (1986a). Comparison of calculated and measured height profiles of transverse electric VLF signals across the daytime earth-ionosphere waveguide, *Radio Sci.* 21, 141.

Field, E.C., Warber, C.R., and Joiner, R.G. (1986b). Focusing and shadowing of ELF signals, *Radio Sci.* 21, 511.

Frisius, J. (1970). Observations of diurnal amplitude and phase variations on 16 kHz transmission path and interpretation by a simple propagation model, in Davies, K. (Ed.): *Phase and Frequency Instabilities,* AGARD Conference Proc. No 33, Advisory Group on Aerospace Research and Development, NATO, p. 76.

Furutsu, K. (1982). A systematic theory of wave propagation over irregular terrain, *Radio Sci.* 17, 1037.

Galej, J. (1972). *Terrestrial Propagation of Long Electromagnetic Waves,* Pergamon Press, Oxford.

Golde, R.H. (1977). Lightning and related phenomena, in Golde, R.H. (Ed.): *Lightning, Vol. I,* Academic Press, London, 309.

Grandt, C. (1992). Thunderstorm monitoring in South Africa and Europe by means of VLF sferics, *J. Geophys. Res.* 97, 18215.

Greifinger, C., and Greifinger, P. (1978). Approximate method for determining ELF eigenvalues in the earth-ionosphere waveguide, *Radio Sci.* 13, 831.

Greifinger, C., and Greifinger, P. (1979). On the ionospheric parameters which govern high latitude ELF propagation in the earth-ionosphere waveguide, *Radio Sci.* 14, 889.

Grosskopf, J. (1970). *Wave Propagation I* (Wellenausbreitung I), Bibliogr. Institut, Mannheim.

Hargreaves, J.K. (1979). *The Upper Atmosphere and Solar-Terrestrial Relations,* van Nostrand Reinhold Co., New York.

Harth, W. (1982). Theory of low frequency wave propagation, in Volland, H. (Ed.): *Handbook of Atmospherics, Vol. II,* CRC Press, Boca Raton, FL, p. 133.

Hildebrand, V. (1993). Investigations of equatorial ionosphere nighttime mode conversion at VLF, in Belrose, J. (Ed.): *ELF/VLF/LF Radio Propagation and System Aspects,* AGARD Conference Proc. No 529, Advisory Group on Aerospace Research and Development, NATO, p. 7-1.

Inan, U.S., Bell, T.F., and Rodriguez, J.V. (1991). Heating and ionization of the lower ionosphere by lightning, *Geophys. Res. Lett.* 18, 705.

Johler, J.R., and Harper, J.F. (1962). Reflection and transmission of radio waves at a continuously stratified plasma with arbitrary magnetic inclination, *J. Res. NBS* 66, 81.

Kikuchi, T., and Ohtani, A. (1984). Anomalous interference in OMEGA VLF wave propagation on east-to-west equatorial paths, *J. Atmos. Terr. Phys.* 46, 697.

Kossey, P.A., and Lewis, E.A. (1992). Propagation characteristics of the ionospheric transmission window relating to long wave radio location issues, AGARD Conference Proceedings 528, *Radiolocation Tech.*, p. 1–2.

Krider, E.P., Weidman, C.D., and Noggle, R.C. (1977). The electric fields produced by lightning stepped leaders, *J. Geophys. Res.* 82, 951.

Labson, V.F., Becker, A., Morrison, H.F., and Conti, U. (1985). Geophysical exploration with audiofrequency natural magnetic fields, *Geophysics* 50, 656.

Lin, Y.T., Uman, M.A., Tiller, J.A., Brantley, R.D., Beasley, W.H., Krider, E.P., and Weidman, C.D. (1979). Characterization of lightning return stroke electric and magnetic fields from simultaneous two-station measurements, *J. Geophys. Res.* 84, 6307.

McDonald, T.B., Uman, M.A., Tiller, J.A., and Beasley, W.H. (1979). Lightning location and lower-ionospheric height determination from two-station magnetic measurements, *J. Geophys. Res.* 84, 1727.

Mclean, T.S.M., and Wu, Z.P. (1993). *Radiowave Propagation over Ground*, Chapman & Hall, London.

Morfitt, D.G. (1978). *Simplified VLF/LF Mode Conversion Computer Programs GRNNDMC and ARBNMC*, Technical Report 514, Ocean Systems Center, San Diego.

Morfitt D.G., and Shellman, C.H. (1976). *MODESRCH, an Improved Computer Program for Obtaining ELF/VLF/LF Mode Constants in an Earth-Ionosphere Waveguide*, Interim Report 77T, Naval Electr. Lab. Center, San Diego.

Pappert, R.A. (1985). Calculated effects of traveling sporadic E on nocturnal ELF propagation: comparison with measurements, *Radio Sci.* 20, 229.

Pappert, R.A., and Ferguson, J.A. (1986). VLF/LF mode conversion model calculations for air to air transmissions in the earth-ionosphere waveguide, *Radio Sci.* 21, 551.

Pappert, R.A., and Hitney, L.R. (1988). Empirical modelling of nighttime easterly and westerly VLF propagation in the earth-ionosphere waveguide, *Radio Sci.* 23, 599.

Pierce, E.T. (1965). Nuclear explosion phenomena and their bearing on radio detection of the explosions, *Proc. IEEE* 53, 1994.

Reder, F., and Westerlund, S. (1970). VLF signal phase instabilities produced by propagation medium: Experimental results, in Davies, K. (Ed.): *Phase and Frequency Instabilities*, AGARD Conference Proc. No. 33, Advisory Group on Aerospace Research and Development, NATO, 103.

Richter, J.H. (Ed.) (1990). *Radio Wave Propagation Modeling, Prediction and Assessment*, AGARDograph No. 326, Loughton, Essex.

Sao, K., and Jindoh, H. (1974). Real time location of atmospherics by single station techniques and preliminary results, *J. Atm. Terr. Phys.* 36, 261.

Sao, K., Yamashita, M., Tanahashi, S., Jindoh, H., and Ohta, K. (1970). Sudden enhancements (SEA) and decreases (SDA) of atmospherics, *J. Atm. Terr. Phys.* 32, 1467.

Sauer, H.H., Spjeldvik, W.N., and Steele, F.K. (1987). Relationship between long term phase advance in high-latitude VLF wave propagation and solar energetic particle fluxes, *Radio Sci.* 22, 405.

Serhan, G.I., Uman, M.A., Childers, D.G., and Lin, Y.T. (1980). The RF spectra of first and subsequent lightning return strokes in the 1–200 km range, *Radio Sci.* 15, 1089.

Shellman, C.H. (1986). *A New Version of MODESRCH using Interpolated Values of the Magnetoionic Reflection Coefficients*, Technical Report 1473, Naval Ocean Systems Center, San Diego.

Sommerfeld, A. (1952). *Lectures in Theoretical Physics, Vol. VI*, Academic Press, New York.

Stein, S., and Jones, J.J. (1967). *Modern Communication Principles*, McGraw-Hill, New York.

Straker, T.W. (1955). The ionospheric reflection of radio waves at frequencies 16 kc/s over short distances, *Proc. IEE* 102C, 396.

Sukhorukov, A.I. (1992). On the excitation of the earth-ionosphere waveguide by pulsed ELF sources, *J. Atm. Terr. Phys.* 54, 1337.

Sukhorukov, A.I., Shimakura, S. and Hayakawa, M. (1992). Approximate solution for the VLF eigenvalues near cut-off frequencies in the nocturnal inhomogeneous earth-ionosphere waveguide, *Planet. Space Sci.* 40, 1363.

Swanson, E.R. (1982). OMEGA, in Belrose, J.S. (Ed.): *Medium, Long and Very Long Wave Propagation*, AGARD Conference Proc. No. 305, Advisory Group on Aerospace Research and Development, NATO, 36–1.

Taylor, W.L., and Sao, K. (1970). ELF attenuation rates and phase velocities from slow-tail components of atmospherics, *Radio Sci.* 5, 1453.

Thomson, N.R. (1993). Experimental daytime VLF ionospheric parameters, *J. Atm. Terr. Phys.* 55, 173.

Uman, M.A. (1987). *The Lightning Discharge*, Academic Press, New York.

Volland, H. (1968). *The Propagation of Long Waves* (Die Ausbreitung langer Wellen), Verlag Friedr. Vieweg & Sohn, Braunschweig.

Volland, H. (1981). A waveguide model of lightning current, *J. Atm. Terr. Phys.* 43, 191.

Volland, H. (1982). Low frequency radio noise, in Volland, H. (Ed.): *Handbook of Atmospherics, Vol. I*, CRC Press, Boca Raton, FL, 179.

Volland, H. (1984). *Atmospheric Electrodynamics*, Springer-Verlag, Heidelberg.

Wait, J.R. (1970). *Electromagnetic Waves in Stratified Media*, McMillan, New York.

Wait, J.R. (1981). *Wave Propagation Theory*, Pergamon Press, New York.

Wait, J.R. (1991). VLF radio wave mode conversion for ionospheric depressions, *Radio Sci.* 26, 1261.

Wait, J.R. (1992). Lateral deviation of VLF radio waves due to diffraction and scattering from coast-lines, mountain ranges and polar ice caps, AGARD Conference Proceedings 528, *Radiolocation Tech.*, p. 1–1.

Wait, J.R., and Spies, K.P. (1964). *Characteristics of the Earth-Ionosphere Waveguide for VLF Radio Waves*, NBS Techn. Note 300.

Watt, A.D. (1967). *VLF Radio Engineering*, Pergamon Press, Oxford.

Weidman, C.D., and Krider, E.P. (1979). The radiation field wave forms produced by intracloud lightning discharge processes, *J. Geophys. Res.* 84, 3159.

Westerlund, S., Reder, F.H., and Abom, C. (1969). Effects of polar cap absorption events on VLF transmissions, *Planet. Space Sci.* 17, 1329.

Yano, S., Ogawa, T., and Hagino, H. (1989). Waveform analysis of tweek atmospherics, *Res. Lett. Atmos. Electr.* 9, 31.

Chapter 4

Electromagnetic Noise Due to Earthquakes

Michel Parrot

CONTENTS

1. INTRODUCTION

Electric and magnetic perturbations in connection with earthquakes have been known for a very long time. Old literature describing such events commonly reports the observation of electric sparks, light in the sky or on mountain tops, deviation of compasses, etc. For example, Milne (1890) presented many observations of such phenomena made by different persons. He also compared the record of atmospheric electricity, made during one year at the Imperial Meteorological Observatory in Tokyo, with the records of earthquakes in various parts of Japan; this showed that the correlation depended on the magnitude of the earthquake and varied with the distance of the epicenter from Tokyo.

Until recently, these effects were considered with much caution, as they were largely unsupported by measurements; but 10 years ago, two papers (Gokhberg et al., 1982; Warwick et al., 1982) boosted the study of electromagnetic emissions (EM) related to earthquakes. They presented observations made with instruments that were very well known, but used for other purposes, and introduced hypotheses concerning the EM wave emissions during earthquakes. Such observations are of great interest, because they start a few hours before the shock and can be considered as short-term precursors. However, not all earthquakes produce such emissions.

This chapter only considers the observations of EM waves from the ULF range (a few hertz) up to the HF range (several megahertz). Many other phenomena observed before and during an earthquake are closely related, but will not be discussed here. For example, Honkura (1981) considered four types of anomalies in the electric and magnetic approach to earthquake prediction: the geomagnetic field, the electric (telluric) field, ground resistivity associated with strain, and the resistivity in and around the focal region. The magnetic and electric field effects near seismic regions and at frequencies lower than the ULF range have been reviewed by Johnston (1989). Japanese data were analyzed by Rikitake (1987). Zlotnicki and Le Mouel (1988, 1990) have reported such events during volcanic eruptions. Anomalous behavior of the telluric field was studied by Corwin and Morrison (1977), Varotsos and Alexopoulos (1984a,b), Varotsos et al. (1986), Ralchovsky and Komarov (1988), and Varotsos and Lazaridou (1991). Preseismic changes in Earth resistivity are described by Rikitake and Yamazaki (1985). Considering all four anomalies and the atmospheric electric-potential gradient, Zubkov and Migunov (1975) investigated the relationship between the energy of an earthquake and the time of appearance of precursor effects. They found that among this group of EM precursors of earthquakes, the atmospheric electric-potential gradient is the latest precursor.

Global seismo-electromagnetic effects are discussed by Dmowska (1977) and Parrot et al. (1993); thus, here we will not describe in detail the observations of perturbations of ionospheric layers before or after earthquakes. Sobolev and Husamiddinov (1985) found local increase of the critical frequencies f_oE and f_oF_2 before earthquakes. Increase of f_oF_2 2 days before the shock and decrease 1 day before were shown to occur by Fatkullin et al. (1989). Perturbations of the sporadic E_s layer during the night have been reported by Alimov et al. (1989), 2 days before five earthquakes of magnitudes between 4.5 and 6. Such ionospheric anomalies also perturb the propagation of waves in the Earth-ionosphere guide. Gokhberg et al. (1989) analyzed data concerning the propagation of the VLF Omega ground transmitter signal through different paths; they presented cases of phase variation when an earthquake occurs on one path, while other paths were used as tests for detecting anomalies. These ionospheric perturbations are similar to the Trimpi effect (Carpenter et al., 1984).

The second section of this paper is separated into four parts. The first subsection presents the ground observations of anomalous perturbations of the EM field, recorded at the time of seismic events, and the second investigates the ionospheric EM field observed by satellites flying over seismic regions. In Section 2.3, some reports of earthquake lights ("this darkest domain of seismology") are presented. Except for these phenomena that can occur before an earthquake, it is well known since the great Alaskan earthquake in 1964 (Leonard and Barnes, 1965) that strong perturbations may also be observed in the ionosphere just after the shock (Tanaka et al., 1984; Wolcott et al., 1984; Kelley et al., 1985). These are due to the propagation of acoustic-gravity waves, which are generated by powerful earthquakes or explosions (Row, 1967). This phenomenon will be discussed, because acoustic-gravity waves can interact in the ionosphere to produce waves at higher frequencies (Al'perovich et al., 1979). Direct observations of acoustic-gravity waves by the HF Doppler technique (Yuen et al., 1969; Weaver et al., 1970; Liu et al., 1982; Wolcott et al., 1984) will not be described in detail, but Liu et al. (1982) presented data recorded during the volcanic eruption of Mount St. Helens. Electromagnetic noise related to volcanoes is slightly different from that generated by earthquakes and will be discussed in Section 2.4.

The third section discusses all hypotheses that have been proposed for explaining EM noise. Many of these models also provide a theoretical analysis of the conditions required to obtain earthquake lights. The end of the chapter is devoted to laboratory experiments, where EM emissions are observed when applying stress to rocks.

The problems raised to explain this phenomenon and the resulting conclusions are summarized in the last chapter.

2. OBSERVATIONS

On the ground as well as in the ionosphere, EM noise is generated by different natural and artificial sources covering a very large frequency spectrum. The main natural source is lightning discharge of thunderstorms, whereas the artificial field can be due to industrial noise and VLF transmitter signals.

In order to discriminate EM disturbances caused by earthquakes, most observations of EM noise that will be described below, were recorded during low magnetic activity and in frequency bands where no artificial radio signals are radiated. In some experiments, underground fields are measured to reduce external noise caused by man-made or natural sources. Satellite observations are useful in places that usually have a low-level background noise. Reports of events are made before and after earthquakes.

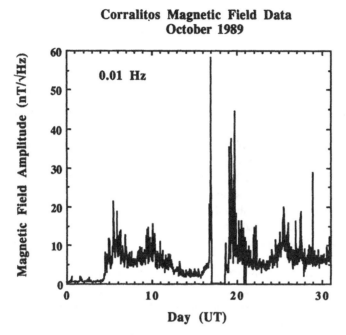

Figure 4.2.1 ULF variations of the magnetic field recorded around the time of the Loma Prieta earthquake, which started on October 18, 1989. Two perturbations can be observed: a small one a few days before the shock and a large one 3 h before. There is a gap in the data due to a power failure at the time of the shock. The experiment was located at 7 km from the epicenter. (From Fraser-Smith, A.C., Bernardi, A., McGill, P.R., Bowen, M.M., Ladd, M.E., Helliwell, R.A., and Villard Jr., O.G., 1990, *Geophys. Res. Lett.,* 17, 1465. With permission.)

2.1. GROUND-BASED OBSERVATIONS

The observations of EM waves will be classified in the order of increasing frequency. At ULF frequencies magnetometers record EM waves, but they are also sensitive to seismic waves when the earthquake is close. A good example was given by Belov et al. (1974). During an $M_s6.0$ earthquake that occurred at 70 km from the station, they recorded a perturbation 10 s before the arrival of the seismic shock. The frequency characteristics of the measuring channel were uniform in the range of 0.1 to 10 Hz, and the first perturbation was most probably due to an EM wave generated at the time of the shock. Fraser-Smith et al. (1990) observed large amplitude increases in ULF signals (0.01 to 10 Hz) 3 h before the Loma Prieta earthquake ($M_s7.1$). Figure 4.2.1 presents the data recorded several days before the quake. The ULF sensors were located very close to the epicenter (7 km), but the ELF/VLF system (10 Hz to 32 kHz), located 52 km from the epicenter, revealed no precursor activity. A more complete study of this event was carried out by Bernardi et al. (1991). Molchanov et al. (1992) compared the characteristics of the ULF emissions recorded during the $M_s6.9$ Spitak and the $M_s7.1$ Loma Prieta earthquakes. Their maximum intensity occurred a few hours before the two quakes (4 h prior to the main shock in the Spitak case, 3 h for Loma Prieta), and their ULF background activity started increasing 3 to 5 days before the Spitak event and 12 days before the Loma Prieta one, ULF activity remained high several weeks after both earthquakes. The only difference that was observed is the much lower amplitude in the Spitak case, which could be explained by the difference in distance between epicenters and sensors (128 km in the Spitak case). Kopytenko et al. (1993) compared other examples of ULF waves before the Spitak earthquake and before an earthquake in Japan ($M_s6.4$). They used different ground stations to show that these events were not due to magnetospheric activity. Dea et al. (1991, 1993) monitored the 0.1 to 10 Hz ULF and 10 to 40 Hz ELF signals, observing broadband ULF signals before and during an $M_s4.6$ earthquake located at 200 km. One day before the earthquake, the power spectra in the 0.1 to 5 Hz band was 7 dB larger than during quiet periods. Two hours after the quake, ULF activity had begun to decrease.

Adams (1990) reported signal increases at frequencies below 1 kHz, 10 days to a month prior to three earthquakes in California that included the Loma Prieta one.

Electromagnetic VLF emissions have been observed at the Kerguelen station during three moderate earthquakes ($M_s \approx 4.7$), which occurred about 100 km away (Parrot et al., 1985). In two cases, an increase in wave intensity between 500 and 3600 Hz was observed 1.5 h before the shock. These emissions decreased after the shock. The main difference with the natural noise usually observed at Kerguelen was that the starting time of the increases was not the same in the different filters, being detected first at the lowest frequency.

Fujinawa and Takahashi (1990) found anomalous EM signals several hours before an $M_s4.9$ earthquake located at 150 km from the receiver. The observations were carried out in three frequency bands: DC (up to 12 Hz), 0.01 to 12 Hz, and 1 to 9 kHz. Anomalous radiations in the third band were characterized by sporadic bursts 6 and 4 h before the shock. Duration of the pulses (50 ms) was much longer than the 5 ms pulses caused by atmospheric activity.

Low frequency, broadband, radio receivers were deployed along the San Andreas fault system by Tate and Daily (1989) between 1983 and 1986. The largest earthquakes occurring during this period were an $M_s6.2$ event located at 40 km from a station and an $M_s5.6$ event located 24 km away. Three channels were used: 0.2 to 1, 1 to 10, and 10 to 100 kHz. They observed two types of EM anomalies just before earthquakes: (1) interruption of radio waves that lasted several hours and (2) a short duration broadband radio emission that is similar to other reported observations.

Ralchovski and Christokov (1985) continuously surveyed the 5- and 9.6-kHz frequencies over a 10% bandwidth. During a minor ($M_s = 3.5$) earthquake that occurred only 10 km away from their recording station, they found that the EM noise level in the 5-kHz channel increased 8 h before the shock and lasted 2 h, whereas in the 9-kHz channel the EM noise level increase started only 5 h before the shock.

Gokhberg et al. (1982) presented several observations of EM emissions before earthquakes in Japan. Their data were collected with a narrowband receiver centered at 81 kHz, and with a wideband receiver whose maximum frequency was 8 kHz. The 81-kHz frequency was chosen to avoid perturbations due to man-made noise and to be far from the maximum of natural noise. Emissions at this frequency increased about three quarters of an hour before an earthquake and decreased abruptly just after it (Figure 4.2.2). In the 8-kHz wideband receiver, which was normally used for the study of whistlers, EM emissions with impulsive noises were seen at 1.5 kHz, about 30 min prior to the shock. Yoshino et al. (1993) have summarized the results obtained over five years, giving a statistical analysis of emission characteristics using the data of 29 events. In Japan too, but at the slightly higher frequency of 183 kHz, Oike and Ogawa (1986) recorded the anomalous low frequency noises observed before and after earthquakes for 2 years. They counted the number of pulses when the signal amplitude exceeded a given threshold. An example of one

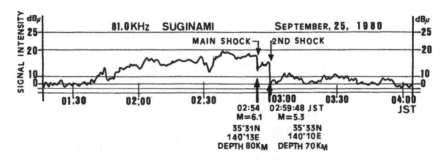

Figure 4.2.2 Electromagnetic radiation levels at a frequency of 81 kHz plotted against time. The arrows indicate the two shocks. The signal returns to the background level only after the second shock. The distance between the receiver and epicenter was around 55 km. (From Gokhberg, M.B., Morgunov, V.A., Yoshino, T., and Tomizawa, I., 1982, *J. Geophys. Res.*, 87, 7824. Copyright by American Geophysical Union. With permission.)

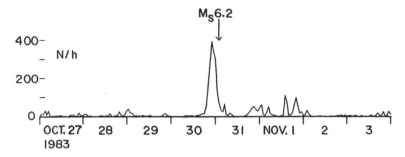

Figure 4.2.3 Variation of the hourly number of low frequency pulses before and after an $M_s6.2$ earthquake at a depth of 15 km. The distance from the receiver was about 200 km. (From Oike, K. and Ogawa, T., 1986, *J. Geomagn. Geoelectr.*, 38, 1031. With permission.)

observation is given in Figure 4.2.3. The magnitudes of the earthquakes were between 5 and 7, and they were located at distances up to 1000 km. No significant noise increment was observed for earthquakes whose epicenters were located in deep sea regions. These authors also showed a relationship between the magnitudes of the main shocks and the number of pulses received 3 days before and 3 days after these shocks. Gufel'd et. al. (1992) have studied radio wave disturbances prior to Rudbar and Rachinsk earthquakes.

In the HF frequency range, Warwick et al. (1982) have shown EM emissions at 18 MHz, 6 days prior to the great Chilean earthquake of May 22, 1960. The observation point was very far (about 10,000 km) from the epicenter. The radio receivers they used are part of a worldwide network to study cosmic radio noise. It must be noted that Gokhberg et al. (1982) also reported intensification of EM radiation at frequencies of 27 kHz and 1.63 MHz, about 30 min before an earthquake in Iran ($M_s7.4$). This was observed in the Caucasus, with detectors placed in a tunnel 50 m below ground level.

Ground observations of the perturbation of natural magnetospheric waves also were published by Hayakawa et al. (1993a) who studied the influence of seismic activity on the propagation of magnetospheric whistlers at low latitudes. Their data came from observations at Sugadeira (Japan) during several years of study. Anomalous whistlers with dispersion values greater than 70 $s^{1/2}$, i.e., more than twice the typical value of 25 to 30 $s^{1/2}$, occur during seismic activity in the longitude sector 100 to 160°E. This ratio is even larger during summer. The dispersion value of whistlers was estimated by using the curve-fitting method in the frequency range of 1 to 8 kHz and its accuracy was about 1 $s^{1/2}$. The anomalous whistlers appear before and after earthquakes occurring in the relevant longitude range.

Finally, it must be noted that two recent papers (Gershenzon and Gokhberg, 1992; Koryavov, 1992) have discussed the possibility of tsunami prediction with EM fields.

2.2. IONOSPHERIC OBSERVATIONS

Gokhberg et al. (1983) have shown disturbances in the upper ionosphere above the epicentral region of imminent earthquakes. With the data from the OGO-6 satellite, they reported an increase of the integrated noise in different filters between 100 and 500 Hz, when the subsatellite point was at 480 km from the epicenter of an $M_s5.4$ earthquake, 14 h before the shock. They also presented variations of ionic and electronic densities observed by the AE-C and ISIS-2 satellites when the distances between subsatellite points and epicenters were between 100 and 200 km; the earthquakes ($M_s \approx 6$) occurred between 2 and 14 h afterward.

Anomalous increase of the intensity of low frequency (0.1 to 15 kHz) radio wave emissions was detected by the Intercosmos-19 satellite when it was over a seismic zone (Larkina et al., 1983, 1989). Studying several cases, Larkina et al. (1983) drew the following conclusions: the increase of ELF/VLF emissions occurs between tens of minutes to hours before and after the quakes; the regions of increase extend in longitude (they suggested that this could be related to the longitudinal drift of ionospheric plasma); the effect is observed more often at 15 kHz (the maximum frequency observed), but the closer the satellite is to the epicenter, the more the effect is observed at lower frequencies (down to 140 Hz); and the emissions are mainly electrostatic. Larkina et al. (1989) presented a study of several orbits around an $M_s5.9$ earthquake, and found that low frequency emissions extended along $\pm 60°$ of geographic longitude and $\pm 2°$ of geographic latitude.

Parrot and Lefeuvre (1985) published a statistical study on the data recorded by the geostationary satellite GEOS-2. Although a wave emitted at the Earth's surface must travel over a longer distance to reach this altitude, a geostationary satellite enables the study as a function of time, whereas in the case of an orbiting satellite the distance between epicenters and the orbital position is constantly changing. Two GEOS-2 data sets were used: one at the time of selected earthquakes with magnitudes larger than 4.7 and with epicenters close to the satellite longitude and a second with random data taken during the lifetime of GEOS-2. With a rough criterion based on the relative intensity of waves at frequencies between 0.3 and 10 kHz, recorded over short (15 min) and long (1.5 h) intervals before and after earthquakes, they obtained a positive correlation of 44%. The same analysis performed with the random data set gave 41%, which is very similar. However, when they decreased the distance between the satellite and the earthquake longitudes, the percentage increased to 51%; when the epicenters are all land based, it goes up to 54%. The studies as a function of frequencies also showed some differences between the two data sets; for the random data set, the maximum of positive correlation was around 1 kHz (the frequency where the natural noise usually is observed), whereas for the earthquake data set the maximum was at lower frequencies.

No characteristic seismic-related spectral feature was found by Matthews and Lebreton (1985) in the micropulsations and ranges (0.3 to 11 Hz) with the same GEOS-2 satellite. The event identification was made by a visual inspection of wave records. However, this result may not be significant, as they used a data set of only 3 months, during which time all earthquake magnitudes were less than 5. They also studied two special events in the case of an earthquake of $M_s6.1$, but during a high level of geomagnetic activity.

Most of the reported events observed by low altitude satellites take place just above the seismic zone or at the same L values. Recorded natural noise has a high level, mainly in the electron slot region ($2 < L < 3$) and at high latitudes. However, electrostatic turbulence could be observed during the night around the equator (Kelley and Mozer, 1972; Holtet et al., 1977).

Cases studies with the data of the low altitude AUREOL-3 satellite (apogee 2012 km, perigee 408 km) were given by Parrot and Mogilevsky (1989) and Parrot (1990). The observations were made with magnetic and electric antennae in the frequency range 10 Hz to 15 kHz (Figure 4.2.4). The three panels represent, respectively, the projection of the AUREOL-3 orbit, the variations of the electric signal recorded at the output of a 72-Hz filter, and at the output of a 150-Hz filter as a function of the time. The star indicates the epicenter of an earthquake ($M_s5.1$) that occurred roughly 20 min after the satellite passed over the epicentral area. In the two filters, a signal enhancement is seen around 19.22 UT when the satellite passed over the same latitude as the epicenter. The signal increase after 19.35 UT was due to the fact that the satellite entered the auroral zone.

With the same AUREOL-3 satellite, Gal'perin et al. (1992) made simultaneous records of ELF waves and increases in precipitation of high energy electrons and protons. Observations were made when satellite passes intersected the L-shell of epicenters. Chmyrev et al. (1989) presented data on quasistatic electric fields and hydromagnetic waves in the frequency range of 0.1 to 8 Hz, detected by the low altitude Intercosmos-Bulgaria-1300 satellite a few minutes before an $M_s4.8$ earthquake. Enhanced electric fields were observed in two conjugated regions. Recently, the data of the COSMOS-1809 satellite were analyzed for numerous aftershocks of the Spitak earthquake (Serebryakova et al., 1992). The satellite orbit was near circular (970 km, i = 82.5°), and the central analyzer frequencies were between 140 Hz and 15 kHz. Intense EM radiation at

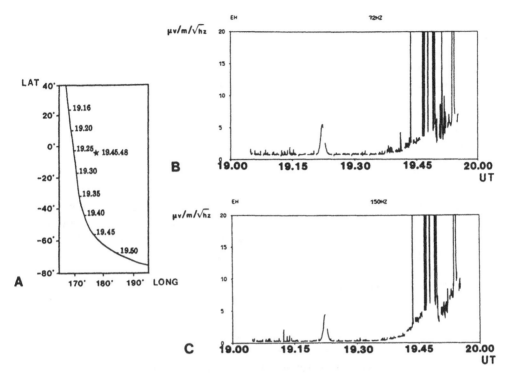

Figure 4.2.4 (A) Orbit of the Aureol-3 satellite between 19.16 and 19.50 UT on March 17, 1982. (B) Time variation of the signal recorded by an electric component in a filter centered around 72 Hz. (C) Same as in (B), for a filter centered around 150 Hz. (From Parrot, M. and Mogilevsky, M.M., 1989, *Phys. Earth Planet. Inter.*, 57, 86. With permission.)

frequencies below 450 Hz were observed at the L-shells of the earthquake when the orbits passed less than 10° from the longitude of the epicenter. Using events recorded by AUREOL-3, they have shown that the power of the waves associated with earthquakes is higher than the power of the natural noise at those latitudes, and also that the amplitudes observed during different events by two low altitude satellites are very similar.

An examination of ELF/VLF waves recorded by the low altitude satellite DE-2 (apogee 1300 km, perigee 300 km) failed to indicate signatures associated with earthquakes (Henderson et al., 1993). They used electric field intensities at the output of 20 filters between 4 Hz and 512 kHz and selected 63 parts of satellite orbits, called earthquake orbits, as DE-2 passed within ±20° geographic longitude of epicenters of earthquakes with $M_s \geq 5.0$. They compared these data with 61 parts of the so-called control orbits, where no earthquakes occurred within ±20° geographic longitude, but which had similar latitudes, longitudes, local times, and magnetic activity, as the earthquake orbits. Statistical analysis of these two data sets, showed that 40 of 63 earthquake orbits had an emission with amplitude larger than 10 µV/m in at least one filter, whereas 38 of 61 control orbits showed an emission in at least one filter in similar conditions. Thus, no differences were found between earthquake and control orbits. They also showed that the average emission strengths of emissions measured on earthquake orbits are similar to those measured on control orbits. They conclude that these results do not eliminate the possibility of EM emissions by earthquakes, but that they point out the difficulties in distinguishing such phenomena, and that more satellite data are needed in order to work with larger sample size. However, when more selections are introduced to obtain the best conditions for observing EM waves, generated by earthquakes under land and at depths of less than 100 km, the number of cases decreases in such a way that the statistics are no longer meaningful.

Anomalous numbers of whistlers with no ELF component in their spectra and excitation of LHR noise have been reported by Mikhaylova et al. (1991), during aftershocks of the Iranian earthquake of June 20, 1990. The satellite was above the seismic region 9 h after the main shock and 3 h before a strong aftershock ($M_s6.1$, depth 3 km). They explained that the anomalous whistlers were certainly due to the formation of large scale horizontal heterogeneities (increases in electron concentration) above the seismically active region.

The satellite observations described above have shown that:

1. Authors disagree about the extension in longitude of the seismic perturbation
2. No significant signature of the events was found
3. Observations are best made at low (<100 Hz) or high frequencies (>10 kHz).

2.3. EARTHQUAKE LIGHTS

Earthquake lights (EQL) are a great source of controversy for different reasons. First of all, their nature is not well defined: electric sparks, chemical luminescence (Hedervari and Noszticzius, 1985), ball lightning, aurora-like beams, etc. Secondly, EQL locations are very different: just above the ground, on top of mountains, in the sky, and even above the sea, which means that they certainly have different origins. Thirdly, except for some photographs, most reports have come from witnesses that certainly were psychologically shocked by the quake itself (Raleigh et al., 1977). A well-documented catalog compiled by Corliss (1982) contains EQL reports from the 13th century until today. Many reported events, like the spontaneous ignition of natural gas, are not relevant to the emission of EM waves, but if some EQL are of an electrostatic nature, they will produce EM noise. An important point, which supports the reality of the phenomena, is that many luminous phenomena have been observed during laboratory measurements (see Section 3.2). EQL may be observed before, during, and after earthquakes.

Finkelstein and Powell (1970) and Finkelstein et al. (1973) have described some observations and have suggested that EQL could be caused by the piezoelectric field produced in surface rocks. More complete reports have been given by Yasui (1973) and Derr (1973). In the paper by Yasui, a correlation of luminous phenomena with earthquakes provides the following main points:

- EQL locations: mostly on hill tops or high mountains
- Geological configuration: rocks with radon or quartz
- Altitude of EQL: low and near earth atmospheric layer
- Hour of EQL: in winter and at the early dawn
- Relations between EQL and sferics: sometimes
- Duration of EQL: several seconds up to several tens of seconds, which is longer than common thunderlights
- Relation between EQL and earthquakes: close to the epicenter

The observations by Yasui were discussed by Derr (1973). He noted that sferics generally follow the luminescence and are strongest in the 10 to 20 kHz range, but no example is given. As EQL have been observed at sea, he has put some restrictions on their generation mechanism, providing that these lights have the same cause as those observed on land. At the time of the paper, two theories had been advanced: (1) violent low level air oscillation (which does not account for precursor observations) and (2) piezoelectric effects generated by quartz-bearing rocks. More theories to explain EQL will be given in Section 3.

Other luminous phenomena (LP), mostly nocturnal lights, have been reviewed by Derr (1986). He postulated that these lights are associated with tectonic strain in active seismic areas and are correlated with earthquakes with a time scale different from the EQL. Luminous phenomena may result from small rock fractures as tectonic strain accumulates over a large region, whereas EQL are the result of strain release close to the epicenters. Derr and Persinger (1986) determined a correlation between LP and microseismicity in southern Washington State.

The only report that results from a long period of observations with known equipment was carried out by Fishkova et al. (1985). They presented photometric observations of the night airglow made at the Abastumani Observatory in the Caucasus. They showed different individual cases and made a statistical analysis of 400 small local earthquakes ($M_s < 4$) located at less than 200 km from the observatory. They found that the intensity of the green oxygen line at 0.5577 μm increases several hours before the earthquakes (6% on average, but sometimes more than 30%).

2.4. PHENOMENA RELATED TO VOLCANOES

The observations of events during volcanic eruptions can be classified into two types: observations during eruptions, and observations before that are similar to those made during earthquakes, but less numerous. However, volcano plumes could be an important source of EM noise. Anderson et al. (1965) and Brook et al. (1974) reported observations of lightning in volcanic clouds. Impressive photographic observations were provided by the first authors, who estimated that the energy release in one of the observed discharges was about 10^6 J, i.e., about 1000 times less than the energy estimated for a lightning discharge in a thunderstorm. During a volcanic eruption near the sea, Brook et al. (1974) have also measured a potential gradient of 7000 V/m at 100 m of the lava-seawater interface.

During the eruption of Mt. Mihara on Izu-Oshima island, Ondoh (1990) recorded atmospherics of 160 kHz at 100 km from the eruption and 80 kHz at 200 km. Such atmospherics were detected only during the night and have time variations correlated with the repeated eruptions of the volcano. He concluded that LF atmospherics were generated by lightning discharge in eruption smoke and ascending warm air currents.

For the same event, Yoshino and Tomizawa (1989) made observations with sensors similar to those used for earthquake studies. They recorded impulsive noise at 82 kHz recorded several days before the volcanic eruption. Fujinawa et al. (1992) reported variations in the ULF (0.01 to 0.6 Hz) and VLF (1. to 3. kHz) bands a few days before and after a volcanic eruption. These authors believe that the observed emissions were produced by magma flow in the mountain, which caused electrokinetic phenomena, or variations in crack density.

3. GENERATION MECHANISMS

The main point concerning the generation mechanisms of EM emissions in seismic areas is related to the large accumulation of energy before the shock. Due to the physical properties of the rocks that are subjected to stress, part of this energy is transformed into EM radiation either directly, or through processes involving the displacement of electric charges. To explain these EM emissions several hypotheses have been put forward. They will be explained by considering all theoretical models that explain the observations before an earthquake and also afterward, when the Earth's surface layers regain their equilibrium. Effects of acoustic emissions at the time of the shock are presented, and in the second part of this section we will review the laboratory measurements where physical properties of rocks are studied.

3.1. THEORETICAL MODELS

Scholz et al. (1973) underlined the importance of rock dilation and water diffusion, which may explain a large class of precursor phenomena to earthquakes. In their dilation model, the rocks undergo an inelastic volumetric increase prior to failure. The dilation is produced by the formation and propagation of cracks within the rocks. Dilation affects many physical properties of rocks, in particular the electric resistivity, which depends on the amount of water contained in the rocks. The rate of water flow increases in the dilatant zone.

Quantitative evaluations of this dilation-diffusion earthquake model have been investigated by several workers (Mizutani et al., 1976; Fitterman, 1978; Ishido and Mizutani, 1981). The consequence of fluid motion into a dilatant zone prior to an earthquake is the generation of an electric-potential anomaly due to electrokinetic phenomena. An electric double layer is formed at a

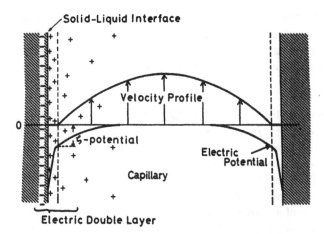

Figure 4.3.1 Simplified diagram of an electric double layer and profiles of the electric potential and liquid velocity in a rock pore. The electric double layer is formed by negative ions linked to the rock, and positive ions inside the fluid. Due to an electrical potential between the solid-liquid interface and the liquid, and the movement of groundwater, a current is induced. (From Mizutani, H., Ishido, T., Yokokura, T., Ohnishi, S., 1976, *Geophys. Res. Lett.*, 3, 365. With permission.)

solid-liquid interface (Figure 4.3.1). According to Mizutani et al. (1976), the electric current induced I is proportional to the gradient of the pore pressure P:

$$I = \frac{\phi \varepsilon \zeta}{\eta} \nabla P \qquad (1)$$

where ϕ is the porosity of the porous medium (10^{-2}), ε is the dielectric constant of the fluid ($80/(4\pi\ 9 \cdot 10^9)$ F/m^2), ζ is the zeta potential ($10^{-2} \sim 10^{-1}$V), and η is the viscosity of the fluid (10^{-4} Nsm^{-2}). With the values indicated in parentheses before and a value of ∇P between 10^5 and 10^7 Nm^{-2}/km (which may correspond to a M$_s$6.4 earthquake), they estimated that I = 0.7 ($10^{-7} \sim 10^{-4}$) A/m^2, and that the electric streaming potential gradient is between 1 and 100 V/km. Ishido and Mizutani (1981) measured the zeta potentials of mineral and rocks in aqueous solutions, under various physicochemical conditions and different temperatures.

Using a friction-vaporization theory, Lockner et al. (1983) and Lockner and Byerlee (1985) explained how a large charge concentration can be generated and maintained in a highly conductive earth. The electric field E created by the generation of an electrostatic charge will decrease with time t as:

$$E = E_0 \exp\left(\frac{-\sigma t}{\varepsilon_0 \kappa}\right) \qquad (2)$$

where σ is the conductivity, ε_0 is the permittivity of free space, and κ the relative permittivity of the material. For common wet rocks, σ is about 10^{-2} Sm^{-1}, and κ is about 10, which leads to a microsecond relaxation time, too short to allow for the accumulation of a significant charge. The general idea of Lockner et al. (1983) is the following: during a large earthquake, frictional heating of the shear zone will occur, which will lead to vaporization of water near the shear zone and thus to a decrease in the electrical conductivity σ for saturated or partially saturated rocks (from about 10^{-1} to less than 10^{-10} Sm^{-1}). Continuous frictional heating produces increasing σ (10^{-5} Sm^{-1} at 500°C; 10^{-4} Sm^{-1} at 650°C) and charge separation in the shear zone. It results in a thin central conductor of a few centimeters wide along the fault, surrounded by a low σ sheath of rocks with pores containing vaporized water. This conductor will collect the charge and because of its dimensions (hundreds of meters deep by a few centimeters wide), it will concentrate the charges along its edges. The charge concentrated at the top will produce an intense electric field at the Earth's surface, enhanced by the normal atmospheric potential gradient. This potential gradient will be strong enough to induce a coronal discharge in the atmosphere above the fault. They have quantified their model, using formulas for: the shear stress as a function of depth; the temperature distribution as a function of shear stress; and the variation of σ as a function of temperature, fault width, length and duration of the slip, and average sliding velocity. Their model shows that an $M_s7.0$ earthquake is the smallest that could produce EQL under the conditions presented. In their theory, EQL are not expected to precede an earthquake. They concluded that, even if EQL do not occur, short term transient electric and magnetic disturbances would result from both the generation and movement of charges and the modification of conductivity around the fault zone.

These results were discussed by Lee and Delaney (1987), who examined the conditions of elevated temperatures and vaporization arising from the frictional generation of heat during an earthquake. They estimated that, with a vapor zone at a depth of 130 m for an event with 1-m displacement, the temperature may exceed 500°C, but only if the duration is less than 0.14 s; such short duration requires an unusually large slip rate of 7 ms^{-1} or more.

With samples of rocks taken near the San Andreas fault, Lockner and Byerlee (1985) have shown that κ is generally frequency dependent. Inverting these data into the time domain they found that, while much of the charge quickly dissipates after an earthquake, as much as 0.5 to 1% can remain for several seconds. This could explain the discharges observed.

A recent study by Draganov et al. (1991) also considered water diffusion; in order to explain the Loma Prieta event in the ULF range, they have shown that it may result from the motion of groundwater. The magnetic field response to fluid motion is expressed using the MHD approximation. They assume the existence of a highly conductive layer containing water at a depth of 4 to 4.5 km with $\sigma = 0.1 \sim 4$ Sm^{-1}. Magnetic field fluctuations at the Earth's surface of $\Delta B/B = 10^{-5}$ at f $<$ 1 Hz are shown to result from the motion of groundwater with a velocity of $\sim$4 cm/s.

Gokhberg et al. (1984) gave estimations of seismic electric field at the Earth's surface of $10^3 - 10^4$ V/m, in a frequency range between 1 and 10^3 Hz and discussed the penetration of such electric fields of seismic origin into the ionosphere. In their upward propagation at altitudes of 70 km, Park and Dejnakarintra (1973) have shown that these fields decrease by 5 or 6 orders of magnitude when the ionosphere is considered isotropic. Gokhberg et al. (1984) expected values of the electric field $\sim$10 to 100 mV/m in the lower ionosphere, but with the creation of an oscillating circuit. They examined the possible instabilities that might be excited by quasivariable electric currents with frequencies between 1 and 10^3 Hz. They restricted their discussion to the E region of the ionosphere (100 to 120 km), where the generated electric currents are maximum. Instabilities are of the Farley-Buneman and the gradient-drift types, which could lead to excitation of electrostatic turbulence that is related to perturbations on the ionograms in the E and F layers. They also discussed the transformation of electrostatic waves into EM waves by inhomogeneities of the ionospheric plasma.

This model of the oscillatory seismic-ionospheric circuit was further developed by Gokhberg et al. (1985a). Here, the circuit is excited by current sources in the Earth's crust during seismic events. They considered the case of a source in the near equatorial region: the disturbed volume is given by L × a × b, with L = 100 km in the direction of the Earth's magnetic field, a = 100 km, and the thickness of the ionospheric layer b = 20 km. They also considered the distance from the Earth's surface to the ionosphere d = 100 km, the resistances of the ionosphere section and the Earth's crust, an electrical source of seismic origin, and an inductance loop. Ground and ionosphere are considered as the two plates of a condenser. They found that the natural frequencies of local oscillation in the circuit are around 300 Hz for the parameter values given above. The generation of natural oscillations in this local Earth-ionosphere circuit may lead to secondary effects in the ionosphere: ionic-acoustic turbulence and heating. They expected that additional ionization and turbulence in the ionosphere could cause light effects and EM radiation in the radiofrequency range.

Another plasma instability was investigated by Martelli and Cerroni (1985). They suggested that this instability is triggered by the non-Maxwellian plasma produced during the failure of rocks under stress. They demonstrated that an EM emission could be generated in the plasma by a beam-plasma interaction, resulting in a Doppler shifted ion-acoustic instability.

A more complete investigation of different models for the sources of EM earthquake precursors was carried out by Gokhberg et al. (1985b). The electric fields are calculated for current sources in the Earth's crust, the atmosphere, and the ionosphere. They first considered that the possibility of recording EM radiations on the Earth's surface from deep local sources was controlled by rock characteristics and frequency range. The high frequency component of the spectrum is attenuated during passage through rock and is proportional to

$$e^{-(\pi f \mu \sigma)^{\frac{1}{2}} x} \tag{3}$$

where x is the distance from the hypocenter, f the wave frequency, and μ the magnetic permeability. With σ between 10^{-1} and 10^{-3} Sm^{-1}, and frequencies between 10^4 and 10^6 Hz, the attenuation is between 10^2 and 10^3 dB/km. The second point they considered is that the stability of the separated charges in samples is determined by the relaxation time of free charge carriers. For rock samples well insulated from external surroundings, the relaxation time is 10^{-1} to 10^{-3} s, but this time constant for relaxation of charges in the Earth's crust lies in the range of 10^{-2} to 10^{-7} s (see Equation 2). Therefore, they concluded that the EM fields can only be caused by large scale sources. The main point of their model is that during the excitation of large scale nonstationary current sources in the rupture phase, conduction currents are created in the Earth's crust and electric charges are induced at its surface. The disturbances of electrotelluric and geomagnetic fields are related to conduction currents, and the disturbance of the atmospheric electrical potential is related to surface charges. At frequencies of 10 to 1000 Hz, the electric fields of surface charges attenuate slowly in the atmosphere. Their calculations will now be described.

The nature of local electric currents is related to the shift of charged dislocations during a seismic event. The charges, carried by charged dislocation, depend on the displacement velocity v and their density. The current caused by dislocation is estimated to be between 10^{-3} and 10^3 A/m^2. The movement of charged dislocations and fractures (electrical charges are carried by fractures as well) is not steady, for which reason EM radiation could be emitted from this excited current system. The excitation period of local dislocation sources is compared to the speed v of fracture propagation. For fractures with dimensions 10^{-5} to 1 m, and v $\sim$ 10 to 100 m/s, the period of a single excitation is $\sim 10^{-7}$ to 10^{-1} s.

They estimated the electric field of a quasi-steady current source located in an infinite half space with uniform conductivity σ. The source excites a current in the crust and induces a charge at the Earth's surface. The spaced and the induced charges create an electric field in the atmosphere. The values of the vertical component Ez at the Earth's surface are estimated for different distances and current strengths. Surface currents cause disturbances of the electrotelluric field. With a size of the focus L = $\sim 10^3 - 10^5$ m, a lateral dimension l between 10 and 10^3 m, a conductivity σ = $10^{-2} - 10^{-5}$ Sm^{-1}, and a depth h between 10^3 and 10^4 m, the maximum disturbance at distances r < h is found between 10 and 10^5 V/m. Yet, when h or r are increased up to 10^5 m, the disturbance will not exceed the background level. Concerning now the electric field in the atmosphere, with h $\sim$ 10^3 to 5.10^4 m and just above the source, Ez lies between 10^{-5} to 10^4 V/m. With increasing r, the field quickly diminishes: at r $\sim$ 10^4 m, Ez $\approx$ 1 to 100 V/m and at r $\sim$ 5 10^4 m, Ez $\approx$ 10^{-3} to 10^{-1} V/m. The calculation of the EM field at the Earth's surface near the source has been given in a table as functions of σ and the frequency f. As an example, for L = 3 10^3 m, r = 5 10^2 m, l = 10^2 m, h = 10^4 m, σ = 10^{-5} Sm^{-1}, and f = 10^3 Hz, they found E = 30 V/m. The magnetic field at frequencies of 10 to 100 Hz in the epicentral region for r < 10^4 m reaches 10^2 to 10^4 nT, and at distances r ~ 10^5 m, it decreases to a fraction of nT. At 10^3 to 10^5 Hz and r ~ 10^5 m, the magnetic component is less than 10^{-2} to 10^{-4} nT.

The last point of the paper by Gokhberg et al. (1985b) concerns the electric field of seismic origin in the atmosphere and ionosphere. They estimated the attenuation of this field with the same method as Park and Dejnakarintra (1973) and considered the possibility of electric break-down in the upper atmosphere, whose field is less than the breakdown values for the normal atmosphere (3.10^6 V/m) because the travel distance of charged particles increases. They also estimated the minimum electric field at the Earth's surface which is necessary to reach the break-down values at higher altitudes. Curves are given as functions of altitude and frequency. The frequency range of the disturbance should correspond to the frequency range of the excitation of the primary source of current in the Earth's crust, i.e., 10 to 10^3 Hz. When the electric field exceeds the breakdown value in the upper layer of the atmosphere, it will be a source of luminous effects and therefore another source of wideband radiation in the radiofrequency range.

In order to solve the problem of large attenuation of EM waves in the ground (see Equation 3), Gershenzon et al. (1987, 1989) postulated that EM emissions can be caused by the cracking of surficial crustal layers in an earthquake area. They evaluated the intensity of the emissions as a function of magnitude. First, using results from Nitsan (1977), they estimated that the total energy emitted by a microcrack produced during failure in quartz-bearing rocks is $7 \cdot 10^{-13}$ J. This value is obtained with a time of crack opening of $2 \cdot 10^{-7}$ s, which corresponds to an emission frequency of 5 MHz. It should be noted that Nitsan (1977) and Warwick et al. (1982) obtained very different estimations. Their results will be discussed in the next section. Second, Gershenzon et al. (1989) calculated with a model the number of microcracks as a function of strain. The radius of the zone of precursory strain phenomena for crustal earthquakes is estimated to be $10^{0.43M}$ km, where M is the earthquake magnitude. The strain is assumed to decrease from the epicenter according to a cubic law. They took into account only the near surface layer, down to 400 m, which is the skin layer thickness for a wave at frequencies between 10 and 10^4 kHz. They obtained a simple formula for the energy:

$$W = 10^{0.86M+1} \tag{4}$$

which gives, for example, $2 \cdot 10^5$ J for an $M_s 5.0$ earthquake and 10^7 J for an $M_s 7.0$ earthquake. Another model of EM signal generation by cracks was given by Mastov and Lasukov (1989), who also attributed the emissions to the formation and variation of the dipole moment of cracks.

In the paper by Gershenzon et al. (1987), a formula is given for calculating the amplitude of EM emissions. They deduced from 22 different experimental measurements, a relationship between the level of the emitted signal, the magnitude M of the earthquake, and the distance R between the reception and the epicenter (in kilometers):

$$\mathrm{Log}\left(\frac{AR}{A_0}\right)^2 = 0.77M + 0.98 \tag{5}$$

where A/A_0 is the ratio between the maximum level A of the EM emissions and the background level A_0.

Assuming only quartz exists at a certain depth and that the pressure generates piezoelectricity phenomena, Cutolo (1988) attributed the creation of an electric field and potential difference to the polarization of dielectrics constituting the materials. He gave a table with or without the presence of electric discharge, the generated frequency, and the electromagnetic energy, as a function of pressure, focal depth, and distances between the sides of the double stratum. When there are emissions, the frequencies are consistently higher than 10 kHz. Kingsley (1989) cal-culated the signal-to-noise ratio expected at HF frequencies from an earthquake at 10 km from a

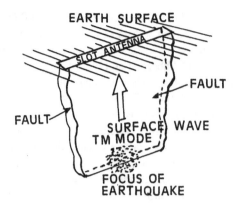

Figure 4.3.2 Schematic representation of a fault to explain how a wave emitted at the hypocenter of an earthquake could be observed at the surface. (From Yoshino, T. and Tomizawa, I., 1988, The Technical Report of the Institute of Electronic Information and Communications, Tech. Rep., EMCJ 88–64. With permission.)

receiver using the results of Cutolo (1988), but the propagation of HF signals through rock is very attenuated, and he suggested that propagation along fault lines might give propagation conditions that are somewhat closer to those of free space. Such a model of EM wave propagation (Figure 4.3.2) was built by Yoshino and Tomizawa (1988). As earthquake foci commonly are located inside a fault, the waves emitted at this place are supposed to propagate toward the surface in the waveguide formed by the fault.

Gershenzon et al. (1993) have considered three mechanisms of mechanoelectrical transformation: piezomagnetic, electrokinetic, and induction (electric field caused by a conducting crust moving in the geomagnetic field), which can produce electromagnetic impulses. The basic equations are Maxwell's equations, which led to estimates of the field intensities observable at the surface at different distances from the epicenter and with earthquakes of different magnitude. They found amplitudes of 1 to 10 nT around the focus, decreasing as r^{-3}. The time scale of the impulses is 1 to 100 s.

A complete model of EM field transmission from seismic sources in the Earth's crust to the upper ionosphere was analyzed by Molchanov (1991). Also using Maxwell's equations, the different conductivities of ground, atmosphere, ionosphere, and magnetosphere, and the equations of continuity at the edges of these different media, he evaluated a transmission coefficient of the EM energy for different depths of the source and for different ground conductivities. He found that this coefficient is maximum between 1 and 10 Hz. This model is also discussed by Kopytenko et al. (1993) to explain the Spitak ULF observations.

Geochemical releases in the atmosphere can also be considered as a perturbation susceptible to change electrical properties. Thomas (1988) and Wakita et al. (1988) reviewed geochemical precursors observed during the past decades. Concentrations of hydrogen and radon can increase a few days before and after earthquakes. The changes observed a few days before are very sharp with spike-like patterns. The rapid increase of radon after earthquakes was attributed to vibration caused by the main shock and aftershocks. The importance of the geochemical precursors was earlier underlined by Pierce (1976), who showed that a release of radon from the ground would increase the atmospheric conductivity and decrease the electric field. He adopted a simple model to estimate atmospheric electrical parameters taking into account the following hypotheses: currents and fields are vertical, and the atmosphere is separated in two layers (a mixed layer just above the ground and an undisturbed layer). The rate of ion production is considered to evaluate the conductivity. A decrease in the fair weather field prior to an earthquake of about 20% by day and 40% at night was estimated, which agrees with Kondo's (1968) report of variations in the atmospheric electric field near the epicenter of an earthquake at the same time as radon emission.

The effects of the acoustic-gravity waves, which occur after a large earthquake, have also been theoretically studied. Reviews of different mechanisms were published by Yeh and Liu (1974),

Blanc (1985), and Al'perovich et al. (1985). The last paper is an introduction to a special issue of *Izvestiya Russian Academy of Sciences, Physics of the Solid Earth* (11, 1985) devoted to the study of the effects of a powerful explosion (the MASSA experiment). Due to the ion-neutral collisions, the acoustic-gravity waves propagating in the lower ionosphere induce fluctuations of plasma density and electric currents and produce magnetohydrodynamic waves (Rietveld, 1985). This changes the plasma wave refractive index, and the intensity of the VLF waves can be modified above the seismic zone. Krechetov (1988) investigated the mechanism of VLF wave generation as the result of the interaction of an acoustic disturbance with the charged plasma components. Infra-acoustic oscillations were considered by Gal'perin et al. (1992) as a possible method of energy transfer to the upper atmosphere. It was proposed that they stimulate MHD waves in the ionospheric region and trigger ELF noise, primarily in the 0.1 to 10 Hz range.

3.2. LABORATORY EXPERIMENTS

In most laboratory experiments described below, a rock sample is submitted to large stress with a press. The experiments differ in the nature of the rock sample, the conditions of observation (external temperature, surrounding gas or liquid), and the frequency range studied.

Nitsan (1977) has reported radiofrequency emissions, associated with the fracturing of quartz-bearing rocks and other piezoelectric materials. The volume of the sample ranged from about 1 to 10^4 cm^3 and the estimated area of cracks from about 0.01 to 100 cm^2. The effective bandwidth of the coil (located between 0.05 and 0.5 m from the sample) was $\sim$1 to 10 MHz. He observed emissions between 1 and 7 MHz and estimated the radiated power as about $2 \cdot 10^{-15}$ W at 5 MHz. This corresponds to an EM energy release of about 10^{-19} J for a transient associated with a single crack. No signal was observed for rocks that did not contain quartz. He explained that the most probable mechanism for this emission is the rapid drop in the piezoelectric field, accompanying the sudden stress release when fracturing occurs.

To support their EM observations during the Chilean earthquake, Warwick et al. (1982) studied the emission processes during fracturing of granite and quartz rocks. They simultaneously recorded electric, magnetic, and acoustic fields. The electric and magnetic signals were analyzed in the range of 0.5 to 250 MHz with coils located at some centimeters from the samples. During application of the stress, all electric and magnetic transients were accompanied by acoustic emissions. The records were dominated by a 50-kHz component of the electric field. The maximum power of the magnetic field is between 1 and 2 MHz. They found that the energy for a single microfracture is $2.5\ 10^{-23}$ J, which is much less than the estimate by Nitsan.

Wideband (10 Hz to 100 kHz) electric fields were digitally recorded by Ogawa et al. (1985) using granite samples fractured by a bending moment. Four kinds of signal were observed: maxima at 30 kHz, 5 kHz, 10 Hz, and intermittent pulses. The charge-generation mechanism in rocks was discussed in terms of electrification that might be due to either contact or separation of the rocks and of the piezoelectrification.

Brady and Rowell (1986) and Cress et al. (1987) found that rocks, when fractured, emit electrical signals with a power spectrum maximum in the band 900 Hz to 5 kHz (Figure 4.3.3). Quartz-free basalt radiates both light and low frequency electric signals as do quartz-bearing rocks, but with an order of magnitude less. They drew the conclusion that the piezoelectric effect of quartz is at most a minor contributor to the total power radiated. Granite samples were broken at ambient atmospheric pressure in a gas consisting of air, argon, nitrogen, and helium, in a vacuum, and under water. Light spectra produced when rocks are broken (even under water) are characteristic of the ambient fluid. They suggested that exoelectrons, emitted from fresh fracture surfaces and bombarding the ambient fluid surrounding the samples, are the excitation source for the light emitted at the fracture, and that it is not due to a piezoelectric discharge. They postulated that electric signals are caused by turbulent motion of rock fragments with a charge distribution. Pyrite, a conductive mineral, produces no detectable electric signals because the conductive fracture surfaces of pyrite cannot maintain a charge distribution. Thus, both light and charged surfaces are the result of exoelectron expulsion from surfaces created as the rock fractures.

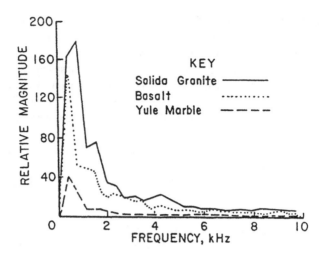

Figure 4.3.3 Power density spectra vs. frequency for different rocks. They result from Fourier analysis of the electric signals recorded when the rock samples are stressed. Salida granite, which contains quartz, produces the largest amplitudes. (From Cress, G.O., Brady, B.T., and Rowell, G.A., 1987, *Geophys. Res. Lett.,* 14, 331. With permission.)

A complementary study has been independently published by Bykova et al. (1987). They observed electron emissions during the formation of freshly broken surfaces in quartz-bearing rocks (quartzite, granite, and limestone) in a vacuum. The experimental apparatus measured the number of emitted electrons with an energy larger than 100 eV. Limestone gave the highest intensity of emission (15,000 pulses per second) at the time of rupture. Emission surges were also observed during the fissuring that preceded the main rupture. Electron emissions during sample fracturing appears to be caused by an accumulation of charges on the sides of cracks.

Enomoto and Hashimoto (1990, 1992) measured emissions of charged particles from different rocks (granite, quartz, feldspar) undergoing indentation fracturing under atmospheric conditions. During maximum load, acoustic emissions were also detected. Their estimation of the production rate of charge is ~ 0.6 C m^{-3} s^{-1}. They noted that, if a failure occurs during ~ 1 s, over an area of some meters, the charge generated is comparable to the total electric charge produced by one lightning event (~ 1 C).

Light emissions have been observed by Schloessin (1985) and Martelli et al. (1989). During tests performed both in air and in vacuum, the latter authors observed a light simultaneously with radio frequency emissions. The presence of ions has also been detected with a Langmuir probe.

Khatiashvili (1988) and Khatiashvili and Perel'man (1989) presented electric signals recorded during crystal fracture. They also theoretically studied the mechanisms of EM emission generation during crystal shearing and the mechanism of EM emissions passing through rocks. They considered the formation of double electric layers (DEL) at media interfaces. A single DEL is modeled by capacitors or dipoles, and the oscillations or variations of their parameters (space between charges and density of charges) lead to generation of EM emissions.

Laboratory experiments were performed by Yamada et al. (1989) on electromagnetic and acoustic emissions from granite loaded at a constant strain rate; 10 to 20% of the acoustic emissions detected were associated with EM emissions. These EM emissions are observed with peaks in the frequency range 0.5 to 1 MHz. The total range measured was 80 kHz to 1.2 MHz. They agree that a possible mechanism of EM emissions is the electrification of a fresh surface, created by cracking in the rock.

Atkinson and Rawlings (1981) have studied the stress corrosion of gabbro and granite in air and water at 20°C. The acoustic response was monitored during the experiment, and they found that the acoustic emission rates and their amplitude distributions are a sensitive indicator of the crack-propagation mechanism.

Table 4.3.1 **Frequencies of the EM Emissions Recorded during Different Laboratory Experiments that Applied Stress to Rocks**

Rock type	Observed frequencies	Authors
Granite, quartzite	1–10 MHz	Nitsan, 1977
Granite	>0.5 MHz	Yamada et al., 1989
Westerly granite and smokey quartz	500 kHz–10 MHz	Warwick et al., 1982
Granite	0.3–300 kHz	Martelli et al., 1989
Granite	10 Hz–30 kHz	Ogawa et al., 1985
Granite and basalt	0–5 kHz	Cress et al., 1987
LiF and KCl crystals	1–100 kHz	Khatiashvili and Perel'man, 1989

An important point, which could explain the different behaviors of earthquakes, even of earthquakes occurring at the same place, has been discussed by Wakita et al. (1988). Results of a rock-fracturing experiment showed that the number of acoustic-emission events in rocks differs between the first loading and the second loading; the amplitude of the signal recorded during the first being larger than that of the second. However, when increasing the interval between the two loadings, the difference gradually disappears.

A summary of all EM emission frequencies of the recorded during laboratory measurements is given in Table 4.3.1.

In the general study of the physical properties of rocks, parameter changes have been determined when the rocks are submitted to different strain and temperature. The variation of electric resistivity of rocks under stress was first investigated by Brace and Orange (1968a,b), who studied the effect of pressure up to 10^9 Nm^{-2} on different crystalline rocks. A compilation of other work, also including the effect of temperature has been published by Parkhomenko (1982). In the experiments, the range of temperature variations is 20 to 1000°C and 10^8 Nm^{-2} to $2 \cdot 10^9$ Nm^{-2} for pressure. His presentation was divided into water-saturated and dry rocks. The electrical resistivity variation of water-saturated rocks with temperature is a function of electrolyte concentration, mineral composition, and porosity. The resistivity of rocks could be reduced by a factor of 4 to 20 in the temperature interval of 20 to 250°C. The porosity in rocks has an important effect on their resistivity. Various rocks (granites, gneisses, basalts) show reduced resistivity when they are not fully saturated with electrolyte. A general conclusion is that the resistivity decreases as pressure and temperature increase.

The experiment reported by O'Keefe and Thiel (1991) is not a proper laboratory experiment because the data were recorded during quarry blasting; but they have observed radio emissions at frequencies up to 5 kHz with horizontal monopole antennas located within 60 m of the blast. One of the mechanisms they invoked to explain such observations is that it is due to microfracturing after pressure adjustment in the remaining rock wall after the explosion.

Laboratory measurements thus indicate that rock samples under stress can radiate EM energy in a wideband spectrum. The recorded frequencies are different because the various authors did not survey the same frequency bands. An important point is that light has been observed in most experiments. An unanswered question concerns the application of such experiments on rock samples to phenomena that occur in strongly seismic areas.

4. DISCUSSIONS AND CONCLUSIONS

This chapter has described the characteristics of EM waves that are related to seismic and volcanic activity. Understanding the generation mechanisms of these EM waves is a difficult problem because very different topics like plasma waves, wave propagation, rock physics, and seismology are concerned. However, it is a rewarding exercise as it can be related to intriguing processes such as the EQL or the anomalous animal behavior preceding earthquakes (Rikitake, 1981).

The study and detection of EM emissions emitted by seismic zones require their isolation from the EM field due to other sources. In the radiofrequency range, EM field sources of artificial origin are industrial noise (mainly power line harmonic radiations) and the pulses of ground-based VLF transmitters. The main sources of natural EM emissions are lightning discharges during thunderstorms. Concerning satellite measurements, low altitude observations at low latitudes and around the equator are more appropriate in searching for seismic emissions, because the electric and magnetic field levels are lower than in the mid-latitude zone.

Large uncertainties exist concerning the EM energy release by an earthquake because it is influenced by many parameters. All earthquakes are different, even those occurring in the same seismic zone. Thus, numerous observations, with a maximum number of parameters measured at the same time, are needed for a better understanding of such EM emissions. In the future, many systems should give new information on EM emissions in seismic areas. Bella et al. (1987) have described equipment installed in a deep cave in the Central Apennines in Italy, which can measure electric signals in a frequency band between 0.3 and 300 kHz (see also Bella et al., 1992). In California, Dea et al. (1993) use a permanent system to survey the ULF frequency range, and an interdisciplinary group monitors several parameters (Park, 1992). The problem of direction finding of precursory radio emissions has been discussed by Hayakawa et al. (1993b). It is based on the simultaneous measurement of waveforms of the electric and magnetic field components in the ELF (1.5 kHz) and ULF (0 to 10 Hz) ranges, at a few stations installed in Japan. Mori et al. (1993) presented a new utilization of the telephone network. Providing that other disturbances (magnetic storms, man-made noise) could be eliminated, anomalous changes in the electric field can be detected using underground telephone cables.

Compared with such ground campaigns as described above, satellite measurements are complementary, because they allow the surveying of large seismic zones.

REFERENCES

Adams, M.H. (1990). Some observations of electromagnetic signals prior to California earthquakes, *J. Sci. Explor.*, 4, 137.

Alimov, O.A., Gokhberg, M.B., Liperovskaia, E.V., Gufeld, I.L., Liperovsky, V.A., and Roubtov, L.N. (1989). Anomalous characteristics of the middle latitude Es layer before earthquakes, *Phys. Earth Planet. Inter.*, 57, 76.

Al'perovich, L.S., Gokhberg, M.B., Sorokin, V.M., and Fedorovich, G.V. (1979). The production of geomagnetic variations by acoustic vibrations during earthquakes, *Izv. Russ. Acad. Sci. Phys. Solid Earth*, 15, 192.

Al'perovich, L.S., Ponomarev, E.A., and Fedorovich, G.V. (1985). Geophysical phenomena modeled by an explosion: a review, *Izv. Russ. Acad. Sci. Phys. Solid Earth*, 21, 816.

Anderson, R., Björnsson, S., Blanchard, D.C., Gathman, S., Hughes, J., Jonasson, S., Moore, C.B., Survilas, H.J., and Vonnegut, B. (1965). Electricity in volcanic clouds, *Science*, 148, 1179.

Atkinson, B.K., and Rawlings, R.D. (1981). Acoustic emission during stress corrosion cracking in rocks. In: D.W. Simpson and P.G. Richard (Eds.), (pp. 605–616). *Earthquake Prediction — An International Review*, Maurice Ewing Series, American Geophysic Union, Washington, D.C.

Bella, F., Della Monica, G., Ermini, A., Sgrigna, V., Biagi, P.F., Manjgaladze, P., and Zilpimiani, D. (1987). Underground monitoring system of electromagnetic emissions, *Il Nuovo Cimento*, 10, 495.

Bella, F., Biadji, P.F., Della Monica, J., Zilpimiani, D.O., Mandzhgaladzhe, P.V., Pokhotelov, O.A., Sgrinya, V., Ermini, A., and Liperovskiy, V.A. (1992). Observations of natural electromagnetic radiation during earthquakes in Central Italy, *Izv. Russ. Acad. Sci. Solid Earth*, 1, 88.

Belov, S.V., Migunov, N.I., and Sobolev, G.A. (1974). Magnetic effects accompanying strong earthquakes on Kamchatka, *Geomagn. Aeron.*, 14, 321.

Bernardi, A., Fraser-Smith, A.C., McGill, P.R., and Villard Jr., O.G. (1991). ULF magnetic field measurements near the epicenter of the $M_s 7.1$ Loma Prieta earthquake, *Phys. Earth Planet. Inter.*, 68, 45.

Blanc, E. (1985). Observations in the upper atmosphere of infrasonic waves from natural or artificial sources: A summary, *Ann. Geophys.*, 3, 673.

Brace, W.F., and Orange, A.S. (1968a). Further studies of the effects of pressure on electrical resistivity of rocks, *J. Geophys. Res.*, 73, 5407.

Brace, W.F., and Orange, A.S. (1968b). Electrical resistivity changes in saturated rocks during fracture and frictional sliding, *J. Geophys. Res.,* 73, 1433.

Brady, B.T., and Rowell, G.A. (1986). Laboratory investigation of the electrodynamics of rock fracture, *Nature,* 321, 488.

Brook, M., Moore, C.B., and Sigurgeirsson (1974). Lightning in volcanic clouds, *J. Geophys. Res.,* 79, 472.

Bykova, V.V., Stakhovskiy, I.R., Fedorova, T.S., Khrustalev, Yu.A. Deryagin, B.V., and Toporov, Yu.P. (1987). Electron emission during fracturing of rocks, *Izv. Russ. Acad. Sci. Phys. Solid Earth,* 23, 690.

Carpenter, D.L., Inan, U.S., Trimpi, M.L., Helliwell, R.A., and Katsufrakis, J.P. (1984). Perturbations of subionospheric LF and MF signals due to whistler-induced electron precipitation bursts, *J. Geophys. Res.,* 89, 9857.

Chmyrev, V.M., Isaev, N.V., Bilichenko, S.V., and Stanev, G. (1989). Observation by spaceborne detectors of electric fields and hydromagnetic waves in the ionosphere over an earthquake centre, *Phys. Earth Planet. Inter.,* 57, 110.

Corliss, W.R. (1982). *Lightning, Auroras, Nocturnal Lights, and Related Luminous Phenomena, A Catalog of Geophysical Anomalies,* (pp. 110–115), The Sourcebook Project, Glen Arm.

Corwin, R.F., and Morrison, H.F. (1977). Self-potential variations preceding earthquakes in central California, *Geophys. Res. Lett.,* 4, 171.

Cress, G.O., Brady, B.T., and Rowell, G.A. (1987). Sources of electromagnetic radiation from fracture of rock samples in the laboratory, *Geophys. Res. Lett.,* 14, 331.

Cutolo, M. (1988). On a new general theory of earthquakes, *Il Nuovo Cimento,* 11, 209.

Dea, J.Y., Richman, C.I., and Boerner, W.M. (1991). Observations of seismo-electromagnetic earthquake precursor radiation signatures along Southern California fault zones: Evidence of long-distance precursor ultra-low frequency signals observed before a moderate Southern California earthquake episode, *Can. J. Phys.,* 69, 1138.

Dea, J.Y., Hansen, P.M., and Boerner, W.M. (1993). Long-term ELF background noise measurements, the existence of window regions, and applications to earthquake precursor emission studies, *Phys. Earth Planet. Inter.,* 77, 109.

Derr, J.S. (1973). Earthquake lights: a review of observations and present theories, *Bull. Seismol. Soc. Am.,* 63, 2177.

Derr, J.S. (1986). Luminous phenomena and their relationship to rock fracture, *Nature,* 321, 470.

Derr, J.S., and Persinger, M.A. (1986). Luminous phenomena and earthquakes in southern Washington, *Experientia,* 42, 991.

Dmowska, R. (1977). Electromechanical phenomena associated with earthquakes, *Geophys. Sur.,* 3, 157.

Draganov, A.B., Inan, U.S., and Taranenko, Yu.N. (1991). ULF magnetic signatures at the earth surface due to ground water flow: a possible precursor to earthquakes, *Geophys. Res. Lett.,* 18, 1127.

Enomoto, Y., and Hashimoto, H. (1990). Emission of charged particles from indentation fracture of rocks, *Nature,* 346, 641.

Enomoto, Y., and Hashimoto, H. (1992). Transient electrical activity accompanying rock under indentation loadings, *Tectonophysics,* 211, 337.

Fatkullin, M.N., Zelenova, T.I., and Legenka, A.D. (1989). On the ionospheric effects of asthenospheric earthquakes, *Phys. Earth Planet. Inter.,* 57, 82.

Finkelstein, D., and Powell, J. (1970). Earthquake lightning, *Nature,* 228, 759.

Finkelstein, D., Hill, R.D., and Powell, J.R. (1973). The piezoelectric theory of earthquake lightning, *J. Geophys. Res.,* 78, 992.

Fishkova, L.M., Gokhberg, M.B., and Pilipenko, V.A. (1985). Relationship between night airglow and seismic activity, *Ann. Geophys.,* 3, 689.

Fitterman, D.V. (1978). Electrokinetic and magnetic anomalies associated with dilatant regions in a layered earth, *J. Geophys. Res.,* 83, 5923.

Fraser-Smith, A.C., Bernardi, A., McGill, P.R., Bowen, M.M., Ladd, M.E., Helliwell, R.A., and Villard Jr., O.G. (1990). Low-frequency magnetic field measurements near the epicenter of the M_s7.1 Loma Prieta earthquake, *Geophys. Res. Lett.,* 17, 1465.

Fujinawa, Y., and Takahashi, K. (1990). Emission of electromagnetic radiation preceding the Ito seismic swarm of 1989, *Nature,* 347, 376.

Fujinawa, Y., Kumagai, T., and Takahashi, K. (1992). A study of anomalous underground electric field variations associated with a volcanic eruption, *Geophys. Res. Lett.,* 19, 9.

Gal'perin, Yu.I., Gladyshev, V.A., Dzhordzhio, N.V., Larkina, V.I., and Mogilevskii, M.M. (1992). Precipitation of high-energy captured particles in the magnetosphere above the epicenter of an incipient earthquake, *Cosmic Res.,* 30, 89.

Gershenzon, N.I., Gokhberg, M.B., Morgunov, V.A., and Nikolayevskiy, V.N. (1987). Sources of electromagnetic emissions preceding seismic events, *Izv. Russ. Acad. Sci. Phys. Solid Earth,* 23, 96.

Gershenzon, N.I., Gokhberg, M.B., Karakin, A.V., Petviashvili, N.V., and Rykunov, A.L. (1989). Modelling the connection between earthquake preparation processes and crustal electromagnetic emission, *Phys. Earth Planet. Inter.,* 57, 129.

Gershenzon, N.I., and Gokhberg, M.B. (1992). Electromagnetic tsunami prediction, *Izv. Russ. Acad. Sci., Phys. Solid Earth,* 2, 130.

Gershenzon, N.I., Gokhberg, M.B., and Yunga, S.L. (1993). On the electromagnetic field of an earthquake focus, *Phys. Earth Planet. Inter.,* 77, 13.

Gokhberg, M.B., Morgunov, V.A., Yoshino, T., and Tomizawa, I. (1982). Experimental measurement of electromagnetic emissions possibly related to earthquakes in Japan, *J. Geophys. Res.,* 87, 7824.

Gokhberg, M.B., Pilipenko, V.A., and Pokhotelov, O.A. (1983). Seismic precursors in the ionosphere, *Izv. Russ. Acad. Sci. Phys. Solid Earth*, 19, 762.

Gokhberg, M.B., Gershenzon, N.I., Gufel'd, I.L., Kustov, A.V., Liperovskiy, V.A., and Khusameddimov, S.S. (1984). Possible effects of the action of electric fields of seismic origin on the ionosphere, *Geomagn. Aeron.*, 24, 183.

Gokhberg, M.B., Buloshnikov, A.M., Gufel'd, I.L., and Liperovskiy, V.A. (1985a). Resonant phenomena accompanying seismic-ionospheric interaction, *Izv. Russ. Acad. Sci. Phys. Solid Earth*, 21, 413.

Gokhberg, M.B., Gufel'd, I.L., Gershenzon, N.I., and Pilipenko, V.A. (1985b). Electromagnetic effects during rupture of the Earth's crust, *Izv. Russ. Acad. Sci. Phys. Solid Earth*, 21, 52.

Gokhberg, M.B., Gufel'd, I.L., Rozhnoy, A.A., Marenko, V.F., Yampolsky, V.S., and Ponomarev, E.A. (1989). Study of seismic influence on the ionosphere by super long-wave probing of the Earth-ionosphere wave guide, *Phys. Earth Planet. Inter.*, 57, 64.

Gufel'd, I.L., Roznoy, A.A., Tyumentsev, S.N., Sherstyuk, S.V., and Yampol'skiy, V.S. (1992). Radio wave field disturbances prior to Rudbar and Rachinsk earthquakes, *Izv. Russ. Acad. Sci. Phys. Solid Earth*, 3, 267.

Hayakawa, M., Yoshino, T., and Morgounov, V.A. (1993a). On the possible influence of seismic activity on the propagation of magnetospheric whistlers at low latitudes, *Phys. Earth Planet. Inter.*, 77, 97.

Hayakawa, M., Tomizawa, I., Ohta, K., Shimakura, S., Fujinawa, Y., Takahashi, K., and Yoshino, T. (1993b). Direction finding of precursory radio emissions associated with earthquakes: a proposal, *Phys. Earth Planet. Inter.*, 77, 127.

Hedervari, P., and Noszticzius, Z. (1985). Recent results concerning earthquake lights, *Ann. Geophys.*, 3, 705.

Henderson, T.R., Sonwalkar, V.S., Helliwell, R.A., Inan, U.S., and Fraser-Smith, A.C. (1993). A Search for ELF/VLF emissions induced by earthquakes as observed in the ionosphere by the DE-2 satellite, *J. Geophys. Res.*, 98, 9503.

Holtet, J.A., Maynard, N.C., and Heppner, J.P. (1977). Variational electric field at low latitudes and their relation to spread-F and plasma irregularities, *J. Atmos. Terr. Phys.*, 39, 247.

Honkura, Y. (1981). Electric and magnetic approach to earthquake prediction. In: T. Rikitake (Ed.), (pp. 301–383). *Current Research in Earthquake Prediction I*, D. Reidel, Dordrecht.

Ishido, T., and Mizutani, H. (1981). Experimental and theoretical basis of electrokinetic phenomena in rock-water systems and its applications to geophysics, *J. Geophys. Res.*, 86, 1763.

Johnston, M.J.S. (1989). Review of magnetic and electric field effects near active faults and volcanos in the USA, *Phys. Earth Planet. Inter.*, 57, 47.

Kelley, M.C., and Mozer, F.S. (1972). A satellite survey of vector electric fields in the ionosphere at frequencies of 10 to 500 Hertz, 3. Low-frequency equatorial emissions and their relationship to ionospheric turbulence, *J. Geophys. Res.*, 77, 4183.

Kelley, M.C., Livingston, R., and McCready, M. (1985). Large amplitude thermospheric oscillations induced by an earthquake, *Geophys. Res. Lett.*, 12, 577.

Khatiashvili, N.G. (1988). Investigation of the results of electromagnetic emission at shear, *Gerlands Beitr. Geophys.*, 97, 287.

Khatiashvili, N.G., and Perel'man, M.E. (1989). On the mechanism of seismo-electromagnetic phenomena and their possible role in the electromagnetic radiation during periods of earthquakes, foreshocks and aftershocks, *Phys. Earth Planet. Inter.*, 57, 169.

Kingsley, S.P. (1989). On the possibilities for detecting radio emissions from earthquakes, *Il Nuovo Cimento*, 12, 117.

Kondo, G. (1968). The variation of the atmospheric electric field at the time of earthquake, *Mem. Kakioka Magn. Obs.*, 13, 11.

Kopytenko, Yu.A., Matiashvili, T.G., Voronov, P.M., Kopytenko, E.A., and Molchanov, O.A. (1993). Detection of ULF emissions connected with the Spitak earthquake and its aftershock activity, based on geomagnetic pulsation data at Dusheti and Vardzia observatories, *Phys. Earth Planet. Inter.*, 77, 85.

Koryavov, V.P. (1992). Electromagnetic field associated with tsunami propagation, *Izv. Russ. Acad. Sci. Phys. Solid Earth*, 2, 142.

Krechetov, V.V. (1988). Generation of VLF emissions in connection with acoustic disturbances in the atmosphere, *Geomagn. Aeron.*, 28, 725.

Larkina, V.I., Nalivayko, A.V., Gershenzon, N.I., Gokhberg, M.B., Liperovskiy, V.A., and Shalimov, S.L. (1983). Observations of VLF emissions related with seismic activity, on the Interkosmos-19 satellite, *Geomagn. Aeron.*, 23, 684.

Larkina, V.I., Migulin, V.V., Molchanov, O.A., Kharkov, I.P., Inchin, A.S., and Schvetcova, V.B. (1989). Some statistical results on very low frequency radiowave emissions in the upper ionosphere over earthquake zones, *Phys. Earth Planet. Inter.*, 57, 100.

Lee, T.C., and Delaney, P.T. (1987). Frictional heating and pore pressure rise due to a fault slip, *Geophys. J. R. Astron. Soc.*, 88, 569.

Leonard, R.S., and Barnes Jr., R.A. (1965). Observation of ionospheric disturbances following the Alaska earthquake, *J. Geophys. Res.*, 70, 1250.

Liu, C.H., Klostermeyer, J., Yeh, K.C., Jones, T.B., Robinson, T., Holt, O., Leitinger, R., Ogawa, T., Sinno, K., Kato, S., Ogawa, T., Bedard, A.J., and Kersley, L. (1982). Global dynamic responses of the atmosphere to the eruption of Mount St. Helens on May 18, 1980, *J. Geophys. Res.*, 87, 6281.

Lockner, D.A., Johnston, M.J.S., and Byerlee, J.D. (1983). A mechanism to explain the generation of earthquake lights, *Nature*, 302, 28.

Lockner, D.A., and Byerlee, J.D. (1985). Complex resistivity of fault gouge and its significance for earthquake lights and induced polarization, *Geophys. Res. Lett.*, 12, 211.

Martelli, G., and Cerroni, P. (1985). On the theory of radio frequency emission from macroscopic hypervelocity impacts and rock fracturing, *Phys. Earth Planet. Inter.*, 40, 316.

Martelli, G., Smith, P.N., and Woodward, A.J. (1989). Light, radiofrequency emission and ionization effects associated with rock fracture, *Geophys. J. Inter.*, 98, 397.

Mastov, S.R., and Lasukov, V.V. (1989). A theoretical model of generation of an electromagnetic signal in brittle failure, *Izv. Russ. Acad. Sci. Phys. Solid Earth*, 25, 478.

Matthews, J.P., and Lebreton, J.P. (1985). A search for seismic related wave activity in the micropulsation and ULF frequency ranges using GEOS-2 data, *Ann. Geophys.*, 3, 749.

Milne, J. (1890). Earthquakes in connection with electric and magnetic phenomena, *Trans. Seismol. Soc. Jpn.*, 15, 135.

Mikhaylova, G.A., Golyavin, A.M., and Mikhaylov, Yu.M. (1991). Dynamic spectra of VLF-radiation in the outer ionosphere associated with the Iranian earthquake of June 21, 1990 (Interkosmos-24 satellite), *Geomagn. Aeron.*, 31, 647.

Mizutani, H., Ishido, T., Yokokura, T., and Ohnishi, S. (1976). Electrokinetic phenomena associated with earthquakes, *Geophys. Res. Lett.*, 3, 365.

Molchanov, O.A. (1991). Transmission of electromagnetic fields from seismic sources to the upper ionosphere, *Geomagn. Aeron.*, 31, 80.

Molchanov, O.A., Kopytenko, Yu.A., Voronov, P.M., Kopytenko, E.A., Matiashvili, T.G., Fraser-Smith, A.C., and Bernardi, A. (1992). Results of ULF magnetic field measurements near the epicenters of the Spitak (M_s = 6.9) and Loma Prieta (M_s = 7.1) earthquakes: comparative analysis, *Geophys. Res. Lett.*, 19, 1495.

Mori, T., Ozima, M., and Takayama, H. (1993). Real-time detection of anomalous geoelectric changes, *Phys. Earth Planet. Inter.*, 77, 1.

Nitsan, U. (1977). Electromagnetic emission accompanying fracture of quartz-bearing rocks, *Geophys. Res. Lett.*, 4, 333.

Ogawa, T., Oike, K., and Miura, T. (1985). Electromagnetic radiations from rocks, *J. Geophys. Res.*, 90, 6245.

Oike, K., and Ogawa, T. (1986). Electromagnetic radiations from shallow earthquakes observed in the LF range, *J. Geomagn. Geoelectr.*, 38, 1031.

O'Keefe, S.G., and Thiel, D.V. (1991). Electromagnetic emissions during rock blasting, *Geophys. Res. Lett.*, 18, 889.

Ondoh, T. (1990). Unusual intensity enhancements of low frequency atmospherics associated with great eruptions of Izu-Oshima volcano in November 1986, *J. Geomagn. Geoelectr.*, 42, 237.

Park, S.K. (1992). Workshop on low-frequency electrical precursors to earthquakes, *EOS Trans. Am. Geophys. Union*, 73, 491.

Park, C.G., and Dejnakarintra, M. (1973). Penetration of thunder clouds electric fields into the ionosphere and magnetosphere, *J. Geophys. Res.*, 78, 6623.

Parkhomenko, E.I. (1982). Electrical resistivity of minerals and rocks at high temperature and pressure, *Rev. Geophys. Space Phys.*, 20, 193.

Parrot, M., and Lefeuvre, F. (1985). Correlation between GEOS VLF emissions and earthquakes, *Ann. Geophys.*, 3, 737.

Parrot, M., Lefeuvre, F., Corcuff, Y., and Godefroy, P. (1985). Observations of VLF emissions at the time of earthquakes in the Kerguelen Islands, *Ann. Geophys.*, 3, 731.

Parrot, M., and Mogilevsky, M.M. (1989). VLF emissions associated with earthquakes and observed in the ionosphere and the magnetosphere, *Phys. Earth Planet. Inter.*, 57, 86.

Parrot, M. (1990). Electromagnetic disturbances associated with earthquakes: An analysis of ground-based and satellite data, *J. Sci. Explor.*, 4, 203.

Parrot, M., Achache, J., Berthelier, J.J., Blanc, E., Deschamps, A., Lefeuvre, F., Menvielle, M., Plantet, J.L., Tarits, P., and Villain, J.P. (1993). High-frequency seismo-electromagnetic effects, *Phys. Earth Planet. Inter.*, 77, 65.

Pierce, E.T. (1976). Atmospheric electricity and earthquake prediction, *Geophys. Res. Lett.*, 3, 185.

Ralchovski, T.M., and Christokov, L.V. (1985). On low-frequency radio emission during earthquakes, *C. R. Acad. Bulg. Sci.*, 38, 863.

Ralchovsky, T.M., and Komarov, L.V. (1988). Periodicity of the Earth electric precursors before strong earthquakes, *Tectonophysics*, 145, 325.

Raleigh, C.B., Bennet, G., Craig, H., Hanks, T., Molnar, P., Nur, A., Savage, J., Scholz, C., Turner, R., and Wu, F. (1977). Prediction of the Haicheng earthquake, *EOS Trans. Am. Geophys. Union*, 58, 236.

Rietveld, M.T. (1985). Ground and in situ excitation of waves in the ionospheric plasma, *J. Atmos. Terr. Phys.*, 47, 1283.

Rikitake, T. (1981). Anomalous animal behaviour preceding the 1978 earthquake of magnitude 7.0 that occurred near Izu-Oshima island, Japan. In: T. Rikitake (Ed.), (pp. 69–80). *Current Research in Earthquake Prediction I*, D. Reidel, Dordrecht.

Rikitake, T., and Yamazaki, Y. (1985). The nature of resistivity precursor, *Earthquake Prediction Res.*, 3, 559.

Rikitake, T. (1987). Magnetic and electric signals precursory to earthquakes: an analysis of Japanese data, *J. Geomagn. Geoelectr.*, 39, 47.

Row, R.V. (1967). Acoustic-gravity waves in the upper atmosphere due to a nuclear detonation and an earthquake, *J. Geophys. Res.,* 72, 1599.

Schloessin, H.H. (1985). Experiments on the electrification and luminescence of minerals and possible origins of EQLs and sferics, *Ann. Geophys.,* 3, 709.

Scholz, C.H., Sykes, L.R., and Aggarwal, Y.P. (1973). Earthquake prediction: A physical basis, *Science,* 181, 803.

Serebryakova, O.N., Bilichenko, S.V., Chmyrev, V.M., Parrot, M., Rauch, J.L., Lefeuvre, F., and Pokhotelov, O.A. (1992). Electromagnetic ELF radiation from earthquakes regions as observed by low-altitude satellites, *Geophys. Res. Lett.,* 19, 91.

Sobolev, G.A., and Husamiddinov, S.S. (1985). Pulsed electromagnetic earth and ionosphere field disturbances accompanying strong earthquakes, *Earthquake Prediction Res.,* 3, 33.

Tanaka, T., Ichinose, I., Okuzawa, T., Shibata, T., Sato, Y., Nagasawa, O., and Ogawa, T. (1984). HF Doppler observation of acoustic waves excited by the Urakawa Oki earthquake on March 21, 1982, *J. Atmos. Terr. Phys.,* 46, 233.

Tate, T., and Daily, W. (1989). Evidence of electro-seismic phenomena, *Phys. Earth Planet. Inter.,* 57, 1.

Thomas, D. (1988). Geochemical precursors to seismic activity, *Pageophysics,* 126, 241.

Varotsos, P., and Alexopoulos, K. (1984a). Physical properties of the variations of the electric field of the earth preceding earthquake. I, *Tectonophysics,* 110, 73.

Varotsos, P., and Alexopoulos, K. (1984b). Physical properties of the variations of the electric field of the earth preceding earthquake. II. Determination of epicenter and magnitude, *Tectonophysics,* 110, 99.

Varotsos, P., Alexopoulos, K., Nominos, K., and Lazaridou, M. (1986). Earthquake prediction and electric signals, *Nature,* 322, 120.

Varotsos, P., and Lazaridou, M. (1991). Latest aspects of earthquake prediction in Greece based on seismic electric signals, *Tectonophysics,* 188, 321.

Wakita, H., Nakamura, Y., and Sano, Y. (1988). Short-term and intermediate-term geochemical precursors, *Pageophysics,* 126, 267.

Warwick, J.W., Stoker, C., and Meyer, T.R. (1982). Radio emission associated with rock fracture: possible application to the great Chilean earthquake of May 22, 1960, *J. Geophys. Res.,* 87, 2851.

Weaver, P.F., Yuen, P.C., Prolss, G.W., and Furumoto, A.S. (1970). Acoustic coupling in the ionosphere from seismic waves of the earthquake at Kurile islands on August 11, 1969, *Nature,* 226, 1239.

Wolcott, J.H., Simons, D.J., Lee, D.D., and Nelson, R.A. (1984). Observations of an ionospheric perturbation arising from the Coalinga earthquake of 2 May 1983, *J. Geophys. Res.,* 89, 6835.

Yamada, I., Masuda, K., and Mizutani, H. (1989). Electromagnetic and acoustic emission associated with rock fracture, *Phys. Earth Planet. Inter.,* 57, 157.

Yasui, Y. (1973). A summary of studies on luminous phenomena accompanied with earthquakes, *Mem. Kakioka Magn. Obs.,* 15, 127.

Yeh, K.C., and Liu, C.H. (1974). Acoustic gravity waves in the upper atmosphere, *Rev. Geophys. Space Phys.,* 12, 193.

Yoshino, T., and Tomizawa, I. (1988). LF seismogenic emissions and its application on the earthquake prediction, *The Technical Report of Institute of Electronic Information and Communications,* Tech. Rep., EMCJ 88–64.

Yoshino, T., and Tomizawa, I. (1989). Observation of low frequency electromagnetic emissions as precursors to the volcanic eruption at Mt. Mihara during November, 1986, *Phys. Earth Planet. Inter.,* 57, 32.

Yoshino, T., Tomizawa, I., and Sugimoto, T. (1993). Results of statistical analysis of low-frequency seismogenic EM emissions as precursors to earthquakes and volcanic eruptions, *Phys. Earth Planet. Inter.,* 77, 21.

Yuen, P.C., Weaver, P.F., Suzuki, R.K., and Furumoto, A.S. (1969). Continuous traveling coupling between seismic waves and the ionosphere evident in May 1968 Japan earthquake data, *J. Geophys. Res.,* 74, 2256.

Zlotnicki, J., and Le Mouel, J.L. (1988). Volcanomagnetic effects on Piton de la Fournaise volcano (Reunion Island), *J. Geophys. Res.,* 93, 9157.

Zlotnicki, J., and Le Mouel, J.L. (1990). Possible electrokinetic origin of large magnetic variations at La Fournaise volcano, *Nature,* 343, 633.

Zubkov, S.I., and Migunov, N.I. (1975). Time of appearance of electromagnetic precursors of earthquakes, *Geomagn. Aeron.,* 15, 751.

Chapter 5

Biological Effects of Electromagnetic Man-Made Noise, Atmospherics, and Small Ions

Reinhold Reiter

CONTENTS

1. INTRODUCTION AND HISTORY

Visible and audible electric phenomena have always fascinated man and induced fear as well as vague imaginations about obscure influences. Among people of certain persuasions, lightning and thunder have been (and still are) revered as deities. Electricity in the real sense, as a common phenomenon in our environment, was described long ago by Wall (1707) who concluded that the "cracklings and light" of a spark discharge from rubbed amber in some degree resemble the phenomena of thunder and lightning. From this time on, electricity became widely popular and was enthusiastically studied by many scientists, initially during the baroque period, in "electricity cabinets," and the opinion was disseminated that electric (and magnetic) forces could cause effects in both the body and the mind and even conferred healing powers. The famous researchers A. Volta and L. Galvani, the philosopher E. Kant, the natural scientists C.W. Hufeland and A.V. Humboldt (who postulated a vital "electric matter" in the air), and many other scholars were definitely convinced that atmospheric electricity had a direct influence on human well-being (for more details see Reiter, 1992).

During the first half of the 19th century another fascination was aroused by Geißler's gas-discharge tubes, the "electric light" that has also been applied in medicine. However, these tubes ended up after the turn of the century in flea markets, a fate symbolic of so many fallacies in the application of electric and magnetic fields in therapy that propagated electricity as a new blessing considered to be beneficial to mankind.

On the other hand, after the turn of the century, the increasingly taller and smoking chimney stacks as well as the impressive pylons of high power lines and the growing networks for electrical energy supplies were clear evidence for the promising industrial development of that period. Today, however, both smoking stacks as well as high power lines have become objects of concern in view of health hazards.

Of course, emissions from chimney stacks can easily be inspected and classified with regard to their effects on health. In contrast, the postulation that electric and/or magnetic fields near electric power supply systems can be harmful to mankind is still not generally confirmed. Even very recently this fear is still being instilled in people by certain lines. Whether, and under what conditions, harm or even danger really do exist will be demonstrated in this article based on our present state of knowledge.

Another consideration involves the hypothesis that atmospherics in the ELF and VLF range could be the direct cause of "weather disease," and a final reflection concerns the postulation of direct effects of small ions on man.

2. PARAMETERS: NATURE, ENVIRONMENT, AND PHYSICAL PROPERTIES

2.1. MOST IMPORTANT QUALITIES

Qualities of high importance to be considered are: (1) static and alternating electric fields (electric field strength E in V/m), (2) static and alternating magnetic fields (magnetic flux density B in Tesla (T), the formerly used unit was Gauss = 10^{-4} T), (3) electromagnetic radiation, and (4) atmospheric small ions of both polarities (n_+, n_-).

Atmospheric small ions are submicron particles consisting of one monomolecular nucleus which carries one positive or negative elementary charge. This nucleus is surrounded by about 3 to 6 water molecule dipoles. The great importance of small ions depends on their high mobility by which they determine the air-earth current in the atmosphere. Furthermore, they show a certain respirability. For more details in atmospheric ions see Chapter I/1 of this volume and Reiter (1992).

2.2. CATEGORIES OF ENVIRONMENT

The following types must be taken into consideration: (1) open air in contrast to closed rooms, (2) pure natural in contrast to technical environment, and (3) materials of special physical nature in the environment, like clothing.

2.3. PROPERTIES, FEATURES, AND EMERGENCE OF THE PHYSICAL PARAMETERS
2.3.1. Natural Environment
2.3.1.1. Atmospheric Electric Fields
The so-called fair weather electric field (Ef) which is maintained by the global thunderstorm activity exists around the Earth, but it is globally modified by the topography (Makino and Ogawa, 1985, see also Chapter I/10). In small scale dimensions, local orographical conditions (valleys, hills) and all vertically extended objects (trees, houses, poles, etc.) modify and determine Ef. It shows a representative value only on a plain, far from such objects in a low altitude and is on the order of 100 to 150 V/m. It decreases with height above ground and shows a typical diurnal variation, which is linked to universal time.

Ef can be superimposed by local and varying fields originating from charges on clouds, on precipitation, and on aerosol particles. In such cases Ef is superimposed by a locally disturbed E which can approach many thousands of volts per meter with fast changes of its strength and polarity within fractions of minutes to seconds (for more details see Reiter, 1992). It is important to point out that Ef and slowly varying E are totally screened by every material with an electrical conductivity significantly higher than that of air, e.g., glass, wood, ceramics, stone, bricks, humus, water, etc. Consequently, the electric outdoor fields do not penetrate into closed rooms independent of the construction material and, e.g., not even inside a dense forest.

2.3.1.2. Electrostatics
In our common environment, electric charges are produced on special materials by manifold processes. Apart from rubbing surfaces of a fur and of carpets and furniture, the friction of different textiles on our body causes extremely high charges and electric fields causing tiny flashes to appear.

2.3.1.3. Natural Electromagnetic Radiation
Electromagnetic (EM) radiation of natural origin in the troposphere (called sferics, see Chapters I/10–I/12) come into being by dipole-like electric discharges from or inside highly electrically charged volumes of clouds. By this way sferics allow a global thunderstorm monitoring (Grandt, 1992). Generation and propagation of such (damped) EM waves occur exclusively as expressed by Maxwell's equations and not only by pure friction and counter-movements of different air masses without electrical discharges, as postulated by Eichmeier and Baumer (1987, 1990). EMs of the VLF range (3 to 30 kHz) or at higher frequencies are able to penetrate into closed rooms as long as too much metal is not used for construction. The amount of penetration depends on the frequency of the respective EM and on the electrical conductivity of the material to be penetrated. In this article only EMs which do not induce any thermal effects in biological tissue are considered. The electric field of spherics has values from small fractions of volts per meter to some volts per meter and the magnetic vector is on the order of at most microTesla.

2.3.1.4. Magnetic Fields
The overall existing almost continuous magnetic field is the geomagnetic field which cannot be shielded without an enormous effort (e.g., by use of the expensive μ-metal). Magnetic fields in the ELF range are typical parameters of the technical environment (see Section 2.3.2.).

2.3.1.5. Small Atmospheric Ions
They exist everywhere in the air. The concentration (number density = ions per cubic meter) outdoors as well as inside rooms depends on the balance of production and decay. Responsible for the production is, apart from cosmic rays, the total of radioactive material in the air (radon + daughters) and in every concrete material and soil (uranium, radium + daughters). The decay of the small ions results from recombination and attachment to aerosol particles and surfaces. The outdoor small ion number density is on the order of some hundred or thousand per cubic meter (for details see Reiter, 1992).

2.3.2. Technical Environment

Technical sources of electric and magnetic fields in our common environment are (1) high power lines, switch yards, and transformer stations and (2) the electrical installations and appliances in houses, apartments, and workrooms. It is remarkable that the strongest concern of the public is directed toward category (1), the most important outward establishments of any aspect.

2.3.2.1. Fields Near Equipment of Electrical Energy Supply

Technical sources of electric and magnetic fields in our common environment are high power lines, switch yards, and transformer stations. In Europe the high power lines are designed for up to 380 kV, in the U.S. and in Russia up to 765 kV. Figure 5.2.1 shows mean values of the magnetic flux density (in microTesla) vs. distance (in meters) rectangular from the centerline of a 2 × 3 110 kV line with 1-kA current load. In a distance of 40 m from the centerline, the magnetic flux density already drops to 20% of the maximum. Some corresponding E values which are of lower biological interest are: at the centerline, 5 kV/m; at 20 m distance, 3 kV/m; and at 60 m distance, 0.8 kV/m.

Here the question may arise as to how fields from a high power line are able to penetrate into a building. Figure 5.2.2 gives a schematic imagination, albeit houses will not be build so near a line. E fields with 50 Hz do not penetrate (maximum about 1% depending on the electrical conductivity of the construction material) in contrast to the magnetic field, which practically penetrates the building material totally.

Other objects of concern are transformer stations because of the generated noise. However, significant electric and magnetic fields do not exist in the vicinity of the transformer body. Of more interest are the high energy current cables leaving the body.

2.3.2.2. Fields Near Appliances

Electric fields are insignificant in contrast to magnetic fields. Some examples (Carstensen, 1987) for B values are: 1000 µT near welders and soldering guns; 2 m off an air furnace, 300 µT; 5 cm off a massager, 200 µT; in contact with a heating pad, 17 µT, and 25 cm off a television set, 1 µT, a number equal to that at a distance of 50 m off the high power line in Figure 5.2.1. These figures already vividly demonstrate how subjective assessments may lead to wrong ideas. Furthermore, one has to consider the duration of exposition which determines the dose (see television performance consumption).

3. REQUIREMENTS FOR BIOLOGICAL EFFECTIVENESS

Any biological effectiveness requires penetrative ability. Without that, the respective parameter being discussed is of no importance.

3.1. CONTINUOUS ELECTRIC FIELDS

Figure 5.3.1a shows the electric lines of force and the electric equipotential surfaces of a static homogeneous and ELF (with 60 Hz) electric field in the environment of a human body. The body is completely shielded from the field due to its high electric conductivity.

Electric fields have always existed in the free atmosphere and in its lower layers above ground. Therefore, it has been and still is speculated that man has adapted to this quality during evolution, all the more considering that in ancient times human beings mainly lived and worked in the open air. Today, we live in buildings (and partly in cars) in which electric fields do not exist. Must, consequently, the agent electric field not be artificially maintained inside rooms? Because E cannot penetrate into bodies, this question is to be disregarded. Nonetheless, until now, equipment was offered commercially with the claim that it produces a necessary and sound "fair weather electric field" inside rooms.

Here, some thoughts should be added concerning clothes on our body. As mentioned above, the electric charges on them are remarkably high. Electric fields near a moving person can be on

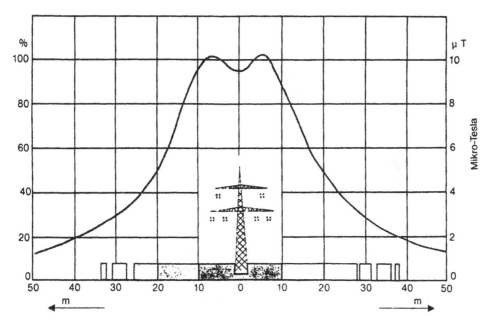

Figure 5.2.1 Magnetic flux density (in Tesla) below and rectangular to the cables of a 110 kV/1 kA transmission line. (Data from WHO, 1984 and Carstensen, 1987.)

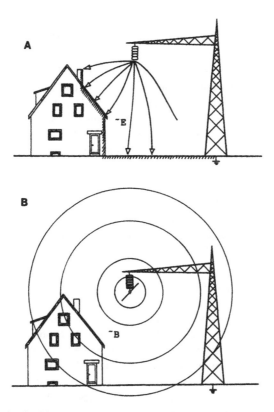

Figure 5.2.2 Schematic sketch of the electric (a) and magnetic field lines (b) of a high power line near a house. (From Reiter, R., 1992, *Phenomena in Atmospheric and Environmental Electricity*, Elsevier, Amsterdam. With permission.)

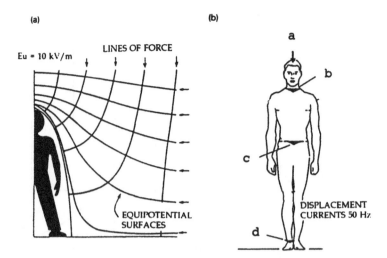

Figure 5.3.1   (a) Deformation of an electric constant or 50/60-Hz field by a human body (From Reiter, R., 1992, *Phenomena in Atmospheric and Environmental Electricity,* Elsevier, Amsterdam. With permission.) (b) Cross sections (a)–(d) showing those parts of the body for which the currents and the internal 50/60-Hz electric fields have been calculated. Results are given in Table 5.3.1.

the order, or even higher than, in the vicinity of a thundercloud (Reiter 1982, 1992), obviously without positive or negative biological effect. It must also be pointed out that charges on clothes have incorrectly never been taken into account in the case of laboratory experiments with persons in artificial electric fields, which may lead to gross errors in the results.

3.2.   VARIABLE NATURAL ELECTRIC FIELDS, ELECTROMAGNETIC WAVES
3.2.1.   Induced Electric Currents
Electromagnetic fields of 50 to 60 Hz can penetrate into biological tissues, inducing electric currents in it. The strength of these currents increases with increasing frequency of the external magnetic field component of the electromagnetic wave. Under natural conditions, sudden changes of B can occur nearby lightning return strokes. The current density peaks induced in the tissue are on the order of only some $nAcm^{-2}$ and remain weaker than omnipresent biological currents in the body. On the other hand, the pulses from thunderstorms are relatively rare events under special weather conditions.

3.2.2.   Atmospherics (Sferics)
The energy of sferics pulses is as low as common LF and VLF radio waves used for telecommunication and signal transmission, from which no significant biological effect is confirmed. The mean power flux of sferics is on the order of some 10^{-11} Wcm^{-2}.

3.3.   ALTERNATING TECHNICAL ELECTRIC FIELDS
The effect of power line electric and magnetic fields on the human body is outlined in Figure 5.3.1. Figure 5.3.1a shows the distorted electric lines put forth and the equipotential surfaces near a human body of an originally homogeneous electric field ("unperturbed" field Eu) of 10 kV/m. This field is two times larger than the strength of the field below the high power line in Figure 5.2.1. The deformation of ambient fields at 50 to 60 Hz is almost identical with that of Figure 5.3.1a. While the body is completely shielded in the case of a static field, an extremely small portion of the ambient alternating ELF field penetrates into the body. The numerical values of electric fields and axial current densities inside the body, depending on the external field amplitude, are listed in Table 5.3.1. Figure 5.3.1b indicates the cross sections of the body related

Table 5.3.1 **Current Densities I (in $\mu A/cm^2$) and Electric Field Strengths Ei (mV/m) Inside the Human Body Below 50/60-Hz High Power Lines of Different Voltage**

Tension on the cable in kV	380		380		220	
External field strength in kV/m	3.6		2.5		1	
Orientation to the HV line	Center		± 30 m from center		Center	
Magnitude	*I*	*Ei*	*I*	*Ei*	*I*	*Ei*
Part of the body						
Head	(a) 0.02	1.1	0.015	0.8	0.006	0.3
Neck	(b) 0.22	11	0.15	8	0.05	2.9
Center of the body	(c) 0.07	3.5	0.05	2.5	0.02	1.1
Ankle	(d) 0.72	36	0.5	27	0.2	10

Note: Characters (a)–(d) are related to Figure 5.3.1b. Data are after Carstensen, 1987; WHO, 1984; Polk and Postow, 1986; and Krause, 1989.

From Reiter, R. (1992). *Phenomena in Atmospheric and Environmental Electricity.* Elsevier, Amsterdam. With permission.

to the corresponding symbols in Table 5.3.1. A biological assessment of such current densities will be given in Section 4.2.

3.4. MAGNETIC FIELDS
Every magnetic field, whether constant or alternating, penetrates biological tissues totally.

3.5. SMALL ATMOSPHERIC IONS
Here only small atmospheric ions with one elementary charge are of interest in biometeorology and medicine. We will not consider so-called "electroaerosols" (highly charged aerosol particles and droplets used for therapy, see Wehner, 1983). The size is about 3×10^{-4} to 10^{-3} μm radius and the average mobility of n_+ amounts to 1.14, that of n_- to 1.24×10^{-4} m²/V. For details on small ion physics and chemistry as well as for small ions in the troposphere, see Chapter I/1 and Reiter (1992). Apart from an application of small ions for healing open wounds (Minehart, 1958), they have been and still are considered to influence the body after deposition in the alveoli. However, do inhaled small ions actually penetrate the respiratory tract down to the alveoli? This has been assumed by some people (Krueger et al., 1958; Sulman, 1971) who postulated effects based on different experiments and who speculated erroneously (Krueger, 1968) that small ions penetrate the alveolar barrier and enter the bloodstream in the capillaries. However, this supposition is a fallacy. First, only about 0.04% of inhaled small ions reach the alveoli (Bailey, 1982), and secondly they decay the moment they touch the inside surface of the alveoli. Consequently, biological effects of inhaled small ions cannot be expected. Notwithstanding, many questionable experiments have been carried out and published. These results have caused a lot of trouble since then and will be reviewed critically in Section 5.

4. BIOLOGICAL EFFECTS OF ELECTRIC AND MAGNETIC QUALITIES

4.1. ATMOSPHERICS (SFERICS)
It is well known that the appearance of a significantly increased impulse frequency of sferics of different wavelengths is directly linked to synoptic scale meteorological conditions (for details see Reiter, 1992). The amount of tropospheric instability correlates with the sferics' impulse frequency. Thus, generation of sferics by thunderstorms and showers are attributes of biotropical weather situations. Consequently, the pulse frequency of sferics can be used as an indicator for

Table 5.4.1   **Threshold Values of Current Densities in Tissues Triggering Certain Excitations**
(in a 50/60 Hz electric field)

Kind of excitation, organ	Brain and nerve cells excitation[a]	Nerve to muscle excitation[b]	Excitation by transmembrane current[c]	Limitation by thermal burden[d]
Threshold current density	$0.1\ \mu A/cm^2$	$50\ \mu A/cm^2$	$100\ \mu A/cm^2$	$1000\ \mu A/cm^2$

[a] Polk, 1986; Krause, 1989; Bernhardt et al., 1983
[b] In Bernhardt et al., 1983
[c] WHO, 1984; Schwan, 1982
[d] Krause, 1989; Bernhardt et al., 1983

From Reiter, R. (1992). *Phenomena in Atmospheric and Environmental Electricity*, Elsevier, Amsterdam. With permission.

a large range biotropical weather condition, as applied by Reiter (1960). Based on about 1 million single biological facts, Reiter (1960) could demonstrate how well synchronous reactions of ill and sound persons are linked to weather processes in large areas. These findings, however, do not allow the deduction of a direct biological effect of electromagnetic waves like sferics. Recently, Baumer and Eichmeier (1980, 1981) observed a correlation between the pulse rate of sferics and the diffusion time of ions in gelatine films, from which they hypothesized (Baumer and Eichmeier, 1983) a mechanism by which sferics are absorbed in gelatine and trigger the pore size and ion diffusion in it. The authors speculate that a direct effect of sferics on biological tissue seems to be plausible. However, neither the results of those experiments nor the theoretical supposition have been confirmed until now. A correlation between sferics and the onset of epileptic seizures has been reported by Ruhenstroth-Bauer et al. (1984). The observation period included only 6 months. The correlation with 28-kHz sferics was positive, with 10 kHz, however, it was negative. This questions the results because a selective and opposite direct effect of sferics in these nearby frequency ranges appears to be unlikely, because sferics are impulsive waveforms rather than harmonic oscillations. The above-mentioned frequencies are only filtered from a complex mixture of a very noisy signal. Recently, Ruhenstroth-Bauer et al. (1988) claimed that a good correlation exists between 8- as well as 10-kHz sferic pulse frequencies and the inflammatory reaction of rats after carrageenan injection, although they used an observation period of only 6 months. However, even a statistically significant correlation does not allow us to conclude that the sferics' pulses directly cause the observed biological effect. Sferics' pulse density and weather processes are directly linked, consequently, certain primary weather components may be the cause (see Reiter, 1960).

4.2.   BIOLOGICAL EFFECTS OF FIELDS CAUSED BY POWER SUPPLY EQUIPMENT
4.2.1.   Results of Theoretical Considerations

A number of authors (for details, see Reiter, 1992) have calculated the induced axial current density I (in $\mu A/cm^2$) and the inner field strengths Ei (in mV/m) in certain cross sections (defined in Figure 5.3.1a) of a man standing upright in the E field of high power lines. The results are given in Table 5.3.1.

Data recently published by different research groups allow us to provide a survey of characteristic physiologic thresholds of perception (see Table 5.4.1). Here a general perception of cells should be understood and not a distinctive one of special sensitive organs and glands. The thresholds in Table 5.3.1 are now to be compared with the currents to be expected in the E fields given in Table 5.4.1.

Even within the highest E fields (centerline, 380 kV) the induced currents are lower by orders of magnitude compared to the thresholds for excitation. This is true indeed when one regards the

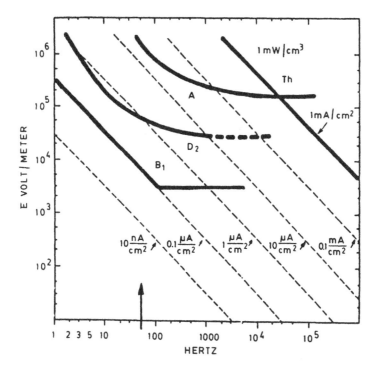

Figure 5.4.1 Model calculation by Bernhardt et al. (1983). Induced electric current densities in the living tissue (head) as function of E in volts per meter and frequency (in hertz). (From Reiter, R., 1992, *Phenomena in Atmospheric and Environmental Electricity,* Elsevier, Amsterdam. With permission.)

low perception threshold in the brain tissue. Obviously, fields directly below or in the nearest vicinity to common high power lines do not induce harm. The same conclusion follows when the fields inside rooms are taken into account.

Similar but rather striking results can be obtained by applying the diagrams in Figures 5.4.1 and 5.4.2 based on model calculations by Bernhardt et al. (1983), which are generally accepted (Tenforde, 1986). They present the induced current densities in the brain in steps of one order of magnitude (dashed lines) as a function of frequency and strength of an ambient alternating electric field E (Figure 5.4.1) or magnetic flux density B (in Tesla) of an ambient magnetic field (Figure 5.4.2). The heavy lines represent different thresholds: B_1 for complete security, D_2 for a beginning but weak perception, and A for general danger. Line Th marks danger by thermal effect. Below 0.1 μA/cm^2 (curve B_1) no effects are to be expected, in accordance with Tenforde (1986). When setting (see arrow) a frequency of 60 Hz and reading the field according to B_1, it then appears that E = 6 kV/m and B = 0.5 mT are the threshold figures. These values are, however, not present in our environment including the technical installations open to the public, neither near the wiring inside buildings nor near common appliances (for more details see Carstensen, 1987; Reiter, 1992).

4.2.2. Results of Laboratory Experiments

Apart from the fact that, seen from theoretical aspects and numerical modeling, significant effects of common environmental fields in man and mammals are rather unlikely, it remains for us to consider the results of experimental and epidemiological studies. Up to the present day numerous papers have reported on results of experimental investigations when objects were exposed to electric or magnetic fields. However, many of those do not meet the requirements which must be observed for the set up of such experiments (see Reiter 1982, 1992; Carstensen, 1987). Consequently, only the results of some reliable working groups should be mentioned here (a total of about 250 relevant publications has been critically reviewed by Carstensen, 1987).

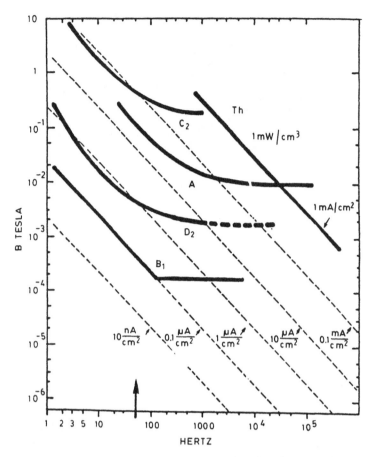

Figure 5.4.2 Similar to Figure 5.4.1 but for the magnetic flux density B (in Tesla). (From Reiter, R., 1992, *Phenomena in Atmospheric and Environmental Electricity,* Elsevier, Amsterdam. With permission.)

Hauf (1974, 1976, 1981, 1986) has performed many series of experiments with volunteers in ambient E (1 to 20 kV/m) and B (0.1 to 0.3 mT) ELF fields. Many physiological functions (e.g., heart function, blood circulation, reaction time) have been observed in conjunction with laboratory tests (e.g., blood picture, serotonin level). Hauf found only some very slight reactions within the physiological frame but no pathological departures in 50-Hz fields when E was <10 kV/m and B was <0.3 mT. Similar results were obtained by Brinkmann and Schaefer (1984) and Schaefer (1986) where more physiological tests have been executed. They confirmed that no harmful reactions occur in 50-Hz fields with 10 to 20 kV/m. Silm (1983) found some weak reactions and sensations in magnetic flux density of B > 5 mT. A summary of all published results obtained in different countries has been given by Reiter (1992). In consideration of that international experience, the following thresholds have been proposed for 50/60 Hz E/B fields: International Radiation Protection Association, 10 kV/0.5 mT; (1) Russia, 15 kV/m/1.8 mT; (2) U.K., 12 kV/m/2 mT; and (3) Germany, 20 kV/m/5 mT for an unlimited exposure (DIN VDE 0848 part 4). In the U.S. the values are on the same order but very different from state to state.

4.2.3. Results of Epidemiological Studies

Results of epidemiological studies based on observations on persons who have been involuntarily exposed to high fields are of special interest. Results of such studies have been published recently, but they need a careful reassessment. Here, mainly the flux-density values of the alternating magnetic fields are of relevance. The greatest concern linked with a field exposure over a long period of time is the question of whether magnetic alternating fields can induce cancer. This topic, indeed, has been heavily discussed in recent years, and this discussion is still continued, unfortunately, not always with unbiased and necessary criticism. A representative study was published by Milham (1983). He analyzed mortality data from 1950 to 1979, a total of 438,000 cases, and calculated the leukemia proportionate mortality ratios (PMR) = observed deaths/expected deaths × 100 for different classes of occupation. Milham's data, however, show that there is no apparent relationship between the kind of occupation linked to field exposure and mortality from leukemia (details in Reiter, 1992).

Miller (1986) has reviewed 10 extensive epidemiological studies based on occupational exposure to fields from transmission lines. Only two studies (carried out in Russia) showed positive results, all others were negative. To date there are no significant results confirming that a steady professional exposure to technical high power fields would cause any kind of cancer which could not have been induced by other influences or agents. One has to take into account that a profession in the field of electrotechnics is linked with the use of organic and also lead vapors. Other carcinogenic parameters, such as smoking, are difficult to verify in epidemiological studies.

One of the most controversial issues related to the interaction between 50/60-Hz magnetic fields and humans is the postulated link between cancer risk and residential exposure to fields of very low flux values. The first report was by Wertheimer and Leeper (1979). They claimed that deaths from cancer (mainly leukemia) in children in the Denver, Colorado region could be correlated to the presence of high current primary and secondary wiring configuration in the vicinity of their residences (344 cases in 1950 to 1973). The possible importance of the power lines in these cases has only been estimated visually and by the investigators themselves. In a subsequent study, Wertheimer and Leeper (1982) reported that a similar association could exist for the incidence of adult cancer in Colorado between 1967 and 1977. Fulton et al. (1980) tried to reproduce this result by matching the methodology of Wertheimer and Leeper (1982) as closely as possible, but no statistically significant correlation was found. Another study of the incidence of childhood leukemia was conducted by Tomenius (1986) in Sweden, with a positive result only being shown when B was >0.3 µT in front of the entrance door of the residences. However, in contrast to Wertheimer and Leeper and Tomenius, (1986) found no relationship between the risk of childhood cancer and overhead lines near residences. Sawitz et al. (1988) and Sawitz (1988) also studied the cancer risk in children in Denver. No significant association was observed between measured residential B values and childhood cancer, although the B fields have been directly measured. Other recent studies in England (McDowall, 1986; Stevens et al., 1986) failed to find any significant association between power fields and cancer risk. It has been confirmed recently by Lovely (1988) and Norris and Bonnell (1988) that persons exposed to 50/60-Hz electric fields in the common environment are without hazards. It is necessary to present this sequence of comparable claims in order to show that the general public concern about possible hazardous influences caused by electric power lines and installations is not justified.

4.3. RECENT NEW DISCOVERIES
4.3.1. The Importance of the Pineal Gland
Very recently, unique observations and perceptions have opened doors to new concepts regarding neuroendocrine effects of ELF fields in tissues. This very rapid approach was perhaps initiated by two concerns. The first effort derived from U.S. congressional hearings on the large naval communication facility "Sanguine", later called "Seafarer". They were intended to use extremely large antenna systems, transmitting at ELF and high power levels to provide communication with submarines. However, it incidentally appeared that high power lines create fields with orders of magnitudes greater than the navy projects. Another was the "Moscow Signal" (Wilson 1990), when the U.S. State Department became aware of extreme microwave flux densities in the U.S. embassy in Moscow. Although microwave effects are not considered in this article, it must be pointed out that a microwave emission that is modulated with extremely low frequencies

can act like a pure ELF field (Lyle et al., 1983, 1990), allowing a deep penetration into highly conductive materials.

Here two new main branches are to be considered, both possibly linked to some extent with Seafarer and Moscow Signal. One is the discovery of Wilson et al. (1981) that ELF fields influence the pineal gland function in such a way that the concentrations of neuroendocrines in the gland are altered. The other is the fascinating outcome that light cations in tissues undergo a cyclotron resonance when an ELF magnetic field is superimposed by a weak constant magnetic field such as the geomagnetic field. From intense work by R.J. Reiter (Reiter, 1973; Reiter et al., 1988) it became apparent that the pineal gland undergoes a series of environmental stresses whereby the melatonin synthesis is affected. This is of importance because the gland is a ubiquitously acting and active organ of internal secretion, and its function is to synchronize the physiology of the organisms with the prevailing environmental conditions. For example, the ambient-light intensity controls the highly important melatonin production of the gland. Consequently, Wilson et al. (1981, 1986) investigated whether the pineal gland reacts in an analogous way to ELF exposure and found that such fields do actually alter the neuroendocrine production in the gland significantly, a fact which was confirmed by Reiter et al. (1988). Nighttime levels of melatonin and associated neurohormones appeared to be significantly changed in rats which were exposed for one month to ambient E fields of 2 kV/m and higher. Melatonin and associated biosynthetic enzymes showed a dark-phase decrease in E-exposed animals (see also Wilson and Anderson, 1986; Wilson et al., 1990). It must be pointed out that (Blask, 1990) one hallmark of the pineal gland function is the pineal gland's ability to inhibit the growth of reproductive organs in pre-pubertal animals, as well as inducing a marked regression in the size of reproductive organs in adults (Blask, 1990). As an important consequence, investigators initiated an application of these principles to explore whether the pineal gland and its primary hormone, melatonin, influence the growth of malignant neoplasms (Blask, 1984; Blask and Hill, 1988). One of the many results is that breast cancer and melanoma are perhaps the most compelling with respect to pineal and melatonin influences. Further, recent work (Wilson et al., 1988) with healthy human subjects suggests that ELF exposure to electric and magnetic fields may alter the pineal function in certain individuals, and this is the case when the current density in the tissues is 0.1 to 0.37 $\mu A/cm^2$, a range which had been considered as safe by Bernhardt et al. (1983).

All of these findings, with essential reserves, suggest that under certain conditions and in individual persons, cancer can be promoted or even triggered by ELF fields, a suggestion, however, that up to now does not seem completely convincing. In particular, it must be pointed out with emphasis that the number of other carcinogenic influences is very high.

4.3.2. The Cyclotron Resonance

Polk (1984) was the first to suggest that free (anydrated) Ca ions in a biological tissue would exhibit cyclotron resonances of about 10 Hz in the natural geomagnetic field. Recent experimental investigations have actually demonstrated that a superimposition of a weak constant magnetic field on the order of the geomagnetic field and an ELF magnetic field produce a resonance interaction which influences the cation movements through membrane tunnels (Liboff et al., 1987; McLeod and Liboff, 1986; Liboff and McLeod, 1988; Durney et al., 1988; Liboff and Parkinson, 1991). The physical process is a resonant transfer of energy from the ELF magnetic field when its frequency corresponds to the cyclotron resonance frequency of a cation moving in a constant magnetic field. This frequency is defined by

$$f_c = q \cdot B/2\pi m_1$$

where f_c is the cyclotron frequency, q the ion charge, m the mass of the cation, and B the magnetic flux density of the constant field. The average geomagnetic B flux is about 30 to 50 μT. In such a field, f_c is in the range of ELF frequencies for biologically important cations like K^+, Na^+, and

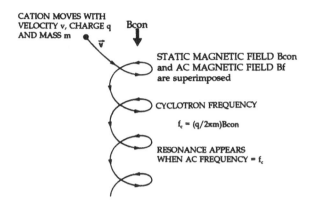

CATION MOVES WITH
VELOCITY v, CHARGE q
AND MASS m

Bcon

STATIC MAGNETIC FIELD Bcon
and AC MAGNETIC FIELD Bf
are superimposed

CYCLOTRON FREQUENCY

$f_c = (q/2\pi m)Bcon$

RESONANCE APPEARS
WHEN AC FREQUENCY = f_c

Figure 5.4.3 Schematic demonstration of the appearance of the cyclotron resonance for a Ca^{2+} ion in a biological tissue. (Adapted from Tenforde, T.S., 1990, *Extremely Low Frequency Electromagnetic Fields, the Question of Cancer*, Battelle Press, Richland, OH and Reiter, R., 1992, *Phenomena in Atmospheric and Environmental Electricity*, Elsevier, Amsterdam. With permission.)

Ca^{2+}. Such cations move as shown schematically in Figure 5.4.3. From experiments evidence appeared that cyclotron resonance interactions can influence cell biological processes in the following way (see Tenforde, 1990): (1) the rate of Ca ions released from brain tissue (Blackman et al., 1985, 1990), (2) the operant behavior of rats in a timing discrimination task (Thomas et al., 1986), and (3) effects on calcium uptake by human lymphocytes (Liboff et al., 1987). Of course, some major problems remain related to the cyclotron-resonance theory which will not be discussed here (for details see Reiter, 1992). However, well-established experimental contributions demonstrating the effectiveness of the cyclotron resonance exist (see Blackman et al., 1985; Blackman, 1988; and Blackman et al., 1990). They showed that the Ca^{2+} release from the brain tissue clearly responds to the strength of the superimposed 16-Hz electromagnetic field.

Adey (1986, 1988a,b) demonstrated that nonionizing electromagnetic fields interact with tetradecanoyl phorbol acetate (TPA) at cell membranes such as to modulate its actions on both inward and outward signal streams; but they do not cause gene modulation. This is of importance because TPA is known to promote cancer. Adey (1990) confirms the importance of weak imposed magnetic fields with ELF B fields below 100 Hz and that cell membranes are primary sites of interaction with such fields. Already Reiter (1960) could demonstrate the importance of the cell membrane in the case of application of rectangular E fields (2 to 20 kHz, 10 V/m), which caused a shift in the pH of the subcutaneous connective tissue.

A good summary is given by Tenforde (1990) who states that time-varying B fields which induce tissue current densities of <1 mA/cm^2 have been found to produce very few, if any, biological effects. This, after Tenforde, is not surprising because the endogenous current densities present in many organs are in the range of about 0.1 to 10 μA/cm^2.

Of course, some contradictions concerning the cyclotron-resonance effect have been published. Halle (1988) shows that the cyclotron resonance model violates the laws of classical mechanics and argues that the ubiquitous presence of dynamic friction in fluid media precludes significant magnetic effects on membrane ion transport. Wilson et al. (1990) published a comprehensive statement on the question of ELF electromagnetic fields being related to cancer induction. In this connection Liboff, McLeod, and Smith (1990) point out that the energy associated with any cyclotron resonance in the cell membrane is far below kiloTesla from where some serious theoretical problems arise.

It has been shown that in view of the conflicting results, any proof of a convincing correlation between cancer risk and residential exposure to ELF fields is still lacking, although a growing body of evidence indicates that processes connected with cell physiologic mechanisms, on the one hand, and endocrinological reactions on the other, suggest a certain possibility of ELF effects. The results of epidemiological investigations, however, demonstrated that causal effects of ELF fields play *de facto* a very subsidiary role in daily life.

More details on the present subject can be found in Reiter (1992).

5. EFFECTS OF SMALL ATMOSPHERIC IONS

First, it should be pointed out that in most of the laboratory experiments extremely high ion number densities have been applied which are unrealistic and do not exist in the natural environment. Effects of small ions in plants (Kotada et al., 1961; Barthakur and Arnold, 1988) and in small animals (Bhartendu and Menon, 1978) have been observed. In most cases the small ions are produced in "ionizers" by point discharge. Apart from the possibility that nitrous gases and/ or ozone are produced by point discharges — a question which is normally not mentioned by the producers. It has been shown recently (Goldstein et al., 1992) that the very reactive peroxide anion is produced by point discharge that also questions most of the "small ion effects". A biological effect of small atmospheric ions in vertebrates requires a large penetration into the respiratory tract. This is very questionable, as has been mentioned in Section 3.5 (see also Bailey, 1982). Nevertheless, some errors should be briefly discussed because they caused avalanches of fallacies which were "warmed up" until today. After Krueger and Smith (1960) (see details in Reiter, 1992) had postulated a direct biological effect of small ions (release of serotonin; 5HT), Sulman (1971) tried to explain the occurrence of disease in warm and dry winds (Foehn, Sharav in Israel) exclusively by inhaling positive small ions. This story was initiated by Czermak (1902) who found in 1900 a surplus of positive small ions in Innsbruck (Austria) during Foehn conditions immediately after the discovery that air is conductive (Schuster, 1884) and that small ions are in the air (Elster and Geitel, 1899). Czermak (1902, 1904), Huber (1915), and Schorer (1928, 1931) assumed that the positive small ions in excess over the negative ions may be of biological importance. Consequently, "Foehn disease", frequently observed in the Foehn valleys of Innsbruck, has been linked to the idea of a strong excess of positive small ions based on unconfirmed assumptions. Later Sulman (1971) was prompted by these assumptions to make comparable measurements in hot and dry winds in Israel, which, of course, are not comparable with Foehn air currents. He carried out measurements of small ions and other atmospheric electric parameters in Israel in an incorrect manner and performed tests on serotonin (5HT and 5HIAA) with patients. In the meantime it appeared that all of this work was of questionable value and could not be confirmed (see Anderson, 1971, 1972; Bailey, 1982). One may forget this story, however, since Sulman distributed his theories in public magazines. Consequently, many companies jumped on the bandwagon and began to produce and sell useless "ionizers" exhausting negative ions in order to overcompensate the "unhealthy" positive ions in rooms. Their appealing cry was and still is: produce a sound climate in your room like it exists in the mountains or over a lake (where actually no surplus of negative small ions exists).

REFERENCES

Adey, W.R. (1986). The sequence and energetics of cell membrane transductive coupling to intercellular enzyme systems, *Bioelectrochem. Bioenergy*, 15, 447.

Adey, W.R. (1988a). Cell membranes: the electromagnetic environment and cancer promotion, *Neurochem, Res.*, 13, 671.

Adey, W.R. (1988b). Physiological signalling across cell membranes and cooperative influences of extremely low frequency electromagnetic fields. In: *Biological Coherence and Response to External Stimuli*, Springer-Verlag, Heidelberg.

Adey, W.R. (1990). Electromagnetic fields and the essence of living systems. In: *Modern Radio Science*, U.R.S.I., Oxford University Press, Oxford, England.

Anderson, I. (1971). Mucocilliary function in trachea exposed to ionized and nonionized air. Ph.D. thesis. Akademical Boghandel. Arhus, Denmark.

Anderson, I. (1972). Effects of natural and artificially generated air ions on mammals, *Int. J. Biometeor.*, 5, 229 (Part II, Supplement to Vol. 16).

Bailey, W.H. (1982). Biological effects of air ions: facts and fancy. In: *Conference on Environmental Ions and Related Biological Effects*, American Institute of Medical Climatology, Philadelphia.

Barthakur, N.N. and Arnold, N.P. (1988). Growth and certain chemical constituents of tobacco plants exposed to air ions, *Int. J. Biometeor.* 32, 78.

Baumer, H. and Eichmeier, J. (1980). Relationship between the pulse rate of atmospherics and the diffusion time of ions in gelatine films, *Int. J. Biometeorol.* 24, 271.

Baumer, H. and Eichmeier, J. (1981). Relationship between the atmospherics pulse rate in the 10 and 27 kHz range, air mass movements and the diffusion time of ions in gelatine films, *Int. J. Biometeorol.* 25, 263.

Baumer, H. and Eichmeier, J. (1983). A hypothesis concerning the absorption of atmospherics in photogelatine films, *Int. J. Biometeorol.* 17, 125.

Bernhardt, J.H., Dahme, M., and Rothe, F.K. (1983). Gefährdung von Personen durch elektromagnetische Felder. STH-Berichte 2, 1, Dietrich Reimer Verlag, Darmstadt.

Bhartendu, S. and Menon, A.I. (1978). Effects of atmospheric small negative ions on the oxygen consumption of mouse liver cells, *Int. J. Biometeorol.* 22, 43.

Blackman, C.F. (1988). Stimulation of brain tissue *in vitro* by extremely low frequency, low intensity, sinusoidal electromagnetic fields. In: *Electromagnetic Waves and Neurobehavioral Functions*, Alan R. Liss, New York.

Blackman, C.F., Bennane, S.G., Rabinowitz, J.R., House, D.E., and Joines, W.T. (1985). A role for the magnetic field in the radiation-induced efflux of calcium ions from brain tissues *in vitro*, *Bioelectromagnetics*, 6, 327.

Blackman, C.F., Benane, S.G., House, D.E., and Elliott, D.J. (1990). Importance of alignment between local DC magnetic field and oscillating magnetic field in responses of brain tissue *in vitro* and *in vivo*, *Bioelectromagnetics*, 11, 159.

Blask, D.E. (1984). The pineal: an oncostatic gland? In: *The Pineal Gland*, Raven Press, New York.

Blask, D.E. (1990). The emerging role of the pineal gland and melatonin in oncogenesis. In: *Extremely Low Frequency Electromagnetic Fields, the Question of Cancer*, Battelle Press, Richland, OH.

Blask, D.E. and Hill, S.M. (1988). Melatonin and cancer: basic clinical aspects. In: *Melatonin: Clinical Perspectives*, Oxford University Press, Oxford, England.

Brinkmann, K. and Schaefer, H. (1984). Der Einfluß von elsktrischen und magnetischen Feldern auf den Menschen, *Med. Klin.*, 79, 49.

Carstensen, E.L. (1987). *Biological Effects of Transmission Line Fields*. Elsevier, Amsterdam.

Czermak, P. (1902). Über die Elektrizitätszerstreuung während Föhn, *Meteorol. Z.*, 19, 75.

Czermak, P. (1904). Über die Elektrizitätszerstreuung in der Atmosphäre. *Denkschr.* Akad. Wiss. Wien, No. 64.

Dumey, C.H., Rushforth, C.K., and Anderson, A.A. (1988). Resonant AC-DC magnetic fields: calculated response, *Bioelectromagnetics*, 9, 315.

Eichmeier, J. and Baumer, H. (1987). Atmospherics emission computer tomography and its importance for biometeorology, Abstracts, 11th Congr. of Biometeorology, S.P.B. Academic Publishing, The Hague, 1987.

Eichmeier, J. and Baumer, H. (1990). A model for the generation of atmospherics and the correlation between spectral atmospherics pulses and behavior of cell DNA, Proc. 12th Int. Congr. of Biometeorol., S.P.B. Academic Publishing, The Hague, 1990.

Elster, J. und Geitel, H. (1989). Über die Existenz elektrischer Ionen in der Atmosphäre, *Terr. Magn. Atmos. Electr.*, 4, 213.

Fulton, J.P., Cobb, S., Preble, L., Leone L., and Forman, E. (1980). Electric wiring configurations and childhood leukemia in Rhode Island, *A. J. Epidemiol.*, 111, 292.

Goldstein, N.I., Goldstein, R.N., and Merzlyak, M.N. (1992). Negative air ions as a source of superoxide, *Int. J. Biometeorol.* 36, 118.

Grandt, Ch. (1992). Thunderstorm monitoring in South Africa and Europe by means of very low frequency sferics, *J. Geophys. Res.*, 97, 18.215.

Halle, B. (1988). On the cyclotron resonance mechanism for magnetic field effects on transmembrane ion conductivity, *Bioelectromagnetics*, 9, 381.

Hauf, R. (1974). Effects of 50 Hz alternating fields on man, *Elektrotech. Z.*, 12, 318.

Hauf, R. (1976). Einfluss elektromagnetischer Felder auf den Menschen, *Elektrotech. Z.*, 23, 181.

Hauf, R. (1981). Untersuchungen zur Wirkung energietechnischer Felder auf den Menschen, Forschungsstelle für Elektropathologie, Freiburg, Germany.

Hauf, R. (1986). Bericht von der wissenschaflichen Tagung über Elektropathologie. Forschungsstelle für Elektropathologie, Freiburg, Germany.

Huber, P.B. (1915). Luftelektrische Beobachtungen und Messungen bei Föhn, *Z. Meteorol.*, 32, 512.

Kotaka, S., Krueger, A.P., and Andriese, P.C. (1961). Effects of air ions on IAA content of barley seedings, *Plant Cell Physiol.*, 6, 711.

Krause, N. (1989). Grenzwerte für elektrische und magnetische Felder im Bereich von 0 bis 30 kHz, *ETZ Elektrotech. Z.*, 110, 280.

Krueger, A.P. and Smith, R.F. (1958). Effects of gaseous ions on tracheal ciliary rate, *Proc. Soc. Exp. Biol. Med.*, 98, 421.

Krueger, A.P. and Smith, R.F. (1960). The biological mechanism of air ions action: 5-hydroxytryptamine as the endogenous mediator of positive air ions effects on the mammalian trachea, *J. Gen. Physiol.*, 43, 533.

Krueger, A.P. (1968). Air ion action on animal and man, in (pp. 42–50) *Aeroionotherapy*, Carlo Erba Foundation, Milan, Italy.

Liboff, A.E., Rozek, R.J., Sherman, M.L., McLeod, B.R., and Smith, S.D. (1987). Ca^{++}-45 cyclotron resonance in human lymphocytes, *J. Bioelectr.*, 6, 13.

Liboff, A.R. and McLeod, B.R. (1988). Kinetics of channelized membrane ions in magnetic fields, *Bioelectromagnetics*, 9, 39.

Liboff, A.R., McLeod, B.R., and Smith, S.D. (1990). *Extremely Low Frequency EFL Fields: The Question of Cancer*, Battelle Press, Richland, OH, p. 290.

Liboff, A.R. and Parkinson, W.C. (1991). Search for ion-cyclotron resonanes in an Na^+-transport system, *Bioelectromagnetics*, 12, 77.

Lovely, R.H. (1988). Recent studies in the behavioral toxicology of ELF electric and magnetic fields. In: *Electromagnetic Fields and Neurobehavioral Function*. Alan R. Liss, New York.

Lyle, D.B., Schlechter, P., Adey, W.R., and Lundak, R.L. (1983). Suppression of T-lymphocyte cytotoxicity following exposure to sinusoidally amplitude-modulated fields, *Bioelectromagnetics*, 4, 281.

Lyle, D.B., Wang, X., Ayotte, R.D., Sheppard, A.R., and Adey, W.R. (1990). Calcium uptake by leukemia and normal T-lymphocytes to ELF magnetic fields, *Bioelectromagnetics*, 12(3), 145, 1991.

Makino, M. and Ogawa. T. (1985). Quantitative estimation of global circuit, *J. Geophys. Res.*, 90, 5861.

McDowall, M.E. (1986). Mortality of persons resident in the vicinity of electricity transmission facilities, *Br. J. Cancer*, 53, 271.

McLeod, B.R. and Liboff, A.R. (1986). Dynamic characteristics of membrane ions in multified configurations of low-frequency electromagnetic radiation, *Bioelectromagnetics*, 7, 177.

Milham, S. Jr. (1983). Occupational mortality in Washington State 1950–1979. U.S. Department of Health and Human Services, Cincinnati, OH.

Miller, M.W. (1986). Extremely low frequency (ELF) electric fields: experimental work on biological effects. In: *CRC Handbook of Biological Effects in Electromagnetic Fields*, CRC Press, Boca Raton, FL.

Minehart, J.R., Davis, T.A., and Kornblueh, I.H. (1958). Artificial ionization and the burned patient, *Med. Sci.*, 3, 363.

Norris, W.T. and Bonnell, J.A. (1988). People in 50 Hz electric and magnetic fields: studies in the United Kingdom. In: *Electromagnetic Fields and Neurobehavioral Function*, Alan R. Liss Inc., New York.

Polk, C. (1984). Time-varying magnetic fields and DNA synthesis: magnetic forces due to magnetic fields on surface-bound counter-ions. In: *Proc. 6th Annu. Meet. Bioelectromagnetics Soc.*, Permagon Press, New York.

Polk, C. (1986). Introduction. In: *CRC Handbook of Biological Effects of Electromagnetic Fields*, CRC Press, Boca Raton, FL.

Polk, C. and Postow, W. (1986). *CRC Handbook of Biological Effects of Electromagnetic Fields*, CRC Press, Boca Raton, FL.

Reiter, R.J. (1973). Pineal control of a seasonal reproductive rhanthm in male golden-hamsters exposed to natural daylight and temperature, *Endocrinology*, 92, 423.

Reiter, R.J., Anderson, L.E., Buschbom, R.L., and Wilson, B.W. (1988). Reduction of the nocturnal rise in pineal melatonin levels in rats exposed to 60-Hz electric fields *in utero* and for 23 days after birth, *Life Sci.*, 42, 2203.

Reiter, R.J., Li, K., Gonzalez-Brito, A., Tannanbaum, M.G., Vaughan, M.K., Vaughan, G.M., and Viluana, M. (1988). Elevated environmental temperature alters the response of the reproductive and thyroid asxes of female Syrian hamsters to afternoon melatonin injections, *J. Pineal Res.*, 5, 301.

Reiter, R. (1960). *Meteorobiologie und Elektrizität der Atmosphäre*, Akademische Verlagsges, Leipzig.

Reiter, R. (1982). Biological effects of atmospherics and man-made noise. In: *CRC Handbook of Atmospherics*, CRC Press, Boca Raton, FL.

Reiter, R. (1992). *Phenomena in Atmospheric and Environmental Electricity*, Elsevier, Amsterdam.

Ruhenstroth-Bauer, G., Baumer, H., Kugler, J., Spatz, R., Sönning, W., and Filipiak, B. (1984). Epilepsy and weather: a significant correlation between the onset of epileptic seizures and specific atmospherics — a pilot study, *Int. J. Biometeorol.*, 28, 333.

Ruhenstroth-Bauer, G., Rösing, O., Baumer, H., Sönning, W., and Lehmacher, W. (1988). Demonstration of correlations between the 8 and 10 kHz atmospherics and the inflammatory reaction of rats after carrageenan injection, *Int. J. Biometeorol.*, 32, 201.

Sawitz, A. (1988). Case-control study of childhood cancer and exposure to 60-Hz-magnetic fields, *Am. J. Epidemiol.*, 128, 21.

Sawitz, D.A., Wachtel, H., Barner, F.A., John, E.M., and Tvrdik, J.G. (1988). Case-control study of childhood cancer and exposure to 60-Hz-magnetic fields, *Am. J. Epidemiol.*, 128, 38.

Schaefer, H. (1986). Neue Untersuchungen über die biologische Wirkung von Magnetfeldern. In: *Berichte von der wissenschaftlichen Tagung über Elektropathologie*, Freiburg, Germany.

Schorer, G. (1928). Über den Elektrizitätsgehalt der Luft und dessen Einfluß auf wetteremfindliche Menschen, *Schweiz. Med. Wochenschr.*, 58, 431.

Schorer, G. (1931). Über die Wirkung der Luftelektrizität auf gesunde und kranke Menschen, *Schweiz. Med. Wochenschr.*, 61, 417.

Schuster, A. (1884). Experiments on the discharge of electricity through gases, Proc. R. Soc. London, 37, 317.

Schwan, H.P. (1982). Biophysics of the interaction of electro-magnetic energy with cells and membranes. In: *Biological Effects and Dosimetry of Non-Ionizing Radiation*, Plenum Press, New York.

Silm, R. (1983). Biologische Wirkungen magnetischer 50-Hz-Felder. Medizinisch-Technischer Bericht, Institut z. Erforschung Elektr. Unfälle, Köln, Germany.

Stevens, R.G., Severson, R.K., Kaune, W.T., and Thomas, D.B. (1986). Epidemiological study of residential exposure to ELF electric and magnetic fields and risk of non-lymphocytic leukemia, DOE/EPRI/NY State Power Lines Project Contractors Review, Denver, CO.

Sulman, F.G. (1971). Meteorologische Frontverschiebung und Wetterfühligkeit — Föhn — Chamissin-Sharav, *Ärztliche Praxis*, 23, 998.

Tenforde, T.S. (1986). Interaction of ELF magnetic fields with living matter. In: *CRC Handbook of Biological Effects of Electromagnetic Fields*, CRC Press, Boca Raton, FL.

Tenforde, T.S. (1990). Biological interactions and human health effects of extremely low frequency magnetic fields. In: *Extremely Low Frequency Electromagnetic Fields, the Question of Cancer*, Battelle Press, Richland, OH.

Thomas, J.R., Schrot, J., and Liboff, R. (1986). Low-intensity magnetic fields alter operant behavior in rats, *Bioelectromagnetics*, 7, 349.

Tomenius, L. (1986). 50 Hz electromagnetic environment and the incidence of childhood tumors in Stockholm county, *Bioelectromagnetics*, 7, 191.

Wall, W. (1707). Experiment of the luminous qualities of amber, diamond and gum lac, *Philos. Trans.*, 26, 69.

Wehner, A.P. (1983). Effects of negatively charged aerosol on blood and cerebrospinal fluid parameters in rat, *J. Biometeorol.*, 27, 259.

Wertheimer, N. and Leeper, E. (1979). Electrical wiring configurations and childhood cancer, *Am. J. Epidemiol.*, 109, 273.

Wertheimer, N. and Leeper, E. (1982). Adult cancer related to electrical wires near the home, *Int. J. Epidemiol.*, 11, 345.

WHO (1984). Extremely Low Frequencies (ELF) Fields, Environmental Health Criteria 35, WHO, Geneva.

Wilson, B.W. (1990). The Moscow signal. In: *Extremely Low Frequency Electromagnetic Fields*, Battelle Press, Richland, OH.

Wilson, B.W., Chess, E.K., and Andersen, L.E. (1986). 60-Hz-electric-field effects on pineal melatonin rhythms, *Bioelectromagnetics*, 7, 239.

Wilson, B.W., Anderson, L.E., Hilton, D.I., and Phillips, R.D. (1981). Chronic exposure to 60-Hz fields; effects on pineal function in the rat, *Bioelectromagnetics*, 2, 371.

Wilson, B.W. and Andersen, L.E. (1986). 60-Hz electric field effects on pineal melatonin rhythms, *Bioelectromagnetics*, 7, 239.

Wilson, B.W., Wright, C.W., Morris, J.A., Stevens, R.G., and Andersen, L.E. (1988). Effects of electric blanket use on human pineal gland function: a preliminary report. In: Proc. of DOE/EPRI/Contractors Review, Washington D.C., U.S. Department of Energy.

Wilson, B.W., Stevens, R.G., and Andersen, L.E. (1990). *Extremely Low Frequency Electromagnetic Fields, the Question of Cancer*, Battelle Press, Richland, OH.

Chapter 6

Electromagnetic Effects of Nuclear Explosions

Conrad L. Longmire

CONTENTS

1. NUCLEAR EXPLOSIVES

Nuclear physics became an active experimental and theoretical science in the 1930s. By 1939 it was known that the nuclei with the largest binding energy per nucleon were those of intermediate mass, like iron. Thus, energy is released when two lighter nuclei are made to join together. The Coulomb repulsion of the two nuclei prevents such *fusion* from occurring naturally, except at the extremely high temperatures and densities near the centers of stars. Even there, the *thermonuclear* reaction lifetime of a given light nucleus is millions of years. In the laboratory, nuclei artificially accelerated to energies of the order of a million electron volts (MeV) may react within 10^{-8} s upon entering a target of normal solid density. However, the energy released is far less than that expended in the accelerator, especially since most of the accelerated nuclei are slowed by atomic rather than nuclear collisions.

By 1939 it was also known, first theoretically and then experimentally, that the heaviest nuclei such as $_{92}U^{235}$ are on the verge of splitting into two lighter nuclei because of the Coulomb repulsion of the 92 protons for each other, and that this force gives the separating fragments a total kinetic energy of about 180 MeV. It was expected and found that *fission* could be triggered by even a low energy neutron entering the U nucleus, for which there is no repulsion. Further, since the ratio of neutrons to protons in stable nuclei increases with atomic number Z (143/92 in U, 60/46 for Z = 46), there are too many neutrons in U to be accommodated in the two fragments. It was therefore expected that a few neutrons would be released immediately following fission, within a time interval of the order of 10^{-14} s or less. The measured number of prompt neutrons released per fission is $\nu \approx 2.5$ for U^{235}. The possibility of an exponentially increasing chain reaction thus became obvious. This possibility was pursued in several countries during World War II, most vigorously in the U.S. where the first nuclear explosion was produced on July 16, 1945, releasing energy equivalent to that of 20,000 tons of TNT (20 kton).

The advantage of a nuclear explosive is its much larger energy release (or *yield*) per mass than chemical explosives. The difference arises from the fact that chemical binding energies are electron volts per atom while nuclear energies are millions of electron volts per atom. One metric ton (1000 kg) of TNT releases 10^9 calories = 4.19×10^9 J. At 180 MeV per fission, 1 kg of U^{235} if completely fissioned would release 7.37×10^{13} J, or 17.6 kton. The actual fraction of U nuclei that undergo fission (the *efficiency*) depends on the rate of growth of the chain reaction.

In a very large sphere of U^{235}, where neutron leakage is negligible, the reaction grows with time t as $\exp(\alpha t)$, where the growth rate α is

$$\alpha = N_u \sigma_f \upsilon (\nu - 1) \approx 1.0 \times 10^8 / \text{sec} \tag{1.1}$$

Here $N_u \approx 5 \times 10^{22}/\text{cm}^3$ is the density of U nuclei in solid uranium, $\sigma_f \approx 1 \times 10^{-24}$ cm^2 is the fission cross section per nucleus, $\upsilon \approx 1.4 \times 10^9$ cm/s is the average speed of the fission neutrons, and $\nu = 2.5$ as given above. Neutron leakage in a finite sphere will reduce α, but this can be offset by use of a neutron reflector and by compression of the uranium by chemical explosives. We will use Equation 1.1 as a typical growth rate.

A crude estimate of the energy released can be made as follows. The fission energy heats the uranium, raising its internal energy ε (per gram) and pressure p. The order of magnitude of the pressure is $p \approx \rho \varepsilon$, where ρ is the mass density. If the radius of the sphere is a there is a pressure gradient $\approx \rho \varepsilon / a$ tending to make the sphere expand. According to Newton's law, the increase in radius δa obeys $\rho \ddot{\delta a} \approx \rho \varepsilon / a$. For exponentially growing ε this gives

$$\delta a \approx \varepsilon / a \alpha^2 \tag{1.2}$$

When δa exceeds some fraction f of a, increased neutron leakage causes the system to go sub-critical ($\alpha < 0$) and energy production dies away. Thus, the final energy per gram is of order of magnitude

$$\varepsilon \approx (\alpha a)^2 f \tag{1.3}$$

Of course, internal energy is eventually converted to kinetic energy of explosion. As a hypothetical example (not necessarily self-consistent), taking α from Equation 1.1, $a = 10$ cm and $f = 0.1$ gives $\varepsilon \approx 1 \times 10^{13}$ J/kg = 2.4 kton/kg, which would correspond to an efficiency of about 14%.

At such high internal energies the state of matter is very different from what we normally see. The matter becomes an ionized gas of free electrons and highly charged nuclei. In collisions with the nuclei, the electrons emit X-rays (photons) so rapidly that the X-radiation comes into thermal equilibrium with the particles in a time short compared with 10^{-8} s. Much of the internal energy is then in the X-rays. For example, in normal density uranium fissioned to 1%, the temperature is about $\theta \approx 5$ keV, radiation and particle energies per cubic centimeter are approximately equal, and the number of electrons remaining bound to each nucleus is between 4 and 10. (These results come from simple atomic physics and the Stefan-Boltzmann law of radiation.) At higher temperatures, the radiation energy dominates since it is proportional to θ^4, while the particle energy is approximately proportional to θ.

The temperatures reached in fission explosives are sufficient to ignite thermonuclear reactions. The most common thermonuclear fuels are the two isotopes of hydrogen, deuterium $D = H^2$ and tritium $T = H^3$, and the isotope of lithium Li^6 in the convenient form of LiH. Most prominent reactions and the energies they release are (n stands for neutron):

$$D + D \rightarrow H^1 + T + 4.0 \text{ MeV} \tag{1.4}$$

$$n + Li^6 \rightarrow He^4 + T + 4.8 \text{ MeV} \tag{1.5}$$

$$D + T \rightarrow He^4 + n + 17.6 \text{ MeV} \tag{1.6}$$

Reaction 1.6 goes much faster than 1.4, and may be used in kiloton range explosives. The neutron in Reaction 1.6 carries off 14.1 MeV, which stands out as a strong peak against the continuous fission neutron spectrum. The other reactions are useful in megaton range explosives. D and Li^6

are plentiful and cheap. T is unstable with a half life of about 12 years and therefore must be continually produced in reactors (by reaction 1.5).

Most of the nuclear physics involved in nuclear explosives can be found, for example, in the book by Kaplan (1955). Equation 1.3, with a numerical coefficient that makes it more accurate than order of magnitude, is known as the Bethe Feynman formula.

2. EFFECTS OF NUCLEAR EXPLOSIONS

The original intent was to detonate the nuclear explosive in the air several hundred meters above the ground. The principal military effect would be the destruction of structures by the blast wave in air. Secondary effects would be heat radiation, which might start fires, and radioactivity. The theory of the blast wave and heat radiation were worked out during the war, but published years later. Original works by Taylor (1950), Bethe et al. (1958), and Sedov (1959) are available. In addition, a detailed monograph on these effects with much test data was edited by Glasstone (1977). We will need only a little of this information.

The energy is transferred from the nuclear device to the air primarily by two mechanisms. First, the hot device materials radiate X-rays, which may carry out 75% of the energy for a megaton range device. These X-rays are absorbed in a few meters of air at sea level. Most of the remaining energy is in kinetic energy of the exploding device debris. When the debris has encountered an air mass that is a few times the debris mass, most of the debris energy has been transferred to air. Debris speeds are of the order of 1 m/μs. By 10 μs or so at sea level, both X-ray and debris energy are driving the expansion of a hot *fireball* in (mostly) air. Until 100 μs or so, the fireball grows by radiation diffusion. By that time a strong outward going hydrodynamic shock wave has formed at the outer edge of the fireball, which outruns diffusion. Subsequently, the surface of the growing fireball is very sharp, with approximately ambient temperature outside and thermally ionized air inside as far out in radius as the shock wave remains strong. The ionized air is a good electrical conductor. Eventually the shock weakens, the shocked-air temperature decreases, and the ionization and electrical conductivity fall rather abruptly for larger radius.

We can define the *conducting fireball* as the sphere in which the conductivity σ exceeds 1 mho/m. Choosing 0.1 or 10 mho/m would make very little difference. The radius R of this sphere depends on the time t approximately as

$$R = R_0 \left(\frac{t}{t_0} \right)^{2/5} \text{ for } t < t_0 \tag{2.1}$$

and $R = R_0$ for $t > t_0$ until the fireball rises after a few seconds due to buoyancy. In this equation

$$R_0 = 1.1 \left(\frac{Y}{Y_0} \frac{\rho_0}{\rho_a} \right)^{1/3} \text{ km}, \quad t_0 = 0.6 \left(\frac{Y}{Y_0} \frac{\rho_0}{\rho_a} \right)^{1/3} \text{ s} \tag{2.2}$$

where Y is the yield in megatons, $Y_0 = 1$ for a surface burst, $Y_0 = 2$ for an air burst, ρ_a is the density of ambient air in gram per liter, and $\rho_0 = 1.23$, the density at sea level. These scaling rules are valid only at those lower burst heights for which $R_0 \leq 3$ km or about half an atmospheric scale height. For $Y = 1$ Mt, the burst height should be less than 30 km.

The visible fireball, which is nearly the same as the conducting fireball, radiates heat in the form of photons with wavelengths longer than about 0.2 μ, i.e., in the ultraviolet, visible, and infrared. These photons travel great distances through dry air at standard temperature and pressure (STP). About one third of the energy initially in the fireball is radiated away within a few seconds. The rest of the energy is either acoustically radiated away in a pressure wave, lifts ambient air against gravity, or remains temporarily in ionization, dissociation, vibration, and rotation of molecules (forms of internal energy not contributing to pressure).

The radioactivity effects are conveniently divided into two groups. The first includes the effects of the *initial* nuclear radiation, principally neutrons and gamma rays, that are emitted by the nuclear device in, e.g., the first second. The second group includes the *residual* nuclear radiation, principally beta and gamma rays, that are emitted by fission fragments and activated device materials over times that range from seconds to many years with a decay rate roughly proportional to $t^{-1.2}$. For an air burst, the vaporized radioactive materials rise with the heated air and tend to disperse widely before condensing and falling to earth. If the fireball reaches the ground, soil and water sucked into the rising fireball lead to earlier condensation with intense local radioactive *fallout*. This is hazardous to personnel, photographic film, and some sensitive electronic equipment.

The initial nuclear radiation is the cause of some of the electromagnetic effects discussed in this paper. Because the fraction of total yield contained in this radiation is generally only about 1%, these effects were not identified and studied in advance. However, starting with the first nuclear test in 1945, electrical damage often occurred to electronic equipment fielded to measure and record physical parameters relating to the explosion. Because the close-in equipment was bunkered to shield against air blast, heat, and nuclear radiation, it was conjectured that electromagnetic effects were driving large currents in the wires entering the bunkers. A characteristic transient electromagnetic pulse (EMP) was indeed routinely observed by placing vertical HF antennas at large distances. Some large currents were measured in close-in radial wires in 1956. The EMP radiated to great distances by several high altitude explosions in 1962 was measured. The detailed and quantitative explanation of these effects was developed in the early 1960s. While atmospheric nuclear testing was stopping at that time, there was the new stimulus of making sure that the ICBM farms would function after receiving a first strike.

With the recent most welcome end of the cold war, military scientists look forward to documenting their sciences and turning to other interesting and important problems. However, the world is not yet free of conflicts and secrecy still shrouds much of the subject of this paper. We are restricted to discussing the known basic physics involved in the effects and magnitudes that can be estimated therein, using nominal outputs of nuclear explosions. Very likely that will be sufficient for the purposes of this handbook.

3. ELECTROMAGNETIC EFFECTS OF GAMMA RAYS

3.1. SOURCE CURRENTS AND AIR CONDUCTIVITY

Typically, about 0.3% of the energy released by a nuclear explosive leaves its outer surface in the form of prompt gamma rays (Glasstone and Dolan, 1977). These are made either in the fission process or in inelastic scatter or capture of neutrons in the device materials. They may be mostly emitted in a time interval of about 30 ns and have average quantum energy of about 1.5 MeV. In traveling through the air, the gammas collide with electrons (Compton effect). The struck electrons stream predominantly in the same direction as the gammas, stopping in a few meters in sea level air as a result of collisions with other (secondary) electrons. The aggregate of streaming Compton electrons in unit volume constitutes an electric current density, which is the source of electromagnetic fields. Of the order of 10^4 secondary electrons are freed by each Compton electron and these, plus the positive ions they left, make the air electrically conducting. The effect of this conductivity is generally to oppose the generation of the fields.

The physical equations that govern the formation of the Compton current and air conductivity and the generation of fields have been reviewed by the present author, Longmire (1978), along with approximate methods for solving them. We make use here of results developed in that paper, which we denote by the letter A. For example, A-Section II shows that for a burst in the air just above the ground (*surface burst*) there are additional sources of gammas due to inelastic scatter and capture of neutrons in the soil and air near the burst; see A-Figure 1. For an isotropic point source of gammas in uniform air, A-Equation 1 gives the approximate dependence of the radial gamma flux on distance from the source. For a time-dependent source strength $(S(t)\ \gamma - \text{MeV/}$ s, the radial flux at distance r is

$$F_\gamma(r,\ t) = S\left(t - \frac{r}{c}\right)e^{-r/\lambda}/4\pi r^2 \quad \gamma - \text{MeV/m}^2\ \text{s} \tag{3.1}$$

where c is the speed of light. This equation is exact for the flux of *unscattered* gammas if $\lambda = \lambda_s$, the mean free path for Compton scattering. It is approximately correct for the total flux if $\lambda = \lambda_a$, the energy absorption length, which is given in A-Table 1. We express S in energy units $\gamma - \text{MeV}$ because the effectiveness of a quantum in making EMP is roughly proportional to its energy.

The time delay in $S(t - r/c)$ is important for the faster parts of the source. If the prompt pulse width is 30 ns, then at any given time the prompt gammas occupy a spherical shell with radial thickness about 10 m, and this shell moves outward with time at the speed of light. Scattered gammas tend to fall behind this advancing shell; thus, the total gamma flux is spread somewhat to later times at larger distances.

The relation of the Compton current to the gamma flux is discussed in A-Section III. Note the simple relation A-Equation 3 between F_γ and the flux F_e of Compton electrons. This equation is appropriate at low altitude where deflection of the Compton electrons by the geomagnetic field may be neglected because the electron range is only about 0.05 of the gyroradius. The range increases with altitude, and is comparable with the gyroradius at an altitude of about 30 km. Here the Compton current has both a radial component J_r and a comparable *transverse* component J_t in the direction perpendicular to both J_r and the geomagnetic field. These currents are graphed in A-Figure 2 for a propagating impulse of gammas, as a function of time after passage of the impulse.

The air conductivity is discussed in A-Section IV in terms of the three-species model which follows the density N_e of free electrons, N_+ of their positive ions and N_- of the negative ions formed by the attachment of an electron to O_2. Rate constants k_1 for attachment, k_2 for recombination of electrons with positive ions, and k_3 for positive and negative ion neutralization are given. Simple approximate solutions of the rate equations are given for various regimes of the ionization source density. Because the mobility (response to an applied electric field) of electrons is much larger than that of ions, electrons tend to dominate the conductivity σ at early times. At late times the electrons are nearly all attached and σ becomes mostly ionic.

For an air burst at low altitude the Compton current $\vec{J}$ is approximately radial, and J and σ are nearly spherically symmetric. If symmetry were exact, no magnetic field would be generated and there would be no radiated signal (A-Section V). A radial electric field would build up due to radial separation of charge, according to the Maxwell equation (MKS units)

$$\varepsilon_0\frac{\partial E}{\partial t} + \sigma E = -J \tag{3.2}$$

After an initial charging phase in which $E \approx -\dfrac{1}{\varepsilon_0}\int J dt$, E limits at a value such that the conduction current σE cancels the Compton current. This value is called the *saturated field*,

$$E_s = -J/\sigma \tag{3.3}$$

and generally (A-Equation 24) has the order of magnitude 3×10^4 V/m.

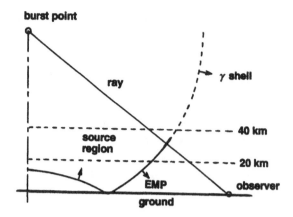

Figure 6.3.1 For a burst at 100-km altitude, the prompt gamma shell, which is about 10 m in radial thickness, is shown at about 0.4 ms after the burst. The gammas are converted into an EM pulse in the source region by means of Compton recoil electrons that are deflected by the geomagnetic field, forming a transverse current. The EMP lies on the same spherical shell as the gammas until it is reflected by the ground (shown here as flat). The EMP seen by the observer is produced near the ray in the source region.

Departures from spherical symmetry that lead to the generation of magnetic fields and radiated signals include the vertical gradient of air density, presence of the ground, and the geomagnetic field.

3.2. HIGH ALTITUDE BURSTS

For bursts at altitudes of the order of 100 km or more, the gammas in A-Figure 1 due to interactions of neutrons with air and ground are much delayed or absent. A 14-MeV neutron travels 50 km/ms and interacts significantly with air only at altitudes below 50 km. Only the prompt gammas contribute significantly to the EMP at times earlier than 1 ms. They occupy an expanding spherical shell, centered about the burst point, with radial thickness of the order of 10 m.

The gammas begin to undergo significant Compton scattering at altitudes below 50 km. Most of the gammas are scattered between 35 and 25 km altitude. A scattered gamma soon falls behind the shell of unscattered gammas and makes no further contribution to the first microsecond of the EMP. The fraction of gammas that remain unscattered falls rapidly below 20 km.

The average velocity of the Compton electrons is radial at their birth and only slightly smaller than c in magnitude, thus, they almost keep up with the gamma shell. As a result, the density of Compton electrons is initially about 10 times the density of their birthplaces. In addition, the retarded time $\tau = t - r/c$ of a Compton electron born at $t_b = r_b/c$ is much smaller than the time $t - t_b$ since birth of that electron. (If the Compton electrons moved outward with speed c, their retarded time would remain at $\tau = 0$.) This is the reason the transverse Compton current in A-Figure 2 peaks at about $\tau = 25$ ns, while the time $t - t_b$ for a Compton electron to be turned through 1 rad by the geomagnetic field is about 250 ns.

A current element radiates no fields in the direction of the current element, but radiates predominately in directions perpendicular to that direction. The transverse Compton current radiates both outgoing (in r) and ingoing waves. The outgoing waves maintain coincidence with the gamma shell, and fields radiated at all r's are superposed. Ingoing fields and those radiated by the radial current do not have that advantage and therefore reach much smaller amplitudes. In addition they are usually much attenuated by the developing air conductivity.

Figure 6.3.1 shows the thin gamma shell being converted into a coherent electromagnetic (EM) pulse in the *source region* between 20 and 40 km altitude, and a ray running from the burst point to an observer point on the ground. The EMP seen by the observer at early retarded times must be generated interior to a slender ellipsoid having these two points as foci. In fact, to good approximation, the EMP can be determined by performing an integration along that part of the ray that lies within the source region. By neglecting the ingoing waves and the curvature of the gamma shell within the source region, it is shown in A-Section VII that the transverse electric field E_t obeys the equation (MKS units)

$$2 \frac{\partial}{\partial r} E_t + Z_0 \, \sigma E_t = -Z_0 \, J_t \qquad (3.4)$$

where $Z_0 = \sqrt{\mu_0/\varepsilon_0} \approx 377$ ohms and the derivative $\partial/\partial r$ is taken at fixed retarded time τ. J_t and σ are to be regarded as functions of r and τ. This equation is similar to Equation 3.2, except that it is to be integrated in r rather than in t. Saturation also occurs here when the conduction current σE_t cancels J_t. Thus E_t is limited by the saturated field E_S,

$$E_t(r, \tau) \leq E_S(r, \tau) = -J_t(r, \tau)/\sigma(r, \tau) \qquad (3.5)$$

after the wave has run far enough in the source region to build E_t up to this value. At the lower edge of the source region, both J_t and σ go to zero when all of the gammas have been scattered. Equation 3.4 then indicates that E_t (τ) becomes constant in r. However, if the curvature of the gamma shell is included, the amplitude falls proportionally to $1/r$, as for any spherically diverging wave when r is much larger than the wavelengths involved. If the operator $\dfrac{\partial}{\partial r}$ in Equation 3.4 is replaced by $\dfrac{1}{r} \dfrac{\partial}{\partial r} r$, this result is automatically achieved, because rE_t becomes constant in r.

The secondary electrons do not attach significantly in the first microsecond in the high altitude source region. As indicated in A-Figure 2, their density continues to increase with retarded time, as does the conductivity σ. The saturated field $E_s(r, \tau)$ therefore decreases with increasing τ. The sooner E_t saturates, the larger will be its peak value, and this result is enhanced by having a gamma flux that rises very rapidly to a very high level of intensity.

The rise rates of actual nuclear devices are secret. In order to provide unclassified information of some relevance for scientists and engineers, Longmire et al. (1987) assumed an isotropic gamma output rate that rises instantaneously from zero at $t = 0$ to

$$S(t) = S_0 e^{-\beta t} \text{ for } t \geq 0 \qquad (3.6)$$

where

$$\beta = 10^8/\text{s}, \, S_0 = 2.6 \times 10^{34} \, \gamma - \text{MeV/s} \qquad (3.7)$$

The total gamma energy output is 10 kton, and the quantum energy was taken as 2 MeV. The burst was at 400 km over a point on the earth where the geomagnetic field is 20 degrees from the vertical and has magnitude 0.56 G. A dipole model was used to find the geomagnetic field at other locations. The EMP was calculated by the computer code CHAP, which is described in the report. The electric field components are graphed as functions of time for observers at various azimuths and ground ranges out to the burst horizon at 2200 km, and contour plots of peak amplitude are provided. The results are justified by analytic means and energy conservation is verified.

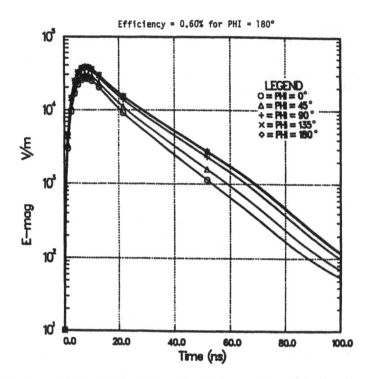

Figure 6.3.2 For burst at 400-km altitude and observer at 501-km ground range at various magnetic azimuths, amplitude of electric field vs. time and efficiency of conversion of gamma energy to EMP energy, per unit solid angle. (From Longmire et al., 1987, *A Nominal Set of High-Altitude EMP Environments,* Oak Ridge Laboratory Report ORNL/Sub/86-18417/1, National Technical Information Service, U.S. Department of Commerce.)

Some results are reproduced here in Figures 6.3.2 through 6.3.5, which show the magnitude of the electric field. The direction of this field is approximately that of $-\vec{r} \times \vec{B}_a$, where $\vec{B}_a$ is the ambient geomagnetic field in the source region for a particular observer. It can be seen that the peak amplitude decreases and the pulse length increases with increasing ground range. The maximum efficiency of conversion of gamma energy per steradian to EMP energy per steradian is 6.0% and occurs for the observer on the horizon. At smaller ranges, stronger saturation leads to more energy loss from the wave to joule heating.

3.3. LOW ALTITUDE BURSTS, NUCLEAR LIGHTNING

There is a large amount of interesting scientific information on the gamma-induced EMP from low altitude air and surface bursts that is still labeled secret, although the military advantage of having it is insignificant. Examples include the EM signals radiated to large distances by nuclear explosions with nominal outputs, due to such asymmetries as the air-ground interface, the air density gradient, and the geomagnetic field. A terse review of some of the underlying physics is in A-Section VI, and the present author hopes to push more information through the declassification process.

An interesting phenomenon that can be discussed in some detail is that of *nuclear lightning*, which was seen on many large yield surface bursts set off by the U.S. at Eniwetok and Bikini Atolls. Figure 6.3.6 shows this phenomenon on the Mike event of Operation IVY in 1952. It was made from a photograph originally taken by EG&G for Los Alamos that was first published by Uman et al. (1972). A sequence of four actual photographs taken at times from 1.4 to 10.4 ms appear in a book of lightning photographs by Salanave (1980). The burst point was on the reef at the center of the hemispherical fireball. The camera was looking across the lagoon. The discharges are believed to have formed on sharp metallic structures also located on the reef at the

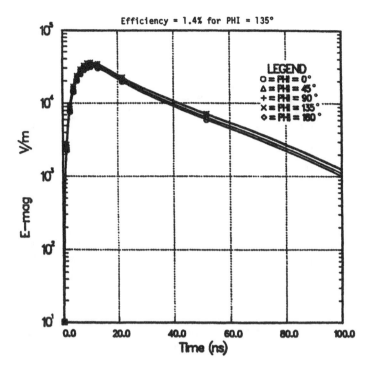

Figure 6.3.3 For burst at 400-km altitude and observer at 777-km ground range and various magnetic azimuths, amplitude of electric field vs. time and efficiency of conversion of gamma energy to EMP energy, per unit solid angle. (From Longmire et al., 1987, *A Nominal Set of High-Altitude EMP Environments,* Oak Ridge Laboratory Report ORNL/Sub/86-18417/1, National Technical Information Service, U.S. Department of Commerce.)

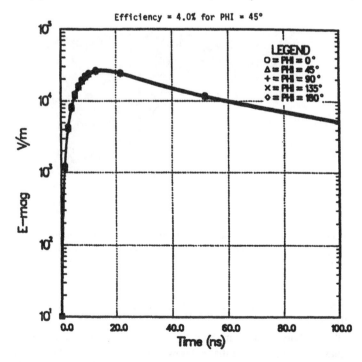

Figure 6.3.4 For burst at 400-km altitude and observer at 1356-km ground range and various magnetic azimuths, amplitude of electric field vs. time and efficiency of conversion of gamma energy to EMP energy, per unit solid angle. (From Longmire et al., 1987, *A Nominal Set of High-Altitude EMP Environments,* Oak Ridge Laboratory Report ORNL/Sub/86-18417/1, National Technical Information Service, U.S. Department of Commerce.)

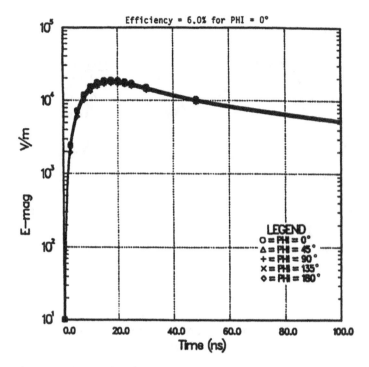

Figure 6.3.5 For burst at 400-km altitude and observer at 2201-km ground range (on horizon) and various magnetic azimuths, amplitude of electric field vs. time and efficiency of conversion of gamma energy to EMP energy, per unit solid angle. (From Longmire et al., 1987, *A Nominal Set of High-Altitude EMP Environments*, Oak Ridge Laboratory Report ORNL/Sub/86-18417/1, National Technical Information Service, U.S. Department of Commerce.)

Figure 6.3.6 Made from photograph of IVY Mike event taken at about 40 ms after burst by EG&G, showing nuclear lightning discharges at ground ranges indicated in meters. Yield of device was 10.4 Mt. Channels were mostly formed before the fireball reached half the size shown. (From Gardner et al., 1984, *Phys. Fluids*, 27, 2694. With permission.)

distances (in meters) indicated from the burst point. They were seen to grow upward at an apparent speed of about 100 m/ms.

The shape of the discharges, circular about the burst point, indicates that they are driven by the EMP in the quasistatic phase, since that is known to be the shape of the electric field lines in that phase (A-Section VI). However, the quasistatic E_θ at the relevant ranges was only about 30 kV/m, while 2 MV/m is required for uniform field breakdown. The solution to this puzzle was found by Gardner et al. (1984).

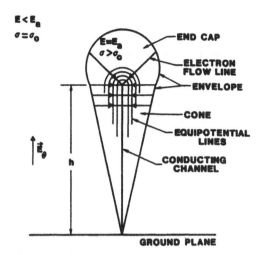

Figure 6.3.7 Conducting channel with tip at height h. Envelope is the boundary of the region in which avalanching occurs. The region, called the cone, is not exactly conical, and the end cap is not exactly spherical. The electric field inside the envelope is almost exactly radial from the tip in the end cap and from the channel in the cone. (From Gardner et al., 1984, *Phys. Fluids,* 27, 2694. With permission.)

The feature that distinguishes nuclear lightning from ordinary lightning is that there is a steady source of ionization, an ambient air conductivity $\sigma_0 \approx 3 \times 10^{-5}$ mho/m and a conduction current driven by E_θ that is returning to the ground negative charge deposited in the air by the divergence of the Compton current. If there is at some time a highly conducting channel from the ground up to a height h, then the electric field and conduction current in its vicinity will have shifted so as to feed current into the channel, especially near its upper end. Because of the increased electric field and current density near the channel tip, joule heating will raise the air temperature there. The idea of the model was that this heating would be sufficient to thermally ionize the air, thus increasing the height of the channel.

It was found that linear electrostatics, i.e., air conductivity assumed constant, did not give enough field and current concentration to produce sufficient joule heating. However, the air conductivity is not constant. On some surface enclosing the channel, called the envelope in Figure 6.3.7, the concentrated electric field reaches the avalanche value $E_a \approx 2$ MV/m, at which the attachment and ionization rates of electrons are equal. If E exceeds E_a in some region, the electron density and the air conductivity increase in times of the order of 10^{-8} s. The resulting increase in current moves charge in such a way as to decrease E there. Thus the electric field is effectively clamped at E_a in magnitude inside the envelope. This means that each equipotential surface is at a constant distance from the highly conducting channel, i.e., it is a hemisphere above the tip and a cylinder below it, as shown in Figure 6.3.7. The direction of the electric field and conduction current is therefore spherically radial in the end cap and cylindrically radial in the cone. Let r denote either radius.

A radial electric field of constant magnitude E_a has a positive divergence $2E_a/r$ in the end cap and E_a/r in the cone. Thus, there is a positive charge density inside the envelope, which can adjust itself in a time $\varepsilon_0/\sigma \le 3 \times 10^{-7}$ s in such a way as to keep E about equal to E_a everywhere. This positive charge has flowed up the channel and into the air (or negative charge flowed the other way) in order to get near the negative Compton electron charge. It is responsible for the modification and concentration of the electric field in the vicinity of the channel.

The radial conduction current J_r in the end cap and the conductivity σ are both proportional to $1/r^2$. At the outer radius of the end cap of about 1 m, they are

$$J_r \approx 60 \text{ A/m}^2, \ \sigma \approx 3 \times 10^{-5} \text{ mho/m} \tag{3.8}$$

At $r = 1$ mm, they are increased by a factor of about 10^6. However, a conductivity of 30 mho/m requires electron and positive ion densities so high ($\approx 3 \times 10^{15}$/cm³) that electron-ion recombination removes electrons faster than the normal attachment rate. Then E has to rise above the original E_a to supply enough electrons to pass the current, and E lines no longer focus on a point. This phenomenon effectively sets the radius a of the region of maximum joule heating rate, which becomes the radius of the channel near its tip. This maximum heating rate determines the time δt for the air temperature in the region to rise to about 1 eV, at which thermal ionization is about 10^{17}/cm³ and increasing rapidly with temperature. Thus in a time increment δt the conducting channel can advance upward into cold air by a distance $\delta h \approx a$, and the velocity V_C of channel growth is determined.

$$V_C = \frac{dh}{dt} \approx \frac{a}{\delta t} \tag{3.9}$$

A detailed computer simulation of the spatial and temporal development of such a discharge by Gardner et al. (1984) gave the results

$$a \approx 0.3 \text{ mm}, V_C = 115 \pm 35 \text{ m/ms} \tag{3.10}$$

It is important to note that this growth is steady and deterministic because of the plentiful ambient ionization ($\approx 10^9$ electrons/cm³). This is in contrast with the growth of streamers in natural lightning, which is nonsteady and seemingly more stochastic. Gardner et al. (1984) hypothesize that the small scale meanderings of the discharge path in Figure 6.3.6 result from the interaction of the growing tip with particulates or water droplets.

From Equation 3.8, the total current collected by a *single* channel 300 m in height is estimated to be 60 kA. However, Figure 6.3.6 shows that either bifurcation or development of side channels occurs. J.L. Gilbert (private communication) argues that a bump on the side of a channel is unstable and grows in a direction perpendicular to the original channel until it reaches the region of relatively undisturbed ambient field, where it turns and follows that field. Nearby channels repel each other because of the electric fields made by their positive charges, and they also rob some current from each other, but do collect current from a larger volume than a single channel. Colvin et al. (1987) have estimated the total current near the base of one of the Mike channels to be 250 kA on the basis of its light emission per unit length of channel. That is about one half the (radial) Compton current per steradian at the range of the discharge.

In the model of Gardner et al. (1984) the channel grows approximately in the direction of the electric field at its tip. This field is determined by all of the charges present, including those deposited by the Compton current, drained away by the ambient conduction current, and those that flowed in discharges. Conversely, if the field is given, the charge density is determined by taking the divergence of the field. In this model at least, it is not logical to say that the growth of the discharge depends more on the charge distribution than on the field, as Williams et al. (1988) seem to say.

The ambient field E_θ vanishes at $\theta = 0$, directly over the burst point. However, a channel may reach this point, if it is the first one to do so. After draining negative charge from the vicinity, the electric field at its tip will point toward the negative charge that remains in regions beyond the zenith. A channel that starts from a taller, sharper structure on the ground can get there first. These observations can explain the Bravo discharges in Figure 6 of Williams et al. (1988).

4. MAGNETOHYDRODYNAMIC EFFECTS

For a nuclear explosion in the atmosphere at any altitude, the fireball and the strongly shocked air are electrically conducting. Hydrodynamic motion of the air in the geomagnetic field therefore generates electric fields, currents, and magnetic perturbations. These effects are called MHD effects or MHD EMP for short.

For low altitude bursts the effects are fairly simple. In much of the fireball volume while it is still expanding, the conductivity σ is greater than 10^2 mho/m. The skin depth, or distance a magnetic field can diffuse into a conducting medium in time t, is

$$\delta(m) \approx \sqrt{2t/\mu_0\sigma} = 1.26 \times 10^3 \sqrt{t/\sigma} \tag{4.1}$$

If $\sigma > 10^2$ mho/m, then $\delta < 126$ m for $t = 1$ s. Thus, the geomagnetic field is mostly pushed out by the expanding fireball, particularly for megaton range yields, and remains out for several seconds. The change in the magnetic field outside the fireball is well represented by that of a magnetic dipole (for an air burst),

$$\Delta B_r = B_0 \cos\theta(a/r)^3, \ \Delta B_\theta = \frac{1}{2} B_0 \sin\theta(a/r)^3 \tag{4.2}$$

where B_0 is the ambient field, a is the fireball radius, and r and θ are spherical coordinates about an axis that is antiparallel to the ambient field. From Equation 4.2 the electric field induced by the increasing magnetic field as a increases can be found by use of Faraday's law of induction. The electric field is generally less than 1 V/m, but for closed wire loops the effect can be significant. For the hemispherical fireballs of surface bursts, exact analysis is complicated, but the spherical case can be used for estimates. Because ground conductivities are typically less than 10^{-2} mho/m, the skin depth in the ground is typically greater than 12 km at 1 s. Thus the magnetic perturbations easily diffuse into the ground. In the ocean $\sigma \approx 4$ mho/m, and diffusion is not so fast.

The fraction of the yield put into geomagnetic distortions for a low altitude burst is about 10^{-9}. For increasing burst altitude, this fraction increases inversely proportional to the ambient air density. For bursts above 100 km, the X-ray energy is not deposited locally, thus, only the kinetic energy yield drives the geomagnetic perturbations.

Those X-rays that go upward from bursts above 100 km mostly escape the earth. Those that go generally downward are mostly absorbed in a layer of air between 70 and 100 km, the ionospheric D region, where they increase the ionization and conductivity. The conductivity is mostly electronic because ions have too much friction with neutral atoms and molecules. Magnetic perturbations diffuse in this region by skin effect. The time T_D needed to diffuse through this region is

$$T_D(\text{sec}) \approx 4 \times 10^{-7} N_e(\text{cm}^{-3}) \tag{4.3}$$

where N_e is the electron density. For a megaton burst at 400 km (e.g., the 1962 Starfish event), $N_e \approx 10^{10}$ in the D region below the burst just after X-ray deposition. Dissociative recombination of electrons with molecular ions causes N_e to fall from its initial value N_0 as

$$N_e \approx N_0/(1 + 2 \times 10^{-7} N_0 t) \tag{4.4}$$
$$\approx 5 \times 10^6/t \text{ if } N_0 >> N_e(t)$$

independent of the initial value if the inequality is satisfied. In that case,

$$T_D(\text{sec}) \approx 2/t \tag{4.5}$$

Thus, magnetic perturbations can begin to diffuse through the D region a few seconds after the burst, and this is true out to the D region horizon from the burst point. In the natural D region beyond that horizon $N_e \leq 10^5$, so that T_D is very short.

The mass of air in the D region is about 10^5 kg/km^2. Only very small motions are induced here by Starfish-like explosions.

In the E region, which we take as extending from 100 to 150 km, there are some new features. The conductivity $\sigma_\parallel$ for an electric field $E_\parallel$ parallel to the magnetic field $\vec{B}$ is very large; for most purposes it can be taken as infinite, so that $E_\parallel = 0$ even if there is a $J_\parallel$. An electric field $\vec{E}_\perp$ perpendicular to $\vec{B}$ drives a current $\vec{J}_p = \sigma_p \vec{E}_\perp$, where σ_p is the *Pedersen* conductivity, and a *Hall* current $J_H = \sigma_H E_\perp$ in the direction of $\vec{E}_\perp \times \vec{B}$. The Hall effect is non-dissipative, but tends to rotate the plane of polarization; it is maximum between 90 and 110 km, near the interface of the D and E regions. The Pedersen conductivity plays the role of ordinary conductivity in magnetic diffusion. It has a maximum in the E region, which is mostly ionic rather than electronic. The ionization density is initially smaller in the E region than in the D region because there are fewer atoms to be ionized by the X-ray flux. However, the recombination is a little slower in the E region because the atomic ions O^+, which are about 20% of the initial ionization, have to react with N_2 or O_2 to form NO^+ or O_2^+ before they can dissociatively recombine. By the time of a few seconds, N_e in the E region is 1.5 to 2.0 times the value given in Equation 4.4. The distance to be diffused in the E region is about twice that in the D region, so that the diffusion time T_E is about four times the expression in Equation 4.3. Altogether,

$$T_E(\text{s}) \approx 14/t \qquad (4.6)$$

A weakness of this analysis is that σ changes during the diffusion process. While this variation can easily be taken into account in the diffusion formulae, it is already clear that a pulse with a 1-s rise time cannot be passed through the E region without substantial time smearing before $t = 14$ s. In the natural ionosphere $T_E < < 1$ s at night but can approach 1 s in the daytime.

The D and E region ionization caused by X-rays is called the *X-ray patch*. For Starfish it extended out to a ground range of about 2200 km. Its conductivity shields observers on the ground from signals at early times that could otherwise travel directly from burst to observer. The X-rays also make an EMP through the photoelectrons that they produce, but this is mainly a collisionally damped plasma oscillation in the X-ray patch that does not radiate. The principal EM effect, other than the high altitude gamma EMP, is the MHD effect arising from the kinetic energy of the exploding device debris.

At the 400-km altitude of Starfish the collisional mean free path of a fast moving debris atom with air atoms is thousands of kilometers. It would appear, therefore, that the expanding debris atoms can simply stream through the ambient ions and neutral atoms, and this is correct for the motion parallel to the geomagnetic field. However, the debris is mostly ionized and the gyroradius of the debris ions in the geomagnetic field is of the order of 10 km. As the ions start to move *across* the field, their magnetic deflection forms a current density that pushes the field ahead. The moving magnetic field causes the air ions to gyrate and move with the field on the average. For motion perpendicular to the field, the gyroradius acts like a momentum-transfer mean free path. On the scale of hundreds of kilometers the perpendicular motion looks like hydrodynamics, with magnetic pressures and tensions added, i.e., MHD.

The effects of these facts on the developing blast wave are illustrated in Figure 6.4.1. The magnetic field has been pushed out of a region, called the *bubble*, that is elongated in the direction of the ambient field. An MHD shock wave runs ahead of the expanding bubble in the perpendicular directions. Hot debris and shocked-air ions are leaking along the field lines and depositing their energy in a *hot ion patch* located in the E region near the central field line, both north of the burst and in the southern *conjugate* region. Beta rays emitted by fission fragments are depositing energy in and below the D region in a *beta patch*. All of these features and the X-ray patch emit enough light to be seen by eyewitnesses and to be photographed. The upward-going shock wave moves faster than the downward one because the plasma density is lower at the higher altitudes. The initial speed of the outermost debris was about 3000 km/s, while the ambient

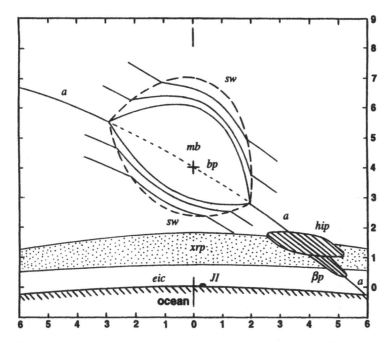

Figure 6.4.1 Sketch of phenomena observed in Starfish event. *aaa*, central magnetic field line; *bp*, burst point; *mb*, magnetic bubble; *sw*, MHD shock wave; *xrp*, X-ray patch; *hip*, hot ion patch; β*p*, beta patch; *eic*, Earth-ionosphere cavity; and *Jl*, Johnston Island. The numbers indicate distances in units of 100 km.

Alfven speed in the vicinity of the burst point was 300 km/s. Thus, the shock is initially very strong; air ions are heated to thermal speeds about equal to the debris speed, or a temperature of 500 keV for an O^+ ion.

Some of the experimental data taken following the Starfish event are shown in Figures 6.4.2 and 6.4.3, selected here because it appears to have good time resolution. Dyal's report is still labeled secret but the data shown have been released. Using rocket-borne magnetometers, Dyal also verified the existence of the magnetic bubble. Bomke et al. (1964) also made measurements at greater distances than the 1500 km distance from Johnston Island to Hawaii. The quantity graphed in the figures is the change in the magnitude of the magnetic field. Since the change δB was small compared with the ambient field B, the measurement gave only the component of $\vec{\delta B}$ that was parallel to $\vec{B}$.

The first MHD signal peaked at about 4 s at both Johnston Island and Hawaii. This signal cannot have diffused through the X-ray patch, which covers both locations. Instead, it propagated high above the F region where the Alfven speed is several thousand kilometers per second, out to horizontal distances beyond the X-ray patch, then down through the natural ionosphere to the earth-ionosphere cavity. In the cavity the signal propagates at the speed of light, and it therefore arrived at Johnston Island only 5 ms later than at Hawaii — not enough to be noticeable in Figure 6.4.3. The more distant measurements of Bomke et al. (1964; in the eastern U.S.) also showed simultaneity of the same quality. This fact proves that the first signal got into the cavity beginning at about 2 s and then propagated in the cavity to all observers. The more distant observers saw a pronounced ringing following the first pulse, which probably arose from local vertical oscillations in the F region. At Johnston Island the ringing was presumably damped by additional conductivity in the E region and below. The fact that the signal at Hawaii started going negative and then reversed is evidence of Hall rotation of the higher frequencies contained in the pulse.

Starting at about 20 s the blast wave began to diffuse through the X-ray patch over Johnston Island. The signal at this time was driven by the magnetic moment due to the plasma thermal energy that had been deposited hydrodynamically at altitudes between 200 and 300 km in the

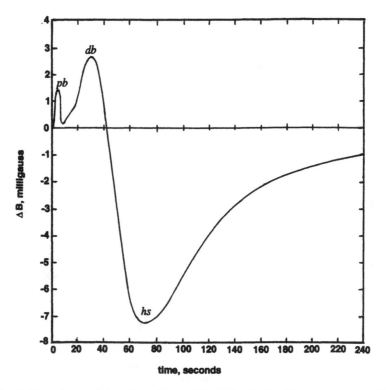

Figure 6.4.2   Variation of magnetic intensity at Johnston Island following Starfish event, as measured by P. Dyal and redrawn by author. *pb,* propagated blast wave; *db,* diffused blast wave; and *hs,* heave signal.

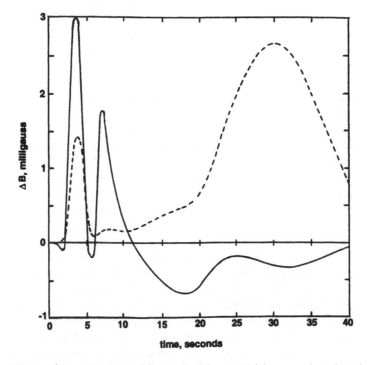

Figure 6.4.3   Variation of magnetic intensity following Starfish event. Solid curve is data of Bomke et al. (1964) taken at Hawaii. Dashed curve is early part of Dyal's Johnston Island data.

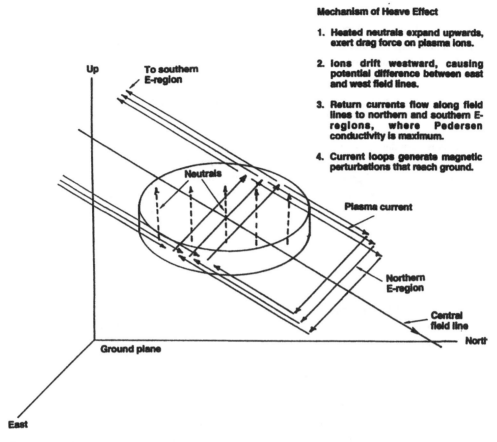

Mechanism of Heave Effect

1. Heated neutrals expand upwards, exert drag force on plasma ions.

2. Ions drift westward, causing potential difference between east and west field lines.

3. Return currents flow along field lines to northern and southern E-regions, where Pedersen conductivity is maximum.

4. Current loops generate magnetic perturbations that reach ground.

Figure 6.4.4 Sketch illustrating the mechanism that produced the heave signal in the case of the Starfish event.

general region underneath the burst point. (The upper part of the magnetic bubble and its magnetic moment had collapsed in a few seconds.) This signal was not seen at Hawaii because of the $1/r^3$ falloff of the static field of a dipole.

Starting at about 30 s, a strong negative-going signal was observed at Johnston Island. This was caused by the upward heave of the air in the E region, which had been heated by the blast wave and by deposition there of the kinetic energy of hot ions. Because there was more atmospheric mass below it than above it, this air converted its internal energy into upward velocities of a few kilometers per second, which persisted until reversed by gravity. By 30 s the geomagnetic field was not far from its original configuration. Ions and electrons would have been stuck on field lines except for the drag force exerted on them by the heaving neutral atoms and molecules. The ions balanced this force by drifting westward across the field lines, electrons eastward. Figure 6.4.4 shows how the charge thus separated was neutralized by flowing along the field lines to the north and south E regions where the conductivity (Pedersen) across field lines was maximum. The magnetic field of the loop currents opposed the ambient field at Johnston Island, but made a small increase at Hawaii. There are similarities between phenomena in the heave effect and those in magnetic storms.

Accurate calculations of the MHD effects are made difficult by the tensor character of the conductivity and by the long collision mean free paths at the higher altitudes. However, calculations using models based on two interacting fluids (neutrals and plasma) are in fair agreement with the time dependence and amplitudes of the experimental data. They support the description of phenomena presented in this section.

The magnetic perturbations at ground level diffuse into the ground, giving rise to electric fields and current densities in the ground. For high altitude bursts the diffusion into the ground can be approximated as a one-dimensional skin effect problem because the fields vary more rapidly with depth than with horizontal position. The magnetic and electric fields are both approximately horizontal and perpendicular to each other. If $B(t)$ is known at the ground surface, then $E(t)$ at the surface can be found from the relation

$$E(t) = \frac{1}{\sqrt{\pi\mu_0\sigma}} \int_{-\infty}^{t} \frac{1}{\sqrt{t - t'}} \frac{\partial B(t')}{\partial t'} \, dt' \quad \text{(MKS units)} \tag{4.7}$$

where σ is the conductivity of the ground averaged over one skin depth. Over continental land masses, $\sigma \approx 10^{-3}$ mho/m, although there are regions of lower conductivity. Thus the same magnetic perturbations would produce much larger electric fields over continents than over oceans.

In the simple case in which B varies linearly with t, changing by an amount ΔB between $t = 0$ and $t = T$ s, and for $\sigma = 10^{-3}$ mho/m, one finds $E(t) = \sqrt{t/T} \, E_p$, where the peak value E_p is reached at $t = T$ and is

$$E_p(V/km) = \frac{10}{\pi\sqrt{T}} \Delta B \text{ (mG)} \tag{4.8}$$

(Note: 1 G = 10^{-4} T.) Application of this simple formula to the data of Figure 6.4.2 gives peak electric fields of about 5 V/km for both the propagated blast wave and the heave signal. This value is comparable with that achieved during the most rapid magnetic fluctuations in magnetic storms. Thus the service outages experienced by long wire communication and electric power transmission systems during magnetic storms are directly relevant to the potential effects of the MHD EMP. In contrast, the interaction of the gamma induced EMP with systems must be investigated chiefly by calculation (in the absence of atmospheric nuclear tests).

The published data on magnetic storm effects on long wire systems are not as extensive as is scientifically desirable. However, the collapse of the entire electric power system of the province of Quebec due to a severe magnetic storm was reported in the July 1989 issue of *Power Engineering Review* (published by IEEE), with two articles on the nature of the effects.

REFERENCES

Bethe, H.A., K. Fuchs, J.O. Hirschfelder, J.L. Magre, R.E. Peierls, and J. von Neumann (1958), *Blast Wave*, Los Alamos Scientific Laboratory, Los Alamos, NM.

Bomke, H.A., I.A. Balton, H.H. Grote, and A.K. Harris (1964), Near and distant observations of the 1962 Johnston Island high-altitude nuclear tests, *J. Geophys. Res.*, 69, 3125.

Colvin, J.D., C.K. Mitchell, J.R. Greig, D.P. Murphy, R.E. Pechacek, and M. Raleigh (1987), An empirical study of the nuclear explosion induced lightning on IVY-Mike, *J. Geophys. Res.*, 92, 5696.

Gardner, R.L., M.H. Frese, J.L. Gilbert, and C.L. Longmire (1984), A physical model of nuclear lightning, *Phys. Fluids*, 27, 2694.

Glasstone, S., and P.I. Dolan (1977) *The Effects of Nuclear Weapons*, U.S. Atomic Energy Commission, Washington, D.C.

Kaplan, I. (1955), *Nuclear Physics*, Addison-Wesley, Cambridge.

Longmire, C.L. (1978), On the electromagnetic pulse produced by nuclear explosions, *IEEE Trans. Antennas Propag.*, AP-26, 3. Also in *IEEE Trans. Electromagn. Compat.*, EMC-20.

Longmire, C.L., R.M. Hamilton, and J.M. Hahn (1987), *A Nominal Set of High-Altitude EMP Environments*, Oak Ridge National Laboratory Report ORNL/Sub/86-18417/1, National Technical Information Service, U.S. Department of Commerce.

Salanave, L.E. (1980), *Lightning and Its Spectrum*, University of Arizona Press, Tucson.

Sedov, L.I. (1959), *Similarity and Dimensional Methods in Mechanics*, Academic Press, New York.

Taylor, G.I. (1950), The formation of a blast wave by a very intense explosion, *Proc. R. Soc.*, A 210, 159.

Uman, M.A., D.F. Seacord, G.H. Price, and E.T. Pierce (1972), Lightning induced by thermonuclear detonations, *J. Geophys. Res.*, 77, 1591.

Williams, E.R., C.M. Cooke, and K.A. Wright (1988), The role of electric space charge in nuclear lightning, *J. Geophys. Res.*, 93, 1679.

Chapter 7

Whistlers

Masashi Hayakawa

CONTENTS

0-8493-2520-X/95/$0.00+$.50
© 1995 by CRC Press

1.   INTRODUCTION

The first unambiguous report of whistlers was made by Barkhausen during World War I when it was common practice to eavesdrop on enemy telephone conversations at the front. Barkhausen carried out the systematic studies of whistlers and suggested that whistlers originated in lightning discharges and that their long descending tone was the result of propagation within a dispersive medium (Barkhausen, 1919, 1930). However, magnetoionic theory was not known at that time, so quantitative explanation was not possible. Eckersley (1931) developed the wave propagation theory in a magnetoactive plasma like the ionosphere and magnetosphere, which could lead to the whistler dispersion law in the low frequency limit. However, the observed dispersion required a puzzle. Eckersley and his colleagues also carried out a VLF observational program and they found many other examples of VLF noises now known as "VLF/ELF emissions (chorus, hiss, etc.)".

At the time of Eckersley's work, Burton and Boardman (1933) investigated whistlers and VLF emissions that were picked up by submarine cables. They carried out detailed studies of whistler spectrograms, which probably showed different kinds of whistlers (including nose whistlers, whistler-triggered emissions, etc.).

A long period of inactivity followed in whistler research until the pioneering work by Storey in 1953. This paper represented dramatic breakthroughs that had far reaching implications (Storey, 1953). He showed conclusive evidence that whistlers originated in lightning discharges in the opposite hemisphere and then propagated in the magnetosphere along the geomagnetic field lines to the hemisphere of the observer. Furthermore, he suggested on the basis of dispersion consideration that the electron density of several hundred per cubic centimeter existed in the remotest parts of the whistler paths (i.e., a few Earth radii). This is the first evidence that plasma exists far beyond the ionospheric layer and opened a vast new region of space for exploration. Whistlers quickly became an important part of space physics, and then many whistler stations were established around the world in the International Geophysical Year (1957 to 1958) (Helliwell, 1965).

With the advent of the space age, it became possible to place VLF receivers on spacecrafts, which revealed many new kinds of whistlers that could not be received on the ground by the nature of their propagation mode. Satellite observations of these whistlers helped us to understand many subtle features of whistler propagation and to deduce important magnetospheric plasma parameters.

During the past 3 decades, whistler research had progressed; it is impossible to cite here the many individual contributions that have been made. Thus, we advise the readers to refer to several earlier reviews for a more comprehensive biography (Helliwell, 1965; Walker, 1976; Hayakawa and Tanaka, 1978; Park, 1982; Al'pert, 1990; Hayakawa and Ohta, 1992).

1.1.   THE WHISTLER PHENOMENON

A lightning discharge radiates electromagnetic waves over a wide frequency range, and radiation in a relatively low frequency range, ~ 0.5 to 30 kHz, is the source of "whistlers". This mode of propagation is possible only in a magnetized plasma and at frequencies below both the electron plasma- and gyrofrequency. Propagation in this mode is strongly affected by the static magnetic field and is characterized by low propagation speeds that vary with frequency.

If whistlers are guided through the magnetosphere by field-aligned irregularities (whistler "ducts"), they can then propagate from one hemisphere to another, as illustrated in Figure 7.1.1. Radiation from a lightning discharge in the southern hemisphere propagates along a duct to the northern hemisphere, and because different frequency components travel at different speeds, the received signal has a frequency-time signature as shown by curve 1 and the propagation time delay is of the order of 1 s. At the lower edge of the ionosphere in the northern hemisphere, a part of the wave energy may be reflected back to the southern hemisphere, thus producing a "two-hop" whistler as illustrated by curve 2. Such partial reflection may be repeated many times to produce whistler echo trains in both hemispheres. When a lightning discharge is composed of multiple flashes and they propagate in the same duct, such a whistler is called "multi-flash." When a lightning discharge illuminates more than one duct in the magnetosphere, then the resulting whistler consists of several discrete components separated in time due to the differences in travel time through different ducts, and such whistlers are called multi-path whistlers.

VLF/ELF wideband signals are picked up by magnetic (loop) or electric (monopole) antenna and recorded on magnetic tapes. Because whistlers are in the audio frequency range, they can be heard with the aid of an earphone, although an amplifier is helpful. When they are played back and displayed as sonograms, whistlers exhibit a falling tone spectrum lasting for a second or more.

When whistlers were explained as radiation from lightning that had traveled several Earth radii out into space, it became clear that they contained useful information about the medium through which they had propagated. Methods were developed to use the whistler frequency-time characteristics to infer plasma density and electric field distributions in the magnetosphere. Applications of these methods led to discoveries that have significantly advanced our understanding of the ionosphere and magnetosphere. If a receiver is placed in space on board satellites or rockets, whistlers whose paths deviate significantly from the Earth's magnetic field lines can be detected. Such unducted whistlers show a wide variety of frequency-time signatures, depending on the location of the receiver and on the electron density distribution and ion composition of the medium.

There is evidence that whistlers interact with energetic electrons in the radiation belts during their traversal through the magnetosphere. Such interactions may result in the amplification of the whistlers, triggering of emissions at new frequencies, and precipitation of some of the interacting energetic electrons in the radiation belts. Precipitating energetic electrons, in turn, produce enhanced ionization and optical emissions in the lower ionosphere, as well as X-rays detectable down to about a 30-km altitude.

Although lightning has been known to be the source of whistlers for several decades, the detailed relationship between lightning characteristics and whistlers is poorly understood. For example, only a small fraction, perhaps 1% or less, of lightning discharges produces detectable whistlers, and it is not clear what is unique about these whistler sources.

2. WAVE PROPAGATION IN A MAGNETOACTIVE PLASMA

Many important properties of whistlers received on the ground and on spacecraft are known to be explained by the classical magnetoionic theory (Ratcliffe, 1959). This theory accounts for the motion of electrons under the influence of wavefields, but ignores their thermal motion as well as ions. Collisions between electrons and heavy neutral particles can be easily included in the theory, and such collisions are unimportant above the lower ionospheric D region ($\gtrsim 90$ km). In the following sections we will use the more generalized theory which can easily include the

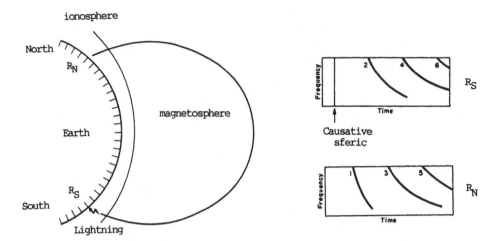

Figure 7.1.1 A lightning discharge produces a whistler that echoes from hemisphere to hemisphere along a geomagnetic field line. The panels on the right indicate idealized spectra of whistler echo-trains observed at the conjugate points R_N and R_S. The causative sferic is indicated by an arrow on the time axis.

presence of multiple ion species (Stix, 1962), because the inclusion of ions plays an important role in the ion whistlers.

2.1. DISPERSION RELATION

The plasma is assumed to be cold and the dispersion relation for the magnetoactive plasma is generally obtained from the combination of Maxwell's equations and the equation of motion of particles. The final solution for the dispersion relation is given as follows (Stix, 1962).

$$An^4 - Bn^2 + C = 0 \tag{1}$$

where

$$\left.\begin{array}{l} A = S\sin^2\theta + P\cos^2\theta \\ B = RL\sin^2\theta + PS(1 + \cos^2\theta) \\ C = PRL \end{array}\right\} \tag{2}$$

In Figure 7.2.1 we indicate the coordinate system in which the static Earth's magnetic field ($\mathbf{B_o}$) is taken to be parallel to the z-axis and the wave normal direction ($\mathbf{k}$) is assumed to be in the xz-plane and to make an angle θ with $\mathbf{B_o}$. The quantities in Equation 2 are defined in the following way

$$\left.\begin{array}{l} R \equiv 1 - \sum_k (f_{pk}^2/f^2)[f/(f + \varepsilon_k f_{Hk})] \\ L \equiv 1 - \sum_k (f_{pk}^2/f^2)[f/(f - \varepsilon_k f_{Hk})] \\ P \equiv 1 - \sum_k (f_{pk}^2/f^2) \\ S = \dfrac{1}{2}(R + L), \ D = \dfrac{1}{2}(R - L) \end{array}\right\} \tag{3}$$

where f_{pk} and f_{Hk} are the plasma- and gyrofrequencies of the particle k, and are expressed by,

$$\omega_{pk}^2 = (2\pi f_{pk})^2 \equiv \frac{e^2 n_k}{m_k \varepsilon_0}; \ \omega_{Hk} = 2\pi f_{Hk} = \frac{Z_k e B_0}{m_k} \tag{4}$$

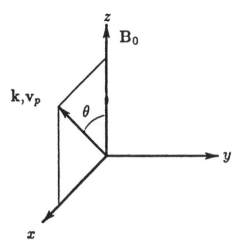

Figure 7.2.1 Propagation in a magnetoactive plasma. The static magnetic field (B_0) is assumed to be parallel to the z-axis, and the wave normal direction (k) is assumed to lay in the xz-plane (without loss of generality) and to make an angle θ with B_0.

Here, n_k and m_k are the number density and mass of k particles with charge of magnitude $Z_k e$ ($e > 0$); f_{Hk} (and ω_{Hk}) is taken to be positive and the sign of the charge, ± 1 is given by ε_k. The refractive indices of two characteristic mode waves in the plasma are given as the solution of Equation 1.

$$n^2 = (-B \pm \sqrt{B^2 - 4AC})/2A \tag{5}$$

The dispersion curve in the form of $\omega - k$ diagram is illustrated in Figure 7.2.2 for two different plasma conditions: (a) $\omega_{pe} = 2\omega_{He}$ (ω_{pe} and ω_{He} are the electron plasma- and gyrofrequencies) and (b) $\omega_{pe} = 0.5\omega_{He}$. The plasma is assumed to be composed of electrons and ions of one species. Figure 7.2.2 suggests that there are three characteristic frequency ranges:

(1) The wave modes at the frequency above ω_L (cut-off frequency of left-handed polarized wave) are called "quasi-free-space mode waves."

(2) When the frequency becomes smaller than ω_{He} but above ω_{Hi} (ion gyrofrequency), there is only one possible mode of propagation (that is, the so-called "whistler-mode" wave), and this frequency range can be called "whistler-mode" frequency range.

(3) Two modes of propagation are possible at frequency below ω_{Hi}, which are hydromagnetic waves.

In this chapter we will deal with the latter two frequency ranges, and the first quasi-free-space mode waves will be treated in another chapter of this handbook.

The wave polarization of electric field components transverse to B_0, is given as follows:

$$iE_x/E_y = (n^2 - S)/D \tag{6}$$

In Equation 6 right-handed circular polarization is given by $iE_x/E_y = 1$, while left-handed circular polarization, $iE_x/E_y = -1$.

2.2. WHISTLER-MODE PROPAGATION (GENERAL THEORY)

As seen from Figure 7.2.2, the whistler mode is characterized by large values of refractive index or slow propagation speeds. The polarization of whistler-mode waves at $\theta = 0°$ is exactly right-handed circular because $iE_z/E_y = 1$ with $n^2 = R$ so that the wavefield vectors rotate around B_0

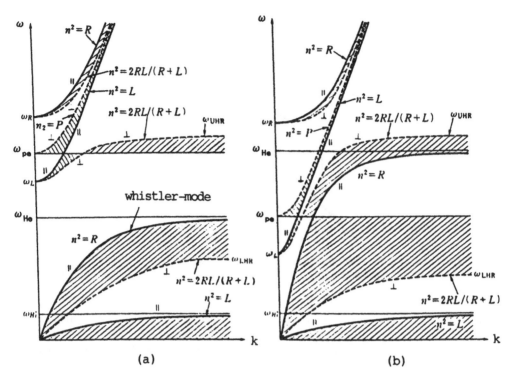

Figure 7.2.2 Dispersion curve in the form of $\omega - k$ diagram for an electron-ion plasma for the two cases; (a) $\omega_{pe} = 2\omega_{He}$ and (b) $\omega_{pe} = 0.5\omega_{He}$. $\parallel$ and $\perp$ mean $\theta = 0°$ and $\theta = 90°$, respectively, and the dispersion curve for an oblique θ must lay in the shaded region. ω_R and ω_L are the cut-off frequencies of right- and left-handed polarized mode waves. While ω_{LHR} and ω_{UHR} are the lower and upper hybrid resonance frequencies obtained from $S = 0$, and whose expressions will be given later. $n^2 = R$ and $n^2 = L$ for $\theta = 0$ indicate the right- and left-handed circularly polarized waves, respectively.

in the same sense as electrons do in their gyration. In the following sections we will indicate the important characteristics of whistler-mode propagation.

2.2.1. Quasilongitudinal (QL) Approximation

The general expression for whistler-mode refractive index given by Equation 5 is difficult to use because of their complexity. However, if the wave normal angle is not too large, considerable simplification is possible. In this process Appleton-Hartree's equations (Ratcliffe, 1959; Helliwell, 1965; Park, 1982) are rather useful, and the expression for QL approximation will become Equation 7 by ignoring the transverse term compared with the longitudinal term.

$$n^2 = 1 + \frac{f_{pe}^2}{f(f_{He} \cos\theta - f)} \approx \frac{f_{pe}^2}{f(f_{He} \cos\theta - f)} \tag{7}$$

Here, we have assumed $f < f_{He} \ll f_{pe}$ is satisfied almost in any part of the magnetosphere and we consider only electrons and neglect the collisional effect.

The shaded area in Figure 7.2.3 as a function of $f_{pe}^2/f\,f_{He}$ indicates where the QL approximation is valid. This figure suggests that the QL approximation is found to extend even to large θ angles for the larger $f_{pe}^2/f\,f_{He}$ values.

In general, the wave's magnetic field is perpendicular to the wave normal direction (**k**), but the electric field of whistler-mode waves has a component in the **k** direction (electrostatic component). There is no electrostatic component for the particular case of $\theta = 0°$, and so the whistler-mode wave for $\theta = 0°$ is a purely electromagnetic wave. In the case of the QL approximation,

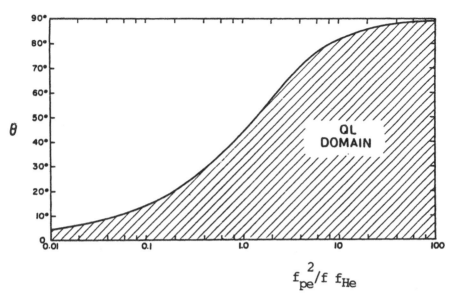

Figure 7.2.3 The conditions under which the QL approximation is valid. The QL domain is given as a function of f_{pe}^2/f_{He}. (From Helliwell, R. A., *Whistlers and Related Ionospheric Phenomena*, Stanford University Press, Stanford, CA, 1965. With permission.)

the wave is found to be right-handed circularly polarized, and it is proved that the plane of polarization of the wave's electric field is normal to $\mathbf{B_0}$ (Helliwell, 1965).

2.2.2. Refractive Index Surface

When we consider the wave propagation in an anisotropic medium, the concept of refractive index surface is of great potential in the study. The refractive index of whistler-mode waves as given by Equation 5 or Equation 7 is a function of the wave normal angle θ, and Figure 7.2.4 illustrates the plot of n vs. θ for three different frequencies. Because there is an azimuthal symmetry around $\mathbf{B_0}$, the refractive index surfaces are generated by the revolution of the corresponding n vs. θ curves around $\mathbf{B_0}$. The angle when n $\rightarrow \infty$ is called the "oblique resonance angle (θ_{res})", which can be obtained as the pole of Equation 7 as follows.

$$\cos\theta_{res} = f/f_{He} \tag{8}$$

Of course, the more accurate expression can be estimated easily from Equation 5 without approximation (Park, 1982). As seen from Figure 7.2.4, we notice a topological change at $f = f_{He}/2$, because the curvature of the refractive index surface at $\theta = 0°$ changes at this frequency. A more accurate expression for a tenuous plasma is also available (Burtis and Helliwell, 1976; Ishikawa et al., 1990). This topological change would result in a significant effect of ray focusing as discussed below.

2.2.3. Ray Direction (Energy Propagation)

In an anisotropic medium a wave packet travels in a direction different from the wave normal direction ($\mathbf{k}$). The direction of the wavepacket is the direction of wave energy propagation and is indicated by the group velocity (v_g) as in Figure 7.2.5a. This direction is called "ray direction". It can easily be proved that the ray direction is always normal to the refractive index surface, as illustrated in Figure 7.2.5b. If the wave normal angle θ is defined as the angle of $\mathbf{k}$ measured

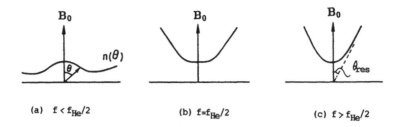

(a) $f < f_{He}/2$ (b) $f = f_{He}/2$ (c) $f > f_{He}/2$

Figure 7.2.4   The refractive index surfaces of whistler-mode waves for three different frequencies. Electrons are only considered.

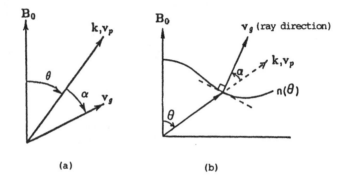

(a)   (b)

Figure 7.2.5   The general relationship between wave normal direction (**k**) (or phase velocity v_p) and group velocity (v_g) (or ray direction). The angles of θ and α are measured in the directions in (a) and the clockwise directions are taken as positive.

from B_0 and α is the angle of v_g measured from **k** as in Figure 7.2.5a (the angle is taken as positive in the clockwise direction), the angle α is defined as follows (Stix, 1962).

$$\tan\alpha = -\frac{1}{n}\frac{\partial n}{\partial \theta} \tag{9}$$

This formula, together with the angle definition, is universal for any kind of plasma waves. As shown in Figure 7.2.5b in the case of whistler-mode propagation at lower frequencies, the angle α takes a negative value, and the ray direction (v_g) is lying between the directions of B_0 and **k**. In the zero-frequency limit, the ray direction (with respect to B_0) is easily estimated to be less than $19°\ 29'$ (Storey, 1953; Helliwell, 1965), which means that the anisotropy of the medium provides a certain amount of wave guiding.

The group velocity of a wavepacket can be expressed as follows. The group velocity in the **k** direction is given by

$$\left.\begin{array}{l} v_{gk} = \dfrac{c}{n_g} \\[2ex] n_g = \dfrac{\partial}{\partial f}(nf) \end{array}\right\} \tag{10}$$

and so the absolute value of the group velocity (v_g) is given by

$$v_g = \frac{v_{gk}}{\cos\alpha} \tag{11}$$

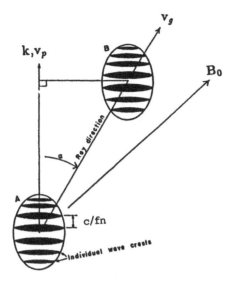

Figure 7.2.6 Schematic illustration of phase velocity and group velocity of whistler-mode waves in the lower frequency approximation. (From Helliwell, R. A., *Whistlers and Related Ionospheric Phenomena,* Stanford University Press, Stanford, CA, 1965. With permission.)

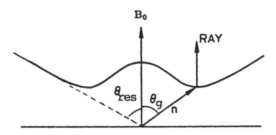

Figure 7.2.7 The definition of Gendrin angle (θ_g), together with the oblique resonance angle (θ_{res}).

where α is the angle between the wave normal and ray directions as in Figure 7.2.5. Figure 7.2.6 illustrates the schematic picture of the propagation of the lower frequency whistler wavepacket in the plasma.

For $f < f_{He}/2$ the refractive index surface has a shape illustrated in Figure 7.2.7. The characteristic angle θ_g where $n \cos\theta$ is minimum (i.e., $\cos\theta_g = 2f/f_{He}$) is called the Gendrin angle (Gendrin, 1960). Propagation at the Gendrin angle has some interesting properties (Helliwell, 1965; Park, 1982). The wavepacket travels in the direction of B_0 at a speed that is independent of frequency, and the corresponding wave velocity is such that its component in the direction of B_0 is equal to that of the wavepacket.

2.2.4. Group Velocity

It is interesting to compare the variations of the group velocity v_g and the phase velocity v_p as a function of wave frequency. Figure 7.2.8 shows the results for $f_{pe}/f_{He} = 3$ typical for the outer magnetosphere and for $\theta = 0°$, in which $v_p = c/n$ (c: speed of light) and $v_g = c/n_g$ (in Equation 10) are used for the calculation. The phase velocity has a maximum at $f/f_{He} = 0.5$, whereas the group velocity is maximized at $f/f_{He} = 0.25$. At frequencies below 0.5 f_{He}, the group velocity exceeds the phase velocity (as in Figure 7.2.8b), and at higher frequencies it is lower than the phase velocity. Depending on whether the wavefrequency was ≤ 0.5 f_{He}, the individual waves would appear to an observer to be coming with the wave envelope and to be moving backward or forward, respectively.

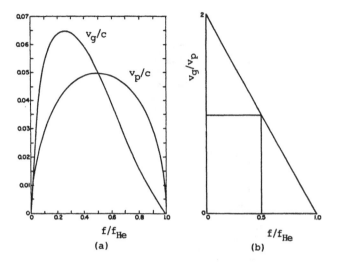

Figure 7.2.8 (a) Group velocity and phase velocity for $\theta = 0$ as a function of wavefrequency (f/f_{He}) and (b) the corresponding ratio of group velocity to phase velocity as a function of wavefrequency. The plasma condition is $f_{pe}/f_{He} = 3$. (From Helliwell, R. A., *Whistlers and Related Ionospheric Phenomena*, Stanford University Press, Stanford, CA, 1965. With permission.)

2.2.5. Time Delay and Dispersion

For the purely longitudinal propagation ($\theta = 0°$), the group delay of a whistler-mode signal traveling over a given path, is simply given by

$$T = \int_{path} \frac{ds}{v_g} \tag{12}$$

By substituting the expression of v_g in Equation 12, we obtain

$$T(f) = \frac{1}{2c} \int_{path} \frac{f_{pe}\, f_{He}}{f^{1/2}\, (f_{He} - f)^{3/2}}\, ds \tag{13}$$

When the wavefrequency is much smaller than the electron gyrofrequency ($f/f_{He} \ll 1$), Equation 13 reduces to

$$T(f) = \frac{1}{2c} \int_{path} \frac{f_{pe}}{\sqrt{f\, f_{He}}}\, ds \tag{14}$$

The frequency-time spectrograms expected from Equations 13 and 14 are illustrated in Figure 7.2.8, and Equations 13 and 14 are the low and high frequency approximations, respectively. The group delay in the low frequency approximation is conveniently described by a quantity called the dispersion D (in $s^{1/2}$) given by

$$D = T\sqrt{f} = \frac{1}{2c} \int_{path} \frac{f_{pe}}{\sqrt{f_{He}}}\, ds \tag{15}$$

which is a constant determined by the parameters of the medium through which whistlers propagate. Equation 15 is sometimes called "Eckersley's law". In the high frequency approximation the spectrogram in Figure 7.2.9a looks like a nose, thus, this kind of whistler is called a nose whistler. The frequency of the minimum time delay is the nose frequency at $f = f_{He}/4$ for a

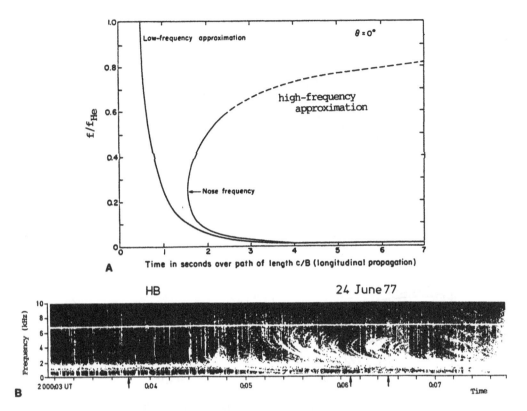

Figure 7.2.9 (a) Normalized frequency vs. time delay for longitudinal propagation ($\theta = 0°$) over a path of c/B_0. (From Helliwell, R. A., *Whistlers and Related Ionospheric Phenomena,* Stanford University Press, Stanford, CA, 1965. With permission.) (b) An example of observed spectrograms of nose whistlers. (From Sazhin, S. S., Smith, A. J., and Sazhin, E. M., *Ann. Geophys.,* 8, 273, 1990. With permission.) The left arrow at the bottom indicates the causative sferic for the two whistlers indicated by the right two arrows.

homogeneous plasma (see Figures 7.2.8 and 7.2.9). Figure 7.2.9b illustrates a series of nose whistlers actually observed at middle latitude.

3. PROPAGATION THROUGH THE IONOSPHERE

In the VLF frequency range of 1 to 10 kHz, the free-space wavelength ranges from 300 to 30 km. Because the plasma properties of the lower ionosphere change significantly with vertical distance in a wavelength, then the assumption that the medium is "slowly varying" (i.e., WKB approximation), which is valid in the outer magnetosphere is not valid. A proper treatment of VLF whistler-mode propagation through the ionosphere requires the exact full-wave solutions to be discussed later. However, certain aspects of whistler propagation through the ionosphere can be studied with the aid of the following basic concepts.

3.1. WAVE PENETRATION THROUGH THE IONOSPHERE
3.1.1. General Behavior of Refractive Index
Wave penetration can be studied by the general behavior of the variation of the refractive indices of two characteristic mode waves with ω_{pe}^2/ω^2 (ω_{pe}^2 is proportional to electron density n_e, and then the abscissa correspondingly indicates the height) as illustrated in Figure 7.3.1. In the figure the collision is neglected and we take $\omega = \omega_{He}/2$. For the general oblique θ case, the right-handed polarized wave possibly propagates between $\omega_{pe}^2/\omega^2 = 0$ and unity. However, in the range of $\omega_{pe}^2/\omega^2 = 1$ and the oblique resonance, ($\omega_{pe}^2/\omega^2 = (\omega_{He}^2 - \omega^2)/(\omega_{He}^2\cos^2\theta - \omega^2)$), the refractive

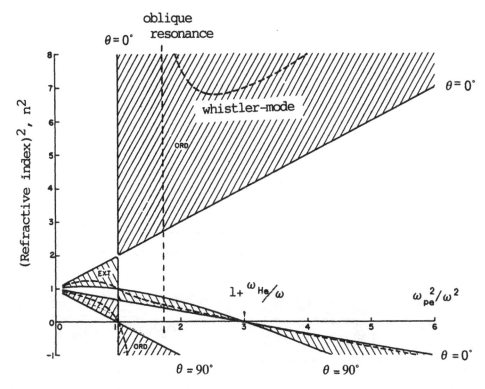

Figure 7.3.1 Variations of n^2 with ω_{pe}^2/ω^2 (proportional to electron density (n_e)) with $\omega_{He}/\omega = 2$. Dashed curves correspond to the situation of $\theta = 30°$. The same shadings indicate the same wave polarization. The resonance takes place at $\omega_{pe}^2/\omega^2 = (\omega_{He}^2 - \omega^2)/(\omega_{He}^2 \cos^2\theta - \omega^2)$. (From Helliwell, R. A., *Whistlers and Related Ionospheric Phenomena,* Stanford University Press, Stanford, CA, 1965. With permission.)

index becomes imaginary, indicating that this mode is evanescent. At higher values of ω_{pe}^2/ω^2, the mode again possibly propagates. For the conditions in Figure 7.3.1, coupling to the whistler mode from free space is difficult because of the presence of the evanescent region between $\omega_{pe}^2/\omega^2 = 1$ and the pole. One factor that promotes coupling is a steep gradient of refractive index, which reduces the thickness of the evanescent region. If the effect of collisions is included, it is found that coupling can occur over a finite range of θ (Ratcliffe, 1959; Budden, 1961, 1985). Together with the gradient of refractive index, this coupling accounts for the excitation of whistler mode waves in the ionosphere.

3.1.2. Snell's Law

Consider a wave propagating across the boundary of two horizontally stratified media with different refractive indices n_1 and n_2, as illustrated in Figure 7.3.2a. In this case Snell's law is expressed as follows

$$n_1 \sin\theta_1 = n_2 \sin\theta_2 \qquad (16)$$

and the corresponding graphical means of determining the wave normal direction of the refracted (or transmitted) wave is given in Figure 7.3.2b. This law can be used for estimating the wave normal direction not only for isotropic but also anisotropic media. Figure 7.3.3 illustrates the refractive index surfaces for the atmosphere and ionosphere in a magnetic meridian plane, when we consider the situation that radiation from lightning discharges penetrates into the ionosphere. In order to satisfy Snell's law, all waves that enter the ionosphere from below (with different incident angles), have their wave normals bent sharply toward the vertical, as illustrated as the

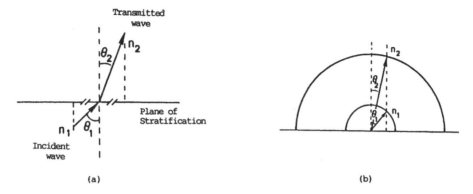

Figure 7.3.2 (a) Refraction (transmission) at a boundary between two media with different refractive indices and (b) a graphical solution by means of Snell's law.

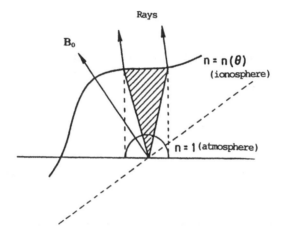

Figure 7.3.3 Application of Snell's law to the wave penetration through the atmosphere-ionosphere boundary.

shaded region in Figure 7.3.3 called the "transmission cone". At the ionospheric layer where the electron density is $\sim 10^6$ cm^{-3}, n is of the order of 100 for a wave frequency of 5 kHz. This means that the transmission cone is extremely narrow, with the maximum allowed θ of $\sim 0.5°$. Thus, we conclude that any whistler mode originating in the atmosphere enters the ionosphere with a vertical wave normal angle, and this in turn provides a validity of using two-dimensional whistler-mode ray tracing in the magnetosphere.

After a whistler wave propagates in an either ducted or unducted mode (to be discussed in the next section) through the magnetosphere and reaches the conjugate ionosphere, the reverse problem appears. This problem would be essentially important for detecting whistlers on the ground. If the wave normal angle of a downgoing whistler is accidentally laid within the narrow transmission cone, it can penetrate through the ionosphere down to the ground. However, when it is outside the transmission cone, the wave will suffer from total reflection and will reflect back to the magnetosphere. The rocket observation of the wave normal direction of whistlers in the ionosphere, together with the simultaneous ground observation, experimentally confirmed the presence of this transmission cone, which was found to be slightly larger than the theoretical value (Iwai et al., 1974; Hayakawa and Tanaka, 1978).

3.1.3. Collisional Absorption

Collisions of electrons with neutral particles are important only in the lower ionosphere such as D and E layers, and they result in the loss (or absorption) of wave energy. If such collisional

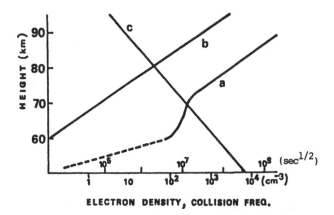

Figure 7.3.4. The profiles of electron density (a) day, (b) night, and (c) collision frequency used for the full-wave calculations.

effects are taken into account, the refractive index for whistler-mode waves becomes complex, with its imaginary part indicating the attenuation factor. In the QL approximation which is valid under a wide range of conditions as seen in Figure 7.3.4, the attenuation factor γ is well approximated by Equation 17 with υ as the collision frequency.

$$\gamma = \frac{f_{pe}\, \upsilon f^{1/2}}{2c(f_{He}\, \cos\theta)} \tag{17}$$

As was found before, the wave normal angle of the upgoing whistler-mode waves in the ionosphere is supposed to be reasonably vertical, thus Equation 17 suggests a few important implications: (1) the attenuation increases with the increase of frequency and (2) the attenuation also increases with increasing θ value with decreasing geomagnetic latitude.

3.2. FULL-WAVE ANALYSIS

Although the fundamental aspects of wave propagation in the ionosphere can be obtained by the important concepts in the previous sections, the detailed quantitative studies require the use of full-wave analyses, because the refractive index varies very rapidly within a wavelength at very low frequencies (in other words, the WKB or ray theory is not valid). Under such circumstances, the waves undergo partial reflection and mode coupling, whose analysis is only possible by those full-wave solutions.

Several numerical techniques to find the full-wave solutions have been proposed (Budden, 1961, 1985; Pitteway, 1965; Altman and Cory, 1969; Wait, 1970; Tsuruda, 1973; Nagano et al., 1975), but the general approach to full-wave solutions is to approximate the inhomogeneous ionosphere by a number of thin slabs and solve the wave equations within each slab subject to the upper and lower boundary conditions. Figure 7.3.5 illustrates an example of the full-wave computations for the profiles of electron density and collision frequency in Figure 7.3.4. This figure suggests the latitudinal dependence of the transmission coefficient of upgoing whistlers, which is roughly consistent with the consideration based on Equation 17 and which also indicates a large day-night asymmetry of the transmission coefficient at lower latitudes.

4. PROPAGATION IN THE MAGNETOSPHERE

Generally speaking, there are two different kinds of propagation modes: ducted and unducted. The unducted propagation is based simply on the geomagnetic guiding effect due to the anisotropic medium, as discussed in Section 2.2. Originally, it was thought that this geomagnetic guiding

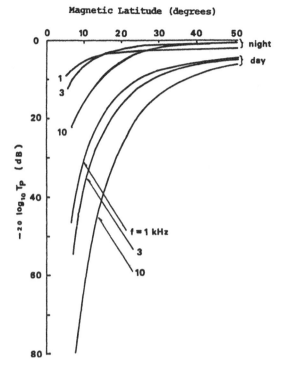

Figure 7.3.5 The latitudinal dependence of the ionospheric transmission loss of upgoing whistler waves with wave frequency as a parameter. (From Hayakawa, M., *J. Geomagn. Geoelectr.*, 41, 573, 1989. With permission.)

was sufficient to account for many characteristics of whistlers. However, recent evidence has shown that some ducting is involved in the propagation of whistlers, and it is postulated that there are numerous field-aligned irregularities of ionization (so-called "ducts") that extend between both hemispheres and trap whistler energy along the magnetic field line in the manner of a metallic waveguide. This is called "ducted propagation".

4.1. DUCTED PROPAGATION
4.1.1. Theory of Ducting
We consider a simple duct in which the static magnetic field is constant in magnitude and direction (z-axis) and the electron density varies only in the x-direction normal to B_o as in Figure 7.4.1 (Smith et al., 1960; Smith, 1961; Helliwell, 1965). There are two types of ducts; crests (increased density) and troughs (decreased density). Because the refractive index surface suffers a topological change at $f_{He}/2$ (see Figure 7.2.4), conditions for ducting also change at this frequency.

Figure 7.4.1 illustrates an example of a ducting mechanism for very low frequency whistlers ($f/f_{He} \ll 1$) for a crest duct. Figure 7.4.1a indicates the variation of wave normal angle of the whistler injected at the crest center with its initial wave normal angle θ_o and the corresponding construction of ray direction based on the concept in Figure 7.2.5. The resulting ray path is given in Figure 7.4.1b, with arrows indicating the wave normal direction at each point. In the figure we plot only one half a cycle, and we have a snake-like ray path. As one may see from Figure 7.4.1, the density gradients on both sides of the crest duct have a function that rotates the wave normal angle toward the magnetic field direction and thus forces the ray to propagate along the magnetic field. The wave normal angle is also confined within a certain limit around the magnetic field direction (called "trapping cone"). As is easily understood from Figure 7.4.1a, the density change required for wave trapping is dependent on the initial wave normal angle θ_o, and Figure 7.4.2 is the result of the minimum enhancement factor required for ducting vs. initial wave normal direction θ_o and wavefrequency, where the enhancement factor (Ec) is defined as the ratio of ΔN to

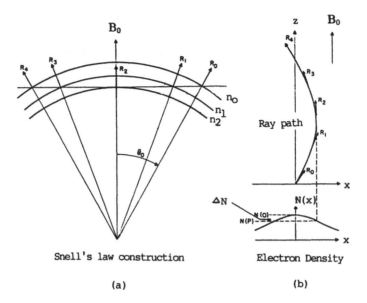

Figure 7.4.1   (a) Application of Snell's law to the crest duct at very low frequencies ($f/f_{He} \ll 1$). θ_o is the initial wave normal angle of the wave injected at the duct center. (b) The resulting snake-like ray path. (From Helliwell, R. A., *Whistlers and Related Ionospheric Phenomena,* Stanford University Press, Stanford, CA, 1965. With permission.)

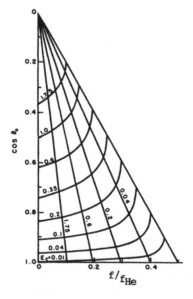

Figure 7.4.2   Minimum enhancement factor (E_c) required for ducting whistler waves vs. wave normal direction (θ_o) and wavefrequency normalized by the electron gyrofrequency. For example, $E_c = 0.01$ means the crest of 1% enhancement factor. (From Helliwell, R. A., *Whistlers and Related Ionospheric Phenomena,* Stanford University Press, Stanford, CA, 1965. With permission.)

the density at the outermost excursion of the ray path from the axis, as given in Figure 7.4.1b. Figure 7.4.2 is the result only for the crest duct, which would be useful for the actual magnetospheric situations. As mentioned in Section 3.1., the wave normal direction of whistlers after entering the ionosphere is supposed to be reasonably vertical, and the initial wave normal angle θ_o in Figures 7.4.1 and 7.4.2 reflects the latitudinal dependence of the minimum enhancement factor for wave trapping. At higher latitudes corresponding to smaller θ_o (i.e., $\cos\theta_o \sim 1.0$), the enhancement factor of ducts of the order of a few percent is only necessary for trapping, whereas it must be increased greatly with decreasing geomagnetic latitude.

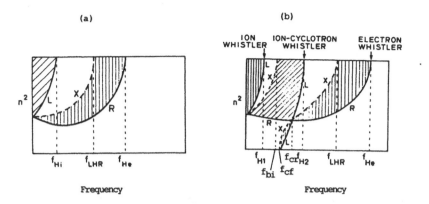

Figure 7.4.3 Dispersion relations for electron-ion plasma. (a) The case of one ion species and (b) two ion species. The longitudinal (θ = 0°) and transverse (θ = 90°) propagations are indicated by solid and broken lines, respectively, and the shaded areas show where the curves would be for intermediate propagation directions. (From Park, C. G., *Handbook of Atmospherics,* Vol. 2, Holland, V., Ed., p. 21, CRC Press, Boca Raton, FL, 1982. With permission.)

The conditions for different frequency ranges and different types of ducts are discussed completely in the book of Helliwell (1965), and the readers are asked to consult his book for the extensive trapping theory.

4.1.2. One-Sided Duct Propagation (Whispering Gallery Mode Propagation)

Recent studies have shown that the wave trapping does not necessitate density gradients on both sides. Because of the curvature of the magnetic field lines of the Earth, the wave normal tends to bend toward the direction of the larger L value (L is McIlwain's L value, and it is used to indicate a particular magnetic field line whose equatorial geocentric distance is given by the unit of the Earth's radius), and a density decrease toward the larger L has a function to counter this tendency. However, if the wave normal does not turn toward lower L, no density gradient is needed on the lower L side. This kind of trapping is called "gradient trapping" (Helliwell, 1965), and this propagation mode is similar to the whispering gallery mode (Budden, 1985). A good example of this kind of wave trapping is found to take place at the plasmapause boundary where the density decreases sharply with increasing L (see Section 7.1.) (Inan and Bell, 1977), and another example is the equatorial anomaly (Hasegawa and Hayakawa, 1980).

4.2. UNDUCTED PROPAGATION

This unducted propagation is based on the anisotropic nature of the plasma as mentioned in Section 2.2. Both the wave normal and ray directions may deviate significantly from the magnetic field direction, and we cannot use QL expression for the refractive index. Hence, the calculation of ray path and travel time requires the use of a ray-tracing technique in a model magnetosphere by using the full expression of the refractive index.

4.2.1. Effect of Ions

In the unducted propagation and at lower frequencies the motion of ions is known to strongly influence the whistler-mode propagation characteristics in the magnetosphere. In the magnetosphere the plasma is composed of electrons and ions of a few species. The dispersion curves for the cases including ions are illustrated in Figure 7.4.3 in the form of n^2 vs. frequency, which are based on Equations 1 through 5. Figure 7.4.3a refers to the case of one ion species which is typical for the magnetospheric plasma (electrons and protons), and Figure 7.4.3b, the case of two ion species which is usually encountered in the ionosphere and lower exosphere. First, we consider the simpler case of Figure 7.4.3a for one ion species. Important effects of including ions are seen at low frequencies. When $f \to 0$, the refractive indices are found to take finite values instead of becoming infinitely large when electrons are only considered, as seen in Equation 7. In addition,

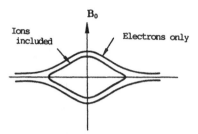

Figure 7.4.4 Refractive index surfaces with and without ion effects at the wavefrequency below the lower hybrid resonance frequency.

a new mode of propagation is made possible by ions below the ion gyrofrequency, f_{Hi}. Waves propagating in this mode are called hydromagnetic Alfven waves or ion cyclotron waves. Characteristic frequencies are discussed here. The infinity of the refractive index (resonance) takes place at $f = f_{He}$ for the right-handed circularly polarized whistler-mode wave ($n^2 = R \rightarrow \infty$). On the other hand, the lower hybrid resonance frequency (f_{LHR}) is obtained by $S = 0$ in Equation 3 (or $n^2 = RL/S \rightarrow \infty$). Because the condition, $f_{pe} \gg f_{He} \gg f_{Hi}$ is usually satisfied in the ionosphere and magnetosphere, f_{LHR} can be approximated by,

$$\frac{1}{f_{LHR}^2} = \frac{1}{f_{He} f_{Hi}} + \frac{f_{He}}{f_{Hi} f_{pe}^2} \tag{18}$$

For a relatively dense plasma, f_{LHR} is further given by a simple formula.

$$f_{LHR} \sim \sqrt{f_{He} f_{Hi}} \tag{19}$$

The most notable feature in Figure 7.4.3a is the fact that the ion motion allows transverse propagation of the whistler-mode waves below the f_{LHR}. Recall that electron motion alone does not support the transverse whistler mode. In order to show this, Figure 7.4.4 illustrates the refractive index surfaces with and without ion effects when the wavefrequency is assumed to be below the f_{LHR}. Without ions, the refractive index goes to infinity at the oblique resonance angle (θ_{res}) given by Equation 8, and the refractive index is open. While, if ion effects are included, the refractive index surface becomes closed, and the propagation is possible for all wave normal angles and correspondingly the ray direction is perpendicular to the magnetic field direction at $\theta \sim 90°$ as seen from Figure 7.4.4 (Hines, 1957; Kimura, 1966). Of course, when the wavefrequency is greater than the f_{LHR}, the refractive index surface does not close even when ions are included and thus transverse whistler-mode propagation is not possible.

If more than one ion species is present as in the ionosphere, additional zeros (cutoffs) and infinities are introduced in the refractive index curve in Figure 7.4.3b. The LHR frequency (f_{LHR}) in this case is obtained by $S = 0$, and it includes the information of the effective ion mass and fractional abundances of the two ion species. An additional cutoff is indicated by f_{cf} and an additional resonance is the bi-ion resonance frequency (f_{bi}). The inclusion of the second ion species is found to cause the refractive index curves to cross at the cross-over frequency (f_{cr}). If two modes have the same refractive index and polarization, there is a possibility of mode coupling whereby the wave energy in one mode is transferred to another mode. For purely longitudinal propagation the two branches crossing at f_{cr} have different polarization, but at small wave normal angles, the two branches have similar polarization, and we can expect the excitation of an "ion-cyclotron whistler" by an electron whistler, which will be discussed in Section 5.2.

4.2.2. Unducted Ray Paths

Figure 7.4.5 illustrates an example of ray paths (or trajectories) at a frequency of 10 kHz, and these have been calculated for a plasma containing three species of ions (H^+, He^+, O^+) (Aikyo

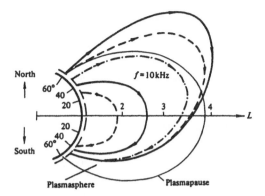

Figure 7.4.5 Ray trajectories at the frequency of 10 kHz, which is higher than the lower hybrid resonance frequency f_{LHR} throughout the entire propagation path. The effects of ions are taken into account. The waves are injected at different initial latitudes with vertical initial wave normal angle. (From Aikyo, K. and Ondoh, T., *J. Radio Res. Lab. Jpn.*, 18, 153, 1971. With permission.)

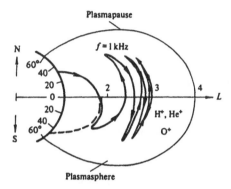

Figure 7.4.6 A ray trajectory at a frequency of f = 1 kHz, in the case where f $<$ f_{LHR} over the entire outer ionosphere. The dashed ray path was calculated neglecting the effect of ions. (From Kimura, I., *Radio Sci.*, 1, 269, 1966. With permission.)

and Ondoh, 1971). However, for the ionosphere-magnetosphere model used, the frequency of 10 kHz is higher everywhere than the local f_{LHR}, thus the nature of the trajectory would be the same if the effect of the ions were neglected.

4.2.3. Ion Effects in Unducted Ray Paths

In Figure 7.4.5 with the approach to f_{LHR} at frequencies f $\lesssim$ f_{LHR} the nature of a ray trajectory becomes greatly altered because of the change in the refractive index surface in Figure 7.4.4. Figure 7.4.6 illustrates an example of ray trajectory at the frequency of f = 1 kHz (Kimura, 1966) where f $<$ f_{LHR} over the entire outer ionosphere. The initial wave normal is directed vertically upward and the height of the start of the trajectory is 300 km. The dashed curve was calculated neglecting the ion effects for the comparison. As seen from Figure 7.4.6, the ray can cross the magnetic field line at certain points of the magnetosphere. The whistlers observed on board a satellite are generally unducted, except when the satellite is inside a duct, which is a rather rare occurrence. Ray-tracing studies indicate that whistler ducts have leaky surfaces, allowing waves to be ducted along only a part of their length. Unducted and partially ducted whistlers are strongly affected by the presence of ions, electron density gradients in a variety of modes depending on the wavefrequency, electron density gradients, and ion composition. They have been given names that are descriptive of their mode of propagation and will be described in Section 5.2.

5.   OBSERVATIONS OF WHISTLERS ON THE GROUND AND IN SPACE

Ground-based whistlers are generally considered to be attributed to ducted propagation (Helliwell, 1965; Walker, 1976; Hayakawa and Tanaka, 1978; Hayakawa and Ohta, 1992), while most whistlers detected in the ionosphere and magnetosphere are propagating in an unducted mode. This section deals with the experimental characteristics of whistlers.

5.1.   GROUND-BASED OBSERVATIONS
5.1.1.   Evidence and Characteristics of Ducts
A lot of indirect evidence for the presence of ducts for whistler phenomena has been accumulated, especially the detection of long-lasting echo-train whistlers which are strongly indicative of such ducts. Whistler observations indicate that ground-based whistlers are attributed to the trapping within enhanced ducts and the ducting should be effective up to one half of the minimum electron gyrofrequency along the path. The latter prediction has bee confirmed by both ground- (Carpenter, 1968) and satellite-based observations of whistler upper cut-off frequencies. Angerami (1970) applied the ray-tracing method to a number of whistlers observed on a high-altitude satellite and deduced that they were ducted up to the satellite. It is recently considered that the ducted propagation is prevalent at middle latitudes (L = ∼ 2 ∼ 6) (Walker, 1976; Thomson and Dowden, 1977a), and also the recent review by Hayakawa and Ohta (1992) has indicated that low latitude whistlers at geomagnetic latitudes greater than 20° are also ducted. Furthermore, very low latitude whistlers at geomagnetic latitudes around 10° are also found to have propagated along the magnetic field lines (but probably by the different field-aligned propagation) (Hayakawa et al., 1990).

Although whistler ducts are essential for whistler phenomena, their characteristics are not well understood. Angerami (1970) obtained the enhancement factor of the order of ∼10 to 30% at middle latitudes and the average diameter of ducts was estimated to be ∼50km at the inospheric heights. Low latitude observational results have yielded the value of more than 100% of the enhancement factor (Hayakawa and Tanaka, 1978), which is much larger than that predicted from Figure 7.4.2 for the homogeneous background and is supported by theoretical ray-tracing studies (Hasegawa et al., 1978). The satellite reception of echo-train whistlers enabled Ondoh (1976) to deduce the latitudinal variation of the duct dimension as shown in Figure 7.5.1. Their duct sizes

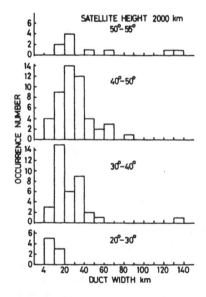

Figure 7.5.1   The latitudinal variation of whistler duct size at 2000 km based on the satellite observation of echo-train whistlers. (From Ondoh, T., *J. Radio Res. Lab. Tokyo,* 23, 139, 1976. With permission.)

(or diameter) at middle latitudes seem to be consistent with Angerami's result. The observed duct size is found to decrease with decreasing geomagnetic latitude. The values at lower latitudes have a lot of support from the extensive ground-based, direction-finding measurements (Ohta et al., 1984; Hayakawa and Ohta, 1992). However, there exists a controversy about the termination altitude of whistler ducts. Satellite data of whistler ducts indicate that the ducts at middle latitudes may terminate at altitudes as high as a few thousand kilometers (Bernhardt and Park, 1977; Aubry, 1968; Cerisier, 1974; James, 1972; Thomson and Dowden, 1977a; Tixier et al., 1984). It is then possible for the emerging waves to spread laterally as observed by low altitude satellites (Thomson and Dowden, 1977a; Tixier et al., 1984).

The lifetime of ducts is reported to vary from a few minutes to many hours. However, recent results may suggest that the duct formation and decay are kinds of cyclic phenomena with the time scale of an hour (Hansen et al., 1983, Hayakawa et al., 1983). The mechanism of formation of ducts is poorly understood, but a few theories have been proposed including flux tube interchange instability (Park and Helliwell, 1971; Thomson, 1978; Walker, 1978). However, further investigation is strongly required in this field.

5.1.2.   Coupling of Downgoing Whistlers to Earth-Ionosphere Waveguide Mode

If we confine our attention to the wave components trapped in ducts, then we are interested only in the wave normals within the trapping cone which lay within the transmission cone for ground reception. Figure 7.5.2 indicates the situation of the excitation of the Earth-ionosphere waveguide mode waves by a downgoing whistler in different latitudes. At very low latitudes (like Figure 7.5.2a) there is no overlapping between the trapping cone of the ducted propagation and the transmission cone, and correspondingly we have only the evanescent waves. With the increase of latitude (Figure 7.5.2b), a small overlapping between the two cones will yield a wave propagating only poleward in the Earth-ionosphere waveguide. This kind of tendency in the subionospheric propagation has been verified by the network observation in Japan (Hayakawa and Ohtsu, 1973), which gives a strong indication of ducted propagation, even at lower latitudes. At higher latitude (Figure 7.5.2d) the excitation of the waveguide modes is easily realized so that the excited wave can propagate in any azimuthal direction.

5.1.3.   Pro-Longitudinal (PL) Propagation

Another possible unducted mode of propagation has been proposed for the whistlers observed on the ground and an example of these PL propagations is illustrated in Figure 7.5.3 (Singh, 1976). This mode requires the existence of significant latitudinal gradients in electron density in the low altitude magnetosphere. In Figure 7.5.3 a negative horizontal gradient around the equator is included, and the wave normal angle remains relatively small with respect to the Earth's magnetic field and also the final wave normal angle suitably lays within the transmission cone for the PL whistler to be observed on the ground (Singh, 1976). The simultaneous observation of whistlers on board the satellite and on the ground at middle latitudes has suggested the presence of PL mode whistlers even on the ground (Thomson and Dowden, 1977b). As one may see from Figure 7.5.3a, the ray paths of the PL and ordinary ducted whistlers are different. However, when they are observed on the ground, both the PL and ducted whistlers are found to follow the Eckersley's dispersion law, and it is rather difficult to distinguish between the two.

5.1.4.   Whistler Characteristics

The frequency spectrum of whistlers is peaked around 5 kHz, and its comparison with that of lightning may yield that the intensity variations of most whistlers, except those that have propagated in the Earth-ionosphere waveguide over great distances, are controlled by the spectrum of the causative lightning discharges.

The amplitude of a whistler can be determined approximately in terms of the apparent field strength received by the antenna, and the peak intensity of whistlers was found to be a few

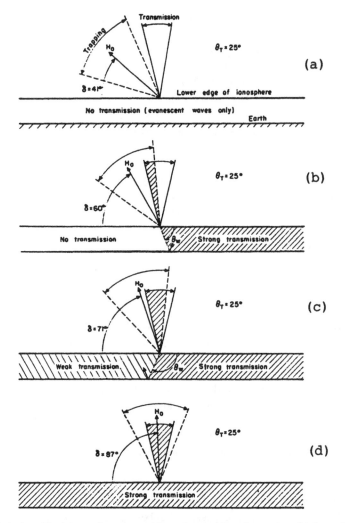

Figure 7.5.2 Excitation of Earth-ionosphere waveguide mode waves by a downgoing whistler at different latitudes. Note the overlapping of the trapping cone with the transmission cone. (From Helliwell, R. A., *Whistlers and Related Ionospheric Phenomena,* Stanford University Press, Stanford, CA, 1965. With permission.)

millivolts per meter at middle latitudes (Helliwell, 1965). The amplitude of whistlers is likely to be dependent on latitude. The direction finding (to be discussed in Section 8.1) at lower latitudes made it possible to estimate the accurate absolute intensities of whistlers, which has yielded that the maximum intensity of daytime strong whistlers at a geomagnetic latitude of 25° is ~250 μV/m and that at 35° it is ~600 μV/m (Hayakawa et al., 1986). This latitudinal dependence is then interpreted satisfactorily in terms of the latitudinal variation of the ionospheric absorption loss as given in Figure 7.3.5 (Hayakawa et al., 1986).

The occurrence rate of whistlers is obviously known to be controlled by the factors of (1) source and (2) propagation in the ionosphere and magnetosphere, with the second factor being further classified into two effects (D-region ionospheric absorption and duct formation). However, the relative importance of these two factors in the whistler occurrence rate is poorly understood, though it is a very fundamental subject. The diurnal variation of whistler occurrence rate shows an enhancement during nighttime at middle and high latitudes (Helliwell, 1965), while that at lower latitudes (~20 to −30°) has an additional sharp peak around sunset (Hayakawa and Tanaka, 1978). The nighttime enhancement is mainly attributed to the lower ionospheric absorption, but the additional sunset peak at lower latitudes is considered to be due to the enhanced occurrence

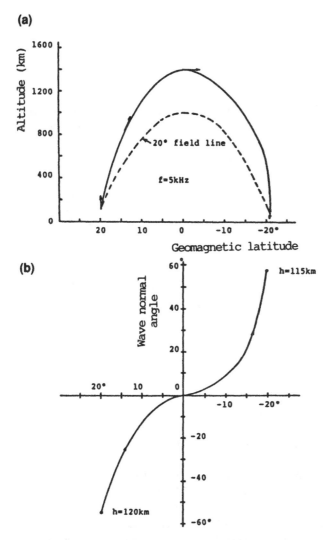

Figure 7.5.3 An example of pro-longitudinal (PL) propagation in the magnetosphere. (a) Ray path of PL mode at 5 kHz and the arrows along the path indicate the wave normal directions. (b) The detailed variation of wave normal directions. The ray is computed in an ionospheric model consisting of a negative horizontal gradient around the equator. (From Singh, B., *J. Geophys. Res.*, 81, 2429, 1976. Copyright by the American Geophysical Union. With permission.)

of ducts at this time. A marked seasonal variation is observed at every latitude, obviously due to the seasonal asymmetry of lightning activity in the conjugate hemisphere. The solar-cycle variation exhibits a depletion during high solar activity and an increase during lower solar activity, in which the D-region absorption might play a primary role.

The latitudinal dependence of whistler occurrence rate is found to show a maximum at the geomagnetic latitude ~45° (L ~ 2.0). This might be explained by the combined effects: (1) the ionospheric absorption loss increases rapidly when we go to lower latitudes (see Figure 7.3.5) and (2) the duct formation may be effective at middle latitudes. The lower latitude cutoff of whistlers is recently found to be 10° ~ 13° (Hayakawa et al., 1990).

Whistler occurrence rate is known to exhibit a strong dependence on geomagnetic activity (Allcock, 1966; Helliwell, 1965; Hayakawa and Tanaka, 1978). Whistler activity reaches a maximum at a certain value K_{opt} of the magnetic index K_p where K_{opt} varies progressively with latitude, being about 4 at 45° latitude and 1 at 68° (Allcock, 1966). This is interpreted in terms

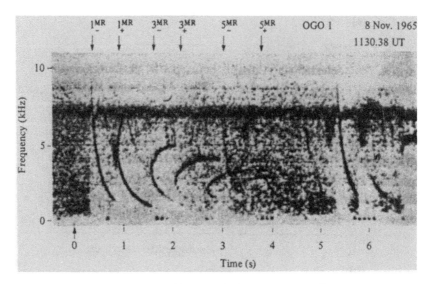

Figure 7.5.4 An example of magnetospherically reflected (MR) whistlers received on a satellite in the outer magnetosphere. (From Smith, R. L. and Angerami, J. J., *J. Geophys. Res.,* 73, 1, 1968. Copyright by the American Geophysical Union. With permission.)

of the relationship of the observing station and the latitudinal movement of the plasmapause with the K_p index.

5.2. SATELLITE OBSERVATIONS OF UNDUCTED WHISTLERS

Most of whistlers observed in space are propagating in the unducted mode, and in the following section we will explain only some interesting unducted whistlers. Other examples of unducted whistlers are described by Park (1982) and Al'pert (1990).

5.2.1. Magnetospherically Reflected (MR) Whistlers

Figure 7.5.4 illustrates a typical example of MR whistlers observed aboard satellites in the deep magnetosphere, which are composed of multiple discrete components (Smith and Angerami, 1968). The propagation of MR whistlers is described adequately by the trajectories depicted in Figure 7.5.5. The important point of MR whistlers is that a lightning discharge illuminates the ionosphere over a wide latitude range, as shown in Figure 7.5.5 with the reasonable assumption of a nearly vertical wave normal angle of each component as the wave enters the ionosphere. At a given frequency there are usually a number of unducted paths from a lightning source to a satellite, as given in Figure 7.5.5. The paths vary with frequency, and the resulting frequency-time spectrograms of received signals are shown on the right in Figure 7.5.5. Each discrete trace is designated by the number of hops N and a subscript (+ or −) indicates the signal moving upward and downward. This MR is based on the nature of the closed refractive index surface in Figure 7.4.4. As the ray travels past the equator and to lower altitudes, the local LHR frequency increases. When the ray reaches a point where the LHR frequency equals the wavefrequency, the refractive index surface becomes closed, making it possible for the wave normal angle to go through 90° and correspondingly the ray direction to reverse. Note the reflection points for the ray paths (B, C in Figure 7.5.5a and B' and C' in Figure 7.5.5c). This is called MR and whistlers undergoing such reflection in the magnetosphere are called MR whistlers. The spectrograms of MR whistlers are often utilized to deduce the electron density profile of the magnetosphere and field-aligned irregularities within it (Edgar, 1976).

5.2.2. Pro-Longitudinal (PL) and Pro-Resonance (PR) Whistlers

We consider the unducted mode of propagation at a higher frequency such that the wave does not undergo LHR reflection. As indicated in Figure 7.5.6a, when the wave propagates downward,

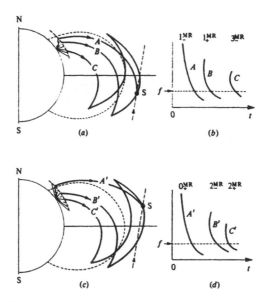

Figure 7.5.5 Possible propagation trajectories of MR whistlers. (From Smith, R. L. and Angerami, J. J., *J. Geophys. Res.*, 73, 1, 1968. Copyright by the American Geophysical Union. With permission.)

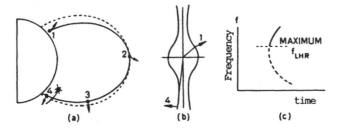

Figure 7.5.6 Pro-resonance (PR) whistlers. (a) Arrows indicate the wave normal directions at several points along the ray path. The dashed curve indicates the magnetic field line. (b) Refractive index near the beginning and the end of the path. (c) The dynamic spectrum of a pro-resonance whistler received on a low altitude satellite. (From Park, C. G., *Handbook of Atmospherics*, Vol. 2, Volland, H., Ed., p. 21, CRC Press, Boca Raton, FL, 1982. With permission.)

the wave normal becomes nearly perpendicular to B_o, but the corresponding ray remains nearly parallel to B_o, as is easily understood from Figure 7.5.6b. If a whistler propagating in this mode is detected on a low altitude satellite, it shows a low cut-off frequency corresponding to the maximum LHR frequency above the satellite altitude. The cut-off frequency is usually ~5 kHz or greater so that only the rising tone part of the whistler is recognized, as illustrated in Figure 7.5.6c. This whistler is called "walking trace" whistler (Walter and Angerami, 1969), and this kind of propagation is called pro-resonance (PR) mode.

In Section 5.1. we have already discussed the PL mode of propagation. The wave normal angle of whistlers propagating in this mode remains relatively small, thus they do not suffer MR even when the wavefrequency is below the LHR frequency. Whistlers propagating in this mode are eventually received on low altitude satellites (Scarabucci, 1969; Morgan, 1980), and the characteristics of these PL whistlers are considerably different from PR whistlers.

5.2.3. Ion Cyclotron Whistlers

Other important ion effects appear in the form of ion-cyclotron whistlers (Gurnett et al., 1965), and one example is illustrated in Figure 7.5.7 (Gurnett and Brice, 1966). As already described in Figure 7.4.3, propagation in the ion-cyclotron mode is possible only if more than two ion species are present, and the frequency is limited in the range between the ion gyrofrequency and the

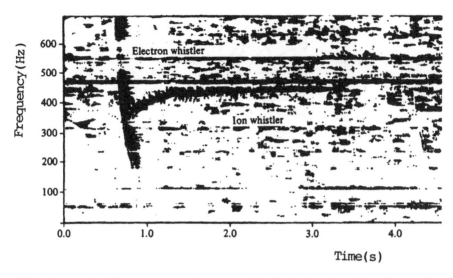

Figure 7.5.7 An example of the spectrogram of a proton whistler in the ionosphere. (From Gurnett, D. A. and Brice, N. M., *J. Geophys. Res.*, 71, 3639, 1966. Copyright by the American Geophysical Union. With permission.)

cross-over frequency characterized by the frequency where the polarization becomes linear during the transition between being right- and left-handed. These whistlers are called "ion whistlers", whereas the conventional whistlers are called "electron whistlers", as in Figure 7.5.7. If the plasma consists of O^+ and H^+ ions, the ion-cyclotron whistler may also be called "a proton whistler". If H_e^+ ions are additionally present, a second additional ion-cyclotron mode exists below the helium gyrofrequency, which can be called "a helium whistler" (Barrington et al., 1966).

The formation of ion whistlers is explained by using Figure 7.5.8 (Gurnett et al., 1965; Ratcliffe, 1972). The two curves in Figure 7.5.8a indicate the height variations of the cross-over frequency and the proton gyrofrequency. The propagation of ion whistlers is allowed between these two curves. The light arrows represent the waves from a lightning below the spacecraft propagating upward in the usual electron whistler mode. This component forms the frequency-time spectrogram of an electron whistler as in Figure 7.5.8b, which is the conventional short-fractional-hop whistler. When an upgoing wave reaches the height where its wavefrequency is equal to the cross-over frequency, some of the wave energy goes into exciting an ion cyclotron wave through mode coupling, which then propagates upward, as indicated by heavy arrows. The ion cyclotron mode is, however, much slower than the electron whistler as easily understood from Figure 7.4.3. Then the corresponding spectrogram of the ion whistler takes the form as shown in Figure 7.5.8b. Ion cyclotron whistlers are extensively used to provide the information on local ion gyrofrequency and cross-over frequency, which then yields much information on ions (ion composition, ion temperature, ion relative concentration, etc.) (Gurnett and Brice, 1966; Gurnett and Shawhan, 1966; Shawhan and Gurnett, 1966).

6.   WHISTLER-ASSOCIATED PHENOMENA AND WAVE-PARTICLE INTERACTIONS

Thus far we have been concerned with the propagation of whistlers, but we can expect that whistlers may interact with the energetic particles present in the magnetosphere. In the following we provide a few examples of evidence of the interaction between whistlers and energetic electrons.

6.1.   WHISTLER-TRIGGERED EMISSIONS

Some whistlers are found to show evidence of amplification in the magnetosphere. Amplification usually occurs in a limited frequency range that presumably depends on the parameters of the

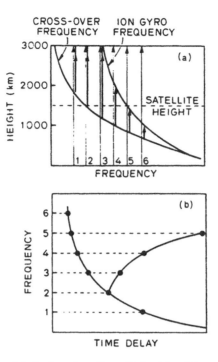

Figure 7.5.8 An illustration to explain the formation of ion whistlers (a) and the corresponding dynamic spectrum (b). (From Ratcliffe, J. A., *An Introduction to the Ionosphere and Magnetosphere,* Cambridge University Press, Cambridge, U.K., 1972. With permission.)

energetic electrons in the magnetosphere. Amplified whistlers are sometimes able to trigger free-running emissions (new emissions), as illustrated in Figure 7.6.1. Figure 7.6.1a illustrates an example of the excitation of several whistler components at the upper cut-off frequency, which is closely related with $f_{He}/2$ along the whistler path (Park and Carpenter, 1978), and Figure 7.6.1b indicates an example of how emissions are triggered by the low frequency tail of a whistler (Hayakawa, 1991). The statistics of the triggering frequency indicate that there is a strong preference for triggering at $f_{He}/2$ (Matsumoto and Kimura, 1971).

Recent studies (Hayakawa, 1991) on the whistler-triggered emissions have indicated that they tend to be highly concentrated in the L value of 2.0 ~ 2.3 (i.e., the electron slot region which is the slot between the outer and inner radiation belts) and also in the inner radiation belt (L = 1.5 ~ 1.8). The former L value seems to be consistent with the latitude where ground-based whistlers are most numerous.

6.2. WHISTLER-INDUCED ELECTRON PRECIPITATION

Wave-particle interactions that amplify whistlers and trigger new emissions also result in the change in the energetic electrons as the consequence of wave-particle interactions. The cyclotron resonance instability is known to lead to the pitch-angle diffusion in the energetic electrons, which then result in the precipitation of energetic electrons into the lower ionosphere as the loss mechanism of magnetospheric energetic particles (Cornwall, 1964; Dungey, 1963; Lyons et al., 1972; Inan et al., 1978). Precipitating particles produce optical emissions, Bremsstrahlung X-rays, and enhanced ionization in the lower ionosphere (see Park, 1982; Rycroft, 1991). The perturbations in VLF subionospheric propagation are known to be extremely sensitive means of detecting particle precipitation (Helliwell et al., 1973; Lohrey and Kaiser, 1979; Inan et al., 1988; Carpenter and LaBelle, 1982; Dowden and Adams, 1989; Smith and Cotton, 1990).

Many types of VLF/ELF waves are generated by naturally occurring plasma processes, and these waves propagate in the whistler mode and interact with energetic electrons the same way

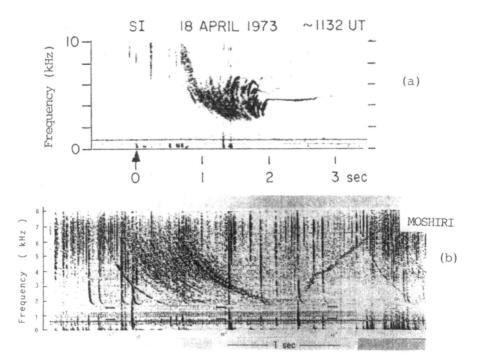

Figure 7.6.1   Examples of whistler-triggered emissions. (a) The triggering takes place at the frequency at one half of the minimum electron gyrofrequency along the whistler path (From Park, C. G. and Carpenter, D. L., *Antarctic Research Series*, Vol. 29, Lanzerotti, L. J. and Park, C. G., Eds., American Geophysical Union, Washington, D.C., 1978. With permission.) and (b) the triggering of emission occurs at the low frequency tail of a whistler. (From Hayakawa, M., *J. Geomagn. Geoelectr.*, 43, 267, 1991. With permission.)

as whistlers (Sazhin and Hayakawa, 1992; Hayakawa and Sazhin, 1992; Sazhin et al., 1993; Sazhin and Hayakawa, 1994), and future problems on wave-particle interactions to be solved are indicated in the review papers (see also Chapter II/13).

7.   WHISTLER PROBING OF THE MAGNETOSPHERIC PLASMA PARAMETERS

The time delay or dispersion of whistlers is defined by Equation 14, which is valid for low frequency waves, and the corresponding high frequency alternative is given by Equation 13. In Equation 13 we have already found the presence of nose frequency. When we observe the whistlers in the lower frequency approximation, their dispersion values would provide us with the information on the electron density of the magnetosphere if we know their path latitude. It is possible for us to infer the whistler path latitude from the observed whistler characteristics, but this can definitely be done with the aid of the ground-based direction finding, which will be discussed in Section 8. Whereas, the situation is completely different for nose whistlers; that is, the minimum time delay or the "nose delay" (denoted by t_n) and the corresponding nose frequency, f_n are related with the path latitude (or its L value) and the electron density along the path.

7.1.   PATH LOCATION AND ELECTRON DENSITY MEASUREMENT BY MEANS OF NOSE WHISTLERS

An approximate correction to the group delay time due to propagation through the ionosphere (up to the altitude of 1000 km), can be obtained by the following equation (Park, 1972),

$$\Delta t_{ion} = 1.4 \, \overline{f_o \, F_2} / \sqrt{f} \tag{20}$$

where Δt_{ion} is given in seconds, $\overline{f_o \, F_2}$ is the critical frequency of the F_2 layer of the ionosphere in megahertz, averaged for both hemispheres, and f is the wavefrequency in hertz. By taking into account the above ionospheric contribution, we can use the following empirical formulas to relate $(f_n, \, t_n)$ to $(f'_n, \, t'_n)$ where f'_n and t'_n are the nose frequency and minimum group delay for the magnetospheric paths (above the altitude of 1000 km),

$$f'_n = \frac{f_n}{1 + (\alpha D_i / t_n f_n^{1/3})} \tag{21}$$

$$t'_n = t_n - D_i \left(\frac{f_n + f'_n}{2} \right)^{-1/2} \tag{22}$$

where the sum of ionospheric dispersion in both conjugate ionospheres, D_i is approximately equal to $1.4 \, \overline{f_o \, F_2}$. The constant α is 0.17 for the diffusive equilibrium model and 0.15 for the collisionless model, with these models being discussed later. We mention that Equations 21 and 22 appear to not be very sensitive to the choice of model (Sazhin et al., 1992), and uncertainties in the measured values are discussed extensively in Park (1972).

Several models of the magnetospheric electron density have been proposed and among them the following two are representative: (1) a diffusive equilibrium (DE) model (Angerami and Thomas, 1964) and (2) a collisionless (CL) model (Eviator et al., 1964; Angerami, 1966). The former is known to be applicable for the plasmasphere where electron densities are of the order of 100 cm^{-3} or greater, whereas the latter model may be more appropriate in the plasmatrough region outside the plasmapause where the equatorial densities are only a few electrons per cubic centimeter.

The relationship of f'_n vs. L for the two magnetospheric electron density models (DE and CL models, the parameters being very typical) is given in Park (1972), and n_{eq} (equatorial electron density of the propagation path) can be determined from a combination of f'_n and t'_n (Park, 1972).

The most outstanding finding by means of nose whistlers is the discovery of the plasmapause. Figure 7.7.1 illustrates how to use the ground-based whistlers in estimating the magnetospheric electron density profile (Park and Carpenter, 1978). The spectrograms on the two top panels are found to be composed of two whistler trains excited by a single lightning discharge (indicated by an arrow on the time axis). The bottom panel indicates the electron density profiles of the magnetosphere at two different times, using the procedure mentioned above. The solid dots show a smooth density profile, reflecting the smooth locus of $f_n - t_n$ in the first top spectrogram. On the other hand, the second top panel exhibits a number of low f_n traces with sharply reduced t_n. The corresponding density profile is indicated by open dots, which indicate the sharp density gradient, so-called "plasmapause".

7.2. NOSE EXTENSION METHOD

Some whistlers, especially those having propagated along lower latitude field lines, do not exhibit a detectable nose frequency in the spectrograms. In such cases, we have to extrapolate f_n and t_n from the available portion of the whistler traces. A number of different extrapolation methods have been proposed for this purpose (Smith and Carpenter, 1966; Likhter and Molchanov, 1968; Dowden and Allcock, 1971; Bernard, 1973; Corcuff and Corcuff, 1973; Rycroft and Mathur, 1973; Smith et al., 1975; Tarcsai, 1975; Stuart, 1977), and readers are advised to consult a recent

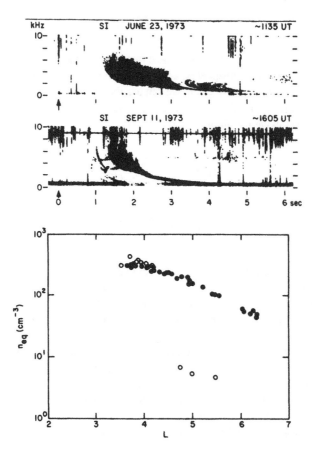

Figure 7.7.1 Two examples of multicomponent whistlers recorded at Siple, Antarctica (76°S, 84°W) and equatorial electron density profiles deduced from them. The solid dots correspond to the whistlers in the first of the top panels, and the open dots, to the second of the top panels. (From Park, C. G. and Carpenter, D. L., *Antarctica Research Series,* Vol. 29, Lanzerotti, L. J. and Park, C. G., Eds., American Geophysical Union, Washington, D.C., 1978. With permission.)

comprehensive review paper on this nose extension method by Sazhin et al. (1992) for details. This review deals with a comparison of different methods proposed thus far.

7.3. ESTIMATION OF OTHER MAGNETOSPHERIC PLASMA PARAMETERS

Other applications of the whistler technique are the measurements of electric field and temperature of the magnetosphere (e.g., Sazhin et al., 1992). By measuring f_n for successively recorded whistlers, we can obtain the changes with time of the L-shell of the duct along which a whistler propagates. The rate of change of L can be related to magnetospheric plasma drift caused by a large scale electric field **E**.

$$v_d = \frac{d(R_e L)}{dt} = c \frac{|E \times B_{eq}|}{|B_{eq}^2|} \tag{23}$$

where B_{eq} is the magnetic induction at the equator and R_e is the Earth's radius. Assuming a dipole model for the magnetospheric magnetic field and assuming E is perpendicular to B_0, we have (Bernard, 1973),

$$|E| = a \frac{d(f_n^{2/3})}{dt} \tag{24}$$

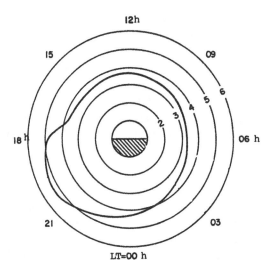

Figure 7.7.2 Equatorial cross section of the inner magnetosphere showing (solid curve) an estimate of the average position of the plasmapause during periods of moderate, steady magnetic agitation (K_p = 2–4). (From Carpenter, D. L., *J. Geophys. Res.,* 71, 693, 1966. Copyright by American Geophysical Union. With permission.)

If **E** is measured in volts per meter and f_n is measured in hertz, then a can be taken as 2.07 × 10^{-2} (Park, 1982). If df_n/dt is positive (a whistler duct drifts inward), then the electric field is directed from east to west. Electric fields as small as 10^{-5} V/m in the equatorial plane can be measured by this technique.

Other parameters that can be determined using the measurements of f_n and t_n are the total electron content N_T in the magnetic field tube with the cross section 1 cm^2 at the reference level (see Park, 1972; Sazhin et al., 1992) and the downward depletion flux or upward refilling flux of electrons (Sazhin et al., 1992).

Beside the above-mentioned traditional methods of diagnostics of electron density, attempts were made to use whistlers as a diagnostic of the electron temperature therein. The method used by Sazhin et al. (1990) is essentially based on the fact that the thermal corrections to whistler group velocity, and correspondingly, to whistler group delay time, are the largest at frequencies close to the upper cut-off frequency and can almost always be neglected elsewhere. See further details in Sazhin et al. (1990, 1992).

Sazhin et al. (1992) have summarized the main results of magnetospheric plasma parameters based on the methods considered in the previous and present sections. Readers can find there the latest results on the large scale electric field (E), electron density in the equatorial plane, duct location, equatorial magnetospheric magnetic field, and electron temperature in the equatorial magnetosphere.

7.4. CHARACTERISTICS OF DISPERSION AND OF MAGNETOSPHERIC ELECTRON DENSITY

7.4.1. Magnetospheric Electron Density Profile (Spatial Variations)

The whistler analyses, as mentioned in the previous sections, have yielded that the electron density in the quiet magnetosphere remains almost constant in the inner plasmasphere within L ~ 4 and decreases with geocentric distance roughly as R^{-4} beyond the plasmapause (Carpenter, 1966; Corcuff, 1975). This kind of spatial variation is seen in Figure 7.7.1. The position of the plasmapause is known to be strongly dependent on local time, and Figure 7.7.2 illustrates the average structure of the plasmapause (Carpenter, 1966), which exhibits the existence of a duskside "bulge".

It is well known that the plasmapause moves inward during the periods of increased magnetic activities. In regard to this storm-like reduction in plasmasphere size, it is widely believed that an erosion process occurs in association with enhanced convection (Carpenter et al., 1992), but the plasmapause structure exhibits a rather complicated dependence on the storm phase, which needs further study.

Because whistler propagation beyond the plasmapause is rather infrequent, the information on the electron density in the plasmatrough region beyond the plasmapause is lacking. Just inside the plasmapause (at $L \sim 4$), the density (~ 300 el/cm^3) varies only slightly from day to night, while just outside the plasmapause the nighttime values (~ 5 el/cm^3) are significantly lower than the dayside values (~ 5 el/cm^3). This effect reflects the filling of the protonosphere during the day by diffusion of plasma upward from the underlying ionosphere.

7.4.2. Temporal Variations

The analyses of whistlers (nose and non-nose) at low, middle, and high latitudes have shown that the electron density in the plasmasphere exhibits large changes during magnetic storms, diurnal, annual (seasonal), and solar-cycle variations.

During the main phases of magnetic storms, electron density levels are often rapidly depressed, and in contrast, the recovery is slow, in general, much slower than that of the underlying ionosphere (Corcuff, 1975). However, the tube contents at low latitudes (geomagnetosphere latitude less than 25°) seem to increase during storm times (Hayakawa and Tanaka, 1978). Further study is required in order to investigate the dynamics of magnetospheric plasma by using whistler dispersions.

The annual variation in magnetospheric electron density exhibits a December maximum and a June minimum at middle latitudes (Helliwell, 1965; Park, 1972; Corcuff, 1975), but it is found that a semi-annual variation is more important than the annual one at low latitudes with maxima near the equinoxes and minima near the solstices (Hayakawa and Tanaka, 1978).

Based on the whistler dispersion data at middle latitudes, it is found that the magnetosphere is less sensitive than the F region of the ionosphere to the long-term variations of solar activity (Helliwell, 1965; Corcuff, 1975). At $L \sim 1.6 - 2$ the electron density diminishes during the solar minimum by about 35%, compared with that of the solar maximum (Hayakawa and Tanaka, 1978).

The diurnal variation of electron density is not easily determined because of the scarcity of daytime whistlers. The dispersion of whistlers at low latitudes is found to be enhanced in the afternoon and decreased toward midnight, closely following the variations in f$_o$F$_2$ (Hayakawa and Tanaka, 1978). Data from middle latitudes also show a definite diurnal variation in dispersion with an afternoon maximum and a postmidnight minimum. The magnitude of this variation is about 25%, corresponding to a change in the electron density over the entire path of 50%, assuming no change in the form of distribution (Helliwell, 1965).

8.   DIRECTION FINDING OF VLF/ELF RADIO WAVES

As mentioned in Section 7, it is very important to locate the region where whistlers emerge from in the ionosphere to know their path latitude; this is called "direction-finding" measurement. Even for other magnetospheric VLF/ELF emissions (Sazhin and Hayakawa, 1994; Hayakawa and Sazhin, 1992; Sazhin et al., 1993), the ground-based, direction-finding measurement would be of essential importance in the study of their generation and propagation mechanism.

8.1.   GROUND-BASED DIRECTION FINDING

Figure 7.8.1 illustrates the wave incidence in which a generally elliptically polarized VLF wave is incident onto an observing point A with an incident angle i and an azimuthal angle φ (measured eastward from the north). The elliptically polarized wave is decomposed into transverse electric (TE) and transverse magnetic (TM) mode components; the fields ($E_\perp$, $H_\parallel$) correspond to the TE

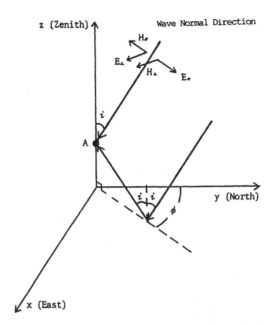

Figure 7.8.1 The principle of direction finding of VLF/ELF waves. A generally elliptically polarized wave is incident onto the direction finder located at a point A. The incident and azimuthal angles of the incident wave are i and ϕ, respectively.

mode wave, while $(E_\parallel, H_\perp)$, correspond to the TM mode wave. The wave polarization (p) is defined as the ratio of the magnetic field components of the TE and TM modes as follows.

$$p = \frac{H_\parallel}{H_\perp} = u - jv \tag{25}$$

Here, an exactly right-handed circular polarization is given by $(u, v) = (0, 1)$. The electromagnetic fields observed at point A are considered to be superpositions of a direct ray and a ray reflected from the ground. When we assume that the height of the observing point A is much smaller than a wavelength, the resultant field components are given as follows:

$$\left.\begin{array}{l} H_x = 2A_\perp \, e^{j\omega t}\{(\cos\phi - u\cos i \sin\phi + jv\cos i \sin\phi\} \\ H_y = 2A_\perp \, e^{j\omega t} \{-(\sin\phi + u\cos i \cos\phi) + jv\cos i \cos\phi\} \\ E_z = -2A_\perp Z_0 \, e^{j\omega t} \sin i \end{array}\right\} \tag{26}$$

Where Z_0 is the characteristic impedance of free space, $A_\perp$ is the amplitude of $H_\perp$ and the time factor $(e^{j\omega t})$ is assumed. Furthermore, the ground is taken to be a perfect conductor, and hence, other field components $(H_z, E_x, \text{and } E_y)$ vanish.

The horizontal magnetic fields H_x and H_y are picked up by the loop aerials whose planes are in the NS and EW directions, respectively. The voltages induced in the NS and EW loops are denoted by V_x and V_y (e.g., $V_x \propto H_y$), while V_z refers to the voltage by a monopole antenna. By making the amplitude and phase characteristics between the loop and vertical antennas the same over the relevant frequency range by inserting different antenna coupling circuits for different antenna systems, we obtain,

$$\left.\begin{array}{l} \dfrac{V_x}{V_z} = -\dfrac{\sin\phi + u\cos i \cos\phi}{\sin i} + j\dfrac{v\cos i \cos\phi}{\sin i} \\[2mm] \dfrac{V_y}{V_z} = \dfrac{\cos\phi - u\cos i \sin\phi}{\sin i} + j\dfrac{v\cos i \sin\phi}{\sin i} \end{array}\right\} \tag{27}$$

When we separate the above two equations into real and imaginary parts ($V_x/V_z = a_1 + jb_1$ and $V_y/V_z = a_2 + jb_2$), we obtain the following relations.

$$
\left.
\begin{aligned}
a_1 &= -\frac{\sin\phi + u \cos i \cos\phi}{\sin i} \\
b_1 &= \frac{v \cos i \cos\phi}{\sin i} \\
a_2 &= \frac{\cos\phi - u \cos i \sin\phi}{\sin i} \\
b_2 &= \frac{v \cos i \sin\phi}{\sin i}
\end{aligned}
\right\}
\tag{28}
$$

These four equations may be solved to find the four unknown wave parameters (i, ϕ) and (u, v) in terms of quantities derivable from relative amplitudes and phase differences (a_1, a_2, b_1, and b_2).

The direction-finding methods thus far proposed (Watts, 1959; Delloue, 1960; Crary, 1961; Cousins, 1972; Bullough and Sagredo, 1973; Tsuruda and Hayashi, 1975; Leavitt, 1975; Tanaka et al., 1976; Okada et al., 1977, 1981) can be classified into three cases (Cousins, 1972), but the two important cases are given as follows.

Case 1 (The "B" method). The quantities b_1 and b_2 are not both zero, and hence i is not 90° and v is nonzero. Therefore, the effective plane wave is incident from above the horizon and has complex (generally elliptical) polarization. The incident angle i and azimuthal angle ϕ are estimated by using Equation 28 in the following way.

$$
\left.
\begin{aligned}
\phi &= \tan^{-1}(b_2/b_1) \\
i &= \sin^{-1}\left(\frac{1}{a_2 \cos\phi - a_1 \sin\phi}\right)
\end{aligned}
\right\}
\tag{29}
$$

We can have the corresponding relationships for estimating the wave polarization (u, v). This determination of the arrival direction will be called the B method because we utilize the imaginary parts (b_1 and b_2) for the determination of ϕ.

Case 2 (The "A" method). The quantities b_1 and b_2 are both zero, implying that either i = 90° or v is zero (linear polarization). Consider first the case with i = 90° (horizontal incidence). In this case of horizontal incidence, we have to determine only the azimuthal direction (ϕ) by using Equation 28.

$$
\phi = \tan^{-1}(-a_1/a_2)
\tag{30}
$$

Since b_1 and b_2 are zero, the magnitudes of the loop-to-vertical relative voltages (V_x/V_z and V_y/V_z) are equal to the real parts. The quadrant of the azimuth may be determined from the signs of a_1 and a_2. The arrival angle determination in this way designated the A method, approximates the results of simple crossed-loop direction finding. The latter is referred to as the "loops-and-goniometer" technique.

Only two direction-finding methods [(1) goniometer and (2) field-analysis method] among the many, are being used extensively in the studies of whistlers and VLF/ELF emissions (Sagredo and Bullough, 1973; Tanaka et al., 1976; Hayakawa et al., 1981; Smith and Carpenter, 1982; Ohta et al., 1984; Hayakawa et al., 1986). The latter field-analysis method falls in the B method, and it is effective for nearby ionospheric exit regions of VLF waves. The details of its instrumentation are described in Okada et al. (1977, 1981). Recent improvements associated with this system are an automatic direction-finding measurement at one single frequency and an extension from a single frequency measurement to a wideband measurement (see Hayakawa et al., 1992).

The A method is a special case of the goniometer technique, and the semi-automated goniometer system has been developed by Smith et al. (1979). The goniometer is effective for linearly polarized waves which have propagated over great distances from the ionospheric exit region, and at least two stations are necessary for the triangulation to locate the ionospheric exit region of VLF waves, because the goniometer is able to determine only the signal azimuth.

The accuracy of the measurement of arrival directions is the most important subject to be investigated in the direction findings (Crary, 1961; Cousins, 1972; Bullough and Sagredo, 1973; Leavitt, 1975; Tanaka et al., 1976). Strangeways (1980) and Strangeways and Rycroft (1980) have discussed the relative merits of different methods by evaluating the systematic error in the azimuthal bearing due to multipath propagation in the Earth-ionosphere waveguide and wave polarization (when applicable). Their assumption of a point source at the bottom of the lower ionosphere is probably acceptable at higher latitudes, but it is uncertain whether this supposition is still valid at low latitudes. Theoretical estimation of the measuring error for a more realistic ionospheric transmission model is required, and also the multistationed direction-finding measurement is highly necessary in order to experimentally know the measuring error.

In all of the previous direction-finding methods, the incident wave is assumed to be a single plane wave, however, on some occasions, either when the whistler duct is rather wide or when a few ducts are present, that supposition breaks down so that the results of direction findings are questionable. The new method, which is effective even for such cases is developed by Shimakura et al. (1992) and yields the distribution of wave energy of magnetospheric VLF waves at the ionospheric base. The idea of wave distribution function was originally proposed by Lefeuvre et al. (1981) for the satellite-based direction finding to be presented in the next section.

8.2. SPACECRAFT-BASED DIRECTION FINDING

Contrary to the situation of ground-based direction finding, one can simultaneously measure multiple electromagnetic field components (three magnetic and three electric) on spacecrafts (rockets and satellites) (Shawhan, 1970; Lefeuvre et al., 1982). When the incident wave is supposed to be a single whistler-mode plane wave, the fact that the wave normal direction is perpendicular to the wave magnetic field, as already described in Section 2.2., is utilized for a direction-finding method on board spacecrafts. In this case, the use of three-dimensional orthogonal loop aerials is adopted by some workers (Iwai et al., 1974; Cerisier, 1974).

Next when the wavefield is supposed to be plane or to be formed by a finite number of plane waves, but the signal is nondeterministic. The wave normal directions are determined from the $n \times n$ spectral matrix of the n magnetic and electric wave field components ($n \leq 6$), whose components are the auto- and cross-power spectra. Means' (1972) method utilizes only the imaginary parts of the spectral matrix to determine the wave normal direction of a single plane wave, while Buchalet and Lefeuvre (1981) have presented a way of estimating the wave normal directions of a few simultaneous waves.

In another method the wavefield is supposed to be random and the signal is nondeterministic. In this case the wavefield is characterized by a statistical quantity like a wave distribution function, specifying how the wave energy at a given frequency is distributed with respect to the wave normal direction (Lefeuvre et al., 1981).

How the spacecraft-based, direction-finding measurements contributed to the study of the generation and propagation mechanism of magnetospheric VLF/ELF emissions, has been recently summarized by Hayakawa (1993).

REFERENCES

Aikyo, K. and Ondoh, T., Propagation of nonducted VLF waves in the vicinity of the plasmapause, *J. Radio Res. Lab. Jpn.*, 18, 153, 1971.

Allcock, G. Mck., Whistler propagation and geomagnetic activity, *J. Inst. Telecomm. Eng.*, 12, 158, 1966.

Al'pert, Ya. L., *Space Plasma,* Cambridge University Press, Cambridge, U.K., 1990.

Altman, C., and Cory, H., The generalized thin film optical method in electromagnetic wave propagation, *Radio Sci.,* 4, 459, 1969.

Angerami, J. J., A whistler study of distribution of thermal electrons in the magnetosphere, Tech. Rep. 3412-7, Radioscience Laboratory, Stanford University, Stanford, CA, 1966.

Angerami, J. J., Whistler duct properties deduced from VLF observations made with OGO-3 satellite near the magnetic equator, *J. Geophys. Res.,* 75, 6115, 1970.

Angerami, J. J. and Thomas, J. O., Studies of planetary atmosphere. I. The distribution of electrons and ions in the earth's exosphere, *J. Geophys. Res.,* 69, 4537, 1964.

Aubry, M. P., Influence des irregularities de densite electronique sur la propagation des ondes TBF dans l'ionosphere, *Ann. Geophys.,* 24, 39, 1968.

Barkhausen, H., Zwei mit Hilfe der neuen Varstarker entdeckte Erscheimongen, *Physik A.,* 20, 401, 1919.

Barkhausen, H., Whistling tones from the earth, *Proc. Inst. Radio Eng.,* 18, 1155, 1930.

Barrington, R. E., Belrose, J. S., and Mather, W. E., A helium whistler observed in the Canadian Satellite Alouette, II., *Nature,* 210, 80, 1966.

Bernard, L. P., A new nose extension method for whistlers, *J. Atmos. Terr. Phys.,* 35, 871, 1973.

Bernhardt, P. A. and Park, C. G., Protonospheric-ionospheric modeling of VLF ducts, *J. Geophys. Res.,* 82, 5222, 1977.

Buchalet, L. J. and Lefeuvre, F., One- and two-direction models for VLF electromagnetic waves observed on board GEOS1, *J. Geophys. Res.,* 86, 2377, 1981.

Budden, K. G., *Radio Waves in the Ionosphere,* Cambridge University Press, Cambridge, U.K., 1961.

Budden, K. G., *The Propagation of Radio Waves (The Theory of Radio Waves of Low Power in the Ionosphere and Magnetosphere),* Cambridge University Press, Cambridge, U.K., 1985.

Bullough, K. and Sagredo, J. L., VLF goniometer observations at Halley Bay, Antarctica-I. The equipment and the measurement of signal bearing, *Planet. Space Sci.,* 21, 899, 1973.

Burtis, W. J. and Helliwell, R. A., Magnetospheric chorus: occurrence patterns and normalized frequency, *Planet. Space Sci.,* 24, 1007, 1976.

Burton, E. T. and Boardman, E. M., Audio-frequency atmospherics, *Proc. Inst. Radio Eng.,* 21, 1476, 1933.

Carpenter, D. L., Whistler studies of the plasmapause in the magnetosphere-1. Temporal variations in the position of the knee and some evidence on plasma motions near the knee, *J. Geophys. Res.,* 71, 693, 1966.

Carpenter, D. L., Ducted whistler mode propagation in the magnetosphere: a half-gyrofrequency upper intensity cutoff and some associated wave growth phenomena, *J. Geophys. Res.,* 73, 2919, 1968.

Carpenter, D. L. and LaBelle, J. W., A study of whistlers correlated with bursts of electron precipitation near L = 2., *J. Geophys. Res.,* 87, 4427, 1982.

Carpenter, D. L., Smith, A. J., Gilles, B. L., Chappell, C. R., and Decreau, P. M. E., A case study of plasma structure in the dusk sector associated with enhanced magnetospheric convection, *J. Geophys. Res.,* 97, 1157, 1992.

Cerisier, J. C., Ducted and partly ducted propagation of VLF waves through the magnetosphere. *J. Atmos. Terr. Phys.,* 36, 1443, 1974.

Corcuff, Y., Probing the plasmapause by whistlers, *Ann. Geophys.,* 31, 53, 1975.

Corcuff, P. and Corcuff, Y., Determination des parametres $f_n - t_n$ caracteristiques des sifflements radio electriques recus an sul, *Ann. Geophys.,* 29, 273, 1973.

Cornwall, J. M., Scattering of energetic trapped electrons by very low frequency waves, *J. Geophys. Res.,* 69, 1251, 1964.

Cousins, M. D., Direction finding on whistlers and related VLF signals, Tech. Rep. No. 3432-2, Radioscience Laboratory, Stanford University, Stanford, CA, 1972.

Crary, J. H., The effect of the earth-ionosphere waveguide on whistlers, Tech. Rep. No. 9, Radioscience Laboratory, Stanford University, Stanford, CA, 1961.

Delloue, J., La determination de la direction d'arrivee et de la polarisation des atmospheriques siffleurs, premere partie, *J. Phys. Radium,* 6, 514, 1960.

Dowden, R. L. and Adams, C. D. D., Phase and amplitude perturbations or the NWC signal at Dunedin from lightning induced precipitation, *J. Geophys. Res.,* 94, 497, 1989.

Dowden, R. L. and Allcock, G. Mck., Determination of nose frequency of non-nose whistlers, *J. Atmos. Terr. Phys.,* 33, 1125, 1971.

Dungey, J. W., Loss of Van Allen electrons due to whistlers, *Planet. Space Sci.,* 11, 591, 1963.

Eckersley, T. L., 1929–1930 developments in the study of radio wave propagation, *Marconi Rev.,* 5, 1, 1931.

Edgar, B. C., The upper- and lower-frequency cutoffs of magnetospherically reflected whistlers, *J. Geophys. Res.,* 81, 205, 1976.

Eviator, A., Lenchek, A. M., and Singer, S. G., Distribution of density in an ion-exosphere of a non-rotating planet, *Phys. Fluids,* 7, 1775, 1964.

Gendrin, R., Guidage des sifflements radio electriques parle champ magnetique terrestre, *Compt. Rend.,* 251, 1085, 1960.

Gurnett, D. A. and Brice, N. M., Ion temperature in the ionosphere obtained from cyclotron damping of proton whistler, *J. Geophys. Res.,* 71, 3639, 1966.

Gurnett, D. A. and Shawhan, S. D., Determination of hydrogen ion concentration, electron density, and proton gyrofrequency from the dispersion of proton whistlers, *J. Geophys. Res.*, 71, 741, 1966.

Gurnett, D. A., Shawhan, S. D., Smith, R. L., and Brice, N. M., Ion cyclotron whistlers, *J. Geophys. Res.*, 70, 1665, 1965.

Hansen, H. J., Scourfield, M. W. J., and Rash, J. P. S., Whistler duct lifetimes, *J. Atmos. Terr. Phys.*, 45, 789, 1983.

Hasegawa, M. and Hayakawa, M., The influence of the equatorial anomaly on the ground reception of whistlers at low latitudes, *Planet. Space Sci.*, 28, 17, 1980.

Hasegawa, M., Hayakawa, M., and Ohtsu, J., On the conditions of duct trapping of low latitude whistlers, *Ann. Geophys.*, 34, 317, 1978.

Hayakawa, M., Satellite observation of low-latitude VLF radio noises and their association with thunderstorms, *J. Geomagn. Geoelectr.*, 41, 573, 1989.

Hayakawa, M., Observation at Moshiri (L = 1.6) of whistler-triggered VLF emissions in the electron slot and inner radiation belt regions, *J. Geomagn. Geoelectr.*, 43, 267, 1991.

Hayakawa, M., Study of generation mechanisms of magnetospheric VLF/ELF emissions based on the direction findings, *Proc. Natl Inst. Polar Res. Symp. Upper Atmos. Phys.*, 6, 117, 1993.

Hayakawa, M. and Ohta, K., The propagation of low-latitude whistlers: a review, *Planet. Space Sci.*, 40, 1339, 1992.

Hayakawa, M. and Ohtsu, J., Ducted propagation of low latitude whistlers deduced from the simultaneous observations at multi-stations, *J. Atmos. Terr. Phys.*, 35, 1685, 1973.

Hayakawa, M., Ohta, K., and Shimakura, S., Spaced direction finding of nighttime whistlers at low and equatorial latitudes and their propagation mechanism, *J. Geophys. Res.*, 95, 15091, 1990.

Hayakawa, M., Ohta, K., and Shimakura, S., Direction finding techniques for magnetospheric VLF waves: recent achievements, *Trends Geophys. Res.*, 1, 157, 1992.

Hayakawa, M., Okada, T., and Iwai, A., Direction findings of medium-latitude whistlers and their propagation characteristics, *J. Geophys. Res.*, 86, 6939, 1981.

Hayakawa, M. and Sazhin, S. S., Mid-latitude and plasmaspheric hiss: a review, *Planet. Space Sci.*, 40, 1325, 1992.

Hayakawa, M. and Tanaka, Y., On the propagation of low-latitude whistlers, *Rev. Geophys. Space Phys.*, 16, 111, 1978.

Hayakawa, M., Tanaka, Y., and Ohtsu, J., Time scales of formation, lifetime and decay of low-latitude whistler ducts, *Ann. Geophys.*, 1, 515, 1983.

Hayakawa, M., Tanaka, Y., Ohta, K., and Okada, T., Absolute intensity of daytime whistlers at low and middle latitudes and its latitudinal variation, *J. Geophys.*, 59, 67, 1986.

Hayakawa, M., Tanaka, T., Sazhin, S. S., Okada, T., and Kurita, K., Characteristics of dawnside mid-latitude VLF emissions associated with substorms as deduced from the two-stationed direction finding measurement, *Planet. Space Sci.*, 34, 225, 1986.

Helliwell, R. A., *Whistlers and Related Ionospheric Phenomena*, Stanford University Press, Stanford, CA, 1965.

Helliwell, R. A., Katsufrakis, J. P., and Trimpi, M. L., Whistler-induced amplitude perturbations in VLF propagation, *J. Geophys. Res.*, 78, 4679, 1973.

Hines, C. O., Heavy-ion effects in audio-frequency radio propagation, *J. Atmos. Terr. Phys.*, 11, 36, 1957.

Inan, U. S. and Bell, T. F., The plasmapause as a VLF wave guide, *J. Geophys. Res.*, 82, 2819, 1977.

Inan, U. S., Bell, T. F., and Helliwell, R. A., Nonlinear pitch angle scattering of energetic electrons by coherent VLF waves in the magnetosphere, *J. Geophys. Res.*, 83, 3235, 1978.

Inan, U. S., Shafter, D. C., Yip, W. Y., and Orville, R. E., Subionospheric VLF signatures of nighttime D region perturbations in the vicinity of lightning discharges, *J. Geophys. Res.*, 93, 11455, 1988.

Ishikawa, K., Hattori, K., and Hayakawa, M., A study of ray focusing of whistler-mode waves in the magnetosphere, *Trans. Inst. Electr. Inform. Comm. Eng. Jpn*, E73, 149, 1990.

Iwai, A., Okada, T., and Hayakawa, M., Rocket measurement of wave normal directions of low-latitude sunset whistlers, *J. Geophys. Res.*, 79, 3870, 1974.

James, H. G., Refraction of whistler-mode waves by large-scale gradients in the middle-latitude ionosphere, *Ann. Geophys.*, 28, 301, 1972.

Kimura, I., Effects of ions on whistler-mode ray tracing, *Radio Sci.*, 1, 269, 1966.

Leavitt, M. K., A frequency-tracking direction finding for whistlers and other VLF signals, Tech. Rep. No. 3456-2, Radioscience Laboratory, Stanford University, Stanford, CA, 1975.

Lefeuvre, F., Neubert, T., and Parrot, M., Wave normal directions and wave distribution functions for ground-based transmitter signals observed on GEOS-1, *J. Geophys. Res.*, 87, 6203, 1982.

Lefeuvre, F., Parrot, M., and Delannoy, C., Wave distribution functions estimation of VLF electromagnetic waves observed onboard GEOS 2, *J. Geophys. Res.*, 86, 2359, 1981.

Likhter, Ya. I. and Molchanov, O. A., Changing in whistler characteristics in the disturbed magnetic field of the magnetosphere, *Geomagn. Aeron.*, 8, 899, 1968 (in Russian).

Lohrey, B. and Kaiser, A. B., Whistler induced anomalies in VLF propagation, *J. Geophys. Res.*, 84, 5122, 1979.

Lyons, L. R., Thorne, R. N., and Kennel, C. F., Pitch-angle diffusion of radiation belt electrons within the plasmasphere, *J. Geophys. Res.*, 77, 3455, 1972.

Matsumoto, H. and Kimura, I., Linear and nonlinear cyclotron instability and VLF emissions in the magnetosphere, *Planet. Space Sci.*, 19, 567, 1971.

Means, J. D., The use of the three-dimensional covariance matrix in analyzing the properties of plane waves, *J. Geophys. Res.*, 27, 5551, 1972.

Morgan, M. G., Some features of pararesonance (PR) whistlers, *J. Geophys. Res.*, 85, 130, 1980.

Nagano, I., Mambo, M., and Hutatsuishi, G., Numerical calculation of electromagnetic waves in an anisotropic multilayered medium, *Radio Sci.*, 10, 6611, 1975.

Ohta, K., Hayakawa, M., and Tanaka, Y., Ducted propagation of daytime whistlers at low latitudes as deduced from the ground direction finding, *J. Geophys. Res.*, 89, 7557, 1984.

Okada, T., Iwai, A., and Hayakawa, M., The measurement of incident and azimuthal angles and the polarization of whistlers at low latitudes, *Planet. Space Sci.*, 25, 233, 1977.

Okada, T., Iwai, A., and Hayakawa, M., A new whistler direction finder, *J. Atmos. Terr. Phys.*, 43, 679, 1981.

Ondoh, T., Magnetospheric whistler ducts observed by ISIS satellites, *J. Radio Res. Lab. Tokyo*, 23, 139, 1976.

Park, C. G., Methods of determining electron concentrations in the magnetosphere from nose whistlers, Tech. Rep. 3454-1, Radioscience Laboratory, Stanford University, Stanford, CA, 1972.

Park, C. G. and Helliwell, R. L., The formation by electric fields of field-aligned irregularities in the magnetosphere, *Radio Sci.*, 6, 299, 1971.

Park, C. G., Whistlers, in *Handbook of Atmospherics*, Vol. 2, Volland, V., Ed., p. 21, CRC Press, Boca Raton, FL, 1982.

Park, C. G. and Carpenter, D. L., Very low frequency radio waves in the magnetosphere, in upper atmosphere research in Antarctica, *Antarctic Research Series*, Vol. 29, Lanzerotti, L. J. and Park, C. G., Eds., American Geophysical Union, Washington, D.C., 1978.

Pitteway, M. L. V., The numerical calculation of wave-fields, reflection coefficients and polarizations for long radio waves, *Philos. Trans. R. Soc. London*, A257, 219, 1965.

Ratcliffe, J. A., *The Magneto-Ionic Theory and its Applications to the Ionosphere*, Cambridge University Press, Cambridge, U.K., 1959.

Ratcliffe, J. A., *An Introduction to the Ionosphere and Magnetosphere*, Cambridge University Press, Cambridge, U.K., 1972.

Rycroft, M. J., Interactions between whistler-mode waves and energetic electrons in the coupled system formed by the magnetosphere, ionosphere and atmosphere, *J. Atmos. Terr. Phys.*, 53, 849, 1991.

Rycroft, M. J. and Mathur, A., The determination of the minimum group delay of a non-nose whistler, *J. Atmos. Terr. Phys.*, 35, 2177, 1973.

Sagredo, J. L. and Bullough, K., VLF goniometer observations at Halley Bay, Antarctica, II. Magnetospheric structure deduced from whistler observations, *Planet. Space Sci.*, 21, 913, 1973.

Sazhin, S. S. and Hayakawa, M., Magnetospheric chorus emissions: a review, *Planet. Space Sci.*, 40, 681, 1992.

Sazhin, S. S. and Hayakawa, M., Periodic and quasiperiodic emissions, *J. Atmos. Terr. Phys.*, 56, 735, 1994.

Sazhin, S. S., Bullough, K., and Hayakawa, M., Auroral hiss: a review, *Planet. Space Sci.*, 41, 153, 1993.

Sazhin, S. S., Hayakawa, M., and Bullough, K., Whistler diagnostics of magnetospheric parameters: a review, *Ann. Geophys.*, 10, 293, 1992.

Sazhin, S. S., Smith, A. J., and Sazhin, E. M., Can magnetospheric electron temperature be inferred from whistler dispersion measurements?, *Ann. Geophys.*, 8, 273, 1990.

Scarabucci, R. R., Interpretation of VLF signals observed in the OGO-4 satellite, Tech. Rep. 3418-2, Radioscience Laboratory, Stanford Electronics Laboratory, Stanford University, Stanford, CA, 1969.

Shawhan, S. D., The use of multiple receivers to measure the wave characteristics of very low frequency noise in space, *Space Sci. Rev.*, 10, 689, 1970.

Shawhan, S. D. and Gurnett, D. A., Fractional concentration of hydrogen ions in the ionosphere from VLF proton whistler measurement, *J. Geophys. Res.*, 71, 46, 1966.

Shimakura, S., Hayakawa, M., Lefeuvre, F., and Lagoutte, D., On the estimation of wave energy distribution of magnetospheric VLF waves at the ionospheric base with ground-based multiple electromagnetic field components, *J. Geomagn. Geoelectr.*, 44, 573, 1992.

Singh, B., On the ground observation of whistlers at low latitudes, *J. Geophys. Res.*, 81, 2429, 1976.

Smith, A. J. and Carpenter, D. L., Echoing mixed path whistlers near the dawn plasmapause, observed by direction-finding receivers at two Antarctic stations, *J. Atmos. Terr. Phys.*, 44, 973, 1982.

Smith, A. J. and Cotton, P. D., The Trimpi effect in Antarctica: observations and models, *J. Atmos. Terr. Phys.*, 52, 341, 1990.

Smith, A. J., Smith, I. D., and Bullough, K., Methods of determining whistler nose-frequency and minimum group delay, *J. Atmos. Terr. Phys.*, 37, 1179, 1975.

Smith, A. J., Smith, I., Deeley, A. M., and Bullough, K., A semi-automated whistler analyser, *J. Atmos. Terr. Phys.*, 41, 578, 1979.

Smith, R. L., Propagation characteristics of whistlers trapped in field-aligned columns of enhanced ionization, *J. Geophys. Res.*, 66, 3699, 1961.

Smith, R. L. and Angerami, J. J., Magnetospheric properties deduced from OGO-1 observations of ducted and nonducted whistlers, *J. Geophys. Res.*, 73, 1, 1968.

Smith, R. L. and Carpenter, D. L., Extension of nose whistler analysis, *J. Geophys. Res.*, 71, 3755, 1966.

Smith, R. L., Helliwell, R. A., and Yabroff, I. W., A theory of trapping of whistlers in field-aligned columns of enhanced ionization, *J. Geophys. Res.*, 65, 815, 1960.

Stix, T. H., *The Theory of Plasma Waves*, McGraw-Hill, New York, 1962.

Storey, L. R. O., An investigation of whistling atmospherics, *Philos. Trans. R. Soc., London A*, 246, 113, 1953.

Strangeways, H. J., Systematic errors in VLF direction-finding of whistler ducts-I, *J. Atmos. Terr. Phys.*, 42, 995, 1980.

Strangeways, H. J. and Rycroft, M. J., Systematic errors in VLF direction-finding of whistler ducts-II, *J. Atmos. Terr. Phys.*, 42, 1009, 1980.

Stuart, G. F., Systematic errors in whistler extrapolation. 2. Comparison of methods, *J. Atmos. Terr. Phys.*, 39, 427, 1977.

Tanaka, Y., Hayakawa, M., and Nishino, M., Study of auroral VLF hiss observed at Syowa Station, Antarctica, *Mem. Natl. Inst. Polar Res., Ser. A*, No. 13, p. 58, 1976.

Tarcsai, G., Routine whistler analysis by means of accurate curve fitting, *J. Atmos. Terr. Phys.*, 37, 1447, 1975.

Thomson, R. J., The formation and lifetime of whistler ducts, *Planet. Space Sci.*, 26, 423, 1978.

Thomson, R. J. and Dowden, R. L., Simultaneous ground and satellite reception of whistlers. 1. Ducted whistlers, *J. Atmos. Terr. Phys.*, 39, 869, 1977a.

Thomson, R. J. and Dowden, R. L., Simultaneous ground and satellite reception of whistlers. 2. PL whistlers, *J. Atmos. Terr. Phys.*, 39, 879, 1977b.

Tixier, M., Charcosset, G., Corcuff, Y., and Okada, T., Propagation modes of whistlers received aboard satellites over Europe, *Ann. Geophys.* 2, 211, 1984.

Tsuruda, K., Penetration and reflection of VLF waves through the ionosphere: fullwave calculations with ground effect, *J. Atmos. Terr. Phys.*, 35, 1377, 1973.

Tsuruda, K. and Hayashi, K., Direction finding technique for elliptically polarized VLF electromagnetic waves and its application to the low latitude whistlers, *J. Atmos. Terr. Phys.*, 37, 1193, 1975.

Wait, J. R., *Electromagnetic Waves in Stratified Media*, Pergamon Press, Elmsford, New York, 1970.

Walker, A. D. M., The theory of whistler propagation, *Rev. Geophys. Space Phys.*, 14, 629, 1976.

Walker, A. D. M., Formation of whistler ducts, *Planet. Space Sci.*, 26, 375, 1978.

Walter, F. and Angerami, J. J., Nonducted mode of VLF propagation between conjugate hemispheres: observations on OGO's-2 and -4 of the walking-trace whistler and of Doppler shifts in fixed frequency transmissions, *J. Geophys. Res.*, 74, 6352, 1969.

Watts, J. M., Direction finding on whistlers, *J. Geophys. Res.*, 64, 2029, 1959.

Chapter 8

Ionospheric F-Region Storms

Gerd W. Prölss

CONTENTS

1. INTRODUCTION

Perturbations of the ionosphere in association with the increased dissipation of solar wind energy remains one of the most challenging topics of upper atmosphere physics. Such disturbances have a profound influence on the global morphology of the ionosphere and constitute an important link in the complex chain of solar-terrestrial relations. They are also of practical interest since they may severely degrade transionospheric radio communications.

The disturbance effects observed on these occasions are numerous, and all ionospheric parameters are affected. Here we are concerned exclusively with perturbations of the ionization density at F-region heights. These changes will be called "ionospheric disturbances" or "ionospheric storms''. The former term is used because the temporal and spatial variations involved are partly irregular and because changes at subauroral latitudes are essentially of a transient nature. The latter term is used to indicate that the observed perturbations may be rather severe. In the following sections both terms will be used, depending on the situation.

Since they were first discovered (Anderson, 1928; Hafstad and Tuve, 1929; Appleton and Ingram, 1935; Kirby et al., 1935)*, ionospheric disturbances and storms have been studied extensively. Even a cursory inspection of some of the review literature in this field (Obayashi, 1964; Matuura, 1972; Prölss, 1980; Danilov and Morozova, 1985; Rishbeth, 1986; Prölss et al., 1991; Abdu, 1991) yields more than 250 papers on this subject. In spite of this immense effort, many properties of ionospheric disturbances remain incompletely documented and poorly understood, testifying to the complexity of this phenomenon. This chapter attempts to summarize what is presently known about the physics of this effect. The material is organized in the following way. Section 2 briefly describes the energy source responsible for the disturbance effects and its parameterization by geomagnetic activity indices. It is followed by a discussion of ionospheric storm effects at middle latitudes (Section 3), which begins with a brief description of the basic storm morphology and a summary of possible disturbance mechanisms. It continues with the outline of a more recent thermospheric-ionospheric storm model and with a compilation of ionospheric storm simulation studies. Special conditions are encountered at equatorial and polar latitudes. Therefore, separate sections deal with storm effects in these regions (Sections 4 and 5). In spite of considerable progress made in our understanding of ionospheric storms, many open questions remain. Some of these are summarized in Section 6.

2. SOLAR WIND ENERGY SOURCE

It is well known that solar energy is transferred to the Earth's upper atmosphere in two very different ways. First, solar radiation in the UV range is directly absorbed in the sunlit upper atmosphere. There it is responsible for the formation of the undisturbed ionosphere. Secondly, solar wind energy is captured by the magnetosphere, transformed, and dissipated in the polar upper atmosphere. It is this latter energy source which is responsible for the ionospheric disturbance effects to be discussed in this contribution.

There exist numerous excellent review articles and books on the formation of the ionosphere by solar radiation (Rishbeth and Garriot, 1969; Ratcliffe, 1972; Banks and Kockarts, 1973; Bauer, 1973; Giraud and Petit, 1978; Schunk, 1983; Ivanov-Kholodny and Mikhailov, 1986; Rees, 1989; Kelley, 1989; Hargreaves, 1992 and many more). Therefore, this process and the associated energy source do not need to be discussed here. It is sufficient to recall some basic facts about the F region of the undisturbed ionosphere as they are summarized in Table 8.2.1. Note that we consider the F region to constitute that part of the ionosphere where atomic oxygen ions dominate. The upper and lower boundaries of this region vary considerably with solar and geophysical conditions. Therefore, the altitudes listed should be considered only as approximate values. The same applies to the density and height of the ionization maximum and the temperatures.

As for the disturbed ionosphere, it is the intensity of the solar wind energy source which is of primary interest. This intensity is nearly exclusively described by means of geomagnetic activity indices. First, these indices correlate reasonably well with the observed dissipation effects, and secondly, these indices are easily accessible. It should be kept in mind, however, that these indicators are only indirectly related to the actual source strength. For example, the Pedersen currents responsible for the electrodynamic heating of the polar atmosphere are associated with a torroidal magnetic field which is virtually undetectable on the ground (Volland, 1984). What are actually observed in the heating region are magnetic perturbations produced by ionospheric

---

*   Please note that all references cited should be considered as representative examples only.

Table 8.2.1   **Typical Daytime Ionospheric F-Region Parameters at Middle Latitudes and for Low Solar Activity**[a]

Height interval	$\approx$170–1800 km
Dominant ion species	O^+ (O_2^+, NO^+, H^+)
Dominant neutral species	O (N_2, O_2)
Maximum ionization density and height	$\approx 4 \times 10^{11} m^{-3}$ at 280 km
Neutral density (280 km)	$\approx 9 \times 10^{14} m^{-3}$
Production	Photoionization of O by EUV radiation (ca. $17 \le \lambda \le 91$ nm)
Loss	Charge transfer reactions with N_2 and O_2 and subsequent dissociative recombination
Transport	Ambipolar diffusion, neutral wind, and electric field induced drifts
Temperatures	$T_{ion} \simeq T_{neutral} \simeq 950$ K
	$T_{electron} \simeq 2200$ K

[a]   12 h local time, 45° latitude, F10.7 = 84. (Köhnlein, 1989a,b.)

Table 8.2.2   **Planetary Geomagnetic Activity Indices**

Index	Monitored current system	Type	Magnetic latitude coverage	Number of stations	Time resolution	Availability
Kp	Polar electrojet, field-	Average disturbance	45–62	11	3 h	Prompt
ap	aligned currents, ring	range			3 h	
Ap	current				24 h	
Km	Same	Same	28–59	22	Same	Same
am						
Am						
AU	Eastward	Maximum positive	60–71	10–12	1 min	Delayed
AL	Westward Electrojet	deviation, maximum negative deviation with respect to station ensemble			(2.5 min)	
AE	Combined	AU-AL				
Dst	Ring current, field-aligned currents, magnetopause currents	Average deviation from mean	21–33	4	1 h	Delayed

Hall currents. Although both Pedersen and Hall currents are driven by the same electric field, their respective conductivities are only loosely coupled (Baumjohann, 1983; Brekke, 1983). At lower latitudes, magnetic ground perturbations are primarily due to the dissipation-free ring current encircling the Earth at a geocentric distance of several Earth radii. Additional magnetic disturbance effects are produced by magnetopause currents and currents which flow along magnetic field lines and connect different regions of the magnetosphere to the polar ionosphere.

All these current systems are presumably related to each other, as has been discussed, for example, by Vasyliunas (1972), Nisbet et al. (1978), Kamide et al. (1981), Harel et al. (1981), Stern (1983), Volland (1984), Akasofu (1984), and many others. For this reason, geomagnetic indices based on different current systems (and even mixtures of these current systems) have been successfully used to describe the intensity of the solar wind energy source. Some of the more commonly used planetary indices are listed in Table 8.2.2 together with some pertinent information. At present, the auroral electrojet indices appear to be best suited to monitor the energy injection at high latitudes. They have a superior time resolution, a fairly well-defined physical meaning, and a relatively close connection to the current system actually responsible for the electrodynamic heating at polar latitudes. Note that through changes in the conductivity, these indices will also depend on the amount of energy deposited by precipitating particles (Ahn et al., 1983). Limitations of these indices arise for different reasons and have been discussed, for example, by Wei et al. (1985) and Feldstein (1992).

The major assets of the 3-h range indices are that they are promptly available, easily accessible, and that they indicate the degree of magnetospheric activity on a level which is sufficient for many purposes. Of these indices, Km is probably the better one, and Kp certainly the more popular one. Both represent a quasi-logarithmic measure of the disturbance range. Frequently, a linear disturbance measure is desired (especially if averaging is required), and in this case the ap and am indices are preferred. Kp and Km values are converted to ap and am values, respectively, using conversion tables (Svalgaard, 1976; Mayaud, 1980). These latter studies also discuss the conversion of Kp into Km indices. The daily indices Ap and Am are obtained by simply averaging the corresponding eight consecutive ap and am indices, respectively.

Whereas much of the absorbed solar wind energy is injected into the ring current ($\approx 10^{16}$ J), only a small fraction of this energy is eventually deposited into the Earth's upper atmosphere (Noël and Prölss, 1993). Nevertheless, it is common practice to use the Dst index to define the onset, the different phases, and the general magnitude of major energy dissipation events. This index is less suited for correlation studies because of the long recovery time constant of the ring current.

Besides geomagnetic indices, solar wind parameters can be used directly to indicate the intensity of the solar wind energy source. Although good results are obtained at times (Burkard, 1970; Prölss, 1985; Sastri et al., 1992), difficulties arise from the limited availability of these parameters.

A more sophisticated specification of the solar wind energy source is required when it comes to simulating ionospheric storms by means of Thermosphere-Ionosphere General Circulation Models (TIGCMs). Here, the ionospheric electric field potential and the particle precipitation rate are used as suitable indicators of the strength, spatial distribution, and temporal variation of the energy dissipation. The electric field potential and the particle precipitation pattern may be derived from models (Volland, 1978; Heelis et al., 1982; Spiro et al., 1982; Hardy et al., 1985; Foster et al., 1986; Heppner and Maynard, 1987) or from a self-consistent combination of measurements using the AMIE technique (Richmond and Kamide, 1988).

An essential characteristic of the solar wind energy source is its highly irregular temporal behavior. It is for this reason that solar wind energy dissipation is considered a disturbance rather than a regular feature. Accordingly, terms like "magnetic (sub)storm" and "magnetic activity" are used to denote periods of increased energy deposition. Conservative estimates indicate that within an hour or two the global energy injection rate may increase by a factor of five or more.

These large, irregular changes also contain smaller, systematic variations. Regular variations are due to systematic changes in the conductivity of the polar ionosphere and also to periodicities in the solar wind properties. Conductivity-associated changes include diurnal, seasonal, and solar cycle variations. Diurnal and universal time-dependent variations arise from the fact that the heating oval and the polar cap regions rotate with the geomagnetic pole around the geographic pole. This way various fractions of the polar oval and polar cap regions are exposed to solar EUV radiation, which produces regions of high conductivity (Rees et al., 1987). Similarly, higher conductivities and heat input rates are expected during summertime (Foster et al., 1983) and also during solar cycle maximum conditions.

Quasiregular variations are also introduced by the solar wind. A solar-rotation (27-day) periodicity is frequently observed in the magnetospheric activity and is attributed to high speed solar wind streams originating in long-lived coronal holes. Likewise, solar cycle variations are evident and can be decomposed into coronal-mass-ejection- and coronal-hole-associated components (Feynman, 1982).

Geomagnetic disturbances (D) have often been analyzed in terms of storm time (Dst) and disturbance local time (DS) variations (Chapman and Bartels, 1962, p. 272):

$$D(ST,LT) = Dst(ST) + DS(ST,LT) + \Delta$$

Here, ST and LT denote the storm time and the local time, respectively, and Δ is an irregular contribution assumed to be small enough to be neglected. Later, this analyzing scheme was also adopted in statistical studies of ionospheric storms (Martyn, 1953a,b; Lewis and McIntosh, 1953; Matsushita, 1959, 1963; Rajaram and Rastogi, 1969, 1970; Mendillo et al., 1972; Hearn, 1974; Huang et al., 1974; Basu et al., 1975; Hargreaves and Bagenal, 1977; Inoue et al., 1978; Jakowski et al., 1990; Balan and Rao, 1990). In these studies it is assumed that the beginning of the disturbance is defined by the onset of the magnetic storm. Now the onset of a magnetic storm is an all but well-defined concept. In the past, the so-called *sudden storm commencement* (SSC) often served as a reference time. By this, we mean an impulse-like disturbance of the magnetic field which is followed by a magnetic storm within 24 h (Joselyn and Tsurutani, 1990; Kamide and Joselyn, 1991). Note that these impulse-like disturbances (generally classified as *sudden impulses* or SI) are not associated with any significant energy deposition, that they may occur in groups, and that they are also observed *after* the onset of a magnetic storm, as indicated, for example, by the decrease in the Dst index (Akasofu, 1970). Evidently, SSCs are a poor choice when defining the beginning of a magnetic storm. Not only are many magnetic storms not preceded by sudden impulses, also the time resolution of this time marker — which according to the definition above is 24 h — is much too low. Accordingly, all statistical studies (magnetic or ionospheric) based on SSC-defined onset times should be considered with due caution.

Alternatively, the main phase onset (MPO) has been used to fix the beginning of magnetic and ionospheric storms (Thomas and Venables, 1966; Jones, 1971a; Kane, 1973a,b). Here, the onset of the growth phase of the magnetospheric ring current (as indicated by the decrease in the horizontal magnetic field component at a low latitude station or by the decrease in the Dst index) serves as a reference time. This is certainly a better choice because the growth of the ring current is related, directly or indirectly, to the energy dissipation at polar latitudes (Davis and Parthasarathy, 1967; Feldstein, 1992). In practice, however, this definition may also lead to some difficulties. Thus, the main phase decrease may proceed gradually, or in steps, or may follow a series of smaller perturbations. Fixing the baseline value for prestorm conditions also may not be an easy task. All these problems will render a determination of the exact onset time difficult.

A third way to fix the beginning of a magnetic-ionospheric disturbance is via the onset of major substorm activity (Akasofu, 1970; Tanaka, 1986; Prölss, 1993a,b). Naturally, the timing of a magnetospheric substorm with its transient energy injection can be determined with much greater precision. The auroral electrojet indices represent a convenient (albeit, far from perfect) means to monitor the occurrence and intensity of these events. As mentioned above, this index is also closely related to the energy dissipation in the polar upper atmosphere which is mainly responsible for ionospheric storm effects. Difficulties arise again if the substorm activity builds up slowly or in steps or if the auroral electrojet indices fail to indicate the true strength of a disturbance event.

3. IONOSPHERIC STORMS AT MIDDLE LATITUDES

3.1. BASIC MORPHOLOGY

Although all parameters are affected during ionospheric storms, only changes in the ionization density and, to a lesser degree, in the ionization temperature have been studied in greater detail (Matuura, 1972 and the references therein for publications before 1970. See also, Evans, 1970a–c; Mendillo et al., 1970; Rajaram and Rastogi, 1970; Thomas, 1970; Jones, 1971a; Klobuchar et al., 1971; Nelson and Cogger, 1971; Rajaram et al., 1971; Somayajulu et al., 1971; Katz and Papagiannis, 1972; Mendillo et al., 1972; Obayashi, 1972; Rush, 1972; Tulunay, 1972; Fatkullin 1972, 1973; Gondhalekar, 1973; Kane, 1973a,b; Low and Roelofs, 1973; Mendillo, 1973; Spurling and Jones, 1973; Hearn, 1974; Huang et al., 1974; Mendillo and Klobuchar, 1974; Park, 1974; Prölss and von Zahn, 1974; Schödel et al, 1974; Brace et al., 1974; Stening and Chance, 1974; Basu et al., 1975; Kane, 1975; Lanzerotti et al., 1975; Mendillo and Klobuchar, 1975; Prölss and Najita, 1975; Spurling and Jones, 1975; Papagiannis et al., 1975; Marubashi et al., 1976; Matsushita, 1976; Gogoshev et al., 1976; Park and Meng, 1976; Soicher, 1976; Hargreaves and Bagenal, 1977; Paul et al., 1977; Prölss, 1977; Sastri and Titheridge, 1977; Inoue et al., 1978; Essex, 1979; Buonsanto et al., 1979; Miller et al., 1979; Tanaka, 1979; Prölss, 1980; Goel and Rao, 1980; Essex et al., 1981; Huang, 1985; Danilov and Morozova, 1985; Wrenn et al., 1987; Kilifarska, 1988; Leitinger et al., 1988; Titheridge and Buonsanto, 1988; Oliver et al., 1988; Forbes et al., 1988; Ezquer and de Adler, 1989; Jakowski et al., 1990; Balan and Rao, 1990; Walker et al., 1991; Oliver et al., 1991; Prölss et al., 1991; Kilifarska et al., 1991; Buonsanto et al., 1992; Rich and Deng, 1992; Yeh et al., 1992; Förster et al., 1992; Jakowski et al., 1992; Lal, 1992; Foster, 1993; Prölss, 1993a,b; Cander, 1993; Pi et al., 1993; Skoblin and Förster, 1993; Yeh et al., 1994 and references therein.). This is because the electron density is the most easily accessible parameter of the ionosphere. Even with this limitation and in spite of the great effort documented in the above studies, no generally accepted concept of the morphology and origin of ionospheric storms has emerged. This is partly due to the great variability of this phenomenon and also to the many different processes at work. Basically, the ionization density can either increase or decrease during disturbed conditions. Traditionally, these changes are denoted as *positive* and *negative* ionospheric storms, respectively. Both disturbance effects are illustrated in Figure 8.3.1. In association with a magnetic storm (indicated here by the sudden and large increase in the ap index), the maximum electron density of the ionosphere above Pt. Arguello exhibits a large enhancement (i.e., positive storm effects), whereas the ionosphere above Brisbane shows a considerable depletion of ionization (i.e., negative storm effects). Because the total electron content of the ionosphere exhibits very similar variations (Taylor, 1961; Mendillo and Klobuchar, 1974), the observed changes in the maximum electron density cannot simply be attributed to a distortion of the height distribution. This is documented in Figure 8.3.2 for a severe negative ionospheric storm. Here, the ionization density in the entire peak region is drastically reduced. In fact, the F2-layer density is observed to drop below that of the F1 region, a situation which has become known as *G condition* (King, 1962).

Another feature of interest documented in Figure 8.3.2 is the changeover from negative to positive storm effects in the upper F region (see also Figure 8.4.4). Evidently, the decrease in the ionization density near the F-region peak is associated with a substantial increase in the plasma temperature and scale height. This anticorrelation between the plasma density and the plasma temperature is a regularly observed and well-understood phenomenon (Prölss et al., 1975; Schunk and Nagy, 1978). In any case, such a changeover from one type of storm effect to the other makes a systematic description of the disturbed topside ionosphere rather complicated (Reddy et al., 1967; Lakshmi and Reddy, 1970; Fatkullin, 1972; Gondhalekar, 1973).

The completely different disturbance behavior of the ionosphere above two comparable locations (see Figure 8.3.1) indicates that the global morphology of ionospheric storms is complex. Thus strong longitudinal and latitudinal asymmetries are frequently observed. In addition, the distribution of storm effects may vary considerably from one event to the next (Kane, 1973a,b; 1975). In spite of this seemingly irregular behavior, certain systematic trends can be recognized. For example, the north-south asymmetry illustrated in Figure 8.3.1 may be attributed to seasonal variations of ionospheric storm effects, with negative storms extending to much lower latitudes in the summer hemisphere. This and similar kinds of systematic changes will be discussed further in Section 3.5.

The distinction between positive and negative ionospheric storms represents the simplest possible classification. A more sophisticated description is based on the different forms and origins of storm effects. For example, at subauroral latitudes and during winter conditions positive storms may be caused by: (1) traveling atmospheric disturbances; (2) large-scale changes in the wind circulation; and (3) an expansion of the polar ionization enhancement. Negative ionospheric storms may be caused by (1) changes in the neutral gas composition and (2) the equatorward displacement of the trough region (Prölss et al., 1991). Figure 8.3.3 summarizes these different classes of ionospheric storm effects, emphasizing their local time behavior. It is understood that this list is by no means complete, and other processes are expected to operate during disturbed

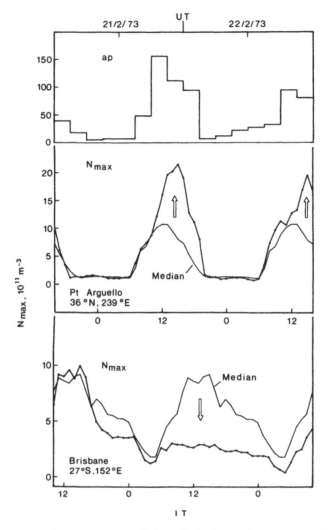

Figure 8.3.1 Magnetic storm-induced changes in the ionospheric electron density. Measured F2-layer maximum electron densities (N_{max}, heavy dotted lines) obtained at two mid-latitude stations during the magnetic disturbance event of February 21 and 22 of 1973 (upper panel) illustrate positive and negative ionospheric storm effects (middle and lower panels). The monthly median serves as a quiet-time reference (thin lines). (From Prölss, G.W., *Rev. Geophys. Space Phys.*, 18, 183–202, 1980. Copyright by the American Geophysical Union. With permission.)

conditions. It is also clear that these effects may be superimposed on one another and that they may partly cancel each other. Therefore, great care should be exercised when averaging storm data.

3.2.   DISTURBANCE MECHANISMS

The point of departure for any explanation of storm-induced changes in the ionospheric plasma density is the continuity equation (Rishbeth, 1986)

$$\frac{\partial N}{\partial t} = q - l(N) - \text{div}(N \cdot \bar{u}) \tag{3.1}$$

Here, N denotes the ionization density (in our case the atomic oxygen ion density or electron density), q the rate of production, l(N) the rate of loss, and $\bar{u}$ the transport velocity. In principle,

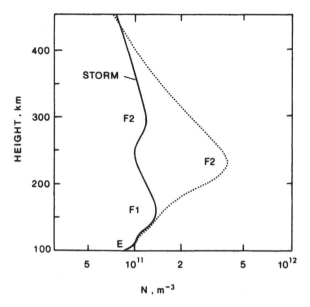

Figure 8.3.2 Height distribution of the electron density during a negative ionospheric storm. The storm profile (solid line) was obtained on April 18, 1965. The densities above the F2 peak were deduced from Alouette I ionograms. The densities below the F2 peak were obtained for ionograms recorded by the St. John's ground-based ionosonde. For comparison, an undisturbed height profile is also shown (dotted line). It is based on data obtained on April 14, 17, and 21 of 1965. (From Norton, R.B., *Proc. IEEE,* 57, 1147–1149, 1969. © 1969 IEEE. With permission.)

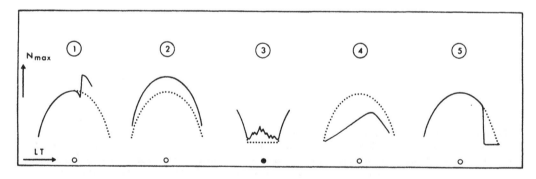

Figure 8.3.3 Classification of ionospheric storm effects observed in the winter hemisphere at subauroral latitudes. In each case, the dotted line indicates the "normal" behavior and the solid line the storm-time variation. The disturbance effects are ordered in a local time frame with an open circle marking local noon and a solid circle marking local midnight. The following five disturbance mechanisms are distinguished: (1) traveling atmospheric disturbances; (2) changes in the large-scale wind circulation; (3) expansion of the polar ionization enhancement; (4) neutral composition changes; and (5) displacement of the ionospheric trough region. (From Prölss, G.W., *J. Geophys. Res.,* 96, 1275–1288, 1991. Copyright by the American Geophysical Union. With permission.)

perturbations of any of the three terms on the right-hand side of this equation may be responsible for ionospheric storm effects and, indeed, all terms and combinations thereof have been considered in possible explanations. A compilation of these explanations, as far as they have become known to the author, is given in Table 8.3.1. Fifteen of the mechanisms refer to positive (P) and sixteen to negative (N) storm effects.

In the meantime, some of these explanations have become obsolete, either because they rest on incorrect assumptions (e.g., mechanisms P1, P3, P6, P7, N2, N3, N6, and N12) or because they were shown to be insufficient (e.g., mechanisms P2, N1, N4, N8, and N10). Others may

Table 8.3.1 **Disturbance Mechanisms**

Positive Phase (P)

1. Height-dependent downward drift due to westward-directed electric fields (Martyn, 1953a,b)
2. Height-independent downward drift (westward-directed electric fields) in the presence of a positive atmospheric temperature gradient (Kamiyama, 1956)
3. Dumping of radiation belt electrons into the upper atmosphere (Muldrew, 1963)
4. Ionization inflow from plasmasphere due to magnetospheric compression and meridional plasma drift (Piddington, 1964)
5. Increase in atomic oxygen density at lower latitudes due to storm-induced convection (Duncan, 1969)
6. Ionization "pile up" in dusk sector caused by a reduction of the corotation electric field (Papagiannis et al., 1971)
7. Upward drift due to internal electric polarization fields (Rajaram et al., 1971)
8. Ionization inflow from nightside plasmasphere caused by westward-directed electric fields (Park, 1971)
9. Upward drift due to equatorward-directed thermospheric winds associated with changes in the large scale circulation (Jones and Rishbeth, 1971)
10. Upward drift due to eastward-directed electric fields (Tanaka and Hirao, 1973)
11. Soft electron precipitation from the storm ring current (Zevakina et al., 1974)
12. Upward drift due to equatorward-directed winds associated with traveling atmospheric disturbances (Prölss and Jung, 1978)
13. Upward drift due to zonal thermospheric winds at locations with a significant magnetic field declination (Kilifarska, 1988)
14. Ionization convergence due to negative vertical gradients in meridional thermospheric winds (Burnside et al., 1991)
15. Convection of high density plasma into evening sector (Foster, 1993)

Negative Phase (N)

1. Thermal expansion of neutral atmosphere (Appleton and Ingram, 1935; Kirby et al., 1935; Matuura, 1963)
2. Localized zonal drifts due to particle precipitation induced vertical polarization fields (Eckersley, 1942)
3. Height-dependent upward drift due to eastward-directed electric fields (Martyn, 1953a,b)
4. Height-independent upward drift (eastward-directed electric fields) in the presence of a positive atmospheric temperature gradient (Kamiyama, 1956)
5. Increase in molecular oxygen density due to turbulent mixing of thermosphere (Seaton, 1956)
6. Lack of ionization production by radiation belt electrons after their depletion (Muldrew, 1963)
7. Acceleration of chemical loss reactions by an increase in the neutral gas temperature (Yonezawa, 1963)
8. Ionization outflow due to refilling of plasmasphere and hydrodynamic pumping (Piddington, 1964)
9. Acceleration of chemical loss reactions by vibrationally excited reactants (e.g., N_2 excitation by high electron temperatures and/or particle precipitation, Thomas and Norton, 1966)
10. Ionization outflow due to inflation of magnetosphere and energization of thermal plasma in ring current region (Bauer and Krishnamurthy, 1968)
11. Decrease of atomic oxygen density at turbopause altitudes (Chandra and Herman, 1969)
12. Downward drift due to internal electric polarization fields (Rajaram et al., 1971)
13. Equatorward motion/expansion of trough region (Mendillo et al., 1974)
14. Weakening of the corotation drift at the dusk terminator (Anderson, 1976)
15. Ionization divergence due to a positive vertical gradient in the meridional thermospheric wind (Taieb and Poinsard, 1984)
16. Downward drift due to zonal thermospheric winds at locations with a significant magnetic field declination (Kilifarska, 1988)

become important only under certain conditions but are not considered to be the main causes of ionospheric storms at middle latitudes (e.g., mechanisms P4, P13, N7, N9, N13, N14, and N16). Finally, there are a number of mechanisms whose significance has yet to be established (e.g., mechanisms P5, P8, P11, P14, and N15).

In what follows, a disturbance scenario will be described which is based on a close coupling between thermospheric and ionospheric storm effects. In this model, daytime positive storm effects are attributed to traveling atmospheric disturbances and changes in the large-scale wind circulation (mechanisms P9 and P12), and negative storm effects are caused by changes in the neutral gas composition (modified versions of mechanisms N5 and N11). Because positive storms are frequently also attributed to electric field perturbations (e.g., P10 and P15), these mechanisms will be considered in a separate section.

3.3. NEGATIVE IONOSPHERIC STORMS CAUSED BY NEUTRAL COMPOSITION CHANGES

The dissipation of solar wind energy continuously affects the density structure of the polar upper atmosphere. Even during magnetically quiet conditions, this energy addition is sufficiently large to generate a permanent disturbance zone. Characteristic changes observed in this region are an increase in the heavier gases and a decrease in the lighter gases (Prölss et al., 1988).

During disturbed conditions these composition changes are no longer restricted to the polar region but expand toward middle latitudes. This is illustrated schematically in Figure 8.3.4. Coming from the equator, a polar-orbiting satellite will penetrate into the composition disturbance zone and survey its latitudinal structure. A latitudinal profile recorded this way is reproduced in the middle panel of Figure 8.3.5. Here, relative changes in the molecular nitrogen and atomic oxygen densities are plotted as functions of magnetic latitude. R(n) is defined as the stormtime value of the gas constituent n divided by the respective quiet-time value, and R(n) = 1 serves as a reference, meaning no change with respect to quiet times. Note that the densities have been adjusted to a common pressure level of 8×10^{-6} Pa. During quiet conditions, this corresponds to a geodetic height of about 290 km. A constant pressure height (rather than a constant geodetic height) was chosen because ionospheric effects are best described in this reference frame (Rishbeth and Edwards, 1989).

The most striking feature of this presentation is the well-developed composition disturbance zone extending from high to middle latitudes. It is marked by a significant increase in the molecular nitrogen density and a concurrent depletion of the atomic oxygen density. Both density changes have important implications for the ionosphere. Consider, for example, the production of ionization. At F2-region heights it is based on the photoionization of atomic oxygen

$$O + \text{Solar radiation } (\lambda \leq 91 \text{ nm}) \rightarrow O^+ + e^-$$

If it is assumed, to a good approximation, that the atmosphere is optically thin in this region, then the production rate is directly proportional to the atomic oxygen density [O]

$$q = J_o [O] \tag{3.2}$$

where J_o denotes the ionization frequency of this constituent. Consequently, a decrease in the atomic oxygen density will directly reduce the ionization production rate.

Next, consider the loss of ionization. At F-region heights it is primarily due to charge transfer reactions of the type

$$O^+ + N_2 \xrightarrow{k_1} NO^+ + N \tag{3.3}$$

$$O^+ + O_2 \xrightarrow{k_2} O_2^+ + O \tag{3.4}$$

where k_1 and k_2 are the associated reaction rate constants. The resulting molecular ions NO^+ and O_2^+ are quickly destroyed by dissociative recombination. Therefore, the loss rate reduces to the simple form

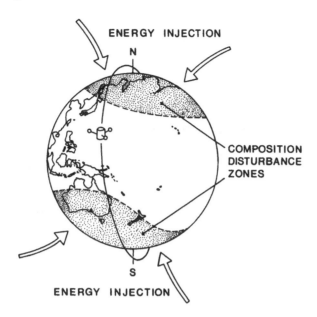

Figure 8.3.4 Energy injection and the formation of neutral composition disturbance zones during a magnetic storm. A polar-orbiting satellite surveys the latitudinal structure of the atmospheric perturbation and provides measurements similar to those presented in Figure 8.3.5. (From Prölss, G.W., *Rev. Geophys. Space Phys.*, 18, 183–202, 1980. Copyright by the American Geophysical Union. With permission.)

$$l(N) = \beta \, [O^+] \tag{3.5}$$

where β depends on the density of the molecular gases N_2 and O_2,

$$\beta = k_1[N_2] + k_2[O_2] \tag{3.6}$$

Consequently, an increase in these gases will directly increase the loss rate of ionization.

From the above consideration it is clear that the decrease in the atomic oxygen density and the increase in the molecular nitrogen density which are illustrated in Figure 8.3.5 combine to reduce the ionization density at F-region heights. Therefore, any ionosonde station located within the composition disturbance zone should observe negative storm effects, and this is indeed the case. The lower part of Figure 8.3.5 shows the local time variation of the maximum electron density as observed at two mid-latitude stations, Pt. Arguello and Boulder. Both are located within the neutral composition disturbance zone, and both exhibit pronounced negative storm effects. This is by no means a special case, and similar correlations between neutral composition changes and negative ionospheric storm effects have been observed on numerous occasions (Prölss and von Zahn, 1974; Chandra and Spencer, 1976; Hedin et al., 1977; Prölss, 1980; Miller et al., 1984; Deng Wei and Förster, 1989; Mikhailov et al., 1992; Skoblin and Förster, 1993).

A simple estimate confirms that the measured composition changes are also sufficient to explain the observed storm effects. Assume, to a first approximation, that photochemical equilibrium conditions prevail up to the F-region peak. In this case the maximum ionization density is given by

$$N_{max} \simeq q/\beta \tag{3.7}$$

The ratio of the disturbed to the undisturbed value of this quantity, therefore, reduces to the simple form

$$R(N_{max}) \simeq R(O)/R(N_2) = R(O/N_2) \tag{3.8}$$

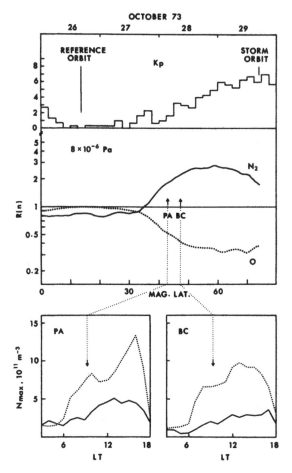

Figure 8.3.5 Magnetic storm-associated changes in the neutral gas composition and negative ionospheric storm effects. The upper panel shows the development of the magnetic activity during a 4-day interval in October 1973. It also indicates the times at which the storm data and the quiet-time reference data were measured. The panel below presents storm-associated changes in the molecular nitrogen and in the atomic oxygen densities. Relative changes are plotted, and R(n) = 1 serves as a reference, meaning no change with respect to quiet times. The data refer to a common pressure level of 8 × 10⁻⁶ Pa (≈290 km during quiet times). Local time and longitude of measurement are approximately 9 SLT and 245°E, respectively. The lower part of the figure shows the local time variations of the maximum electron density of the F2 layer as observed at the two ionosonde stations Boulder, CO. (40.0 geographic latitude, 254.7 geographic longitude E) and Pt. Arguello (35.6 geographic latitude, 239.4 geographic longitude E) whose relative positions with respect to the atmospheric disturbance zone are indicated by arrows. Storm-time data (October 29, solid lines) are compared to quiet-time reference data (October 26, dotted lines). The times of the satellite measurements are indicated by arrows. (From Prölss, G.W., *J. Geomagn. Geoelectr.*, 43, 537–549, 1991. With permission.)

where we have assumed that the molecular oxygen density increases by the same factor as the molecular nitrogen density (both have approximately the same molecular weight). For the specific case illustrated in Figure 8.3.5 we have R(O) ≈ 0.5 and R(N₂) ≈ 2. Accordingly, R(N$_{max}$) should be of the order of 0.25, which is close to the measured values. The validity of Equation 3.8 has been checked using a larger number of storm data. As is demonstrated in Figure 8.3.6, it is reasonably well confirmed by the observations. Note that in the upper thermosphere, R(O/N₂) is approximately independent of height and nearly the same in a geodetic and in a constant pressure coordinate system (Prölss, 1992).

More accurate results can be obtained if the observed composition changes are used as input into a numerical model of the ionosphere. The outcome of such a simulation is illustrated in

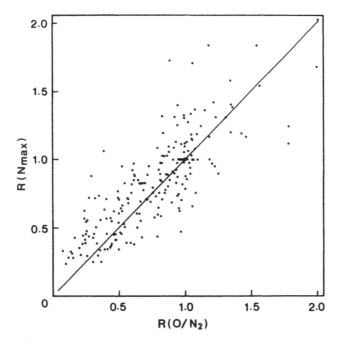

Figure 8.3.6 Connection between neutral composition changes and negative ionospheric storm effects. Relative changes in the maximum electron density of the F2 layer, $R(N_{max})$, are plotted vs. relative changes in the atomic oxygen to molecular nitrogen concentration ratio, $R(O/N_2)$. The data were obtained at 30 different locations during the February and October 1973 storm events. The relationship $R(N_{max}) = R(O/N_2)$ to be expected from a simple estimate is also shown for comparison. (From Prölss, G.W., *Rev. Geophys. Space Phys.,* 18, 183–202, 1980. Copyright by the American Geophysical Union. With permission.)

Figure 8.3.7. The upper part shows the variation of the maximum electron density as calculated for average quiet-time conditions in October 1973 and for the disturbed day October 29, 1973 (see Figure 8.3.5). For comparison, the variations actually observed are shown in the lower part of this figure. Apart from details, the observed and calculated negative storm effects are of comparable magnitude. This confirms that the observed composition changes are entirely sufficient to explain the observed reduction in ionization density, at least during periods of low solar activity.

It should be emphasized that both the increase in molecular gases and the decrease in atomic oxygen density contribute to the decrease in the ionization density. Therefore, the original suggestion by Seaton (1956) that negative storm effects are caused by an increase in the molecular gas O_2 and the suggestion by Chandra and Herman (1969) that negative storm effects are caused by a reduction of the atomic oxygen density are incomplete.

Additional evidence for the close coupling between neutral composition changes and negative ionospheric storm effects is derived from the fact that both phenomena exhibit the same kind of regular variations. These include systematic changes with the disturbance intensity, with magnetic position, with local time, and with season (Prölss, 1980). The latter two variations will be discussed in more detail in Section 3.5.

3.4. POSITIVE IONOSPHERIC STORMS CAUSED BY MERIDIONAL WINDS

It has been repeatedly suggested that positive ionospheric storms are also caused by neutral composition changes (Duncan, 1969; Chandra and Stubbe, 1971; Obayashi and Matuura, 1972; Danilov et al., 1987; Rodger et al., 1989; Rishbeth, 1991). Thus far, however, no firm evidence has been presented to support this supposition for more severely disturbed conditions. Thus the increase in the O/N_2 density ratio frequently observed equatorward of the main disturbance zone

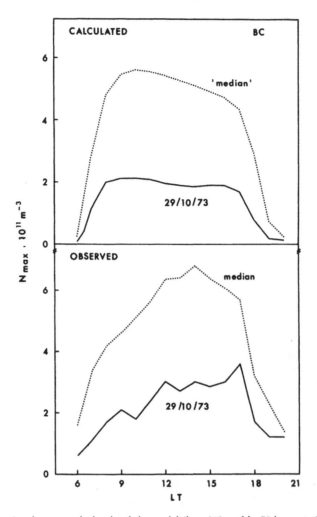

Figure 8.3.7 Comparison between calculated and observed daily variation of the F2-layer maximum electron density during quiet and disturbed conditions. The observed storm data were recorded at Boulder (BC) on October 29, 1973 (lower panel, see also Figure 8.3.5). The monthly median of October 1973 serves as a quiet-time reference. The calculated "median" (upper panel) corresponds to the daily variation on October 15, 1973, and median values of solar and magnetic activity. (From Jung, M.J. and G.W. Prölss, *J. Atmos. Terr. Phys.,* 40, 1347–1350, 1978. With permission.)

is certainly not sufficient to explain the rather large storm effects sometimes observed in this region.

In contrast, the idea that positive ionospheric storms are caused by the *transport* of ionization is well supported by observations. Consider, for example, the data presented in Figure 8.3.8. In response to a sudden energy injection at polar latitudes (as indicated by the AE index), a well-defined positive ionospheric storm develops at middle latitudes. The immediate cause of this density increase is a sudden uplifting of the F layer, as demonstrated in the middle part of this figure. The interplay between layer height and density becomes even more evident if one considers the associated density height profiles. Figure 8.3.9 shows that initially the upward motion of the F layer is associated with a decrease in the ionization density. This decrease is attributed here to a positive height gradient in the drift velocity. It is only after reaching the maximum layer height that an increase in the density is observed (Tanaka and Hirao, 1973; Spurling and Jones, 1976; Prölss and Jung, 1978).

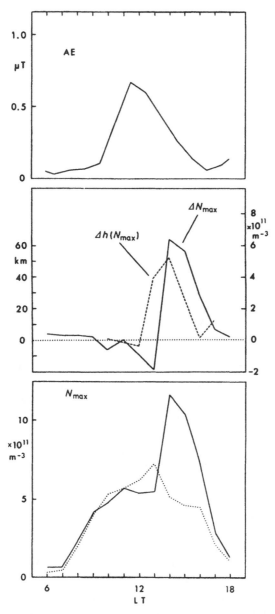

Figure 8.3.8 Short-duration positive ionospheric storm in the local afternoon sector. In response to an isolated burst of substorm activity on January 23, 1973 (hourly averaged AE index, upper panel), the ionosonde at Slough (50° invariant magnetic latitude) first observes an impulse-like uplifting of the layer height ($\Delta h(N_{max})$, middle panel) and subsequently an impulse-like increase in the ionization density (ΔN_{max}, middle panel, and N_{max}, lower panel). Data recorded on January 22 serve as a quiet-time reference (dotted lines). (From Prölss, G.W., *Ann. Geophys.*, 11, 1–9, 1993a. With permission.)

That an increase in layer height will lead to positive storm effects is easily understood if the height dependences of the ionization production and loss rates are considered. According to Equations 3.5 and 3.6, the loss rate is proportional to the molecular nitrogen and molecular oxygen densities. Therefore, it decreases much faster with height than the production rate, which is proportional to the atomic oxygen density (see Equation 3.2). An upward displacement of the F layer will therefore lead to an overall increase in the ionization density. To evaluate the efficiency

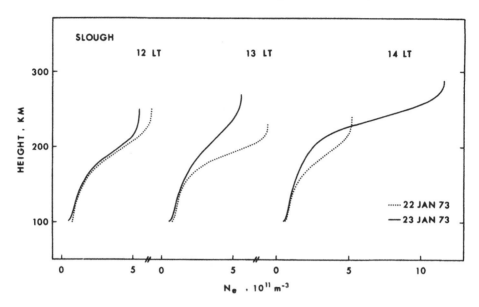

Figure 8.3.9 Changes in the height distribution of the electron density during the positive ionospheric storm of January 23, 1973 (see Figure 8.3.8). The dotted lines describe the undisturbed behavior (January 22, 1973) and the solid lines the stormtime variations.

of this mechanism we assume again that the peak ionization density may be calculated using the photochemical equilibrium formula, Equation 3.7. We further assume that at peak altitudes the thermosphere is nearly isothermal and that the molecular species N_2 and O_2 have roughly the same scale heights. Then the ratio of the disturbed to the undisturbed values of the maximum ionization density is given by

$$R(N_{max}) \simeq e^{\Delta h(1/H_{N_2,O_2} - 1/H_o)} \qquad (3.9)$$

where Δh is the displacement height, and H_{N_2,O_2} and H_o are the pressure scale heights of N_2, O_2 and O, respectively. For the case illustrated in Figure 8.3.8, the increase in layer height is of the order of 50 km. If an exospheric temperature of 900 K (low solar activity, winter conditions) is assumed, Equation 3.9 predicts a factor 2.3 increase in the ionization density. This is close to the observed value. Of course, the situation is highly dynamic, and reliable results are obtained only if time-dependent numerical models of the ionosphere are used in the simulation.

The observed upward motion of the F layer may be caused either by electric fields or by thermospheric winds. The electric field mechanism will be discussed further in Section 3.6. Here, it is assumed that winds are responsible for the upward drift. The specific disturbance scenario envisaged is illustrated in Figure 8.3.10. During a magnetic substorm (indicated by an increase in the AE index) a significant amount of energy is injected into the polar upper atmosphere. This sudden energy addition launches a so-called traveling atmospheric disturbance (TAD). By this we mean a pulse-like atmospheric perturbation formed by a superposition of gravity waves which propagates with high velocity (here, 600 m/s) toward the equator (Testud et al., 1975; Richmond and Matsushita, 1975; Fuller-Rowell and Rees, 1981; Forbes et al., 1987; Crowley et al., 1989a; Fesen et al., 1989; Chang and St. Maurice, 1991; Fuller-Rowell et al., 1994; Millward et al., 1993a). An essential feature of such a TAD is that it carries along equatorward-directed winds of moderate magnitude (here, 150 m/s). At middle latitudes these meridional winds cause an uplifting of the F2 layer which in turn leads to an increase in the ionization density (Prölss and Jung, 1978; Roble et al; 1978; Zevakina et al., 1978; Crowley et al., 1989b; Prölss et al., 1991; Prölss, 1993a,b; Millward et al., 1993b). Note that this mechanism is applicable only to the daytime ionosphere. During night, lack of ionization production will not allow the formation of

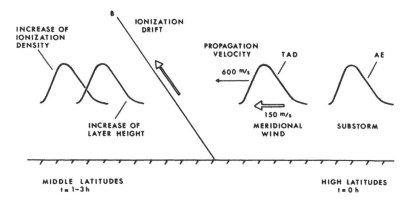

Figure 8.3.10 Daytime positive ionospheric storms caused by traveling atmospheric disturbances. See text for details. The explicit values for the propagation velocity and the meridional wind velocity of the TAD only serve to illustrate the order of magnitude of these quantities. (From Prölss, G.W., *Ann. Geophys.*, 11, 1–9, 1993a. With permission.)

larger positive storm effects. TAD-associated changes in layer height, on the other hand, will be observable irrespective of local time.

Although the above disturbance scenario is fully consistent with the data set presented in Figure 8.3.8, an unambiguous verification is still lacking. Suitable incoherent backscatter radar (ISR) measurements, for example, are scarce and mostly incomplete. All that can be said at the moment is that the few ISR data sets available are compatible with the TAD hypothesis (Evans, 1970a, 1973; Testud et al., 1975; Roble et al., 1978; Goel and Jain, 1988; Hagan, 1988; Buonsanto et al., 1989, 1990; Reddy et al., 1990a). Indirect support for the above model is derived from the properties of positive storm effects. For example, increases in layer height and ionization density are observed with a certain time delay with respect to substorm activity (Becker, 1961; Rüster, 1965; Bowman, 1978; Prölss and Jung, 1978; Hajkowicz, 1990, 1991a,b, 1992). Also, these enhancements do not occur simultaneously but are increasingly delayed with decreasing latitude (Wright, 1961; Chan and Villard, 1962; Thome, 1968; Buonsanto et al., 1989; Prölss et al., 1991; Prölss, 1993b; Walker and Wong, 1993). Lastly, they exhibit dispersion effects which are typical for TAD-associated perturbations.

Besides the TAD-produced positive storms of short duration, longer lasting increases in the ionization density are observed; see Figure 8.3.11. These long-duration events are also associated with an increase in layer height (Alcaydé et al., 1972; Evans, 1973; Prölss et al., 1991; Prölss, 1993a). This is documented in Figure 8.3.11 for the February 21, 1973, event. In response to the continuing energy injection at polar latitudes (AE index), the layer height remains elevated for more than 6 h.

These long-duration upward drifts are attributed again to meridional winds. Thus, it appears less probable that the magnetosphere can maintain a large-scale electric field which extends to low latitudes and lasts for many hours. On the other hand, it is easy to visualize that a continuing energy injection at high latitudes leads to changes in the global wind circulation. Thereby, it is sufficient that the polar high-pressure zone reduces the intensity of the poleward-directed daytime winds. This disturbance scenario is supported by many observations and calculations (Jones and Rishbeth, 1971; Volland and Mayr, 1971; Jones, 1973; Anderson, 1976; Davies and Rüster, 1976; Mayr et al., 1978; Miller et al., 1979; Volland, 1979, 1983; Mazaudier et al., 1985; McCormac et al., 1987; Rees et al., 1987; Salah et al., 1987; Roble et al., 1988; Hernandez and Killeen, 1988; Hagan, 1988; Forbes, 1989; Buonsanto et al., 1990; Codrescu et al., 1992; Miller et al., 1993 and references therein).

There remains the question as to the relationship between positive storm effects produced by TADs and changes in the large-scale wind circulation. Here it is argued that positive storms are *initiated* by TADs. How well developed these disturbance fronts will be in each case depends on

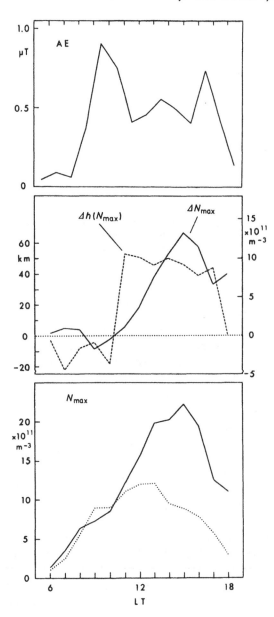

Figure 8.3.11 Long-duration positive ionospheric storm which begins in the local forenoon sector. The form of presentation corresponds to that of Figure 8.3.8. The storm data were recorded at Pt. Arguello (41° invariant magnetic latitude) on February 21, 1973; see also Figure 8.3.1. Measurements obtained on February 18 and 19 serve as a quiet-time reference (dotted lines). (From Prölss, G.W., *Ann Geophys.,* 11, 1–9, 1993a. With permission.)

the specific properties of the excitation source (Hunsucker, 1982). It is only in the wake of these transient perturbations that the large-scale wind system changes, causing longer lasting positive storm effects.

3.5. LOCAL TIME AND SEASONAL VARIATIONS OF IONOSPHERIC STORM EFFECTS

3.5.1. Local Time Variations

Local time variations are one of the more prominent features of ionospheric storms (Obayashi, 1964; Matuura, 1972; Prölss, 1980; Danilov and Morozova, 1985). A first, if incomplete, description of this effect was given by Kirby et al. (1936), who stated that severe magnetic storms beginning during the daytime may show little correlation with radio data (in this case, negative storm effects) while a severe magnetic disturbance before sunrise is accompanied by disturbed radio conditions during the entire day. Subsequent studies confirmed that negative ionospheric storms at middle latitudes are usually observed to follow magnetic activity which occurred during the preceding night. They also showed that positive ionospheric storms are generally associated with magnetic activity beginning in the local daytime sector (Appleton and Piggot, 1952; Martyn, 1953b; Thomas and Venables, 1966; Jones, 1971a; Mendillo, 1973; Hargreaves and Bagenal, 1977; Inoue et al., 1978; Titheridge and Buonsanto, 1988; Jakowski et al., 1990; Balan and Rao, 1990; Prölss, 1993a). Independent support for this pattern comes from the observation that negative ionospheric storms commence most frequently in the early morning and very rarely in the noon and afternoon sectors (Jones, 1971a; Prölss and von Zahn, 1978; Danilov and Morozova, 1985).

Recently, a possible explanation of this local time variation of ionospheric disturbance effects was given by Prölss (1993a). This explanation is based on the assumptions that positive ionospheric storms are caused by meridional winds and negative ionospheric storms by changes in the neutral gas composition, as described in the previous sections. The storm scenario envisaged is illustrated in Figure 8.3.12.

This figure presents a view from above onto the northern hemisphere. Three concentric circles indicate the locations of 80, 60, and 40° magnetic latitude. A fixed local time frame is used, with midnight at the bottom and morning to the right. The dotted area indicates regions in the upper atmosphere where the neutral gas composition is disturbed and anomalously high values of the N_2/O density ratio are observed. To illustrate the ionospheric storm development, observations at two fictitious ionosonde stations will be discussed. Initially, ionosonde station No. 1 is situated in the noon sector and ionosonde station No. 2 in the midnight sector.

Prestorm Conditions

In the first plot (Figure 8.3.12a) moderately quiet conditions just prior to storm onset are considered. Heating by joule dissipation and particle precipitation maximizes along an annular region at polar latitudes. Accordingly, the composition changes produced by this heating are also restricted to the higher latitude region except for the midnight/early morning sector. Here winds of moderate magnitude, designated as midnight surge (Bates and Roberts, 1977; Babcock and Evans, 1979; Killeen and Roble, 1986), transport composition perturbations toward subauroral latitudes. These composition changes, however, are relatively small and quickly dissolve as they rotate with the Earth into the morning sector (Prölss, 1981).

Initial Phase

Next consider storm conditions (Figure 8.3.12b). Here, it is assumed that they are initiated by a burst of substorm activity. This sudden energy addition launches a traveling atmospheric disturbance which moves in the form of a global and circumpolar disturbance front toward lower latitudes. As it passes ionosonde station No. 1, positive ionospheric storm effects are observed at this location. This is not the case at ionosonde station No. 2, which will register only a sudden rise in layer height but no significant ionization density increase.

Following the first major substorm, the magnetic activity may temporarily subside. In this case, any new burst of activity may generate another TAD and a further positive storm effect. Such a case is documented in Figure 8.3.13. Alternatively, magnetic activity may remain at a high level. In this case, the continuing energy injection will lead to prolonged changes in the global wind circulation. During daytime the poleward-directed winds will be reduced or even reversed. This in turn will sustain the uplifting of the ionospheric F layer and cause long-duration positive storm effects, as discussed in Section 3.4.

Important characteristics of the initial phase of a storm are the expansion of the polar heating zone and of the associated composition perturbations, as indicated in Figure 8.3.12b. Except during large magnetic storms (Miller et al., 1990; Yeh et al., 1991; Oliver et al., 1991; Buonsanto

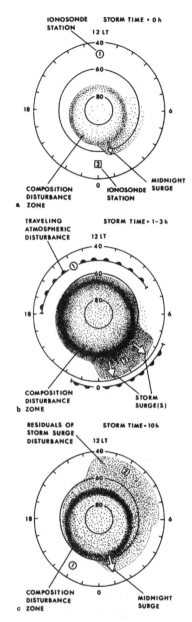

Figure 8.3.12 Time sequence of thermospheric-ionospheric storm effects. The schematic illustration presents a view from above onto the northern hemisphere. Three concentric circles indicate the location of 80, 60, and 40° magnetic latitude. Local time is given on the outer latitude circle. The dotted area identifies regions with an anomalously high N_2/O density ratio. The locations of two fictitious ionosonde stations at middle latitudes are indicated by a circle (ionosonde station No. 1) and by a square (ionosonde station No. 2). (a) Illustrates prestorm conditions, (b) the expansion phase, and (c) the later phase of a storm. To show the expansion of the composition disturbance toward middle latitudes, the situation in the night sector of part b refers to a somewhat later time than that in the associated day sector (note the progression of the traveling atmospheric disturbance front). (From Prölss, G.W., *Ann. Geophys.*, 11, 1–9, 1993a. With permission.)

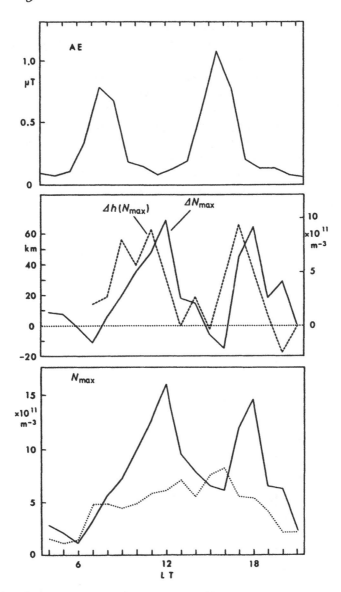

Figure 8.3.13 Short-duration positive ionospheric storms caused by two consecutive magnetospheric substorms. The presentation is similar to that of Figure 8.3.8. The storm was observed at Rome (37° invariant magnetic latitude) on March 6, 1973. Data recorded on March 4 serve as a quiet-time reference (dotted lines). (From Prölss, G.W., *Ann. Geophys.*, 11, 1–9, 1993a. With permission.)

et al., 1992), this disturbance expansion does not extend to middle latitudes. Accordingly, ionosonde station No. 1 will not observe composition disturbance effects.

This is not the case for ionosonde station No. 2, which is located in the early morning sector. Here, strong winds ("storm surges", Hays and Roble, 1971; Hernandez and Roble, 1976; Sipler and Biondi, 1979; Volland, 1979; Rees et al., 1979; Hagan, 1988; Buonsanto et al., 1990, 1992; Burns et al., 1991) carry air of disturbed composition out of the heating region toward middle latitudes. Thus this region is swamped with air of strongly enhanced molecular content. Figure 8.3.14 documents the asymmetric extent of the composition disturbance zone in the afternoon/evening and early morning sectors.

In response to these composition perturbations, ionosonde station No. 2 will observe the onset of a negative ionospheric storm. On actual records, these night effects may not seem to be as spectacular as their daytime counterparts. This mostly has to do with the lower ionization density at night. Thus *relative* changes may be quite impressive (Wrenn et al., 1987). Also note that strong equatorward-directed winds may temporarily slow down the ionization losses.

In the above scenario, it is tacitly assumed that the disturbance transport in the midnight/early morning sector is affected by advection. There is now considerable theoretical evidence in support of this assumption (Mayr and Volland, 1972, 1973; Volland, 1988; Burns et al., 1991; Fuller-Rowell et al., 1991b; Kersken et al., 1992; Burrage et al, 1992).

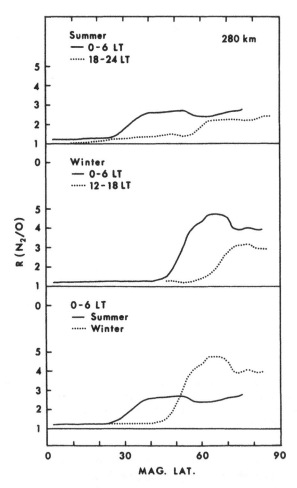

Figure 8.3.14 Local time and seasonal variations in the extent of the neutral atmospheric composition disturbance zone. Relative changes in the molecular nitrogen to atomic oxygen density ratio, $R(N_2/O)$, are plotted as a function of magnetic latitude. $R(N_2/O) = 1$ serves as a reference, meaning no change with respect to the quiet-time ESRO 4 model (Ap set equal to zero). All data refer to a common altitude of 280 km. The magnetic activity corresponds to more strongly disturbed conditions (Kp* $\geq$ 4+, with Kp* a modified Kp index which takes the development of the disturbance into account). Note that the curves represent the *median of superimposed latitudinal profiles.* These profiles have been superimposed in such a way that the equatorward boundary of the composition disturbance serves as a common reference location. This way the typical latitudinal structure of the disturbance is preserved. (From Prölss, G.W., *Ann. Geophys.* 11, 1–9, 1993a. With permission.)

Later Phase

Lastly, consider a later phase of the storm when magnetic activity has already subsided (Figure 8.3.12c). The heating oval and the associated composition perturbations have already retreated to higher latitudes, and the storm surges have lost their intensity. Ionosonde stations Nos. 1 and 2 are now located in the premidnight and prenoon sector, respectively.

The prominent feature of this situation is that the composition disturbance generated in the early morning sector is now located in the forenoon sector. It has simply rotated there with the Earth, and although a partial recovery has taken place, the perturbations are still large enough to produce substantial daytime negative ionospheric storm effects which will dominate the ionospheric behavior observed at station No. 2.

In contrast, ionosonde station No. 1 has yet to encounter a region of disturbed composition. It will do so while passing through the night sector, provided magnetic activity continues. In this case, it is only in the following morning that this station will observe "delayed" daytime negative storm effects.

The above model does remarkably well in explaining the distinct local time variations of ionospheric storm effects. It explains why a station in middle latitudes and during daytime either observes positive storm effects or negative storm effects or a sequence of positive and delayed negative storm effects. This is an improvement over previous thermospheric-ionospheric storm models which suggest a much closer coupling between positive and negative disturbance effects (Davies, 1974a,b; Rishbeth, 1974, 1975; Mayr et al., 1978; Namgaladze et al., 1981; Volland, 1983). It is also consistent with the observation that positive storm effects are chiefly observed in the local daytime sector and during the initial phase of a storm and with observations that negative storms most frequently begin in the early morning sector, and that these effects may be considerably delayed with respect to positive storm effects.

Of course, the idea that composition perturbations, once they have been generated, "rotate" into the daytime sector, presents only a first-order description. Actually, the disturbance bulge will be pushed around by winds and may move back and forth in latitude, as has been demonstrated in the storm simulation by Fuller-Rowell et al. (1993). This wind-induced oscillation in the latitudinal extent of the composition disturbance zone may well explain the rapid recovery of some storms observed in the winter season (Skoblin and Förster, 1993).

3.5.2. Seasonal Variations

Beside the local time-dependent changes, seasonal variations belong to the more prominent features of ionospheric storms. They were discovered early (Appleton et al., 1937; Kirby et al., 1937) and have since been investigated in numerous studies (Berkner and Seaton, 1940b; Appleton and Piggot, 1952; Sato, 1957a; Matsushita, 1959; Duncan, 1969; Rajaram et al., 1971; Fatkullin, 1973; Spurling and Jones, 1973; Prölss, 1977; Wrenn et al., 1987; Titheridge and Buonsanto, 1988; Jakowski et al., 1990). One important aspect of this variation is that during summer, negative storm effects are observed to extend all the way from the polar region to the subtropics, whereas during winter they are restricted to the higher latitude region. Figure 8.3.15 illustrates this asymmetric distribution for summer conditions in the southern hemisphere.

Given the cause-effect relationship between neutral composition changes and negative ionospheric storm effects, the thermospheric perturbation should exhibit similar variations. This is indeed the case, and Figure 8.3.14 documents that in summer the composition disturbance zone extends much further equatorward (by about $20°$ on the average) than in winter. Today it is generally believed that these variations arise from the interaction between seasonal and storm-induced winds. In summer, both types of winds support each other; in winter they are out of phase (Duncan, 1969; Roble et al., 1977; Mayr et al., 1978; Volland, 1979, 1988). The gradual rise of the N_2/O disturbance in the summer hemisphere and the relatively steep rise in the winter hemisphere are consistent with this explanation.

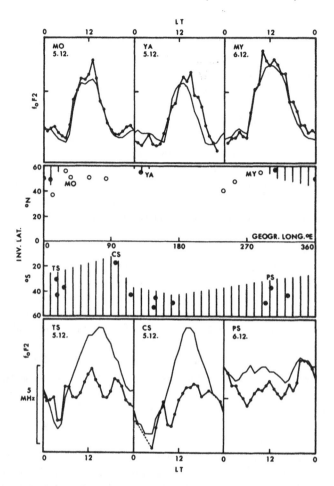

Figure 8.3.15 Global distribution of negative ionospheric storm effects. Using a geographic longitude/magnetic invariant latitude coordinate system, the middle panel shows the extent of negative ionospheric storm effects during the magnetic activity period of December 4 to 6, 1973 (hatched area). The solid, semisolid, and open circles indicate that significant, small, and no negative storm effects, respectively, were observed at these locations. The upper and lower panels directly illustrate the magnetic-storm-induced changes in the critical frequency of the F2 layer ($f_oF2[Hz]$ $\simeq 9\sqrt{N_{max}[m^{-3}]}$) as observed at six representative stations (heavy dotted lines). The monthly medians for December 1973 (thin lines) serve as quiet-time references. A tick mark on the ordinate indicates the 5-MHz level (for each station separately), and the bar on the lower left-hand side shows the scale used for all stations. (From Prölss, G.W., *J. Geophys. Res.,* 82, 1635–1640, 1977. Copyright by the American Geophysical Union. With permission.)

Another important aspect of seasonal variations is that positive ionospheric storms are mainly observed in winter. This may largely be explained by the limited extent of the composition disturbance zone in this season. Thus, a much larger proportion of the mid-latitude region will be exposed only to wind perturbations. In addition, the winter ionosphere may be especially sensitive to wind perturbations because of the much smaller vertical scale heights encountered in this season.

3.6. ELECTRIC FIELD EFFECTS

The idea that externally applied electric fields are responsible for ionospheric storm effects has a long tradition (Martyn, 1951, 1953a,b; Sato, 1957a,b, 1957a,b, 1959; Maeda and Sato, 1959; Rüster, 1965, 1969; Thomas, 1968; Evans, 1970a, 1973; VanZandt et al., 1971; Park, 1971, 1976; Park and Meng, 1971, 1973; Papagiannis et al., 1971; Tanaka and Hirao, 1973; Lanzerotti et al.,

1975; Anderson, 1976; Hargreaves and Bagenal, 1977; Tanaka, 1979, 1986; Ben'kova et al., 1981; Nazerets and Troschichev, 1984; Jakowski et al., 1990, 1992; Mendillo et al., 1992). Only the most recent variants of this hypothesis will be addressed here. The first is based on the assumption that sudden increases in layer height, similar to those illustrated in Figures 8.3.8 and 8.3.11, are generated by zonal electric fields of polar origin (Reddy et al., 1990a,b; Reddy and Nishida, 1992). A second one (Pi et al., 1993) concludes that the same kinds of electric fields are also responsible for the diurnal double-maxima pattern, similar to that illustrated in Figure 8.3.13. The main arguments put forward by these authors in support of their theory are that the ionospheric perturbations are closely associated with the rise and fall of substorm activity and that the observed changes occur nearly simultaneously at all latitudes. The validity of these suppositions has recently been challenged by Prölss et al. (1991) and Prölss (1993b), who maintain that ionospheric per-turbations are frequently delayed with respect to substorm activity and that this delay increases with decreasing latitude. These latter authors also note that electric fields of polar origin should produce height changes in opposite directions in the day and night sectors which are not observed. Also, they point out that the considerable dispersion effects of positive storms are not consistent with the electric field hypothesis. Because of these difficulties, Prölss et al. consider it more likely that both sudden height rises and twin-peak phenomena are caused by traveling atmospheric disturbances.

Here we note that a final evaluation of both theories will be possible only on the basis of simultaneous global electric field and wind measurements of sufficient temporal and spatial res-olutions. Such measurements are only beginning to become available (Buonsanto and Foster, 1993). With regard to the interpretation of these measurements, a clear distinction should be made between *external* electric fields (Nopper and Carovillano, 1978; Kikuchi et al., 1978; Kamide and Matsushita, 1981; Senior and Blanc, 1987; Spiro et al., 1988; Fejer et al., 1990a,b; Denisenko and Zamay, 1992 and references therein) and *dynamo* electric fields (Blanc and Richmond, 1980; Prakash and Pandey, 1985; Mazaudier et al., 1987), the latter being a secondary effect of the atmospheric perturbation. We also note that the TAD hypothesis by no means excludes the possibility that significant electric fields penetrate to middle latitudes during more severely dis-turbed conditions.

The third study to be addressed promotes the idea that a special type of positive storm observed in the North American dusk sector is caused by the convection electric field (Foster, 1993). Using ISR measurements, this study identifies a subauroral density enhancement that is well separated from the positive storm effects observed at middle latitudes. This density enhancement is several degrees wide and extends along the equatorward boundary of the trough region. It is attributed to the transport of plasma from lower to higher latitudes by convection electric fields. Any mid-latitude station rotating into this region in the afternoon/evening sector will observe conspicuous positive storm effects.

The above disturbance scenario is certainly interesting and deserves further study. Here we note that a similar distinction between the "normal" positive phase and the dusk enhancement in the American sector was also made by Tanaka (1979). Evidently such a distinction can be made only if global maps of the ionization density are available. This is because positive storm effects very similar to those observed in the evening sector are also observed in the afternoon sector and therefore, presumably, well equatorward of the trough region.

3.7. STORM SIMULATIONS

The simulation of ionospheric storms using first-principle models is one of the most challenging tasks of ionospheric physics. Accordingly, much effort has gone into the solution of this problem, and Table 8.3.2 contains a selection of nearly 40 publications dealing with this subject. Most of these studies rely on local, one-dimensional models of increasing sophistication to identify and test possible disturbance mechanisms. As a result, all the basic physical processes that may be

Table 8.3.2 **Numerical Simulation of F-Region Storms**

Lepechinsky (1951)
Investigates thermal expansion theory of negative ionospheric storms using Chapman layer model

Martyn (1953a)
Simulates positive and negative ionospheric storms using height-dependent $E \times B$ drifts; considers only transport effects

Kamiyama (1956)
Studies electrodynamic drift effects on the layer height and ionization density in the presence of strong thermospheric temperature gradients (diffusion effects not included)

Sato (1957a, 1959)
Models positive and negative ionospheric storms using height-independent $E \times B$ drifts. Includes production and loss processes (the latter in an unrealistic way) but no diffusion

Yonezawa (1963)
Simulates negative ionospheric storms by an increase in the thermospheric temperature using temperature-dependent loss coefficients

Matuura (1963)
Simulates transient daytime negative storm effects by means of a sudden rise in the thermospheric temperature which increases the loss rate (via an enhancement of the molecular species) and decreases the diffusion velocity (via an enhancement of the total density)

Rüster (1965, 1969)
Simulates magnetic activity associated uplifting of the nighttime F layer using eastward-directed electric fields

Chandra and Herman (1969)
Reproduce negative ionospheric storms by reducing the O concentration at homopause altitudes

Jones and Rishbeth (1971)
Reproduce long-lasting positive ionospheric storms by introducing large scale pole to equator-directed meridional winds (servo model)

King (1971)
Tests different storm theories, including thermal expansion, protonospheric fluxes, electric field drift, composition changes, and vibrationally excited N_2, using simplified F-region model

Jones (1971b)
Investigates positive ionospheric storm effects produced by eastward-directed electric fields (servo model)

Chandra and Stubbe (1971)
Reproduce negative ionospheric storms by changing O concentration at homopause level (similar to Chandra and Herman, 1969, study but time-dependent model).

Moffet and Murphy (1973)
Investigates Park (1971) hypothesis that nighttime positive ionospheric storm effects are caused by downward-directed $E \times B$ drifts (coupled ionosphere-plasmasphere model)

Tanaka and Hirao (1973)
Reproduce daytime short duration positive ionospheric storms using eastward-directed electric fields

Jones (1973)
Tests wind, electric field, and composition theories of ionospheric storms (servo model is replaced by a more complete ionospheric model)

Rishbeth and Hanson (1974)
Criticizes plasma "pile up" theory of positive ionospheric storms

Park and Banks (1974)
Simulates nighttime positive ionospheric storm effects using westward-directed electric fields, and inward convection of plasma tubes

Moffett et al. (1975)
Simulates nighttime positive ionospheric storm effects using westward-directed electric fields, inward convection of plasma tubes and local time drifts

Anderson (1976)
Simulates an evening positive ionospheric storm and a subsequent rapid transition to a negative ionospheric storm using meridional winds and east-west drifts (northward-directed electric fields)

Davies and Rüster (1976)
Test wind, electric field, and composition theories of ionospheric total electron content storms

Roble et al. (1978)
Simulate ionospheric response to traveling atmospheric disturbance

Table 8.3.2 **Continued**

Prölss and Jung (1978)
Simulate a short-duration positive ionospheric storm using a traveling atmospheric disturbance

Zevakina et al. (1978)
Simulate response of total electron content to traveling atmospheric disturbance

Jung and Prölss (1978)
Reproduce observed daytime negative ionospheric storms using observed composition changes

Miller et al. (1979)
Simulate positive and negative phases of ionospheric storms using meridional winds and composition changes predicted by the Mayr-Volland general circulation model

Zakharov and Namgaladze (1979)
Investigate response of ionosphere to particle influx from plasmasphere

Goel and Rao (1981)
Test various storm mechanisms including winds, temperature changes, composition changes, and electric fields (Stubbe model)

Serebryakov (1982)
Simulates seasonal variations of ionospheric storms using a thermosphere-ionosphere model.

Taieb and Poinsard (1984)
Reproduce observed ionospheric storm effects using input parameters constrained by ISR observations

Goel and Jain (1988)
Simulate observed ionospheric response to major substorm activity using winds, electric fields, and protonospheric fluxes (servo model)

Richards et al. (1989)
Model negative ionospheric storm using input data constrained by ISR observations (FLIP model)

Mendillo et al. (1992)
Model latitude and longitude dependence of ionospheric storm effects produced by winds and dawn-to-dusk electric fields

Codrescu et al. (1992)
Simulate ionospheric storm effects observed at a latitudinal chain of stations using a global interactive thermosphere-ionosphere model (NCAR-TIGCM)

Pi et al. (1993)
Model magnetic activity associated diurnal double maxima patterns using zonal electric fields and meridional winds

Pavlov (1994)
Models negative ionospheric storm effects produced by vibrationally excited N_2.

Fuller-Rowell et al. (1994)
Simulate global distribution of ionospheric storm effects as a function of UT onset of magnetospheric perturbation using a fully coupled thermosphere-ionosphere model (UCL-TIGCM)

important in explaining ionospheric storm behavior are thought to the known. This, however, is not a closed matter, and new ideas are advanced every now and then, challenging our present understanding of ionospheric storm physics (see Table 8.3.1).

To run a storm simulation code, various input parameters are needed. These include the intensity of the solar radiation, the thermospheric density and composition, the thermospheric winds, the electric field distribution, the particle precipitation pattern, and the protonospheric heat and particle fluxes. Unfortunately, none of these parameters are available on a larger scale, and all simulations have to rely on *ad hoc* assumptions. Therefore, the validation of the proposed concepts remains incomplete.

Recently, attempts were made to *calculate* some of the needed input information. Thus, TIGCMs were used to self-consistently specify the thermospheric density, composition, and winds. Going one step further, an extended version of these models also calculates the dynamo electric field produced by the disturbance winds [thermosphere-ionosphere-electrodynamic general circulation model (TIEGCM), Richmond et al., 1992]. As it turns out, even these rather sophisticated and demanding algorithms are only partly successful in reproducing the observed ionospheric storm effects. One reason for this is the lack of suitable input information. To run the models, the global distribution and temporal variation of magnetospheric electric fields and particle precipitation have to be specified, and this is by no means a trivial task.

Even as these problems and other shortcomings of large-scale models are being overcome, there will be a continuing need for simpler models that isolate and focus on specific physical mechanisms and processes, perhaps with a numerical resolution unattainable with three-dimensional models. Therefore, local and global models are expected to coexist for some time to come and to contribute to our understanding of ionospheric storm physics.

4. IONOSPHERIC STORMS AT LOW LATITUDES

The direction of ionization transport by diffusion, winds, and electrodynamic drift depends on the inclination (*dip*) of the Earth's magnetic field. A special condition is encountered near the dip equator where the magnetic field is horizontal. Here, a number of anomalies are observed which justify a separate treatment of this region. For example, near the dip equator, meridional winds will not lead to a rise in the F-layer height. Therefore, mechanisms P9 and P12 in Table 8.3.1 cannot be responsible for positive storm effects in this region. Also, electrodynamic (E × B) upward drifts are not expected to produce positive storm effects. This is because once lifted, the plasma will diffuse down the magnetic field lines and away from the equator. In fact, it is this combination of electrodynamic upward drift and subsequent downward diffusion along magnetic field lines which is invoked to explain the anomalous depletion of the F-region ionization density observed at the dip equator during quiet conditions (Rishbeth and Garriot, 1969; Stening, 1992). This *fountain effect* also produces two maxima (*crests*) in the ionization density near 15° *dip latitude* on both sides of the equator. Together these ionization maxima at subequatorial latitudes and the ionization minimum at the equator constitute the *equatorial* (or Appleton) anomaly in the latitudinal distribution of F-region plasma (see, for example, Figure 8.4.5). The fountain effect also produces an anomaly in the daily variation of the ionization density at the equator with a depression (*bite out*) near noon, and this is illustrated in Figure 8.4.3. Likewise, the ionization density at subequatorial latitudes is anomalously enhanced during the afternoon/evening hours.

In what follows, the special properties of ionospheric storm effects at low latitudes will be discussed. After a brief description of the storm morphology, various disturbance scenarios will be considered. More recent reviews of this subject have been presented by Matuura (1972), Rajaram (1977), and Abdu et al (1991).

4.1. BASIC MORPHOLOGY

As in middle latitudes, both positive and negative ionospheric storm effects are observed in the equatorial region (Rastogi and Rajaram, 1965; Raghavarao and Sivaraman, 1973; Jones and Davies, 1974). Depletions are primarily registered during the initial phase of a storm or during particularly strong substorms (Turunen and Rao, 1980; Alamelu et al., 1982; Adeniyi, 1986). The duration of these perturbations is of the order of some tens of minutes to a few hours. Depending on the intensity of the storm, veritable ionospheric holes may be created on such occasions (Berkner et al., 1939; Berkner and Seaton, 1940a; Kotadia, 1962; Batista et al., 1991; Greenspan et al., 1991; Lakshmi et al., 1991). Figure 8.4.1 illustrates such a case using DMSP satellite data.

Positive ionospheric disturbance effects are primarily observed during the main phase of a magnetic storm and in association with moderately strong activity (Berkner and Seaton, 1940b; Appleton and Piggot, 1952; Martyn, 1953a,b; Skinner and Wright, 1955; Sato, 1956; Rastogi, 1962; Matsushita, 1963; Kotadia, 1965; Olatunji, 1966; Kotadia and Jani, 1967; Rajaram and Rastogi, 1969; Yeboah-Amankwah, 1976). Figure 8.4.2 summarizes the results of the most recent statistical study of this subject (Adeniyi, 1986). The histogram indicates the percentage of iono-spheric storms which exhibit only a positive phase (+), or only a negative phase (−), or both a positive and negative phase (+/−), or no significant disturbance effects (0). The ionospheric response is considered separately for the initial and main phases of a magnetic storm and also

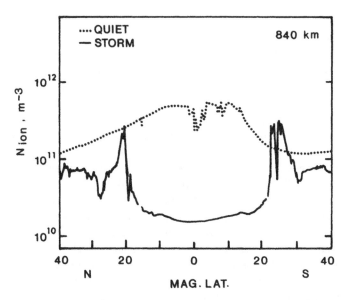

Figure 8.4.1 Severe negative storm effects in the topside ionosphere at low latitudes. The ion density measured by the DMSP F9 satellite at an altitude of 840 km is plotted as a function of magnetic latitude. The storm data (solid line) were obtained during the great magnetic storm of March 14, 1989. They refer to a local time of 21:30 and a geographic longitude of 284°E. Undisturbed conditions are illustrated with the help of data obtained on March 11, 1989 (dotted line). These data also refer to an altitude of 840 km and a local time near 21:30 and were also obtained above the South American continent (294°E). (From Greenspan, M.E. et al., *J. Geophys. Res.*, 96, 13931–13942, 1991. Copyright by the American Geophysical Union. With permission.)

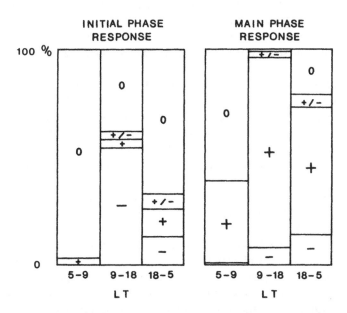

Figure 8.4.2 Response of the equatorial ionosphere to magnetic storms. Considered are changes in the maximum electron density of the F2 layer as recorded at the Central African station Ibadan (7.4°N, 3.9°E, 6°S magnetic dip). The following cases are distinguished: (1) positive storms (+); (2) negative storms (−); (3) positive and negative storm effects (+/−); and (4) no significant response (0). The above diagram indicates the relative frequency of these cases for two different storm phases and for three different local time sectors (based on a statistic published by Adeniyi, 1986).

for morning (5 to 9 LT), daytime (9 to 18 LT), and nighttime (18 to 5 LT) conditions. The results are based on maximum electron density data recorded at Ibadan (7.4°N, 3.9°E, 6°S magnetic dip) during 94 magnetic storms ($Kp_{max} \geq 5$ and $Ap \geq 26$). The magnetic storm phases were identified using the magnetic H component measured at the same location. As is evident, positive ionospheric storm effects are primarily observed during the (local) main phase of a magnetic storm and in particular during the daytime. Here, 88% of the storms exhibit an increase in the ionization density. An example of such a positive storm is illustrated in Figure 8.4.3. Typically, the noon bite out produced by the fountain effect is reduced or even disappears during disturbed conditions.

The height and latitude distributions of ionospheric storm effects at low latitudes have been investigated with the help of topside sounder data (King et al., 1967; Basu and Das Gupta, 1968; Sato, 1968; Rush et al., 1969; Sharma and Hewens, 1976). Figure 8.4.4 illustrates the kind of results obtained in the study by Sato. Positive storm effects are observed in a dome-shaped region at the magnetic equator. The dome-shaped region is surrounded by a belt-shaped region of reduced ionization density. Evidently, positive storm effects are recorded at the minimum of the equatorial anomaly and negative storm effects along magnetic field lines which are connected to anomaly crest regions. This general picture is in agreement with density measurements at a constant altitude such as those presented in Figure 8.4.5 (King et al., 1967). It is also consistent with the observed anticorrelation between ionospheric disturbance effects at the equator and at subequatorial latitudes (Rajaram and Rastogi, 1969; Walker, 1973; Huang and Cheng, 1993), and with the disappearance of the equatorial anomaly in the total electron content data (Deshpande et al., 1977).

4.2. DISTURBANCE MECHANISMS

Initially, it was thought that ionospheric storm effects at low latitudes are produced the same way as those at higher latitudes. Thus, they were attributed to a thermal expansion of the upper atmosphere, zonal electrodynamic drifts caused by the precipitation of energetic particles, height-dependent upward and downward electrodynamic drifts, and neutral composition changes (mechanisms N1 to N3, N5, and P1 in Table 8.3.1). Today these explanations, at least in their original form, are only of historical interest. Modern theories are based on a modification of the equatorial anomaly by electric fields and winds, on wind-induced transport, and on neutral composition changes. In the following sections these mechanisms will be briefly discussed.

4.2.1. Perturbations of the Equatorial Anomaly by Electric Fields

The observations presented in the previous section indicate that significant elements of the ionospheric disturbance effects at low latitudes are caused by modifications of the equatorial anomaly. Basically, these modifications may be brought about by changes in the electric fields and in the thermospheric winds. Changes in the electric fields in turn may be caused by (1) a superposition of external electric fields of polar and magnetospheric origin and (2) modifications of the electric dynamo field by storm winds.

Details on the morphology and theory of these disturbance electric fields can be found, for example, in Volland (1984), Kelley (1989), Fejer (1991), and Richmond (Ionospheric S_q- and L-currents, Chapter II/9).

Because electric field perturbations of *external* origin are primarily associated with the initial phase of a magnetic storm and with very intense substorm activity, they may well be responsible for the *negative* phase of the equatorial disturbance. Consider, for example, the following scenario. Sudden changes in the polar cap and/or ring current potential cause a temporary enhancement of the eastward-directed zonal electric field. This enhancement will increase the upward drift and the subsequent drainage of ionization from the equatorial region. Accordingly, negative storm effects will be observed at these latitudes. This scenario is supported by observations of significant F-region height changes and upward drifts during severely disturbed conditions (Berkner et al., 1939; Berkner and Seaton, 1940a; Somayajulu, 1963; Woodman et al., 1972; Fejer et al., 1990a,b; Somayajulu et al., 1991; Greenspan et al., 1991; Batista et al., 1991; Lakshimi et al., 1991; Sastri et al., 1992; Rasmussen and Greenspan, 1993). It is also consistent with the coincidence of

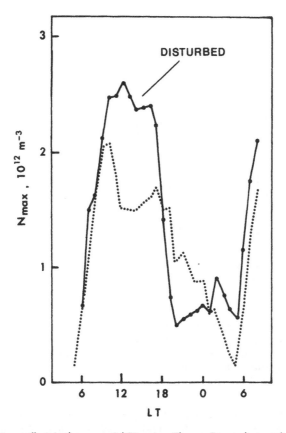

Figure 8.4.3   Positive storm effects in the equatorial F2 region. The maximum electron density of the F2 layer as recorded at Ibadan (7.4°N, 3.9°N, 6°S magnetic dip) is plotted as a function of local time. The continuous line indicates the variation observed during the magnetic storm of May 6, 1960. The average local time variation observed on five undisturbed days serves as a reference (dotted line). Note the disappearance of the noon "bite out" during disturbed conditions. (From Adeniyi, J.O., *J. Atmos. Terr. Phys.,* 48, 695–702, 1986. With permission.)

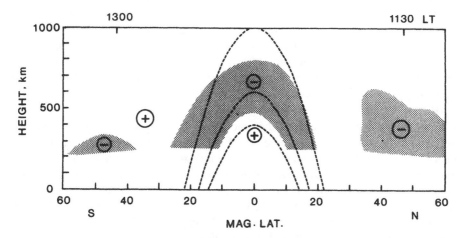

Figure 8.4.4   Height-latitude variation of storm effects in the topside ionosphere. Shaded parts represent regions of reduced ionization density, and nonshaded parts indicate enhancements or no changes. Three representative dipole field lines are shown for comparison. The storm data were recorded by the Alouette 1 satellite on December 18, 1962 (Kp = 5−). Measurements obtained on December 16, 1962, (Kp = 1+) serve as a quiet-time reference. Note that the different scales used for the height and latitude variations distort the inclination of the magnetic field lines. (From Sato, T., *J. Geophys. Res.,* 73, 6225–6241, 1968. Copyright by the American Geophysical Union. With permission.)

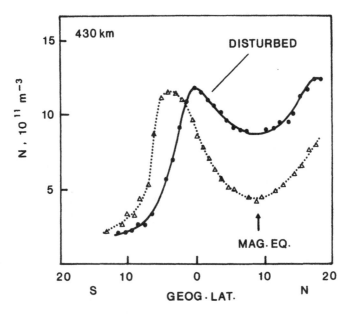

Figure 8.4.5 The equatorial anomaly during disturbed conditions. The electron concentration at a height of 430 km is plotted as a function of geographic latitude. The location of the magnetic equator is indicated by an arrow. The storm data (continuous line) were obtained by the Alouette 1 satellite on September 15, 1963 (Ap = 43). Measurements recorded on September 10, 1963, (Ap = 9) serve as a quiet-time reference (dotted line). Local time of observation is roughly 1300 h. (From King, J.W. et al., *J. Atmos. Terr. Phys.*, 29, 1355–1363, 1967. With permission.)

negative storm effects at equatorial latitudes and positive storm effects at subequatorial latitudes observed on such occasions (Tanaka, 1981; Forbes, 1989; Huang and Cheng, 1993).

Electric field perturbations may also arise from a modification of the dynamo winds (Blanc and Richmond, 1980). These changes are expected to set in with a certain time delay. Thus, it takes a traveling atmospheric disturbance about 3 to 4 hours to propagate from high to low latitudes, and changes in the large-scale wind circulation are induced on a similar time scale. Therefore, this mechanism can explain only ionospheric disturbance effects which occur during the main and recovery phases of a magnetic storm. As indicated in the study by Blanc and Richmond (1980), the storm winds blow in such a way that the quiet-time electric dynamo field is reduced. This weakens the fountain effect and leads to daytime positive ionospheric storm effects at equatorial latitudes and negative effects at subequatorial latitudes, as is observed in the majority of cases. Efforts are being undertaken to test this mechanism using a thermospheric-ionospheric general circulation model in which both regions are coupled self-consistently to each other (Richmond et al., 1992).

Note that the inhibition of the electrodynamic upward drift may lead to negative storm effects in the premidnight sector. This is because the relatively large densities observed at this time are thought to be produced by pronounced prereversal upward drifts (Anderson, 1973) which may be reduced during disturbed conditions (Namboothiri et al., 1989). In addition, it should be kept in mind that the inhibition of the equatorial anomaly has also been attributed to westward-directed electric fields of magnetospheric origin (Abdu et al., 1993). Finally, it should be noted that the equatorial anomaly tends to be more pronounced *following* a storm (Rush et al., 1969), indicating an intensification of the dynamo winds during this time.

4.2.2. Meridional Wind-Induced Perturbations

Another way to modify the equatorial anomaly is through wind-induced drifts. As has been pointed out by Burge et al. (1973), equatorward-directed winds will oppose the poleward transport of ionization along the magnetic field lines. This will hinder the formation of the equatorial anomaly and generate negative storm effects in the anomaly crest regions and positive storm effects near the equator. Model calculations by Rüster and King (1976) confirm the validity of this concept. They note that without the fountain effect, equatorward-directed winds will increase the ionization density, as is the case at middle latitudes. This tendency prevails down to about 20° magnetic latitude where a changeover takes place.

The complex interplay between storm-associated traveling atmospheric disturbances and the ionospheric F region at low latitudes has recently been studied using a thermospheric general circulation code coupled to a low-latitude ionospheric model (Fesen et al., 1989). It is found that the ionospheric variations can largely be explained in terms of TAD-associated winds. Also, the observed anticorrelation between the ionization density and the layer height in the anomaly crest region is reproduced by the calculation. Winds alone, however, cannot explain the significant height variations observed at the equator (see also Abdu et al., 1990).

4.2.3. Composition Changes

Positive ionospheric storms at low latitudes have also been attributed to changes in the neutral gas composition (Duncan, 1969; Chandra and Stubbe, 1971; Obayashi and Matuura, 1972; Mayr et al., 1978; Rishbeth, 1991). The idea is that the storm-induced, large-scale thermospheric circulation transports air rich in atomic oxygen toward lower latitudes. This enhanced oxygen density will affect both the ionization production and diffusion, leading to positive storm effects.

While this mechanism may well contribute to the general increase in the ionization density at lower latitudes (Mikhailov et al., 1994), its significance is all but established. Thus far, composition changes in this region may largely be attributed to an increase in the gas temperature, although nonthermal effects are also observed at times (Prölss, 1982; Burrage et al., 1992; Burns and Killeen, 1992). Because temperature-induced composition changes will hardly affect the ionization density, there remains only the nonthermal component, which appears to be of secondary importance.

Composition changes may also be responsible for low-latitude *negative* ionospheric storm effects. Thus, during the very largest storms, the polar disturbance zone marked by an increase in the molecular species and a decrease in the atomic oxygen density may extend all the way down to low and even equatorial latitudes, at least in one hemisphere. This kind of composition change will lead to long-duration negative storm effects such as those observed on a few occasions (Berkner et al., 1939; Somayajulu, 1963; Kotadia, 1965; Sastri, 1980; Huang and Cheng, 1991; Walker and Wong, 1993). Direct confirmation of this mechanism is still lacking, although increases in the N_2/O density ratio at fixed pressure heights are regularly observed to extend as far down as 30° magnetic latitude (Prölss, 1987).

5. IONOSPHERIC STORMS AT HIGH LATITUDES

Several factors combine to make the polar ionosphere a special and rather complicated region. It is here where most of the solar wind energy is dissipated and where this dissipation often dominates the atmospheric energy budget. Accordingly, changes induced by this energy source are not of a transient nature; rather, there is a continuous transition between more and less active conditions.

Of special importance are the substantial electric fields encountered in this region. Combined with the almost vertical magnetic field, they cause large-scale horizontal motions ($E \times B$ drifts) which have become known as *convection* (see, e.g., Chapter 12). Therefore, the ionosphere in this region is strongly affected by nonlocal processes. This motion also causes considerable frictional heating which affects the ionosphere in different ways.

Another special feature is that particle precipitation constitutes an important source of ionization. In fact, this source will dominate the polar ionosphere during winter conditions. Finally, magnetic field lines in the polar caps are "open", allowing ionization to escape continuously

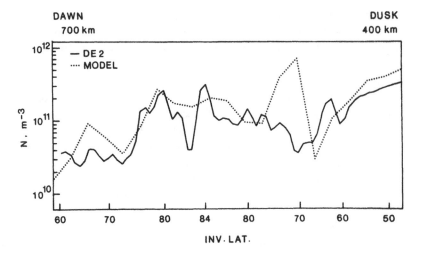

Figure 8.5.1 Typical electron density variations along a satellite pass across the polar region (continuous line). The data were recorded by the DE 2 satellite in the northern hemisphere on November 22, 1981, at approximately 9:30 UT. At the beginning of the traverse, the spacecraft was in the drawn sector (≈0800 MLT) at an altitude of about 700 km, at the end it was in the dusk sector (≈1800 MLT) at an altitude of about 400 km. The conditions were moderately disturbed (Kp = 3+), with a substorm (AE$_{max}$ ≈ 750 nT) preceding the measurements. Note that Rodger et al. (1992) identify the depression at about 83° invariant latitude on the dawn side and the depression at about 70° invariant latitude on the dusk side as high latitude troughs, and the depression at about 60° invariant latitude on the dusk side as the main trough. Using the USU time-dependent ionospheric model, an attempt was made to reproduce the satellite measurements, and the result is indicated by the dotted line. (From Sojka, J.J. et al., *J. Geophys. Res.*, 97, 1245–1256, 1992a. Copyright by the American Geophysical Union. With permission.)

from this region. This *polar wind* has important implications for the vertical structure of the outer ionosphere.

5.1 MORPHOLOGICAL ASPECTS

Early attempts to identify storm effects in the polar ionosphere were hampered by the strong absorption encountered in this region. During less disturbed conditions it was found that with increasing magnetic activity the F-region ionization density decreased during the day and some-times increased during the night (Appleton et al., 1937; Harang, 1951; Martyn, 1953b; Meek, 1953; Sato, 1957b; Maehlum, 1958). These studies also indicated that nonlocal effects might play an important role (Knecht, 1959).

A more complete if rather complicated picture of the polar ionosphere was provided by satellite-borne topside sounders (Muldrew, 1965; Sato and Colin, 1969; Thomas and Andrews, 1969; Nelms and Chapman, 1970). Measurements of this kind show that the ionization density in this region is highly structured, particularly in the absence of sunlight. Typically, a satellite pass will show a number of peaks and troughs as it makes a traverse across the polar region; this is illustrated in Figure 8.5.1. These nonstationary features render the determination of a quiet-time pattern, which could be used as a reference in storm studies, rather difficult. Therefore, only a few attempts were made to establish storm-associated changes (Nishida, 1967; Sato and Chan, 1969; Tulunay and Grebowsky, 1987). Figure 8.5.2 summarizes the results obtained from a comparison of individual quiet and disturbed satellite passes.

Because of the complex overall structure of the polar ionosphere, more recent investigations have concentrated on the study of specific morphological features identified in the satellite data. Prominent among these structures is the *main ionospheric trough*. By this we mean a latitudinally narrow and circumpolar region of reduced ionization density which is located adjacent and equa-torward of the auroral oval; see Figure 8.5.3. The copious literature on this subject has recently been reviewed by Moffett and Quegan (1983) and Rodger et al. (1992). Here we are interested

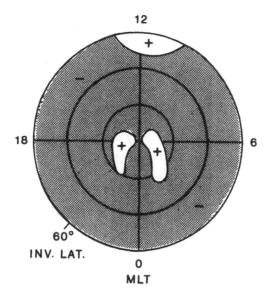

Figure 8.5.2 Distribution of ionospheric storm effects in the polar region. The coordinates are invariant magnetic latitude and magnetic local time. White areas indicate regions of enhanced electron densities and shaded areas those of reduced concentrations. The drawing summarizes the results of a comparison between quiet and disturbed Alouette 1 satellite passes across the polar region. The distribution refers to an altitude of 300 km. (From Sato, T. and K.L. Chan, *J. Geophys. Res.,* 74, 2208–2216, 1969. Copyright by the American Geophysical Union. With permission.)

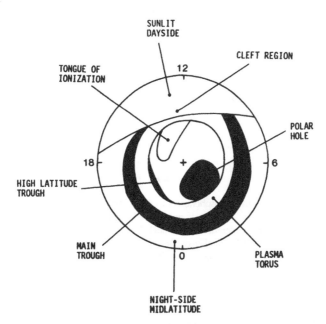

Figure 8.5.3 Schematic diagram showing the spatial distribution of important morphological phenomena in the polar region. (From Sojka, J.J. et al., *Adv. Space Res.,* 12, No. 6, 89–92, 1992b, supplement. With permission.)

only in magnetic-activity-associated changes. It is found that during disturbed conditions the trough is increasingly displaced toward lower latitudes. In fact, during storm conditions it may occupy regions that are normally considered middle latitudes. An ionosonde station rotating into such a displaced trough structure will observe severe negative storm effects (Mendillo et al., 1974; Prölss et al., 1991). Such a case is documented in Figure 8.5.4.

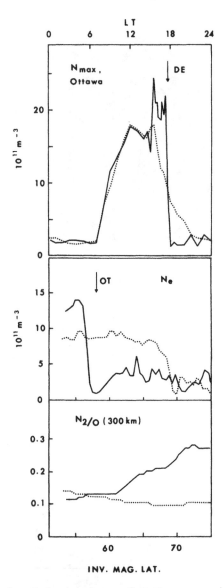

Figure 8.5.4 Trough-associated negative ionospheric storm effects. The upper panel shows the (local) time variation of the maximum electron density of the F layer, N_{max}, as observed at Ottawa on December 7, 1982. Data obtained on December 6 serve as the quiet-time reference (dotted line). The middle panel shows the latitudinal structure of the electron density as recorded by the DE 2 satellite on December 7 at approximately 17:40 LT and at a geographic longitude (287°E) close to Ottawa (284°E). The measurements refer to altitudes between about 270 km (53° invariant latitude) and 300 km (75° invariant latitude). The time of the satellite measurement is indicated in the upper panel and the relative position of Ottawa in the middle panel. A comparison of the two data sets demonstrates that the low ionization densities observed at Ottawa after 18 LT are a trough-associated phenomenon. The bottom panel shows the neutral composition changes observed during the same satellite pass. The lack of correlation between the latitudinal variation of the N_2/O density ratio and the trough structure is evident. (From Prölss, G.W. et al., *J. Geophys. Res.*, 96, 1275–1288, 1991. Copyright by the American Geophysical Union. With permission.)

Latitudinally narrow and longitudinally elongated regions of low ionization density also exist poleward of the auroral oval and have become known as *high-latitude troughs* (Rodger et al., 1992 and references therein). Although being a much less studied feature, the equatorward boundary of the high-latitude trough also seems to move equatorward with increasing magnetic activity (Grebowsky et al., 1983).

A third type of density depressions are so-called *polar holes*. By this we mean long-lived and extended areas of low ionization density which are chiefly observed in the winter polar cap region

(Brinton et al., 1978). The magnetic activity dependence of this feature has recently been studied by Hoegy and Grebowsky (1991). They found that Kp is not an important parameter when it comes to the quantitative description of the polar hole density.

Beside depletions, conspicuous increases in the ionization density are observed. In fact, the poleward side of the main trough is formed by a "wall" of enhanced ionization density (Andrews and Thomas, 1969; Mikkelsen, 1975; Leitinger et al., 1982; Rodger et al., 1986). This rise constitutes the equatorward boundary of an annular density enhancement variously called the *auroral peak,* the *auroral cliffs,* the *plasma ring,* or the *polar plasma torus.* It is roughly colocated with the auroral oval and shows similar variations with magnetic activity. Thus during storms it broadens and expands toward lower latitudes. Any mid-latitude station affected by this expansion will observe positive storm effects (Buonsanto et al., 1979; Prölss et al., 1991).

Another prominent type of density enhancement is the so-called *tongue of ionization.* By this we mean an elongated density structure which extends from the cleft into the polar cap region (Sato and Rourke, 1964; Nishida, 1967; Foster and Doupnik, 1984). There are indications that this tongue of ionization is not always a coherent structure but may break apart into patches of ionization which drift with considerable velocity across the polar cap (Weber et al., 1984, 1986; Foster, 1993). If this disintegration into patches is controlled by magnetic reconnection, as suggested by Lockwood (1992), the tongue of ionization should be much more structured during disturbed conditions.

5.2. DISTURBANCE MECHANISMS
5.2.1. Composition Changes
Some of the disturbance mechanisms listed in Table 8.3.1 are also applicable to the polar ionosphere. This is especially true for perturbations caused by neutral composition changes. Thus anomalous increases in the N_2/O density ratio are a permanent feature of the polar thermosphere (Prölss et al., 1988). As illustrated in Figure 8.3.12, these perturbations maximize along an annular region which is roughly colocated with the auroral oval. During more active conditions, these composition perturbations intensify and expand, thus leading to a general depletion of the polar ionization density.

Spectacular as they may be, these composition changes are not sufficient to explain the rather narrow and large decreases observed in trough structures (Raitt et al., 1975; Prölss et al., 1991); this is documented in Figure 8.5.4. Therefore additional disturbance mechanisms must be operative in these regions.

5.2.2. Convection
Figure 8.5.5 illustrates a steady convection pattern based on the superposition of a typical polar electric field distribution and the corotation electric field. As can be seen, some of the drift trajectories are confined to the polar cap region, which during winter is in total darkness (path 1). As the plasma decays, very low ionization densities may be achieved. This kind of plasma confinement can therefore explain the formation of polar holes (Sojka et al., 1981a, 1991). On other trajectories the flow stagnates such that the plasma spends many hours away from production sources (path 5). This is the case in the evening sector where steady decay may lead to very pronounced ionization depletions. Model calculations confirm that such stagnant flows are most likely responsible for the development of the main trough (Knudsen, 1974; Watkins, 1978; Spiro et al., 1978; Sojka et al., 1981a; Quegan et al., 1982; Fuller-Rowell et al., 1991a). Plasma tubes of low content may subsequently be convected into other regions, extending the trough structure into other local time sectors (Sojka et al., 1990).

In addition, high density plasma from the sunlit daytime region is convected into the polar cap region (paths 3 and 4), forming a tongue of ionization (Knudsen, 1974; Sojka et al., 1981a, 1992a; Fuller-Rowell et al., 1987). This higher density plasma may subsequently drift back into the dusk and dawn sectors, contributing to the poleward wall of the main trough (Robinson et al., 1985; Rodger et al., 1986; Senior et al., 1987). Evidently, the transport of ionization and

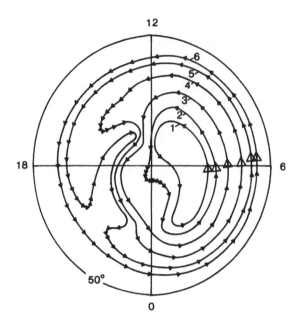

Figure 8.5.5 Examples of convection paths of plasma at 300-km altitude in the northern hemisphere under the combined influence of the magnetospheric and corotation electric fields, depicted in the fixed sun-Earth frame. The large open triangles indicate the starting points used in the calculation of the paths. The time between successive solid triangle is 1 h. The polar cap boundary (not marked) is a circle of radius 15° and centered 5° toward midnight from the geomagnetic pole. (From Quegan, S. et al., *J. Atmos. Terr. Phys.,* 44, 619–640, 1982. With permission.)

nonlocal processes play preeminent roles in the polar ionosphere. Note that in regions where the magnetic field lines are sufficiently inclined, large convection velocities will also cause considerable upward and downward drifts.

Stationary flow patterns like that depicted in Figure 8.5.5 are the exception rather than the rule because the electric field intensity and distribution are constantly adjusting to the varying conditions in interplanetary space. During more active conditions, the convection pattern expands and the drift speeds increase. This leads to the development of new troughs and density enhancements, changing the plasma distribution in the polar ionosphere (Sojka et al., 1981b).

5.2.3. Electrodynamic Heating

Beside affecting the transport of ionization, electric fields also act to increase the loss of ionization. This is because the reaction rate constants k_1 and k_2 in Equations 3.3, 3.4, and 3.6 strongly depend on both the ion temperature and the relative flow speeds of the interacting gases (Schunk et al., 1975; St. Maurice and Torr, 1978). At high temperatures, for example, the rate coefficient k_1 is approximately proportional to the square of the effective temperature. Via electrodynamic heating, this effective temperature increases roughly with the square of the effective electric field strength. Therefore, the reaction rate constant k_1 will change with the fourth power of the electric field intensity. To give an example, in a stationary neutral atmosphere an electric field of the order of 100 mV/m may increase the loss rate by an order of magnitude (Rodger et al., 1992). Such drastic increases will create veritable holes in the plasma density and may explain the formation of additional troughs both at sub- and transauroral latitudes (Schunk et al., 1976; Spiro et al., 1979; Schlegel, 1984; Jones et al., 1990; Häggström and Collis, 1990; Collis and Häggström, 1991; Anderson et al., 1991; Moffett et al., 1992; Quegan et al., 1992; Buonsanto et al., 1992; Millward et al., 1993). Electrodynamic heating will also cause thermal expansion and large upward flows which amplify the decrease in the ionization density (Winser et al., 1986; Yeh and Foster, 1990; Moffett et al., 1992).

5.2.4. Particle Precipitation

Particle precipitation constitutes an important ionization source of the polar ionosphere, especially during solar eclipse conditions. A prominent F-region feature associated with this ionization source is the auroral oval density enhancement, which has been denoted as *plasma torus* in Figure 8.5.3. Thus, it is generally believed that at least part of this feature is produced by low energy electrons (<1 keV) which precipitate from the plasma sheet, from the magnetosheath, and from the magnetopause boundary layer (Turunen and Liszka, 1972; Whitteker et al., 1972; Watkins and Richards, 1979; Labelle et al., 1989; Sojka et al., 1989; Schumaker et al., 1989; Newell et al., 1991). Because the plasma torus forms the poleward wall of the main trough, particle precipitation will also affect the trough structure (Raitt et al., 1975; Mikkelsen, 1975; Rodger et al., 1986). During more active conditions, particle precipitation will intensify and expand and may cause positive storm effects at higher and lower latitudes (Essex and Watkins, 1973; Buonsanto et al., 1979; Mendillo et al., 1987; Providakes et al., 1989; Prölss et al., 1991 and references therein).

Note that soft particle precipitation will also increase the vibrational excitation of molecular nitrogen. This in turn will increase the rate coefficient k_1 of Equations 3.3 and 3.6 and therefore the loss of ionization (Newton and Walker, 1975; Richards et al., 1989; Buonsanto et al., 1992). To what extent this mechanism is able to counteract the production of ionization is not clear at present.

5.2.5. Storm Simulation

If the significance of nonlocal processes is considered, large-scale models are required to simulate the polar ionosphere. More recent versions of these models include almost all physical processes thought to be important in this region, including photo and particle ionization, temperature-dependent loss reactions, vertical diffusion, horizontal convection, neutral air drag, and vertical heat flow (Sojka, 1989 and references therein). To run these models a number of input parameters have to be specified. These include the solar flux, the particle precipitation, the convection pattern, and the thermospheric conditions. Recently, lower resolution thermosphere-ionosphere models have been developed, which eliminate the need to specify the thermospheric conditions as an external input parameter (Fuller-Rowell et al., 1987; Rees et al., 1987; Roble et al., 1987).

Thus far, only a few attempts have been made to apply these first principal models to disturbed conditions. The main problem here is to specify the appropriate precipitation and electric field distributions. In a first exploratory study, Sojka et al. (1981b) investigated the response of their model to an increased polar cap potential, to an asymmetric electric field distribution with a smaller dawn convection cell, and to an extended auroral oval. Only stationary conditions were considered, and all changes were kept constant for 24 h. They found that under these circumstances the tongue of ionization became more pronounced, the main trough became more shallow, and the normal polar hole disappeared with a new hole developing at a different location.

In a follow-up study, a moderate storm of a few hours' duration was simulated (Sojka and Schunk, 1983). The storm perturbed the quiet-time ionosphere, and then the ionosphere relaxed back to a quiet-time situation. The effects of the storm were manifested in time-dependent changes in the magnetospheric convection electric field, the auroral oval size, and the auroral particle precipitation flux. One of the more important results to emerge from this study was that the upper ionosphere responds rather sluggishly to time-varying changes. Consequently, near and above the F-region peak, density variations are not correlated with the morphology of the storm auroral precipitation or the temporal variation of the storm electric field pattern. In fact, the upper ionosphere may not feel the full effect of the storm until after storm activity has ceased.

One of the difficulties with the above storm simulations is that they are based on statistically averaged electric field and particle precipitation patterns which may not represent the actual situation. In an effort to partly overcome this deficiency, a recent study by Sojka et al. (1992a) constrains the particle precipitation model with the help of satellite measurements. The period investigated was moderately disturbed and contained several substorms. Figure 8.5.1 illustrates the degree of agreement between data and model predictions. As is evident, there are regions where the observed densities are reasonably well reproduced by the simulation, and there are areas where data and model values differ by an order of magnitude. The discrepancies are probably due to the fact that the electric field pattern could not be tuned to the substorm situation. Consequently, very sophisticated input information is needed to successfully simulate the disturbed polar ionosphere.

6. UNSOLVED PROBLEMS

More than 60 years after their discovery, ionospheric storms remain a most fascinating and challenging topic of upper atmospheric physics. This is because so many aspects of this striking phenomenon are still incompletely documented and understood. Some of these problems have already been addressed in the previous sections. A more succinct summary is given here. It is understood that this compilation is far from complete and that it reflects somewhat the personal preferences of the author.

One of the most important points to be clarified is the significance of external electric fields in explaining ionospheric disturbance effects:

1. Are external electric fields of secondary importance when it comes to explaining short-duration positive storm effects at middle latitudes, as has been suggested in this review?
2. Is the assumption correct that TADs are a more likely explanation for sudden height rises in the day- and nightside ionospheres?
3. How important are convection electric fields in explaining positive storm effects at subauroral latitudes?
4. Is it correct to assume that it is only during the largest magnetic storms and substorms that significant external electric fields penetrate all the way to equatorial latitudes? Are dynamo electric fields more important during less severely disturbed conditions?

Neutral composition associated disturbance effects are another problem area.

5. Is it correct to assume that neutral composition changes (i.e., increases in the O/N_2 density ratio) play a secondary role when it comes to explaining positive ionospheric storm effects at middle and lower latitudes, as has been done in the present review?
6. Are neutral composition changes at higher and middle latitudes (i.e., decreases in the O/N_2 density ratio) sufficient to explain the rather large negative storm effects observed during periods of high solar activity or are additional mechanisms (e.g., vibrationally excited species) required?

Other open questions concern ionospheric disturbance effects in the night sector, a topic which has not been treated adequately in this review.

7. How important are the inflow and outflow of plasma from and into the protonosphere for the explanation of nighttime positive and negative storm effects at middle latitudes?
8. What role do height gradients of meridional winds (e.g., shears) play in explaining both kinds of disturbance effects?
9. Is particle precipitation from the storm ring current an important ionization source for the nighttime F region?
10. How important are the compression and dilution of plasma by meridional winds during night- (and day-) time at equatorial latitudes?

In order to simulate ionospheric storm effects it is of paramount interest to know how reliable first principal models are.

11. Is it correct to assume that the physics of the ionosphere is basically understood and that it is primarily the lack of suitable input which limits the accuracy of ionospheric models?
12. If this is the case, how well will we be able in the foreseeable future to specify this input with sufficient accuracy to make reliable predictions?

Admittedly, this is but a small selection of open questions concerning ionospheric storm effects, but answering only these questions will require considerable effort and concerted action from ionospheric experimentalists, data analysts, and theoreticians.

ACKNOWLEDGMENTS

The author would like to thank Hans Volland for his invitation to write this review. Thanks are also due to all who participated in collecting the data discussed in this study.

REFERENCES

Abdu, M.A., G.O. Walker, B.M. Reddy, J.H.A. Sobral, B.G. Fejer, T. Kikuchi, N.B. Trivedi, and E.P. Szuszczewicz, Electric field versus neutral wind control of the equatorial anomaly under quiet and disturbed condition: A global perspective from SUNDIAL 86, *Ann. Geophys.*, 8, 419–430, 1990.

Abdu, M.A., J.H.A. Sobral, E.R. de Paula, and I.S. Batista, Magnetospheric disturbance effects on the Equatorial Ionization Anomaly (EIA): an overview, *J. Atmos. Terr. Phys.*, 53, 757–771, 1991.

Abdu, M.A., G.O. Walker, B.M. Reddy, E.R. de Paula, J.H.A. Sobral, B.G. Fejer, and E.P. Szuszczewicz, Global scale equatorial ionization anomaly (EIA) response to magnetospheric disturbances based on the May–June 1987 SUNDIAL-coordinated observations, *Ann. Geophys.*, 11, 585–594, 1993.

Adeniyi, J.O., Magnetic storm effects on the morphology of the equatorial F2-layer, *J. Atmos. Terr. Phys.*, 48, 695–702, 1986.

Ahn, B.-H., S.-I. Akasofu, and Y. Kamide, The joule heating production rate and the particle injection rate as a function of the geomagnetic indices AE and AL, *J. Geophys. Res.*, 88, 6275–6287, 1983.

Akasofu, S.-I., Diagnostics of the magnetosphere using geomagnetic, auroral and airglow phenomena, *Ann. Geophys.*, 26, 443–457, 1970.

Akasofu, S.-I., The magnetospheric currents: an introduction, in *Magnetospheric Currents* (T.A. Potemra, Ed.), American Geophysical Union, Washington D.C. 29–48, 1984.

Alamelu, V., M.M. Rao, K.G.S. Charya, and R. Sethuraman, Effect of severe auroral disturbances on the equatorial ionosphere, *J. Atmos. Terr. Phys.*, 44, 161–171, 1982.

Alcaydé, D., J. Testud, G. Vasseur, and P. Waldteufel, L'orage magnetique du 11 fevrier 1969: Étude par diffusion incoherente, *J. Atmos. Terr. Phys.*, 34, 1037–1052, 1972.

Anderson, C.N., Correlation of long wave transatlantic radio transmission with other factors affected by solar activity, *Proc. Inst. Radio Eng.*, 16, 297–347, 1928.

Anderson, D.N., A theoretical study of the ionospheric F region equatorial anomaly. II. Results in the American and Asian sectors, *Planet. Space Sci.*, 21, 421–442, 1973.

Anderson, D.N., Modeling the midlatitude F-region ionospheric storm using east-west drift and a meridional wind, *Planet. Space Sci.*, 24, 69–77, 1976.

Anderson, P.C., R.A. Heelis, and W.B. Hanson, The ionospheric signatures of rapid subauroral ion drifts, *J. Geophys. Res.*, 96, 5785–5792, 1991.

Andrews, M.K. and J.O. Thomas, Electron density distribution above the winter pole, *Nature*, 221, 223–227, 1969.

Appleton, E.V. and L.J. Ingram, Magnetic storms and upper-atmospheric ionisation, *Nature*, 136, 548–549, 1935.

Appleton, E.V. and W.R. Piggot, The morphology of storms in the F_2 layer of the ionosphere I. Some statistical relationships, *J. Atmos. Terr. Phys.*, 2, 236–252, 1952.

Appleton, E.V., R. Naismith, and L.J. Ingram, British radio observations during the second international polar year 1932–33, *Philos. Trans., R. Soc.*, A236, 191–259, 1937.

Babcock, R.R. and J.V. Evans, Effects of geomagnetic disturbances on neutral winds and temperatures in the thermosphere observed over Millstone Hill, *J. Geophys. Res.*, 84, 5349–5354, 1979.

Balan, N. and P.B. Rao, Dependence of ionospheric response on the local time of sudden commencement and the intensity of geomagnetic storms, *J. Atmos. Terr. Phys.*, 52, 269–275, 1990.

Banks, P.M. and G. Kockarts, *Aeronomy, A, B*, Academic Press, New York 1973.

Basu, S. and A. Das Gupta, Latitude variation of electron content in equatorial region under magnetically quiet and active conditions, *J. Geophys. Res.*, 73, 5599–5602, 1968.

Basu, S., B.K. Guhathakurta, and S. Basu, Ionospheric response to geomagnetic storms at low midlatitudes, *Ann. Geophys.*, 31, 497–505, 1975.

Bates, F.B. and T.D. Roberts, The southward midnight surge in F-layer wind observed with the Chatanika incoherent scatter radar, *J. Atmos. Terr. Phys.*, 39, 87–93, 1977.

Batista, I.S., E.R. de Paula, M.A. Abdu, N.B. Trivedi, and M.E. Greenspan, Ionospheric effects of the March 13, 1989, magnetic storm at low and equatorial latitudes, *J. Geophys. Res.*, 96, 13943–13952, 1991.

Bauer, S.J., *Physics of Planetary Ionospheres*, Springer-Verlag, Heidelberg 1973.

Bauer, S.J. and B.V. Krishnamurthy, Behavior of the topside ionosphere during a great magnetic storm, *Planet. Space Sci.*, 16, 653–663, 1968.

Baumjohann, W., Ionospheric and field-aligned current systems in the auroral zone: a concise review, *Adv. Space Res.*, 2, No. 10, 55–62, 1983.

Becker, W., The varying electron density profile of the F-region during magnetically quiet nights, *J. Atmos. Terr. Phys.*, 22, 275–289, 1961.

Ben'kova, N.P., Yu.M. Berezin, G.V. Bukin, and N.K. Osipov, Global structure of ionosphere disturbances of convective origin, *Geomagn. Aeron.*, 21, 547–549, 1981.

Berkner, L.V. and S.L. Seaton, Ionospheric changes associated with the magnetic storm of March 24, 1940, *Terr. Magn. Atmos. Electr.*, 45, 393–418, 1940a.

Berkner, L.V. and S.L. Seaton, Systematic ionospheric changes associated with geomagnetic activity, *Terr. Magn. Atmos. Electr.*, 45, 419–423, 1940b.

Berkner, L.V., H.W. Wells, and S.L. Seaton, Ionospheric effects associated with magnetic disturbances, *Terr. Magn. Atmos. Electr.*, 44, 283–311, 1939.

Blanc, M. and A.D. Richmond, The ionospheric disturbance dynamo, *J. Geophys. Res.*, 85, 1669–1686, 1980.

Bowman, G.G., A relationship between polar magnetic substorms, ionospheric height rises and the occurrence of spread-F, *J. Atmos. Terr. Phys.*, 40, 713–722, 1978.

Brace, L.H., E.J. Maier, J.H. Hoffman, J. Whitteker, and G.G. Shepherd, Deformation of the night side plasmasphere and ionosphere during the August 1972 geomagnetic storm, *J. Geophys. Res.*, 79, 5211–5218, 1974.

Brekke, A., Joule heating and particle precipitation, *Adv. Space Res.*, 2, No. 10, 45–53, 1983.

Brinton, H.C., J.M. Grebowsky, and L.H. Brace, The high-latitude winter F region at 300 km: thermal plasma observations from AE-C, *J. Geophys. Res.*, 83, 4767–4776, 1978.

Buonsanto, M.J. and J.C. Foster, Effects of magnetospheric electric fields and neutral winds on the low-middle latitude ionosphere during the March 20–21, 1990, storm, *J. Geophys. Res.*, 98, 19133–19140, 1993.

Buonsanto, M.J., M. Mendillo, and J.A. Klobuchar, The ionosphere at L = 4: average behavior and the response to geomagnetic storms, *Ann. Geophys.*, 35, 15–26, 1979.

Buonsanto, M.J., J.E. Salah, K.L. Miller, W.L. Oliver, R.G. Burnside, and P.G. Richards, Observations of neutral circulation at mid-latitudes during the equinox transition study, *J. Geophys. Res.*, 94, 16987–16997, 1989.

Buonsanto, M.J., J.C. Foster, A.D. Galasso, D.P. Sipler, and J.M. Holt, Neutral winds and thermosphere/ionosphere coupling and energetics during the geomagnetic disturbances of March 6–10, 1989, *J. Geophys. Res.*, 95, 21033–21050, 1990.

Buonsanto, M.J., J.C. Foster, and D.P. Sipler, Observations from Millstone Hill during the geomagnetic disturbances of March and April 1990, *J. Geophys. Res.*, 97, 1225–1243, 1992.

Burge, J.D., D. Eccles, J.W. King, and R. Rüster, The effects of thermospheric winds on the ionosphere at low and middle latitudes during magnetic disturbances, *J. Atmos. Terr. Phys.*, 35, 617–623, 1973.

Burkard, O.M., A coupling between solar wind and ionosphere, *Planet. Space Sci.*, 18, 1832–1834, 1970.

Burns, A.G. and T.L. Killeen, The equatorial neutral thermospheric response to geomagnetic forcing, *Geophys. Res. Lett.*, 19, 977–980, 1992.

Burns, A.G., T.L. Killeen, and R.G. Roble, Processes responsible for the compositional structure of the thermosphere, *J. Geophys. Res.*, 94, 3670–3686, 1989.

Burns, A.G., T.L. Killeen, and R.G. Roble, A theoretical study of thermospheric composition perturbations during an impulsive geomagnetic storm, *J. Geophys. Res.*, 96, 14153–14167, 1991.

Burnside, R.G., C.A. Tepley, M.P. Sulzer, T.J. Fuller-Rowell, D.G. Torr, and R.G. Roble, The neutral thermosphere at Arecibo during geomagnetic storms, *J. Geophys. Res.*, 96, 1289–1301, 1991.

Burrage, M.D., V.J. Abreu, N. Orsini, C.G. Fesen, and R.G. Roble, Geomagnetic activity effects on the equatorial neutral thermosphere, *J. Geophys. Res.*, 97, 4177–4187, 1992.

Cander, Lj.R., On the global and regional behaviour of the mid-latitude ionosphere, *J. Atmos. Terr. Phys.*, 55, 1543–1551, 1993.

Chan, K.L. and O.G. Villard, Observations of large-scale traveling ionospheric disturbances by spaced-path high-frequency instantaneous frequency measurements, *J. Geophys. Res.*, 67, 973–988, 1962.

Chandra, S. and J.R. Herman, F-region ionization and heating during magnetic storms, *Planet. Space Sci.*, 17, 841–851, 1969.

Chandra, S. and N.W. Spencer, Thermospheric storms and related ionospheric effects, *J. Geophys. Res.*, 81, 5018–5026, 1976.

Chandra, S. and P. Stubbe, Ion and neutral composition changes in the thermospheric region during magnetic storms, *Planet. Space Sci.*, 19, 491–502, 1971.

Chang, C.A. and J.-P. St. Maurice, Two dimensional high latitude thermospheric modeling: a comparison between moderate and extremely disturbed conditions, *Can. J. Phys.*, 69, 1007–1031, 1991.

Chapman, S. and J. Bartels, *Geomagnetism, I*, Clarendon Press, Oxford, 1962.

Codrescu, M.V., R.G. Roble, and J.M. Forbes, Interactive ionosphere modeling: a comparison between TIGCM and ionosonde data, *J. Geophys. Res.*, 97, 8591–8600, 1992.

Collis, P.N. and I. Häggström, High-latitude ionospheric response to a geomagnetic sudden commencement, *J. Atmos. Terr. Phys.*, 53, 241–248, 1991.

Crowley, G., B.A. Emery, R.G. Roble, H.C. Carlson, Jr., and D.J. Knipp, Thermospheric dynamics during September 18–19, 1984. 1. Model simulations, *J. Geophys. Res.*, 94, 16925–16944, 1989a.

Crowley, G., B.A. Emery, R.G. Roble, H.C. Carlson, Jr., J.E. Salah, V.B. Wickwar, K.L. Miller, W.L. Oliver, R.G. Burnside, and F.A. Marcos, Thermospheric dynamics during September 18–19, 1984. 2. Validation of the NCAR Thermospheric General Circulation Model, *J. Geophys. Res.*, 94, 16945–16959, 1989b.

Danilov, A.D. and L.D. Morozova, Ionospheric storms in the F_2 region. Morphology and physics (Review), *Geomagn. Aeron.*, 25, 593–605, 1985.

Danilov, A.D., L.D. Morozova, Tc. Dachev, and I. Kutiev, Positive phase of ionospheric storms and its connection with the dayside cusp, *Adv. Space Res.*, 7, No. 8, 81–88, 1987.

Davies, K., Studies of ionospheric storms using a simple model, *J. Geophys. Res.*, 79, 605–613, 1974a.

Davies, K., A model of ionospheric F-2 region storms in middle latitudes, *Planet. Space Sci.*, 22, 237–253, 1974b.

Davies, K. and R. Rüster, Theoretical studies of storm effects in ionospheric total electron content, *Planet. Space Sci.*, 24, 867–872, 1976.

Davis, T.N. and R. Parthasarathy, The relationship between polar magnetic activity DP and growth of the geomagnetic ring current, *J. Geophys. Res.*, 72, 5825–5836, 1967.

Deng Wei and M. Förster, Changes of thermospheric composition and the response of the ionosphere during the magnetic storm of January 1974, *Gerlands Beitr. Geophys.*, 98, 240–250, 1989.

Denisenko, V.V. and S.S. Zamay, Electric field in the equatorial ionosphere, *Planet. Space Sci.*, 40, 941–952, 1992.

Deshpande, M.R., R.G. Rastogi, H.O. Vats, J.A. Klobuchar, G. Sethia, A.R. Jain, B.S. Subarao, V.M. Patwari, A.V. Janve, R.K. Rai, M. Singh, H.S. Gurm, and B.S. Murthy, Effect of electrojet on the total electron content of the ionosphere over the Indian subcontinent, *Nature*, 267, 599–600, 1977.

Duncan, R.A., F-region seasonal and magnetic-storm behaviour, *J. Atmos. Terr. Phys.*, 31, 59–70, 1969.

Eckersley, T.L., Holes in the ionosphere and magnetic storms, *Nature*, 150, 177, 1942.

Essex, E.A., The effects of geomagnetic activity on the F-region of the ionosphere, *J. Atmos. Terr. Phys.*, 41, 951–960, 1979.

Essex, E.A. and B.J. Watkins, Enhancements of ionospheric total electron content in the southern auroral zone associated with magnetospheric substorms, *J. Atmos. Terr. Phys.*, 35, 1015–1018, 1973.

Essex, E.A., M. Mendillo, J.P. Schödel, J.A. Klobuchar, A.V. da Rosa, K.C. Yeh, R.B. Fritz, F.H. Hibberd, L. Kersley, J.R. Koster, D.A. Matsoukas, Y. Nakata, and T.H. Roelofs, A global response of the total electron content of the ionosphere to the magnetic storm of 17 and 18 June 1972, *J. Atmos. Terr. Phys.*, 43, 293–306, 1981.

Evans, J.V., The June 1965 magnetic storm: Millstone Hill observations, *J. Atmos. Terr. Phys.*, 32, 1629–1640, 1970a.

Evans, J.V., Midlatitude ionospheric temperatures during three magnetic storms in 1965, *J. Geophys. Res.*, 75, 4803–4813, 1970b.

Evans, J.V., F-region heating observed during the main phase of magnetic storms, *J. Geophys. Res.*, 75, 4815–4823, 1970c.

Evans, J.V., The cause of storm-time increases of the F-layer at mid-latitudes, *J. Atmos. Terr. Phys.*, 35, 593–616, 1973.

Ezquer, R.G. and N.O. de Adler, Electron content over Tucuman, *J. Geophys. Res.*, 94, 9029–9032, 1989.

Fatkullin, M.N., Topside ionosphere disturbance effects during different phases of two successive magnetic storms in September, 1963, *Planet. Space Sci.*, 627–636, 1972.

Fatkullin, M.N., Storms and the seasonal anomaly in the topside ionosphere, *J. Atmos. Terr. Phys.*, 35, 453–468, 1973.

Fejer, B.G., Low latitude electrodynamic drifts: a review, *J. Atmos. Terr. Phys.*, 53, 677–693, 1991.

Fejer, B.G., R.W. Spiro, R.A. Wolf, and J.C. Foster, Latitudinal variations of perturbation electric fields during magnetically disturbed periods: 1986 SUNDIAL observations and model results, *Ann. Geophys.*, 8, 441–454, 1990a.

Fejer, B.G., M.C. Kelley, C. Senior, O. de la Beaujardiere, J.A. Holt, C.A. Tepley, R. Burnside, M.A. Abdu, J.H.A. Sobral, R.F. Woodman, Y. Kamide, and R. Lepping, Low- and mid-latitude ionospheric electric fields during the January 1984 GISMOS campaign, *J. Geophys. Res.*, 95, 2367–2377, 1990b.

Feldstein, Y.I., Modelling of the magnetic field of magnetospheric ring current as a function of interplanetary medium parameters, *Space Sci. Rev.*, 59, 83–165, 1992.

Fesen, C.G., G. Crowley, and R.G. Roble, Ionospheric effects at low latitudes during the March 22, 1979, geomagnetic storm, *J. Geophys. Res.*, 94, 5405–5417, 1989.

Feynman, J., Geomagnetic and solar wind cycles, 1900–1975, *J. Geophys. Res.*, 87, 6153–6162, 1982.

Förster, M., N. Jakowski, A. Best, and J. Smilauer, Plasmaspheric response to the geomagnetic storm period March 20–23, 1990, observed by the ACTIVNY (MAGION-2) satellite, *Can. J. Phys.*, 70, 569–574, 1992.

Forbes, J.M., Evidence for the equatorward penetration of electric fields, winds, and compositional effects in the Asian/Pacific sector during the September 17–24, 1984, ETS interval, *J. Geophys. Res.*, 94, 16999–17007, 1989.

Forbes, J.M., R.G. Roble, and F.A. Marcos, Thermospheric dynamics during the March 22, 1979, magnetic storm. 2. Comparison of model predictions with observations, *J. Geophys. Res.*, 92, 6069–6081, 1987.

Forbes, J.M., M. Codrescu, and T.J. Hall, On the utilization of ionosonde data to analyze the latitudinal penetration of ionospheric storm effects, *Geophys. Res. Lett.*, 15, 249–252, 1988.

Foster, J.C., Storm time plasma transport at middle and high latitudes, *J. Geophys. Res.*, 98, 1675–1689, 1993.

Foster, J.C. and J.R. Doupnik, Plasma convection in the vicinity of the dayside cleft, *J. Geophys. Res.*, 89, 9107–9113, 1984.

Foster, J.C., J.-P. St. Maurice, and V.J. Abreu, Joule heating at high latitudes, *J. Geophys. Res.*, 88, 4885–4896, 1983.

Foster, J.C., J.M. Holt, R.E. Musgrove, and D.S. Evans, Ionospheric convection associated with discrete levels of particle precipitation, *Geophys. Res. Lett.*, 13, 656–659, 1986.

Fuller-Rowell, T.J. and D. Rees, A three-dimensional, time dependent simulation of the global dynamical response of the thermosphere to a geomagnetic substorm, *J. Atmos. Terr. Phys.*, 43, 701–721, 1981.

Fuller-Rowell, T.J., D. Rees, S. Quegan, R.J. Moffett, and G.J. Bailey, Interactions between neutral thermospheric composition and the polar ionosphere using a coupled ionosphere-thermosphere model, *J. Geophys. Res.*, 92, 7744–7748, 1987.

Fuller-Rowell, T.J., D. Rees, S. Quegan, and F.J. Moffett, Numerical simulations of the sub-auroral F-region trough, *J. Atmos. Terr. Phys.*, 53, 529–540, 1991a.

Fuller-Rowell, T.J., D. Rees, H. Rishbeth, A.G. Burns, T.L. Killeen, and R.G. Roble, Modelling of composition changes during F-region storms: a reassessment, *J. Atmos. Terr. Phys.*, 53, 541–550, 1991b.

Fuller-Rowell, T.J., M.V. Codrescu, R.J. Moffett, and S. Quegan, Response of the thermosphere and ionosphere to geomagnetic storms, *J. Geophys. Res.*, 99, 3893–3914, 1994.

Giraud, A. and M. Petit, *Ionospheric Techniques and Phenomena*, D. Reidel, Dordrecht, 1978.

Goel, M.K. and B.C.N. Rao, Ionospheric variations during the 13 September 1967 storm: Incoherent scatter radar results, *J. Geomagn. Geoelectr.*, 32, 561–565, 1980.

Goel, M.K. and B.C.N. Rao, Theoretical calculations on the changes in the ionospheric composition and temperature profiles during storms, *J. Atmos. Terr. Phys.*, 43, 111–120, 1981.

Goel, M.K. and A.R. Jain, Changes in mid-latitude F_2 region associated with a major substorm event, *J. Atmos. Terr. Phys.*, 50, 579–584, 1988.

Gogoshev, M.M., K.B. Serafimov, Ts.N. Gogosheva, and K.M. Kazakov, The behaviour of the midlatitude F-region at the time of four magnetospheric substorms during the night of 29/30 October 1973, *Planet. Space Sci.*, 24, 293–298, 1976.

Gondhalekar, P.M., The behaviour of the topside ionosphere during magnetically disturbed conditions, *J. Atmos. Terr. Phys.*, 35, 1293–1298, 1973.

Grebowsky, J.M., H.A. Taylor, Jr., and J.M. Lindsay, Location and source of ionospheric high latitude troughs, *Planet. Space Sci.*, 31, 99–105, 1983.

Greenspan, M.E., C.E. Rasmussen, W.J. Burke, and M.A. Abdu, Equatorial density depletions observed at 840 km during the great magnetic storm of March 1989, *J. Geophys. Res.*, 96, 13931–13942, 1991.

Häggström, I. and P.N. Collis, Ion composition changes during F-region density depletions in the presence of electric fields at auroral latitudes, *J. Atmos. Terr. Phys.*, 52, 519–529, 1990.

Hafstad, L.R. and M.A. Tuve, Note on Kennelly-Heaviside layer observations during a magnetic storm, *Terr. Magn. Atmos. Electr.*, 34, 39–43, 1929.

Hagan, M.E., Effects of geomagnetic activity in the winter thermosphere. 2. Magnetically disturbed conditions, *J. Geophys. Res.*, 93, 9937–9944, 1988.

Hajkowicz, L.A., A global study of large scale travelling ionospheric disturbances (TIDs) following a step-like onset of auroral substorms in both hemispheres, *Planet. Space Sci.*, 38, 913–923, 1990.

Hajkowicz, L.A., Global onset and propagation of large-scale travelling ionospheric disturbances as a result of the great storm of 13 March 1989, *Planet. Space Sci.*, 39, 583–593, 1991a.

Hajkowicz, L.A., Auroral electrojet effect on the global occurrence pattern of large scale travelling ionospheric disturbances, *Planet. Space Sci.*, 39, 1189–1196, 1991b.

Hajkowicz, L.A., Universal time effect in the occurrences of large-scale ionospheric disturbances, *Planet. Space Sci.*, 40, 1093–1099, 1992.

Harang, L., *The Aurorae*, Chapman and Hall, London, 1951.

Hardy, D.A., M.S. Gussenhoven, and E. Holeman, A statistical model of auroral electron precipitation, *J. Geophys. Res.*, 90, 4229–4248, 1985.

Harel, M., R.A. Wolf, P.H. Reiff, R.W. Spiro, W.J. Burke, F.J. Rich, and M. Smiddy, Quantitative simulation of a magnetospheric substorm. 1. Model logic and overview, *J. Geophys. Res.*, 86, 2217–2241, 1981.

Hargreaves, J.K., *The Solar-Terrestrial Environment*, Cambridge University Press, Cambridge, U.K., 1992.

Hargreaves, J.K. and F. Bagenal, The behavior of the electron content during ionospheric storms: a new method of presentation and comments on the positive phase, *J. Geophys. Res.*, 82, 731–733, 1977.

Hays, P.B. and R.G. Roble, Direct observations of thermospheric winds during geomagnetic storms, *J. Geophys. Res.*, 76, 5316–5321, 1971.

Hearn, A.L., Longitudinal effect in the response of total electron content to magnetic storms, *Nature*, 249, 133, 1974.

Hedin, A.E., P. Bauer, H.G. Mayr, G.R. Carignan, L.H. Brace, H.C. Brinton, A.D. Parks, and D.T. Pelz, Observations of neutral composition and related ionospheric variations during a magnetic storm in February 1974, *J. Geophys. Res.*, 82, 3183–3189, 1977.

Heelis, R.A., J.K. Lowell, and R.W. Spiro, A model of the high-latitude ionospheric convection pattern, *J. Geophys. Res.*, 87, 6339–6345, 1982.

Heppner, J.P. and N.C. Maynard, Empirical high-latitude electric field models, *J. Geophys. Res.*, 92, 4467–4489, 1987.

Hernandez, G. and T. Killeen, Optical measurements of winds and kinetic temperatures in the upper atmosphere, *Adv. Space Res.*, 8, No. 5, 149–213, 1988.

Hernandez, G. and R.G. Roble, Direct measurements of nighttime thermospheric winds and temperatures. 2. Geomagnetic storms, *J. Geophys. Res.*, 81, 5173–5181, 1976.

Hoegy, W.R. and J.M. Grebowsky, Dependence of polar hole density on magnetic and solar conditions, *J. Geophys. Res.*, 96, 5737–5755, 1991.

Huang, T.-X. Major ionospheric storms during July–September 1982 at lower latitudes in East Asia, *J. Atmos. Terr. Phys.*, 47, 1031–1036, 1985.

Huang, Y.-N. and K. Cheng, Ionospheric disturbances at the equatorial anomaly crest region during the March 1989 magnetic storms, *J. Geophys. Res.*, 96, 13953–13965, 1991.

Huang, Y.-N. and K. Cheng, Ionospheric disturbances around East Asian region during the 20 October 1989 magnetic storm, *J. Atmos. Terr. Phys.*, 55, 1009–1020, 1993.

Huang, Y.-N., K. Najita, T.H. Roelofs, and P.C. Yuen, Dst and SD variations of electron content at low latitude, *J. Atmos. Terr. Phys.*, 36, 9–28, 1974.

Hunsucker, R.D., Atmospheric gravity waves generated in the high-latitude ionosphere: a review, *Rev. Geophys. Space Phys.*, 20, 293–315, 1982.

Inoue, T., Sh. Handa, and S. Maeda, Ionospheric F2-region storms at mid-latitudes starting in the daytime and the nighttime, *J. Atmos. Terr. Phys.*, 40, 1081–1084, 1978.

Ivanov-Kholodny, G.S. and A.V. Mikhailov, *The Prediction of Ionospheric Conditions*, D. Reidel, Dordrecht, 1986.

Jakowski, N., E. Putz, and P. Spalla, Ionospheric storm characteristics deduced from satellite radio beacon observations at three European stations, *Ann. Geophys.*, 8, 343–352, 1990.

Jakowski, N., A. Jungstand, K. Schlegel, H. Kohl, and K. Rinnert, The ionospheric response to perturbation electric fields during the onset phase of geomagnetic storms, *Can. J. Phys.*, 70, 575–581, 1992.

Jones, G.O.L., P.J.S. Williams, K.J. Winser, and M. Lockwood, Characteristics of the high-latitude trough, *Adv. Space Res.*, 10, No. 6, 191–196, 1990.

Jones, K.L., Storm time variation of F2-layer electron concentration, *J. Atmos. Terr. Phys.*, 33, 379–389, 1971a.

Jones, K.L., Electrodynamic drift effects in mid-latitude F-region storm phenomena, *J. Atmos. Terr. Phys.*, 33, 1311–1319, 1971b.

Jones, K.L., Wind, electric field and composition perturbations of the mid-latitude F-region during magnetic storms, *J. Atmos. Terr. Phys.*, 35, 1515–1527, 1973.

Jones, K.L. and H. Rishbeth, The origin of storm increases of mid-latitude F-layer electron concentration, *J. Atmos. Terr. Phys.*, 33, 391–401, 1971.

Jones, T.B. and K. Davies, Electron density and temperature changes in the equatorial ionosphere during magnetic storms, *J. Atmos. Terr. Phys.*, 36, 1071–1078, 1974.

Joselyn, J.A. and B.T. Tsurutani, Geomagnetic sudden impulses and storm sudden commencements, a note on terminology, *Eos Trans. AGU*, 71, 1808–1809, 1990.

Jung, M.J. and G.W. Prölss, Numerical simulation of negative ionospheric storms using observed neutral composition data, *J. Atmos. Terr. Phys.*, 40, 1347–1350, 1978.

Kamide, Y. and J.A. Joselyn, Toward a standardized definition of geomagnetic sudden impulses and storm sudden commencements, *EOS Trans. AGU*, 72, 300, 1991.

Kamide, Y. and S. Matsushita, Penetration of high-latitude electric fields into low latitudes, *J. Atmos. Terr. Phys.*, 43, 411–425, 1981.

Kamide, Y., A.D. Richmond, and S. Matsushita, Estimation of ionospheric electric fields, ionospheric currents, and field-aligned currents from ground magnetic records, *J. Geophys. Res.*, 86, 801–813, 1981.

Kamiyama, H., Ionospheric changes associated with geomagnetic bays, *Sci. Rep. Tohoku Univ., Geophys.*, 7, 125–135, 1956.

Kane, R.P., Storm-time variation of F2 region, *Ann. Geophys.*, 29, 25–42, 1973a.

Kane, R.P., Global evolution of F2-region storms, *J. Atmos. Terr. Phys.*, 35, 1953–1966, 1973b.

Kane, R.P., Global evolution of the ionospheric electron content during some geomagnetic storms, *J. Atmos. Terr. Phys.*, 37, 601–611, 1975.

Katz, A.H. and M.D. Papagiannis, Nighttime changes of the ionosphere during geomagnetic storms, *J. Atmos. Terr. Phys.*, 34, 525–530, 1972.

Kelley, M.C., *The Earth's Ionosphere*, Academic Press, New York, 1989.

Kersken, H.-P., G.W. Prölss, and M. Roemer, Convective transport of composition perturbations, *Adv. Space Res.*, 12, No. 10, 261–264, 1992.

Kikuchi, T., T. Araki, H. Maeda, and K. Maekawa, Transmission of polar electric fields to the equator, *Nature*, 273, 650–651, 1978.

Kilifarska, N.A., Longitudinal effects in the ionosphere during geomagnetic storms, *Adv. Space Res.*, 8, No. 4, 23–26, 1988.

Kilifarska, N.A., Ts.P. Dachev, A.S. Besprozvannaya, and T.I. Schuka, An influence of IMF-By on the ion density planetary distribution during disturbed period 20–23 October 1981: satellite and ground-based results, *Adv. Space Res.*, 11, No. 10, 47–50, 1991.

Killeen, T.L. and R.G. Roble, An analysis of the high-latitude thermospheric wind pattern calculated by a thermospheric general circulation model. 2. Neutral parcel transport, *J. Geophys. Res.*, 91, 11291–11307, 1986.

King, G.A.M., The ionospheric F region during a storm, *Planet. Space Sci.*, 9, 95–100, 1962.

King, G.A.M., The ionospheric F-region storm, *J. Atmos. Terr. Phys.*, 33, 1223–1240, 1971.

King, J.W., K.C. Reed, E.O. Olatunji, and A.J. Legg, The behaviour of the topside ionosphere during storm conditions, *J. Atmos. Terr. Phys.*, 29, 1355–1363, 1967.

Kirby, S.S., T.R. Gilliland, E.B. Judson, and N. Smith, The ionosphere, sunspots, and magnetic storms, *Phys. Rev.*, 48, 849, 1935.

Kirby, S.S., T.R. Gilliland, N. Smith, and S.E. Reymer, The ionosphere, solar eclipse and magnetic storm, *Phys. Rev.*, 50, 258–259, 1936.

Kirby, S.S., N. Smith, T.R. Gilliland, and S.E. Reymer, The ionosphere and magnetic storms, *Phys. Rev.*, 51, 992–993, 1937.

Klobuchar, J.A., M. Mendillo, F.L. Smith, III, R.B. Fritz, A.V. da Rosa, M.J. Davis, P.C. Yuen, T.H. Roelofs, K.C. Yeh, and B.J. Flaherty, Ionospheric storm of March 8, 1970, *J. Geophys. Res.*, 76, 6202–6207, 1971.

Knecht, R.W., Observations of the ionosphere over the South Geographic Pole, *J. Geophys. Res.*, 64, 1243–1250, 1959.

Knudsen, W.C., Magnetospheric convection and the high-latitude F_2 ionosphere, *J. Geophys. Res.*, 79, 1046–1055, 1974.

Köhnlein, W., A model of the terrestrial ionosphere in the altitude interval 50–4000 km I. Atomic ions (H^+, He^+, N^+, O^+), *Earth, Moon and Planets*, 45, 53–100, 1989a.

Köhnlein, W., A model of the terrestrial ionosphere in the altitude interval 50–4000 km. II. Molecular ions (N_2^+, NO^+, O_2^+) and electron density, *Earth, Moon and Planets*, 47, 109–163, 1989b.

Kotadia, K.M., The great magnetic storm of 11 February 1958 and associated changes in the F2-layer of the ionosphere in low and middle latitudes, *J. Atmos. Terr. Phys.*, 24, 975–988, 1962.

Kotadia, K.M., Variations in critical frequency of the F2-layer of the ionosphere associated with geomagnetic storms at equatorial stations, *J. Atmos. Terr. Phys.*, 27, 723–733, 1965.

Kotadia, K.M. and K.G. Jani, Effects of magnetic storms on the F2-layer of the ionosphere near the boundary of the equatorial zone, *J. Atmos. Terr. Phys.*, 29, 661–672, 1967.

Labelle, J., R.J. Sica, C. Kletzing, G.D. Earle, M.C. Kelley, D. Lummerzheim, R.B. Torbert, K.D. Baker, and G. Berg, Ionization from soft electron precipitation in the auroral F region, *J. Geophys. Res.*, 94, 3791–3798, 1989.

Lakshmi, D.R. and B.M. Reddy, Nighttime response of the topside ionosphere to magnetic storms, *J. Geophys. Res.*, 75, 4335–4338, 1970.

Lakshmi, D.R., B.C.N. Rao, A.R. Jain, M.K. Goel, and B.M. Reddy, Response of equatorial and low latitude F-region to the great magnetic storm of 13 March 1989, *Ann. Geophys.*, 9, 286–290, 1991.

Lal, C., Global F2 layer ionization and geomagnetic activity, *J. Geophys. Res.*, 97, 12153–12159, 1992.

Lanzerotti, L.J., L.L. Cogger, and M. Mendillo, Latitude dependence of ionosphere total electron content: Observations during sudden commencement storms, *J. Geophys. Res.*, 80, 1287–1306, 1975.

Leitinger, R., G.K. Hartmann, W. Degenhardt, A. Hedberg, and P. Tanskanen, The electron content of the ionosphere and the southern boundary of diffuse aurora, *J. Atmos. Terr. Phys.*, 44, 369–374, 1982.

Leitinger, R., P. Wilkinson, and R. Hanbaba, The ionosphere in mid-latitudes during the SUNDIAL campaign, *Ann. Geophys.*, 6, 59–68, 1988.

Lepechinsky, D., Effects of temperature variations of the upper atmosphere on the formation of ionospheric layers, *J. Atmos. Terr. Phys.*, 1, 278–285, 1951.

Lewis, R.P.W. and D.H. McIntosh, Diurnal and storm-time variations of geomagnetic and ionospheric disturbance, *J. Atmos. Terr. Phys.*, 3, 186–193, 1953.

Lockwood, M., Incoherent scatter radar measurements of the cusp, in *Proceedings of ESA Cluster Workshop*, ESA, Nevilly, France, 1992.

Low, N.C. and T.H. Roelofs, On the large scale vertical movements of the F-layer and its effects on the total electron content over low latitude during the magnetic storm of 25 May 1967, *Planet. Space Sci.*, 21, 1805–1810, 1973.

Maeda, K.I. and T. Sato, The F region during magnetic storms, *Proc. Inst. Rad. Eng.*, 47, 232–239, 1959.

Maehlum, B., The diurnal variation of f_oF2 near the auroral zone during magnetic disturbances, *J. Atmos. Terr. Phys.*, 13, 187–190, 1958.

Martyn, D.F., The theory of magnetic storms and auroras, *Nature*, 167, 92–94, 1951.

Martyn, D.F., The morphology of the ionospheric variations associated with magnetic disturbance. I. Variations at moderately low latitudes, *Proc. R. Soc.*, A218, 1–18, 1953a.

Martyn, D.F., Geo-morphology of F_2-region ionospheric storms, *Nature,* 171, 14–16, 1953b.

Marubashi, K., C.A. Reber, and H.A. Taylor, Jr., Geomagnetic storm effects on the thermosphere and the ionosphere revealed by in situ measurements from OGO 6, *Planet. Space Sci.*, 24, 1031–1041, 1976.

Matsushita, S., A study of the morphology of ionospheric storms, *J. Geophys. Res.*, 64, 305–321, 1959.

Matsushita, S., Equatorial ionospheric variations during geomagnetic storms, *J. Geophys. Res.*, 68, 2595–2601, 1963.

Matsushita, S., Ionospheric and thermospheric responses during August 1972 storms — a review, *Space Sci. Rev.*, 19, 713–737, 1976.

Matuura, N., Thermal effect on the ionospheric F region disturbance, *J. Rad. Res. Lab. Jpn.* 10, 1–35, 1963.

Matuura, N., Theoretical models of ionospheric storms, *Space Sci. Rev.*, 13, 124–189, 1972.

Mayaud, P.N., Derivation, meaning, and use of geomagnetic indices, *Geophys. Monogr.* 22, AGU, Washington, D.C., 1980.

Mayr, H.G. and H. Volland, Magnetic storm effects in the neutral composition, *Planet. Space Sci.*, 20, 379–393, 1972.

Mayr, H.G. and H. Volland, Magnetic storm characteristics of the thermosphere, *J. Geophys. Res.*, 78, 2251–2264, 1973.

Mayr, H.G., I. Harris, and N.W. Spencer, Some properties of upper atmosphere dynamics, *Rev. Geophys. Space Sci.*, 16, 539–565, 1978.

Mazaudier, C., R. Bernard, and S.V. Venkateswaran, Saint-Santin radar observations of lower thermospheric storms, *J. Geophys. Res.*, 90, 6685–6686, 1985.

Mazaudier, C., A.D. Richmond, and D. Brinkman, On thermospheric winds produced by auroral heating during magnetic storms and associated dynamo electric fields, *Ann. Geophys.*, 5, 443–448, 1987.

McCormac, F.G., T.L. Killeen, J.P. Thayer, G. Hernandez, C.R. Tschan, and J.-J. Ponthieu, Circulation of the polar thermosphere during geomagnetically quiet and active times as observed by Dynamics Explorer 2, *J. Geophys. Res.*, 92, 10133–10139, 1987.

Meek, J.H., Correlation of magnetic, auroral, and ionospheric variations at Saskatoon, *J. Geophys. Res.*, 58, 445–456, 1953.

Mendillo, M., A study of the relationship between geomagnetic storms and ionospheric disturbances at mid-latitudes, *Planet. Space Sci.*, 21, 349–358, 1973.

Mendillo, M. and J.A. Klobuchar, An Atlas of the Midlatitude F-Region Response to Geomagnetic Storms, Tech. Rep. 74-0065, Air Force Cambridge Research Lab, Cambridge, MA, 267 pp., 1974.

Mendillo, M. and J.A. Klobuchar, Investigations of the ionospheric F region using multistation total electron content observations, *J. Geophys. Res.*, 80, 643–650, 1975.

Mendillo, M., M.D. Papagiannis, and J.A. Klobuchar, Ionospheric storms at midlatitudes, *Radio Sci.*, 5, 895–898, 1970.

Mendillo, M., M.D. Papagiannis, and J.A. Klobuchar, Average behavior of the midlatitude F-region parameters N_T, N_{max}, and τ during geomagnetic storms, *J. Geophys. Res.*, 77, 4891–4895, 1972.

Mendillo, M., J.A. Klobuchar, and H. Hajeb-Hosseinieh, Ionospheric disturbances: evidence for the contraction of the plasmasphere during severe geomagnetic storms, *Planet. Space Sci.*, 22, 223–236, 1974.

Mendillo, M., J. Baumgardner, J. Aarons, J. Foster, and J. Klobuchar, Coordinated optical and radio studies of ionospheric disturbances: initial results from Millstone Hill, *Ann. Geophys.*, 5A, 543–550, 1987.

Mendillo, M., X.-Q. He, and H. Rishbeth, How the effects of winds and electric fields in F2-layer storms vary with latitude and longitude: a theoretical study, *Planet. Space Sci.*, 40, 595–606, 1992.

Mikhailov, A.V., Yu. L. Terekhin, M.G. Skoblin, and V.V. Mikhailov, On the physical mechanism of the ionospheric storms in the F2-layer, *Adv. Space Res.*, 12, No. 10, 269–272, 1992.

Mikhailov, A.V., M. Förster, and M.G. Skoblin, Neutral gas composition changes and ExB vertical plasma drift contribution to the daytime equatorial F2-region storm effects, *Ann. Geophys.*, 12, 226–231, 1994.

Mikkelsen, I.S., Enhancements of the auroral zone ionization during substorms, *Planet. Space Sci.*, 23, 619–626, 1975.

Miller, K.L., P.G. Richards, and H.Y. Wu, A global-scale study of meridional winds and electron densities in the F-region during the SUNDIAL 1987 campaign, *Ann. Geophys.*, 11, 572–584, 1993.

Miller, N.J., J.M. Grebowsky, H.G. Mayr, I. Harris, and Y.K. Tulunay, F layer positive response to a geomagnetic storm — June 1972, *J. Geophys. Res.*, 84, 6493–6500, 1979.

Miller, N.J., H.G. Mayr, N.W. Spencer, L.H. Brace, and G.R. Carignan, Observations relating changes in thermospheric composition to depletions in topside ionization during the geomagnetic storm of September 1982, *J. Geophys. Res.*, 89, 2389–2394, 1984.

Miller, N.J., L.H. Brace, N.W. Spencer, and G.R. Carignan, DE 2 observations of disturbances in the upper atmosphere during a geomagnetic storm, *J. Geophys. Res.*, 95, 21017–21031, 1990.

Millward, G.H., S. Quegan, R.J. Moffett, T.J. Fuller-Rowell, and D. Rees, A modelling study of the coupled ionospheric and thermospheric response to an enhanced high-latitude electric field event, *Planet. Space Sci.*, 41, 45–56, 1993a.

Millward, G.H., R.J. Moffett, S. Quegan, and T.J. Fuller-Rowell, Effects of an atmospheric gravity wave on the midlatitude ionospheric F layer, *J. Geophys. Res.*, 98, 19173–19179, 1993b.

Moffett, R.J. and J.A. Murphy, Coupling between the F-region and protonosphere: numerical solution of the time-dependent equations, *Planet. Space Sci.*, 21, 43–52, 1973.

Moffett, R.J. and S. Quegan, The mid-latitude trough in the electron concentration of the ionospheric F-layer: A review of observations and modelling, *J. Atmos. Terr. Phys.*, 45, 315–343, 1983.

Moffett, R.J., J.A. Murphy, and G.J. Bailey, Storm-time increases in the ionospheric total electron content, *Nature*, 253, 330–331, 1975.

Moffett, R.J., R.A. Heelis, R. Sellek, and G.J. Bailey, The temporal evolution of the ionospheric signatures of subauroral ion drifts, *Planet. Space Sci.*, 40, 663–670, 1992.

Muldrew, D.B., The relationship of F-layer critical frequencies to the intensity of the outer Van Allen belt, *Can. J. Phys.*, 41, 199–202, 1963.

Muldrew, D.B., F-layer ionization troughs deduced from Alouette data, *J. Geophys. Res.*, 70, 2635–2650, 1965.

Namboothiri, S.P., N. Balan, and P.B. Rao, Vertical plasma drifts in the F region at the magnetic equator, *J. Geophys. Res.*, 94, 12055–12060, 1989.

Namgaladze, A.A., L.P. Zakharov, and A.N. Namgaladze, Numerical simulation of ionospheric storms, *Geomagn. Aeron.*, 21, 184–187, 1981.

Nazerets, V.P. and O.A. Troshichev, Effect of magnetic disturbances on the altitude of the F layer in the midlatitude ionosphere, *Geomagn. Aeron.*, 24, 108–109, 1984.

Nelms, G.L. and J.H. Chapman, The high latitude ionosphere: results from the Alouette/ISIS topside sounders, in *The Polar Ionosphere and Magnetospheric Processes* (G. Skovli, Ed.), 233–269, Gordon and Breach, London, England, 1970.

Nelson, G.J. and L.L. Cogger, Enhancements in electron content at Arecibo during geomagnetic storms, *Planet. Space Sci.*, 19, 761–775, 1971.

Newell, P.T., W.J. Burke, C.-I. Meng, E.R. Sanchez, and M.E. Greenspan, Identification and observations of the plasma mantle at low altitude, *J. Geophys. Res.*, 96, 35–45, 1991.

Newton, G.P. and J.C.G. Walker, Electron density decrease in SAR arcs resulting from vibrationally excited nitrogen, *J. Geophys. Res.*, 80, 1325–1327, 1975.

Nisbet, J.S., M.J. Miller, and L.A. Carpenter, Currents and electric fields in the ionosphere due to field-aligned auroral currents, *J. Geophys. Res.*, 83, 2647–2657, 1978.

Nishida, A., Average structure and storm-time change of the polar topside ionosphere at sunspot minimum, *J. Geophys. Res.*, 72, 6051–6061, 1967.

Noël, S. and G.W. Prölss, Heating and radiation production by neutralized ring current particles, *J. Geophys. Res.*, 98, 17317–17325, 1993.

Nopper, Jr., R.W. and R.L. Carovillano, Polar-equatorial coupling during magnetically active periods, *Geophys. Res. Lett.*, 5, 699–702, 1978.

Norton, R.B., The middle-latitude F region during some severe ionospheric storms, *Proc. IEEE*, 57, 1147–1149, 1969.

Obayashi, T., Morphology of storms in the ionosphere, *Research in Geophysics*, Vol. 1 (H. Odishaw, Ed.), MIT Press, Cambridge, MA, 335–366, 1964.

Obayashi, T., World-wide electron density changes and associated thermospheric winds during an ionospheric storm, *Planet. Space Sci.*, 20, 511–520, 1972.

Obayashi, T. and N. Matuura, Theoretical model of F-region storms, in *Solar-Terrestrial Physics IV* (Dyer, E.R., Ed.), D. Reidel, Dordrecht, 199–211, 1972.

Olatunji, E.O., Some features of equatorial ionospheric storms, *Ann. Geophys.*, 22, 485–491, 1966.

Oliver, W.L., S. Fukao, T. Sato, T. Tsuda, S. Kato, I. Kimura, A. Ito, T. Saryou, and T. Araki, Ionospheric incoherent scatter measurements with the middle and upper atmosphere radar: observations during the large magnetic storm of February 6–8, 1986, *J. Geophys. Res.*, 93, 14649–14655, 1988.

Oliver, W.L., S. Fukao, T. Takami, T. Tsuda, and S. Kato, Four-beam measurements of ionospheric structure with the MU radar during the low-latitude auroral event of 20–23 October 1989, *Geophys. Res. Lett.*, 18, 1975–1978, 1991.

Papagiannis, M.D., M. Mendillo, and J.A. Klobuchar, Simultaneous storm-time increases of the ionospheric total electron content and the geomagnetic field in the dusk sector, *Planet. Space Sci.*, 19, 503–511, 1971.

Papagiannis, M.D., H. Hajeb-Hosseinieh, and M. Mendillo, Changes in the ionospheric profile and the Faraday factor $\overline{M}$ with Kp, *Planet. Space Sci.*, 23, 107–113, 1975.

Park, C.G., Westward electric fields as the cause of nighttime enhancements in electron concentrations in midlatitude F region, *J. Geophys. Res.*, 76, 4560–4568, 1971.

Park, C.G., A morphological study of substorm-associated disturbances in the ionosphere, *J. Geophys. Res.*, 79, 2821–2827, 1974.

Park, C.G., Substorm electric fields in the evening plasmasphere and their effects on the underlying F layer, *J. Geophys. Res.*, 81, 2283–2288, 1976.

Park, C.G. and P.M. Banks, Influence of thermal plasma flow on the mid-latitude nighttime F_2 layer: Effects of electric fields and neutral winds inside the plasmasphere, *J. Geophys. Res.*, 79, 4661–4668, 1974.

Park, C.G. and C.-I. Meng, Vertical motions of the midlatitude F_2 layer during magnetospheric substorms, *J. Geophys. Res.*, 76, 8326–8332, 1971.

Park, C.G. and C.-I. Meng, Distortions of the nightside ionosphere during magnetospheric substorms, *J. Geophys. Res.*, 78, 3828–3840, 1973.

Park, C.G. and C.-I. Meng, Aftereffects of isolated magnetospheric substorm activity on the mid-latitude ionosphere: localized depressions in F layer electron densities, *J. Geophys. Res.*, 81, 4571–4578, 1976.

Paul, M.P., S. Matsushita, and A.D. Richmond, Ionospheric storm of 4–5 August 1972 in the Asia-Australia-Pacific sector, *J. Atmos. Terr. Phys.*, 39, 43–50, 1977.

Pavlov, A.V., The role of vibrationally excited nitrogen in the formation of the mid-latitude negative ionospheric storms, *Ann. Geophys.*, 12, 554–564, 1994.

Pi, X., M. Mendillo, M.W. Fox, and D.N. Anderson, Diurnal double maxima patterns in the F region ionosphere: substorm-related aspects, *J. Geophys. Res.*, 98, 13677–13691, 1993.

Piddington, J.H., Ionospheric and magnetospheric anomalies and disturbances, *Planet. Space Sci.*, 12, 553–566, 1964.

Prakash, S. and R. Pandey, Generation of electric fields due to the gravity wave winds and their transmission to other ionospheric regions, *J. Atmos. Terr. Phys.*, 47, 363–374, 1985.

Prölss, G.W., Seasonal variations of atmospheric-ionospheric disturbances, *J. Geophys. Res.*, 82, 1635–1640, 1977.

Prölss, G.W., Magnetic storm associated perturbations of the upper atmosphere: recent results obtained by satellite-borne gas analyzers, *Rev. Geophys. Space Phys.*, 18, 183–202, 1980.

Prölss, G.W., Latitudinal structure and extension of the polar atmospheric disturbance, *J. Geophys. Res.*, 86, 2385–2396, 1981.

Prölss, G.W., Perturbation of the low-latitude upper atmosphere during magnetic substorm activity, *J. Geophys. Res.*, 87, 5260–5266, 1982.

Prölss, G.W., Correlation between upper atmospheric temperature and solar wind conditions, *J. Geophys. Res.*, 90, 11096–11100, 1985.

Prölss, G.W., Storm-induced changes in the thermospheric composition at middle latitudes, *Planet. Space Sci.*, 35, 807–811, 1987.

Prölss, G.W., Thermosphere-ionosphere coupling during disturbed conditions, *J. Geomagn. Geoelectr.*, 43, 537–549, 1991.

Prölss, G.W., Satellite mass spectrometer measurements of composition changes, *Adv. Space Res.*, 12, No. 10, 241–251, 1992.

Prölss, G.W., On explaining the local time variation of ionospheric storm effects, *Ann. Geophys.*, 11, 1–9, 1993a.

Prölss, G.W., Common origin of positive ionospheric storms at middle latitudes and the geomagnetic activity effect at low latitudes, *J. Geophys. Res.*, 98, 5981–5991, 1993b.

Prölss, G.W. and K. Najita, Magnetic storm associated changes in the electron content at low latitudes, *J. Atmos. Terr. Phys.*, 37, 635–643, 1975.

Prölss, G.W. and M.J. Jung, Travelling atmospheric disturbances as a possible explanation for daytime positive storm effects of moderate duration at middle latitudes, *J. Atmos. Terr. Phys.*, 40, 1351–1354, 1978.

Prölss, G.W. and U. von Zahn, Esro 4 gas analyzer results. 2. Direct measurements of changes in the neutral composition during an ionospheric storm, *J. Geophys. Res.*, 79, 2535–2539, 1974.

Prölss, G.W. and U. von Zahn, On the local time variation of atmospheric-ionospheric disturbances, *Space Res.*, 18, 159–162, 1978.

Prölss, G.W., U. von Zahn, and W.J. Raitt, Neutral atmospheric composition, plasma density, and electron temperature at F region heights, *J. Geophys. Res.*, 80, 3715–3718, 1975.

Prölss, G.W., M. Roemer, and J.W. Slowey, Dissipation of solar wind energy in the earth's upper atmosphere: The geomagnetic activity effect, CIRA 1986, *Adv. Space Res.*, 8, No. 5, 215–261, 1988.

Prölss, G.W., L.H. Brace, H.G. Mayr, G.R. Carignan, T.L. Killeen and J.A. Klobuchar, Ionospheric storm effects at subauroral latitudes: a case study, *J. Geophys. Res.*, 96, 1275–1288, 1991.

Providakes, J.F., M.C. Kelley, W.E. Swartz, M. Mendillo, and J. Holt, Radar and optical measurements of ionospheric processes associated with intense subauroral electric fields, *J. Geophys. Res.*, 94, 5350–5366, 1989.

Quegan, S., G.J. Bailey, R.J. Moffett, R.A. Heelis, T.J. Fuller-Rowell, D. Rees, and R.W. Spiro, A theoretical study of the distribution of ionization in the high-latitude ionosphere and the plasmasphere: first results on the mid-latitude trough and the light-ion trough, *J. Atmos. Terr. Phys.*, 44, 619–640, 1982.

Quegan, S., G.H. Millward, and T.J. Fuller-Rowell, A study of the evolution of a high latitude trough using a coupled ionosphere/thermosphere model, *Adv. Space Res.*, 12, No. 6, 161–169, 1992.

Raitt, W.J., U. von Zahn, and P. Christophersen, A comparison of thermospheric neutral gas heating and related thermal and energetic plasma phenomena at high latitudes during geomagnetic disturbances, *J. Geophys. Res.*, 80, 2277–2288, 1975.

Raghavarao, R. and M.P. Sivaraman, Enhancement of the equatorial anomaly in the topside ionosphere during magnetic storms, *J. Atmos. Terr. Phys.*, 35, 2091–2095, 1973.

Rajaram, G., Structure of the equatorial F-region, topside and bottomside — a review, *J. Atmos. Terr. Phys.*, 39, 1125–1144, 1977.

Rajaram, G. and R.G. Rastogi, A synoptic study of the disturbed ionosphere during IGY-IGC — (1) the Asian zone, *Ann. Geophys.*, 25, 795–805, 1969.

Rajaram, G. and R.G. Rastogi, North-South asymmetry of ionospheric storms — dependence on longitude and season, *J. Atmos. Terr. Phys.*, 32, 113–118, 1970.

Rajaram, G., A.C. Das, and R.G. Rastogi, Ionospheric F-region disturbances and their possible mechanisms, *Ann. Geophys.*, 27, 469–475, 1971.

Rasmussen, C.E. and M.E. Greenspan, Plasma transport in the equatorial ionosphere during the great magnetic storm of March 1989, *J. Geophys. Res.*, 98, 285–292, 1993.

Rastogi, R.G., The effect of geomagnetic activity on the F_2 region over Central Africa, *J. Geophys. Res.*, 67, 1367–1374, 1962.

Rastogi, R.G. and G. Rajaram, Abnormal behaviour of f_oF2 at Huancayo in magnetically active periods of IGY-IGC, *J. Atmos. Terr. Phys.*, 27, 1097–1103, 1965.

Ratcliffe, J.A., *An Introduction to the Ionosphere and Magnetosphere*, Cambridge University Press, Cambridge, U.K., 1972.

Reddy, B.M., L.H. Brace, and J.A. Findlay, The ionosphere at 640 kilometers on quiet and disturbed days, *J. Geophys. Res.*, 72, 2709–2727, 1967.

Reddy, C.A. and A. Nishida, Magnetospheric substorms and nighttime height changes of the F2 region at middle and low latitudes, *J. Geophys. Res.*, 97, 3039–3061, 1992.

Reddy, C.A., S. Fukao, T. Takami, M. Yamamoto, T. Tsuda, T. Nakamura, and S. Kato, A MU radar-based study of mid-latitude F region response to a geomagnetic disturbance, *J. Geophys. Res.*, 95, 21077–21094, 1990a.

Reddy, C.A., A. Nishida, S. Fukao, and V.V. Somayajulu, Magnetospheric substorm-related electric fields in the ionosphere: discrepancy of an observation with model predictions, *Geophys. Res. Lett.*, 17, 2333–2336, 1990b.

Rees, D., P.A. Rounce, G.T. Best, and A.F. Quesada, Midlatitude measurements of the thermospheric neutral wind during the Aladdin programme, *J. Atmos. Terr. Phys.*, 41, 1171–1178, 1979.

Rees, D., T.J. Fuller-Rowell, S. Quegan, R.J. Moffett, and G.J. Bailey, Thermospheric dynamics: understanding the unusual disturbances by means of simulations with a fully-coupled global thermospheric/high-latitude ionosphere model, *Ann. Geophys.*, 5A, 303–328, 1987.

Rich, F.J., and W.F. Denig, The major magnetic storm of March 13–14, 1989 and associated ionospheric effects, *Can. J. Phys.*, 70, 510–525, 1992.

Rees, M.H., *Physics and Chemistry of the Upper Atmosphere*, Cambridge University Press, Cambridge, U.K., 1989.

Richards, P.G., D.G. Torr, M.J. Buonsanto, and K.L. Miller, The behavior of the electron density and temperature at Millstone Hill during the equinox transition study September 1984, *J. Geophys. Res.*, 94, 16969–16975, 1989.

Richmond, A.D. and Y. Kamide, Mapping electrodynamic features of the high-latitude ionosphere from localized observations: technique, *J. Geophys. Res.*, 93, 5741–5759, 1988.

Richmond, A.D. and S. Matsushita, Thermospheric response to a magnetic substorm, *J. Geophys. Res.*, 80, 2839–2850, 1975.

Richmond, A.D., E.C. Ridley, and R.G. Roble, A thermosphere/ionosphere general circulation model with coupled electrodynamics, *Geophys. Res. Lett.*, 19, 601–604, 1992.

Rishbeth, H., Some problems of the F region, *Radio Sci.*, 9, 183–187, 1974.

Rishbeth, H., F-region storms and thermospheric circulation, *J. Atmos. Terr. Phys.*, 37, 1055–1064, 1975.

Rishbeth, H., On the F2-layer continuity equation, *J. Atmos. Terr. Phys.*, 48, 511–519, 1986.

Rishbeth, H., F-region storms and thermospheric dynamics, *J. Geomagn. Geoelectr.*, 43, Suppl., 513–524, 1991.

Rishbeth, H. and R. Edwards, The isobaric F2-layer, *J. Atmos., Terr. Phys.*, 51, 321–338, 1989.

Rishbeth, H. and O.K. Garriot, *Introduction to Ionospheric Physics*, Academic Press, New York, 1969.

Rishbeth, H. and W.B. Hanson, A comment on plasma "pile up" in the F-region, *J. Atmos. Terr. Phys.*, 36, 703–706, 1974.

Robinson, R.M., R.T. Tsunoda, J.F. Vickrey, and L. Guerin, Sources of F region ionization enhancements in the nighttime auroral zone, *J. Geophys. Res.*, 90, 7533–7546, 1985.

Roble, R.G., R.E. Dickinson, and E.C. Ridley, Seasonal and solar cycle variations of the zonal mean circulation in the thermosphere, *J. Geophys. Res.*, 82, 5493–5504, 1977.

Roble, R.G., A.D. Richmond, W.L. Oliver, and R.M. Harper, Ionospheric effects of the gravity wave launched by the September 18, 1974, sudden commencement, *J. Geophys. Res.*, 83, 999–1009, 1978.

Roble, R.G., J.M. Forbes, and F.A. Marcos, Thermospheric dynamics during the March 22, 1979, magnetic storm. 1. Model simulations, *J. Geophys. Res.*, 92, 6045–6068, 1987.

Roble, R.G., T.L. Killeen, N.W. Spencer, R.A. Heelis, P.H. Reiff, and J.D. Winningham, Thermospheric dynamics during November 21–22, 1981: dynamics Explorer measurements and thermospheric general circulation model predictions, *J. Geophys. Res.*, 93, 209–225, 1988.

Rodger, A.S., L.H. Brace, W.R. Hoegy, and J.D. Winningham, The poleward edge of the mid-latitude trough — its formation, orientation and dynamics, *J. Atmos. Terr. Phys.*, 48, 715–728, 1986.

Rodger, A.S., G.L. Wrenn, and H. Rishbeth, Geomagnetic storms in the Antarctic F-region. II. Physical interpretation, *J. Atmos. Terr. Phys.*, 51, 851–866, 1989.

Rodger, A.S., R.J. Moffett, and S. Quegan, The role of ion drift in the formation of ionization troughs in the mid- and high-latitude ionosphere — a review, *J. Atmos. Terr. Phys.*, 54, 1–30, 1992.

Rush, C.M., F-region behavior above North America during the magnetic disturbance of May 28, 1970, *J. Geophys. Res.*, 77, 757–760, 1972.

Rush, C.M., S.V. Rush, L.R. Lyons, and S.V. Venkateswaran, Equatorial anomaly during a period of declining solar activity, *Radio Sci.*, 4, 829–841, 1969.

Rüster, R., Height variations of the F2-layer above Tsumeb during geomagnetic bay-disturbances, *J. Atmos. Terr. Phys.*, 27, 1229–1245, 1965.

Rüster, R., Theoretical treatment of the dynamical behaviour of the F-region during geomagnetic bay disturbances, *J. Atmos. Terr. Phys.*, 31, 765–780, 1969.

Rüster, R. and J.W. King, Negative ionospheric storms caused by thermospheric winds, *J. Atmos. Terr. Phys.*, 38, 593–598, 1976.

Salah, J.E., M.-L. Duboin, and C. Mazaudier, Ionospheric electrodynamics over Saint-Satin and Millstone Hill during 26–28 June 1984, *Ann. Geophys.*, 5A, 351–358, 1987.

Sastri, J.H., Ionospheric storm of 4–6 December 1958 in the Indian equatorial region, *Indian J. Radio Space Phys.*, 9, 209–213, 1980.

Sastri, J.H. and J.E. Titheridge, Depressions in midlatitude F-region under relatively quiet geomagnetic conditions, *J. Atmos. Terr. Phys.*, 39, 1307–1316, 1977.

Sastri, J.H., K.B. Ramesh, and H.N.R. Rao, Transient composite electric field disturbances near dip equator associated with auroral substorms, *Geophys. Res. Lett.*, 19, 1451–1454, 1992.

Sastri, J.H., H.N.R. Rao, and K.B. Ramesh, Response of equatorial ionosphere to the transit of interplanetary magnetic cloud of January 13–15, 1967. Transient disturbance in F region, *Planet. Space Sci.*, 40, 519–534, 1992.

Sato, T., Disturbances in the ionospheric F2 region associated with geomagnetic storms. I. Equatorial zone, *J. Geomagn. Geoelectr.*, 8, 131–137, 1956.

Sato, T., Disturbances in the ionospheric F2 region associated with geomagnetic storms. II. Middle latitudes, *J. Geomagn. Geoelectr.*, 9, 1–22, 1957a.

Sato, T., Disturbances in the ionospheric F2 region associated with geomagnetic storms III. Auroral latitudes, *J. Geomagn. Geoelectr.*, 9, 94–106, 1957b.

Sato, T., Ionospheric F2-disturbances associated with geomagnetic storms, *J. Atmos. Terr. Phys.*, 15, 116–120, 1959.

Sato, T., Electron concentration variations in the topside ionosphere between 60°N and 60°S geomagnetic latitude associated with geomagnetic disturbances, *J. Geophys. Res.*, 73, 6225–6241, 1968.

Sato, T. and K.L. Chan, Storm-time variations of the electron concentration in the polar topside ionosphere, *J. Geophys. Res.*, 74, 2208–2216, 1969.

Sato, T. and L. Colin, Morphology of electron concentration enhancement at a height of 1000 kilometers at polar latitudes, *J. Geophys. Res.*, 74, 2193–2207, 1969.

Sato, T. and G.F. Rourke, F-region enhancements in the Antarctic, *J. Geophys. Res.*, 69, 4591–4607, 1964.

Schlegel, K., A case study of a high latitude electron density depletion, *J. Atmos. Terr. Phys.*, 46, 517–520, 1984.

Schödel, J.P., A.V. da Rosa, M. Mendillo, J.A. Klobuchar, T.H. Roelofs, R.B. Fritz, E.A. Essex, B.J. Flaherty, K.C. Yeh, F.H. Hibberd, L. Kersley, J.R. Koster, L. Liszka, and Y. Nakata, A global description of the F-region during the ionospheric storm of 17 December 1971, *J. Atmos. Terr. Phys.*, 36, 1121–1134, 1974.

Schumaker, T.L., M.S. Gussenhoven, D.A. Hardy, and R.L. Carovillano, The relationship between diffuse auroral and plasma sheet electron distributions near local midnight, *J. Geophys. Res.*, 94, 10061–10078, 1989.

Schunk, R.W., The terrestrial ionosphere, in *Solar-Terrestrial Physics* (R.L. Carovillano and J.M. Forbes, Eds.), D. Reidel, Dordrecht, 609–676, 1983.

Schunk, R.W. and A.F. Nagy, Electron temperature in the F region of the ionosphere: theory and observations, *Rev. Geophys.*, 16, 355–399, 1978.

Schunk, R.W., W.J. Raitt, and P.M. Banks, Effect of electric fields on the daytime high-latitude E and F regions, *J. Geophys. Res.*, 80, 3121–3130, 1975.

Schunk, R.W., P.M. Banks, and W.J. Raitt, Effects of electric fields and other processes upon the nighttime high-latitude F layer, *J. Geophys. Res.*, 81, 3271–3282, 1976.

Seaton, M.J., A possible explanation of the drop in F-region critical densities accompanying major ionospheric storms, *J. Atmos. Terr. Phys.*, 8, 122–124, 1956.

Senior, C. and M. Blanc, Convection in the inner magnetosphere: model predictions and data, *Ann. Geophys.*, 5, 405–420, 1987.

Senior, C., J.R. Sharber, O. de la Beaujardière, R.A. Heelis, D.S. Evans, J.D. Winningham, M. Sugiura, and W.R. Hoegy, E and F region study of the evening sector auroral oval: A Chatanika/Dynamics Explorer 2/NOAA 6 comparison, *J. Geophys. Res.*, 92, 2477–2494, 1987.

Serebryakov, B.Ye., Seasonal variations in the manifestations of the negative phase of disturbances in the mid-latitude F_2 region, *Geomagn. Aeron.*, 22, 410–411, 1982.

Sharma, R.P., and E.J. Hewens, A study of the equatorial anomaly at American longitudes during sunspot minimum, *J. Atmos. Terr. Phys.*, 38, 475–484, 1976.

Sipler, D.P. and M.A. Biondi, Midlatitude F region neutral winds and temperatures during the geomagnetic storm of March 26, 1976, *J. Geophys. Res.*, 84, 37–40, 1979.

Skinner, N.J. and R.W. Wright, Some geomagnetic effects in the equatorial F_2-region, *J. Atmos. Terr. Phys.*, 6, 177–188, 1955.

Skoblin, M.G. and M. Förster, An alternative explanation of ionization depletions in the winter night-time storm perturbed F2-layer, *Ann. Geophys.*, 11, 1026–1032, 1993.

Soicher, H., Response of electrons in ionosphere and plasmasphere to magnetic storms, *Nature*, 259, 33–35, 1976.

Sojka, J.J., Global scale, physical models of the F region ionosphere, *Rev. Geophys.*, 27, 371–403, 1989.

Sojka, J.J. and R.W. Schunk, A theoretical study of the high latitude F region's response to magnetospheric storm inputs, *J. Geophys. Res.*, 88, 2112–2122, 1983.

Sojka, J.J., W.J. Raitt, and R.W. Schunk, A theoretical study of the high-latitude winter F region at solar minimum for low magnetic activity, *J. Geophys. Res.*, 86, 609–621, 1981a.

Sojka, J.J., W.J. Raitt, and R.W. Schunk, Plasma density features associated with strong convection in the winter high-latitude F region, *J. Geophys. Res.*, 86, 6908–6916, 1981b.

Sojka, J.J., R.W. Schunk, J.D. Craven, L.A. Frank, S. Sharber, and J.D. Winningham, Modeled F region response to auroral dynamics based upon Dynamics Explorer auroral observations, *J. Geophys. Res.*, 94, 8993–9008, 1989.

Sojka, J.J., R.W. Schunk, and J.A. Whalen, The longitude dependence of the dayside F region trough: a detailed model-observation comparison, *J. Geophys. Res.*, 95, 15275–15280, 1990.

Sojka, J.J., R.W. Schunk, W.R. Hoegy, and J.M. Grebowsky, Model and observation comparison of the universal time and IMF By dependence of the ionospheric polar hole, *Adv. Space Res.*, 11, No. 10, 39–42, 1991.

Sojka, J.J., M. Bowline, R.W. Schunk, J.D. Craven, L.A. Frank, J.R. Sharber, J.D. Winningham, and L.H. Brace, Ionospheric simulation compared with Dynamics Explorer observations for November 22, 1981, *J. Geophys. Res.*, 97, 1245–1256, 1992a.

Sojka, J.J., R.W. Schunk, D. Rees, T.J. Fuller-Rowell, R.J. Moffett, and S. Quegan, Comparison of the USU ionospheric model with the UCL-Sheffield coupled thermospheric-ionospheric model, *Adv. Space Res.*, 12, No. 6, 89–92, 1992b.

Somayajulu, Y.V., Changes in the F region during magnetic storms, *J. Geophys. Res.*, 68, 1899–1922, 1963.

Somayajulu, Y.V., A.K. Sehgal, T.R. Tyagi, and N.K. Negi, Changes in electron content during the 25–26 May 1967 magnetic storm event, *Ind. J. Pure Appl. Phys.*, 9, 548–552, 1971.

Somayajulu, V.V., B.V. Krishna Murthy, and K.S.V. Subbarao, Response of night-time equatorial F-region to magnetic disturbances, *J. Atmos. Terr. Phys.*, 53, 965–976, 1991.

Spiro, R.W., R.A. Heelis, and W.B. Hanson, Ion convection and the formation of the mid-latitude F region ionization trough, *J. Geophys. Res.*, 83, 4255–4264, 1978.

Spiro, R.W., R.A. Heelis, and W.B. Hanson, Rapid subauroral ion drifts observed by Atmosphere Explorer C, *Geophys. Res. Lett.*, 6, 657–660, 1979.

Spiro, R.W., P.H. Reiff, and L.J. Maher, Precipitating electron energy flux and auroral zone conductances: an empirical model, *J. Geophys. Res.*, 87, 8215–8227, 1982.

Spiro, R.W., R.A. Wolf, and B.G. Fejer, Penetration of high-latitude-electric-field effects to low latitudes during SUNDIAL 1984, *Ann. Geophys.*, 6, 39–50, 1988.

Spurling, P.H. and K.L. Jones, The nature of seasonal changes in the effects of magnetic storms on mid-latitude F-layer electron concentration, *J. Atmos. Terr. Phys.*, 35, 921–927, 1973.

Spurling, P.H. and K.L. Jones, The observation of storm induced positive and negative processes in the mid-latitude F-region using N-h profiles, *J. Atmos. Terr. Phys.*, 37, 1385–1389, 1975.

Spurling, P.H. and K.L. Jones, The observation of related F-region height and electron content changes at mid-latitudes during magnetic storms and their comparison with a numerical model, *J. Atmos. Terr. Phys.*, 38, 1237–1244, 1976.

Stening, R.J., Modelling the low latitude F region, *J. Atmos. Terr. Phys.*, 54, 1387–1412, 1992.

Stening, R.J. and M.P. Chance, Geomagnetic effects on the F-region of the ionosphere, *J. Atmos. Terr. Phys.*, 36, 1663–1673, 1974.

Stern, D.P., The origins of Birkeland currents, *Rev. Geophys. Space Phys.*, 21, 125–138, 1983.

St. Maurice, J.-P. and D.G. Torr, Nonthermal rate coefficients in the ionosphere: the reactions of O^+ with N_2, O_2, and NO, *J. Geophys. Res.*, 83, 969–977, 1978.

Svalgaard, L., Recalibration of Bartel's geomagnetic activity indices Kp and ap to include universal time variations, *J. Geophys. Res.*, 81, 5182–5188, 1976.

Taieb, C. and P. Poinsard, Modelling of the mid-latitude ionosphere: application to storm effects, II, *Ann. Geophys.*, 2, 359–368, 1984.

Tanaka, T., The worldwide distribution of positive ionospheric storms, *J. Atmos. Terr. Phys.,* 41, 103–110, 1979.

Tanaka, T., Severe ionospheric disturbances caused by the sudden response of evening subequatorial ionospheres to geomagnetic storms, *J. Geophys. Res.,* 86, 11335–11349, 1981.

Tanaka, T., Low-latitude ionospheric disturbances: results for March 22, 1979, and their general characteristics, *Geophys. Res. Lett.,* 13, 1399–1402, 1986.

Tanaka, T. and K. Hirao, Effects of an electric field on the dynamical behavior of the ionospheres and its application to the storm time disturbance of the F-layer, *J. Atmos. Terr. Phys.,* 35, 1443–1452, 1973.

Taylor, G.N., The total electron content of the ionosphere during the magnetic disturbance of November 12–13, 1960, *Nature,* 189, 740–741, 1961.

Testud, J., P. Amayenc, and M. Blanc, Middle and low latitude effects of auroral disturbances from incoherent-scatter, *J. Atmos. Terr. Phys.,* 37, 989–1009, 1975.

Thomas, J.O. and M.K. Andrews, The trans-polar exospheric plasma, 3: a unified picture, *Planet. Space Sci.,* 17, 433–446, 1969.

Thomas, L., World-wide disturbances in the F-region accompanying the onset of the main phase of severe magnetic storms, *J. Atmos. Terr. Phys.,* 30, 1623–1630, 1968.

Thomas, L., F2-region disturbances associated with major magnetic storms, *Planet. Space Sci.,* 18, 917–928, 1970.

Thomas, L. and R.B. Norton, Possible importance of internal excitation in ion-molecule reactions in the F region, *J. Geophys. Res.,* 71, 227–230, 1966.

Thomas, L. and F.H. Venables, The onset of the F-region disturbance at middle latitudes during magnetic storms, *J. Atmos. Terr. Phys.,* 28, 599–605, 1966.

Thome, G., Long-period waves generated in the polar ionosphere during the onset of magnetic storms, *J. Geophys. Res.,* 73, 6319–6336, 1968.

Titheridge, J.E. and M.J. Buonsanto, A comparison of northern and southern hemisphere TEC storm behaviour, *J. Atmos. Terr. Phys.,* 50, 763–780, 1988.

Tulunay, Y.K., Some topside electron density measurements from Ariel III satellite during the geomagnetic storm of 25–27 May 1967, *Planet. Space Sci.,* 20, 1299–1307, 1972.

Tulunay, Y.K. and J.M. Grebowsky, Hemispheric differences in the morphology of the high latitude ionosphere measured at ≈500 km, *Planet. Space Sci.,* 35, 821–826, 1987.

Turunen, T. and L. Liszka, Comparison of simultaneous satellite measurements of auroral particle precipitation with bottomside ionosonde measurements of the electron density in the F-region, *J. Atmos. Terr. Phys.,* 34, 365–372, 1972.

Turunen, T. and M.M. Rao, Examples of the influence of strong magnetic storms on the equatorial F-layer, *J. Atmos. Terr. Phys.,* 42, 323–330, 1980.

VanZandt, T.E., V.L. Peterson, and A.R. Laird, Electrodynamic drift of the midlatitude F_2 layer during a storm, *J. Geophys. Res.,* 76, 278–281, 1971.

Vasyliunas, V.M., The interrelationship of magnetospheric processes, in: *Earth's Magnetospheric Processes,* (B. Mc-Cormac, Ed.) p. 29, D. Reidel, Hingham, MA 1972.

Volland, H., A model of the magnetospheric electric convection field, *J. Geophys. Res.,* 83, 2695–2699, 1978.

Volland, H., Magnetospheric electric fields and currents and their influence on large scale thermospheric circulation and composition, *J. Atmos. Terr. Phys.,* 41, 853–866, 1979.

Volland, H., Dynamics of the disturbed ionosphere, *Space Sci. Rev.,* 34, 327–335, 1983.

Volland, H., *Atmospheric Electrodynamics,* Springer, New York, 1984.

Volland, H., *Atmospheric Tidal and Planetary Waves,* Kluwer Academic Publishing, Dordrecht, 1988.

Volland, H. and H.G. Mayr, Response of the thermospheric density to auroral heating during geomagnetic disturbances, *J. Geophys. Res.,* 76, 3764–3776, 1971.

Walker, G.O., Observations of ionospheric storms at low latitudes and their correlation with magnetic field changes near the magnetic equator, *J. Atmos. Terr. Phys.,* 35, 1573–1582, 1973.

Walker, G.O. and Y.W. Wong, Ionospheric effects observed throughout East Asia of the large magnetic storm of 13–15 March 1989, *J. Atmos. Terr. Phys.,* 55, 995–1008, 1993.

Walker, G.O., W.Y. Wong, Y.N. Huang, T. Kikuchi, J. Soegiyo, V. Badillo, and E.P. Szuszczewicz, Periodic behaviour of the ionosphere in South East Asia on storm and quiet days during the May/June SUNDIAL campaign, 1987, *J. Atmos. Terr. Phys.,* 53, 627–641, 1991.

Watkins, B.J., A numerical computer investigation of the polar F-region ionosphere, *Planet. Space Sci.,* 26, 559–569, 1978.

Watkins, B.J. and P.G. Richards, A theoretical investigation of the role of neutral winds and particle precipitation in the formation of the auroral F-region ionosphere, *J. Atmos. Terr. Phys.,* 41, 179–187, 1979.

Weber, E.J., J. Buchau, J.G. Moore, J.R. Sharber, R.C. Livingston, J.D. Winningham, and B.W. Reinisch, *J. Geophys. Res.,* 89, 1683–1694, 1984.

Weber, E.J., J.A. Klobuchar, J. Buchau, H.C. Carlson, Jr., R.C. Livingston, O. de la Beaujardiere, M. McCready, J.G. Moore, and G.J. Bishop, Polar cap F layer patches: structure and dynamics, *J. Geophys. Res.,* 91, 12121–12129, 1986.

Wei, S., B.H. Ahn, and S.-I. Akasofu, The global joule heat production rate and the AE index, *Planet. Space Sci.,* 33, 279–281, 1985.

Whitteker, J.H., L.H. Brace, J.R. Burrows, T.R. Hartz, W.J. Heikkila, R.C. Sagalyn, and D.M. Thomas, Isis 1 observations of the high-latitude ionosphere during a geomagnetic storm, *J. Geophys. Res.*, 77, 6121–6128, 1972.

Winser, K.J., G.O.L. Jones, and P.J.S. Williams, A quantitative study of the high latitude ionospheric trough using EIS-CAT's common programmes, *J. Atmos. Terr. Phys.*, 48, 893–904, 1986.

Woodman, R.F., D.L. Sterling, and W.B. Hanson, Synthesis of Jicamarca data during the great storm of March 8, 1970, *Radio Sci.*, 7, 739–746, 1972.

Wrenn, G.L., A.S. Rodger, and H. Rishbeth, Geomagnetic storms in the Antarctic F-region. I. Diurnal and seasonal patterns for main phase effects, *J. Atmos. Terr. Phys.*, 49, 901–913, 1987.

Wright, M.D., Possible identification of atmospheric waves associated with ionospheric storms, *Nature*, 190, 898–899, 1961.

Yeboah-Amankwah, D., Increases of equatorial total electron content (TEC) during magnetic storms, *J. Atmos. Terr. Phys.*, 38, 45–50, 1976.

Yeh, H.-C. and J.C. Foster, Storm time heavy ion outflow at mid-latitudes, *J. Geophys. Res.*, 95, 7881–7891, 1990.

Yeh, H.-C., J.C. Foster, F.J. Rich, and W. Swider, Storm time electric field penetration observed at mid-latitude, *J. Geophys. Res.*, 96, 5707–5721, 1991.

Yeh, K.C., K.H. Lin, and R.O. Conkright, The global behavior of the March 1989 ionospheric storm, *Can. J. Phys.*, 70, 532–543, 1992.

Yeh, K.C., S.Y. Ma, K.H. Lin, and R.O. Conkright, Global ionospheric effects of the October 1989 geomagnetic storm, *J. Geophys. Res.*, 99, 6201–6218, 1994.

Yonezawa, T., The characteristic behavior of the F2 layer during severe magnetic storms, *Proc. Conf. on Ionosphere*, Institute of Physics and the Physics Society, London, 128–133, 1963.

Zakharov, L.P. and A.A. Namgaladze, Formation of positive disturbances of the F_2 region by protonospheric plasma fluxes, *Geomagn. Aeron.*, 19, 491–492, 1979.

Zevakina, R.A., Ye.Ye. Goncharova, and L.A. Yudovich, Ionospheric and geomagnetic effects of solar activity in August 1972, *Geomagn. Aeron.*, 14, 698–701, 1974.

Zevakina, R.A., A.A. Namgaladze, and V.M. Smertin, Interpretation of positive disturbances of the F_2 region, *Geomagn. Aeron.*, 18, 708–710, 1978.

Chapter 9

Ionospheric Electrodynamics

Arthur D. Richmond

CONTENTS

1. INTRODUCTION

The free electrons and ions in the Earth's ionosphere make it electrically conducting. Currents flow that are connected with the magnetosphere above, and to a much weaker extent with the poorly conducting atmosphere below. One of the important generators of the ionospheric current is the *ionospheric wind dynamo* (or *ionospheric dynamo* for short). The ionospheric dynamo operates when upper atmospheric winds move the electrically conducting medium through the Earth's magnetic field, creating an electromotive force that drives currents and causes electric polarization charges and electric fields to develop. Other current generation mechanisms exist associated with the interaction of the solar wind with the magnetosphere and, to a much less significant extent, with electrical-storm activity in the troposphere. In the daytime ionosphere the largest currents flow between 90 and 200 km; this general region is sometimes called the *dynamo region*. The currents and electric fields interact with the dynamics of the ionospheric plasma and neutral air in and above the dynamo region. We call the electrical phenomena and their interacting dynamical effects *ionospheric electrodynamics*. Some useful general references relating to ionospheric electrodynamics are the works of Matsushita and Campbell (1967), Maeda (1968), Akasofu and Chapman (1972), Kato (1980), Volland (1984), Kelley (1989), Richmond (1989, 1994), Roble (1991), and Hargreaves (1992).

The currents in the ionosphere and above produce magnetic perturbations that can be sensed at the ground and in space. The currents associated with the ionospheric dynamo have regular, smooth, daily variations. The main part of these currents is often referred to as S_q (solar quiet) or S_R (solar regular); there is also a smaller component related to lunar periods that is sometimes referred to as L. By contrast, the currents associated with solar-wind/magnetosphere interactions are highly variable: when they are weak one speaks of a *magnetically quiet* period, and when they are strong, one speaks of a *magnetically disturbed* period, the most dramatic manifestation of which is the magnetic storm. During disturbed periods the magnetospherically produced electric fields and currents can dominate over those produced by the ionospheric wind dynamo. It is therefore important to distinguish between magnetically quiet and disturbed periods when analyzing ionospheric electrodynamics.

Ionospheric electrodynamics depends, among other things, on the conductivity of the ionosphere and on the strength of thermospheric winds. Both of these depend on the flux of solar UV

0-8493-2520-X/95/$0.00+$.50
© 1995 by CRC Press

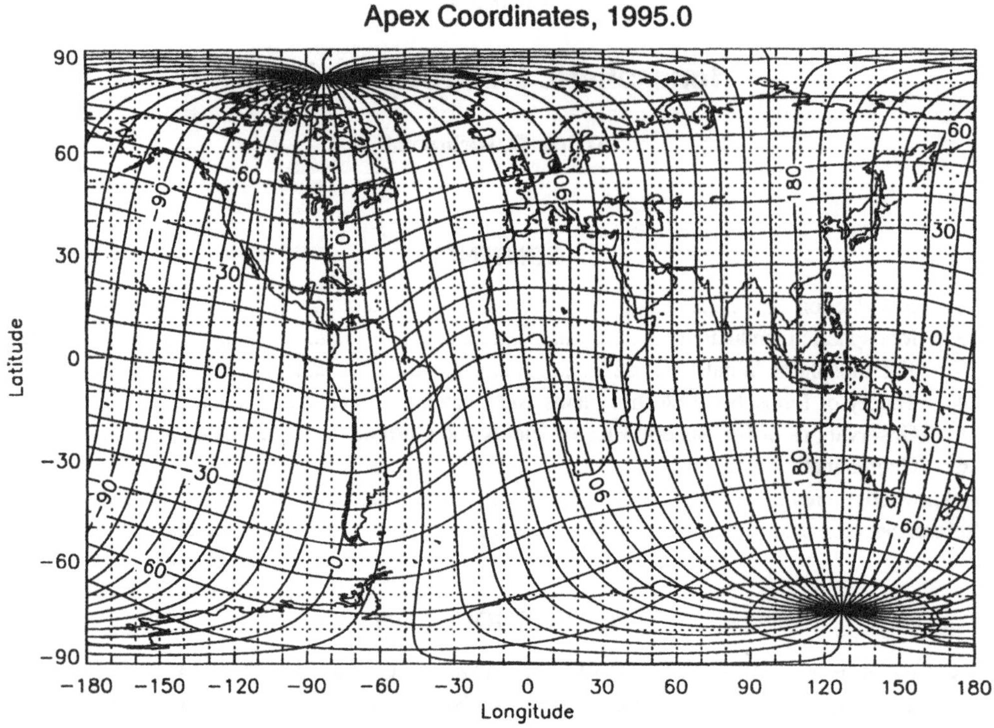

Apex Coordinates, 1995.0

Figure 9.1.1 Magnetic apex coordinates at ground level, derived from the International Geomagnetic Reference Field (IGRF, Langel, 1992) for epoch 1995.0. The dashed grid represents geographic coordinates, while lines of constant apex latitude and longitude are drawn at 10° intervals.

radiation absorbed in the upper atmosphere, which varies considerably with solar activity. Major changes in the solar UV irradiance occur over the 11-year solar activity cycle; between solar minimum and solar maximum the total ionizing extreme-UV flux varies by more than a factor of two. Thus, it is natural to expect significant changes in ionospheric electrodynamics with the solar cycle. A measure of solar activity often used is the index S_a, representing the flux of solar radio emissions at a wavelength of 10.7 cm, in units of 10^{-22} Wm^{-2}Hz^{-1}.

The conductivity of the ionosphere and magnetosphere is highly anisotropic, organized by the geomagnetic field. Electrodynamic features are therefore strongly organized with respect to the geomagnetic field, and it is common to use magnetic coordinates to organize the observations and to do model simulations. Several different magnetic-coordinate systems exist. For global electrodynamics, a useful system is magnetic apex coordinates (VanZandt et al., 1972), which are defined in terms of the altitude and geomagnetic longitude of the apex of the geomagnetic field line passing through the point in question, calculated using a geomagnetic-field model that neglects distortions caused by currents external to the Earth. Apex coordinates have well-behaved properties at both high and low latitudes and incorporate the important feature that geomagnetic lines of force have constant values of apex latitude and longitude, so that *magnetically conjugate points* in the northern and southern magnetic hemispheres have the same values of apex coordinates. Figure 9.1.1 is a map of apex coordinates at ground level.

The first suggestions that the rarefied upper atmosphere might conduct electricity came in the 18th century, when it was noticed that there was striking similarity between the appearance of the aurora borealis and the glow created in an evacuated chamber through which electricity is discharged. Benjamin Franklin proposed that atmospheric electricity might flow globally between the polar and equatorial regions, being visible as auroras only at high latitudes where it is most

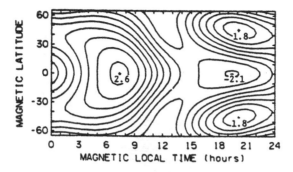

Figure 9.1.2 Yearly average electric potential at 300 km between ±65° apex latitude as a function of magnetic local time. The contour interval is 1 kV and extremum values are in kV. (Adapted from Richmond, A. D. et al., 1980, *J. Geophys. Res.*, 85, 4658. Copyright 1980 by the American Geophysical Union. With permission.)

concentrated. It was also noticed, by Hiorter and Celsius in 1741, that a perturbation in the geomagnetic field occurs simultaneously with an overhead aurora. When it later was discovered that electric currents produce a magnetic perturbation, the link between upper atmospheric aurora and electric currents was strengthened, and 19th century scientists like Carl Friedrich Gauss suspected that nonauroral currents responsible for daily geomagnetic variations at middle latitudes could also be created by upper atmospheric currents. In an *Encyclopaedia Britannica* article, even before the discovery of the ionosphere, Balfour Stewart (1882) suggested that such currents could be produced by the upper atmospheric dynamo mechanism that is today accepted as the primary source of quiet-day magnetic perturbations at middle and low latitudes. The demonstration by Marconi in 1901 that radio waves could propagate across the Atlantic led to suggestions by Heaviside and Kennely that these waves are reflected by a conducting upper atmosphere. In 1924, Appleton and Barnett in England and Breit and Tuve in the U.S. clearly demonstrated the ionosphere's existence by recording echoes from radio waves transmitted nearly vertically. The theory of ionospheric conductivity was developed by Hirono (1952) and Baker and Martyn (1952, 1953). Some sources for further references to the historical developments are Stewart (1882), Chapman and Bartels (1940), Matsushita (1967a), and Brekke and Moen (1993).

Figure 9.1.2 shows the average distribution of mid-latitude ionospheric electric potential as a function of apex latitude and *magnetic local time* (MLT), the latter representing a longitude scale for which 12 MLT is essentially the magnetic longitude of the subsolar point. The potential is symmetric about the magnetic equator, because the extremely high electrical conductivity parallel to the geomagnetic field effectively shorts out any potential difference at opposite ends of a field line in the northern and southern magnetic hemispheres that otherwise might exist. A maximum and minimum exist at the magnetic equator near 7 and 19 MLT, respectively, with a total potential difference of about 5 kV. At mid-latitudes the maximum and minimum occur at earlier local times. These potentials are relatively small in comparison with those commonly present at high latitudes, associated with magnetospheric dynamo action (see Chapter II/12).

To represent the ground-level magnetic perturbations associated with overhead currents, use is often made of the concept of *equivalent* ionospheric currents, defined to be that distribution of horizontal sheet currents flowing overhead that would produce the observed magnetic perturbations. Equivalent current is not necessarily the same as the true height-integrated horizontal ionospheric electric-current density, since the true current system is three-dimensional, with significant amounts of current flowing along geomagnetic field lines above the ionosphere. Figure 9.1.3 shows yearly average distributions of mid-latitude equivalent current functions for S_q (at solar maximum) and L (for average solar activity, at the lunar phase of new or full moon). Equivalent current flows along isocontours of the current function, counterclockwise around maxima and clockwise around minima. The S_q and L current systems superpose, but S_q dominates

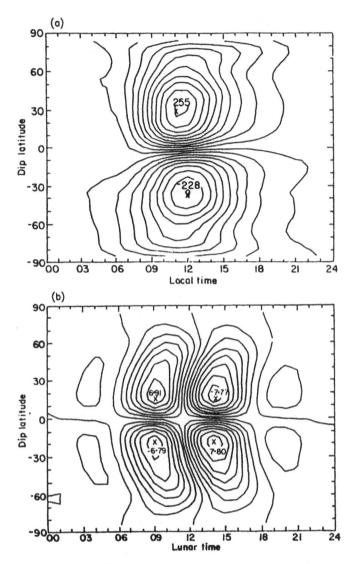

Figure 9.1.3 (a) Yearly average S_q ionospheric equivalent current function as a function of magnetic (dip) latitude and local time, during a period of high solar activity. The high-latitude currents associated with magnetospheric dynamo action have been removed by smooth extrapolation of the function between about 65° and the pole in each hemisphere. The contour interval is 25 kA and extremum values are in kiloamperes. (b) Yearly average L ionospheric equivalent current function for a new or full moon and for average solar activity. The contour interval is 1 kA and extremum values are in kiloamperes. (Adapted from Matsushita, S., 1968, *Geophys. J. R. Astron. Soc.,* 15, 109. With permission from Blackwell Scientific Publishers.)

by far in total intensity. The S_q system is comprised of two large vortices flowing on the dayside of the Earth, counterclockwise in the northern hemisphere and clockwise in the southern hemisphere. On average, the total current in each vortex is around 150 kA, although this varies with solar cycle (being 228 to 255 kA for the solar-maximum conditions in Figure 9.1.3a) and with season (being stronger in summer than winter). Not incorporated in Figure 9.1.3 are the much more intense high-latitude equivalent currents associated with magnetospheric dynamo processes.

2. IONOSPHERIC CONDUCTIVITY

The upper atmosphere is ionized primarily by solar UV and X-radiation and, at higher latitudes, by the precipitation of energetic charged particles from the magnetosphere (see Rees, 1989).

Starlight and cosmic rays are minor ionization sources that have some influence in the nightside ionosphere and at low altitudes. The left side of Figure 9.2.1 shows typical rates of ionization as a function of altitude for the different sources. The solid lines show regular daytime sources, while the dashed lines show sources that are sporadic or highly variable in time. The ionization rate depends on the intensity of ionizing radiation, on the atmospheric density and composition, and on the ionization cross sections of the atmospheric constituents. Above 100 km the densities of all the major atmospheric constituents decrease with increasing altitude. For a radiation source coming from above, absorption causes the intensity to decrease with decreasing altitude. The ionization production rate peaks at an altitude near where the radiation has decreased to 1/e of its intensity at the top of the atmosphere, which depends on the absorption cross sections of the atmospheric constituents to the particular type of radiation (wavelength of electromagnetic radiation or species and energy of precipitating particles). Except for Lyman-α radiation, which can ionize the minor constituent nitric oxide, and for cosmic rays, almost all ionizing radiation is absorbed above 80 km in the atmosphere.

The primary ions produced are N_2^+, O_2^+, N^+, and O^+, but these are reactive with the neutral gases, and the nitrogen ions are rapidly converted to O_2^+ and NO^+. The dominant ions present are therefore NO^+, O_2^+, and O^+, with the molecular ions predominant below 150 km and O^+ predominant above 200 km, and a mixture in between. The right side of Figure 9.2.1 shows typical distributions of the major ion species in the daytime ionosphere. Negative ions and more complex positive ions become important only below about 90 km, where conductivities are relatively small. Between about 90 and 150 km the sum of the densities of NO^+ and O_2^+ is approximately equal to the electron number density N_e. These molecular ions NO^+ and O_2^+ recombine with electrons with a reaction coefficient represented by the symbol α. These ions are usually near photochemical equilibrium, meaning that their production rate is in balance with their chemical loss rate. If Q_e is the electron production rate, then between 90 and 150 km photochemical equilibrium implies that

$$Q_e = \alpha N_e^2 \tag{2.1}$$

or

$$N_e = \sqrt{Q_e/\alpha} \tag{2.2}$$

In reality, the recombination coefficient is somewhat different for NO^+ and O_2^+, so α in Equations 2.1 and 2.2 represents an effective mean value, weighted by the relative densities of the two ions. In addition, α is approximately inversely proportional to the electron temperature. A characteristic value of α is $3 \times 10^{-13}m^3s^{-1}$. Equation 2.2 shows that the electron density in the lower ionosphere is proportional to the square root of the ionization production rate. When sunlight disappears the electron density is greatly reduced, unless there is production at high latitudes by energetic particles precipitating from the magnetosphere.

The O^+ ions at higher altitudes do not directly recombine with electrons, but rather react first with molecular neutral constituents to form molecular ions that can then recombine with electrons. Equations 2.1 and 2.2 become invalid at these heights. As neutral densities become small at high altitudes, the loss of O^+ ions becomes slow, and ions can be transported significant distances before being lost, so that photochemical equilibrium breaks down. Characteristic time scales for ionization loss, diffusion, and advection by winds and electric fields in the mid-latitude daytime ionosphere are given in the center of Figure 9.2.1 at different altitudes. The shorter the time scale is of one process in comparison with the time scales of the other processes, the more important that process is for helping determine the electron density at that altitude.

The primary ionization source at day, solar extreme-UV light, varies significantly with the level of solar activity. The electron density also shows important variations with solar activity. Figure 9.2.2 shows representative vertical profiles of electron density for day and night conditions,

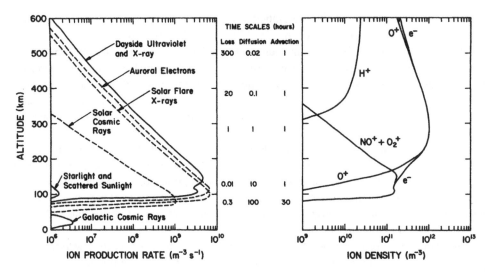

Figure 9.2.1 (Left) Typical rates of ion production by different sources. Solid lines show regular sources. Starlight and sunlight scattered off hydrogen in the geocorona are important for maintaining the nighttime *E* region around 100 km. (Middle) Characteristic time scales for the ion loss, diffusion, and advection at different altitudes in the daytime ionosphere. (Right) Typical daytime profiles of ion and electron densities. (Richmond, A. D. 1987, *The Solar Wind and Earth*, S. I. Akasofu and Y. Kamide, Eds., p. 123, Terra Scientific, Tokyo. With permission.)

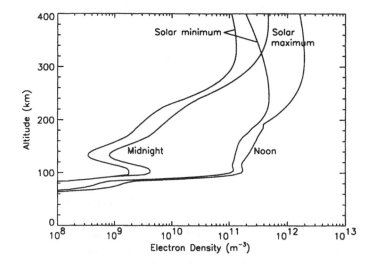

Figure 9.2.2 Electron densities at 44.6°N, 2.2°E from the 1990 International Reference Ionosphere (IRI, Bilitza, 1990) at noon and midnight on March 21, for solar minimum and maximum conditions (annual average sunspot numbers 15 and 160, respectively).

and for low and high levels of solar activity (represented by 12-month average sunspot numbers of 15 and 160, respectively). These are obtained for the location of the St. Santin incoherent scatter radar (44.6°N, 2.2°E) from the 1990 version of the International Reference Ionosphere (IRI, Bilitza, 1990), an empirical model of ionospheric density, composition, and temperature. The ionospheric region around the minor peak in density near 105 km is known as the *E* region, while the larger density region above 150 km is known as the *F* region. The *E* region and lower part of the *F* region undergo relatively greater variations in electron density between day and night than does the upper *F* region.

Theoretical expressions for the conductivities are usually derived from multifluid theory, in which neutrals and the different charged species are treated as separate fluids that interact through collisions. The charged-constituent fluids are assumed to be in force balance, and for the purpose of calculating ionospheric conductivities the pressure-gradient and gravitational forces are neglected. The relevant forces are then the Lorentz force and the frictional forces between fluids moving at different velocities. For calculating ionospheric conductivities above 90 km, where nearly all ions have a single positive charge, it is adequate to consider all ion species together as a single fluid of number density equal to that of the electrons, N_e. The ion mass m_i and velocity $\mathbf{v}_i$, and the collision frequencies of ions with neutrals, ν_{in}, and with electrons, ν_{ie}, then represent mass-density-weighted averages over all ion species. (This approximation is reasonable because of the facts that differences in velocity among the different ion species are normally small, and that their respective collision frequencies with neutrals and electrons are generally similar.) For ions and electrons the force-balance conditions are, respectively:

$$N_e e(\mathbf{E} + \mathbf{v}_i \times \mathbf{B}) - N_e m_i \nu_{in}(\mathbf{v}_i - \mathbf{v}_n) - N_e m_i \nu_{ie}(\mathbf{v}_i - \mathbf{v}_e) = 0 \tag{2.3}$$

$$-N_e e(\mathbf{E} + \mathbf{v}_e \times \mathbf{B}) - N_e m_e \nu_{en}(\mathbf{v}_e - \mathbf{v}_n) + N_e m_e \nu_{ei}(\mathbf{v}_i - \mathbf{v}_e) = 0 \tag{2.4}$$

where e is the magnitude of the electron charge, m_e and $\mathbf{v}_e$ are the electron mass and velocity, $\mathbf{v}_n$ is the velocity of the neutral gas, ν_{en} and ν_{ei} are the electron-neutral and electron-ion collision frequencies, and $\mathbf{E}$ and $\mathbf{B}$ are the electric and magnetic fields. In reality, the collision frequencies are tensors rather than scalars in the presence of a magnetic field (Shkarofsky et al., 1966; Hill and Bowhill, 1977). Above 80 km the electron-neutral collision frequency is about 1.4 times as large for motions perpendicular to $\mathbf{B}$ as for motions along $\mathbf{B}$ (Gagnepain et al., 1977); below this height off-diagonal terms in the collision-frequency tensor can also be significant. ν_{ei} is similarly anisotropic, but we will be concerned only with its value for motions along $\mathbf{B}$. For ions the anisotropy of the collision frequency is unimportant. Table 9.2.1 lists formulas for the collision frequencies as functions of density, composition, and temperature of the colliding species. These expressions are based on a combination of laboratory measurements and theory. There is significant uncertainty in the O^+-O collision frequency (Salah, 1993), which is important for determining the conductivity in the upper ionosphere.

The frictional force that the ions exert on electrons must be equal and opposite to the frictional force that electrons exert on the ions, signifying that

$$m_i \nu_{ie} = m_e \nu_{ei} \tag{2.5}$$

The components of Equations 2.3 and 2.4 parallel to $\mathbf{B}$, after dividing by N_e and rearranging, become

$$m_i \nu_{in}(\upsilon_i - \upsilon_n)_\| + m_i \nu_{ie\|}(\upsilon_i - \upsilon_e)_\| = eE_\| \tag{2.6}$$

$$m_e \nu_{en\|}(\upsilon_e - \upsilon_n)_\| - m_e \nu_{ei\|}(\upsilon_i - \upsilon_e)_\| = -eE_\| \tag{2.7}$$

where the subscript $\|$ denotes the direction along $\mathbf{B}$. The velocity difference between ions and electrons is found by multiplying Equations 2.6 and 2.7 by $m_e \nu_{en\|}$ and $m_i \nu_{in}$, respectively, and subtracting:

$$m_i m_e(\nu_{in}\nu_{en\|} + \nu_{ie\|}\nu_{en\|} + \nu_{ei\|}\nu_{in})(\upsilon_i - \upsilon_e)_\| = (m_i \nu_{in} + m_e \nu_{en\|})eE_\| \tag{2.8}$$

This expression simplifies considerably when we take into account the relative sizes of the masses and collision frequencies. The ratio m_e/m_i depends on ion composition, but is of order 10^{-5}. The ratio ν_{en}/ν_{in} is of the order 10. Taking Equation 2.5 into account, we can drop $\nu_{ie\|}\nu_{en\|}$ on the left-hand side and $m_e \nu_{en\|}$ on the right-hand side of Equation 2.8, so that the velocity difference is found to be

$$(\upsilon_i - \upsilon_e)_\| = \frac{eE_\|}{m_e(\nu_{en\|} + \nu_{ei\|})} \tag{2.9}$$

Table 9.2.1 **Formulas for Collision Frequencies**[a]

$$\nu_{in}(NO^+) = [3.4\,(N_{N_2} + N_{O_2})\,R_i^{-0.16} + 1.9\,N_O\,R_i^{-0.19}] \times 10^{-16}\text{m}^3\text{s}^{-1}$$

$$\nu_{in}(O_2^+) = [3.3\,N_{N_2}\,R_i^{-0.17} + 6.1\,N_{O_2}\,R_i^{0.37} + 1.8\,N_O\,R_i^{-0.19}] \times 10^{-16}\text{m}^3\text{s}^{-1}$$

$$\nu_{in}(O^+) = [5.4\,N_{N_2}\,R_i^{-0.20} + 7.0\,N_{O_2}\,R_i^{0.05} + 8.9\,N_O\,R_i^{0.5}] \times 10^{-16}\text{m}^3\text{s}^{-1}$$

$$\nu_{en\perp} = [7.2\,N_{N_2}\,R_e^{0.95} + 5.2\,N_{O_2}\,R_e^{0.79} + 1.9\,N_O\,R_e^{0.85}] \times 10^{-15}\text{m}^3\text{s}^{-1}$$

$$\nu_{en\|} = [4.6\,N_{N_2}\,R_e^{0.95} + 4.3\,N_{O_2}\,R_e^{0.79} + 1.5\,N_O\,R_e^{0.85}] \times 10^{-15}\text{m}^3\text{s}^{-1}$$

$$\nu_{ei\|} = (1.84 \times 10^{-6}\text{s}^{-1}\text{m}^3\,\text{K}^{3/2})\,(\ln\Lambda)\,N_e\,T_e^{-3/2}$$

Note: $N_{N_2}, N_{O_2}, N_O, N_e$ = number densities of N_2, O_2, O, electrons; $R_i = (T_i + T_n)/1000$ K; $R_e = T_e/300$ K; T_i, T_n, T_e = temperatures of ions, neutrals, electrons; and $\ln\Lambda$ = Coulomb logarithm = $16.33 + 0.5 \ln (T_e^3/N_e)$, with T_e in K and N_e in m^{-3}

[a] Formulas for ν_{in} are based on Table 3 of Mason (1970) except for the contribution of N_O to $\nu_{in}(O^+)$, which is based on Salah (1993). The formula for $\nu_{en\perp}$ is based on Table 2 of Itikawa (1971), as parameterized by Gagnepain et al. (1977). The formula for $\nu_{en\|}$ takes into account the factor $g_a(\alpha)$ of Itikawa, with α estimated from the temperature dependence of ν_{en}. The formula for $\nu_{ei\|}$ is from equation (8.30) of Shkarofsky et al. (1966).

The scaling approximations correspond to the assumption that ions move at the neutral velocity in the direction along the magnetic field, and that only the electrons move in response to the parallel electric field. (In reality, ions do move parallel to **B** in response to the neglected pressure-gradient and gravitational forces, and Equation 2.9 does not represent the complete physics of the plasma motions. In fact, a small parallel electric field tends to be set up so as to nullify differential ion and electron velocities in the presence of plasma pressure gradients and gravity to ensure that plasma diffusion is ambipolar, i.e., that charge neutrality is preserved by maintaining similar ion and electron velocities along **B**. This ambipolar-diffusion electric field is of no importance for ionospheric currents, except that Equation 2.9 is not valid when applied to it.) The electric current parallel to **B** is given by

$$J_\| = N_e e(\upsilon_i - \upsilon_e)_\| = \sigma_\| E_\| \tag{2.10}$$

where $\sigma_\|$ is the *parallel* or *direct* conductivity (also frequently symbolized by σ_0). Comparing Equations 2.9 and 2.10, we readily see that

$$\sigma_\| = \frac{N_e e^2}{m_e(\nu_{en\|} + \nu_{ei\|})} \tag{2.11}$$

To derive expressions for motions and conductivities perpendicular to the magnetic field, we begin by neglecting effects of collisions between ions and electrons in Equations 2.3 and 2.4. This turns out to be an excellent approximation (unlike what we found for motions parallel to **B**), because electron-ion collisions are important in comparison with ion-neutral and electron-neutral collisions only in the upper ionosphere, where, as we will see, the ions and electrons move with almost the same velocity in the direction perpendicular to **B**. Thus, $(\mathbf{v}_i - \mathbf{v}_e)_\perp$ is nearly zero where ν_{ei} is significant; here the subscript $\perp$ signifies the component perpendicular to **B**. With this simplification, the component of Equation 2.3 perpendicular to **B**, upon rearrangement, gives

$$e(\mathbf{v}_i - \mathbf{v}_n) \times \mathbf{B} - m_i\nu_{in}(\mathbf{v}_i - \mathbf{v}_n)_\perp = -e(\mathbf{E}_\perp + \mathbf{v}_n \times \mathbf{B}) \tag{2.12}$$

Taking the cross product of Equation 2.12 with **B** gives

$$-eB^2(\mathbf{v}_i - \mathbf{v}_n)_\perp - m_i\nu_{in}(\mathbf{v}_i - \mathbf{v}_n) \times \mathbf{B} = -e(\mathbf{E}_\perp + \mathbf{v}_n \times \mathbf{B}) \times \mathbf{B} \tag{2.13}$$

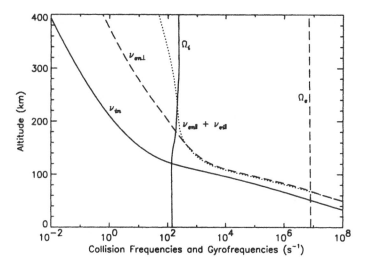

Figure 9.2.3 Collision frequencies v_{in}, $v_{en\perp}$ (for motion perpendicular to **B**) and $v_{en\parallel} + v_{ei\parallel}$ (for motion along **B**), and gyrofrequencies Ω_i and Ω_e at 44.6°N, 2.2°E for solar-minimum conditions ($S_a = 75$) on March 21. Collision frequencies are derived from the formulas in Table 9.2.1, using neutral densities from the MSISE-90 model (Hedin, 1991) and the electron densities and temperatures from the 1990 IRI model (Bilitza, 1990). The magnetic field is from the IGRF (Langel, 1992), epoch 1986.2.

Eliminating $(\mathbf{v}_i - \mathbf{v}_n) \times \mathbf{B}$ between Equations 2.12 and 2.13 yields the following expression for the ion velocity relative to the neutrals

$$(\mathbf{v}_i - \mathbf{v}_n)_\perp = \frac{v_{in}\Omega_i(\mathbf{E}_\perp + \mathbf{v}_n \times \mathbf{B}) - \Omega_i^2\mathbf{b} \times (\mathbf{E}_\perp + \mathbf{v}_n \times \mathbf{B})}{B(v_{in}^2 + \Omega_i^2)} \qquad (2.14)$$

where **b** is a unit vector in the direction of **B** and where

$$\Omega_i = eB/m_i \qquad (2.15)$$

is the definition of the *angular gyrofrequency* for the ions, describing their gyration in the geomagnetic field. Similar equations are obtained for electrons:

$$(\mathbf{v}_e - \mathbf{v}_n)_\perp = \frac{-v_{en\perp}\Omega_e(\mathbf{E}_\perp + \mathbf{v}_n \times \mathbf{B}) - \Omega_e^2\mathbf{b} \times (\mathbf{E}_\perp + \mathbf{v}_n \times \mathbf{B})}{B(v_{en\perp}^2 + \Omega_e^2)} \qquad (2.16)$$

$$\Omega_e = eB/m_e \qquad (2.17)$$

Figure 9.2.3 shows the height variations of the collision frequencies v_{in}, $v_{en\perp}$ and $v_{en\parallel} + v_{ei\parallel}$ for daytime solar-minimum conditions, along with the angular gyrofrequencies of the ions and electrons. The collision frequencies are derived from the MSISE-90 model (Hedin, 1991), an empirical model of atmospheric density, temperature, and composition. The ion-neutral and electron-neutral collision frequencies are proportional to the neutral density, which falls off exponentially with increasing altitude in the upper atmosphere, changing by a factor of the order of 10^5 between 80 and 300 km. The ion-electron and electron-ion collision frequencies, on the other hand, are approximately proportional to the electron number density N_e, and peak near the height of maximum electron density.

Notice that at high altitudes, where $v_{in} \ll \Omega_i$ and $v_{en\perp} \ll \Omega_e$, Equations 2.14 and 2.16 reduce to

$$\mathbf{v}_{i\perp} = \mathbf{v}_{e\perp} = \frac{\mathbf{E} \times \mathbf{b}}{B} \equiv \mathbf{v}_E \qquad (2.18)$$

which represents the so-called $\mathbf{E} \times \mathbf{B}$ (*E-cross-B*) drift velocity of charged particles in crossed electric and magnetic fields. Thus, at high altitudes in the ionosphere the ions and electrons essentially move together in the direction perpendicular to the magnetic field, and the current component perpendicular to $\mathbf{B}$ goes to zero. At lower altitudes the motion of ions relative to neutrals becomes more complex, having components both parallel and perpendicular to $\mathbf{E}$. At the lowest altitudes, where $v_{in} \gg \Omega_i$, (below 110 km), the right-hand side of Equation 2.14 becomes very small, so that the ion velocity is almost the same as the neutral velocity. Because the electron gyrofrequency is much larger than the ion gyrofrequency, the electron velocity approaches the neutral velocity only below 70 km.

By subtracting Equation 2.16 from Equation 2.14 we can obtain the velocity difference $\mathbf{v}_i - \mathbf{v}_e$, which, when multiplied by $N_e e$, gives the electric current density perpendicular to $\mathbf{B}$. An expression for Ohm's Law can then be obtained that expresses the total current density $\mathbf{J}$ in terms of the electric field that exists in the frame of reference of the moving neutral gas, $\mathbf{E}_\perp + \mathbf{v}_n \times \mathbf{B}$ (representing a Lorentz transformation from the Earth-based reference frame to one moving at velocity $\mathbf{v}_n$):

$$\mathbf{J} = \sigma_P(\mathbf{E}_\perp + \mathbf{v}_n \times \mathbf{B}) + \sigma_H \mathbf{b} \times (\mathbf{E}_\perp + \mathbf{v}_n \times \mathbf{B}) + \sigma_\parallel E_\parallel \mathbf{b} \tag{2.19}$$

$$\sigma_P = \frac{N_e e}{B}\left(\frac{v_{in}\Omega_i}{v_{in}^2 + \Omega_i^2} + \frac{v_{en\perp}\Omega_e}{v_{en\perp}^2 + \Omega_e^2}\right) \tag{2.20}$$

$$\sigma_H = \frac{N_e e}{B}\left(\frac{\Omega_e^2}{v_{en\perp}^2 + \Omega_e^2} - \frac{\Omega_i^2}{v_{in}^2 + \Omega_i^2}\right) \tag{2.21}$$

σ_P and σ_H are called the Pedersen and Hall conductivities, respectively (sometimes symbolized by σ_1 and σ_2, respectively). For currents perpendicular to $\mathbf{B}$, the Pedersen conductivity gives the component in the electric-field direction, while the Hall conductivity gives the component perpendicular to both $\mathbf{E}$ and $\mathbf{B}$. The quantity $\mathbf{v}_n \times \mathbf{B}$ is often called the *dynamo electric field*.

It should be noted that the above derivation of ionospheric conductivities rests on the assumption that the collision frequencies are independent of the fluid velocities. This is a reasonable assumption as long as the relative velocities are small with respect to the thermal velocities, of the order of 300 to 1000 m/s for ions and neutrals, depending on composition and temperature of the gases. However, the differential velocities of the ion and neutral gases can sometimes be comparable to the thermal velocities, especially in the auroral regions, where strong electric fields associated with magnetospheric processes can exist. Under such circumstances the above formulas can become inaccurate.

Figure 9.2.4 shows typical mid-latitude vertical profiles of the daytime conductivity components for low solar activity. At all altitudes above 80 km the parallel conductivity is much larger than the Pedersen and Hall conductivities, attaining a value on the order of 100 S/m in the upper ionosphere. This value depends on the electron temperature, varying as $T_e^{3/2}$, but it is practically independent of the electron density. The Pedersen conductivity peaks at an altitude of around 125 km during the day, while the Hall conductivity peaks around 105 to 110 km. At a given altitude, both σ_P and σ_H are essentially proportional to the electron density. The ratio between the Hall and Pedersen conductivities is greater than 1 between about 70 and 125 km, and maximizes a little below 100 km with a value of about 36, implying an angle between the current and electric field of 88.4° in the plane perpendicular to $\mathbf{B}$.

Figure 9.2.5 illustrates the variations of Pedersen conductivity between day and night and between low and high solar-activity levels. These profiles correspond to the respective electron-density profiles in figure 9.2.2. At night, when the *E*-region electron density decays strongly away, the largest Pedersen conductivities can sometimes be in the ionospheric *F* region, above 200 km.

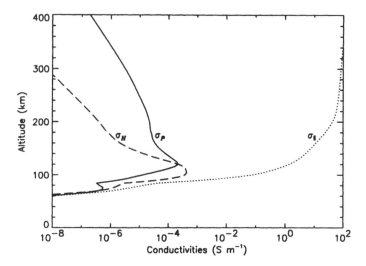

Figure 9.2.4 Noontime parallel ($\sigma_{\parallel}$), Pedersen (σ_P), and Hall (σ_H) conductivities at 44.6°N, 2.2°E for solar-minimum conditions on March 21. The electron density is the noon, solar-minimum profile of Figure 9.2.2, and the collision frequencies are those of Figure 9.2.3.

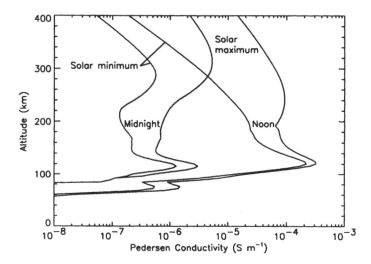

Figure 9.2.5 Pedersen conductivities at 44.6°N, 2.2°E at noon and midnight on March 21, for solar minimum and maximum conditions. The electron densities are those of Figure 9.2.2. Collision frequencies are derived from the formulas in Table 9.2.1, using neutral densities from the MSISE-90 model (Hedin, 1991) for S_a = 75 (solar minimum) or 200 (solar maximum).

This is especially true during high levels of solar activity, when F-region Pedersen conductivities are particularly large, both because of high electron densities and high neutral densities that lead to high ion-neutral collision frequencies. Because the vertical distribution of F-region electron density is highly variable, the F-region conductivity also shows a great deal of variability. The Hall conductivity also varies in proportion to the electron density, but in contrast to σ_P, σ_H always peaks in the E region of the ionosphere.

The very large parallel conductivity prevents the establishment of any significant potential differences along magnetic field lines, thus the field lines are essentially equipotential. Consequently, the perpendicular electric field, which is determined by the electric-potential difference between adjacent field lines, is nearly constant with height over most of the Earth. (An exception to this near height constancy of **E** occurs where geomagnetic field lines are almost horizontal, in

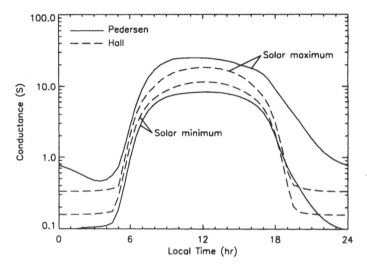

Figure 9.2.6 Height-integrated Pedersen and Hall conductivities at 44.6°N, 2.2°E on March 21, for solar minimum and maximum conditions (defined as for Figures 9.2.2 and 9.2.5).

the vicinity of the magnetic equator.) For this reason it is conceptually useful to consider the height-integrated ionospheric Pedersen and Hall conductivities, or Pedersen and Hall *conductances*, when analyzing the current-carrying capacity of the ionosphere. Figure 9.2.6 shows the modeled diurnal variations of the conductances for the same mid-latitude location as in Figures 9.2.2 to 9.2.5. Between solar minimum and solar maximum the Pedersen conductance changes proportionately more than the Hall conductance, even more so at night than during the day. This can be attributed to the relatively greater contribution of the F region to the Pedersen conductances than to the Hall conductances, and to the fact that the F-region conductances have a large solar-cycle variation. The following formulas are found to give representative daytime values of the Pedersen and Hall conductances at mid-latitudes for values of the solar zenith angle χ less than 80°:

$$\int \sigma_P dz = (11 \text{ S}) \left(\frac{S_a}{S_0}\right)^{1.1} \left(\frac{B}{B_0}\right)^{-1.6} (\cos\chi)^{0.5} \qquad (2.22)$$

$$\int \sigma_H dz = (14 \text{ S}) \left(\frac{S_a}{S_0}\right)^{0.5} \left(\frac{B}{B_0}\right)^{-1.3} (\cos\chi)^{0.8} \qquad (2.23)$$

The values of the normalizing constants are $S_0 = 100$ and $B_0 = 5 \times 10^{-5}$ T. The conductances have a fairly strong dependence on the geomagnetic-field strength B, which for these formulas is assumed to be evaluated at an altitude of 125 km. There are additional seasonal modulations of the conductances; for example, the Pedersen conductance tends to be lower in summer and higher in winter than suggested by Equation 2.22, due to the so-called "winter anomaly" in F-region electron densities, according to which the winter electron densities are greater than the summer densities at middle latitudes (see Chapter II/8).

The conductances on the nightside of the Earth are considerably more complicated than on the dayside, because of the large variability of the electron density in response to movements of the plasma, to changes in thermospheric conditions, and to additional ionization sources. The high-latitude nightside ionosphere displays the greatest degree of variability, due to the highly variable nature of auroral precipitation. Particle precipitation can also influence the low-latitude nighttime ionosphere enough to affect the Pedersen conductance at solar minimum (Rowe and Mathews, 1973).

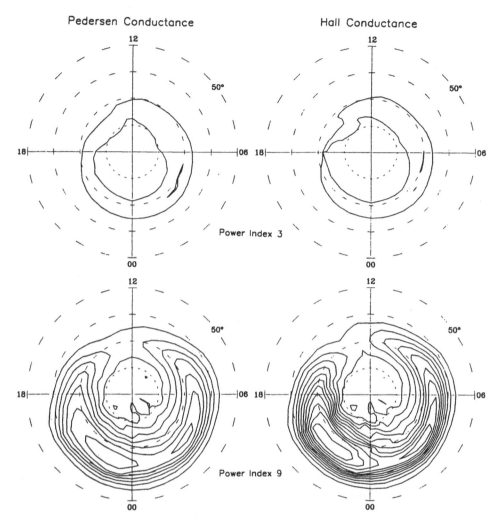

Figure 9.2.7 Average conductances due to auroral electron precipitation for low (Power Index 3) and high (Power Index 9) levels of auroral activity. Coordinates are apex latitude, from 50° to the pole, and magnetic local time. The contour interval is 2 S. (Based on the model of Fuller-Rowell and Evans, 1987)

High-latitude conductivities, although highly variable, tend to be greatest in the auroral zones. Fuller-Rowell and Evans (1987) determined the average fluxes of auroral electron precipitation as a function of magnetic latitude and magnetic local time for ten levels of hemispherically integrated energy fluxes and used these to calculate the auroral component of ionospheric conductance. Figure 9.2.7 illustrates the auroral components of Pedersen and Hall conductances for two levels of the hemispherically integrated fluxes, represented by power indices of 3 (4.6 GW) and 9 (69.5 GW). Whereas the maximum average Pedersen and Hall conductances for Power Index 3 are located in the 3 to 5 MLT sector and have similar magnitudes of about 4 S, the conductance maxima during disturbed periods tend to shift to the 21 to 23 MLT sector, and to have larger Hall values (23 S) than Pedersen values (16 S), owing to the more energetic, deeper penetrating auroral electrons. The aurorally produced conductances are thus comparable in magnitude to the conductances produced by solar UV radiation. When the auroral zone is sunlit, both conductance components contribute to the total conductance. Although the ionization rates associated with auroral electrons and solar UV radiation add in an approximate linear fashion, the conductances due to the two separate sources do not add linearly because of the fact that the electron density tends to scale as the square root of the ionization rate (Equation 2.2). To the extent that the vertical profiles of ion production are similar for both sources, which is often a reasonable rough approximation, the total conductance due to the combined sources will be approximately the square root of the sum of the squares of the individual conductance components.

3. THERMOSPHERIC WINDS

Ohm's Láw, as expressed by Equation 2.19, relates the current density to the electric field as would be measured in a reference frame moving at the wind velocity v_n. It is the wind that drives the dynamo current, leading to the buildup of polarization charges responsible for the electric field **E**. The distribution of electric fields and currents therefore depends very much on the characteristics of global winds in the *thermosphere*, that part of the atmosphere above 90 km. These characteristics and the factors that determine them are discussed in this section. Greater detail may be found in Volland (1988) and Rees (1989).

The dominant effect driving winds in the thermosphere is the diurnal variation in the absorption of solar UV radiation, which heats and expands the dayside thermosphere, creating day-to-night horizontal pressure gradients. Because of the atmosphere's much greater horizontal than vertical extent, the motions are constrained to be predominantly horizontal; vertical winds are typically only of the order of 1% the horizontal wind magnitude. At high latitudes, especially during magnetospheric disturbances, joule heating by strong electric currents and momentum input by the *Ampère force* ($\mathbf{J} \times \mathbf{B}$, also referred to as the *Lorentz force* or *Laplace force*) are also important forcing mechanisms. An additional important effect, especially in the lower thermosphere, is the upward propagation of global wave features from the atmosphere below.

The basic relations governing thermospheric dynamics are the equation of state for a perfect gas and the balance equations for mass, momentum, and energy. The equation of state for a perfect gas is

$$p = \rho kT/m_n = \rho gH \tag{3.1}$$

where p is pressure, ρ is mass density, k is Boltzmann's constant, T is temperature, m_n is the mean molecular mass of the atmospheric constituents, g is gravitational acceleration (9.4 m/s^2 at 130-km altitude), and $H = kT/m_n g$ is the pressure scale height. The equation of continuity is

$$\frac{D\rho}{Dt} + \rho \nabla \cdot \mathbf{v}_n = 0 \tag{3.2}$$

where the operator $D/Dt = \partial/\partial t + \mathbf{v}_n \cdot \nabla$ represents time differentiation following the motion of the fluid. The horizontal momentum equation in the frame of reference rotating with the Earth is

$$\rho \frac{D\mathbf{v}_h}{Dt} + 2\rho\Omega \cos\theta \mathbf{z} \times \mathbf{v}_h + \nabla_h p - \frac{\partial}{\partial z}\left(\mu \frac{\partial \mathbf{v}_h}{\partial z}\right) = (\mathbf{J} \times \mathbf{B})_h \tag{3.3}$$

where $\mathbf{v}_h$ is the horizontal component of $\mathbf{v}_n$; Ω is the angular rotation rate of the Earth (7.29×10^{-5}s^{-1}); θ is colatitude; z is altitude, $\mathbf{z}$ is a unit vertical vector, ∇_h is the horizontal component of the gradient operator, μ is the coefficient of molecular viscosity, and $(\mathbf{J} \times \mathbf{B})_h$ is the horizontal component of the Ampère force. The second term in Equation 3.3 represents the Coriolis force. In the vertical direction, the pressure gradient and gravity are nearly in balance for global-scale motions:

$$\frac{\partial p}{\partial z} = -\rho g = -p/H \tag{3.4}$$

The energy-balance equation comes from the First Law of Thermodynamics, which equates changes in internal energy with work done on the fluid plus heat added:

$$\rho\frac{D(c_v T)}{Dt} = -p\nabla \cdot \mathbf{v}_n + \frac{\partial}{\partial z}\left(\kappa\frac{\partial T}{\partial z}\right) + \rho Q \tag{3.5}$$

where c_v is specific heat at constant volume, κ is the coefficient of molecular heat conduction, and Q is the net heating associated with radiative transfer, joule heating by ionospheric currents, and viscous heating.

The viscous and heat-conduction terms in Equations 3.3 and 3.5, respectively, implicitly assume that horizontal gradients of $\mathbf{v}_h$ and T are negligible in comparison with vertical gradients, as is the case for global-scale atmospheric dynamics. In the lower thermosphere there are additional effective viscous and heat-conduction effects associated with turbulence; these effects can mostly be accommodated by adding turbulent viscous and heat-conduction coefficients to the molecular ones (although there is also an additional downward transport of heat by turbulence in a stably stratified atmosphere).

Because a 1-day periodicity is a dominant feature of thermospheric dynamics, a characteristic time rate of change of the winds is the angular frequency $2\pi/(24\ \text{h})$, of the order of 10^{-4}s^{-1}. The characteristic time rate of change associated with the Coriolis acceleration is $2\Omega\cos\theta$, also of the order of 10^{-4}s^{-1} at mid-latitudes. The viscous acceleration varies strongly with the magnitude of the wind shear and with altitude. A characteristic time rate of change associated with viscosity is $\mu/\rho H^2$, which is about 10^{-4}s^{-1} at an altitude of 200 km, varying exponentially with altitude as the inverse air density for heights above and below. The characteristic time rate of change associated with molecular heat conductivity turns out to be similar to that for molecular viscosity.

The Ampère acceleration $\mathbf{J} \times \mathbf{B}/\rho$ is also an important effect in the thermosphere. To compare it with the other terms in the momentum equation, let us evaluate it with Ohm's Law (Equation 2.19) and determine appropriate characteristic time rates of change to compare with those of the other terms in Equation 3.3. This acceleration can be expressed as

$$\frac{\mathbf{J} \times \mathbf{B}}{\rho} = \frac{\sigma_P B^2}{\rho}(\mathbf{v}_E - \mathbf{v}_n)_\perp + \frac{\sigma_H B^2}{\rho}\,\mathbf{b} \times (\mathbf{v}_E - \mathbf{v}_n) \tag{3.6}$$

The coefficients $\sigma_P B^2/\rho$ and $\sigma_H B^2/\rho$ are called the *ion-drag* coefficients, because the acceleration represents a frictional force resulting from ion-neutral collisions. At middle and low latitudes, where $\mathbf{v}_n$ is usually larger than $\mathbf{v}_E$, the first term on the right-hand side of Equation 3.6 represents a drag force, while the second term represents an acceleration perpendicular to the wind vector, similar in effect but generally opposite in sign to the Coriolis acceleration. At high magnetic latitudes $\mathbf{v}_E$ often exceeds $\mathbf{v}_n$ in magnitude, and there can be strong acceleration of the wind. Above about 150 km the Pedersen ion-drag coefficient $\sigma_P B^2/\rho$ is simply equal to the effective neutral-ion collision frequency for momentum transfer, $\nu_{ni} = (N_e m_i/\rho)\nu_{in} \sim 10^{-15}\text{m}^3\text{s}^{-1}N_e$, and is thus roughly proportional to the electron density N_e. In the daytime F-region ionosphere, where N_e can exceed 10^{12}m^{-3}, the Pedersen ion-drag coefficient can exceed 10^{-3}s^{-1}, which is much greater than the characteristic time rates of changes associated with the inertial and Coriolis terms in Equation 3.3. During the day the Hall ion-drag coefficient $\sigma_H B^2/\rho$ peaks around 125 km, with a value usually less than 10^{-4}s^{-1}, roughly comparable with the Coriolis parameter. It is less important at higher and lower altitudes.

Figure 9.3.1 shows results from a model simulation of horizontal thermospheric winds and temperature variations at heights of approximately 300 km (top) and 125 km (bottom), at 12 UT for equinox solar-minimum conditions. The subsolar point is in the center of these diagrams, i.e., $0°$ latitude, $0°$ longitude. The effects of high-latitude auroral currents were not included in this

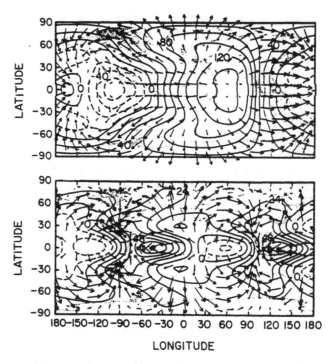

Figure 9.3.1 Temperatures (solid curves) and winds (arrows) at 12 UT for equinox and solar-minimum conditions, at atmospheric pressure levels of 6.8 μPa (approximately 300 km, top) and 2.7 mPa (approximately 125 km, bottom). Temperatures are expressed as departures from the global mean. Contour intervals are 20 (top) and 12 K (bottom). The maximum wind arrows are 166 (top) and 71 m/s (bottom). (Adapted from Fesen, C. G. et al. (1986). *J. Geophys. Res.,* 99, 4471. Copyright by the American Geophysical Union. With permission.)

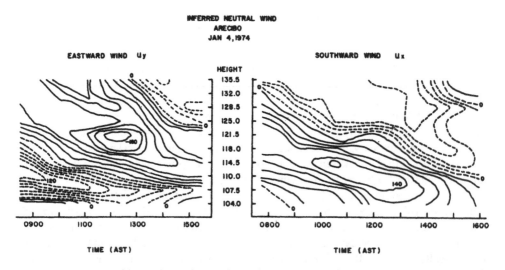

Figure 9.3.2 Variations of the winds over the Arecibo incoherent scatter radar on January 4, 1974, during daylight hours. The time scale is Atlantic Standard Time (AST, or 60° W Mean Time). Solid contours represent westward and southward winds, while dashed contours represent eastward and northward winds. The contour interval is 20 m/s. (From Harper, R. M. et al. (1976). *J. Geophys. Res.,* 81, 25. Copyright by the American Geophysical Union. With permission.)

simulation; as will be seen later, these currents have a very important effect on the dynamics at high latitudes. The wind and temperature patterns are relatively fixed in a nonrotating reference frame, such that a fixed point on the Earth sweeps through these patterns from left to right through the course of a day. In the upper thermosphere, the temperature maximizes in the afternoon, where it is associated with atmospheric expansion and a high-pressure region. This produces a general day-to-night force on the air which is counteracted primarily by ion drag and viscosity as the wind flows from high to low pressure. In the lower thermosphere ion drag and viscosity are not nearly as important, and the wind represents more of a balance among inertia, the Coriolis force, and the pressure-gradient force. In the lower part of Figure 9.3.1 the wind and temperature variations cycle twice in longitude, corresponding to a temporal variation with a 12-h period. This arises from the inclusion in the model of upward-propagating *atmospheric tides*.

Atmospheric tides (Chapman and Lindzen, 1970; Kato, 1980; Volland, 1988) are global oscillations forced by a regular periodic source, particularly at periods of one solar day (*diurnal* tides) and of one half solar or lunar day (*semidiurnal* tides). For tides with lunar-based periods, lunar gravitation is the sole forcing mechanism, but for tides with solar-based periods thermal generation is much more important than gravitational. Tides are capable of propagating vertically as global atmospheric waves, and their associated wind amplitudes tend to grow with increasing altitude as the air density decreases, until strong dissipation of the wave occurs in the thermosphere. The primary tides in the upper atmosphere result from the daily variations of solar-radiation absorption by the major atmospheric constituents in the thermosphere, as well as by ozone in the 30- to 60-km altitude range and by water vapor below 5 km.

Tidal theory was originally developed for a horizontally uniform atmosphere without any background winds; this is commonly referred to as classical tidal theory. In this case the atmospheric response can be described by distinct modes of oscillation, arising from the fact that the wave structure must be continuous around the globe. (The tidal modes are analogous to spherical harmonic functions except that their structure is altered by the latitudinally varying Coriolis parameter.) Because most of the tidal forcing migrates westward around the Earth with the apparent position of the sun or moon, westward-migrating tidal modes are expected to dominate the atmospheric response. Such modes are sometimes labeled $S_{m,n}$ or $L_{m,n}$ for solar and lunar migrating modes, respectively, where m gives the longitudinal wavenumber and where n is an index identifying the latitudinal structure of the mode. Some of the modes that have been discussed in relation to the ionospheric dynamo are as follows.

1. $S_{1,-2}$ is the main diurnal mode, symmetric about the equator, driven by local heating. It does not propagate vertically, and has its largest winds at high latitudes.
2. $S_{1,-1}$ is the main antisymmetric diurnal mode driven by local heating. It does not propagate vertically.
3. $S_{1,1}$ is the main diurnal mode, symmetric about the equator, that is capable of propagating vertically. It has a vertical wavelength around 24 km in the thermosphere and has its largest winds at low latitudes.
4. $S_{2,2}$ is the main semidiurnal mode, symmetric about the equator. It is capable of vertical propagation above about 110 km, where it has a relatively long vertical wavelength, and it has winds that are greater at low latitudes than at high latitudes.
5. $S_{2,3}$ is the main antisymmetric semidiurnal mode. It propagates vertically with a wavelength on the order of 60 km.
6. $S_{2,4}$ is another important symmetric semidiurnal mode. It propagates vertically with a wavelength on the order of 45 km and has winds that are greater at upper mid-latitudes than at low latitudes.
7. $L_{2,2}$ is the main lunar mode, symmetric about the equator, with a period of 12.42 h. It is similar in structure to the $S_{2,2}$ mode.

The combination of different modes can give rise to complex structures of the wind in the thermosphere.

When the true latitudinal structure of the background atmosphere is taken into account, or when dissipative forces are considered, individual tidal modes with distinct horizontal structures no longer exist. Forbes and Vial (1989, 1991) have calculated tidal structures for these more realistic conditions. The atmospheric response is still found to display many of the characteristics of the classical tidal modes, such as phase variations with altitude and latitude characteristic of propagating waves and amplitude growth and decay in altitude characteristic of waves propagating upward into less dense atmospheric regions and then being dissipated. Figure 9.3.2 shows vertical structures of observed winds in the lower thermosphere above the Arecibo incoherent scatter radar (18.3°N, 66.75°W) on January 4, 1974. The descending structures are characteristic of tides.

Planetary waves (Volland, 1988) have periods longer than one day. In the lower atmosphere they are associated with large-scale weather patterns. Planetary waves with periods around 2, 5, and 16 days have been detected in the upper mesosphere and lower thermosphere. Modeling studies (Hagan et al., 1993) indicate that these waves can penetrate well into the dynamo region, though unambiguous observational evidence of planetary waves with significant amplitudes above 100 km has not yet been obtained. The waves seem to be quasiresonant, representing normal modes of atmospheric motion. The quasi-2-day wave appears most often in January and August, with an amplitude that often exceeds the tidal amplitudes around 90 km for an interval of 1 to 2 weeks. Theory and observation are consistent with it having a longitudinal wavenumber (number of sinusoidal oscillations around a circle of latitude) of 3, and to be antisymmetric about the equator. The properties of longer period planetary waves have not been as well defined as those for the 2-day wave.

Waves with periods shorter than tidal periods, but longer than the natural buoyancy period (or *Brunt-Väisälä period*) of the atmosphere (roughly 15 min in the thermosphere) are called *gravity waves*. In fact, tides are just a large-scale, long-period form of gravity wave. Gravity waves are ubiquitous in the atmosphere and have relatively large amplitudes in the upper atmosphere, associated with their tendency to grow exponentially with height. Their upward growth often can lead to nonlinear breakdown of the wave, whereupon momentum and energy are deposited locally in the atmosphere. Although gravity waves are of considerable importance for the dynamical state of the upper atmosphere, they are probably of less importance for global ionospheric electrodynamics.

Above 120 km the dynamics of the thermosphere can be strongly influenced by energy input from the magnetosphere (Fuller-Rowell et al., 1988; Roble, 1992). Rapidly drifting ions at high magnetic latitudes set the neutral atmosphere into motion through ion drag. A strong correlation exists between wind and ion motions above 130 km at magnetic latitudes above 65° (Killeen and Roble, 1988). Figure 9.3.3 shows the high-latitude wind distribution at 145 km, along with the distribution of ion velocities, obtained from a numerical simulation. The influence of the two-cell ion convection on the wind is evident, although the high-latitude, two-cell wind pattern is rotated clockwise by about 2 hours of local time with respect to the ion convection cells, caused by the Earth's rotation and the few hour time scale of ion-drag effects at this altitude. Joule dissipation of electric energy in the high-latitude ionosphere affects not only the local, but also the global thermospheric circulation, especially during magnetic storms. The heated air rises at high latitudes, and there is a general outflow of air above the region of peak heating, centered roughly at an

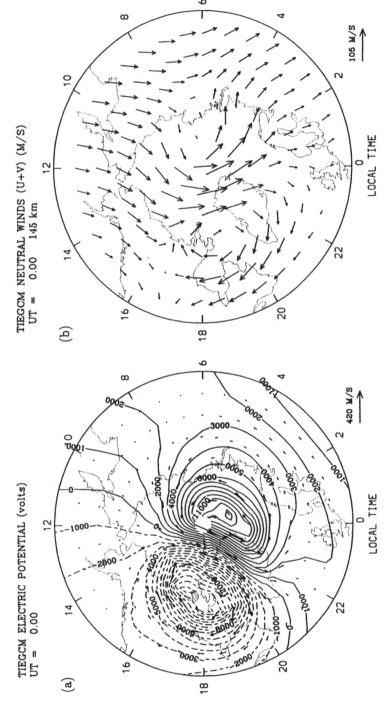

Figure 9.3.3 Results from a numerical simulation with the Thermosphere Ionosphere Electrodynamics General Circulation Model (TIEGCM) for solar minimum equinox conditions at 0 UT. (a) Electric potential and ion convection velocities between 47.5° and the geographic north pole. (b) Pattern of wind velocities at 145 km altitude. (From Richmond, A. D. (1994). *The Upper Mesosphere and Lower Thermosphere*, R. M. Johnson and T. L. Killeen, Eds. in press. Copyright by the American Geophysical Union. With permission.)

altitude of 125 km. The equatorward flow at upper mid-latitudes is acted upon by the Coriolis force, producing a westward motion. Because the energy input during a storm is highly erratic, large-scale gravity waves are also generated that are able to propagate worldwide so that the resultant thermospheric motions are very irregular. The energy deposited at high latitudes is partially redistributed globally by the altered thermospheric circulation, thus the global thermospheric temperature rises by as much as several hundred kelvins. After a storm, the altered thermospheric circulation and structure relax toward the prestorm state on a time scale that is linked to the characteristic viscous and heat-conduction time scale at the lowest altitude where significant alteration took place. This time scale is of the order of 1 day at 130 km.

4. THEORY OF IONOSPHERIC ELECTRIC FIELDS AND CURRENTS

If the distributions of thermospheric winds and ionospheric conductivities are known, the electric fields and currents generated by dynamo action can be calculated. In addition to Ohm's Law (Equation 2.19), the conditions of current continuity and of electrostatic fields must be satisfied:

$$\nabla \cdot \mathbf{J} = 0 \qquad (4.1)$$

$$\mathbf{E} = -\nabla\Phi \qquad (4.2)$$

where Φ is the electrostatic potential. Combining Equations 2.19, 4.1, and 4.2 gives a partial differential equation for Φ:

$$\nabla \cdot [\sigma_P(\nabla\Phi)_\perp + \sigma_H \mathbf{b} \times \nabla\Phi + \sigma_\parallel(\nabla\Phi)_\parallel] = \nabla \cdot [\sigma_P \mathbf{v}_n \times \mathbf{B} + \sigma_H \mathbf{b} \times (\mathbf{v}_n \times \mathbf{B})] \qquad (4.3)$$

Solving Equation 4.3 requires boundary conditions that represent the effects of electric coupling with the magnetosphere above and with the lower atmosphere below. Because the atmosphere below 80 km is a very poor conductor, electric-current coupling between the ionosphere and the lower atmosphere is usually negligible as concerns large-scale ionospheric electric fields and currents. An adequate lower boundary condition for Equation 4.3 at the base of the ionosphere is thus obtained from the requirement that the vertical component of electric current density vanish there.

Coupling with the magnetosphere is not negligible, so that formulating a realistic high-altitude boundary condition for Equation 4.3 can be much more difficult. Ohm's Law becomes invalid in the collisionless energetic plasma of the magnetosphere where currents are more closely related to gradient-curvature drifts of the particles and to other effects, like plasma polarization and magnetic-mirror forces, than to the electric field. Equation 4.3 is adequate on geomagnetic field lines that peak at radial distances below about $4R_E$ (R_E = radius of the Earth) or magnetic latitudes below about 60° (lower during disturbed periods), but the additional magnetospheric sources of electric fields and currents become important for higher-latitude field lines (Kamide and Baumjohann, 1993). Under quiet, steady-state magnetospheric conditions, however, there is a tendency for ionospheric electrodynamics at middle and low latitudes to be significantly decoupled from magnetospheric electrodynamics. Energetic plasma in the middle magnetosphere, at the inner edge of the plasma sheet, tends to adjust itself so as to produce polarization electric fields that counteract the strong magnetospheric electric fields present in the outer magnetosphere. This reaction has the effect of shielding the inner magnetosphere and the middle- and low-latitude ionosphere from magnetospheric effects (Wolf et al., 1986). For this reason, middle- and low-latitude electric fields and currents at quiet times are mainly responsive to ionospheric dynamo effects. An upper boundary condition for Equation 4.3 that approximately reflects this "shielding" effect is to specify that the electric potential is constant along a high-latitude boundary that maps to the inner edge of the plasma sheet. Equation 4.3 would then be solved only for the region of space composed of fields lines inside this boundary.

The fact that the conductivity of the ionosphere is extremely anisotropic necessitates the utilization of a geomagnetic-field-oriented coordinate system in order to carry out numerical solutions of Equation 4.3. One of the coordinates will vary along the geomagnetic field while the other two coordinates must be defined so as to be constant along geomagnetic field lines. For practical purposes, a major simplification can be made to Equation 4.3 by additionally assuming that Φ is constant along field lines and that $\sigma_\parallel$ is infinite. The component of $\mathbf{J}$ along $\mathbf{B}$ is then no longer determined by Ohm's Law, but rather by the condition that it assumes whatever value is necessary in order to ensure that the current density remain divergence free so that Equation 4.1 is satisfied everywhere. The partial differential equation for Φ can then be reduced from three to two dimensions upon multiplying Equation 4.3 by $1/B$ and integrating along field lines from the base of the ionosphere in the southern magnetic hemisphere to the base of the ionosphere in the northern magnetic hemisphere. In doing so, note that

$$\nabla \cdot \mathbf{J}_\parallel = \nabla \cdot (J_\parallel \mathbf{B}/B) = \mathbf{B} \cdot \nabla(J_\parallel/B) = B\frac{\partial}{\partial s}\left(\frac{J_\parallel}{B}\right) \tag{4.4}$$

where s is distance along a field line. The integral of $(\nabla \cdot \mathbf{J}_\parallel)/B$ from one end of the field line to the other thus equals the difference between $(J_\parallel/B)$ at the northern end and $(J_\parallel/B)$ at the southern end. Because $J_\parallel$ goes to zero at both ends, owing to the vanishing conductivity at the base of the ionosphere, the integral also vanishes. If we define the two field perpendicular coordinates as x_1 and x_2 such that $\mathbf{B}$ is in the direction of $\nabla x_2 \times \nabla x_1$, the following two-dimensional partial differential equation is obtained:

$$\frac{\partial}{\partial x_1}\left(S_{11}\frac{\partial\Phi}{\partial x_1} + (S_{12} + \Sigma_H)\frac{\partial\Phi}{\partial x_2}\right) + \frac{\partial}{\partial x_2}\left((S_{21} - \Sigma_H)\frac{\partial\Phi}{\partial x_1} + S_{22}\frac{\partial\Phi}{\partial x_2}\right) = \frac{\partial D_1}{\partial x_1} + \frac{\partial D_2}{\partial x_2} \tag{4.5}$$

where

$$S_{ij} = \int (WB\nabla x_i \cdot \nabla x_j)\sigma_P ds \tag{4.6}$$

$$\Sigma_H = \int \sigma_H ds \tag{4.7}$$

$$D_i = \int WB\nabla x_i \cdot (\sigma_P \mathbf{v}_n \times \mathbf{B} + \sigma_H B\mathbf{v}_n)ds \tag{4.8}$$

$$W = (-\mathbf{B} \cdot \nabla x_1 \times \nabla x_2)^{-1} \tag{4.9}$$

The integrals in Equations 4.6 through 4.8 are taken along the entire field line between the base of the ionosphere in each magnetic hemisphere. For simplified geometries like a purely dipole geomagnetic field, x_1 and x_2 can be selected to be orthogonal so that S_{12} and S_{21} vanish. With application of appropriate boundary conditions, as discussed above, it is possible to solve Equation 4.5 for Φ, from which $\mathbf{E}$ and the component of $\mathbf{J}$ perpendicular to $\mathbf{B}$, $\mathbf{J}_\perp$, can be determined using Equations 4.2 and 2.19. $J_\parallel$ can then be calculated by integrating $-(\nabla \cdot \mathbf{J}_\perp)/B$ with respect to s from the southern end of a field line to the desired point and multiplying the result by the local value of B. Figure 9.4.1 shows an example of the height-integrated current density obtained from a simulation for the December solstice.

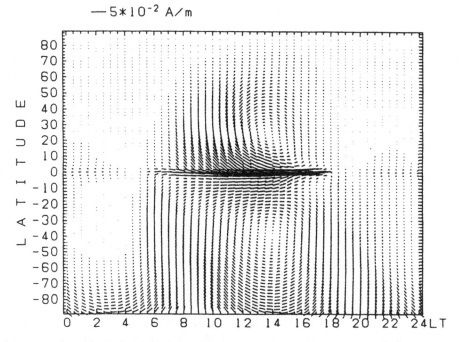

Figure 9.4.1 Global distribution of modeled height-integrated horizontal current density for December solstice, low solar activity. The model assumes a dipolar geomagnetic field aligned with Earth's axis so that the latitude scale is both geographic and geomagnetic. (Reprinted from *J. Atmos. Terr. Phys.*, 52: Takeda, M., Geomagnetic field variation and the equivalent current system generated by an ionospheric dynamo at the solstice, pp. 59–67, Copyright 1990, with kind permission from Pergamon Press Ltd, Headington Hill Hall, Oxford 0X3 0BW, UK.)

Numerical simulation models of the ionospheric dynamo have been used to shed light on many characteristics of ionospheric electrodynamics. (See reviews by Kato, 1980; Wagner et al., 1980; Volland, 1984; and Richmond, 1989, 1994.) Schuster (1908) developed the first model that showed how tidal winds could drive upper atmospheric currents. Hirono (1952), Baker (1953), and J. A. Fejer (1953) developed the first models to use realistic anisotropic conductivities. Maeda (1955, 1957), Kato (1956, 1957), and Schieldge et al. (1973) used dynamo models to deduce wind patterns that could explain observed geomagnetic perturbations at the ground. Richmond et al. (1976), Forbes and Lindzen (1976a,b, 1977), Forbes and Garrett (1979), Takeda and Maeda (1980, 1981), and Richmond and Roble (1987) demonstrated that tidal winds with vertical variations of amplitude and phase consistent with observed lower thermospheric winds produce both electric fields and currents that have a number of similarities with observations.

Dynamo models have demonstrated the importance of winds in the ionospheric E region to explain S_q currents and magnetic perturbations, which are predominantly a daytime phenomenon. However, the F region is also the seat of significant dynamo action under certain conditions: especially around the maximum of the solar cycle, when F-region conductivities are large; at night, when F-region conductances often dominate over E-region conductances; and at low magnetic latitudes, where the curved magnetic field lines can have much longer portions of their length in the F region than in the E region. Rishbeth (1971a, 1981) pointed out that the F-region dynamo should dominate low-latitude electrodynamics at night. Heelis et al. (1974), Matuura (1974), Takeda and Maeda (1980, 1981), Stening (1981), and Takeda et al. (1986) modeled the coupled E- and F-region dynamos and demonstrated how the strong F-region winds can have important effects. Takeda and Maeda (1983) showed the importance of the F-region dynamo for explaining observed magnetic perturbations on the MAGSAT spacecraft. Mayr et al. (1990) found that the F-region dynamo effects can be important even during the day, a finding substantiated by Crain et al. (1993a) for low latitudes. Takeda et al. (1986) and Takeda and Yamada (1987) showed that the much larger F-region conductivities at solar maximum than at solar minimum lead to a strong solar-cycle dependence of dynamo effects.

Because thermospheric winds are influenced by the Ampère force of the electric currents, the neutral dynamics and electrodynamics are strongly coupled, especially in the F region (Rishbeth, 1971b, 1979; Volland, 1976a,b, 1984; Glushakov et al., 1980, 1981; Mayr et al., 1990). Further coupling occurs owing to redistribution of the ionospheric plasma by $\mathbf{E} \times \mathbf{B}$ drifts, which influence the distribution of F-region conductivity (Haerendel and Eccles, 1992; Haerendel et al., 1992; Crain et al., 1993b). Recently, models have been developed that consider the mutual coupling among neutral dynamics, plasma dynamics, and electric currents and fields (Namgaladze et al., 1990, 1991; Richmond et al., 1992).

Dynamo effects with time periodicities different from a solar day have also been modeled. Maeda and Fujiwara (1967) and Tarpley (1970) modeled lunar dynamo effects, showing that the lunar wind amplitudes in the dynamo region must be of the order of 10 m/s. Ito et al. (1986), Takeda and Yamada (1989), and Chen (1992) simulated the dynamo effects associated with planetary waves of a quasi-2-day period, which have been observed in the upper mesosphere, though not yet in the main part of the dynamo region just above. There are not yet sufficient observations of dynamo effects with quasi-2-day periods to be able to verify or refute the validity of the model simulations. Planetary waves of periods other than 2 days have not yet been used in dynamo simulations. At shorter periods, there have been calculations of the electrodynamic effects of atmospheric acoustic-gravity waves (Prakash and Pandey, 1985; Jacobson and Bernhardt, 1985), which can contribute to the short period, local variability in ionospheric electric fields.

Disturbances to the global current system can be produced by changes in ionospheric conductivity. Occasionally, a significant increase in solar ionizing radiation, especially at X-ray wavelengths, occurs during a solar flare. This causes a temporary increase in the ionospheric electron density and conductivity, allowing enhanced currents to flow over the dayside of the Earth. A few modeling studies have been carried out of the ionospheric dynamo under solar-flare conditions (Richmond and Venkateswaran, 1971). The influence of a solar eclipse on ionospheric currents has been modeled by Takeda and Araki (1984).

It is believed that changes in thermospheric winds associated with magnetospheric energy input at high latitudes can produce significant dynamo effects, not only at high latitudes, but also on a global scale. The high-latitude dynamo effects are often referred to as *flywheel* effects (Banks, 1972), because they represent the tendency for the winds accelerated by the Ampère force to maintain the sense of plasma convection originally imposed from the magnetosphere. When such winds extend equatorward of the shielding region, after this region has moved poleward during a relaxation of magnetic activity, their dynamo effect is often referred to as the *fossil-wind* effect (Spiro et al., 1988). Mid-latitude winds produced by high-latitude joule heating that drives an outflow of air from the auroral region, resulting in westward winds that arise from the Coriolis effect, can produce a dynamo effect referred to as the *disturbance dynamo* (Blanc and Richmond, 1980).

Dynamo simulations predict significant asymmetry in the current system about the magnetic equator associated with various effects: the seasonal variation in conductivity; hemispherical asymmetries of thermospheric winds caused by seasonally asymmetric absorption of solar radiation or by asymmetric upward-propagating tides like the semidiurnal $S_{2,3}$ and $S_{2,5}$ modes; and the tilt of the geomagnetic dipole away from the Earth's rotation axis. The asymmetry does not occur in the electric potential, which is constrained to be nearly the same at conjugate points owing to high conductivity along geomagnetic field lines (Dougherty, 1963; Peterson et al., 1977), but the horizontal ionospheric current can be very different between the two hemispheres, and the models find an important flow of electric current along geomagnetic field lines from one hemisphere to the other. Figure 9.4.2 shows the modeled field-aligned current distribution at 1000

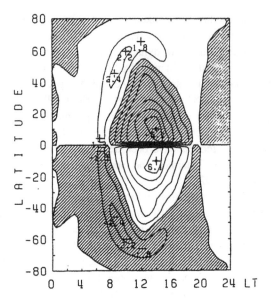

Figure 9.4.2 Global distribution of modeled field-aligned current density at 1000-km altitude, corresponding to the ionospheric current of Figure 9.4.1. Values are in units of nA/m², with a contour value of 1 nA/m². The shading corresponds to current flow out of the ionosphere. (Reprinted from *J. Atmos. Terr. Phys.,* 52: Takeda, M., Geomagnetic field variation and the equivalent current system generated by an ionospheric dynamo at the solstice, pp. 59–67, Copyright 1990, with kind permission from Pergamon Press Ltd, Headington Hill Hall, Oxford 0X3 0BW, UK.)

km corresponding to the horizontal ionospheric current shown in Figure 9.4.1. (Notice that this simulation does not account for magnetospheric dynamo effects and thus has no strong field-aligned currents in the auroral zones.) Typical current magnitudes are found to be of the order of 10^{-9} to 10^{-8} Am^{-2}. The total amount of current flow at mid-latitudes can be of the order of 10^5 A or more, comparable to the total horizontal current flowing in the main S_q vortices. van Sabben (1966, 1968), Richmond (1974), and Takeda (1990) gave procedures for calculating the ground-level magnetic perturbations associated with such currents.

Although the conductivity of the atmosphere below 80 km is very small in comparison with that above, some current does still flow between the ionosphere and the lower atmosphere. A global upward current of the order of 1000 A flows, driven by the strong upward electric fields above the tops of thunderstorms, and an equal downward current flows in fair-weather regions (see Volland, 1984; Roble and Tzur, 1986). Makino and Takeda (1984) have estimated that these currents produce global scale polarization electric fields in the ionosphere of magnitudes up to about 10^{-5} Vm^{-1}, two orders of magnitude smaller than those generated by the ionospheric dynamo. Hays and Roble (1979) modeled a larger, though still relatively modest, effect. Although not important for global ionospheric electrodynamics, thunderstorm-generated ionospheric electric fields are likely to be significant locally in regions directly over the polarized clouds (see Chapter I/10).

Close to the magnetic equator, the combination of the nearly horizontal geomagnetic field and the large difference between σ_P and σ_H in the lower ionosphere gives rise to a peculiar phenomenon known as the *equatorial electrojet* (Forbes, 1981; Reddy, 1989; Rastogi, 1989). During the daytime, when the ionospheric electric field in the equatorial region is usually eastward, a downward component of Hall current is produced. In order to maintain current continuity, a counterbalancing upward component of Pedersen current must flow, necessitating an upward component of electric field. Polarization charge develops until sufficient Pedersen current flows. Because the Pedersen conductivity is small in the lower part of the dynamo region in comparison with the Hall conductivity, the vertical polarization field must be much larger than the original eastward electric field. However, this strong upward electric field also produces a Hall current, in the eastward direction. Consequently, a strong eastward Hall current flows in the lower ionosphere close to the magnetic equator, comprising the equatorial electrojet.

In regions where the electrojet is not changing rapidly with longitude, Equation 4.5 can be solved in an approximate way for the vertical polarization electric field in terms of the eastward field and the winds. Let x_1 be in the magnetic-eastward direction and x_2 in the direction generally upward (with an additional poleward tilt as one moves off the magnetic equator). Let us assume that east-west gradients of the electric field can be neglected. Because the component of electric field in the x_2-direction is proportional to $(\partial\Phi/\partial x_2)$, this means that the x_1-derivative of $(\partial\Phi/\partial x_2)$ is zero, and consequently,

$$\frac{\partial}{\partial x_2}\left(\frac{\partial\Phi}{\partial x_1}\right) = \frac{\partial}{\partial x_1}\left(\frac{\partial\Phi}{\partial x_2}\right) = 0 \tag{4.10}$$

signifying that $(\partial\Phi/\partial x_1)$ is independent of x_2, that is, constant in height and latitude.

To solve Equation 4.5, let us first simplify it by neglecting S_{21}; i.e., we assume that x_1 and x_2 are orthogonal. The lower boundary condition, of vanishing vertical current, corresponds to the condition

$$\Sigma_H\frac{\partial\Phi}{\partial x_1} - S_{22}\frac{\partial\Phi}{\partial x_2} + D_2 = 0 \tag{4.11}$$

Assume that the eastward current varies only slowly in the east-west direction, so that the first terms on the left and right of Equation 4.5 can be neglected. The remaining terms of Equation 4.5 correspond to the condition that the x_2-derivative of Equation 4.11 vanishes, implying that Equation 4.11 must be independent of x_2 and therefore valid at all heights, not only at the lower boundary. Equation 4.11 can readily be solved for $-\partial\Phi/\partial x_2$ to give

$$-\frac{\partial\Phi}{\partial x_2} = -\frac{1}{S_{22}}\left(\Sigma_H\frac{\partial\Phi}{\partial x_1} + D_2\right) \tag{4.12}$$

which relates the polarization electric field to the eastward electric field and the wind. This is seen more clearly if we make the further approximation that within the lower equatorial ionosphere ($\pm10°$ magnetic latitude, 80 to 200 km altitude) the quantities $|\nabla x_1|$, $|\nabla x_2|$, and B can be considered nearly constant along the magnetic field. Compatible with this approximation, x_1 and x_2 can be taken to be actual spatial distances perpendicular to **B** in the eastward and upward/poleward directions, respectively (neglecting the fact that in reality the distance between two field lines varies slightly along the field). Then the electric-field components perpendicular to **B**, which we label E_1 in the eastward direction and E_2 in the x_2-direction, can also be considered to be constant along field lines. With these simplifications, Equation 4.12 becomes

$$E_2 = \frac{\Sigma_H}{\Sigma_P}E_1 - \frac{\int(\sigma_P\upsilon_1 + \sigma_H\upsilon_2)Bds}{\Sigma_P} \tag{4.13}$$

where

$$\Sigma_P = \int\sigma_Pds \tag{4.14}$$

The contribution to E_2 of the term in Equation 4.13 proportional to E_1 is very large for field lines peaking below 110 km, where $\Sigma_H \gg \Sigma_P$. This is the primary source of the strong polarization field responsible for the electrojet current. The contribution to E_2 of the wind terms in Equation 4.13 depends both on the wind velocity and on its variation along the magnetic field line. An eastward wind υ_1 contributes negatively to E_2. In fact, it can be seen that an eastward wind that is constant along the field line would produce an electric field equal to $-\upsilon_1 B$, which in turn would produce an eastward $\mathbf{E} \times \mathbf{B}$ velocity of exactly $\mathbf{v}_1$, and it would be found that the ions and electrons move at the same velocity as the neutral wind. The contribution of the wind component υ_2 to the wind-related integral in Equation 4.13 tends to be less important. Because large scale vertical winds tend to be much smaller than their horizontal counterparts, υ_2 represents primarily the projection of the (magnetically) north-south wind onto the direction perpendicular to $\mathbf{B}$; near the magnetic equator this direction is nearly vertical so that υ_2 tends to be quite small. In addition, for large-scale winds the projections of the north-south wind onto the field-perpendicular direction tend to have opposite signs on opposite sides of the magnetic equator so that a large amount of cancellation tends to occur in the integration of $\sigma_H \upsilon_2 B$ along the field line. Nonetheless, because $\sigma_H \gg \sigma_P$ below 110 km, a north-south wind that is strongly asymmetric about the magnetic equator might contribute significantly to the wind-related integral in Equation 4.13, since the asymmetry would imply a noncanceling variation of $\sigma_H \upsilon_2 B$ along the field line.

The eastward current density, obtained by using Equation 4.13 in Equation 2.19, is

$$ J_1 = \left(\sigma_P + \frac{\Sigma_H}{\Sigma_P} \sigma_H \right) E_1 + \sigma_H \left[\left(\upsilon_1 - \frac{\int \sigma_P \upsilon_1 ds}{\int \sigma_P ds} \right) - \left(\frac{\sigma_P}{\sigma_H} \upsilon_2 + \frac{\int \sigma_H \upsilon_2 ds}{\int \sigma_P ds} \right) \right] B \quad (4.15) $$

The multiplier of E_1 on the right-hand side of Equation 4.15 is sometimes called the *Cowling conductivity*, although the usual expression of Cowling conductivity is derived for a one-dimensional geometry, for which Σ_H / Σ_P is replaced by σ_H / σ_P. (The two expressions give similar results below 105 km at the equator.) It is much larger than either the Pedersen or the Hall conductivity below 110 km at the magnetic equator. It can be seen that the terms dependent on υ_1 in Equation 4.15 would cancel if υ_1 were constant along $\mathbf{B}$; thus only spatially varying east-west winds affect the current.

Figure 9.4.3 shows an example of modeled low-latitude electric currents. A strong electrojet current layer appears at the equator, while only 5° away the current density is much weaker (notice the different scale) and is highly structured in the vertical due to presence of similarly structured winds.

Comparisons of model calculations like that for 0° in Figure 9.4.3 with observations of the electrojet structure have tended to indicate that the models produce a peak polarization electric field and a current layer that are situated too low with respect to the observations when standard models for the Pedersen and Hall conductivities are used. Ronchi et al. (1990) showed that plasma irregularities observed in the lower part of the electrojet are of sufficient magnitude to strongly reduce the vertical polarization field and horizontal current there and to raise the heights of peak field and current, provided it is assumed that the irregularities are very highly aligned with the geomagnetic field. Alternatively, the standard models used to calculate the collision frequencies and conductivities may be incorrect (Stening, 1986). Gagnepain et al. (1977) showed that the discrepancies could be largely removed by arbitrarily increasing the electron-neutral collision frequency $\nu_{en\perp}$ by a factor of around 4; Untiedt (1967, 1968) had adjusted $\nu_{en\perp}$ by a similar factor for his electrojet model.

5. OBSERVATIONS OF IONOSPHERIC ELECTRIC FIELDS AND CURRENTS

Most of the available observations that give us electric-field information are ion-drift measurements in the ionosphere above 200 km, either by radar or by spacecraft. The component of the

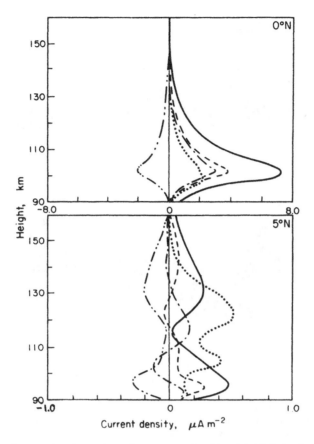

Figure 9.4.3 Eastward current densities at 0°N and 5°N, at 0600 local time (LT) (dash-dot-dash), 0900 LT (dashes), 1200 LT (solid), 1500 LT (dots), and 1800 LT (dash-dot-dot-dash), computed from the $S_{1,-2}$ $S_{1,1}$, $S_{2,2}$, and $S_{2,4}$ tidal modes. Note the difference in scales for 0 and 5°. (Reprinted from *J. Atmos. Terr. Phys.,* 38: Forbes, J. M., and R. S. Lindzen, Atmospheric solar tides and their electrodynamic effects. II. The equatorial electrojet, pp. 911–920, Copyright 1976, with kind permission from Pergamon Press Ltd, Headington Hill Hall, Oxford OX3 0BW, UK.)

electric field perpendicular to the geomagnetic field is readily derived from these drift measurements by inverting Equation 2.18:

$$\mathbf{E}_{\perp} = -\mathbf{v}_i \times \mathbf{B} \tag{5.1}$$

The observations show a high degree of variability (B. G. Fejer, 1991), even on magnetically quiet days. Figure 9.5.1 shows a superposition of several days of quiet-day ion-drift measurements during the months of November to February, at solar-cycle minimum from the Arecibo incoherent scatter radar (see Table 9.5.1). Although the average trend of the drifts throughout the day is discernible, the variability is striking. The variability tends to be greater at night than at day. Only after the data are averaged for many days does a regular pattern emerge.

Figure 9.5.2 shows the daily variations of the average quiet-day drifts for middle- and low-latitude incoherent scatter radars (Table 9.5.1) for different seasons and levels of solar activity. (For St. Santin the solar-cycle dependence of the drifts has not yet been determined.) In order to have sufficient data in each season, only three seasons are used for the year, centered around June, December, and the combined equinoxes: J months (May through August), E months (March, April, September, and October), and D months (November through February). Note that the seasons are ordered differently for the St. Santin and MU radars than for the American sector

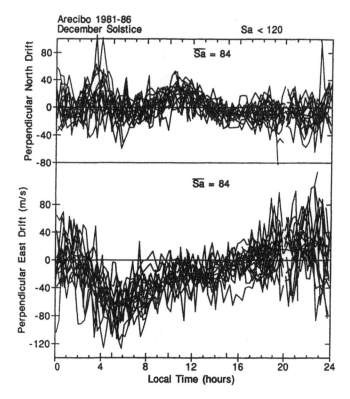

Figure 9.5.1 Scatter plots of the perpendicular/north and east components of the Arecibo *F*-region plasma drifts during magnetically quiet (*K$_p$* < 3) and low solar activity conditions, during November through February. (From Fejer, B. G. (1993). *J. Geophys. Res.,* 98, 13,645. Copyright by the American Geophysical Union. With permission.)

Table 9.5.1 **Middle- and Low-Latitude Incoherent Scatter Radars Observing Ionospheric E × B Drifts**

	Geographic				
Radar	**Latitude (°)**	**Longitude (°)**	**1985.0 Apex Latitude (°)**	**MLT-UT, h**	**Ref.**
Millstone Hill	42.6	−71.5	53.7	−4.4	Buonsanto et al. (1993)
St. Santin	44.6	2.2	39.6	0.5	Blanc and Amayenc (1979)
Arecibo	18.3	−66.75	29.8	−4.2	B. G. Fejer (1993)
MU	34.85	136.1	27.3	9.1	Oliver et al. (1993)
Jicamarca	−11.9	−76.0	1.0	−5.1	B. G. Fejer et al. (1991)

radars in Figure 9.5.2, because geometrical considerations suggest that the solstitial variations might be opposite in the Eurasian and American longitude sectors, owing to the opposite relative placement of the magnetic and geographic equators in these two sectors (see Figure 9.1.1). Because the electric fields equalize along the geomagnetic field between conjugate points in the northern and southern hemispheres, seasonal variations of the electric fields and **E × B** drifts are not great, but differences between the J and D solstices can occur in association with the offset magnetic and geographic equators, among other factors. The upward/poleward drift component (perpendicular to the tilted geomagnetic field lines in the magnetic meridian) tends to be positive in the morning and negative in the afternoon at upper middle latitudes, with the phase shifting somewhat later at low latitudes. The general phase of the diurnal variation of the eastward drift reverses between the magnetic latitudes of Millstone Hill (54°) and St. Santin (40°). Different criteria were used to define low and high activity for the different data sets shown in Figure 9.5.2,

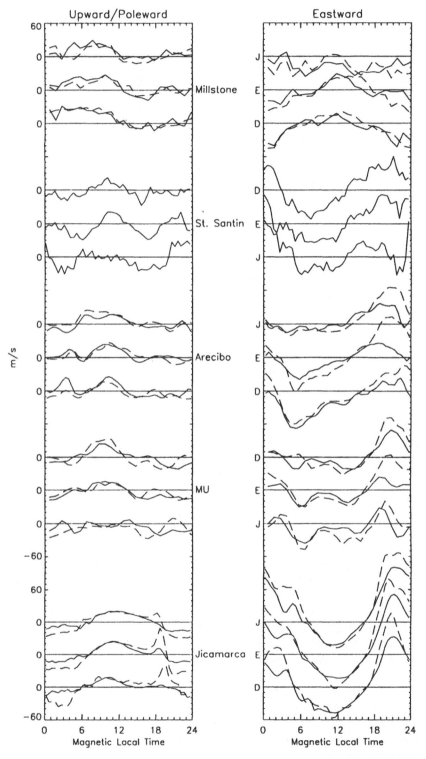

Figure 9.5.2 Average quiet-day **E** × **B** drift components perpendicular to the geomagnetic field over the incoherent scatter radars listed in Table 9.5.1. The mean altitude is 300 km. Tick marks are every 10 m/s. Solid lines are for low solar activity and dashed lines for high solar activity (different criteria were used to define low and high activity for the different data sets). Averages for the months of November through February are denoted by *D* (December solstice); for March, April, September, and October by *E* (equinox); and for May through August by *J* (June solstice). Note that the seasons are ordered differently for the St. Santin and MU radars then for the other radars (see text).

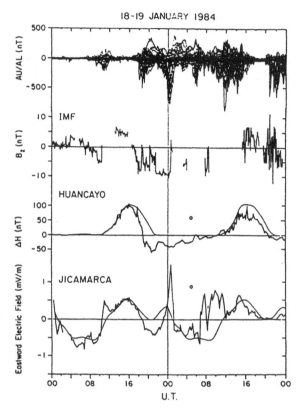

Figure 9.5.3 Auroral electrojet indices AU/AL, northward (B_z) component of the interplanetary magnetic field (IMF), perturbation of the horizontal component of the magnetic field at Huancayo (near Jicamarca), and eastward electric field over Jicamarca during January 18–19, 1984. The smooth curves denote the average quiet-time variations, and the small circles indicate local midnight. (From Fejer, B. G. et al. (1990). *J. Geophys. Res.,* 95, 2367. Copyright by the American Geophysical Union. With permission.)

so quantitative comparisons of the solar-cycle variations among the various curves are not possible. However, some qualitative features are found to be consistent among the curves; for example, the nighttime east-west drifts are generally stronger for high solar activity. At Jicamarca a notable phenomenon in the upward/poleward drift is the strong solar activity amplification of the peak that occurs just after sunset. This feature is generally attributed to the *F*-region dynamo (Heelis et al., 1974; Rishbeth, 1981; Farley et al., 1986; Haerendel and Eccles, 1992; Crain et al., 1993b).

Magnetic disturbances, which are associated with the imposition at high latitudes of strong and variable magnetospheric electric fields and currents, produce fluctuating electric fields over the entire globe. An example of disturbed electric fields at Jicamarca is shown in Figure 9.5.3; together with correlative data: the auroral electrojet indices AU and AL, representing the upper and lower envelopes, respectively, of superposed *H* perturbations from auroral-zone magnetometers (*H* is defined in Section 6); the northward (B_z) component of the interplanetary magnetic field (IMF); and the perturbation in *H* at Huancayo, which is near Jicamarca under the equatorial electrojet. Negative values of IMF B_z tend to produce magnetic activity in the auroral zone and electric-field fluctuations at Jicamarca, which during daylight hours are correlated with perturbations in *H* at Huancayo.

In addition to the fluctuations, the *average* **E** × **B** drifts are altered during disturbed periods, with a general weakening of the average upward/poleward daytime drift and a general average westward shift of the zonal drift at all times, most strongly during the night. Figure 9.5.4 shows clearly the average magnetic activity changes in the eastward drift, changes that become stronger

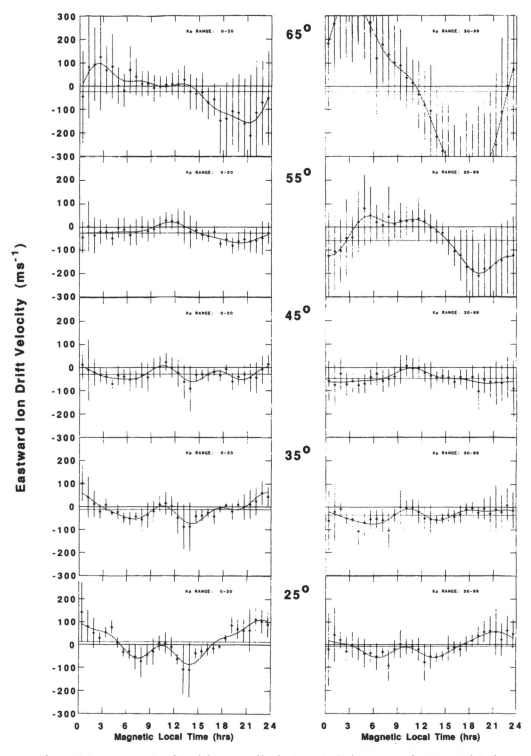

Figure 9.5.4 Average eastward ion drifts measured by the Dynamics Explorer spacecraft at magnetic latitudes of 65 to 25° for quiet (left) and disturbed (right) times. Vertical bars indicate the scatter about the mean values and the solid curve is obtained from a fit to the data with a mean (light dotted line) and four Fourier harmonics. (From Heelis, R. A. and Coley, W. R. (1992). *J. Geophys. Res.,* 97, 19, 461. Copyright by the American Geophysical Union. With permission.)

with increasing magnetic latitude. Observations of equatorial $\mathbf{E} \times \mathbf{B}$ vertical drifts have sometimes shown a decrease in magnitude following magnetic storms (Blanc and Richmond, 1980; B.G. Fejer et al., 1983; Sastri, 1988), suggesting possible disturbance dynamo effects.

Ionospheric currents have been measured by rocket-borne magnetometers (Cahill, 1969; Shuman, 1970; Yabuzaki and Ogawa, 1974; Burrows and Sastry, 1976; Burrows et al., 1977; Sampath and Sastry, 1979) and by incoherent scatter radars (Harper, 1977). Figure 9.5.5 shows typical vertical profiles of current density calculated from radar measurements of electron density, electric field, and ion velocity. At other times and locations the vertical distribution of current can be quite different, reflecting the influence of winds that vary in height and time, as in Figure 9.4.3. Figure 9.5.6 shows a collection of current profiles obtained from rocket-borne magnetometers near the magnetic equator off the coast of South America, showing the equatorial electrojet. The magnetic effects of meridional currents in the equatorial ionosphere have been reported by Musmann and Seiler (1978), Maeda et al. (1982), Takeda and Maeda (1983), and Langel et al. (1993).

6. GEOMAGNETIC VARIATIONS

Perturbations in the geomagnetic field were one of the earliest phenomena of ionospheric electrodynamics to be observed, and magnetic-perturbation data have existed for a long period of time, from points widely spread over the Earth. These data have provided us with some of our most detailed information about global ionospheric electrodynamics and its variability (Chapman and Bartels, 1940; Matsushita and Campbell, 1967; Maeda, 1968; Kane, 1976; Campbell, 1989). The average equivalent-current patterns of Figure 9.1.3 were derived from such data.

In the standard terminology of geomagnetism, the geomagnetic vector is given by three components labeled either H (horizontal intensity), D (declination), and Z (vertical component, defined to be positive downward), or else X (geographically northward), Y (geographically eastward), and Z. The Earth's main field has a strength at the Earth's surface of around 25,000 to 60,000 nT. Typical quiet-day perturbations at mid-latitudes are usually less than 100 nT, while disturbed perturbations in the auroral zone rarely exceed 2000 nT, much smaller than the main field. An increase in H represents a vector perturbation component in the magnetic-northward direction (the direction of a compass needle). An algebraic increase in D represents a vector perturbation component in the magnetic-eastward direction. An algebraic increase in Z represents a vector perturbation component in the downward direction.

Figure 9.6.1 shows the average quiet-day solar variations of the geomagnetic field, S_q, for a chain of magnetometers in the American longitude sector. The three seasons of June solstice, equinox, and December solstice are shown (defined the same as for the electric-field data shown in Figure 9.5.2). The solid and dashed curves represent average data for solar-cycle minimum and maximum conditions, respectively. One can readily see the relation of these variations to the equivalent-current pattern of Figure 9.1.3a by imagining the stations to move from left to right underneath this current pattern through the course of the day and by applying the right-hand rule for the direction of magnetic perturbations produced by a current. In general, the solar-cycle and seasonal variations are what one would expect on the basis of ionospheric-conductivity variations: larger perturbations at solar maximum and in summer. The strong enhancement in the H component at the magnetic equator (Huancayo) is evident, caused by the equatorial electrojet. Global analyses of geomagnetic perturbations like these have been presented by Price and Wilkins (1963), Price and Stone (1964), Matsushita (1967a, 1968), Parkinson (1971), Malin (1973), Suzuki (1973, 1978), Malin and Gupta (1977), and Winch (1981).

An interesting feature of the magnetic perturbations is the asymmetry of their local time variation about the magnetic equator, seen both in Figure 9.6.1 in all seasons, including equinox, and in the yearly average equivalent-current pattern of Figure 9.1.3a. The equinoctial asymmetry has been studied by van Sabben (1964) and Suzuki (1978). Wulf (1963, 1965) pointed out that

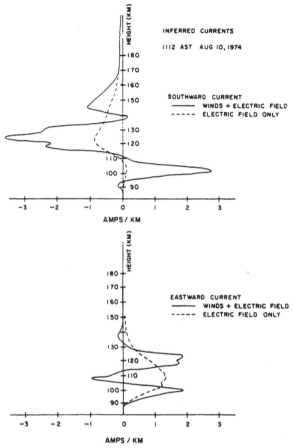

Figure 9.5.5 Typical profile of the horizontal current density vs. height for midday summer conditions over Arecibo. The currents driven by the electrostatic field are also shown. (From Harper, R. M. (1977). *J. Geophys. Res.,* 82, 3233. Copyright by the American Geophysical Union. With permission.)

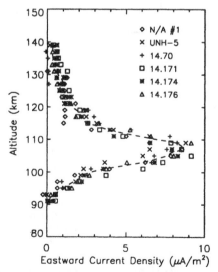

Figure 9.5.6 Electric current density around noon near the magnetic dip equator over Peru measured by rocketborne magnetometers. Values have been normalized to a ground-level magnetic perturbation of 100 nT at the nearby Huancayo observatory. Symbols: current density measured on each of six flights. Dashed line: mean profile of measured current density. The profile measured on flight N/A #1 is from Shuman (1970); that on flight UNH-5 is from Maynard (1967); and those on flights 14.70, 14.171, 14.174, and 14.176 are from Davis et al. (1967).

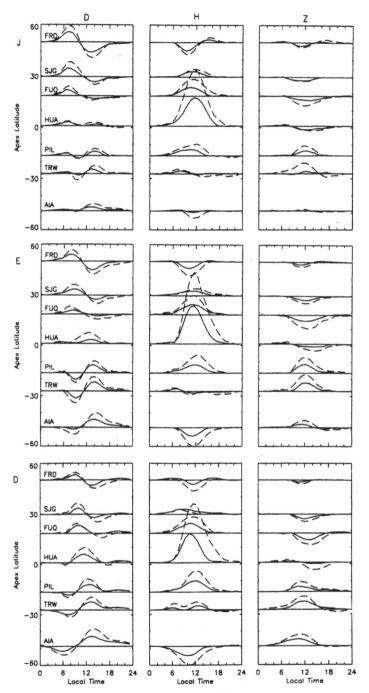

Figure 9.6.1 Average quiet-day magnetic perturbations at several stations in the American longitude sector for J months (May through August; top), E months (March, April, September, October; middle), and D months (November through February; bottom). D = magnetically eastward; H = magnetically northward; Z = downward. Station names and their geographical coordinates from north to south are: Fredericksburg (38.2°N, 282.6°E), San Juan (18.1°N, 293.8°E), Fuquene (5.5°N, 286.3°E), Huancayo (12.0°S, 284.7°E), Pilar (31.7°S, 296.1°E), Trelew (43.2°S, 294.7°E), and Argentine Island (65.2°S, 295.7°E). Solid lines are for solar-cycle minimum (1964 to 1965 or 1985 to 1986) and dashed lines are for solar-cycle maximum (1957 to 1958, 1967 to 1968, or 1979 to 1980). Values are adjusted to a mean solar radio flux value of S_a = 75 (solar minimum) or 200 (solar maximum). The perturbations are measured from a baseline defined to be the average value between 0 and 3 LT, after removal of a linear trend. The scale is 50 nT per 10°. Local time, in hours, is defined as universal time + longitude/(15°h).

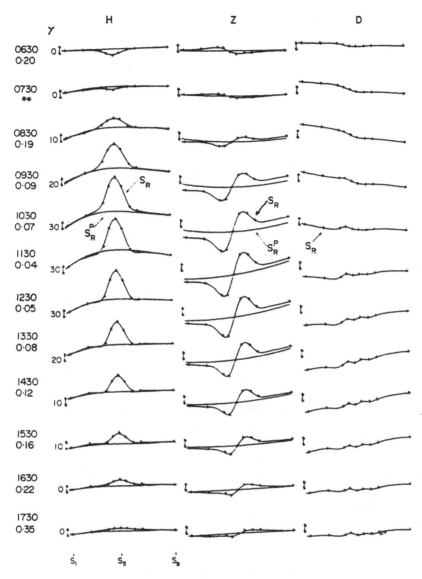

Figure 9.6.2   Yearly averaged latitudinal profiles of the variation S_R in *H*, *Z*, and *D*, from north to south (left to right) across the magnetic equator in Africa at hourly intervals between 0630 and 1730 local time. The total distance from S_1 to S_9 is 3020 km. Data are from the period from January 1969 to March 1970, with a mean S_9 value of 153. Also shown for *H* and *Z* are fitted curves labeled S_R^p representing a smooth interpolation between mid-latitude values to the north and south. Scales correspond to 10 nT, positive upward. The value of the scale base is zero for *Z* and *D* and is the indicated value (in nT) for *H*. The number written below each local hour indicates the goodness of fit to the data by an empirical model of the currents. (Reprinted from *J. Atmos. Terr. Phys.*, 38: Fambitakoye, O., and P. N. Mayaud, Equatorial electrojet and regular daily variations S_R, I. A. determination of the equatorial electrojet parameters, pp. 1–17, Copyright 1976, with kind permission from Pergamon Press Ltd, Headington Hill Hall, Oxford OX3 0BW, UK.)

the seasonal variation of the geomagnetic perturbations is more complex than implied by a simple organization according to solar declination as in Figure 9.6.1, since days equally distant from solstice, preceding and succeeding, have noticeably different variations.

The latitudinal structure of the equatorial electrojet is illustrated in Figure 9.6.2, showing yearly average latitudinal profiles of perturbations in *H*, *Z*, and *D* across the magnetic equator in Africa at hourly intervals throughout daylight hours. The *H* and *Z* perturbations clearly show the effects of a ribbon of current overhead, eastward during most of the day but westward at 0630 and 0730.

Lunar variations are usually less than 10% the magnitude of solar variations (Bartels and Johnston, 1940; Matsushita, 1967a,b, 1968). They depend not only on latitude, solar time, season, and solar cycle, but also on lunar phase. (There is a further dependence on the astronomical declination of the moon and on the Earth-moon distance.) The two oppositely circulating vortices of equivalent current in each hemisphere seen in Figure 9.1.3b slowly migrate toward later solar local times from day to day, as does the apparent lunar position, growing and decaying as they enter and leave daylight. Global analyses of geomagnetic lunar effects (Malin, 1973; Stening and Winch, 1979; Winch, 1981) have also found significant longitudinal variations. The seasonal variations of the lunar magnetic perturbations tend to be greater than those for the solar perturbations, and Stening and Winch (1979) and Gupta (1982) found that the seasonal variations have different forms at different locations: sometimes an annual variation dominates, while at other sites the main seasonal variation is semi-annual. In the equatorial electrojet, the lunar geomagnetic effect is considerably stronger around the December solstice than around the June solstice, on a global average basis (Rastogi and Trivedi, 1970; Gupta, 1973). There is a tendency for a maximum in summer in both hemispheres, but there is an additional tendency for a maximum around the December solstice, so that low-latitude sites in the northern hemisphere tend to have maximum amplitude variations around December rather than June (Schlapp and Malin, 1979; Gupta, 1982; Matsushita and Xu, 1984). In addition, the phase of the lunar variation around January is frequently considerably different from the phase in other seasons (Schlapp and Malin, 1979). The solar-cycle variation is also stronger in December than in June for most sites, a feature noted by Chapman et al. (1971). Stening (1989) found that the amplitude of lunar geomagnetic perturbations is often larger in the morning than in the afternoon.

Variability in the geomagnetic perturbations is caused both by variable magnetospheric activity and by changes in the winds and conductivities in the dynamo region. The auroral electrojets (see Chapter II/12) are the most prominent manifestation of the variability in magnetospheric activity; these are connected by geomagnetic-field-aligned currents with the outer magnetosphere. The field-aligned currents, along with magnetospheric ring currents and currents at the magnetopause and in the magnetotail, contribute to ground-level magnetic perturbations seen at middle and low latitudes and are responsible for the fact that disturbance magnetic perturbations occur even at night in the absence of significant ionospheric conductivity, as is evident in Figure 9.5.3. Daytime geomagnetic disturbances are often amplified in the equatorial electrojet with respect to nearby low-latitude locations (Rastogi, 1989), reflecting the penetration of disturbance electric fields to the equator.

Significant day-to-day changes in the geomagnetic perturbations are observed not only on magnetically disturbed days, but also on quiet days. It is generally believed that these changes are due in large part to changes in the propagation conditions for tides entering the dynamo region from below, as well as to some contribution from planetary waves that are able to propagate to ionospheric heights. Figure 9.6.3 shows an extreme example of day-to-day variability in quiet-day variations of the horizontal magnetic component at Huancayo. The ten magnetically disturbed days of the month have been removed, although some of the remaining variability apparent in Figure 9.6.3 could be due to residual activity, especially the more rapid fluctuations. The day-to-day variability is found to be only weakly correlated between stations widely separated in longitude (Kane, 1972).

The influence of modified ionospheric conductivity on geomagnetic perturbations is often observed during solar flares and eclipses. Flare modifications to the magnetic perturbations are known as *solar-flare effects* or *magnetic crochets,* the latter name arising from the shape of a plot of the magnetic perturbations as a function of time. Figure 9.6.4 shows an example of such plots for two magnetic observatories during a 3-hour interval on July 8, 1968. The dashed traces, subscripted with a q, represent the estimated quiet-day magnetic perturbations unmodified by the solar flare, while the differences between these traces and the solid lines, subscripted with an f, represent the flare modification to the magnetic perturbation. Notice that the flare modification is

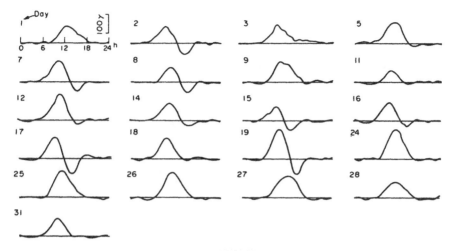

Figure 9.6.3 S_R variations in the magnetic horizontal component H at Huancayo during the month of January 1923. All the quiet days of the month, as classified by the average aa indices for the local day, are displayed. (Reprinted from *J. Atmos. Terr. Phys.,* 39: Mayaud, P. N., The equatorial counter-electrojet — a review of its geomagnetic aspects, pp. 1055–1070, Copyright 1977, with kind permission from Pergamon Press Ltd, Headington Hill Hall, Oxford OX3 0BW, UK.)

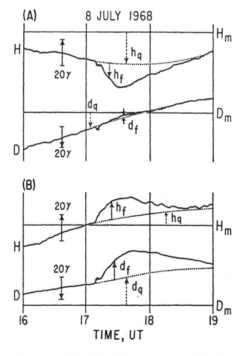

Figure 9.6.4 Geomagnetic crochets recorded at (A) Boulder and at (B) Fredericksburg on July 8, 1968. H_m and D_m are the average quiet-time night values of the magnetic elements, representing a baseline from which to measure the perturbations. h_q and d_q (dotted lined) are the estimated quiet-day variations interpolated through the crochet, while h_f and d_f are the modifications due to the solar flare. Arrows give scales and directions of positive increases for each curve, where 20 γ = 20 nT. (From Richmond, A. D. and Venkateswaran, S. V., 1971, *Radio Sci.,* 6, 139. Copyright by the American Geophysical Union. With permission.)

not necessarily a simple enhancement of the quiet-day perturbation (as around 1730 UT in the D trace for Boulder, when d_q and d_f have opposite signs), and that the temporal evolution of the flare perturbation can be different for the H and D components (for example, the small initial perturbation near 1710 UT does not appear in h_f for Fredericksburg). Solar-eclipse perturbations have been reviewed by Rastogi (1982).

ACKNOWLEDGMENTS

This work was supported by the NASA Space Physics Theory Program. I am grateful to M. Buonsanto, B. Fejer, and W. Oliver for providing digital files of their data shown in Figure 9.5.2, and to G. Lu and R. A. Wolf for helpful comments on a draft manuscript.

REFERENCES

Akasofu, S.-I., and Chapman, S. (1972). *Solar-Terrestrial Physics,* Clarendon, Oxford.

Baker, W. G. (1953). Electric currents in the ionosphere, I. The atmospheric dynamo, *Philos. Trans. R. Soc.,* A246, 295.

Baker, W. G., and Martyn, D. F. (1952). Conductivity of the ionosphere, *Nature,* 170, 1090.

Baker, W. G., and Martyn, D. F. (1953). Electric currents in the ionosphere, I. The conductivity, *Philos. Trans. R. Soc.,* A246, 281.

Banks, P. M. (1972). Magnetospheric processes and the behaviour of the neutral atmosphere, *Space Res.,* 12, 1051.

Bartels, J., and Johnston, H. F. (1940). Geomagnetic tides in horizontal intensity at Huancayo, *Terr. Magn. Atmos. Electr.,* 45, 269.

Bilitza, D., Ed., (1990). *International Reference Ionosphere 1990,* NSSDC 90-22, Greenbelt, Maryland.

Blanc, M., and Amayenc, P. (1979). Seasonal variations of the ionospheric $E \times B$ drifts above Saint-Santin on quiet days, *J. Geophys. Res.,* 84, 2691.

Blanc, M., and Richmond, A. D. (1980). The ionospheric disturbance dynamo, *J. Geophys. Res.,* 85, 1669.

Brekke, A., and Moen, J. (1993). Observations of high latitude ionospheric conductances, *J. Atmos. Terr. Phys.,* 55, 1493.

Buonsanto, M. J., Hagan, M. E., Salah, J. E., and Fejer, B. G. (1993). Solar cycle and seasonal variations in F region electrodynamics at Millstone Hill, *J. Geophys. Res.,* 98, 15677.

Burrows, K., and Sastry, T. S. G. (1976). Rocket measurements of current distribution in a normal and an intense electrojet, *J. Atmos. Terr. Phys.,* 39, 307.

Burrows, K., Sastry, T. S. G., Sampath, Stolarik, J. D., and Usher, M. J. (1977). The storm-time equatorial electrojet, *J. Atmos. Terr. Phys.,* 39, 125.

Cahill, J. R., Jr. (1969). Rocket and satellite studies of the geomagnetic field during the IQSY, in *Annals of the IQSY,* 5, A. C. Stickland, Ed., p. 349, MIT Press, Cambridge, MA.

Campbell, W. H. (1989). The regular geomagnetic-field variations during quiet solar conditions, in *Geomagnetism,* Vol. 3, J. A. Jacobs, Ed., p. 385, Academic Press, San Diego, CA.

Chapman, S., and Bartels, J. (1940). *Geomagnetism,* Clarendon, Oxford.

Chapman, S., and Lindzen, R. S. (1970). *Atmospheric Tides,* D. Reidel, Dordrecht, Holland.

Chapman, S., Gupta, J. C., and Malin, S. R. C. (1971). The sunspot cycle influence on the solar and lunar daily geomagnetic variations, *Proc. R. Soc.,* A324, 1.

Chen, P.-R., (1992). Two-day oscillation of the equatorial ionization anomaly, *J. Geophys. Res.,* 97, 6343.

Crain, D. J., Heelis, R. A., and Bailey, G. J. (1993a). Effects of electrical coupling on equatorial ionospheric plasma motions: when is the F region a dominant driver in the low-latitude dynamo?, *J. Geophys. Res.,* 98, 6033.

Crain, D. J., Heelis, R. A., Bailey, G. J., and Richmond, A. D. (1993b). Low-latitude plasma drifts from a simulation of the global atmospheric dynamo, *J. Geophys. Res.,* 98, 6039.

Davis, T. N., Burrows, K., and Stolarik, J. D. (1967). A latitude survey of the equatorial electrojet with rocket-borne magnetometers, *J. Geophys. Res.,* 72, 1845.

Dougherty, J. P. (1963). Some comments on dynamo theory, *J. Geophys. Res.,* 68, 2383.

Fambitakoye, O., and Mayaud, P. N. (1976). Equatorial electrojet and regular daily variations S_R. I. A determination of the equatorial electrojet parameters, *J. Atmos. Terr. Phys.,* 38, 1.

Farley, D. T., Bonelli, E., Fejer, B. G., and Larsen, M. F. (1986). The prereversal enhancement of the zonal electric fields in the equatorial ionosphere, *J. Geophys. Res.,* 16, 195.

Fejer, B. G. (1991). Low latitude electrodynamic plasma drifts: a review, *J. Atmos. Terr. Phys.,* 53, 677.

Fejer, B. G. (1993). F region plasma drifts over Arecibo: solar cycle, seasonal, and magnetic activity effects, *J. Geophys. Res.*, 98, 13,645.

Fejer, B. G., Larsen, M. F., and Farley, D. T. (1983). Equatorial disturbance dynamo electric fields, *Geophys. Res. Lett.*, 10, 537.

Fejer, B. G., Kelley, M. C., Senior, C., de la Beaujardière, O., Holt, J. A., Tepley, C. A., Burnside, R., Abdu, M. A., Sobral, J. H. A., Woodman, R. F., Kamide, Y., and Lepping, R. (1990). Low- and mid-latitude ionospheric electric fields during the January 1984 GISMOS campaign, *J. Geophys. Res.*, 95, 2367.

Fejer, B. G., de Paula, E. R., González, S. A., and Woodman, R. F. (1991). Average vertical and zonal F region plasma drifts over Jicamarca, *J. Geophys. Res.*, 96, 13901.

Fejer, J. A. (1953). Semidiurnal currents and electron drifts in the ionosphere, *J. Atmos. Terr. Phys.*, 4, 184.

Fesen, C. G., Dickinson, R. E., and Roble, R. G. (1986). Simulation of thermospheric tides at equinox with the National Center for Atmospheric Research Thermospheric General Circulation Model, *J. Geophys. Res.*, 91, 4471.

Forbes, J. M. (1981). The equatorial electrojet, *Rev. Geophys. Space Phys.*, 19, 469.

Forbes, J. M., and Garrett, H. B. (1979). Solar tidal wind structures and the E-region dynamo, *J. Geomagn. Geoelectr.*, 31, 173.

Forbes, J. M., and Leveroni, S. (1992). Quasi 16-day oscillation in the ionosphere, *Geophys. Res. Lett.*, 19, 981.

Forbes, J. M., and Lindzen, R. S. (1976a). Atmospheric solar tides and their electrodynamic effects. I. The global S_q current system, *J. Atmos. Terr. Phys.*, 38, 897.

Forbes, J. M., and Lindzen, R. S. (1976b). Atmospheric solar tides and their electrodynamic effects. II. The equatorial electrojet, *J. Atmos. Terr. Phys.*, 38, 911.

Forbes, J. M., and Lindzen, R. S. (1977). Atmospheric solar tides and their electrodynamic effects. III. The polarization electric field, *J. Atmos. Terr. Phys.*, 39, 1369.

Forbes, J. M., and Vial, F. (1989). Monthly simulations of the solar semidiurnal tide in the mesosphere and lower thermosphere, *J. Atmos. Terr. Phys.*, 51, 649.

Forbes, J. M., and Vial, F. (1991). Semidiurnal tidal climatology of the E region, *J. Geophys. Res.*, 96, 1147.

Fuller-Rowell, T., and Evans, D. S. (1987). Height-integrated Pedersen and Hall conductivity patterns inferred from the TIROS-NOAA satellite data, *J. Geophys. Res.*, 92, 7606.

Fuller-Rowell, T. J., Rees, D., Quegan, S., Moffett, R. J., and Bailey, G. J. (1988). Simulations of the seasonal and Universal Time variations of the high-latitude thermosphere and ionosphere using a coupled, three-dimensional, model, *Pure Appl. Geophys.*, 127, 189.

Gagnepain, J., Crochet, M., and Richmond, A. D. (1977). Comparison of equatorial electrojet models, *J. Atmos. Terr. Phys.*, 39, 1119.

Glushakov, M. L., Dul'kin, V. N., and Ivanovskiy, A. I. (1980). Model of the daily variations of thermospheric characteristics. III. Self-consistent method of accounting for the electrostatic field in the daily variations of thermospheric composition, *Geomagn. Aeron.*, 20, 39 (Engl. trans.).

Glushakov, M. L., Dul'kin, V. N., and Ivanovskiy, A. I. (1981). Model of diurnal variations in the variables of the thermosphere. IV. Results of calculations upon self-consistent correction for the effect of the electrostatic polarization field, *Geomagn. Aeron.*, 21, 629 (Engl. trans.).

Gupta, J. C. (1973). On solar and lunar equatorial electrojets, *Ann. Géophys.*, 29, 49.

Gupta, J. C. (1982). Solar and lunar seasonal variations in the American sector, *Ann. Géophys.*, 38, 255.

Haerendel, G., and Eccles, J. V. (1992). The role of the equatorial electrojet in the evening ionosphere, *J. Geophys. Res.*, 97, 1181.

Haerendel, G., Eccles, J. V., and Çakir, S. (1992). Theory for modeling the equatorial evening ionosphere and the origin of the shear in the horizontal plasma flow, *J. Geophys. Res.*, 97, 1209.

Hagan, M. E., Forbes, J. M., and Vial, F. (1993). Numerical investigation of the propagation of the quasi-two-day wave into the lower thermosphere, *J. Geophys. Res.*, 98, 23193.

Hargreaves, J. K. (1992). *The Solar-Terrestrial Environment: An Introduction to Geospace — The Science of the Terrestrial Upper Atmosphere, Ionosphere, and Magnetosphere*, Cambridge University Press, New York.

Harper, R. M. (1977). A comparison of ionospheric currents, magnetic variations, and electric fields at Arecibo, *J. Geophys. Res.*, 82, 3233.

Harper, R. M., Wand, R. H., Zamlutti, C. J., and Farley, D. T. (1976). E region ion drifts and winds from incoherent scatter measurements at Arecibo, *J. Geophys. Res.*, 81, 25.

Hays, P. B., and Roble, R. G. (1979). A quasi-static model of global atmospheric electricity. 1. The lower atmosphere, *J. Geophys. Res.*, 84, 3291.

Hedin, A. E. (1991). Extension of the MSIS thermosphere model into the middle and lower atmosphere, *J. Geophys. Res.*, 96, 1159.

Heelis, R. A., and Coley, W. R. (1992). East-west ion drifts at mid-latitudes observed by Dynamics Explorer 2, *J. Geophys. Res.*, 97, 19,461.

Heelis, R. A., Kendall, P. C., Moffett, R. J., Windle, D. W., and Rishbeth, H. (1974). Electrical coupling of the E- and F-regions and its effect on F-region drifts and winds, *Planet. Space Sci.*, 22, 743.

Hill, R. J., and Bowhill, S. A. (1977). Collision frequencies for use in the continuum momentum equations applied to the lower ionosphere, *J. Atmos. Terr. Phys.,* 39, 803.

Hirono, M. (1952). A theory of dirunal magnetic variations in equatorial regions and conductivity of the ionospheric E-region, *J. Geomagn. Geoelectr.,* 4, 7.

Hirono, M., and Kitamura, T. (1956). A dynamo theory in the ionosphere, *J. Geomagn. Geoelectr.,* 8, 9.

Itikawa, Y. (1971). Effective collision frequency of electrons in atmospheric gases, *Planet. Space Sci.,* 19, 993.

Ito, R., Kato, S., and Tsuda, T. (1986). Consideration of an ionospheric wind dynamo driven by a planetary wave with a two-day period, *J. Atmos. Terr. Phys.,* 48, 1.

Jacobson, A. R., and Bernhardt, P. A. (1985). Electrostatic effects in the coupling of upper atmospheric waves to ionospheric plasma, *J. Geophys. Res.,* 90, 6533.

Kamide, Y., and Baumjohann, W. (1993). *Magnetosphere-Ionosphere Coupling,* Springer-Verlag, Berlin.

Kane, R. P. (1972). Longitudinal spread of equatorial S_q variability, *J. Atmos. Terr. Phys.,* 34, 1425.

Kane, R. P. (1976). Geomagnetic field variations, *Space Sci. Rev.,* 18, 413.

Kato, S. (1956). Horizontal wind systems in the ionospheric E-region deduced from the dynamo theory of the geomagnetic S_q variations. Part II. Rotating Earth, *J. Geomagn. Geoelectr.,* 8, 24.

Kato, S. (1957). Horizontal wind systems in the ionospheric E-region deduced from the dynamo theory of the geomagnetic S_q variations. Part IV., *J. Geomagn. Geoelectr.,* 9, 107.

Kato, S. (1980). *Dynamics of the Upper Atmosphere,* D. Reidel, Dordrecht, Holland.

Kelley, M. C. (1989). *The Earth's Ionosphere: Plasma Physics and Electrodynamics,* Academic Press, San Diego, CA.

Killeen, T. L., and Roble, R. G. (1988). Thermosphere dynamics: contributions from the first 5 years of the dynamics explorer program, *Rev. Geophys.,* 26, 329.

Langel, R. A. (1992). International Geomagnetic Reference Field: the sixth generation, *J. Geomagn. Geoelectr.,* 44, 679.

Langel, R. A., Purucker, M., and Rajaram, M. (1993). The equatorial electrojet and associated currents as seen in Magsat data, *J. Atmos. Terr. Phys.,* 55, 1233.

Maeda, H. (1955). Horizontal wind systems in the ionospheric E-region deduced from the dynamo theory of the geomagnetic S_q variations. Part I. Non-rotating Earth, *J. Geomagn. Geoelectr.,* 7, 121.

Maeda, H. (1957). Horizontal wind systems in the ionospheric E region deduced from the dynamo theory of the geomagnetic S_q variations. Part III. *J. Geomagn. Geoelectr.,* 9, 86.

Maeda, H. (1968). Variation in geomagnetic field, *Space Sci. Rev.,* 8, 555.

Maeda, H., and Fujiwara, M. (1967). Lunar ionospheric winds deduced from the dynamo theory of geomagnetic variations, *J. Atmos. Terr. Phys.,* 29, 917.

Maeda, H., Iyemori, T., Araki, T., and Kamei, T. (1982). New evidence of a meridional current system in the equatorial ionosphere, *Geophys. Res. Lett.,* 9, 337.

Makino, M., and Takeda, M. (1984). Three-dimensional ionospheric currents and fields generated by the atmospheric global circuit current, *J. Atmos. Terr. Phys.,* 46, 199.

Malin, S. R. C. (1973). Worldwide distribution of geomagnetic tides, *Philos. Trans. R. Soc. London,* A274, 551.

Malin, S. R. C., and Gupta, J. C. (1977). The S_q current system during the International Geophysical Year, *Geophys. J. R. Astron. Soc.,* 49, 515.

Mason, E. A. (1970). Estimated ion mobilities for some air constituents, *Planet. Space Sci.,* 18, 137.

Matsushita, S. (1967a). Solar quiet and lunar daily variation fields, in *Physics of Geomagnetic Phenomena,* S. Matsushita and W. H. Campbell, Eds., p. 301, Academic Press, New York.

Matsushita, S. (1967b). Lunar tides in the ionosphere, *Handbuch der Physik,* 49/2, p. 547, Springer-Verlag, Berlin.

Matsushita, S. (1968). S_q and L current systems in the ionosphere, *Geophys. J. R. Astron. Soc.,* 15, 109.

Matsushita, S., and Campbell, W. H., Eds. (1967). *Physics of Geomagnetic Phenomena,* Academic Press, New York.

Matsushita, S., and Xu, W.-Y. (1984). Seasonal variations of L equivalent current systems, *J. Geophys. Res.,* 89, 285.

Matuura, N. (1974). Electric fields deduced from the thermospheric model, *J. Geophys. Res.,* 79, 4679.

Mayaud, P. N. (1977). The equatorial counter-electrojet — a review of its geomagnetic aspects. *J. Atmos. Terr. Phys.,* 39, 1055.

Maynard, N. C. (1967). Measurements of ionospheric currents off the coast of Peru, *J. Geophys. Res.,* 72, 1863.

Mayr, H. G., Harris, I., and Herrero, F. A. (1990). The dynamo of the diurnal tide and its effect on the thermospheric circulation, *Planet. Space Sci.,* 38, 301.

Musmann, G., and Seiler, E. (1978). Detection of meridional currents in the equatorial ionosphere, *J. Geophys.,* 44, 357.

Namgaladze, A. A., Koren'kov, Yu. N., Klimenko, V. V., Karpov, I. V., Bessarab, F. S., Surotkin, V. A., Glushchenko, T. A., and Naumova, N. M. (1990). Global numerical model of the thermosphere, ionosphere, and protonosphere of the earth, *Geomagn. Aeron.,* 30, 515 (Engl. trans.).

Namgaladze, A. A., Koren'kov, Yu. N., Klimenko, V. V., Karpov, I. V., Surotkin, V. A., and Naumova, N. M. (1991). Numerical modelling of the thermosphere-ionosphere-protonosphere system, *J. Atmos. Terr. Phys.,* 53, 1113.

Oliver, W. L., Yamamoto, Y., Takami, T., Fukao, S., Yamamoto, M., and Tsuda, T. (1993). Middle and Upper atmosphere radar observations of ionospheric electric fields, *J. Geophys. Res.,* 98, 11615.

Parkinson, W. D. (1971). An analysis of the geomagnetic diurnal variation during the international geophysical year, *Gerlands Beitr. Geophys.,* 80, 199.

Peterson, W. K., Doering, J. P., Potemra, T. A., Bostrom, C. O., Brace, L. H., Heelis, R. A., and Hanson, W. B. (1977). Measurement of magnetic field-aligned potential differences using high resolution conjugate photoelectron energy spectra, *Geophys. Res. Lett.*, 4, 373.

Prakash, S., and Pandey, R. (1985). Generation of electric fields due to the gravity wave winds and their transmission to other ionospheric regions, *J. Atmos. Terr. Phys.*, 47, 363.

Price, A. T., and Stone, D. J. (1964). The quiet-day magnetic variations during the IGY, *Ann. IGY*, 35, 63.

Price, A. T., and Wilkins, G. A. (1963). New methods for the analysis of geomagnetic fields and their application to the S_q field of 1932–1933, *Philos. Trans. R. Soc.*, A256, 31.

Rastogi, R. G. (1982). Solar eclipse effects on geomagnetism, *Proc. Indian Natl. Sci. Acad.*, 48A, Suppl. 3, 464.

Rastogi, R. G. (1989). The equatorial electrojet: magnetic and ionospheric effects, in *Geomagnetism*, Vol. 3, J. A. Jacobs, Ed., p. 461, Academic Press, San Diego, CA.

Rastogi, R. G., and Trivedi, N. B. (1970). Luni-solar tides in *H* at stations within the equatorial electrojet, *Planet. Space Sci.*, 18, 367.

Reddy, C. A. (1989). The equatorial electrojet, *Pure Appl. Geophys.*, 131, 485.

Rees, M. H. (1989). *Physics and Chemistry of the Upper Atmosphere*, Cambridge University Press, New York.

Richmond, A. D. (1974). The computation of magnetic effects of field-aligned magnetospheric currents, *J. Atmos. Terr. Phys.*, 36, 245.

Richmond, A. D. (1987). The ionosphere, in *The Solar Wind and the Earth*, S.-I. Akasofu and Y. Kamide, Eds., p. 123, Terra Scientific, Tokyo.

Richmond, A. D. (1989). Modeling the ionospheric wind dynamo: a review, *Pure Appl. Geophys.*, 131, 413.

Richmond, A. D. (1994). The ionospheric wind dynamo: effects of its coupling with different atmospheric regions, in *The Upper Mesosphere and Lower Thermosphere*, edited by R. M. Johnson and T. L. Killeen, in press.

Richmond, A. D., and Roble, R. G. (1987). Electrodynamic effects of thermospheric winds from the NCAR Thermospheric General Circulation Model, *J. Geophys. Res.*, 92, 12365.

Richmond, A. D., and Venkateswaran, S. V. (1971). Geomagnetic crochets and associated ionospheric current systems, *Radio Sci.*, 6, 139.

Richmond, A. D., Matsushita, S., and Tarpley, J. D. (1976). On the production mechanism of electric currents and fields in the ionosphere, *J. Geophys. Res.*, 81, 547.

Richmond, A. D., Blanc, M., Emery, B. A., Wand, R. H., Fejer, B. G., Woodman, R. F., Ganguly, S., Amayenc, P., Behnke, R. A., Calderon, C., and Evans, J. V. (1980). An empirical model of quiet-day ionospheric electric fields at middle and low latitudes, *J. Geophys. Res.*, 85, 4658.

Richmond, A. D., Ridley, E. C., and Roble, R. G. (1992). A thermosphere/ionosphere general circulation model with coupled electrodynamics, *Geophys. Res. Lett.*, 19, 601.

Rishbeth, H. (1971a). The *F*-layer dynamo, *Planet. Space Sci.*, 19, 263.

Rishbeth, H. (1971b). Polarization fields produced by winds in the equatorial *F*-region, *Planet. Space Sci.*, 19, 357.

Rishbeth, H. (1979). Ion-drag effects in the thermosphere, *J. Atmos. Terr. Phys.*, 41, 885.

Rishbeth, H. (1981). The *F*-region dynamo, *J. Atmos. Terr. Phys.*, 43, 387.

Roble, R. G. (1991). On modeling component processes in the Earth's global electric circuit, *J. Atmos. Terr. Phys.*, 53, 831.

Roble, R. G. (1992). The polar lower thermosphere, *Planet. Space Sci.*, 40, 271.

Roble, R. G., and Tzur, I. (1986). The global atmospheric-electrical circuit, in *The Earth's Electrical Environment*, p. 206, National Academy Press, Washington, D.C.

Ronchi, C., Sudan, R. N., and Similon, P. L. (1990). Effect of short-scale turbulence on kilometer wavelength irregularities in the equatorial electrojet, *J. Geophys. Res.*, 95, 189.

Rowe, J. F., Jr., and Mathews, J. D. (1973). Low-latitude nighttime *E* region conductivities, *J. Geophys. Res.*, 78, 7461.

Salah, J. E. (1993). Interim standard for the ion-neutral atomic oxygen collision frequency, *Geophys. Res. Lett.*, 20, 1543.

Sampath, S., and Sastry, T. S. G. (1979). Results from in situ measurements of ionospheric currents in the equatorial region, *J. Geomagn. Geoelectr.*, 31, 373.

Sastri, J. H. (1988). Equatorial electric fields of ionospheric disturbance dynamo origin, *Ann. Geophys.*, 6, 635.

Schieldge, J. P., Venkateswaran, S. V., and Richmond, A. D. (1973). The ionospheric dynamo and equatorial magnetic variations, *J. Atmos. Terr. Phys.*, 35, 1045.

Schlapp, D. M., and Malin, S. R. C. (1979). Some features of the seasonal variation of geomagnetic lunar tides, *Geophys. J. R. Astron. Soc.*, 59, 161.

Schuster, A. (1908). The diurnal variation of terrestrial magnetism, *Philos. Trans. R. Soc.*, A208, 163.

Shkarofsky, I. P., Johnston, T. W., and Bachynski, M. P. (1966). *The Particle Kinetics of Plasmas*, Addison-Wesley, Reading, MA.

Shuman, B. M. (1970). Rocket measurement of the equatorial electrojet, *J. Geophys. Res.*, 75, 3889.

Spiro, R. W., Wolf, R. A., and Fejer, B. G. (1988). Penetration of high-latitude-electric-field effects to low latitudes during SUNDIAL 1984, *Ann. Geophys.*, 6, 39.

Stening, R. J. (1981). A two-layer ionospheric dynamo calculation, *J. Geophys. Res.*, 86, 3543.

Stening, R. J. (1986). Inter-relations between current and electron density profiles in the equatorial electrojet and effects of neutral density changes, *J. Atmos. Terr. Phys.*, 48, 163.

Stening, R. J. (1989). A diurnal modulation of the lunar tide in the upper atmosphere, *Geophys. Res. Lett.*, 16, 307.

Stening, R. J., and Winch, D. E. (1979). Seasonal changes in the global lunar geomagnetic variation, *J. Atmos. Terr. Phys.*, 41, 311.

Stewart, B. (1882). Terrestrial magnetism, in *Encyclopaedia Britannica*, 9th ed., Vol. 16, 159.

Suzuki, A. (1973). A new analysis of the geomagnetic S_q field, *J. Geomagn. Geoelectr.*, 25, 259.

Suzuki, A. (1978). Geomagnetic S_q field at successive Universal Times, *J. Atmos. Terr. Phys.*, 40, 449.

Takeda, M. (1990). Geomagnetic field variation and the equivalent current system generated by an ionospheric dynamo at the solstice, *J. Atmos. Terr. Phys.*, 52, 59.

Takeda, M., and Araki, T. (1984). Ionospheric currents and fields during the solar eclipse, *Planet. Space Sci.*, 32, 1013.

Takeda, M., and Maeda, H. (1980). Three-dimensional structure of ionospheric currents. 1. Currents caused by diurnal tidal winds, *J. Geophys. Res.*, 85, 6895.

Takeda, M., and Maeda, H. (1981). Three-dimensional structure of ionospheric currents. 2. Currents caused by semidiurnal tidal winds, *J. Geophys. Res.*, 86, 5861.

Takeda, M., and Maeda, H. (1983). *F*-region dynamo in the evening — interpretation of equatorial ΔD anomaly found by MAGSAT, *J. Atmos. Terr. Phys.*, 45, 401.

Takeda, M., and Yamada, Y. (1987). Simulation of ionospheric electric fields and geomagnetic field variation by the ionospheric dynamo for different solar activity, *Ann. Geophys.*, 5A, 429.

Takeda, M., and Yamada, Y. (1989). Quasi two-day period of the geomagnetic field, *J. Geomagn. Geoelectr.*, 41, 469.

Takeda, M., Yamada, Y., and Araki, T. (1986). Simulation of ionospheric currents and geomagnetic field variations of S_q for different solar activity, *J. Atmos. Terr. Phys.*, 48, 277.

Tarpley, J. D. (1970). The ionospheric wind dynamo. I. Lunar tide, *Planet. Space Sci.*, 18, 1075.

Untiedt, J. (1967). A model of the equatorial electrojet involving meridional currents, *J. Geophys. Res.*, 72, 5799.

Untiedt, J. (1968). *Der äquatoriale Elektrojet — Stromsystem und Magnetfeld*, Habilitationsschrift, Technische Hochschule Carolo-Wilhelmina zu Braunschwig.

van Sabben, D. (1964). North-south asymmetry of S_q, *J. Atmos. Terr. Phys.*, 26, 1187.

van Sabben, D. (1966). Magnetospheric currents, associated with the N-S asymmetry of S_q, *J. Atmos. Terr. Phys.*, 28, 965.

van Sabben, D. (1968). Errata, *J. Atmos. Terr. Phys.*, 30, 327.

VanZandt, T. E., Clark, W. L., and Warnock, J. M. (1972). Magnetic apex coordinates: a magnetic coordinate system for the ionospheric F_2 layer, *J. Geophys. Res.*, 77, 2406.

Volland, H. (1976a). Coupling between the neutral wind and the ionospheric dynamo current, *J. Geophys. Res.*, 81, 1621.

Volland, H. (1976b). The atmospheric dynamo, *J. Atmos. Terr. Phys.*, 38, 869.

Volland, H. (1984). *Atmospheric Electrodynamics*, Springer-Verlag, Berlin.

Volland, H. (1988). *Atmospheric Tidal and Planetary Waves*, Kluwer Academic Publishers, Dordrecht, Netherlands.

Wagner, C.-U., Möhlmann, D., Schäfer, K., Mishin, V. M., and Matveev, M. I. (1980). Large-scale electric fields and currents and related geomagnetic variations in the quiet plasmasphere, *Space Sci. Rev.*, 26, 391.

Winch, D. E. (1981). Spherical harmonic analyses of geomagnetic tides, 1964–1965, *Philos. Trans. R. Soc.*, A303, 1.

Wolf, R. A., Mantjoukis, G. A., and Spiro, R. W. (1986). Theoretical comments on the nature of the plasmapause, *Adv. Space Res.*, 6, 177.

Wulf, O. R. (1963). A possible effect of atmospheric circulation in the daily variation of the Earth's magnetic field, *Mon. Weather Rev.*, 91, 520.

Wulf, O. R. (1965). A possible effect of atmospheric circulation in the daily variation of the Earth's magnetic field, II, *Mon. Weather Rev.*, 93, 127.

Yabuzaki, T., and Ogawa, T. (1974). Rocket measurement of S_q ionospheric currents over Kagoshima, Japan, *J. Geophys. Res.*, 79, 1999.

Chapter 10

Power Line Harmonic Radiation: Sources and Environmental Effects

Ken Bullough*

CONTENTS

---

* Sadly, since the preparation of this chapter, Dr. Ken Bullough passed away. His colleague, Dr. S. Sazhin
  is to be thanked for his subsequent work on this contribution.

---

1. INTRODUCTION

Radiation from power lines at high (especially odd) harmonics (PLHR) of the fundamental (60, 50 Hz) has long been recognized as an electromagnetic compatibility (EMC) problem for telephonic communication, particularly in industrial areas (see Chapter I/12; Tatnall et al., 1983; Yearby et al., 1983). However, it is only in the last 2 to 3 decades that the important influence of PLHR (and other man-made VLF transmissions) on the environment has become more fully recognized (Boerner et al., 1983; Bullough, Kaiser, and Strangeways, 1985).

This has been due mainly to the extremely small signals radiated; for example, the total power radiated over North America in a 1-kHz band centered on 3.2 kHz is probably less than 100 W (Tatnall et al., 1983) and that from a typical power line in an industrial area (e.g., in Newfoundland) probably less than 1 μW (Yearby et al., 1983). Thus, at least for several years, the suggestion that PLHR might significantly modify the magnetospheric environment was received with skepticism by many colleagues (Tsurutani and Thorne, 1981). Eventually, these objections were overcome by the steady accumulation of evidence from two main sources: (1) the analysis of many years of ground-based recordings, obtained at Eights (from 1965) and Siple (1973–76) in Antarctica (Helliwell et al., 1975; Park and Helliwell, 1978; Park and Miller, 1979), and (2) comprehensive global morphological studies of VLF emissions using satellites with on-board tape recorders (Bullough et al., 1976; Tatnall et al., 1983; Bullough and Cotterill, 1983; Bullough, 1983, 1988, 1990; Parrot, 1990; Molchanov et al., 1991).

A schematic diagram illustrating the effects of PLHR is shown in Figure 10.1.1 and three examples of VLF magnetospheric line radiation (MLR; possibly triggered by PLHR leaking into the magnetosphere) are presented in Figure 10.1.2 (Mathews and Yearby, 1981).

A major discovery of the Ariel 3 and 4 global surveys was that of permanent zones of PLHR-stimulated ELF/VLF emission, particularly prominent at 3.2 kHz, occurring on field lines grounded on major industrial areas in the northeastern U.S. and southern Canada. These were particularly prominent in northern summer (1972) when, due to multihopping and ducting along

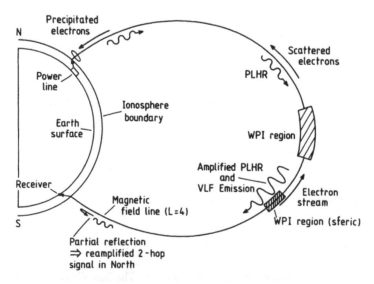

Figure 10.1.1   PLHR propagation and wave-particle interaction (WPI) in the magnetosphere. Harmonic radiation (few kilohertz) from a power line traverses the lower ionosphere boundary and is guided along a field-aligned duct to the equatorial region where it interacts with counterstreaming energetic electrons (according to Yoshida et al., 1983 this region is displaced to the south for a sferic-like signal). This WPI gives rise to stimulation and amplification of VLF emissions of intensity perhaps some 30 dB higher than the input signal. Pitch-angle scattering of the energetic electrons causes some of them to precipitate into the upper atmosphere with energy fluxes that may be more than 10^6 higher than the input wavepower. Often the process may be repeated for south to north propagation (two-hop) of a partially reflected signal in or above the southern ionosphere.

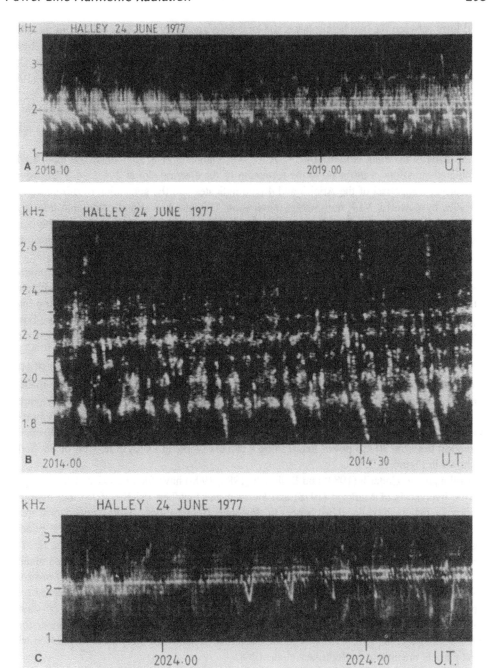

Figure 10.1.2 Examples of magnetospheric VLF line radiation (MLR) observed at Halley, Antarctica, June 24, 1977. (A) The spectrogram shows a well-defined array of magnetospheric lines which are gradually drifting up in frequency at ~80 Hz/min, and lie in the frequency range 1.75 to 2.5 kHz. The lines trigger emission activity and show some evidence of a two-hop modulation in the envelope of the wave spectrum. Local induction lines from Halley station are just visible between 1 and 1.5 kHz and are regularly spaced lines of constant frequency. (B) Spectrogram on an expanded frequency scale showing magnetospheric lines in greater detail (resolution: 2 Hz). Line frequency drifts, small-scale amplitude fluctuations, and triggering of risers and fallers are exhibited. (C) The three magnetospheric lines illustrated here all drift at different rates. The lower line triggers several well-defined hooks, and there is a clear two-hop amplitude modulation. Reprinted from *Planet. Space Sci.*, Vol. 29: Mathews, J. P. and Yearby, K. H., pp. 97–106, Magnetospheric VLF Line Radiation observed at Halley, Antarctica. Copyright (1981), with permission from Pergamon Press Ltd., Headington Hill Hall, Oxford, OX3 0BW, U.K.

field lines between the two hemispheres, the emissions appearing at 500 to 600 km in the north and south, were closely isomorphic in invariant coordinates. The secular increase in 3.2-kHz emission was assumed to be partly related to the increasing electricity consumption in the U.S. and the CIS. Percentage increases in annual electrical energy production were 40.5 and 46 (1967 to 1972) and 53 and 100 (1972 to 1989), respectively.

In Antarctica, at ground stations close to $60°$ Λ (L = 4), PLHR events appear as a subset of naturally occurring MLR events. They are stronger and more frequent at Siple than at Halley with typical line spacings of 120 and 80 Hz, respectively.

At medium invariant latitudes ($40° < \Lambda < 55°$), Tatnall et al. (1983) suggested that PLHR-stimulated emissions contributed to the formation of the electron slot at $2 < L < 3$. However, our current reappraisal of the Ariel 3 and 4 data indicates that the generation of difference frequencies by powerful VLF transmitters (navigation: Omega, Alpha) and communication (e.g., NAA and NSS) may also have been responsible for generation of key frequencies in the range 2 to 4 kHz (see Section 5) during the life time of Ariel 3 and 4. Support for this conclusion is provided by the more recent Aureol 3 global survey of Parrot (1990).

Molchanov et al. (1991) found that a small decrease in the total global electric power production during the weekend (<20%) appears to lead to a large decrease of the electrostatic turbulence intensity (~2x) of their magnetic field-aligned E_z component centered on 72 Hz. They have suggested that this large decrease is not only due to a decrease in total power production, but is mainly caused by the change in power line configuration during the weekend periods (see their Figures 2, 3, and 4). The Sunday (weekend) effect is discussed in Section 5.

At higher invariant latitude ($\Lambda > 55°$, L > 3) the much more spatially localized PLHR-stimulated discrete VLF emissions observed in the northern winter (1971 to 1972) have made it possible to identify cement works (and their power lines) at Edmonton, Winnipeg, and Montreal as prime sources of PLHR (Bullough, 1990). The field line grounded on the 500-km Winnipeg power line appears to be the most active for wave-particle interaction (WPI) in the magnetosphere. Radiation from the 20-km power line feeding a cement works in Derbyshire, west of Sheffield, is shown in Figure 10.2.3.

Bullough and Cotterill (1983) and Bullough (1988, 1990) have found good statistical support for the hypothesis of Markson (1978) and Herman and Goldbert (MHG; 1978) that lightning occurrence may be modulated by changes in atmospheric conductivity above thunderstorm areas. At medium invariant latitudes ($45° < \Lambda < 55°$) the controlling influence is that of the galactic cosmic ray (GCR) flux (Forbush decreases) whereas at high latitudes ($\Lambda > 55°$) energetic electron (>20 keV) precipitation, associated with auroral absorption (AA) events (>1 dB), is dominant.

Direct experimental evidence to support the above hypothesis has been found by Armstrong (1987). Supporting evidence for magnetospheric modulation of atmospheric conductivity over areas comparable to those of the riometer-measured precipitation regions is provided by shuttle observations of "sympathetic" triggering of lightning distributed over neighboring thunderstorm cells (Vonnegut et al., 1985); likewise in the clustering of radio atmospherics in association with burst particle precipitation (Rodriguez et al., 1988).

These PLHR studies have provided an excellent example of the subtle way in which man's activities may interact with, and possibly modify, his environment, and an important insight into the sun-weather relationship.

2. MEASUREMENT OF POWER LINE HARMONIC RADIATION (PLHR)

Ground measurements of the intensity and character of VLF radiation from power lines are essential to the interpretation of both satellite observations of PLHR emissions above the ionosphere and the associated line radiation events recorded at Siple and Halley in Antarctica. Detailed theoretical studies of the radiation properties of horizontal long-wire antennae were first made by

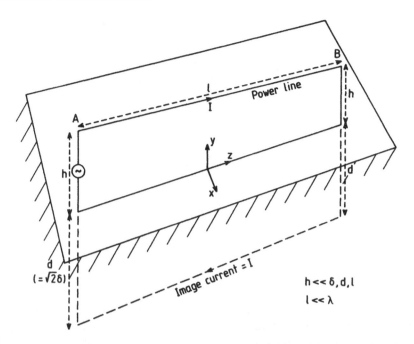

Figure 10.2.1 A simple model for PLHR. The power line AB has length, l, and height h above ground of conductivity, σ, and skin depth, δ, with source at A, and short to ground at B. The unbalanced current, I, flows at the mean height h of the three-phase line with a return (image) current in the ground at depth d $= \sqrt{2}\delta$. h $\ll \delta$, d, $l; l \ll \lambda$.

Carson (1926), more recently, for power lines, by Wedepohl and Efthymiadis (1978), and at higher frequencies (>1 MHz) by Kikuchi (1983). Here we will present the results of simpler modeling, based on Carson's equations. A detailed presentation is given by Yearby et al., (1983); see also Tatnall et al., (1983).

2.1. THEORY

Measurements of radiofrequency interference (RFI) to telephone systems at audiofrequencies [(Whitehead and Radley, (1949); Woodland, (1970)] indicate that such interference is due principally to magnetic induction from unbalanced harmonic currents flowing in the power lines with ground return. This justifies the simple model shown in Figure 10.2.1. The resultant unbalanced current (I) is assumed to flow, from source A, at the mean height (h) of the high-voltage power line (typically three conductors for a three-phase system) for a distance l. The line terminates in a short to ground at B and the return current flows via the ground back to A. The latter may be modeled as a line image current flowing at a distance d $= \sqrt{2}\delta$ below the surface where the skin depth:

$$\delta = (\mu_o \pi f \sigma)^{-1/2} \tag{2.1}$$

and the ground is assumed to have a uniform conductivity σ.

For simplicity we assume that h $\ll \delta$, d, λ and $l \ll \lambda$ where the free-space wavelength, $\lambda = 100$ km at 3 kHz. Thus, the power line behaves like a large vertical loop of area: $l \times$ d. This result may be obtained by applying the antenna reciprocity theorem to Cagniard's (1953) basic theory of the audiomagnetotelluric method of geophysical prospecting.

At a distance $x \gg h$ from the line, the induction components B_x, B_y are given by:

$$B_x = \frac{\mu_0 I}{2\pi} \left[\frac{h}{x^2 + h^2} + \frac{d}{x^2 + d^2} \right] \simeq \frac{\mu_0 I d}{2\pi (x^2 + d^2)} \tag{2.2}$$

where $h \ll x \ll l$:

$$B_y = \frac{\mu_0 I}{2\pi} \left[\frac{x}{x^2 + h^2} - \frac{x}{x^2 + d^2} \right] \simeq \frac{\mu_0 I d^2}{2\pi x (x^2 + d^2)} \tag{2.3}$$

where $x \ll \lambda$.

The magnitude and direction of the resultant field are:

$$B = (B_x^2 + B_y^2)^{1/2} = \mu_0 I d / 2\pi x (x^2 + d^2)^{1/2} \tag{2.4}$$

$$\tan\theta = B_y / B_x = d/x \tag{2.5}$$

From Equations 2.4 and 2.5 and graphs of log B vs. log x and log (B_y/B_x) vs. log x it is possible to determine the skin depth and ground conductivity as a function of frequency; hence, from measurement of the harmonic currents (at $x = 100$ m where the image current contribution may be neglected) the radiated power in a given passband, e.g., 3.2 ± 0.5 kHz.

The radiation field in free space at a height y directly above the power line ($y \gg h$) is given by:

$$B_x = \sqrt{(\mu_0 \varepsilon_0)} E_y = \frac{\sqrt{2} \pi \mu_0 \delta I l}{\lambda^2 y} \exp(2\pi i y / \lambda) \tag{2.6}$$

and the power radiated upward:

$$P = \frac{8\pi^3}{3} I^2 \eta_0 \left[\frac{l \cdot \delta}{\lambda^2} \right]^2 = 4.035 \times 10^{-16} \, I^2 \, l^2 \, \delta^2 \, W \tag{2.7}$$

$$= 3.20 \times 10^{-14} \, I^2 \, l^2 / \sigma \, W \tag{2.8}$$

for $f = 3.2$ kHz, $\lambda = 93.75$ km.

2.2. MEASUREMENTS

A special campaign was mounted by the Sheffield Space Physics Group in Newfoundland in the summer of 1982 to aid in the interpretation of Ariel 4 data and line radiation events recorded at the conjugate Antarctic station, Halley. Related measurements of earth resistivity have been made by McCollor et al. (1983) in the Vancouver area.

These measurements and those of the radiation from a cement works 18 km west of Sheffield indicate that PLHR may vary greatly for different industrial processes, but is particularly large for power lines to cement grinding mills.

2.2.1. Newfoundland

The amplitudes of 60-Hz harmonic currents in the power lines were calculated from the induced vertical magnetic components B_y measured close to the lines ($x \sim 100$ m) where $x \ll d$ and the effect of ground return currents can be neglected. Harmonics were detected up to the highest frequency examined, 4.5 kHz. Skin depths determined as a function of frequency for power lines close to Baie Verte (40°55'N, 56°12'W) and Buchans (48°49'N, 56°53'W) are shown in

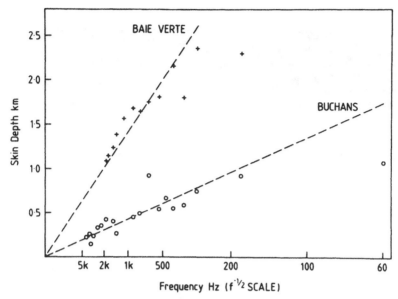

Figure 10.2.2 Skin depths as a function of frequency for the Baie Verte and Buchans lines. Best-fit straight lines through the two sets of points are shown dashed. Reprinted from *J. Atmos. Terr. Phys.*, Vol. 45: Yearby, K. H., Smith, A. J., Kaiser, T. R., and Bullough, K., pp. 409–419, Power line harmonic radiation in Newfoundland. Copyright (1983), with permission from Pergamon Press Ltd., Headington Hill Hall, Oxford, OX3 0BW, U.K.

Table 10.2.1 **Estimated Radiated Power and Magnetic Field at 100-km Altitude for f = 2.7 − 3.7 kHz.**

Line	I (mA)	d (m)	l (km)	P (μW)	B_x (rms) ($\times 10^{-16}$ T)
Buchans	3.2	240	50	0.6	2.4
Baie Verte	0.31	800	50	0.06	0.8
Cooks Harbour	3.0	280	14	0.06	0.7

From Yearby, K. H., Smith, A. J., Kaiser, T. R., and Bullough, K. (1983). *J. Atmos. Terr. Phys.*, 45, 409. With permission.

Figure 10.2.2. An increase in ground conductivity, from 1.2×10^{-4} to 1.4×10^{-3} S/m in the south was found.

The estimated radiated power and magnetic field at 100-km altitude for f = 2.7 to 3.7 kHz are listed in Table 10.2.1 for these power lines and also that at Cooks Harbour, near St. Anthony (51°24′N, 55°37′W).

The signal levels in the Ariel 4 3.2-kHz passband are all less than 1μW and therefore too small for direct detection and unlikely to suffice for triggering of emissions which might be detected on Ariel 4 or in PLHR-stimulated emission at Halley. Helliwell et al. (1980), from experiments with the Siple transmitter, suggested that a threshold of 1-W radiated power may be required for growth and triggering. Park and Chang (1978) used an estimated radiated power of 0.5 W at 3.42 kHz in a simulation of PLHR at Siple. See, however, Sá (1990) and Chapter II/13.

2.2.2. Derbyshire Cement Works
An example of PLHR from one of two 20-km, 66-kV three-phase lines (~4 MW) feeding a cement works in Derbyshire, 18 km west of Sheffield, is shown in Figure 10.2.3. The measured rms field, at x = 100 m from one of the lines in a 300-Hz bandwidth at 3 kHz, corresponds to an unbalanced peak rms current of about 0.1 A in a 1-kHz passband at 3 kHz. The rms magnetic

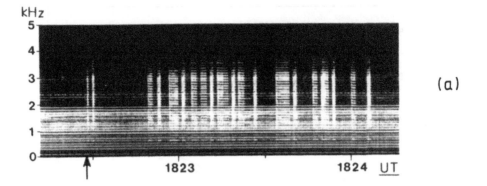

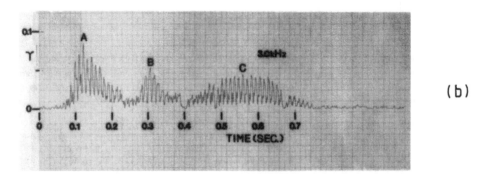

Figure 10.2.3   A typical industrial PLHR event recorded on December 3, 1981, 1822 UT at 100 m from a 66-kV, three-phase line, one of two such lines supplying about 4 MW to a cement works in Derbyshire. (a) Spectrogram of PLHR. (b) rms induction field amplitude (1 λ = 1 nT), measured in a 300-Hz passband centered on 3 kHz, of the PLHR transient arrowed in (a). Reprinted from *Proc. 5th Symposium on EMC, Zurich:* Bullough, K. and Cotterill, A., (1983), pp. 7–12, Ariel 4 observations of power-line harmonic radiation over North America and its effect on the magnetosphere. Permission from ETH Zentrum-IKT, Zurich.

component [$\tilde{B}(80)$] of the electromagnetic (EM) wave at the base of the ionosphere, ~80 km above the lines, is then given by:

$$\tilde{B}(80) = 0.070\,\frac{\tilde{I}\delta}{\lambda^2} = 0.028\delta\ \mu\gamma = 8.14 \text{ to } 81.4\ \mu\gamma \tag{2.9}$$

for $\sigma = 10^{-3}$ to 10^{-5} S/m at 3 kHz; $2,-20\tilde{I} = 0.2$ A rms in a 1-kHz passband; $l = 20$ km; $\lambda = 100$ km.
Therefore,

$$[\tilde{B}(80)]^2\ Hz^{-1} = 18 \text{ to } 38 \text{ dB above } 10^{-15}\gamma^2\ Hz^{-1}$$

At night, loss in transmission through the ionosphere ~4 dB and that due to wavefield divergence ~1 dB. However, at satellite altitude (400 to 600 km), the magnetic component is increased by about 15 dB in the magnetoionic medium. Thus, at the satellite, $B^2/Hz = 28$ to 48 dB above $10^{-15}\gamma^2/Hz$ and would, therefore, for dry ground, be about 10 dB above the Ariel 4 receiver noise level (37 dB $> 10^{-15}\gamma^2/Hz$) at 3.2 kHz. Such a signal would be detectable on the peak reading circuit. Increased D-region absorption (~10 dB) during daytime makes detection unlikely.

Table 10.2.2 **Locations and Capacities of Cement Works: E, W, and M**

City	Geographic coordinates		Invariant coordinates		% Annual cement capacity (100% = 12,805,274 metric tons)
	Longitude W	**Latitude N**	**Longitude E**	**Latitude N**	
Edmonton	113.4°	53.6°	306.7°	61.6°	4.0
Winnipeg	97.2°	49.9°	329.4°	61.0°	6.0
Montreal	73.6°	45.5°	4.5°	58.8°	20.4

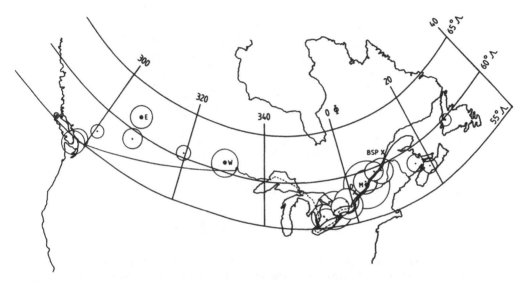

Figure 10.2.4 The location of cement plants in Canada (1970). Annual capacity is proportional to circle area. Total national capacity (1970): 14,572,200 short tons. Map traced from the *Canada Atlas*, 1974: invariant latitude and longitude contours, at 600 km altitude, from Evans, J. E., Newkirk, M. L. and McCormac, B. M., (1969): North Polar, South Polar, world maps of invariant magnetic coordinates for six altitudes: 0, 100, 300, 600, 1000, and 3000 km. Defense Atomic Support Agency No. 2347, Lockheed Palo Alto Research Laboratory, 3251 Hanover Street, Palo Alto, CA 94304, U.S.

Note that the harmonic current in the 3.2-kHz passband (power radiated ~0.05 W for dry ground) is about 100 times that measured in Newfoundland and indicated that power lines to the grinding mills of cement works, similar to but larger than that in Derbyshire, would be particularly powerful sources of PLHR.

2.2.3. PLHR from the Winnipeg Cement Works Power Line

This section will include discussion of Ariel 4 satellite observations and will therefore require prior reading of Section 3.2.

Until our measurements of PLHR from the Derbyshire cement works (Bullough and Cotterill, 1983) we had assumed, incorrectly, that the most important sources were probably arc furnaces. Possible Canadian sources (Canada Atlas, 1974), their locations, and capacities are listed in (Bullough, 1990).

In Table 10.2.2 we have listed the locations and capacities of the major cement works at Edmonton, Winnipeg, and Montreal. These locations, together with a grid of invariant coordinates, are shown superposed on maps of Canada in Figures 10.2.4 and 10.2.5. Note that the power line from Winnipeg extends some 500 km to the northeast and lies within the invariant longitude sector: 330 to 340°.

The percentage occurrences of 3.2-kHz peak and minimum signal intensities (dB > $10^{-15}\gamma^2$/Hz fse) occurring in the 10° longitude sectors on the map (Figure 10.2.5) are plotted in Figure 10.2.6

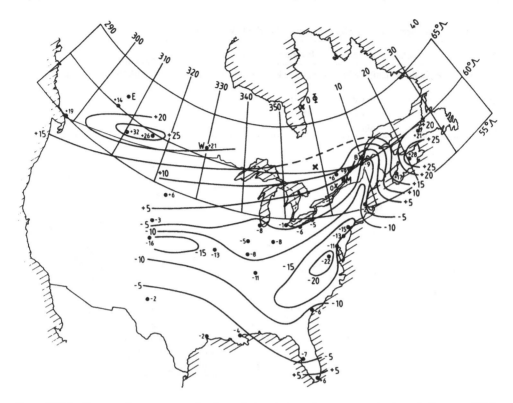

Figure 10.2.5 Contours of percentage secular change in thunderstorm day occurrence for the period between 1936 to 1970 relative to 1901 to 1935. The percentage change for individual stations is shown. Edmonton: E, Winnipeg: W, Montreal: M. Riometer stations (x) at Ottawa: O and Baie St. Paul: B; Val d'Or and Cape Jones also indicated (refer to Figure 10.6.1). Invariant coordinates: 600-km altitude. Contours of percentage change in thunderstorm day occurrence (Changnon, S. A., 1977) taken from *Proc. 5th Int. Conference on Atmospheric Electricity: Electrical Processes in Atmosphere,* Copyright (1977), with permission from Dr. Dietrich Steinkopff Verlag, Darmstadt, Germany.)

for the northern summer and winter periods. We are particularly interested in the northern hemisphere, northern winter in Figure 10.2.6b. Clearly discernible are three peaks corresponding to PLHR emissions originating in cement works at Edmonton (E), Winnipeg (W), and Montreal (M).

The largest plant for copper and nickel (Copper Cliff) lies in the sector immediately west of sector M where the occurrence of the more intense signals in northern winter, in the north, is particularly low. This is probably due to prior removal of eastward-drifting resonant electrons in sector W. Note that the Winnipeg grinding mill normally functions 24 h per day but is routinely closed in January and February (Yakymyshyn, 1990). For full local time coverage the winter period in Figure 10.2.6 has been taken as December 21 to March 19 and the mill is therefore operative for about one third of the period.

The Winnipeg power line is a dedicated 7000-kVA line which extends some 500 km to the northeast over Ontario and the Canadian Shield. Audiofrequency (10 Hz to 10 kHz) magneto-telluric measurements in this region by Koziar and Strangeway (1978) and airborne measurements by Hauser et al., (1969) at higher frequencies (14.7 to 26.1 kHz) lead us to suggest (Bullough, 1990) that a fairly small effective conductivity (10^{-4} S/m) may be possible at 3.2 kHz. McCollor et al. (1983) at 200 km northeast of Vancouver measured a conductivity of 10^{-4} S/m at power line harmonics of 300 and 420 Hz.

An attempt was made to separate out the contribution to strong WPI/AA events by the PLHR in the Winnipeg sector by assuming that, in the absence of PLHR, the seasonal variation in 3.2-kHz emission due to WPIs would be the same as that in the southern hemisphere of the invariant

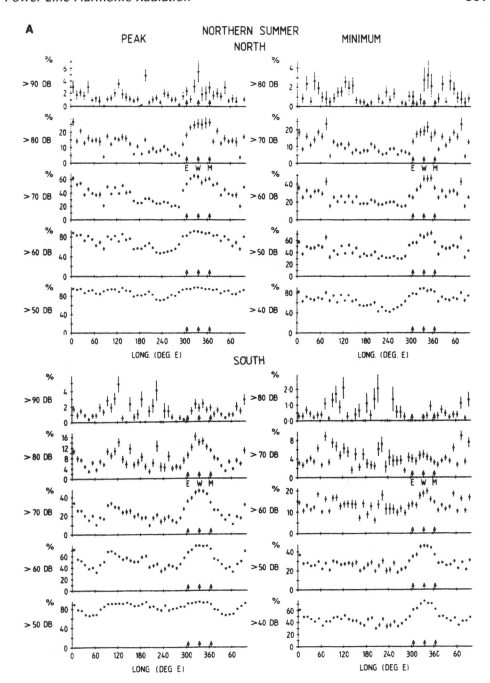

Figure 10.2.6 The percentage occurrence of 3.2-kHz signal intensities above 40, 50, . . . , 90 dB $> 10^{-15}\gamma^2$/Hz in 10° sectors of invariant longitude; invariant latitude interval: 55° to 65°, Kp ≤ 2+. North and south. Edmonton: E, Winnipeg: W, Montreal: M. (A) North summer; (B) North winter. Reprinted from the *Proc. 10th Int. Wroclaw Symposium on EMC*, Bullough, K., (1990), pp. 819–825. More studies in PLHR: cementing a link in the Sun-Weather Relationship? Permission given by Wroclaw Technical University Press, Poland.

longitudinal sector: 150 to 160° (see Figure 10.4.1). We chose this particular sector, which lies over the southern Indian Ocean rather than land as in Canada, because the invariant and geographic latitudes are similar. Thus, diurnal and seasonal dependencies on D-region absorption, magnetospheric whistler propagation, etc. will be similar. Note, however, that the southern summer data

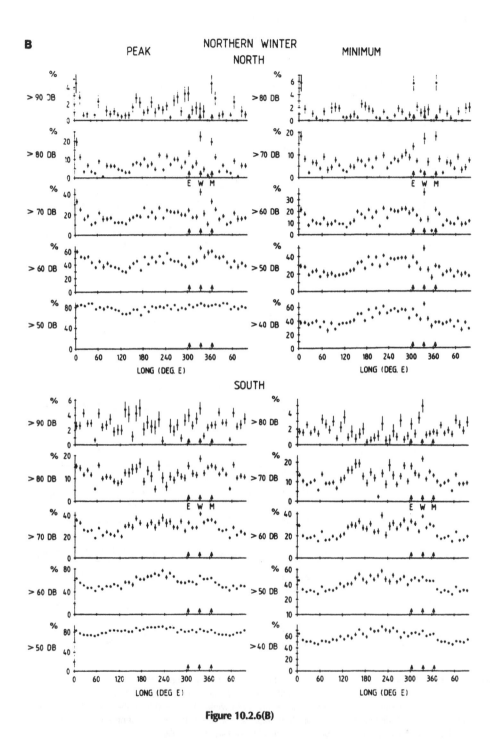

Figure 10.2.6(B)

(December 21 to March 19) is displaced toward autumn. In both locations the winter thunderstorm activity is negligible and, in summer, is probably more active in southern Canada. For example, in Edmonton for the period 1953 to 1988, mean monthly thunderstorm day occurrence, usually in the evening, peaked at 10.5 in July falling to 6 to 7 in June and August, 2 in May, and less than 1 in April and September. In both locations there is negligible activity in the conjugate region

Table 10.2.3   **Estimated Percentage Occurrence of PLHR Signals at Winnipeg**

$dB > 10^{-15}$	Summer (North)						Winter (North)					
γ^2/Hz^{-1}:	>40	>50	>60	>70	>80	>90	>40	>50	>60	>70	>80	>90
Peak	—	86	82	50	13	4.5	—	−19	28	23	12	−0.9
Mean	79	66	22	−1.4	−0.7	—	26	36	28	13	0.8	—
Min	65	32	24	0.4	2	—	38	27	22	13	1	—

Occurrence: Summer (North)/Winter (North)

Peak	—	−4.4	2.9	2.2	1.1	−5.0
Mean	3.1	1.9	0.8	−0.1	−0.9	—
Min	1.7	1.2	1.1	0.0	2.0	—

(geographic latitude too high); also, sferics from distant thunderstorms will be strongly attenuated (~18 dB/Mm) in the Earth-ionosphere waveguide. Thus, in both locations, the observed emissions will be dominated by downgoing whistler-mode signals of nonthunderstorm origin. We also assume closely similar satellite scanning and sampling.

If we now pair the Winnipeg northern summer readings with those from the Indian Ocean southern summer (similarly for winter) we may estimate the percentage occurrences (above given signal levels) of PLHR and PLHR-stimulated emissions. These are listed in Table 10.2.3 where, if the observed percentage occurrence of a signal equals W in the presence of PLHR and I in the absence of PLHR, then the tabulated value:

$$x = (W - I)/[1 - I/100] \qquad (2.10)$$
$$= 80\% \text{ for } W = 96\% \text{ and } I = 80\%$$

The following interpretation, taking account of the many assumptions involved, is very tentative.

First, the Winnipeg grinding mill normally functions 24 h per day but is routinely closed in January and February (Yakymyshyn, 1990). Thus, if there existed a signal intensity level above which upgoing PLHR signals were dominant and their occurrence remained constant over a given area of the lower ionosphere then we might expect a ratio, summer/winter, of ~ 3:1 in occurrence. There is evidence for this at peak readings >60 dB and mean readings >40 dB; however, direct satellite recordings (detailed spectra) of such intense PLHR signals have not thus far been reported (Bullough et al., 1985; Helliwell et al., 1975) even though our ground measurements, as above, indicate that they are feasible. The >60 dB level corresponds to the 1-W (radiated) threshold level that Helliwell et al. (1980) suggested was necessary for emission stimulation in Siple local winter.

Secondly, the most intense events, e.g., mean and minimum >70 and 80 dB, peak >80, 90 dB, appear to be less dependent on PLHR stimulation, in agreement with our earlier work (Bullough et al., 1975).

Thirdly, at the lowest signal levels, background noise of sferic and magnetospheric origin clearly dominates; the peak reading almost never drops below 40 dB.

Fourthly, the extremely high percentage occurrence (>50 dB: 86%; >60 dB: 82%) in the peaks, in the northern summer, is probably due to:

1. A large harmonic current component and the radiation efficiency of the very long power line which extends some 500 km northeast from the western edge of the sector
2. Optimum conditions for PLHR signal entrapment and ducting (hence stronger WPI; see later); this also affects weaker PLHR sources extending across Canada

3. The persistence in gyrophase coherence as the resonant electrons drift eastward away from the power line field (Cornwall and Schultz, 1970)
4. Inward propagation of the stimulating and amplified signal to lower L-shells, within the plasmasphere, where the gyroresonance condition would apply to electrons of too low an energy (keV) to yield an AA (>1 dB) event
5. A greater contribution from local thunderstorms in Canada
6. Contributions (XM), due to mixing and production of difference frequencies in the ionosphere, from lower latitude VLF transmitters (in particular, NAA and NSS) which, since the signals propagate with little loss to higher latitudes, may have helped to fill the winter longitudinal "gaps" between E, W, and M in the summer period; NSS probably did not operate in the winter period. Emissions, recorded in the 3.2-kHz passband on the ground at Halley, were reduced during NAA and NSS maintenance periods on Friday and Thursday (awaiting verification) afternoons (~1500 LT); see Section 5.

Finally, the PLHR contribution to the stronger AA (>1 dB) events appears to be more important in the local winter than in the summer (e.g., mean and minimum >70 dB).

2.2.4.   Other Relevant Studies

Barr (1979) has made measurements of the vertical electric field strength of ELF and VLF radio waves in the south of New Zealand. He found no significant mains harmonic (>1 μV/m) at frequencies greater than 1.5 kHz. The PLHR decreased toward the south of New Zealand until at Stewart Island only a single, but strong (~10 μV/m) harmonic could be detected at 300 Hz. The latter was radiated by a DC power line, not by the standard 220-kV AC power grid.

Dazey and Koons (1982) adopted an available power line for use as a VLF antenna in northern Norway. Their comprehensive report, theoretical and practical, would be of use when making more detailed ELF/VLF measurements on power lines, particularly should the length of the power line (e.g., Winnipeg), impedance, etc., invalidate the simple small loop radiator theory presented above.

Kikuchi (1972) has reported on Japanese PLHR.

3.   SATELLITE VLF EXPERIMENTS

The basic PLHR signal is extremely weak and difficult to identify on satellites (Bell et al., 1982; Koons et al., 1978; Luette et al., 1979) and we normally recognize its existence by the stimulated emissions or wave-wave interactions to which it may rise (Helliwell et al., 1975; Sá, 1990). Thus, to understand and identify the occurrence of PLHR-stimulated emissions it helps to correlate their satellite-determined geographical distribution with more detailed complementary ground-based studies, especially those made in Antarctica (see Section 5).

Parrot (1994) has recently reported on observations of PLHR by the low-altitude Aureol 3 satellite. In 940 real-time (TMF), high-resolution (2.5 Hz) spectrograms (durations ~12 to 15 min; time resolution: 80 ms) only five, recorded at Toulouse, Tromsoe, and Terre Adelie, contained events identified as PLHR. They had the following features:

1. Only the electric field component was observed.
2. The frequency range lay mainly between 3 and 4 kHz.
3. The lines drifted together (mainly 3 to 4 Hz/s) with a roughly constant frequency spacing of 50 Hz.
4. They were only observed during periods of very quiet magnetic activity.

Interestingly, the event locations were on field lines grounded well distant from industrial areas.

3.1. CHOICE OF ORBITAL PARAMETERS

The observed distribution of an emission in space is a convolution of the spatial distribution of that phenomenon and the sampling distribution and caution is required in interpretation, especially when orbital periods are simple multiples or submultiples of a day or year (see discussion by Russell, 1980 of work by Luette et al., 1977 and their reply; Luette et al., 1980). See also the search for evidence of geographic control (based on OGO 5 data) by Tsurutani et al. (1979) and discussions by Thorne and Tsurutani (1981), and Park and Miller (1981) who include, in their reply to "Comment on the Sunday decrease in PLHR events observed at Siple," a discussion of OGO 5 data on chorus events.

The low altitude (4 to 600 km) Ariel 3 and 4 satellites had orbital inclinations (i = 80, 83°) so as to provide (1) high latitude coverage and (2) full local time coverage in each three-monthly season (80 and 96 days, respectively). The orbital period (96 min) was such that the subsatellite paths of successive revolutions were displaced 24° in longitude. The period is a close submultiple of 24 h, and each set of 15 revolutions suffered only very small displacement in longitude ($\sim$1 to 1.2° W) per day. For a basic sampling interval (longitude/latitude) box of $10 \times 2°$ this small daily longitudinal precession leads to gaps in both LT and UT coverage (Tatnall et al., 1983).

Storm-time variations in emission activity (Kp > 2+) and associated LT maxima in emission occurrence are largely filtered out by preselecting only magnetically quiet periods (Kp $\leq$ 2+), however, emission activity is often enhanced in the quiet period following a magnetic storm. Conversely, the 3.2-kHz signal may be diminished by LT-dependent LHR reflection above the satellite (Lefeuvre and Bullough, 1973). This made it necessary to check all zone maxima for the possible effects of both LT and UT (storm-time: Dst) dependence. On Ariel 3 the percentage of quiet 3-h periods were 70 (summer) and 68 (autumn) and on Ariel 4: 64 (winter, 1971 to 1972); 67 (spring) and 72 (summer).

To obtain the maps in Section 4, a count was made of the number of observations in each $10° \times 2°$ longitude-latitude sector and the percentage of such occasions on which the signal intensity exceeded a preselected value found. The number of counts was typically 35 to 55 per sector, large enough to yield smooth computer-drawn contours in the majority of cases, although the effect of the "gaps" in LT, UT coverage (Figure 10.4.2) is clearly seen, especially in maps at ELF (0.75 kHz; see Tatnall et al., 1983, Figure 2(iv)a). Ideally, three such satellites, spaced in longitude and LT, should be used.

Because the latitude interval (2°) is only slightly larger than the sampling interval (28 s $\equiv$ 1.75° along the orbit) most data points in the same box are obtained at least one day apart. This reduces the problem of possible oversampling (Tsurutani et al., 1979).

If we assume that there are n independent observations per sector then the standard error in a measured percentage x will be given by:

$$x = \pm \frac{x(100 - x)}{n} \tag{3.1}$$

3.2. ELF/VLF RECEIVER PARAMETERS

For global morphological studies, the satellites Ariel 3 and 4 had on-board tape recorders. Measurements were made of the peak, mean, and minimum signal intensities (magnetic component parallel to the spin axis and Earth axis) in several narrowband channels each successive 28-s period ($\equiv$200 km) around the orbit. The time constants of the peak and minimum reading circuits (f $\leq$ 9.6 kHz) were 0.01 and 0.1 s, respectively, each sampling circuit being reset immediately after readout. The mean circuit (running mean) had a time constant of 30 s. The ratios peak/mean, mean/minimum, and peak/minimum made it possible to identify the type of emission present in the passband. Ariel 4 had the following center frequencies (kHz) and (passbands): 0.75 (0.5); 1.25 (0.5); 3.2 (1.0); 9.6 (1.0); also VLF transmissions: 16 (1.0, 0.1); 17.8 (0.17). Ariel 3 was similar but lacked the passbands at 0.75, 1.25, and 17.8 kHz. The 3.2-kHz channel was chosen, fortuitously, for ease in internal calibration. It turned out to be well optimized for PLHR (and MLR) phenomena (see Section 5 and Figure 10.5.2). The dynamic range (75 dB) of each receiver channel (log. response) and sensitivity (37 dB above $10^{-15}\gamma^2$/Hz at 3.2 kHz) were such that almost all signals exceeded receiver noise level but almost never exceeded the dynamic range.

Many satellites utilize wideband VLF receivers for signal identification and obtain a wideband spectrum with fast AGC, such that the most intense signal stays within the dynamic range available but the weaker signals (like PLHR) tend to be suppressed (see Luette et al., 1979, p. 2659, 2660). On OGO 5 the ELF experiment employed linear receivers with limited dynamic range, and the receiver gain was adjusted by ground command to try and avoid saturation by strong signals (Park and Miller, 1981, p. 1643). Great care was taken to eliminate spacecraft VLF interference on the simpler spacecraft Ariel 3 and 4. This was more difficult to achieve on the extremely sophisticated OGOs.

The three-axis stabilized satellite, Aureol 3, carried a much more sophisticated ELF/VLF experiment (ARCAD-3: Berthelier et al., 1982) and was placed in a low-altitude orbit (perigee: 408 km, apogee: 2012 km, inclination: 82°, period: 109 min, observing period: October 1981 to December 1983). Three magnetic and two electric components were measured with one electric dipole (E_z) oriented approximately parallel to the local magnetic field at high latitude and the other (E_h) maintained parallel to the satellite's motion (direction: x).

World maps were derived by Parrot (1990) which utilized tape-recorded data of the magnetic component (B_y) selected by narrowband filters at 800, 4500, and 15,000 Hz and the electric field component (E_h) selected by narrowband filters at 725, 4500, and 15,000 Hz. Molchanov et al. (1991) also used tape-recorded measurements of E_h and E_z processed by a bank of 6 filters at 15, 32, 72, 150, 325, and 725 Hz in their discussion of daily variations of ELF data (see Section 5.2 for discussion of the "Sunday (weekend) effect").

The data used in plotting the maps were averaged over a period of 3 min and then an overall mean signal intensity calculated for each basic latitude/longitude blok of 5° × 10°, similarly for each blok of 10° (invariant latitude) × 1 h (MLT). These maps were presented in a format color-coded in 5 dB steps of mean signal intensity over a range of 40 dB, selected such that the maximum and minimum intensities corresponded to 5 x M (red) and M/20 (dark blue), respectively, where M is the average signal amplitude at a given frequency.

On Ariel 3 and 4, the existence of permanent zones of VLF emissions associated with (PLHR + XM) is most clearly shown by means of world maps of VLF emission activity at 1.25, 3.2, and 9.6 kHz during magnetically quiet (Kp ≤ 2+) periods, particularly at 3.2 kHz. (These zones are also identifiable on Aureol 3 at 4.5 and 15 kHz.) These maps are restricted to data from the mean reading circuits of the various channels and were obtained by computer processing (Tatnall et al., 1983) in the following way:

First, the "free-space equivalent" (fse) wave intensity (Wm^{-2} Hz^{-1}) was calculated from the measured magnetic component, B(γ), by assuming that the quasilongitudinal (QL) approximation was valid (Helliwell, 1965; Sazhin, 1993). Thus,

$$B(\gamma) \text{ (fse)} = B(\gamma)/\sqrt{n} \qquad (3.2)$$

where the local refractive index, n, is given by

$$n \sim f_0(ff_H\cos\theta)^{-1/2} \sim f_0(ff_H)^{-1/2} \qquad (3.3)$$

where f_H is the local electron gyrofrequency and f_0 the electron plasma frequency. Equation 3.2 yields B(γ) (fse) with an error of less than 3 dB for a wave-normal angle to the local magnetic field, $\theta < 75°$. Because n >10, B(γ) (fse) is always at least 10 dB less than B(γ) and, for a circularly polarized wave in free space:

$$1\gamma^2/\text{Hz} = 4.8 \times 10^{-4} \text{ Wm}^{-2} \text{ Hz}^{-1}$$

where,

$$1\gamma = 10^{-9} \text{ T in free space}$$

Having obtained a percentage, x, for each $10° \times 2°$ box a smooth latitudinal profile was obtained for each $10°$ longitudinal sector by fitting a $15°$ polynomial, weighted according to the standard error, (Equation 3.1) to each set of latitude ($2°$ interval) points. This was done, in 10-dB steps, for signal intensities (fse) exceeding $4.8 \times 10^{-19} \text{ Wm}^{-2} \text{ Hz}^{-1}$ where the latter reference level is equivalent to $B^2 = 10^{-15}\gamma^2 \text{ Hz}^{-1}$ (fse). Contouring was then performed on these "smoothed" values.

3.3. THE ARIEL 3 AND 4 AND AUREOL 3 WORLD MAPS

When comparing the Ariel 3 and 4 and Aureol 3 world maps, note the following:

1. The higher resolution ($2°$) in latitude on Ariel 3 and 4 relative to Aureol 3 ($5°$) makes it easier to identify invariant latitude-dependent features, e.g., the electron slot, inner and outer radiation belts, and the plasmapause.
2. Ariel 3 and 4 mapping was restricted to quiet periods (Kp < 2+) when we found that great magnetic storms tended to obscure dependence on geographic coordinates.
3. Localized sources, such as VLF transmitters, may be more easily identified and located on the percentage contour maps; also the peak and minimum readings on approaching and leaving the source may better define location and wavefield intensity.
4. The receiver channel at 3.2 kHz was, fortuitously, better chosen than that at 4.5 kHz on Aureol 3: (a) emissions on the Ariel 3.2 kHz channel have maximum occurrence in the key frequency range, 2 to 4 kHz, (b) the "half-gyrofrequency cut-off" for ducting at the geomagnetic equator occurs at a slightly higher L (5.2 for 3.2 kHz; 4.6 for 4.5 kHz), thus improving measurement of the Canadian PLHR sources, and (c) waveguide attenuation of 3.2 kHz (18 dB/Mm) limits obscuration by sferics from distant lightning.

4. ZONES OF ELF/VLF EMISSION

During our reappraisal of the Ariel 3 and 4 studies and, in particular, after intercomparing the Ariel and Aureol 3 world maps, we realized that ground-based VLF transmitters were more important than recognized in earlier papers (Tatnall et al., 1983; Bullough, Strangeways, and Kaiser, 1985). Their probable contribution is discussed below.

4.1. THE EFFECTS OF VLF GROUND-BASED TRANSMITTERS

The International Frequency List (1987) of geographical locations, power, and operating frequencies of the ground-based VLF transmitters is published in Parrot (1990). Figure 10.4.1 is similar to his but includes the earlier temporary Omega transmitters: Forestport (New York: 75°05'W, 43°27'N; which transmitted at low power, ~1 kW) and Tasman (Woodside: 146°56'E, 30°20'S, east coast of Australia, reactivated in August 1982). Both Omega and Alpha transmitters were undergoing changes in location, frequency, power, and operating schedule in their development period during the Ariel life times, but were probably stable in operation during the observing period of Aureol 3.

One important conclusion reached by Parrot was that "enhanced ELF waves occur at the same location (region around transmitter) as the precipitation of electrons anticipated by the model of Inan et al. (1984)." He concluded that the most favorable mechanism for ELF wave generation was that suggested by Chmyrev et al. (1989): a parametric interaction between whistler-mode waves and electromagnetic ion-cyclotron waves which triggers ELF waves.

We propose a second possible explanation. This is the mixing of two VLF signals in the lower ionosphere so as to yield the sum and difference (XM) frequencies in the nonlinear medium

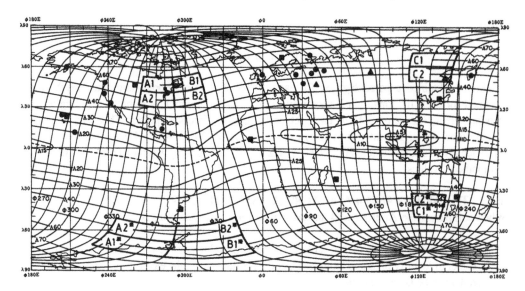

Figure 10.4.1 World map with invariant longitude and latitude contours, at an altitude of 600 km, superimposed. The locations of VLF transmitters: navigation (Omega: ■; Alpha: △) and communication: (●) are shown. Also shown are the three invariant longitude sectors centered on: Eights/Siple (NE USA), Halley (West Atlantic), and Yakutsk with invariant latitude intervals: 45 to 55° and 55 to 65°. [North and South (geomagnetic conjugate)]. Map traced from the *Canada Atlas*, 1974: Invariant latitude and longitude contours, at 600 km altitude, from Evans, J. E., Newkirk, M. L. and McCormac, B. M., (1969) North Polar, South Polar, world maps of invariant magnetic coordinates for six altitudes: 0, 100, 300, 600, 1000, and 3000 km. Defense Atomic Support Agency No. 2347, Lockheed Palo Alto Research Laboratory, 3251 Hanover Street, Palo Alto, CA 94304, U.S.

(Gurevitch, 1978). The difference frequency may often lie in the key 2 to 4 kHz range which not only has a sharp maximum in occurrence for emissions at 3.2 kHz on higher L-shells ($L \sim 4$; see Section 5), but also, as Hayakawa (1991) has shown, for the starting frequencies of emissions in the electron slot ($2.1 < L < 3.2$) and inner radiation belt ($L < 1.8$). The occurrence probability of such emissions increases with Kp. The electron slot region of precipitation is consistent with the theoretical prediction by Chang and Inan (1985) of a region of enhanced whistler-induced higher energy (40 to 300 keV) electron precipitation. Preferential whistler ducting and WPIs may occur on L-shells in this region and frequency range where initially nonducted whistler waves can come to a focus in the conjugate ionosphere. There are two possible ways of generating these difference frequencies.

4.1.1. Omega and Alpha Transmissions

The North Dakota (invariant coordinates: 358°/48°) Omega transmission consists of a sequence of three pulses (20 kW) of frequency: 10.2, 13.6, and 11.33 kHz (each pulse: 0.9 to 1.2 s duration and separated by 0.2 s) every 10 s. The two-hop group travel time at $L \sim 2.5$ is very close to 1 s such that there is a good time overlap between the two-hop 10.2-kHz pulse and the direct (half-hop) 13.6-kHz pulse in the northern ionosphere close to the transmitter. These two frequencies are then mixed in the nonlinear medium to yield the difference frequency: $13.6 - 10.2 = 3.4$ kHz. The latter lies well within the receiver 3.2-kHz passband, is often more appropriate for a gyroresonant WPI at, or near, the equator, and may be amplified, with electron precipitation, to establish a multihop regime between hemispheres (Lefeuvre and Bullough, 1973). The westerly half of the dumbbell-shaped 100% occurrence contour at 3.2 kHz (>30 dB) over North America (Figure 10.4.2A) is probably due to this.

The similar, but more powerful (0.5 MW) Alpha navigation transmitters utilize a format of 0.4-s pulses at frequencies of 14.88, 12.64, and 11.88 kHz with a 0.1 s (?) gap. While the power is much greater, the time overlap will be neglible for adjacent pulses and small for the first and third. Other 0.4-s pulses in the sequence may exist and be used for time information, etc., at

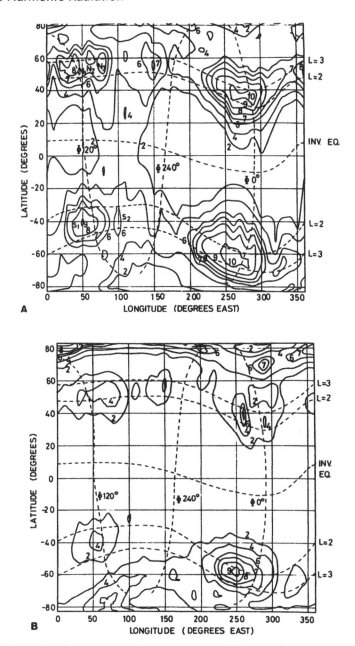

Figure 10.4.2 (A) 1972 summer (days 117 to 213). The percentage occurrence of emissions (3.2 kHz) of intensity greater than 30 dB above 4.8×10^{-19} Wm^{-2} Hz^{-1}; Kp $\leq$ 2+. Contour values 1, 2, 3. . . .10 = 10, 20, 30 . . . . , 100%. Dashed lines represent contours of invariant longitude ϕ and L values at 600 km. (B) As above but for 9.6-kHz emissions of intensity >30 dB. Reprinted from *Space Sci. Rev.*, Vol. 35: Tatnall, A. R. L., Mathews, J. P., Bullough, K., and Kaiser, T. R., pp. 139–173: Power-line harmonic radiation and the electron slot. Copyright (1983) with permission from Kluwer Academic Publishers, P.O. Box 17, 3300 AA Dordrecht, Holland.

frequencies up to 15.625 kHz (additional pulses may also be included in the Omega format). These transmissions may also interact with those of nearby communication transmitters, such as UMS (16.2/17.1 kHz).

4.1.2. The NAA/NSS Difference Frequency

During the Ariel 4 lifetime, the NAA transmission was 17.8 kHz (FSK) and that of NSS: 21.4 kHz, probably FSK, possibly MSK. We monitored NAA [also GBR (16 kHz)] at Sheffield throughout the Ariel 4 lifetime. NSS, unfortunately, was not monitored. However, the difference frequency, $21.4 - 17.8 = 3.6$ kHz, again lay well within the receiver 3.2-kHz passband and could be responsible for the eastern half of the 100% occurrence contour mentioned above. Note that the 10.2-kHz pulse is clearly identifiable in Figure 10.4.2B (9.6 kHz, > 30 dB) for the North Dakota transmission, but absent for the NAA/NSS interaction which would probably maximize near to NSS.

Evidence to support the above hypothesis includes the following:

1. Halley ground recordings in receiver passbands identical to those on Ariel 4 (Yearby et al., 1981) show significant secondary minima in the 3.2-kHz minimum reading recording on Thursday and Friday afternoons (Figure 10.5.6). There were no similar minima in the 0.75-kHz passband (width: 500 Hz).

2. Arnoldy and Kintner (1989), in the WIPP campaign of July 31, 1987, recorded the E and B components (MSK) of NAA and NSS above Wallops Island together with the amplified, Doppler-broadened two-hop wave magnetic component at the difference frequency 2.6 kHz. The peak signal amplitude ($\sim 1 \mathrm{pT}/\sqrt{\mathrm{Hz}}$) of the latter exceeded that of the NAA signal.

3. During the Ariel 3 and 4 life times, Omega and Alpha locations, power, frequency, and scheduling were varied during lengthy trials. This could account for some of the abrupt large changes in the percentage occurrence maps in Tatnall et al. (1983), e.g., compare 1967 summer to 1967 autumn. The latter had high peaks in occurrence (3.2 kHz, >30 dB) centered on the U.S. west coast which could have been due to mixing of NPL, NPG, and NLK transmissions which were absent in the summer. On the east coast, Omega Forestport (New York) may have been active (probably at low power). It now appears less likely that such spatially large, sudden changes in the wavefield above North America could be brought about solely by PLHR from industrial undertakings and domestic habitats. With weekend closing such PLHR sources are unlikely to yield 100% occurrence contours.

4. Alpha Krasnodar may not only have generated difference frequencies like North Dakota Omega, but also interacted with UMS (16.2/17.1 kHz).

Note that the Tasman Omega (identifiable in Tatnall et al., 1983: Figure 2(iv)f: 9.6 kHz, >20 dB), situated on the east coast of Australia, gave rise to nonconjugate one-hop signals in the north which were strongly cross-meridian refracted (Denby et al., 1980) to the west. The direction of refraction was consistent with observations restricted to nighttime passes on that occasion, though the magnitude of the displacement was surprising. The transmitter was probably located at too low an invariant latitude ($40°$) to generate, like North Dakota, conjugate ducted and amplified multihop 3.4 kHz signals.

5. In Parrot (1990), Figure 9, world maps of B_y at 4.5 kHz "all" data and "winter only" data are compared. The two maps are very similar over North America and appear to show a double peak in longitude somewhat reminiscent of that found in Ariel 4 summer (1972) at 3.2 kHz. This might indicate that the strong seasonal variation in the 3.2 kHz mean signal intensity found on Ariel 4 was misleading. Perhaps the North Dakota Omega and the NSS transmitters were inoperative in the 1971/1972 winter. On February 25, 1972, only four Omega stations were operating in a "limited operational status" (Cone, 1972). These were located at Bratland, Norway; Trinidad, West Indies; Haiku Valley, Oahu, Hawaii; and Forestport, New York.

6. Vampola and his colleagues (e.g., Vampola and Adams, 1988) have used a "trace back in longitude" technique to determine that longitude where the observed electron pitch-angle distribution was brought about (for easterly drifting electrons diffused into the drift loss cone) by a WPI initiated by a ground-based transmitter such as UMS (16.2/17.1 kHz). Using this technique they have often found it necessary to postulate nonducted gyroresonant

WPIs at great distances from the equator. For example, see Vampola and Adams (1988), Figure 7, where they present three examples where the L = 1.66 resonance for 195-keV electrons occurs at the equator whereas other resonance locations on L-shells, L = 2.28 and 4.48, occur at altitudes of 5000 to 6000 km. Lower (difference) frequencies would move the resonance condition closer to the equator and render the assumption of a non-ducted resonance less necessary (see Imhof et al., 1986, Figure 1).

7. Imhof and his colleagues (Imhof et al., 1983) have had similar problems when attempting to match sharp, L-dependent peaks in the energy of precipitating electrons, to the equatorial gyroresonance condition for particular transmitter frequencies and have postulated marked decreases in the equatorial electron densities from the standard models normally used. Lower (difference) frequencies may sometimes make this unnecessary.

4.2. SFERIC CONTRIBUTION?

It is unlikely that sferic impulses make a significant contribution to the *mean* signal intensity and occurrence in the northern summer, especially if we note that the Ariel "mean reading" is the "mean of the logarithm of the signal," not the "logarithm of the mean signal," in each 28-s period. This has the effect of more strongly suppressing highly impulsive signals. Analysis of data from the sferic counting experiment on Ariel 4 (Bullough et al., 1975) supports this conclusion.

4.3. EMISSION INTENSITY AND CHARACTER
AT THREE INVARIANT LONGITUDES

Bullough et al. (1985) made a detailed comparison, at three key invariant longitudes, of the percentage occurrence of emission intensities on the 3.2-kHz peak, mean, and minimum sampling circuits for medium and high invariant latitude intervals: $45° < \Lambda < 55°$ and $55° < \Lambda < 65°$, respectively.

The three invariant longitude intervals were:

1. Northeast U.S. (335 to 10°E); centered on the main (PLHR + XM) zone and including the conjugate (Roberval) to Siple, Antarctica.
2. West Atlantic (10 to 40°E); this included Newfoundland and the conjugate to Halley, Antarctica, and is relatively free of significant PLHR and XM sources.
3. Yakutsk (185 to 215°E); this had few, if any, industrial sources of PLHR and a conjugate mainly to the south of Australia. There may have been early testing of a Soviet navigation transmitter (Alpha) at Komsomolskamur (identified on Aureol 3, Parrot, 1990) though this seems unlikely. The geomagnetic field and geographic cords made it comparable to Siple, such that conditions (magnetic field geometry, sunlight) in the Siple winter and summer should be similar to those in the corresponding Yakutsk winter and summer, apart from the presence/lack of strong PLHR sources. Yakutsk has a slightly lower invariant latitude than Siple and Halley.

The stations (Table 10.5.1) and locations of these medium and high invariant latitude zones are shown on the map (Figure 10.4.1) and the percentage occurrences in 5-dB intervals of emission intensity in Figure 10.4.3A (medium invariant latitudes) and Figure 10.4.3B (high invariant latitudes). These analyses make it easier to compare satellite and ground data at Halley, Siple, and Yakutsk.

In Figure 10.4.3 the standard error lies between ±1 and 1.5% for the range 10 to 30%.

4.3.1. Medium Invariant Latitudes ($45° < \Lambda < 55°$)

For zones A2, A2*, but not B2, B2* over the West Atlantic, northern summer peak, mean, and minimum profiles are remarkably similar in north and south, where the 1- to 2-dB displacement

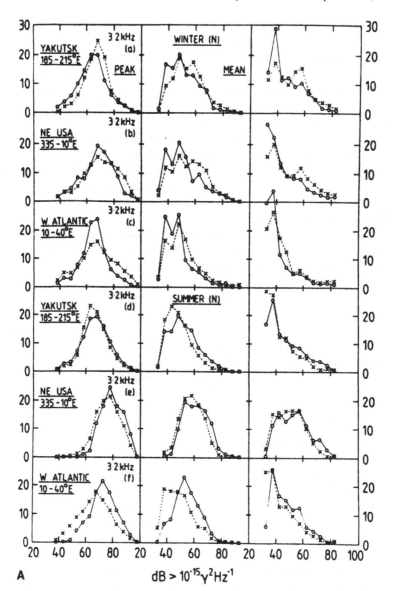

Figure 10.4.3 (A) The percentage occurrence in 5-dB intervals of emission intensity at 3.2 kHz, (45° < Λ < 55°) in the invariant longitude intervals: Yakutsk (185 to 215°E); northeast U.S. (335 to 10°E); West Atlantic (10 to 40°E). Ariel 4: 1971 to 1972 north winter; 1972 north summer; Kp ≤ 2+. North, (o-o); South, (x-x). (B) As above but for higher invariant latitudes: 55° < Λ < 65°. Reprinted from *J. Atmos. Terr. Phys.* Vol. 47: Bullough, K., Kaiser, T. R., and Strangeways, H. J., pp. 1211–1223. Unintentional manmade modification effects in the magnetosphere. Copyright (1985) with permission from Pergamon Press Ltd., Headington Hill Hall, Oxford, OX3 0BW, U.K.

of the summer profile to greater intensities may be attributed to a higher local refractive index. By contrast, in zones B2, B2*, over the West Atlantic, there is a marked signal drop in the south which may be due to LHR reflection. For data taken at all longitudes, the range of latitude and local time (mainly evening) in which the 3.2-kHz intensity is reduced (Lefeuvre and Bullough, 1973), is in accordance with the study by McEwen and Barrington (1967) of lower hybrid resonance noise bands observed by Alouette.

From this we conclude that in 1972 northern summer, a principal role of the PLHR (+XM) may have been to sustain duct structure and multihop propagation which is relatively much rarer over the Atlantic (Bullough, 1983).

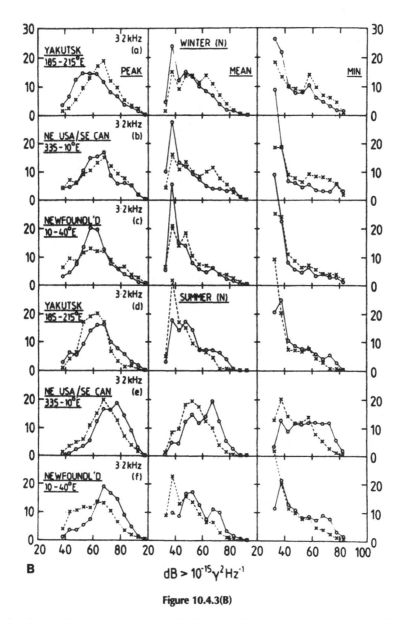

Figure 10.4.3(B)

Note that the minimum reading, even with its short (0.1 s) time constant, rarely falls to noise level in the A2, A2* zones in the northern summer and indicates the greater occurrence of sustained intense emissions.

It is also interesting to compare the northeastern U.S. and western Atlantic to the zone centered on the longitude of Yakutsk. The Yakutsk southern conjugate zone is centered on a low geographic latitude of 36.5°S, to the south of Australia, similar to that (35°N) for the northeastern U.S. If we compare these two zones in their local summer, then high peak and mean signals occur much more frequently in A2, A2*. However, the minimum signals have a somewhat similar double-peaked distribution, that at Yakutsk being particularly marked. The peak in occurrence at B = 55 to 60 dB > 10^{-15} γ^2/Hz^{-1} occurs elsewhere on Earth (see also Figure 10.4.3B) and suggests that once a sustained emission has attained a certain intensity then it has a tendency to continue to grow into a more intense, and/or more sustained, signal. The threshold intensity at which growth to the higher intensity occurs may be related to the threshold intensity (0.5 to 1 W radiated

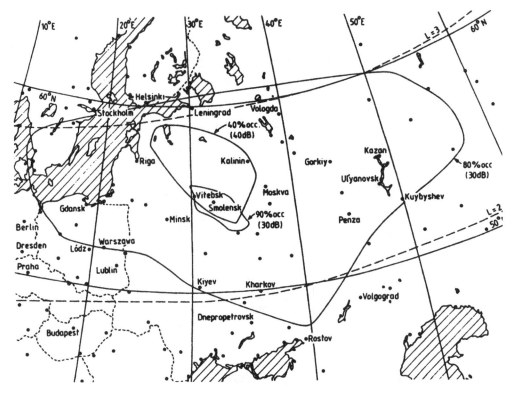

Figure 10.4.4 The PLHR (+XM) zone to the SSE of Madagascar (see Figure 10.4.2A) mapped back along the field lines to an altitude of 80 km over Eastern Europe and the former Soviet Union in the north. The zones correspond to 80 and 90% occurrence of 3.2-kHz emissions of intensity greater than 30 dB above 4.8×10^{-19} Wm^{-2} Hz^{-1} and 40% occurrence for 40 dB above 4.8×10^{-19} Wm^{-2} Hz^{-1}. The L = 2, 3 contours (45°, 55° Λ) at 100 km are shown. Reprinted from *J. Atmos. Terr. Phys.* Vol. 47: Bullough, K., Kaiser, T. R., and Strangeways, H. J., pp. 1211–1223. Unintentional man-made modification effects in the magnetosphere. Copyright (1985) with permission from Pergamon Press Ltd., Headington Hill Hall, Oxford, OX3 0BW, U.K.

power) found by Stanford colleagues to be required for triggering of VLF emissions at Siple (see Sections 2 and 6, II/13). It may also be a propagation effect, whereby the precipitation of energetic electrons leads to an enhancement of ionization in the ionosphere and backward diffusion of cold plasma up the field line. Duct enhancement (see Lefeuvre and Bullough, 1973) may strengthen the WPI (near the equator) by maintaining the wave normal more accurately oriented along the local field line and, simultaneously, reduce or remove LHR reflection of a downcoming whistler wave above the satellite.

4.4. SOUTH OF MADAGASCAR (PLHR + XM), CONJUGATE TO THE CIS AND EASTERN EUROPE

On the map of Figure 10.4.2A, a second PLHR zone (S1, south of Madagascar) which was barely detectable in 1967, may be attributed to the increased PLHR (+XM) (XM = sum and difference frequencies) from sources in the former USSR (now the Commonwealth of Independent States: CIS) and Eastern Europe (Bullough et al., 1985). In Figure 10.4.4 we have mapped the contours of this zone along the geomagnetic field to the mean base of the D region at 80-km altitude in the north. The percentage occurrence contours mapped across are those of 80 and 90% for the intensity level: 30 dB above 4.8×10^{-19} Wm^2 Hz^{-1} and that of 40% for the 40-dB level. The invariant latitude contours $\Lambda = 45$ and 55°, corresponding to L = 2 and 3 at 100-km altitude, are shown.

As in the case of the American zone, there is a tendency for the more intense emissions to occur at greater L. The 30- and 40-dB maxima occur on field lines to the geomagnetic north of the principal Soviet iron ore production region centered on Dnepropetrovsk. There is also extensive heavy industry at the steel production center to the north of the Black Sea, extending from Odessa in the west to Vologograd in the east and Kiyev and Kharkov in the north. The eastern wing of the 80% (30 dB) zone extends over the longitudes 50 to 60°E (Ul'yanovsk, Kazen, and Kirov) where, again, many steel and iron ore production centers are located. The western wing may be attributed in part to northward propagation of PLHR from the East European industrial areas.

It is interesting to note that the 80% (30-dB) zone lies largely over geological structure similar to that in southern Canada. In this zone, thunderstorm activity (May, June, July) in the north and south is relatively low compared to that in the southern summer when the southern zone is well-known as one of high activity (Volland et al., 1987; WMO, 1956). Since the latter is located over the sea where there are no complicating factors (local topology, heat islands, etc.), it will be particularly interesting to monitor its secular change and see to what extent this may be coupled to increasing PLHR (+XM) in the north. Thunderstorms were not reported in this region prior to 1925 (Brooks, 1925). Locations of northern cement works (grinding mills) were not known at the time of this writing.

Generation of difference-frequencies by UMS, Alpha Krasnodar, and other VLF transmitters in the Moscow region, is possible.

4.5. HIGH INVARIANT LATITUDES ($55° < \Lambda < 65°$)

As mentioned earlier, the localization of VLF emission activity over North America in 1972 was most pronounced for weaker emissions during quiet periods (Kp $\leq$ 2+) and at lower invariant latitudes and L-shells ($2 < L < 3$). However, the more intense emissions (in the same quiet periods) while still most frequent over North America, are displaced poleward to L > 3 (>55°Λ). Winter data (Kp $\leq$ 2+) then reveal that power lines to cement works are particularly strong sources of PLHR (Section 2.2.3) where most of the stronger events probably occurred in the quiet aftermath of geomagnetic storms. The scarcity of the stronger events over the Pacific indicates that the onset and stimulation of magnetospheric WPIs are perhaps more dependent on sources of triggering signals in the atmosphere than we have thus far acknowledged (see Section 6). Remember, however, that during great storms the activity becomes world-wide and, in terms of integrated energy, exceeds that occurring during quiet periods.

In Figure 10.2.6, in the northern summer (April 26 to July 31, 1972), there is a broadband of emission extending across southern Canada which is roughly symmetrical about the sectors: Edmonton (E), Winnipeg (W), and Montreal (M). The whistler-mode signal is clearly present in the south and accurately conjugate to that in the north. For the same percentage occurrence, the northern peak and minimum signals are about 10 dB more intense. The ratio peak/minimum, north and south, for the same percentage occurrence, is about 20 dB reducing to 10 dB at greater signal intensity in the north. This indicates in both north and south, a more intense signal superimposed on a weaker, continuous background and several propagation/amplification possibilities which include:

1. A strong ground signal in the north which is weakened after one-hop whistler-mode propagation to the south.
2. A weaker ground signal which is amplified on both the first and the second hop back to the north.
3. The most likely: one- and two-hop propagation with progressive emission stimulation and amplification to stronger signals, to yield the flat-topped distributions in the north: peak reading > 80 dB and minimum reading > 70 dB. These flat-topped maxima suggest that we are approaching the "strong diffusion (pitch-angle) limit" (Chapter II/13) where the energetic electrons, immediately outside the plasmapause, would give rise to the strongest AA (AA > 1 dB) events (see Hargreaves and Bullough, 1972).

The data selected for the winter period (December 21, 1971 to March 19, 1972) have been extended into early spring in order to obtain full local time coverage. Note, however, that the Winnipeg grinding mill (possibly also Edmonton, Montreal) was switched off in January and February. This makes the prominence of Winnipeg even more remarkable (see Section 2). Comparing the summer period, the activity over southern Canada is greatly decreased and there is no longer good conjugacy between north and south. However, sector W (and to a lesser extent, E and M) is an exception, being clearly present in both north and south with a peak/minute ratio decreasing from about 20 dB to 10 dB with increasing intensity. Thus, as in summer, we have good ducting with amplification between the two hemispheres, but the phenomenon is much more spatially localized.

This marked seasonal change is partly due to the greater ease with which the upgoing PLHR signal may be entrapped in field-aligned enhancements of ionization (ducts) in the local summer (Tatnall et al., 1983; Strangeways and Rycroft, 1980; Strangeways, 1981ab; (Chapter II/7)). Secondly, conditions for a WPI and downward ducting of the two-hop signal will also be better in the sunlit northern hemisphere. Thirdly, difference-frequencies generated by the VLF transmitters Omega North Dakota and NAA/NSS in the 1972 summer may have complemented the PLHR signals in the 3.2-kHz channel so as to "fill in" the east-west gaps in the wavefield above the ionosphere between cement works.

Note that, in the winter, in the north and, to a lesser extent, in the south, there is a tendency for signal intensity to rise abruptly (see Figure 10.2.6B, >70, 80 dB peak and >60, 70 dB minimum) at the cement works, especially W, and then fall steeply to a value lower than normal background. This can be attributed to the following process: energetic electrons after substorm injection at local midnight onto these L-shells ($3 < L < 5.6$) will drift eastward around the Earth with a period:

$$P \sim 44/LE \text{ min, where } E = \text{electron energy (MeV)}$$
$$= 4 \text{ h } 35 \text{ min for } E = 0.045 \text{ MeV, } L = 4.$$

Thus, a 45-keV electron will require 7.6 min to traverse a 10° longitudinal sector. Such electrons, during the WPI, have their pitch angles decreased so as to lower their mirror points and precipitate into the atmosphere. These electrons, if removed in the Winnipeg sector, reduce the possibility of a strong WPI to the east and hence the percentage occurrence of such emissions will tend to drop below background. As a result, emission occurrence, while it is greatest in the W sector, is simultaneously least in the longitude sector immediately to the west of Montreal (M). Note the evidence for adiabatic preservation of gyrophase coherence (Cornwall and Schulz, 1970) as the electrons drift eastward from W. As we move eastward into sector M, a further strong PLHR source (or sources), on a lower L-shell, again increase(s) the percentage occurrence abruptly, as at E and W. The electrons participating in the WPI induced by the PLHR from Montreal will be those drifting on an L-shell below that grounded on Winnipeg. The standard errors in Figure 10.2.6 indicate that none of these features can be ascribed to pure chance.

While the most intense "mean reading" signals (>60 dB) occur at these higher latitudes (as one might expect), the minimum readings, by contrast, more frequently fall to receiver noise level, except for northeast U.S./southeast Canada in the summer (see Figure 10.4.3B). There is a greater difference between north and south and multihopping presumably occurs less frequently. The larger magnetic component observed in the summer hemisphere can again be attributed mainly to the larger local refractive index. Note especially the higher occurrence of the most intense and sustained emissions in North America and, for the minimum readings, the similar peak in the profiles at 55 to 60 dB for the southern conjugate of the northeast U.S./southeast Canada longitude in northern summer and both hemispheres at the Yakutsk longitude in the northern winter.

Parrot (1990) found a decrease in ELF (B_y; 800 Hz) emissions to the east of the South Atlantic anomaly (SAA). This was less evident at 750 Hz (Ariel 4, 1972 summer: Tatnall et al., Figure

Table 10.5.1 **Locations of Ground Stations**

Station	Geographic		Invariant	
	Longitude	Latitude	Longitude	Latitude
Halley	26.6°W	75.5°S	31°E	61°S
Eights	77°W	75°S	9°E	61°S
Siple	84°W	76°S	7°E	61°S
Yakutsk	130°E	62°N	198°E	56°N
Faraday	64°W	65°S	13°E	49°S

2(iv)a) and, in 1967 summer, there was preferential amplification at 17.8 kHz (NAA) and 16 kHz (GBR) (see Bullough, Tatnall, and Denby, 1976; Denby et al., 1980) in this longitude sector. This requires further investigation.

5. GROUND-BASED OBSERVATIONS OF MAGNETOSPHERIC LINE RADIATION (MLR), PLHR WHISTLER-MODE SIGNALS

We are initially interested in MLR phenomena, of which PLHR-stimulated emissions are a subset, at the higher invariant latitudes (>55°). Whistler observations at Antarctic stations such as Eights, Siple, Halley, and SANAE, located close to $L = 4$, $\Lambda = 60°$, indicate reception of whistler-mode signals which have propagated along field lines $2.5 < L < 6$, $51° < \Lambda < 66°$ (Chapter II/7). Some locations of ground stations are listed in Table 10.5.1 below.

5.1. GENERAL CHARACTERISTICS OF MLR SIGNALS

Magnetospheric line radiation has the following characteristics (Helliwell et al., 1975; Park, 1977; Park and Helliwell, 1978; Mathews and Yearby, 1980, 1981; Yearby et al., 1981; Vershinin et al., 1983; Adrianova et al., 1977):

1. MLR normally consists of an array of lines (Figures 10.5.1 and 10.1.2) lying in a frequency range from 1 to 8–11 kHz but with a strong, well-defined peak in occurrence at 3 to 3.5 kHz (Figure 10.5.2). The low frequency cutoff is very steep, dropping to zero just above the ELF plasmaspheric hiss and chorus. The line widths are typically 30 Hz, much wider than normal power induction lines, and the lines themselves, though they may originate in PLHR, normally differ in frequency from power line harmonics and often exhibit a slow upward drift in frequency, as seen in Figure 10.5.1. These sustained emissions are usually modulated at the two-hop period indicating excellent ducting between hemispheres. Chorus elements (risers, hooks, etc.) are triggered, or modified, at the line frequencies. Multihop whistlers are also common. Only at Siple, where MLR events occur more frequently (perhaps ~3x) and last longer than at Halley, do the line spacings show a tendency to cluster around 120 Hz, thereby giving support to a PLHR origin. At Halley we have a broad distribution centered on 50 to 90 Hz and at Yakutsk, Vershinin et al. (1977) obtained a very wide spread, from 40 up to 400 to 600 Hz.

2. An outstanding difference between the three stations is that of the diurnal variation in occurrence. This is shown in Figure 10.5.3 for several years' data from Eights (1965) and Siple (1973 to 1976) (E/S), and Yakutsk (1973 to 1982) (Y). The relatively limited data from Halley exhibit the same peak in afternoon occurrence as at Yakutsk. This provides perhaps the clearest evidence of greater PLHR influence on MLR and emission stimulation at E/S invariant longitudes.

Park and Helliwell (1978) suggested that the sharp increase in E/S PLHR activity at 0.5 MLT is due to a sudden increase in power usage at that time and that the gradual decrease in PLHR

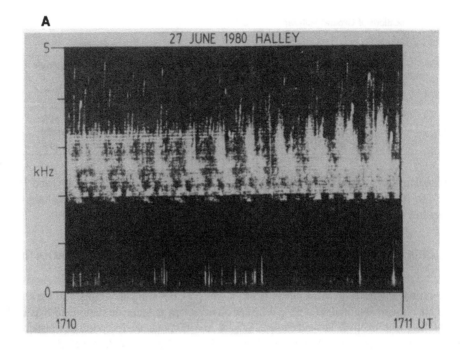

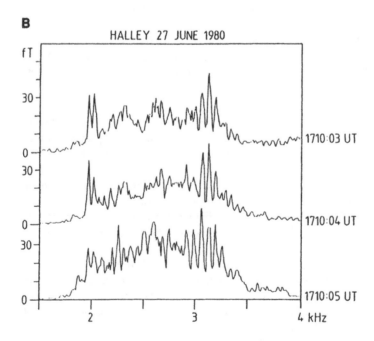

Figure 10.5.1 A spectrogram of an MLR event received at Halley on June 27, 1980. (A) The spectrogram shows a well-defined array of magnetospheric lines which are gradually drifting up in frequency at ~80 Hz/min and lie mainly in the frequency range: 2 to 3.5 kHz. Intensity modulation at the two-hop period is clearly visible. (B) Amplitude (in a 10-Hz bandwidth) against frequency; spectra averaged over three successive 1-s intervals for the above event. (Kindly provided by my colleague Dr. Keith Yearby, Department of Control Engineering, University of Sheffield, S3 7RH, U.K.)

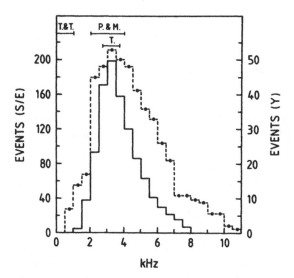

Figure 10.5.2 Frequency distribution of MLR- and PLHR-induced events observed at Siple and Eights, Antarctica. The frequency ranges of studies by Park and Miller, (1979; Sunday decreases), Thorne and Tsurutani, 1981; OGO 5, 6 ELF data) and Ariel 4 are shown. Superimposed are MLR events recorded at Yakutsk, 1973 to 1982 where there is virtually no industrial activity. These figures contrast data from Siple and Eights with those from Yakutsk. The latter data were provided in a short internal report (unpublished) by Vershinin et al. (1983). The histograms of Siple/Eights data are from Park and Helliwell (1978): Magnetospheric effects of power line radiation, *Science*, 200, No. 4343, p. 727 and appear with permission of the AAAS (copyright 1978).

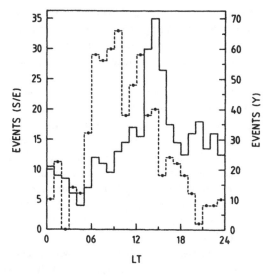

Figure 10.5.3 The diurnal variation in occurrence of MLR/PLHR events at Yakutsk and Siple/Eights. These figures contrast data from Siple and Eights with those from Yakutsk. The latter data were provided in a short internal report (unpublished) by Vershinin et al. (1983). The histograms of Siple/Eights data are from Park and Helliwell (1978): Magnetospheric effects of power line radiation, *Science*, 200, No. 4343, p. 727 and appear with permission of the AAAS (copyright 1978).

activity throughout the afternoon hours is due to "limited accessibility of the afternoon-evening sector to the energetic electrons that resonate with the PLHR waves." However, the sudden increase in power usage is less than the increase (>10 dB) in D-region absorption of the upgoing signal at dawn and therefore this sharp increase at dawn must be due primarily to industrial processes which generate more intense PLHR.

It must be noted that Figure 10.2.6 is for *quiet* periods (Kp ≤ 2+) and the satellite data do not indicate as marked a longitude difference (E/S:H) in signal intensity/occurrence above the ionosphere as the MLR data would lead us to expect (see Section 5.2). Maintenance of permanent duct structure at the Roberval/Siple invariant longitude relative to Newfoundland and Halley (Section 4 and Bullough, 1983) will help to maintain downcoming wave normals within the transmission cone. The similar afternoon peaks in occurrence at Yakutsk and Halley may be due to the greater probability of lightning-induced duct creation in the summer hemisphere as we move into the local evening maximum of thunderstorm activity (Chapter II/7; Park and Helliwell, 1971; Thomson, 1978).

The Ariel 3 and 4 satellite data, while unsuitable for detailed MLR studies at any particular invariant longitude, nevertheless, does support the PLHR (+XM) hypothesis by indicating that the baseline signal detected over the northeastern U.S. varies diurnally in accordance with daytime D-region absorption (Tatnall et al., 1983).

5.2. THE SUNDAY EFFECT AT SIPLE/EIGHTS AND HALLEY

Park and Miller (1979) have identified a significant decrease in the occurrence and mean intensity of VLF emissions in the frequency interval 2 to 4 kHz recorded on Sundays in Eights and Siple, relative to the rest of the week. The tape recordings were analyzed using a minimum reading circuit with time constants of 1 and 0.1 s for positive- and negative-going signals, respectively. This reduced the contribution of sferics to the measured mean amplitude (Figure 10.5.4).

There is an important feature of the S/E Sunday effect which merits more careful examination; this is the strong dependence on magnetic disturbance (Kp) shown in Figure 10.5.4b,c. The Sunday effect, corresponding to a signal decrease in the period 06 to 18 MLT relative to the rest of the week (ROW), is not only missing during *quiet* periods but even has a tendency to reverse in the period 18 to 06 MLT. This dependence on Kp may be explained if we take account of strong industrial PLHR sources at invariant longitudes to the west in southern Canada (e.g., E, W, and M in Figures 10.2.6 and 10.2.4).

Note that it is even possible for such a source to produce an *increase* in Sunday activity at a ground station to the east if it is switched off on Saturday night through Sunday. Let us assume a quiet period of low energetic-electron flux drifting eastward. The threshold signal-level required for a wave-particle interaction (WPI), destabilization, and precipitation will increase with decreasing flux. Thus, if the strong PLHR source is preferentially switched off before Sunday morning, there may be a greater possibility of a low flux of resonant electrons, injected at local magnetic midnight, drifting eastward to the E/S longitude without prior precipitation. On Saturday night and early Sunday morning, before sunrise, upgoing PLHR (+XM) signals will be greater and, therefore, the probability of exceeding the threshold at the Siple longitude may also be greater than for the daytime (Figure 10.5.4c).

The satellite data indicate that PLHR, XM, lightning, or other signals may be less important, perhaps unnecessary, for emission stimulation during great storms (e.g., Kp ≥ 6), but the Siple statistics suggest that during and following moderate activity (3− ≤ Kp ≤ 5+) they retain a key role. During such periods, energetic electrons are being continuously injected during substorm activity to levels above the threshold for precipitation by PLHR sources such as Alcatel (near Roberval) and the Montreal cement works. Absence of a suitable PLHR source at the Siple conjugate will then lead to the observed Sunday decrease.

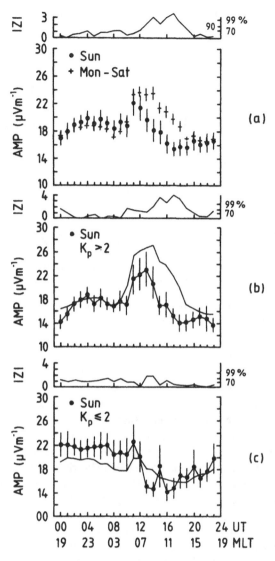

Figure 10.5.4 Daily variations of 2- to 4-kHz wave amplitude showing differences between Sunday and Monday through Saturday. The vertical bars indicate ±1 standard error of the mean. The curve at the top shows the magnitude of the z score, which is related to the confidence level of statistical significance. (a) All data, Eights and Siple. (b, c) The data in (a) are divided into two groups according to the Kp index: (b) Kp > 2, (c) Kp ≤ 2. (This is a composite figure, assembled from figures in Park, C. G. and Miller, T. R., (1979). Sunday decreases in magnetospheric VLF wave activity, *J. Geophys. Res.*, 84(A3), 943. Copyright by the American Geophysical Union. With permission.)

The weaker MLR activity at Halley relative to Siple may be explained by assuming:

1. Strong PLHR (or XM) sources at the Siple longitude and L-shell
2. Greater interception and precipitation of eastward-drifting electrons prior to their arrival at the Halley longitude
3. Halley lies on the edge of the South Atlantic precipitation region of the ''drift loss cone'' where electrons are lost when mirror points (in the south) dip below 100-km altitude
4. More sustained ducting at the S/E longitude (Section 4)

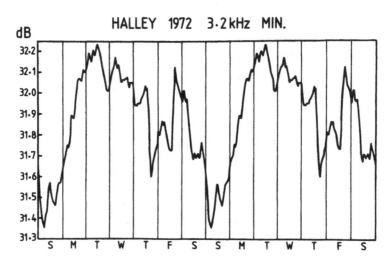

Figure 10.5.5 Average hourly measurement at Halley, for each day of the week, of 3.2-kHz minimum signal intensity from January to February, 1972. Two cycles are shown. In addition to the Sunday minimum, two smaller minima on Thursday and Friday afternoons may be due to the normal weekly maintenance periods of NSS and NAA, respectively. [An original figure, kindly provided by my colleagues, Drs. John P. Mathews and Keith Yearby (formerly members of the Space Physics Group, Sheffield University)].

The Sunday effect was also observed at Halley (Figure 10.5.5) by Yearby et al. (1981). This analysis included all Kp for the period January to November, 1972. A true minimum reading circuit [time constants: ∞ (+), 0.1 s (−)] at 3.2 kHz ($\Delta f = 1$ kHz), similar to that used on the satellites Ariel 3 and 4, sampled the signal level present in a 30-s period, every hour in this period. The decrease (~0.5 dB) in mean signal level is similar to that for Siple (Figure 10.5.4a) though the number and duration of events are greatly reduced. Perhaps the most powerful Newfoundland PLHR source for Halley is a small cement works (Figure 10.2.4); other industrial sources, initially considered more likely, were found to radiate too little PLHR (Section 2). For Halley, the deviation of Sunday intensities from the grand mean was 0.52 dB, whereas the standard error of each mean daily value was only 0.12 dB. The same analysis performed on an ELF channel at 0.75 kHz ($\Delta f = 0.5$ kHz) did not show any Sunday effect.

A key feature found in the Halley data were decreases in the 3.2-kHz minimum VLF intensities on Thursday and Friday afternoons. The latter coincided with the maintenance period of the NAA transmitter and the former probably with that of NSS (awaiting verification). This probably indicates coincident cancellation of the difference-frequency (3.6 kHz) produced by mixing of the NAA (17.8 kHz) and NSS (21.4 kHz) transmissions in the ionosphere (Section 4.1). This would suggest that the NAA/NSS difference-frequency signal intensity generated in the Halley ionosphere was similar to, or possibly larger than, that of any PLHR signal, but not that at Siple.

Molchanov et al. (1991) have found a marked weekly variation in the Aureol 3 electric field components, Ez and Eh. It appears that there is a possible 2:1 variation in the amplitude of Ez at 72 Hz with a maximum (1.1×10^{-4}Vm^{-1}Hz$^{-1/2}$) on Monday and minimum (~0.6 × 10^{-4}Vm^{-1}Hz$^{-1/2}$) on Saturday (their Figure 3). They attribute this large variation not only to lower power consumption by industry but also make the ingenious suggestion that the current distribution at the surface of the Earth must be different at weekends; a ring- and octopus-shaped distribution on weekdays and weekends, respectively. This is used to explain why a small decrease (<20%) in total electric power production may lead to a large decrease (~2 times) of the electrostatic turbulence intensity. The magnetic field strength at the fundamental frequencies, 50 and 60 Hz, has been observed by Tomizawa and Yoshino (1985).

The importance of this electrostatic turbulence which, presumably, acts in parallel with the continuous weak diffusion process of Lyons et al. (1972), remains to be evaluated. The evidence

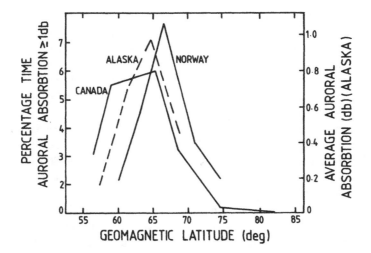

Figure 10.6.1 The location, in geomagnetic latitude, of the AA zone in Alaska, Norway, and Canada. The Alaskan data show the mean magnitude of AA, while the other curves show the percentage of time for which the absorption equaled or exceeded 1.0 dB. Note that, of the three lower latitude riometers in the Canadian chain, Ottawa (45.4°N, 75.9°W) was located in sector M while Val d'Or (48.1°N, 77.8°W) and Cape Jones (54.6°N, 79.8°W) were both located in the invariant sector (350 to 360°E); see Figures 10.2.5 and 10.2.6. Canada: July 1959 to June 1961 (Winnipeg Cement probably began operations in this period); Norway: October 1958 to June 1959; Alaska: three days in April 1958. From Hartz, T. R., Montbriand, L. E., and Vogan, E. L., (1963). A study of auroral absorption at 30 Mc/s, *Can. J. Phys.*, 41, 581. Permission granted by the publishers (copyright 1963).

for transmitter effects, as pointed out by Parrot (1990), would tend to support Vampola's hypothesis (Section 4.1) that pitch-angle scattering is dominated by discrete events.

6. PLHR, THE WAVE-PARTICLE INTERACTION (WPI) AND RELATED PHENOMENA

In this section we examine the relationship between the VLF (3.2 kHz) wave intensity and the flux of precipitating energetic electrons which gives rise to auroral absorption (AA) by increase of electron density in the D region. The stronger events (AA > 1 dB), corresponding to precipitation of electrons($E > 20$ to 40 keV), occur outside the plasmapause which, during disturbed periods ($Kp \geq 5_0$), may be displaced to lower latitudes with a mean location at 1800 LT of <60.1° ± 1.8° decreasing to <57.4° ± 1.8° at 0600 LT (Rycroft and Burnell, 1970). Bremsstrahlung X-rays generated in the D region may then penetrate to lower heights and significantly increase the atmosphere conductivity down to 30 to 40 km altitude (Chapters I/10; I/1; Bullough, 1988; Berger and Selzer, 1972; Bering et al., 1988).

6.1. RIOMETER OBSERVATIONS OF THE AURORAL ABSORPTION (AA) ZONE

An interesting feature of the AA zone over Canada (Figure 10.6.1), almost certainly due to PLHR, is that its maximum occurs at a significantly lower geomagnetic latitude (63°) than in Alaska (65°) or Norway (67°) (Hartz et al., 1963; Hargreaves, 1966: Figure 4). The zone is also broader, extending to lower latitudes in a similar fashion to the stronger VLF emissions.

During the period from July 1959 to June 1961 the absorption exceeded 1 dB in 7.8% of 30-min periods at 60° geomagnetic latitude and for 4% of such periods at Ottawa (56.4°N, geomagnetic).

The AA zone has a sharp lower edge, located just poleward of the plasmapause. Direct evidence for this is provided by satellite observations (Holzer et al., 1974) of high energy (>45 keV) electron precipitation, occurring in conjunction with chorus. This lower edge is displaced equatorward for high magnetic disturbance, the zone maximum being overhead at Baie St. Paul (47°22'N, 70°33'W; 60°Λ) for Kp = 7 (Ecklund and Hargreaves, 1968, Figure 14).

The absorbing region has typical dimensions, E-W and N-S, of 750 and 470 km, respectively, during the day and 1500 and 650 km during the night. Using a four-corner-reflector antenna array, Ecklund and Hargreaves (1968) failed to identify any cross-meridian movement of an AA region; thus the individual event appeared to move with the Earth and had typical durations (for AA > 0.3 dB) of 4 h (day) and 3 h (night).

At near conjugate Baie St. Paul and Eights (L = 4), the maxima in AA median event intensity in 1963 to 1965 occurred at 06 to 08 h and 18 to 20 h UT with a quite significant increase in intensity for Kp ≥ 5_0 (Hargreaves and Cowley, 1967; Figure 6). At Baie St. Paul, LT = UT − 4 h and 42 min. This would yield an AA maximum in the early afternoon, close to that of the MLR at Halley and Yakutsk.

Hargreaves et al. (1987) in a more recent study of AA using a meridional chain of riometers in Finland (from 1972 to 1983), found that the daily maximum occurred between 0500 and 0800 MLT at magnetic latitudes greater than 62°, but at lower latitudes the weaker absorption showed a daily maximum several hours later, between 1000 and 1400 MLT (see their Figure 8). For greater magnetic disturbance (Ap > 20), when the activity moves to lower latitudes, it is reasonable to associate the Yakutsk afternoon peak in MLR occurrence with that of the maximum percentage occurrence of AA > 1 dB [AA(Q1)]. Note the importance of distinguishing the diurnal variation in median event intensity from that of event occurrence where the former tends to precede the latter at all latitudes (Hargreaves and Cowley, 1967: Figures 1 and 3). Note also that the day events tend to be longer and less structured than those at night.

PLHR from the cement works in southern Canada provides the most probable source for stimulation of strong WPIs and hence, AA > 1 dB, at invariant latitudes down to 55°. The latter latitude defines a key boundary between medium and high latitude for sun-weather relationships. The strong WPI localization near Canadian cement works, so clearly evident in the winter of 1971 to 1972, makes necessary a reappraisal of AA studies, particularly those undertaken in Canada.

6.2. THE WAVE-PARTICLE INTERACTION

The condition for gyroresonance is such that, just within the plasmapause, because of the marked increase (10 to 100×) in electron density at the equator, relatively low energy electrons (<10 keV) are precipitated (Chapter I/13; Rycroft, 1976; Inan et al., 1978). These do not yield D-region absorption (AA > 1 dB) or Bremsstrahlung X-ray flux of sufficient strength to modify significantly [relative to galactic cosmic rays (GCR)] the atmosphere conductivity (Chapter I/1, I/10).

PLHR consists of a train of narrow impulses with pulse separation of $(2 fm)^{-1}$ where fm (50, 60 Hz) is the main frequency (Figure 10.2.3). Thus, the conditions for a strong WPI for these "man-made sferics" will be similar to those pertaining to sferics of natural origin. Yoshida et al. (1983) have applied the two resonance conditions (Helliwell, 1967) for phase bunching of electrons by a one-hop whistler originating in a sferic and reach the following conclusion for an L = 4 field-line: the strength of the WPI has a strong maximum in the range 3 to 3.5 kHz for an interaction region located some 15° on the upstream side of the geomagnetic equator.

This is in excellent agreement with the frequency distribution of line radiation events observed at Siple, Eights, Halley, and Yakutsk (Figure 10.5.2).

Siple pulse injection experiments (Helliwell et al., 1990) have demonstrated that optimization of the frequency/time curvature of the input pulse (for first- and second-order resonance) may result in more rapid whistler-mode wave growth. See also Carlson et al. (1990) and earlier work by Chang et al. (1983). Carlson et al. (1990) found that their simulations of the whistler-mode wave growth process, based on test particle trajectory calculations, showed no indication of an amplitude threshold for growth as the applied wave intensity was varied. They conclude that "this suggests that the threshold phenomenon (Helliwell et al., 1980; Section 4) might be associated with the interfering noise that is always present in the magnetosphere."

Tatnall et al. (1983) showed that it is in the critical region between the largest pitch angles where Landau resonance with the ELF hiss is important and the lower, where cyclotron resonance with the ELF hiss dominates pitch angle diffusion, that PLHR (and, since our reappraisal, XM) WPIs may occur.

Following our reappraisal we may modify the conclusions reached by Tatnall et al. (1983) to conclude:

1. XM and PLHR interact with electrons of energy, $E > 100$ keV, to augment the pitch diffusion process and contribute to the formation of the electron slot between the radiation belts.
2. The diffusion coefficient is comparable to that obtained for ELF hiss. Note, however, that the longitude intervals in which PLHR and XM are effective are limited; ELF hiss is ubiquitous.
3. The process becomes relatively more important for lower energies, on lower L-shells, and at lower electron densities.
4. Because the source of these electrons is cross-L diffusion from higher L-shells with high conservation of the first and second invariants (Schultz and Lanzerotti, 1972), then the flux at large pitch angles is increased, favoring a PLHR and/or XM interaction.

More recently, Sá (1990) has attributed the creation of triggered emissions, associated with monochromatic wave transmissions, to an interaction of the wave with PLHR, even though the intensity of the latter may be 50 dB less!

6.3. RELATED SATELLITE STUDIES OF WPIs AND ELECTRON PRECIPITATION IN THE SLOT REGION

Vampola (see Vampola and Adams, 1988) states: "In surveys made by one of us (ALV) of large magnetospheric electron data bases obtained . . . on various satellites, a recurrent observation has been that *electron precipitation occurs in "events" rather than in a slow diffusion process, especially in the slot region*" (author's emphasis). Vampola and Gorney (1983) made a statistical study of energy deposition in the middle atmosphere by precipitating electrons and observed that for "even as high as $L = 4$ the hemispheric pattern of precipitation was that which would occur if electron scattering from stable drift orbits into the loss cone were primarily large-angle scattering rather than a slow pitch angle diffusion." On the other hand, Imhof et al. (1986) found that "the slowly varying electron fluxes observed in the drift loss cone frequently display peaks in the energy spectra which indicate that they are precipitated by plasmaspheric hiss and that this process therefore represents a major loss mechanism."

Thus, there is clear disagreement concerning the relative importance of, on the one hand, discrete and burst-like events (lightning, chorus, triggered emissions, etc.) and, on the other hand, slow diffusion processes (mainly ELF hiss). If the energy loss by electron precipitation is simply related to that of mean wave power then the Ariel 3 and 4 mapping (also, Aureol 3) would indicate that loss mechanisms involving ELF dominate. This can be seen by comparing the global percentage occurrence of signals exceeding 40 dB above 4.8×10^{-19} Wm^{-2} Hz^{-1} at 0.75 ± 0.25; 1.25 ± 0.25; 3.2 ± 0.5; 9.6 ± 0.5 kHz (Tatnall et al., Figures 2(iv) a, b, d, and h). During magnetically quiet (Kp $\leq$ 2+) periods (64 to 72% of the time), the ubiquitous ELF clearly dominates all other emissions.

Our reappraisal of the Ariel 4 data, which indicates the probable generation of difference-frequencies, reinforces the importance of understanding man-made effects on the environment. On the assumption of first-order gyroresonance between wave and particle giving rise to pitch angle diffusion into the loss cone, both Vampola (1983, 1987) and Imhof and his colleagues (Imhof et al., 1983) have often found it difficult to reconcile the peak in electron energy with both transmitter frequency and the standard model of equatorial electron density/L. Imhof states:

> Since there are no known powerful transmitters in regular operation at frequencies below 10 kHz, the need for exceptionally low plasma densities cannot be avoided simply by invoking a lower wave frequency.

It is possible that at least some of the anomalous results obtained by Vampola and Imhof and their colleagues may be explained by the generation of far lower "difference" frequencies as postulated (Section 4.1). This possibility is easily visualized by referring to Figure 1 of Imhof et al. (1986) and Figure 2 of Imhof et al. (1974).

Observations at Moshiri (L = 1.6) of whistler-triggered VLF emissions in the electron slot and inner radiation belt regions (Hayakawa, 1991) would support the possibility that the frequency range of 2 to 4 kHz is more important for WPI, since the starting frequency of VLF emissions often lies in this range. Also, preliminary studies of ducted NAA, NSS signals received at Faraday show little evidence of wave amplification (Smith et al., 1987). WPIs involving NAA (17.8 kHz, FSK) appeared to occur mainly to the east of the SAA (Bullough et al., 1976; Davies et al., 1984).

It now seems probable that, during the summer of 1972, difference frequencies dominated the stimulation of 2 to 4 kHz emissions over North America at medium invariant latitudes (45° < Λ < 55°) and also contributed to the Soviet zone, south of Madagascar. Parrot (1990) has suggested that, during the Aureol 3 observing period (October 1981 to December 1983), ELF emissions could be associated with the Alpha transmitter at Komosomolskamur. Note, however, the nearby transmitter (UBE2: 14.3 to 17.9 kHz) at Petropavlosk which could have interacted with the latter.

6.4. CRITICAL LEVELS OF AURORAL ABSORPTION, 3.2-kHZ SIGNAL INTENSITY, AND ATMOSPHERIC CONDUCTIVITY

The storm-time variation in emission intensity is such that emissions occur first on the morningside of the Earth during the main phase and correlate closely with Kp and Dst. Strong emissions occur on the evening side of the Earth during the recovery phase a few hours later when Kp is often decreasing (Bullough et al., 1969). A statistical analysis by Rycroft and Burnell (1970) of the movement of the plasmapause indicates that, for Kp > 5_0, we would expect a displacement (>5° Λ) to lower latitude with a mean location <60.1° ± 1.8° at 1800 LT decreasing to <57.4° ± 1.8° at 0600 LT. This is consistent with the evidence for the occurrence of AA (> 1 dB) at Λ < 60° when Kp ≥ 5_0.

Hargreaves and Bullough (1972) found that the AA (dB at 30 MHz) at Byrd is proportional to the maximum mid-latitude rms VLF field strength at 3.2 kHz. The flux (J) of precipitated electrons (>40 keV) estimated from the 30-MHz auroral absorption (A) according to the relationship of Jelly et al. (1964):

$$A(dB) = 4 \times 10^{-3} [J (cm^{-2}s^{-1}sr^{-1})]^{1/2}; \tag{6.1}$$
$$= 1 \text{ for } J \sim 10^5;$$

yielding,

$$J = 3.6 \times 10^{12} I(3.2) (\gamma^2/Hz) \tag{6.2}$$

where I (3.2) is the intensity of 3.2 kHz emissions, and,

$$I = 72 \text{ dB} > 10^{-15}\gamma^2/Hz$$

The Kennel and Petschek (1966) trapping limit for >40 keV electrons, corresponding to pitch-angle isotropy and a VLF saturation intensity of $I(3.2) = 90$ dB $>10^{-15}\gamma^2$/Hz, is about 3×10^7 cm^{-2}s^{-1}.

An intense precipitation event, giving rise to significant riometer absorption and Bremsstrahlung X-rays was observed and studied in detail by Foster and Rosenberg (1976). This occurred at Siple (L = 4.2) where bursts of Bremsstrahlung X-rays were correlated with VLF emissions ("risers") in the frequency range 1.5 to 3.8 kHz. They deduced an e-folding energy of 45 keV, a flux ($\sim 10^6$ cm^{-2} s^{-1}) of precipitated electrons (>40 keV) and estimated that the stably trapped flux limit (Chapter I/13) at L = 4.2 was about 2×10^8 cm^{-2} s^{-1}.

From work by Berger and Selzer (1979) (see Chapters I/10, I/1), Bullough (1988) deduced that Bremsstrahlung X-rays from the stronger electron precipitation events (AA > 1 dB) give comparable or greater ion-pair production rates than the GCR background at altitudes above 40 km and, occasionally, down to 30 km. Solar proton events yield higher ion-pair production rates (Herman and Goldberg, 1978; Figure 4) but occur too infrequently; e.g., 40 PCA events in 1949 to 1959 (Jelly et al., 1962), and usually poleward of 60° Λ (Imhof et al., 1971).

The introduction of strong localized PLHR sources (especially cement works) which stimulate WPIs and AA events in southern Canada helps to explain why it has been so difficult to produce a coherent, long-term global picture of AA occurrence and its seasonal and longitudinal dependence; especially from the meridional riometer chains in Canada, Alaska, Finland, Norway, and the former USSR (Basler, 1963, 1966; Berkey et al., 1974; Foppiano and Bradley, 1985 and, most recently, Hargreaves et al., 1987). The Winnipeg power line not only gave rise to the most active field line, at 3.2 kHz, in the magnetosphere, with multihopping PLHR-stimulated emissions (and strong AA events) in both hemispheres (summer and winter) but also, by contrast, by removal of easterly drifting electrons created, at least in northern winter (1971 to 1972), the *least active* invariant longitude sector (350 to 360°); see Figure 10.2.6B.

Hartz et al. (1963) found a marked change in seasonal variation for the period from July 1959 to June 1961 (a strong maximum in autumn) relative to that (annually consistent equinoctial maxima with a marked minimum in *summer*) found at Ottawa by Collins et al. (1961): their Figure 6, for the period 1949 to 1959. This was probably due to the Winnipeg cement works which began operation in the early 1960s (Yakymyshyn, 1990).

Additional evidence for localized PLHR sources in Figure 10.2.6 include an occurrence maximum, in the northern summer, centered on Iceland (invariant longitude: 70 to 80°E) followed by a sharp minimum in the sector 80 to 90°E containing the U.K.; also possibly in Tasmania (220 to 230°E) in northern winter.

A global study by Agy (1975) on the geographic distribution of AA based on 60 "substorm events" in the IQSY (1964 to 1965) and the IASY (1969) identified two key maxima in AA occurrence near longitudes 60°E (Scandinavia) and 210°E (Alaska). This extreme nonuniformity in the longitude distribution, especially the Alaskan peak, is most easily explained if we assume that most of the AA events required ground-based or atmospheric sources to stimulate the WPI responsible for the AA event. Thus, energetic particles injected at local magnetic midnight, during substorms over the Pacific, tended to drift eastward to Alaska before precipitation; similarly for substorm injection over the Atlantic and Scandinavia.

6.5. CAN PLHR INFLUENCE THUNDERSTORM ACTIVITY?
THE SUN-WEATHER RELATIONSHIP

If the thunderstorm activity in southern Canada is linked to magnetospheric phenomena in the manner suggested by Markson (1978) and Herman and Goldbert (1978) (MHG): that is to say, by modulation of atmospheric conductivity and the atmospheric electric field, then (see Section 6.4) the only viable candidate, in terms of the range of invariant latitude involved (55° < Λ < 65°), ion-pair production rate relative to that of the background GCR, and frequency of occurrence, is that of the AA (> 1 dB) event. The latter is associated with planetary magnetic disturbance $K_p > 5_0$ and VLF emission intensity, $I(3.2 \text{ kHz}) > 10^{-7}\gamma^2$/Hz.

Bullough (1990) has shown that the spatial and secular changes (especially over North America: Figure 10.2.4; Changnon, 1977), solar cycle variations, and the correlation with sector boundary passes (SBP) of thunderstorm activity, are all consistent with the MHG hypothesis if we assume that: (1) at high latitudes ($\Lambda > 55°$) AA (>1 dB) events dominate the GCR background in modulating lightning occurrence when the meteorological conditions are appropriate and (2) at medium latitudes ($45° < \Lambda < 55°$) the GCR variations (Forbush decreases) modulate the occurrence of lightning discharges.

Direct evidence in support of the MHG hypothesis is provided by Armstrong (1987) (Chapter I/6) who identified the occurrence of multihopping VLF chorus bursts which were phase locked to the triggering of sferics by the correlated bursts of energetic electrons. Additional evidence is provided by the near simultaneous occurrence of "sympathetic" lightning discharges over mesoscale distances (Vonnegut et al., 1985) and the equivalent occurrence (in time) of bursts of sferics (Rodriguez et al., 1988). On the MHG hypothesis these could have originated in a simultaneous increase in atmospheric conductivity over an area of the stratosphere ≥ 100 km in extent, due to an increase in either AA or GCR.

The secular increase in thunderstorm activity at high invariant latitudes in the period from 1936 to 1970 relative to the period from 1901 to 1935, over North America (Figure 10.2.5) and the corresponding secular decrease in thunderstorm occurrence at lower latitudes may be attributed primarily to the variation in solar activity, which also exhibited a secular increase in the period from 1900 to 1970. The smoothed sunspot numbers at solar maximum, from 1901 to 1935 and 1936 to 1970 were 60, 106, and 79 and 119, 152, and 200, respectively. Thus, increases in numbers of AA events would be expected in southern Canada and, at lower latitudes ($45° < \Lambda < 55°$), the integrated effect of significant Forbush decreases (Δ GCR $> 5\%$), each of which would endure several days, will have decreased lightning activity. Changnon showed that this secular change in thunderstorm activity was part of a world-wide trend (U.S., Europe, South Africa, and Japan) and has argued that observational errors or omissions are unlikely explanations of what he suggests was a natural climatic fluctuation. His study is consistent with all other related studies (Bullough and Cotterill, 1983; Bullough, 1988; Bullough, 1990), especially those of the detailed solar cycle dependence, in latitude, etc., of thunderstorms (Brooks, 1925, 1934; Lethbridge, 1981) and extreme hourly rainfall in the U.K. (May and Hitch, 1989).

In the period following World War II there was a tenfold increase in Canadian cement production. This will have contributed to increases in lightning occurrence in longitude sectors dominated by PLHR stimulated AA (>1 dB) events. Removal of energetic electrons by these stimulated emissions has made the magnetosphere less hostile to spacecraft (e.g., Vampola, 1977, 1983).

ACKNOWLEDGMENTS

The author wishes to thank Miss Dawn Taylor for the typing of this manuscript and his colleague, Sergei Sazhin, and the editor for helpful comments. He is grateful to the authors and publishers cited for permission to reproduce figures.

REFERENCES

Adrianova, N.V., Vershinin, E.F., Trakhtengerts, V.Yu and Shapaev, V.I. (1977). Generation of Lower Hybrid Resonance Waves in outer ionosphere by fluxes of low energy electrons and protons, *Res. Geomagn. Aeron. Solar Phys.*, 43, 101.

Agy, V. (1975). On the geographic distribution of auroral absorption, *J. Atmos. Terr. Phys.*, 37, 681.

Armstrong, W.C. (1987). Lightning triggered from the Earth's magnetosphere as the source of synchronised whistlers, *Nature*, 327, 405.

Arnoldy, R.L. and Kintner, P.M. (1989), Rocket observations of the precipitation of electrons by ground VLF transmitters, *J. Geophys. Res.*, 94, 6825.

Barr, R. (1979). ELF radiation from the New Zealand power system, *Res. Note, Planet. Space Sci.*, 27, 537.

Basler, R.P. (1963). Radio wave absorption in the auroral ionosphere, *J. Geophys. Res.*, 68, 4465.

Basler, R.P. (1966). Annual variation of auroral absorption, *J. Geophys. Res.*, 71, 982.

Bell, T.F., Luette, J.P. and Inan, U.S. (1982). ISEE 1 observations in the Earth's magnetosphere, *J. Geophys. Res.*, 87(A5), 3530.

Berger, M.J. and Selzer, S.M. (1972). Bremmstrahlung in the atmosphere, *J. Atmos. Terr. Phys.*, 34, 85.

Bering, E.A., III, Benbrook, J.R., Leverenz, H., Roeder, J.L., Stansbery, E.G. and Sheldon, W.R. (1988). Longitudinal differences in electron precipitation near L = 4, *J. Geophys. Res.*, 93(A10), 11385.

Berkey, F.T., Driatskiy, V.M., Henriksen, K., Hultqvist, B., Jelly, D.H., Shchuka, T.I., Theander, A. and Yliniemi, J. (1974). A synoptic investigation of particle precipitation dynamics for 60 substorms in IQSY (1964–1965) and IASY (1969), *Planet. Space Sci.*, 22, 255.

Berthelier, J.J., Lefeuvre, F., Mogilevsky, M.M., Molchanov, O.A., Galperin, Yu.I., Karczewski, J.F., Ney, R., Gogly, G., Guerin, C., Leveque, M., Moreau, J.M. and Sene, F.X. (1982). Measurements of the VLF electric and magnetic components of waves and DC electric field on board the AUREOL-3 spacecraft: the TBF-ONCH experiment, *Ann. Geophys.*, 38, 643.

Boerner, W.-M., Cole, J.B., Goddard, W.R., Tarnawecky, M.Z., Shafai, L. and Hall, D.H. (1983). Impacts of solar and auroral storms on power line systems, *Space Sci. Rev.*, 35, 195.

Brooks, C.E.P. (1925). The distribution of thunderstorms over the globe, London, Air Ministry, Met. Office, Geophys, Mem. London, 24.

Brooks, C.E.P. (1934). The variation of the annual frequency of thunderstorms in relation to sunspots, *Q.J.R. Meteorol. Soc.*, 60, 153.

Bullough, K., Hughes, A.R.W. and Kaiser, T.R. (1969). VLF observations on Ariel 3, *Proc. R. Soc.*, A311, 563.

Bullough, K. (1983). Satellite observations of power line harmonic radiation, *Space Sci. Rev.*, 35, 175.

Bullough, K. (1988). An update on PLHR and its effect on the environment, *Proc. 9th Int. Wroclaw Symposium on EMC*, p. 185, Wroclaw Technical University Press, Poland.

Bullough, K. (1990). More studies in PLHR: cementing a link in the Sun-Weather Relationship?, *Proc. 10th Int. Wroclaw Symposium on EMC*, p. 819, Wroclaw Technical University Press, Poland.

Bullough, K., Tatnall, A.R.L. and Denby, M. (1976). Man-made e.l.f./v.l.f. emissions and the radiation belts, *Nature*, 260, 401.

Bullough, K. and Cotterill, A. (1983). Ariel 4 observations of power-line harmonic radiation over North America and its effect on the Magnetosphere, *Proc. 5th Symp. on EMC, Zurich, Switzerland, 1983*, ETH Zentrum-IKT, Zurich, Switzerland.

Bullough, K., Denby, M., Gibbons, W., Hughes, A.R.W., Kaiser, T.R. and Tatnall, A.R.L. (1975). ELF/VLF emissions observed on Ariel 4, *Proc. R. Soc.*, A343, 207.

Bullough, K., Kaiser, T.R. and Strangeways, H.J. (1985). Unintentional man-made modification effects in the Magnetosphere, *J. Atmos. Terr. Phys.*, 47, 1211.

Cagniard, L. (1953). Basic theory of the magneto-telluric method of geophysical prospecting, *Geophysics*, 18, 605.

Carlson, C.R., Helliwell, R.A. and Inan, U.S. (1990). Space-time evolution of whistler-mode wave-growth in the magnetosphere, *J. Geophys. Res.*, 95(A9), 15073.

Carson, J.R. (1926). Wave propagation in overhead lines with ground return, *Bell Syst. Technol. J.*, 5, 539.

Chang, H.C., Inan, U.S. and Bell, T.F. (1983). Energetic electron precipitation due to gyroresonance interactions in the magnetosphere involving coherent VLF waves with slowly varying frequency, *J. Geophys. Res.*, 88(A9), 7037.

Changnon, S.A. (1977). Secular trends in thunderstorm frequencies, *Proc. 5th Int. Conf. on Atmospheric Electricity: Electrical Processes in Atmospheres*, Dr. Dietrich Steinkopff Verlag, Darmstadt, Germany, 1977.

Chmyrev, V.M., Mogilevsky, M.M., Molchanov, O.A., Sobolev, Y.P., Titova, E.E., Yakhnina, T.A., Suncheleev, R.N., Gladyshev, N.V., Baranets, N.V., Jorjio, N.V., Galperin, Y.I. and Streltsov, A.V. (1989). Parametric excitation of ELF waves and acceleration of ions during injection of strong VLF waves into the ionosphere, *Sov. Phys. Space Res.*, 27, 248.

Collins, C., Jelly, D.H. and Mathews, A.G. (1961). High-frequency radiowave blackouts at medium and high latitudes during a solar cycle, *Can. J. Phys.*, 39, 35.

Cone, Capt. W.M. (1972). Private communication, Dept. of the Navy, Office of the Chief of Naval Operations, Washington, D.C.

Cornwall, J.M. and Schulz, M. (1970). Adiabatic preservation of gyrophase coherence in the Earth's Field, *J. Geophys. Res.*, 75, 4339.

Davies, D.J., Strangeways, H.J. and Bullough, K. (1984). The NAA wavefield above the ionosphere, *Proc. 7th Int. Wroclaw Symposium on EMC*, Wroclaw Technical University Press, Poland.

Dazey, M.H. and Koons, H.C. (1982). Characteristics of a power line used as a VLF antenna, Report No. SD-TR-82-22, Space Division, Air Force Systems Command, P.O. Box 92960, Worldway Postal Center, Los Angeles, CA, 90009.

Denby, M., Bullough, K., Alexander, P.D. and Rycroft, M.J. (1980). Observational and theoretical studies of a cross-meridian refraction of VLF waves in the ionosphere and magnetosphere, *J. Atmos. Terr. Phys.*, 42, 51.

Ecklund, W.L. and Hargreaves, J.K. (1968). Some measurements of auroral absorption structure over distances of about 300 km and of absorption correlation between conjugate regions, *J. Atmos. Terr. Phys.*, 30, 265.

Evans, J.E., Newkirk, L.L. and McCormac, B.M. (1969). North polar, South polar, world maps and tables of invariant magnetic coordinates for six altitudes: 0, 100, 300, 600, 1000 and 3000 km, Defense Atomic Support Agency No. 2347, Lockheed Palo Alto Res Lab., 3251 Hanover Street, Palo Alto, CA 94304, USA.

Foppiano, A.J. and Bradley, P.A. (1985). Morphology of background auroral absorption, *J. Atmos. Terr. Phys.*, 47, 663.

Foster, J.C. and Rosenberg, T.J. (1976). Electron precipitation and VLF emissions associated with cyclotron resonance interactions near the plasmapause, *J. Geophys. Res.*, 81, 2183.

Gurevich, A.V. (1978). *Nonlinear Phenomena in the Ionosphere*, Springer-Verlag, New York.

Hargreaves, J.K. (1966). On the variation of auroral radio absorption with geomagnetic activity, *Planet. Space Sci.*, 14, 991.

Hargreaves, J.K. and Bullough, K. (1972). Midlatitude VLF emissions and the mechanism of dayside auroral particle precipitation, *Planet. Space Sci.*, 20, 803.

Hargreaves, J.K. and Cowley, F.C. (1967). Studies of auroral radio absorption events at three magnetic latitudes. 1, *Planet. Space Sci.*, 15, 1571.

Hargreaves, J.K., Feeney, M.T., Ranta, H. and Ranta, A. (1987). On the prediction of auroral absorption on the equatorial side of the absorption zone, *J. Atmos. Terr. Phys.*, 49, 259.

Hartz, T.R., Montbriand, L.E. and Vogan, E.L. (1963). A study of auroral absorption at 30Mc/s, *Can. J. Phys.*, 41, 581.

Hayakawa, M., Bullough, K. and Kaiser, T.R. (1977). Properties of storm-time magnetospheric VLF emissions deduced from the Ariel 3 satellite and ground-based observations, *Planet. Space Sci.*, 25, 353.

Hayakawa, M. (1991). Observation at Moshiri (L = 1.6) of whistler-triggered VLF emissions in the electron slot and inner radiation belt regions, *J. Geomagn. Geoelectr.*, 43, 267.

Helliwell, R.A. (1965). Whistlers and Related Ionospheric Phenomena, Stanford University Press, CA.

Helliwell, R.A. (1967). A theory of discrete VLF emissions from the magnetosphere, *J. Geophys, Res.*, 72, 4773.

Helliwell, R.A., Katsufrakis, J.P., Bell, T.F. and Raghuram, R. (1975). VLF line radiation in the Earth's magnetosphere and its association with power system radiation, *J. Geophys. Res.*, 80, 4249.

Helliwell, R.A., Carpenter, D.L., and Miller, T.R. (1980). Power threshold for growth of coherent VLF signals in the magnetosphere, *J. Geophys. Res.*, 85(A7), 3360.

Helliwell, R.A., Mielke, T. and Inan, U.S. (1990). Rapid whistler-mode wave growth resulting from frequency-time curvature, *J. Geophys. Res.*, 17, 599.

Herman, J.A. and Goldbert, R.A. (1978). Initiation of non-tropical thunderstorms by solar activity, *J. Atmos. Terr. Phys.*, 40, 121.

Holzer, R.E., Farley, T.A. and Burton, R.K. (1974). A correlated study of ELF waves and electron precipitation on OGO 6, *J. Geophys. Res.*, 79, 1007.

Imhof, W.L., Reagan, J.B. and Gaines, E.E. (1971). Solar particle cut-offs as observed at low altitudes, *J. Geophys. Res.*, 76, 4276.

Imhof, W.L., Nightingale, R.W., Reagan, J.B. and Nakano, G.H. (1980). The morphology of widespread electron precipitation at high latitudes, *J. Atmos. Terr. Phys.*, 42, 443.

Imhof, W.L., Reagan, J.B., Gaines, E.E. and Anderson, R.R. (1983). Narrow spectral peaks in electrons precipitating from the Slot region, *J. Geophys. Res.*, 88(A10), 8103.

Imhof, W.L., Voss, H.D., Walt, M., Gaines, E.E., Mobilia, J., Datlowe, D.J., and Reagan, J.B. (1986). Slot region electron precipitation by lightning, VLF chorus, and plasmapheric hiss, *J. Geophys. Res.*, 91, 8883.

Imhof, W.L., Gaines, E.E. and Reagan, J.B. (1974). Evidence for the resonance precipitation of energetic electrons from the slot region of the radiation belt, *J. Geophys. Res.*, 79(22), 3141.

Inan, U.S., Bell, T.F. and Helliwell, R.A. (1978). Nonlinear pitch angle scattering of energetic electrons by coherent VLF waves in the magnetosphere, *J. Geophys. Res.*, 83, 3235.

Inan, U.S., Chang, H.C. and Helliwell, R.A. (1984). Electron precipitation zones around major ground-based VLF signal sources, *J. Geophys. Res.*, 89(A5), 2891.

Jelly, D.H., Mathews, A.G. and Collins, C. (1962). Study of polar cap and auroral absorption at HF and VHF frequencies, *J. Atmos. Terr. Phys.*, 23, 206.

Jelly, D.H., McDiarmid, I.B. and Burrows, J.R. (1964). Correlation between intensities of auroral absorption and precipitated electrons, *Can. J. Phys.*, 42, 2411.

Kennel, C.F. and Petschek, H.E. (1966). Limit on stably trapped particle fluxes, *J. Geophys. Res.*, 71, 1.

Kikuchi, H. (1972). Investigations of electromagnetic noise and interference due to power lines in Japan and some results from the aspect of electromagnetic theory, Proc. Purdue Symposium on Electromagnetic Hazards, Pollution and Environmental Quality, Lafayette, Indiana.

Kikuchi, H. (1983). Power line transmission and radiation, *Space Sci. Rev.*, 35, 59. Special Issue on "Power Line Radiation and Its Coupling to the Ionosphere and Magnetosphere," Ed: H. Kikuchi.

Koons, H.C., Dazy, M.H. and Edgar, B.C. (1978). Satellite observation of discrete VLF line radiation within transmitter-induced amplification bands, *J. Geophys. Res.*, 83(A8), 3887.

Koziar, A. and Strangway, D.W. (1978). Shallow crustal sounding in the superior province by audiofrequency magneto-tellurics, *Can. J. Earth Sci.*, 15, 1701.

Lefeuvre, F. and Bullough K. (1973). Ariel 3 evidence of zones of VLF emission at medium invariant latitudes which co-rotate with the Earth, in *Space Research XIII*, Akademie-Verlag, Berlin, 699.

Lethbridge, M.D. (1981). Cosmic rays and thunderstorm frequency, *Geophys. Res. Lett.*, 8, 521.

Luette, J.P., Park, C.G. and Helliwell, R.A. (1977). Longitudinal variations of very-low-frequency chorus activity in the magnetosphere: evidence of excitation by electrical power transmission lines, *Geophys. Res. Lett.*, 4(7), 275.

Luette, J.P., Park, C.G. and Helliwell, R.A. (1979). The control of the magnetosphere by power line radiation, *J. Geophys. Res.*, 84(A6), 2657.

Luette. J.P., Park, C.G., and Helliwell, R.A. (1980). Comment on 'On possible causes of apparent longitudinal variations in OGO 3 observations of VLF chorus' by C.T. Russell, *J. Geophys. Res.*, 85 (A3), 1343.

Lyons, L.R. and Thorne, R.M. (1973). Equilibrium structure of radiation belt electrons, *J. Geophys. Res.*, 78, 2142.

Lyons, L.R., Thorne, R.M. and Kennel, C.F. (1972). Pitch angle diffusion of radiation belt electrons within the plasmas-phere, *J. Geophys. Res.*, 77, 3455.

Lyons, L.R. and Williams, D.J. (1978). A comment on the effects of man-made VLF waves on the radiation belts, *Geophys. Res. Lett.*, 5(2), 116.

Markson, R. (1978). Solar modulation of atmospheric electrification and possible implications for the sun-weather rela-tionship, *Nature*, 273, 103.

Mathews, J.P. and Yearby, K. (1980). Magnetospheric VLF line radiation observed at Halley, Antarctica, Memoirs of National Institute of Polar Res. Special Issue No. 16, IMS in Antarctica, p. 95; Nat. Inst. of Polar Res., Tokyo, September, 1980.

Mathews, J.P. and Yearby, K. (1981). Magnetospheric VLF line radiation observed at Halley, Antarctica, *Planet. Space Sci.*, 29, 97.

May, B.R. and Hitch, T.S. (1989). Periodic variations in extreme hourly rainfalls in the United Kingdom, *Meteorol. Mag.*, 118, 45.

McCollor, D.C., Watanabe, T., Slawson, W.F. and Shier, R.M. (1983). An E.M. method for earth resistivity measurements using power line harmonic fields, *J. Geomagn. Geoelectr.*, 35, 221.

McEwen, D.J. and Barrington, R.E. (1967). Some characteristics of the lower hybrid resonance noise bands observed by the Alouette 1 satellite, *Can. J. Phys.*, 45, 13.

Molchanov, O.A., Parrot, M., Mogilevsky, M.M. and Lefeuvre, F. (1991). A theory of PLHR emissions to explain the weekly variation of ELF data observed by a low-latitude satellite, *Ann. Geophys.*, 9, 669.

Park, C.G. (1977). VLF wave activity during a magnetic storm: a case study of the role of power line radiation, *J. Geophys. Res.*, 82, 3251.

Park, C.G. and Helliwell, R.A. (1971). The formation by electric fields of field-aligned irregularities in the magnetosphere, *Radio Sci.*, 6, 299.

Park, C.G. and Chang, D.C.D. (1978). Transmitter simulation of power line radiation effects in the magnetosphere, *Geophys. Res. Lett.*, 5, 861.

Park, C.G. and Helliwell, R.A. (1978). Magnetospheric effects of power line radiation, *Science*, 200, No. 4343, 727.

Park, C.G. and Miller, T.R. (1979). Sunday decreases in magnetospheric VLF wave activity, *J. Geophys. Res.*, 84(A3), 943.

Park, C.G. and Miller, T.R. (1981). Reply (to Thorne and Tsurutani, (1981)), *J. Geophys. Res.*, 86(A3), 1642.

Parrot, M. (1990). World map of ELF/VLF emissions as observed by a low-orbiting satellite, *Ann. Geophys.*, 8, 135.

Parrot, M. (1994). Observations of power line harmonic radiation by the low-altitude Aureol-3 satellite, *J. Geophys. Res.*, 99, 3961.

Pursch, A., Kahn, R., Haskins, R. and Granger-Gallegos, S. (1992). New tools for working with spatially non-uniform-sampled data from satellites, *Earth Observer*, 4(5), 19.

Rodriguez, J.V., Inan, U.S., Armstrong, W.C., Yang, C., Smith, A.J., Orville, R.E., McCarthy, M., Holsworth, R.H., Arnoldy, R.L., Kintner, P.M. and Rosenberg, P.J. (1988). Burst particle precipitation and clustered radio atmospherics: an apparent cause and effect relationship, EOS 69(44) Nov. 1st, 1988: Abstract A22C-06, 1515.

Russell, C.T. (1980). On possible causes of apparent longitudinal variations in OGO 3 observations of VLF chorus, *J. Geophys. Res.*, 85(A3), 1341.

Rycroft, M.J. (1976). Gyroresonance interactions in the outer plasmasphere, *J. Atmos. Terr. Phys.*, 38, 1211.

Rycroft, M.J. and Burnell, S.J. (1970). Statistical analysis of movements of the ionospheric trough and the plasmapause, *J. Geophys. Res.*, 75, 5600.

Sá, L.A.D. (1990). A wave-particle-wave interaction as a cause of VLF triggered emissions, *J. Geophys. Res.*, 95(A8), 12277.

Sazhin, S.S. (1993). *Whistler-Mode Waves in a Hot Plasma*, Cambridge University Press, Cambridge.

Schultz, M. and Lanzerotti, L.J. (1972). *Particle Diffusion in the Radiation Belts*, Springer-Verlag, New York.

Smith, A.J., Yearby, K.H., Bullough, K., Saxton, J.M., Strangeways, H.J. and Thomson, N.R. (1987). Whistler mode signals from VLF transmitters, observed at Faraday, Antarctica, *Proc. of the Nagata Symposium on Geomagnetically Conjugate Studies and the Workshop on Antarctica Middle and Upper Atmosphere Physics*, p. 183; Nat. Inst. of Polar Res., Tokyo.

Strangeways, H.J. and Rycroft, M.J. (1980). Trapping of whistler-waves through the side of ducts, *J. Atmos. Terr. Phys.*, 42, 983.

Strangeways, H.J. (1981a). Determination by ray-tracing of the regions where mid-latitude whistlers exit from the lower ionosphere, *J. Atmos. Terr. Phys.*, 43, 231.

Strangeways, H.J. (1981b). Trapping of whistler-mode waves in ducts with tapered ends, *J. Atmos. Terr. Phys.*, 43, 1071.

Tatnall, A.R.L., Mathews, J.P., Bullough, K. and Kaiser, T.R. (1983). Power-line harmonic radiation and the electron slot, *Space Sci. Rev.*, 35, 139.

Thomson, R.J. (1978). The formation and lifetime of whistler ducts, *Planet. Space Sci.*, 26, 423.

Thorne, R.M. and Tsurutani, B.T. (1979). Power-line harmonic radiation: can it significantly affect the Earth's radiation belts?, *Science*, 204, 839.

Thorne, R.M. and Tsurutani, B.T. (1981). Comment on "Sunday decreases in magnetospheric VLF wave activity" by C.G. Park and T.R. Miller, *J. Geophys. Res.*, 86(A3), 1639.

Tomizawa, I. and Yoshino, T. (1985). Power line radiation observed by the satellite "OHZORA," *J. Geomagn. Geolectr.*, 37, 309.

Tsurutani, B.T., Church, S.R. and Thorne, R.M. (1979). A search for geographic control on the occurrence of magnetospheric ELF emissions, *J. Geophys. Res.*, 84(A8), 4116.

Tsurutani, B.T. and Thorne, R.M. (1981). A skeptic's view of PLR effects in the magnetosphere, *Adv. Space Res.*, 1, 439.

Vampola, A.L. (1977). VLF transmission-induced slot electron precipitation, *Geophys. Res. Lett.*, 4, 569.

Vampola, A.L. (1983). Observations of VLF transmitter-induced depletion of inner zone electrons, *Geophys. Res. Lett.*, 10, 619.

Vampola, A.L. (1987). Electron precipitation in the vicinity of a VLF transmitter, *J. Geophys. Res.*, 92, 4525.

Vampola, A.L. and Gorney, D.J. (1983). Electron energy deposition in the middle atmosphere, *J. Geophys. Res.*, 88, 6267.

Vampola, A.L. and Adams, C.D. (1988). Outer zone electron precipitation produced by a VLF transmitter, *J. Geophys. Res.*, 93(A3), 1849.

Vaughan, O.H. and Vonnegut, B. (1982). Lightning to the ionosphere?, Weatherwise, April 1982, p. 70. (Plus Gales, D.M., another account).

Vershinin, E.F., Mullayarov, V.A. and Shapaev, V.I. (1983). Statistical properties of VLF line structured hiss, short report, Institute of Cosmophysics and Aeronomy, Yakutsk, C.I.S.

Volland, H., Schmodlers, M., Prolss, G.W. and Schafer, J. (1987). VLF propagation parameters derived from spherics observations at high southern latitudes, *J. Atmos. Terr. Phys.*, 49, 33.

Vonnegut, B., Vaughan, O.H., Brook, M. and Krehbiel, P. (1985). Mesoscale observations of lightning from Space Shuttle, *Bull. Am. Meteorol. Soc.*, 66, 20.

Wedepohl, L.M. and Efthymiadis, A.E. (1978). Wave propagation in transmission lines over lossy ground: a new complete field solution, *Proc. IEEE*, 125, 505.

Whitehead, S. and Radley, W.G. (1949). Generation and flow of harmonics in transmission systems, *Proc. IEEE*, 96, 29.

WMO (1956). World distribution of thunderstorm days, Report TP 21, II, World Meteorological Organization, Geneva.

Woodland, F. (1970). Electric interference aspects of buried electric and telephone lines, *IEEE Trans.*, PAS-89, 275.

Yakymyshyn, M. (1990). Private communication, Inland Cement Co., Edmonton, Canada.

Yearby, K.H., Mathews, J.P. and Smith, A.J. (1981). VLF line radiation observed at Halley and Siple, *Adv. Space Res.*, 1, 445.

Yearby, K.H., Smith, A.J., Kaiser, T.R. and Bullough, K. (1983). Power line harmonic radiation in Newfoundland, *J. Atmos. Terr. Phys.*, 45, 409.

Yoshida, T., Ohtsu, J. and Hayakawa, M. (1983). A study of the mechanism of whistler-triggered VLF emissions, *J. Geophys. Res.*, 53, 59.

Chapter 11

Magnetospheric Configuration

Gerd-Hannes Voigt

CONTENTS

0-8493-2520-X/95/$0.00+$.50
© 1995 by CRC Press

1. INTRODUCTION

In this Introduction, the main features of Earth's magnetosphere are explained qualitatively. Many details await a more quantitative treatment in the sections that follow. The interaction between the solar wind plasma and Earth's magnetic field will be described largely in terms of magnetohydrodynamics (MHD) and will be restricted to large-scale processes that determine the global configuration of Earth's magnetosphere. Table 11.1.1 contains a list of textbooks and tutorial review articles on the subject of magnetospheric physics.

The formation of Earth's magnetosphere, its global configuration, and its dynamic evolution could, in principle, be understood by answering the following question: "What will happen, when a vacuum dipole field (an approximation of Earth's main field) is exposed to the steady stream of solar wind plasma originating from the Sun?" That seemingly simple question has turned out to be extraordinarily complex. It is not possible to calculate on today's computers the motions of approximately 10^{32} individual solar wind plasma particles and their interaction with Earth's magnetic field. Instead, the solar wind is viewed as a fluid, and its interaction with the magnetic field is treated mathematically in terms of MHD.

A continuous supersonic solar wind plasma stream departs from the solar corona, travels through the interplanetary space, and carries with it the solar magnetic field, called the interplanetary magnetic field (IMF). In analogy to a supersonic neutral gas flow, the supersonic solar wind forms a shock wave (bow shock) in front of Earth's magnetic obstacle. The solar wind also confines Earth's own magnetic field to a limited region in space, called the magnetosphere. The outermost boundary of the magnetosphere is called magnetopause; it distinctly separates the terrestrial magnetic field from the magnetosheath, which is the transition region located between the magnetopause and the bow shock. The magnetopause forms where the solar wind pressure on the outside balances the magnetic field energy density on the inside (more details about pressure balance appear in Section 5).

1.1. CLOSED AND OPEN MAGNETOSPHERES

The concept of a confined magnetosphere follows from a basic theorem of ideal MHD (see Section 10), which states that a highly conductive plasma (the solar wind) cannot penetrate into an existing magnetic field, or vice versa, the magnetic field cannot diffuse into the plasma. That scenario applies if the (scalar) electric conductivity of the plasma, $\sigma \to \infty$, approaches infinity. Then there is no connection between the IMF and the magnetospheric magnetic field, and the magnetosphere is called "closed". If the solar wind carries no magnetic field, then a so-called "teardrop" configuration similar to the one in Figure 11.1.1 might form. If the solar wind is magnetized, then the outside IMF completely drapes around the magnetopause boundary, and the teardrop magnetosphere is surrounded by IMF lines (see Figure 11.1.2).

The closed teardrop model for Earth's magnetosphere was proposed by Johnson (1960). Both Figures 11.1.1 and 11.1.2 show that under ideal MHD conditions the terrestrial magnetic field is completely confined to its magnetospheric cavity. The field is compressed on the dayside by both the solar wind dynamic and thermal pressures; on the nightside it extends into a tail-like configuration. The teardrop magnetosphere is closed on the nightside by the solar wind thermal pressure. The day-night asymmetries of the magnetospheres in Figures 11.1.1 and 11.1.2 can be explained by comparing the dynamic-, IMF-, and thermal solar wind pressure components listed in Table 11.1.2.

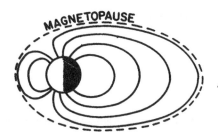

Figure 11.1.1 A closed magnetosphere, the so-called teardrop model, according to Johnson (1960). Such a magnetosphere could form conceptually under the conditions of perfectly ideal MHD. Shown are magnetic field lines in the noon-midnight meridian plane. The solar wind is coming from the left side compressing the dayside of the magnetosphere but creating a tail-like magnetic field configuration on the nightside of the Earth. In this sketch the solar wind is assumed to be unmagnetized. (From Johnson, F.S. (1960). *J. Geophys. Res.,* 65, 3049. 1960 Copyright by the American Geophysical Union. With permission.)

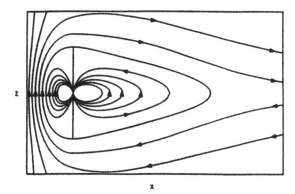

Figure 11.1.2 A closed magnetosphere surrounded by IMF lines. This field line configuration is conceptually similar to the teardrop model in Figure 11.1.1, but here the solar wind is magnetized such that the IMF drapes around the magnetosphere. (From Birn, J. et al. (1992). *J. Geophys. Res.,* 97, 14833. 1992 Copyright by the American Geophysical Union. With permission.)

Table 11.1.1   **Suggested Literature for Space Plasma Physics and Magnetospheric Physics**

Authors	Year	Title	Source	Comment
Parks	1991	*Physics of Space Plasmas*	Textbook	Introductory text
Hasegawa and Sato	1989	*Space Plasma Physics*	Textbook	Advanced text
Kamide and Slavin (Editors)	1986	*Solar Wind-Magnetosphere Coupling*	Proceedings	Individual contributions on various levels
Carovillano and Forbes (Editors)	1983	*Solar-Terrestrial Physics*	Proceedings	Each tutorial starts as an introduction but reaches advanced levels
Nishida (Editor)	1982	*Magnetospheric Plasma Physics*	Textbook	Series of five advanced tutorials by different author teams
Schindler and Birn	1978	*Magnetospheric Physics*	Review paper	Advanced text

Because ideal MHD prevents the solar wind plasma from penetrating into the magnetosphere, there are no currents flowing inside the magnetospheric cavity of the teardrop magnetosphere. Thus, the deformation of the Earth's dipole field, and its shielding against the outside, is caused entirely by surface currents flowing in closed loops on the magnetopause. Those shielding currents are called "Chapman-Ferraro" currents (see Section 4) in reference to Chapman and Ferraro (1931).

Table 11.1.2 **Energy Densities in the Interplanetary Space**

Kinetic energy density of the solar wind	5000 eV/cm³
Thermal pressure of the solar wind	260 eV/cm³
Magnetic energy density of the solar wind	90 eV/cm³
Cosmic radiation	1 eV/cm³

From Kertz, W. (1971). *Einführung in die Geophysik II*, Bibliographisches Institut, Mannheim, Germany. With permission.

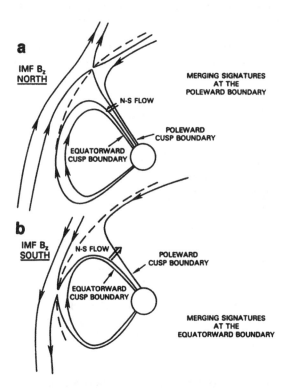

Figure 11.1.3 Schematic representation of field-line interconnection between interplanetary and magnetospheric magnetic field lines near the polar cusp region. (a) When the IMF B_z points northward, interconnection occurs poleward of the cusp. (b) When the IMF B_z points southward, interconnection occurs equatorward of the cusp. (From Basinska, E. M. et al. (1992) *J. Geophys. Res.*, 97, 6369. 1992 Copyright by the American Geophysical Union. With permission.)

If, on the other hand, the electric conductivity, $\sigma < \infty$, is finite at the magnetopause, then the physical configuration is described by nonideal MHD. In that case interplanetary and terrestrial magnetic field lines do interconnect at magnetopause locations where the internal and external fields are anti-parallel. Figure 11.1.3, which displays part of the so-called "polar cusp" region, sketches such an interconnection between anti-parallel fields (Basinska et al., 1992). When the IMF points northward, then interconnection occurs poleward of the cusp. Conversely, when the IMF points southward, then interconnection occurs equatorward of the cusp. It is evident from Figure 11.1.3 why the magnetosphere is called "open" for such interconnected field configurations.

A finite scalar conductivity, σ, or tensor conductivity, $\underline{\sigma}$, causes, in the picture of an MHD fluid, a process of magnetic diffusion. However, it is one of the outstanding problems in magnetospheric physics to derive, from the plasma kinetic point of view, a macroscopic conductivity tensor, $\underline{\sigma}(\mathbf{x}_{mp})$, for magnetopause regions ($\mathbf{x}_{mp}$) where field-line interconnection occurs. The two

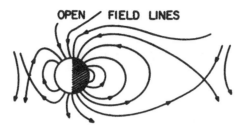

Figure 11.1.4 Magnetic field configuration of an open magnetosphere (Levi et al., 1964). This drawing follows Dungey's (1961) concept of magnetic field-line interconnection between interplanetary and terrestrial field lines at locations where interplanetary and magnetospheric field lines are anti-parallel. Such a magnetosphere forms when ideal MHD conditions become violated locally at the magnetopause boundary. (From Levi, R.H., Petschek, H.E., and Siscoe, G.L., (1964). *AIAA Journal,* 2, 2065. With permission.)

related plasma processes, plasma diffusion and field-line interconnection, still await a satisfactory microphysical explanation for a collisionless plasma.

Figure 11.1.4 shows an open magnetosphere according to Levi et al. (1964). Their sketch refers to Dungey (1961) who postulated that ideal MHD might not hold right at the magnetopause boundary where the guiding-center approximation for plasma particles might not be valid. The guiding-center approximation becomes invalid at MHD discontinuities when magnetic field lines have sharp kinks (see, for example, Figure 11.1.3). A magnetic field line is said to have a sharp kink when the field-line radius of curvature at the kink is in the order of a typical ion gyroradius.

The magnetospheric field points northward at the subsolar stagnation point (i.e., the magnetosphere's outermost point on the Earth-sun line). Hence, the IMF must point southward there to favor interconnection at that very location (see Figure 11.1.4). Part of the solar wind plasma might enter the magnetosphere along interconnected field lines thus partially filling the magnetosphere with mostly thermal plasma. The plasma inside the magnetosphere can now support electric currents within the magnetotail region that flow predominantly in a westward direction on the nightside of the Earth (more details about magnetotail formation appear in Section 6.1).

The two magnetic neutral lines (X-lines) in Figure 11.1.4, namely, the one on the dayside at the magnetopause subsolar point and the other one on the nightside at the center of the tail plasma sheet, complement each other. An unspecified violation of ideal MHD lets field lines break open at the subsolar point. The two newly opened field lines are then carried along with the solar wind toward the nightside of the Earth. On their way, the thermal expansion of the solar wind into the wake of Earth's magnetic obstacle, or simply ($\mathbf{E} \times \mathbf{B}$) drift from the magnetopause toward the plasma sheet, allows tail field lines to move toward the center of the plasma sheet. There, field lines reconnect at the nightside X-line and move toward the Earth thus completing what is called a magnetospheric convection (plasma circulation) cycle (more details about convection appear in Section 8).

1.2. A QUALITATIVE PICTURE OF EARTH'S MAGNETOSPHERE

In the early days of magnetospheric research, it was not clear whether the magnetosphere was open or closed. Figures 11.1.1 and 11.4.4 exemplify the two major magnetosphere-forming processes. Our present understanding of the global magnetospheric configuration is a combination of those two processes. Figure 11.1.5, sketched by Hill and Dessler (1991), combines findings from both observations and MHD computer simulations.

Measured along the Earth-sun line, the above-mentioned bow shock is located about 15 Re (1 Re = 1 earth radius = 6371 km) away from the center of the Earth. The magnetopause subsolar stagnation point is located about 11 Re away from the center of the Earth (see Figure 11.1.6 and Section 5.3). In the transition region between the bow shock and Earth's magnetopause, i.e., in the magnetosheath, both the highly directed solar wind stream and the IMF experience turbulent variations indicating a partial thermalization of the solar wind plasma.

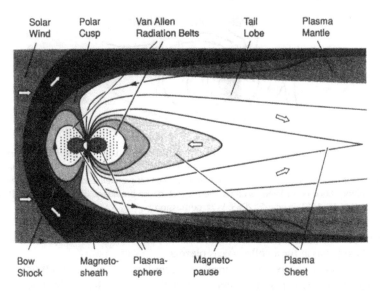

Figure 11.1.5 Modern view of Earth's partially open magnetosphere with a long magnetotail. Drawn are magnetic field lines in the noon-midnight meridian plane. Open arrows indicate the direction of plasma circulation called convection. Gray shades refer to different plasma regimes explained in the text. (From Hill, T. W. and Dessler, A. J. (1991). *Science,* 252, 345. With permission.)

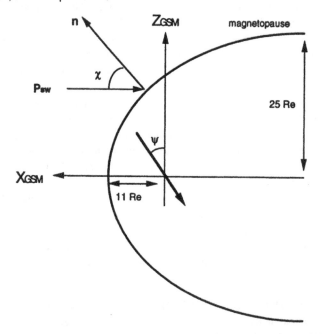

Figure 11.1.6 Sketch of the magnetopause in Geocentric Solar Magnetospheric (GSM) coordinates. Shown are typical scale sizes for both the subsolar stand-off distance (11 R_E) and the magnetotail radius (25 R_E). The angle χ measures the angle between the incoming (idealized) solar wind stream, **v**, and the normal vector, **n**, at the magnetopause surface. The dipole tilt angle ψ measures the angle between the Z_{GSM}-axis and the geomagnetic dipole axis.

In reality, magnetic shielding at the magnetopause is not perfect: about 10% of the solar wind plasma enters the magnetosphere to form the magnetotail plasma sheet. The "openness" of the magnetosphere is described quantitatively in terms of the magnetic normal component, $\mathbf{n} \cdot \mathbf{B} = f(\mathbf{x}_{mp})$, specified on the entire magnetopause surface ($\mathbf{x}_{mp}$). The quantity **n** denotes the surface unit vector pointing outward. Those magnetic boundary conditions determine the amount of magnetic flux entering or leaving the magnetosphere. The function $f(\mathbf{x}_{mp})$ cannot be derived from MHD arguments, and a satisfactory microscopic theory has yet to be established. Recently, Toffoletto and Hill (1989, 1993), and Crooker et al. (1990) have begun to derive semi-empirical functions $f(\mathbf{x}_{mp})$ to construct models for an open magnetosphere (more details about magnetic boundary conditions appear in Section 4).

The compression of Earth's main field on the dayside leads to the formation of two polar cusps which are exposed to the magnetosheath flow (see Figures 11.1.3 and 11.1.5). The cusp geometries allow direct plasma entry into the magnetosphere along field-line bundles that form the polar cusp regions.

On the nightside of the Earth the magnetotail forms owing to the existence of thermal plasma on closed magnetic field lines. Because of distinctly different plasma populations in various tail regions, one distinguishes five different magnetic field domains in the magnetotail.

The *first domain* is the plasma mantle adjacent to the magnetopause. It consists of mainly magnetosheath plasma which is found inside the magnetopause because the magnetosphere is not completely closed. The plasma mantle widens down the tail owing to the thermal expansion of the plasma on the nightside of the Earth (Siscoe and Sanchez, 1987; Sanchez et al., 1990).

The *second domain* are the two magnetic tail lobes where almost no plasma can be found. The lobe field lines are all open and ultimately, far downstream, connect with the IMF. There the plasma, that might have been on lobe field lines, escapes.

The *third domain* comprises the plasma sheet boundary layer. Boundary layer field lines map to the distant neutral line explained above in Figure 11.1.4. The exact location of the distant neutral line varies widely, depending on the overall plasma loading of the magnetotail. Therefore, that location is difficult to observe directly. Indirect evidence, however, suggests that the distant neutral line be located at about somewhere between 80 and 120 Re down the tail (see Section 6.4). The distant neutral line is a site of field-line reconnection and corresponding plasma acceleration. Thus, fast plasma streams are commonly observed along the entire plasma sheet boundary layer.

The *fourth domain* is called the inner plasma sheet. That region is composed of closed magnetotail field lines which have their two ends in the terrestrial northern and southern ionosphere, respectively. Owing to its diamagnetic effect, thermal plasma accumulates in the inner plasma sheet in a plane that is defined by the weakest magnetic field strength. The very center of that plane is often called the neutral sheet; it carries the highest magnetotail current density. The magnetotail currents flow across the tail in a predominantly westward direction and close over the northern and southern tail magnetopause (see Figure 11.1.7). The magnetic field generated by those currents causes a stretching of the entire field configuration into a long magnetotail.

Closed plasma sheet field lines are stretched toward the nightside in response to the amount of thermal plasma loaded on each field line (often the term "flux tube" is used instead of "field line"). The two processes, plasma loading and field-line stretching, correspond to each other in that they manifest a pressure balance between Lorentz forces and plasma pressure gradients (more details about force balance and magnetotail MHD equilibria appear in Section 6.2).

The *fifth domain* comprises the plasma population of the equatorial ring current. Viewed through the glasses of MHD equilibrium and convection theories, there is not really a distinction between the magnetotail currents and the equatorial ring current. Between 5 and 7 Re radial distance from the Earth, the magnetotail currents close in the equatorial plane, thus forming a ring current around the Earth. From the viewpoint of a single particle's motion, the ring current consists of plasma particles that are trapped in the Earth's (nearly axially symmetric) magnetic dipole field and thus circulate around the Earth.

The ring current plasma population has first been measured by Frank (1967). Figure 11.1.8 shows energy density profiles of ring current protons plotted as a function of the radial distance

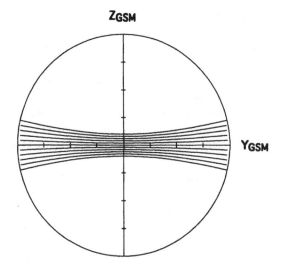

Figure 11.1.7   Cross section of Earth's magnetotail. The tail magnetopause has an idealized circular shape for reasons of computational simplicity. Plotted are tail currents flowing across the tail in westward direction, according to a model by Voigt (1984). The center current stream line in the plane of symmetry carries the highest current density. The tail currents close on the tail magnetopause to satisfy the condition $\nabla \cdot \mathbf{j} = 0$. (From Voigt, G.-H. (1984). *J. Geophys. Res.,* 89, 2169. 1984 Copyright by American Geophysical Union. With permission.)

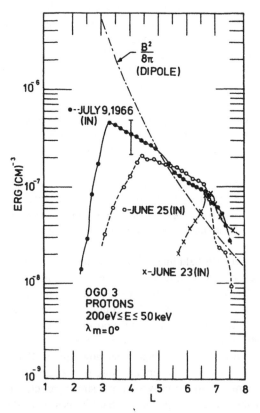

Figure 11.1.8   Energy density profiles of ring current protons plotted as a function of the radial distance L from the center of Earth's magnetic dipole (Frank, 1967). During magnetically quiet periods, the energy density maximum is located between L = 6 R_E and L = 7 R_E. During the progression of a magnetic storm, the energy density maxima shift toward higher proton energies and closer to the Earth. (From Frank, L.A. (1967). *J. Geophys. Res.,* 72, 3753. 1967 Copyright by the American Geophysical Union. With permission.)

L from the center of Earth's magnetic dipole. During magnetically quiet periods (June 23, 1966), the energy density maximum is located between L = 6 Re and L = 7 Re. During the progression of a magnetic storm (caused by a sudden increase in the solar wind pressure and a corresponding compression of the entire magnetosphere), the energy density maxima shift toward higher proton energies and closer to the Earth (June 25 and July 9, 1966).

According to Frank (1967), 20% of the corresponding magnetic field variations at the Earth's surface are due to electrons, the other 80% are due to protons in the energy range 200 eV $\leq$ E $\leq$ 50 keV. Close to the Earth, at around L $\leq$ 7, the energy density of the dipole field is much higher than the energy density of the ring current plasma. Thus the configuration of Earth's main magnetic field remains practically unchanged under the presence of the ring current plasma.

Small changes of Earth's vacuum dipole field configuration, under the influence of measured ring current plasma distributions, have been calculated by Sckopke (1972). He computed, in the gyrotropic approximation, self-consistent axially symmetric MHD equilibria for anisotropic thermal plasma.

1.3. THE PLASMA β-PARAMETER

Whenever a thermal plasma interacts with a magnetic field, the question arises whether at a given location the plasma properties are influenced by the magnetic field, or whether the magnetic field is deformed by the plasma. To quantify the answer, one defines the ratio between the thermal plasma pressure and the magnetic field energy density at a given location in space. That ratio is called the "plasma β-parameter",

$$\beta = \frac{2 \, \mu_0 \, P}{B^2} \tag{1.1}$$

Configurations with $\beta \ll 1$ are called "low-β" regimes where the vacuum magnetic field determines the global plasma motion. The dayside part of Earth's magnetosphere, close to the subsolar point, is a typical low-β regime and can thus be modeled most successfully in terms of vacuum magnetic fields (see Section 4). Low-β, nonvacuum fields are related to force-free magnetic field configurations; those fields exist in the solar corona but are not important in Earth's magnetosphere, except perhaps in the distant magnetotail where substantial field-line twists have been observed (see Section 6.4).

Conversely, configurations with $\beta \gg 1$ are called "high-β" regimes where the magnetic field is deformed by the plasma. The center of the plasma sheet in Earth's magnetotail is a typical example for a high-β regime; values of $\beta \approx 100$ can often be found there.

2. THE EARTH'S MAIN FIELD

The Earth's internal magnetic field is generated deep inside the Earth by an MHD dynamo which describes a convective flow of ionized material near the boundary between the iron core and the mantle. In this section, the magnetic field outside the dynamo region is expressed in terms of a scalar potential $\Phi_D(x)$. For most magnetospheric applications, the complex multipole expansion of $\Phi_D(x)$ reduces to a simple dipole term.

Owing to the Earth's daily and yearly rotations, the axis of the geomagnetic dipole continuously changes its orientation with respect to the solar wind stream. The corresponding time-dependent dipole tilt angle ψ relative to the direction of the solar wind is measured from the z_{GSM} axis of the GSM coordinate system (the GSM system is defined in Section 2.3).

Table 11.2.1 **The First Three Gauss Coefficients of the International Geomagnetic Reference Field (IGRF)**

Coefficient	1950	1960	1970	1980	1990
a_1^0	−30,554	−30,421	−30,220	−29,992	−29,775
a_1^1	−2250	−2169	−2068	−1956	−1851
b_1^1	+5815	+5791	+5737	+5604	+5411

Note: These terms in Equation 2.6 represent, at the Earth's surface, the equatorial magnetic field of a magnetic dipole according to $|\mathbf{B}_D| = [(a_1^0)^2 + (a_1^1)^2 + (b_1^1)^2]^{1/2} = 30,319$ nT for 1990. Time variations of the Gauss coefficients are a manifestation of the so-called secular variations of the geomagnetic main field.

2.1. SCALAR POTENTIAL REPRESENTATION

At the Earth's surface, the geomagnetic main field is represented by a vacuum magnetic field which satisfies the two conditions

$$\nabla \cdot \mathbf{B} = 0 \tag{2.1}$$

$$\nabla \times \mathbf{B} = 0 \tag{2.2}$$

Therefore, Earth's main field, $\mathbf{B}_D$, can be described by a scalar potential $\Phi_D(\mathbf{x})$ according to

$$\mathbf{B}_D = -\nabla \, \Phi_D \tag{2.3}$$

$$\nabla^2 \, \Phi_D = 0 \tag{2.4}$$

In spherical (r, θ, φ) coordinates a separable solution to Laplace's Equation 2.4 can be expressed in terms of spherical harmonics,

$$\Phi_D(r, \theta, \phi) = \sum_{n=1}^{\infty} \sum_{m=0}^{n} (Y_n^m(\theta, \phi) \cdot \frac{1}{r^{n+1}} \tag{2.5}$$

where the spherical harmonics $Y_n^m(\theta, \phi)$ are defined as

$$Y_n^m(\theta, \phi) = [a_n^m \cos(m\phi) + b_n^m \sin(m\phi)] \cdot P_n^m (\cos\theta) \tag{2.6}$$

The radius r in Equation 2.5 is given in units of Earth radii (1 Re = 6371 km). The functions P_n (cosθ) are Legendre polynomials, and P_n^m (cosθ) are associated Legendre polynomials. The coefficients in Equation 2.6 are given in units of nanoTesla (nT) and are called Gauss coefficients. Their values can be obtained from the International Geomagnetic Reference Field (IGRF). Owing to the secular variations of the Earth's magnetic field (see Table 11.2.1), the IGRF is revised about every 5 years by an international commission (Barraclough, 1987; Peddie, 1982).

2.2. THE DIPOLE FIELD

For most magnetospheric applications, the Earth's main field can be approximated by a simple magnetic dipole located at the center of the Earth. That dipole is represented by the first leading term a_1^0 in Equation 2.6. The dipole z_D-axis is anti-parallel to the dipole moment $\mathbf{M}_D$, and the

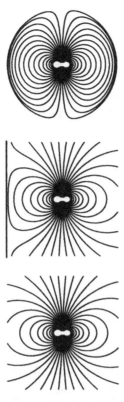

Figure 11.2.1 Shielded and unshielded magnetic dipole configurations. Top: a magnetic dipole is confined by a spherical magnetopause. Chapman-Ferraro currents flow horizontally. The magnetic field configuration results from a superposition of the dipole field in Equations 2.8 and 2.9 and the Chapman-Ferraro field in Equations 4.13 and 4.14. Middle: A magnetic dipole is shielded by an infinitely extended plane. The magnetic field configuration results from a superposition of the dipole field in Equations 2.8 and 2.9 and an exact mirror-image dipole located on the other side of the boundary. Bottom: The pure vacuum dipole field in Equations 2.8 and 2.9.

angle θ is measured from the z_D-axis, i.e., from the northern magnetic north pole. Then the scalar potential Φ_D reads in dipole-aligned (r, θ, ϕ) coordinates

$$\Phi_D(r, \theta) = -\frac{M_D}{r^2} \cos\theta \tag{2.7}$$

Following Equation 2.3 the magnetic dipole components in spherical coordinates read

$$B_r = -\frac{2\, M_D}{r^3} \cos\theta \tag{2.8}$$

$$B_\theta = -\frac{M_D}{r^3} \sin\theta \tag{2.9}$$

$$B_\phi = 0 \tag{2.10}$$

The bottom panel in Figure 11.2.1 shows the magnetic field line configuration of an undisturbed vacuum dipole according to the field components in Equations 2.8 through 2.10. At the Earth's magnetic equator, the geomagnetic dipole field strength is usually approximated by the first three Gauss coefficients in Equation 2.6, namely,

$$|\mathbf{B}_D| = \sqrt{(a_1^0)^2 + (a_1^1)^2 + (b_1^1)^2}$$ (2.11)

This leads to an equatorial magnetic dipole field strength of $|\mathbf{B}_D| = 30,319$ nT $= 0.303$ G for the epoch of 1990. With $r = 1$ Re $= 6371$ km the numerical value of Earth's magnetic dipole moment follows from Equation 2.9, namely,

$$M_D = 7.84 \times 10^{30} \text{ [nT cm}^3\text{]}$$ (2.12)

Table 11.2.1 lists the first three Gauss coefficients of the IGRF for the present and previous epochs. The trend of the secular variations is slightly different for each Gauss coefficient, but is clear that the entire geomagnetic main field gradually decreases with time, at a rate of about 0.15% per year. It is not known whether the main field will vanish completely.

2.3. THE DIPOLE TILT ANGLE

The Earth's rotational axis is tilted by 23.5° against the ecliptic north, and the geomagnetic dipole axis is tilted by 11.4° against the rotational axis. The combination of these two angles leads to a maximum dipole tilt-angle $\psi = \pm 34.9°$. For $\psi = 0°$ the geomagnetic dipole axis is perpendicular to the solar wind stream (see, for example, Figure 11.1.5).

The tilt-angle ψ between the solar wind stream and the Earth's dipole axis is usually measured from the Z_{GSM} axis of the GSM coordinate system. The GSM coordinates are defined such that the X_{GSM}-axis points toward the sun, and the X_{GSM}-Z_{GSM} plane always contains the Earth's dipole (see Figure 11.1.6). The GSM system thus rotates with both a daily and yearly period with respect to inertial coordinates. Detailed definitions of other geophysical coordinate systems can be found in Russell's (1971) review paper.

The tilt-angle ψ is a crucial parameter for the overall configuration of Earth's magnetosphere. Its daily and seasonal time variations may be calculated from the following formula: The day has to be given in days counted from January 1. The time of the day, minute, and hour, has to be given in Universal Time (UT).

$$\psi = \psi \text{ (day, hour, minute)}$$

$$h = \text{hour} + \text{minute}/60.0$$
$$\omega_1 = 2\pi \cdot (\text{day} - 80.)/365.0$$
$$\omega_2 = 2\pi \cdot (h - 10.6667)/24.0$$
$$\psi = 23.5 \cdot \sin(\omega_1) + 11.4 \cdot \sin(\omega_2)$$

3. THE MAGNETOPAUSE AS AN MHD DISCONTINUITY

The equations of an ideal MHD medium (see Section 10) allow for the existence of discontinuous flows. Earth's bow shock and magnetopause can be seen as such discontinuities. Any MHD discontinuity (or shock front) must satisfy certain jump conditions which are derived from the full set of the ideal MHD equations written in conservative form. The closed magnetopause is described by a simple tangential discontinuity, but the open magnetopause must be treated by the full set of jump conditions.

3.1. RANKINE-HUGONIOT JUMP CONDITIONS

In the following notation, $[q] = q_2 - q_1$ denotes an incremental jump of the quantity q at the discontinuity, and the subscripts 1 and 2 refer to the values on the upstream and downstream sides of the discontinuity. The subscripts n and t refer to the respective normal and tangential components of the denoted vector quantities at the discontinuity. The full set of jump conditions for an ideal MHD medium with isotropic scalar pressure is the following:

$$[\rho\, v_n] = 0 \tag{3.1}$$

$$\left[P + \frac{B^2}{2\,\mu_0} - \frac{B_n^2}{\mu_0} + \rho\, v_n^2\right] = 0 \tag{3.2}$$

$$\left[\rho\, v_n\, \mathbf{v_t} - \frac{B_n\, \mathbf{B_t}}{\mu_0}\right] = 0 \tag{3.3}$$

$$\left[\left(P + \frac{B^2}{2\,\mu_0}\right) v_n - \frac{B_n}{\mu_0}\,(\mathbf{v}\cdot\mathbf{B}) + \left(\frac{P}{\gamma - 1} + \frac{1}{2}\,\rho\, v^2 + \frac{B^2}{2\,\mu_0}\right) v_n\right] = 0 \tag{3.4}$$

$$[B_n] = 0 \tag{3.5}$$

$$[v_n\, \mathbf{B_t} - \mathbf{v_t}\, B_n] = 0 \tag{3.6}$$

These conditions follow from the MHD Equations 10.12 through 10.17 written in conservative form. Equation 3.1 follows from the requirement of mass conservation at the boundary, Equations 3.2 and 3.3 follow from the conservation of momentum, and Equation 3.4 follows from the conservation of energy. Equation 3.5 says that the magnetic normal component must be continuous at the boundary which follows from $\nabla \cdot \mathbf{B} = 0$. Finally, Equation 3.6 says that the tangential component of the electric field must be continuous at the boundary; here the electric field is given by $\mathbf{E} = -\mathbf{v} \times \mathbf{B}$ which is the ideal MHD approximation.

Equations 3.1 through 3.6 are called "Rankine-Hugoniot jump conditions". These equations are very general and can be written in a more compact form (c.f., Vol. 8, Sec. 70, of Landau and Lifshitz, 1984).

3.2. TANGENTIAL AND ROTATIONAL DISCONTINUITIES

The closed magnetosphere is described by $B_n = 0$ at the magnetopause, and the flow field $\mathbf{v}$ is parallel to the boundary. The condition $v_n = 0$ means that there is no mass transport across the magnetopause. A boundary with those properties is called a "tangential discontinuity". Under those conditions, five of the equations, 3.1 through 3.6, are satisfied identically. However, the mass density, tangential flow velocity, and tangential magnetic field may be arbitrarily discontinuous but subject to the constraint of Equation 3.2. Hence, the closed magnetosphere must meet the boundary conditions

$$
\begin{array}{ll}
v_n = 0 & [\mathbf{v_t}] \neq 0 \\[2mm]
B_n = 0 & [\mathbf{B_t}] \neq 0 \\[2mm]
\left[P + \dfrac{B_t^2}{2\,\mu_0}\right] = 0 & [\rho] \neq 0
\end{array}
\tag{3.7}
$$

Jumps in $\mathbf{v_t}$, $\mathbf{B_t}$, and ρ are quite frequently observed at the magnetopause, even if the magnetic normal component B_n is not strictly zero. Especially the condition $[\mathbf{B_t}] \neq 0$ at the boundary is measured for the strength and configuration of the Chapman-Ferraro currents (more details about the Chapman-Ferraro currents appear in Section 4).

If $B_n \neq 0$ at the magnetopause, the boundary is open, and the full set of jump conditions, Equations 3.1 through 3.6, must be considered. The open magnetopause is sometimes said to be a "rotational discontinuity". In that case, the magnitude of the total magnetic field $[B_n^2 + B_t^2]$

must be continuous at the boundary, but the condition $[\mathbf{B}_t] \neq 0$ is possible. This means that the magnitude of the field vector remains unchanged, but the field vector might rotate across the boundary. However, rotational discontinuities require that the thermal pressure $[P]$, the normal velocity $[v_n]$, and the mass density $[\rho]$ be continuous as well. Those very stringent conditions are often not observed. Thus the term rotational discontinuity for the open magnetopause should be used with caution. A class of rotational discontinuities for anisotropic pressure in the gyrotropic approximation has been analyzed theoretically by Hudson (1971, 1973).

For most modeling purposes, the magnetopause can be regarded as a slightly disturbed tangential discontinuity with a small magnetic normal component B_n. That deviation from physical and mathematical rigor is painful but convenient, because the shape and location of the magnetopause may now be determined by two boundary conditions which must hold simultaneously at the magnetopause:

1. Magnetic boundary conditions must be met to represent a magnetically closed or open magnetosphere. Only the closed magnetosphere is a direct consequence of ideal MHD. The boundary conditions for the open magnetosphere require additional physical information which the MHD theory does not provide.
2. Pressure boundary conditions must be met such that at each surface point the solar wind pressure balances the magnetic field pressure just inside the boundary.

Magnetic boundary conditions and pressure boundary conditions are discussed, respectively, in Sections 4 and 5.

4. MAGNETIC BOUNDARY CONDITIONS (THE CHAPMAN-FERRARO PROBLEM)

In a visionary paper, Chapman and Ferraro (1931) suspected that occasional outbursts of the solar wind might create a boundary between the solar plasma stream and Earth's magnetic field, shielding the terrestrial field effectively against the solar wind (see the middle panel in Figure 11.2.1). They concluded that currents must flow on that boundary such that their magnetic field exactly cancels the Earth's magnetic field on the solar wind side of the boundary, but those shielding magnetopause currents generate an additional field on the earthward side of the boundary that might explain periods of enhanced magnetic activity at the Earth's surface.

This section contains the mathematical formalism for solving the Chapman-Ferraro problem. The solution leads to the confinement of the terrestrial magnetic field to the space inside the magnetosphere. For simple, prescribed magnetopause shapes, the potential problem with Neumann's boundary values has analytic solutions.

4.1. THE MAGNETIC BOUNDARY VALUE PROBLEM

Instead of explicitly calculating the magnetic field from Chapman-Ferraro currents on the magnetopause (Beard, 1973), it is much more convenient to solve a corresponding magnetic boundary value problem. Inside the magnetospheric cavity, the total magnetic field, $\mathbf{B}_{total}$, is given by

$$\mathbf{B}_{total} = \mathbf{B}_{sources} + \mathbf{B}_{CF} \qquad (4.1)$$

where the field $\mathbf{B}_{sources}$ may contain the geomagnetic main field, the field produced by magnetotail currents, and any other magnetic field that has its sources inside the magnetosphere. The yet unknown field $\mathbf{B}_{CF}$ is an additional magnetic field produced by the Chapman-Ferraro currents which flow on the magnetopause. The field $\mathbf{B}_{CF}$ is determined by the shape of the magnetopause and by the magnetic boundary conditions specified on the entire magnetopause. General boundary conditions $f(\mathbf{x}_{mp})$ at the magnetopause ($\mathbf{x}_{mp}$) can be written as

$$\mathbf{n} \cdot \mathbf{B}_{total} = f(\mathbf{x}_{mp}) \tag{4.2}$$

Because the Chapman-Ferraro currents flow on the boundary, the corresponding magnetic field in Equation 4.1,

$$\mathbf{B}_{CF} = -\nabla \, \Phi_{CF} \tag{4.3}$$

is current free inside the magnetosphere and can thus be derived from a scalar potential Φ_{CF}. The normal derivative of Φ_{CF} represents magnetic boundary conditions at the magnetopause. The combination of Equations 4.1 through 4.3 leads to a scalar Neumann boundary value problem of the potential theory (Voigt, 1981), namely,

$$\frac{\partial \Phi_{CF}}{\partial n} = \mathbf{n} \cdot \mathbf{B}_{sources} - f(\mathbf{x}_{mp}) \tag{4.4}$$

$$\nabla^2 \, \Phi_{CF} = 0 \tag{4.5}$$

Note that the "closed" magnetosphere, $f(\mathbf{x}_{mp}) = 0$ with $\mathbf{n} \cdot \mathbf{B}_{total} = 0$, remains as a special case in a class of more general solutions to the boundary value problems in Equations 4.4 and 4.5. The Chapman-Ferraro field shown in Equation 4.3 depends on the shape of the magnetopause (see Section 4.3).

Realistic magnetopause shapes do not usually represent surfaces of coordinate systems for which Laplace's equation (4.5) has separable solutions. Thus, elegant analytic solutions of the Chapman-Ferraro problem can rarely be found. Exceptions are the analytic models of Stern (1985) and Voigt (1981).

Figure 11.4.1 depicts a magnetospheric field configuration for the three-dimensional Voigt (1981) magnetosphere. Plotted are field lines in the noon-midnight meridian for two different dipole tilt angles ψ. The magnetopause can be regarded as the outermost shell of magnetic field lines. Magnetopause boundary conditions are chosen for a "closed" magnetosphere, that is, $\mathbf{n} \cdot \mathbf{B}_{total} = 0$ in Equation 4.2. The center field lines of the two polar cusps end mathematically in two magnetic "neutral points" at the magnetopause where the total magnetic field vanishes completely.

The Chapman-Ferraro current system $\mathbf{j}_{CF}(\mathbf{x}_{mp})$ on the magnetopause is given by the difference in the tangential components of the total magnetic field on both sides of the magnetopause. If the field outside the magnetosphere is exactly zero, then we obtain the surface current density

$$\mathbf{j}_{CF}(\mathbf{x}_{mp}) = -\frac{1}{\mu_0} \mathbf{n} \times (\mathbf{B}_{sources} + \mathbf{B}_{CF}) \tag{4.6}$$

The magnetopause normal unit vector points outward in the above notation. The Chapman-Ferraro currents seen in Equation 4.6 result from a solution to the potential problem in Equations 4.1 through 4.5; they do not need to be known explicitly in this formalism.

It was already mentioned in the introduction that the surface currents flow entirely in closed loops on the magnetopause. Figure 11.4.2 shows an example of Chapman-Ferraro current stream lines in Equation 4.6 projected into the X_{GSM}-Z_{GSM} plane. The corresponding magnetospheric field configuration (Voigt, 1981) appears in the upper panel of Figure 11.4.1. The surface currents circle around the two singular polar cusp neutral points denoted "N". The surface current density in Equation 4.6 reaches maximum values at the subsolar point and decreases with increasing distance down the tail.

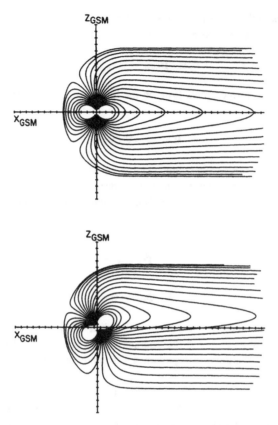

Figure 11.4.1 Magnetic field configuration of the three-dimensional Voigt (1981) magnetosphere in the noon-midnight meridian for the dipole tilt angles $\psi = 0°$ (top) and $\psi = +35°$ (bottom). Field lines are spaced every 2° in magnetic latitude, starting at 66°. Magnetopause boundary conditions are chosen for a "closed" magnetosphere, that is, $\mathbf{n} \cdot \mathbf{B}_{total} = 0$ in Equation 4.2. The center field lines of the two polar cusps end in two magnetic "neutral points" at the magnetopause where the total magnetic field vanishes completely. (From Voigt, G.-H. (1981). *Planet Space Sci.,* 29, 1. With permission.)

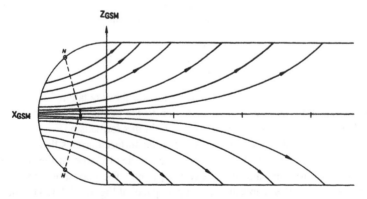

Figure 11.4.2 Chapman-Ferraro current system from Equation 4.6 corresponding to the magnetic field model in the top panel of Figure 10. The surface currents form closed loops on the magnetopause and circle around the two singular polar cusp neutral points denoted "N". The surface current density from Equation 4.6 reaches *maximum* values at the subsolar point and decreases with increasing distance down the tail.

4.2. AN ANALYTIC EXAMPLE: THE DIPOLE IN A SPHERE

As an example of the boundary value problem in Equations 4.1 through 4.5, let us calculate the Chapman-Ferraro field in Equation 4.3 for a dipole located at the center of a spherical magnetopause with radius R (see the top panel in Figure 11.2.1). Thus, the magnetic source term in Equation 4.1 reduces to a dipole field, $\mathbf{B}_{sources} = \mathbf{B}_D$, given by Equations 2.8 through 2.10. For simplicity, let us specify boundary conditions for a closed magnetosphere, that is, $\mathbf{n} \cdot \mathbf{B}_{total} = f$ $(\mathbf{x}_{mp}) = 0$, in Equation 4.2. With these assumptions the boundary conditions in Equation 4.4 at the magnetopause, $r = R$, read

$$\frac{\partial \Phi_{CF}}{\partial n} = \mathbf{n} \cdot \mathbf{B}_D \tag{4.7}$$

$$\frac{\partial \Phi_{CF}}{\partial r} = -\frac{2\,M_D}{R^3} \cos\theta \tag{4.8}$$

The two solutions of Laplace's equation (4.5) for the inner space, $r \leq R$, and for the outer space, $r \geq R$, read in spherical coordinates,

$$\Phi_{CF}^{in} = \sum_{n=1}^{\infty} \sum_{m=0}^{n} (Y_n^m(\theta, \phi)) \frac{1}{n} \frac{r^n}{R^{n-1}} \qquad r \leq R \tag{4.9}$$

$$\Phi_{CF}^{ex} = -\sum_{n=1}^{\infty} \sum_{m=0}^{n} (Y_n^m(\theta, \phi)) \frac{1}{n+1} \frac{R^{n+2}}{r^{n+1}} \qquad r \geq R \tag{4.10}$$

The spherical harmonics $Y_n^m(\theta, \phi)$ are defined by Equation 2.6. Note that the normal derivatives of both the external and internal potentials in Equations 4.9 and 4.10 are the same at $r = R$ which guarantees, in accordance with the condition in Equation 3.5, that the normal component of the magnetic field is continuous at the magnetopause. In comparison with Equation 4.8 only the terms $n = 1$ and $m = 0$ remain in Equations 4.9 and 4.10. Thus we obtain for the dipole at the center of the sphere,

$$\Phi_{CF}^{in} = -\frac{2\,M_D}{R^3} \cos\theta \cdot r \qquad r \leq R \tag{4.11}$$

$$\Phi_{CF}^{ex} = +M_D \cos\theta \cdot \frac{1}{r^2} \qquad r \geq R \tag{4.12}$$

Note that the external solution in Equation 4.12 for the Chapman-Ferraro potential exactly cancels the dipole potential in Equation 2.7 in the region $r \geq R$ outside the magnetosphere which indicates perfect shielding according to the assumed boundary condition $\mathbf{n} \cdot \mathbf{B}_{total} = 0$ in this example. The magnetic field components in Equation 4.3 for the region $r \leq R$ follow from Equation 4.11, namely,

$$B_{CF,\,r} = +\frac{2\,M_D}{R^3} \cos\theta \tag{4.13}$$

$$B_{CF,\,\theta} = -\frac{2\,M_D}{R^3} \sin\theta \tag{4.14}$$

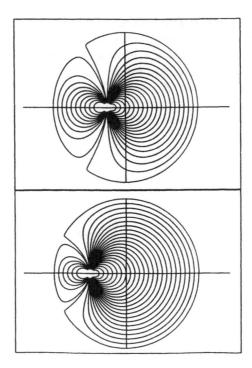

Figure 11.4.3 Magnetic dipole shifted along the x_{GSM}-axis within a spherical magnetopause. Magnetopause boundary conditions are chosen for a "closed" magnetosphere, that is, $\mathbf{n} \cdot \mathbf{B}_{total} = 0$ in Equation 4.2. With increasing eccentricity, the two polar cusps move toward lower magnetic latitudes. This effect corresponds to an increased magnetospheric compression by the solar wind. The two magnetic field configurations represent analytic solutions to the boundary value problem in Equations 4.4 and 4.5 in spherical coordinates.

The field-line configuration in the top panel of Figure 11.2.1 results from a superposition of the dipole field in Equations 2.8 and 2.9 and the Chapman-Ferraro field in Equations 4.13 and 4.14. The field components in Equations 4.13 and 4.14 represent a uniform magnetic field inside the spherical magnetopause, namely,

$$B_{CF, x} = B_{CF, y} = 0 \qquad B_{CF, z} = +\frac{2\,M_D}{R^3} = const \qquad (4.15)$$

Thus, the shielding Chapman-Ferraro field $\mathbf{B}_{CF}$ is anti-parallel to the Earth's dipole moment M_D.

If the magnetic dipole is shifted from the center of the sphere along the x_{GSM}-axis, then the configuration in the top panel of Figure 11.2.1 changes into the configurations plotted in Figure 11.4.3. The potential problem is still analytic in this case, but instead of Equations 4.11 and 4.12 for $n = 1$ and $m = 0$, the general Chapman-Ferraro potentials in Equations 4.9 and 4.10 for $n \geq 1$ and $m = 0$ have to be evaluated (Voigt, 1978). This exercise demonstrates that the two magnetospheric polar cusps in Figure 11.4.3 are, in terms of their topology, nothing other than the shifted magnetic northern and southern poles of the shielded dipole in the top panel of Figure 11.2.1.

The following example is significant for the overall topology of the magnetosphere, because it distinguishes the "closed" from the "open" magnetic field configuration: a simple superposition of the magnetic dipole field in Equations 2.8 and 2.9 and a uniform interplanetary field, $\mathbf{B}_{IMF} = \{0, 0, \pm B_{IMF}\, \mathbf{e}_z\}$, leads to the field line configurations in Figure 11.4.4. There are no shielding Chapman-Ferraro currents in this case, because Equations 4.1 and 4.2 read, respectively,

$$\mathbf{B}_{total} = \mathbf{B}_D + \mathbf{B}_{IMF} + \mathbf{B}_{CF} \qquad (4.16)$$

$$f(\mathbf{x}_{mp}) = \mathbf{n} \cdot (\mathbf{B}_D + \mathbf{B}_{IMF}) \qquad (4.17)$$

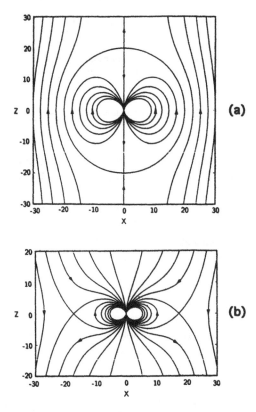

Figure 11.4.4 Simple superposition of a magnetic dipole field in Equations 2.8 and 2.9 and a uniform interplanetary field $\mathbf{B}_{IMF}$ $\{0, 0, \pm B_{IMF}\, \mathbf{e}_z\}$. There is no shielding magnetopause in this case. In (a) the IMF is positive (northward) and anti-parallel to the magnetic dipole moment $\{-M_D\, \mathbf{e}_z\}$. In (b) the IMF is negative (southward) and parallel to the dipole moment. If a realistic solar wind flow was added, streaming from the left-hand side, then the closed configuration would change into Figure 11.1.2, and the open configuration would change into Figure 11.1.4. (From Birn, J. et al. (1992). *J. Geophys. Res.*, 97, 14833. 1992 Copyright by the American Geophysical Union. With permission.)

It is therefore $\mathbf{B}_{CF} = 0$, according to Equations 4.3 and 4.5. In Figure 11.4.4a the IMF is positive (northward) and anti-parallel to the magnetic dipole moment $\{-M_D\, \mathbf{e}_z\}$ which leads to a "closed" magnetosphere. In Figure 11.4.4b the IMF is negative (southward) and parallel to the dipole moment which leads to an "open" magnetosphere.

Birn et al. (1992) pointed out the following consequences: if a realistic solar wind flow, streaming from the left-hand side, was added to the scenarios in Figure 11.4.4, then the closed configuration would change conceptually into Figure 11.1.2, and the open configuration would change into Figure 11.1.4. Although this last example is oversimplified, one can clearly see that the fundamental difference between the "open" and "closed" magnetospheres may be understood in terms of magnetic boundary conditions specified at the magnetopause.

4.3. THE MAGNETOPAUSE FORM FACTOR f_0

The strength of the combined dipole and Chapman-Ferraro fields at the magnetopause subsolar stagnation point depends on the shape of the magnetopause. For many analytic estimates, it is quite useful to define a magnetopause form factor f_0 as follows,

$$f_0 = \frac{|\mathbf{B}_D + \mathbf{B}_{CF}|}{2|\mathbf{B}_D|} \tag{4.18}$$

In the middle panel of Figure 11.2.1, the magnetopause is an infinitely extended plane. The corresponding magnetic field configuration can be computed by placing an identical image dipole

Table 11.4.1   **Magnetopause Form Factor f_0 Defined in Equation 4.18**

$f_0 = 1.00$	Infinite plane magnetopause: shielded dipole
$f_0 = 1.16$	Realistic magnetopause shape with tail field
$f_0 = 1.18$	Voigt (1981): shielded dipole
$f_0 = 1.20$	Stern (1985): shielded dipole
$f_0 = 1.50$	Spherical magnetopause: shielded dipole

on the other side of the plane such that the magnetopause is positioned exactly in the middle between the two dipoles. For that configuration, it is $|\mathbf{B}_{CF}| = |\mathbf{B}_D|$ at the subsolar point, and hence $f_0 = 1.0$ in Equation 4.18.

The subsolar point of the spherical magnetopause is located at $\theta = 90°$, and r = R. Thus, a comparison of Equations 2.9 and 4.14 gives $|\mathbf{B}_{CF}| = 2\,|\mathbf{B}_D|$, hence $f_0 = 1.5$ in Equation 4.18. The plane and the spherical magnetopause shapes represent extreme geometries. Thus, the range of the form factor f_0 can be given as

$$1.0 < f_0 < 1.5 \qquad (4.19)$$

It is $f_0 = 1.16$ for realistic magnetopause geometries which include the effect of an additional magnetotail field (see Table 11.4.1). That value holds for a rather wide range of solar wind pressure conditions. The f_0 values for the analytic models of Voigt (1981) and Stern (1985) are listed in Table 11.4.1 for information.

5.   PRESSURE BOUNDARY CONDITIONS (THE MAGNETOPAUSE GEOMETRY)

This section presents satellite observations of magnetopause crossings which give us an empirical shape of Earth's magnetopause. From a theoretical point of view, the shape and location of the magnetopause should follow from the basic principles of ideal MHD. For mathematical convenience, the MHD momentum equation can be simplified such that the overall shape of the magnetopause results from a balance between the upstream solar wind ram pressure and the magnetospheric magnetic field energy density inside the magnetopause.

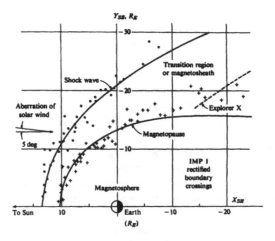

Figure 11.5.1   Position of Earth's bow shock and magnetopause during magnetically quiet conditions, according to IMP 1 satellite observations (Ness et al., 1964). This figure is one of the first reports on magnetosphere boundary crossings. (From Ness, N.F., Scearce, C.S., and Seek, J.B. (1964). *J. Geophys. Res.,* 69, 3531. 1964 Copyright by the American Geophysical Union. With permission.)

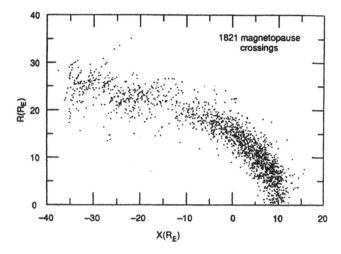

Figure 11.5.2 A total of 1821 magnetopause crossings observed throughout the last couple of decades (Sibeck et al., 1991; Fairfield, 1991). The crossings are plotted in a X_{GSM} − R coordinate system with R = $[Y_{GSM}^2 + Z_{GSM}^2]^{1/2}$. (From Sibeck, D.G., Lopez, R.E., and Roelof, E.C. (1991). *J. Geophys. Res.*, 96, 5489. 1991 Copyright by the American Geophysical Union. With permission.)

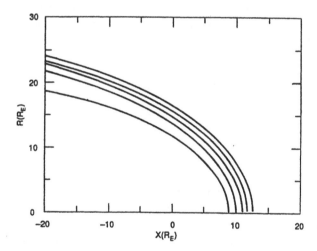

Figure 11.5.3 Average magnetopause curves ordered with respect to five different ranges of the solar wind pressure (Sibeck et al., 1991). These curves are derived from the data set shown in Figure 11.5.2. The shapes of the curves are self-similar in that the ratio between the subsolar stand-off distance and the magnetotail radius remains constant. (From Sibeck, D.G., Lopez, R.E., and Roelof, E.C. (1991). *J. Geophys. Res.*, 96, 5489. 1991 Copyright by the American Geophysical Union. With permission.)

5.1. OBSERVATIONAL FINDINGS

One of the earliest observations of the magnetopause crossings are due to Ness et al. (1964) and are shown in Figure 11.5.1. Fifteen years later, Formisano et al. (1979) were able to analyze nearly 1000 magnetopause crossings from various HEOS, OGO, and IMP satellites. Another 12 years later, Sibeck et al. (1991) and Fairfield (1991) were able to present a total of 1821 boundary crossings which are shown in Figure 11.5.2. That view gives an overall impression of the magnetopause shape and location with respect to the position of the Earth's main dipole. In Figure 11.5.3 the Earth's dipole is located at x_{GSM} = R = 0, and it is R = $[Y_{GSM}^2 + Z_{GSM}^2]^{1/2}$.

If the data set in Figure 11.5.2 is separated into five different solar wind pressure ranges (Sibeck et al., 1991), the curves in Figure 11.5.3 emerge. There is clear observational evidence that the dayside magnetopause moves toward the Earth with increasing solar wind pressure. This effect is explained by a (static) balance between the solar wind dynamic pressure and the energy density of the magnetospheric magnetic field just inside the magnetopause (see Section 5.3).

It is an important issue in magnetospheric physics to derive the shape and location of the magnetopause from the basic principles of ideal MHD. One might think that today's global MHD codes (see Section 10.3) should be able to calculate the magnetopause shape. Unfortunately, even the very best of those codes do not readily satisfy ideal MHD conditions right at the magnetopause discontinuity. Moreover, it is not clear whether or not the magnetosphere reaches a steady state which would resemble average magnetospheric conditions such as the ones displayed in Figures 11.5.2 and 11.5.3. Despite those constraints, one obtains from MHD the shape of the magnetopause under some simplifying assumptions that are presented in Section 5.2.

5.2. SELF-CONSISTENT BOUNDARY SHAPES

When the solar wind passes Earth's bow shock, most thermodynamic quantities and vector fields change in the transition. However, momentum must be conserved in the transition. Therefore, one may use the undisturbed upstream solar wind parameters to derive a pressure balance relation at the magnetopause itself.

Observations from satellite measurements allow us to adopt the following simplifying assumptions right at the magnetopause: inside the boundary, the thermal plasma pressure and the kinetic energy density are observed to be small in comparison with the total magnetic field energy density. In the upstream solar wind, the kinetic energy density is the dominant quantity, as can be seen from Table 11.1.2. In addition, the following approximations are employed right at the magnetopause, namely,

$$|\mathbf{n} \cdot \mathbf{B}_{\text{total}}| \ll |\mathbf{n} \times \mathbf{B}_{\text{total}}| \tag{5.1}$$

$$2(\mathbf{B} \cdot \nabla)\mathbf{B} \ll \nabla B^2 \tag{5.2}$$

$$\rho(\mathbf{v}\,\mathbf{v}) \to \rho\, v_x^2 \tag{5.3}$$

Equation 5.1 says that, at the magnetopause, the normal component of the total magnetic field is much smaller than the tangential component. Thus the magnetosphere is basically closed under this approximation. The magnetic field magnitude $|\mathbf{B}_{\text{total}}|$ is given by Equation 4.1. Equation 5.2 employs the so-called "plane approximation" indicating that magnetic tension is negligible at the boundary in comparison with magnetic field gradients perpendicular to the magnetopause. Finally, Equation 5.3 says that the solar wind is approximated by a cold plasma stream with only one velocity component in the negative x-direction. The approximation in Equation 5.3 is sketched in Figure 11.5.4 and may be compared with a realistic MHD flow which is shown in Figure 11.5.5.

Momentum conservation in Equation 10.13 must hold inside and outside the magnetopause. One obtains from Equation 10.13 under the assumptions in Equations 5.1 through 5.3 the preliminary pressure balance relation for steady-state conditions, namely,

$$\rho\, v_x^2 = \frac{1}{2\,\mu_0}\, (\mathbf{B}_D - \nabla\, \Phi_{\text{CF}})^2 \tag{5.4}$$

which in this form holds at the Earth-sun line. Only the pressure perpendicular to the magnetopause contributes to the pressure balance with the internal magnetic field. Thus the velocity vector v_x in Equation 5.3 and the magnetopause surface normal vector $\mathbf{n}$ define the reflection angle χ according to

$$\cos\chi = -\mathbf{n} \cdot \mathbf{v} \tag{5.5}$$

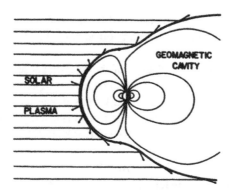

Figure 11.5.4 Sketch of the approximations from Equations 5.3 through 5.5. In the pressure balance relation Equation 5.4 the solar wind is approximated by a cold plasma stream with only one velocity component in the negative x-direction. The effects of an MHD flow in the magnetosheath region have been neglected as well. (This figure was drawn by David B. Beard.)

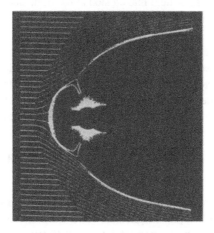

Figure 11.5.5 Flow lines of the undisturbed solar wind outside the bow shock and in the magnetosheath region. The three-dimensional MHD simulation shows that the plasma stream smoothly flows around the magnetosphere which is in contrast to the approximation in Equation 5.3 sketched in Figure 11.5.4. (From Hasegawa, A. and Sato, T. (1989). *Space Plasma Physics,* Springer-Verlag, Tokyo. With permission.)

which is sketched in Figure 11.1.6. The solar wind pressure P_{SW} perpendicular to the magnetopause is then given by

$$P_{SW} = k \, m_p \, n_0 \, v_x^2 \cos^2 (\chi) \tag{5.6}$$

Here, m_p is the proton mass, n_0 and v_0 are the density and velocity of the undisturbed solar wind. The factor $k \approx 0.88$ results from the gas dynamic approximation for the magnetosheath flow (Spreiter and Jones, 1963; Spreiter et al., 1966; Spreiter and Alksne, 1969). The value $k < 1$ compensates the rather drastic approximation in Equation 5.3 and is explained by the fact that in reality the (laminar) solar wind streams smoothly around the magnetospheric obstacle (see Figure 11.5.5). With Equations 5.4 through 5.6 the pressure balance at the magnetopause can now be written as

$$k \, m_p \, n_0 \, (\mathbf{n} \cdot \mathbf{v_x})^2 = \frac{1}{2 \, \mu_0}[\mathbf{n} \times (\mathbf{B}_D - \nabla \, \Phi_{CF})]^2 \tag{5.7}$$

The shape of the magnetopause is then calculated by an iterative numerical procedure between the Neumann boundary value problem as seen in Equations 4.4 and 4.5 and the pressure balance relation in Equation 5.7.

One of the first self-consistent calculations of the magnetopause shape is due to Mead and Beard (1964). Their work is summarized in a tutorial review paper by Beard (1973). The asymmetric shape of the magnetopause for the tilted Earth's dipole has been calculated by Olson (1969) and by Choe et al. (1973). Useful analytic mathematical representations of the magnetopause shape are due to Cap and Leubner (1974) and Kosik (1977).

5.3. THE SUBSOLAR STAND-OFF DISTANCE

The stand-off distance, r_0, is a measure for the overall size of the magnetosphere (see, for example, Figure 11.5.3); it is measured between the center of the Earth and the subsolar stagnation point at the magnetopause. The numerical value of r_0 depends primarily on the upstream solar wind conditions, namely, particle density n_0 and streaming velocity v_x. For a given solar wind pressure, the IMF causes additional small, but physically relevant, changes in the stand-off distance that can be explained by the process of field line erosion at the front-side of the magnetosphere (see, for example, Figure 11.1.3).

5.3.1. Theoretical Analysis

An analytic formula for the stand-off distance can be derived from the pressure balance between the solar wind and the magnetosphere at the Earth-sun line. In contrast to Equation 5.7 the pressure balance relation now includes the interplanetary field $\mathbf{B}_{IMF}$. With the solar wind pressure in Equation 5.6, one obtains from Equation 5.7 for $\cos\chi = 1$ the expression

$$P_{SW} + \frac{1}{2\,\mu_0}\,\mathbf{B}_{IMF}^2 = \frac{1}{2\,\mu_0}\,[(\mathbf{B}_D + \mathbf{B}_{CF} + f_1\,\mathbf{B}_{IMF})]^2 \tag{5.8}$$

The new form factor $f_1 \geq 1$ accounts for the fact that the magnetosphere is open and that, under the influence of a southward IMF, the magnetotail currents increase. Conversely, the tail currents decrease when the IMF points northward (Voigt and Toffoletto, 1993). According to a three-dimensional model for the open magnetosphere developed by Toffoletto and Hill (1993), one finds the approximate value $f_1 = 2.05$ (Toffoletto and Voigt, 1993). A rigorous theoretical justification for f_1 on the level of the MHD theory has yet to be established.

The Chapman-Ferraro field $\mathbf{B}_{CF}$ on the right-hand side of Equation 5.8 does not change under the influence of the field $\mathbf{B}_{IMF}$, because it is assumed that the IMF completely penetrates the magnetopause. Two independent theoretical arguments by Voigt and Fuchs (1979) and by Crooker et al. (1990) suggest the assumption of complete IMF penetration.

On the left-hand side of Equation 5.8 the term $\mathbf{B}_{IMF}^2$ can safely be neglected against the solar wind pressure P_{SW} (see Table 11.1.2). However, the IMF has a non-negligible first-order influence inside the magnetosphere because the field $\mathbf{B}_{IMF}$ is included by a linear superposition of vector fields. With the magnetopause form factor in Equation 4.18, and by neglecting second-order terms in Equation 5.8, one obtains

$$2\,\mu_0\,P_{SW} = \left[2 f_0 \frac{\cos(\psi)\,B_0}{r_0^3}\,\mathbf{e}_z + f_1\,\mathbf{B}_{IMF}\right]^2 \tag{5.9}$$

where r_0 is the stand-off distance in units of Earth radii (1 Re = 6371 km), and $B_0 = 30,319$ nT denotes the Earth's equatorial field strength explained in Section 2.2. The dipole tilt angle ψ is explained in Section 2.3. To derive a handy formula for r_0, it is useful to change into electromagnetic cgs units (emu) for the remainder of this section. In the (emu) system, it is $2\,\mu_0 = 8\pi$,

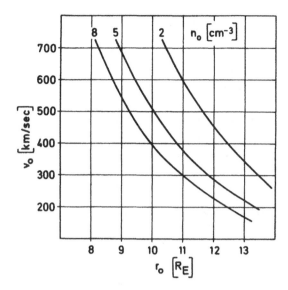

Figure 11.5.6 Magnetopause subsolar stand-off distance r_0 as a function of the solar wind speed $v_0 = v_x$ for various solar wind densities n_0. The curves are obtained from Equation 5.10 assuming an untilted Earth's dipole ($\cos\psi = 1$) and a closed magnetosphere, that is, $B_{IMF} = 0$.

and the magnetic flux density **B** and the magnetic field strength **H** are formally identical. Thus, the combination of Equations 5.6 and 5.9 leads to

$$r_0 = \left(\frac{f_0^2}{k} \cdot \frac{\cos(\psi)^2 \, B_0^2}{2\pi \, m_p \, n_0 \, v_x^2}\right)^{1/6} \cdot \left(1 + \frac{f_1 \, B_{IMF}}{\sqrt{2\pi \, k \, m_p \, n_0 \, v_x^2}}\right)^{1/6} \tag{5.10}$$

It is assumed in Equation 5.10 that the IMF points either in a northward or in a southward direction, namely,

$$\mathbf{B}_{IMF} = \pm B_{IMF} \, \mathbf{e}_z \tag{5.11}$$

The following two sections deal with visualizations of Equation 5.10 with the condition in Equation 5.11. Let us first consider the closed magnetosphere and then the effect of IMF-induced field-line erosion at the subsolar magnetopause.

5.3.2. The Closed Magnetosphere

Figure 11.5.6 shows the stand-off distance r_0 as a function of the solar wind speed v_x for various plasma densities n_0. The curves apply to the untilted dipole, $\cos\psi = 1$. The magnetosphere is assumed to be closed, and the field $\mathbf{B}_{IMF}$ does not appear inside the magnetosphere. Thus, it is $\mathbf{B}_{IMF} = 0$ in Equation 5.10 for the curves in Figure 11.5.6.

Fairfield (1971) quotes $r_0 = 10.8$ Re for magnetic quiet conditions. This result, derived purely from observations, corresponds to average quiet solar wind conditions $n_0 = 5$ cm^{-3} and $v_x = 400$ km/s. In contrast, Sibeck et al. (1991) quote $r_0 = 11.3$ Re for average solar wind conditions without the IMF. This corresponds to a solar wind stream with the parameters $n_0 = 5$ cm^{-3} and $v_x = 350$ km/s.

Figure 11.5.7 shows the effect of the dipole tilt angle ψ on the stand-off distance r_0. For this computation, the closed magnetosphere is balanced by a solar wind stream with the parameters $n_0 = 5$ cm^{-3} and $v_x = 400$ km/s. The stand-off distance decreases with increasing dipole tilt, because the total magnetic field strength at the magnetopause decreases as the polar cusp neutral

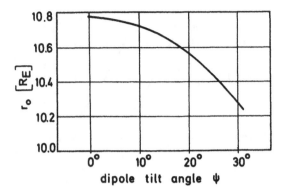

Figure 11.5.7 Effect of the dipole tilt angle ψ on the stand-off distance r_0. For this computation, it is $B_{IMF} = 0$ in Equation 5.10, and the magnetosphere is balanced by a solar wind stream with the parameters $n_0 = 5$ cm^{-3} and $v_x = 400$ km/s.

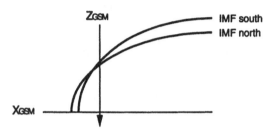

Figure 11.5.8 The effect of magnetic field erosion on the location of the dayside magnetopause. When the IMF has a southward component, magnetic flux is eroded from the subsolar magnetic field, and the magnetopause moves inward. Conversely, when the IMF has a northward component magnetic flux is added, and the magnetopause moves outward (compare with Figure 11.1.3).

point moves closer toward the Earth-sun line (compare, for example, the two field-line configurations in Figure 11.4.1).

5.3.3. Erosion Effects Caused by the IMF

The effect of magnetic field erosion on the location of the dayside magnetopause is sketched in Figure 11.5.8. When the IMF has a southward component, magnetic flux is eroded from the subsolar magnetic field, and the magnetopause moves inward. Conversely, when the IMF has a northward component, magnetic flux is added, and the magnetopause moves outward (see also Figure 11.1.3).

Figure 11.5.9 shows the IMF influence on the stand-off distance for the untilted dipole, $\cos\psi$ = 1. The upper curve indicates quiet solar wind conditions with the parameters $n_0 = 5$ cm^{-3} and $v_x = 350$ km/s. The compressed state of the magnetosphere arises for the solar wind parameters $n_0 = 8$ cm^{-3} and $v_x = 600$ km/s. The nonlinear effect of IMF-induced magnetic field erosion on the front side of the magnetosphere is obvious under quiet solar wind conditions. By inspecting Equation 5.10 one finds that the term that contains B_{IMF} is most prominent for a weak solar wind pressure.

The theoretical results displayed in Figure 11.5.9 are in qualitative agreement with observations reported by Sibeck et al. (1991). Their data suggest that the magnetopause moves inward more rapidly for southward IMF than for northward IMF variations.

The main results obtained from the pressure balance relation in Equation 5.10 are the following: (1) the subsolar stand-off distance is mainly determined by the solar wind ram pressure; (2) for a given solar wind pressure, smaller changes in the stand-off distance are caused by the IMF B_{IMF}

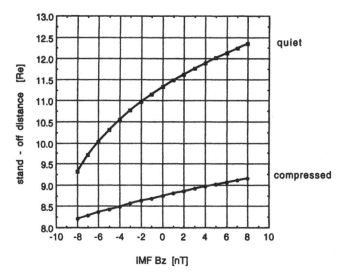

Figure 11.5.9 IMF influence on the stand-off distance r_0 for the untilted dipole, $\cos\psi = 1$. The upper curve indicates quiet solar wind conditions with the parameters $n_0 = 5$ cm^{-3} and $v_x = 350$ km/s. The lower curve arises for the solar wind parameters $n_0 = 8$ cm^{-3} and $v_x = 600$ km/s. The non-linear effect of IMF-induced magnetic field erosion on the front side of the magnetosphere is most obvious under quiet solar wind conditions.

such that the magnetopause moves earthward for southward IMF and moves sunward for northward IMF; and (3) the IMF has a stronger influence on the stand-off distance when the solar wind pressure is weak and vice versa.

6. THE MAGNETOTAIL

This section describes the main features of Earth's magnetotail. The magnetotail has been observed to stretch beyond a distance of 200 Re (1 Re = 6371 km). The magnetic field configuration of the near-Earth tail (earthward of about $x_{GSM} = -30$ Re) may be described in terms of quasistatic MHD equilibria. In contrast, the distant tail (tailward of about $x_{GSM} = -60$ Re) is probably dominated by a balance between plasma flows and Lorentz forces. On the level of the one-fluid MHD theory, no simple approximation exists for modeling the transition region between those two limits.

It is not clear whether or not Earth's magnetotail ever reaches a steady state. Thus, the empirical findings reported in this section represent average magnetospheric conditions which guide us to construct quantitative magnetospheric **B**-field models for average and slowly time-dependent periods.

Field-aligned (Birkeland) currents parallel to the magnetic field, $\mathbf{j}_\parallel = (\mathbf{j} \cdot \mathbf{B}) \mathbf{B}/B^2$, arise near the earthward inner edge of the tail plasma sheet (region 2 currents) and near the equatorial flanks of the tail, at the so-called low-latitude boundary layer (region 1 currents). Field-aligned currents couple the high-latitude ionosphere with the magnetospheric tail region (see Section 7.2 and Chapter II/12).

6.1. MAGNETOTAIL FORMATION

The formation of Earth's magnetotail remains one of the fundamental questions in magnetospheric physics. Pilipp and Morfill (1978) pointed out a mechanism according to which the plasma sheet is filled from a source in the plasma mantle (see Figure 11.6.1). In the plasma mantle, close to the magnetopause, plasma streams tailward, but $\mathbf{E} \times \mathbf{B}$ drift (solar wind convection electric field and magnetotail magnetic field) causes that plasma to eventually convect toward the center of the

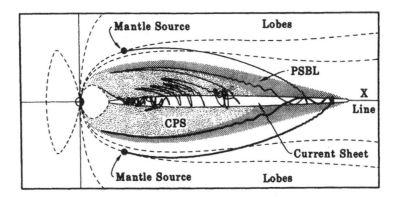

Figure 11.6.1 Schematic view of the mechanism that fills the magnetotail plasma sheet with plasma originating from a source in the plasma mantle. In the plasma mantle, close to the magnetopause, plasma streams tailward, but **E** × **B** drift lets that plasma eventually convect toward the center of the plasma sheet. (From Ashour-Abdalla, M. et al. (1993). *J. Geophys. Res.,* 98, 5651. 1993 Copyright by the American Geophysical Union. With permission.)

plasma sheet. The process sketched in Figure 11.6.1 acts like an energy filter, such that one would expect lower energetic particles to reach the tail plasma sheet at smaller distances from the Earth.

Owing to the energy discrimination of the **E** × **B** drift, the resulting plasma pressure tensor $P(x_{ps})$ in the plasma sheet depends very sensitively on the entry mechanism of magnetosheath plasma into the magnetosphere: Pilipp and Morfill (1978) calculated different plasma energy distribution functions at the center of the plasma sheet, depending on whether the magnetosheath plasma enters the plasma mantle near the dayside polar cusps or along the whole tail magnetopause, following interconnected "open" tail field lines. A kinetic simulation of the above plasma sheet filling mechanism has been carried out more recently by Ashour-Abdalla et al. (1993). The construction of a plasma sheet pressure tensor $P(x_{ps})$ from individual particle trajectories remains a formidable task.

The filling level of the magnetotail, and thus the corresponding pressure tensor $P(x_{ps})$, determine the overall stretching of the entire magnetic field configuration. As the magnetotail stretches with increasing plasma loading, the magnetic field strength decreases at the center of the plasma sheet. Figure 11.6.2 displays that effect in terms of a two-dimensional magnetosphere which satisfies the MHD equilibrium condition $(\nabla \times \mathbf{B}) \times \mathbf{B} = \mu_0 \nabla P$ for isotropic thermal plasma. The plasma β-parameter (Equation 1.1), measured at the center of the plasma sheet, increases with increasing plasma loading (more theoretical details about MHD equilibria are explained in Section 6.2).

It is important to point out that the near Earth part of the tail plasma sheet is populated with plasma from the ionosphere as well (Lennartsson and Shelley, 1986). Thus both the solar wind and the ionosphere are major plasma sources for filling the magnetotail plasma sheet with thermal plasma.

6.2. MAGNETOTAIL MHD EQUILIBRIA

If one wishes to understand the large-scale structure and the stability of Earth's magnetotail plasma sheet, one has to start any analysis from a magnetic field configuration that is in MHD force equilibrium. In order to take into account collective plasma effects, as they result from particle dynamics in a high-β (β ≫ 1) configuration, one commonly constructs equilibria at least on the level of the one-fluid ideal MHD. In other words, one has to solve the equilibrium equations

$$\mathbf{j} \times \mathbf{B} = \nabla P \qquad (6.1)$$

$$\nabla \times \mathbf{B} = \mu_0 \mathbf{j} \qquad (6.2)$$

$$\nabla \cdot \mathbf{B} = 0 \qquad (6.3)$$

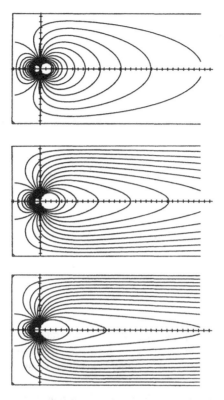

Figure 11.6.2 Two-dimensional MHD equilibrium magnetosphere. The field-line plots demonstrate how the entire magnetotail becomes stretched with increasing plasma loading in the plasma sheet. The magnetic field configurations result from solutions to the Grad-Shafranov Equation 6.4 with the pressure function Equation 6.7. The asymptotic plasma β values from Equation 6.8 at $z = 0$ and $|x| \gg 1$ are $\beta_0 = 10$ (top), 100 (middle), and 1000 (bottom).

with appropriate magnetopause boundary conditions for the magnetic field and plasma sheet boundary conditions for the thermal plasma pressure. Equation 6.1 represents Equation 10.1 in the quasistatic approximation. In that approximation of the MHD, the convection flow velocities are small in comparison with the corresponding Alfvén velocities, and the characteristic travel time for a plasma element is long in comparison with the travel time of an Alfvén wave. Then the flow field is decoupled from the MHD equation of motion. According to observations in the plasma sheet for magnetically quiet periods, the pressure tensor **P** may be replaced by a scalar pressure function P such that $\nabla \cdot \mathbf{P}$ reduces to ∇P in Equation 6.1.

General three-dimensional MHD equilibria for the entire magnetosphere are still not available. Thus it is useful to utilize the two-dimensional equilibrium theory in various degrees of approximations. The two-dimensional model displayed in Figure 11.6.2 is based on linear solutions to the Grad-Shafranov equation

$$\nabla^2 \alpha + \mu_0 \frac{dP}{d\alpha} = -\mu_0 \, j_{dipole} \tag{6.4}$$

where the pressure function is $P = P(\alpha)$. Equation 6.4 is completely equivalent to the three equilibrium equations, 6.1 through 6.3. The function $\alpha(x, z)$ is a magnetic flux function which satisfies the conditions $\mathbf{B} \cdot \nabla \alpha = 0$ and $\nabla \cdot \mathbf{B} = 0$. The variables x and z are used in the (GSM) coordinate system where the x-axis points toward the sun, and the x-z-plane contains the planetary dipole (see Section 2.3). In the isotropic limit, the plasma pressure $P(\alpha)$ is constant on magnetic field lines. The plasma currents are given by

$$j_y\,(\alpha) = \frac{dP}{d\alpha} \tag{6.5}$$

and the magnetization current which describes the planetary magnetic dipole depends on the dipole tilt angle ψ according to

$$\mu_0\,j_{\text{dipole}} = -M_D\!\left(\frac{\partial}{\partial x}\,\delta(x)\,\delta(z)\,\cos\psi - \delta(x)\,\frac{\partial}{\partial z}\,\delta(z)\,\sin\psi\right) \tag{6.6}$$

where M_D is the magnetic dipole moment. The dipole is located at the origin of the (GSM) coordinate system. As long as no thermodynamic conditions are specified, the plasma pressure function $P(\alpha)$ may assume the form

$$P(\alpha) = \frac{1}{2\,\mu_0}\,k^2\,\alpha^2 \tag{6.7}$$

With the pressure function in Equation 6.7 the Grad-Shafranov Equation 6.4 becomes a linear equation which can be solved analytically for a rectangular magnetopause (Fuchs and Voigt, 1979). Further mathematical details can be found in Sections 3.1 and 3.2 of Voigt's (1986) review paper.

As the thermal pressure in Equation 6.7 increases (k increases), owing to an unspecified plasma sheet filling mechanism, magnetotail currents develop and stretch the magnetic field lines accordingly (see Figure 11.6.2). The parameter k in Equation 6.7 can be written in terms of the plasma beta parameter β_0 at the center of the tail plasma sheet at some asymptotic distance down the tail. One obtains for $z = 0$ and $|x| \gg 1$ the expression

$$\beta_0 = \frac{2\,\mu_0\,P}{B^2} = \frac{k^2\,\alpha^2}{B_z^2} = \frac{k^2}{(\pi/2)^2 - k^2} \tag{6.8}$$

The magnetic field configurations in Figure 11.6.2 result from solutions to the Grad-Shafranov Equation 6.4 with the pressure function as seen in Equation 6.7. From top to bottom, the asymptotic plasma β values in Equation 6.8 are $\beta_0 = 10$ (top), 100 (middle), and 1000 (bottom).

The effects of plasma loading and magnetic field line stretching can be viewed, in a global sense, by computing the potential and free energies of the entire magnetosphere. The potential energy W and Grad's (1964) free energy F are expressed by the integrals

$$W = \int_V\!\left[\frac{B^2}{2\,\mu_0} + \frac{P}{(\gamma - 1)}\right] dV \tag{6.9}$$

$$F = \int_V\!\left[\frac{B^2}{2\,\mu_0} - P\right] dV \tag{6.10}$$

The two integrals extend over the entire volume of the magnetospheric cavity. The quantity γ is the polytropic index. It is $\gamma = 5/3$ for a monatomic plasma with three degrees of freedom.

Figure 11.6.3 demonstrates that an increased plasma loading and tail stretching lead to an increase of the potential energy W of the entire magnetosphere. Plotted are the energies in Equations 6.9 and 6.10 as functions of the plasma β-parameter β_0 in Equation 6.8. The two energies, W and F, are computed for the MHD equilibrium magnetosphere depicted in Figure 11.6.2. The

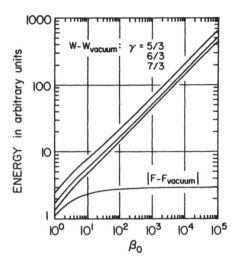

Figure 11.6.3 Potential energy in Equation 6.9 and free energy in Equation 6.10 as functions of the plasma β-parameter β_0 in Equation 6.8. The quantities $W_{vacuum} = F_{vacuum}$ indicate the β-independent energy density of the two-dimensional vacuum dipole field which has been removed from the integration. The plot demonstrates that plasma loading and magnetic field line stretching lead to a storage of potential energy in the entire magnetosphere. The two energies, W and F, are computed for the MHD equilibrium magnetosphere depicted in Figure 11.6.2.

β-independent two-dimensional vacuum dipole field is excluded from the integration in Equations 6.9 and 6.10 in order to avoid the dipole singularities as r approaches zero.

One of the unresolved questions in magnetospheric physics concerns the lower and upper limits of the function W in the real magnetosphere. The minimum potential energy, $W = W_{min}$, constitutes a yet unknown "ground state" of the magnetosphere (Voigt and Wolf, 1988). Beyond the maximum energy, $W > W_{max}$, stable plasma confinement becomes impossible. Then the magnetotail may experience a macroscopic instability which could manifest itself as a "magnetospheric substorm". In a substorm, part of magnetic and plasma energies are released into the Earth's ionosphere where that process leads to the display of magnificent northern lights (more details about substorms appear in Section 9).

6.3. THE NEAR EARTH TAIL

Figure 11.6.4 shows hourly average magnetic field vectors in Earth's magnetotail measured by Explorer 33 (Behannon, 1968). Field vectors are plotted as projections onto the X_{GSM}-Z_{GSM} plane. The sharp magnetic field reversal at $Z_{GSM} = 0$ suggests the existence of a magnetotail current sheet in the equatorial plane with currents flowing across the tail in westward direction. Typical plasma densities in the near Earth tail range is from 0.5 to 10 cm^{-3} with energies from 0.1 to 10 keV.

Figure 11.6.5 shows total magnetic field magnitudes in the magnetotail also measured by Explorer 33 (Behannon, 1968). Plotted are hourly averages for magnetically quiet conditions as a function of the geocentric distance down the tail. Selected are data for which the spacecraft was more than 2 Re away from the neutral sheet. Mihalov and Sonett (1968) and Slavin et al. (1985) match magnetotail magnetic field data with power laws,

$$|\mathbf{B}_{TAIL}| = \frac{A}{|X_{GSM}|^{+B}} \tag{6.11}$$

Useful exponential laws for the magnetic field decrease down the magnetotail are also given by Mihalov and Sonett (1968), namely,

$$|\mathbf{B}_{TAIL}| = C \exp(-|X_{GSM}|/D) \tag{6.12}$$

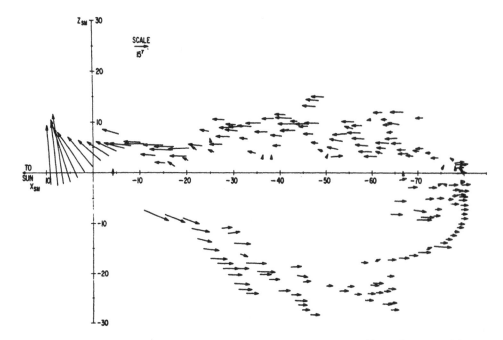

Figure 11.6.4 Average hourly magnetic field vectors in Earth's magnetotail measured by Explorer 33 (Behannon, 1968). Field vectors are plotted as projections onto the X_{GSM}-Z_{GSM} plane. The sharp magnetic field reversal at $Z_{GSM} = 0$ suggests the existence of a magnetotail current sheet in the equatorial plane with currents flowing across the tail in westward direction. (From Behannon, K.W. (1968). *J. Geophys. Res.*, 73, 907. 1968 Copyright by the American Geophysical Union. With permission.)

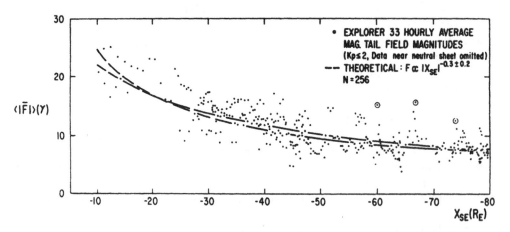

Figure 11.6.5 Magnetic field magnitudes in Earth's magnetotail measured by Explorer 33 (Behannon, 1968). Plotted are hourly averages for magnetically quiet conditions as a function of the geocentric distance down the tail. Selected are data for which the spacecraft was more than 2 R_E away from the neutral sheet. (From Behannon, K.W. (1968). *J. Geophys. Res.*, 73, 907. 1968 Copyright by the American Geophysical Union. With permission.)

The coefficients for both formulas 6.11 and 6.12 are listed in Table 11.6.1. Table 11.6.2 contains values of the equatorial magnetic B_z component at the very center of the tail plasma sheet at local midnight according to Explorer 33 and 35 magnetotail measurements (Behannon, 1970). For this analysis the plasma sheet is assumed to be in the plane of symmetry located at $Y_{GSM} = Z_{GSM} = 0$.

All the information contained in Figure 11.6.5, in Equations 6.11 and 6.12, and in Table 11.6.2 demonstrate that the magnetic field in Earth's tail region decreases slower with geocentric distance than the undisturbed dipole field which possesses an r^{-3} dependency (see Section 2.2). Again

Table 11.6.1 **Coefficients for Equations 6.11 and 6.12 Describing the Down-Tail Decrease of the Total Magnetic Field Strength in Earth's Magnetotail**

A	B	C	D	$|X_{GSM}|$	Ref.
223	0.798	27.4	50.4	10–66 Re	Mihalov and Sonett (1968)
191	0.736	26.8	59.4	10–81 Re	Mihalov and Sonett (1968)
125	0.53			20–220 Re	Slavin et al. (1985)

Table 11.6.2 **Equatorial Magnetic B_z Component at the Center of Earth's Plasma Sheet at Local Midnight According to Explorer 33 and 35 Magnetotail Measurements**

Geocentric distance $|X_{GSM}|$	Equatorial B_z
20–30 Re	3.52 nT
30–40 Re	1.87 nT
40–50 Re	1.01 nT
50–60 Re	0.69 nT
60–70 Re	0.55 nT

From Behannon, K. W. (1970). *J. Geophys. Res.*, 75, 743. Copyright 1970 by the American Geophysical Union. With permission.

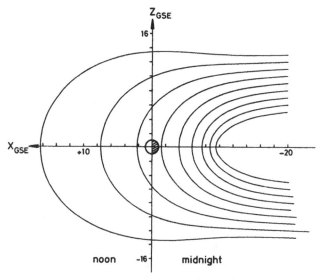

Figure 11.6.6 Magnetic field lines in the X_{GSM}–Z_{GSM} plane generated by the magnetotail current sheet shown in Figure 11.1.7 with plasma currents flowing across the tail in westward direction. For this computation the geomagnetic dipole is assumed to be perpendicular to the ecliptic plane; under that symmetry condition the (GSM) and (GSE) coordinate systems coincide. At $Z_{GSE} = 0$ the magnetic field points southward, i.e., $B_z < 0$. Thus, a strong tail magnetic field can significantly reduce the total magnetic field at the subsolar stagnation point on the dayside.

these findings suggest the existence of a magnetotail current sheet in the equatorial plane with currents flowing across the tail in a westward direction. Figure 11.6.6 shows magnetic field lines in the noon-midnight plane generated by such a tail current sheet with plasma currents flowing westward. At $Z_{GSM} = 0$ the magnetic field points southward, i.e., $B_z < 0$. Thus a strong tail magnetic field can significantly reduce the total magnetic field at the subsolar stagnation point which, in turn, can lead to a decrease of the stand-off distance, even if the solar wind pressure remains unchanged (the subsolar stand-off distance is discussed in Section 5.3).

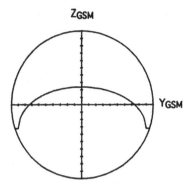

Figure 11.6.7 Shape and position of the tail current sheet shown in a cross section of Earth's magnetotail. The ellipsoidal shape results from Fairfield's (1980) data analysis. For a dipole tilt $\psi > 0$ (summer in the northern hemisphere), the neutral sheet is raised above equatorial plane at local midnight but is depressed below that plane near the flanks of the tail. (From Fairfield, D. H. (1990). *J. Geophys. Res.*, 85, 775. 1980 Copyright by the American Geophysical Union. With permission.)

When the Earth's dipole is tilted with respect to the solar·wind stream (the dipole tilt is explained in Section 2.3) then the tail current sheet is warped. Fairfield (1980) showed through a statistical analysis of IMP 6, 7, and 8 magnetometer data that, for a dipole tilt $\psi > 0$ (summer in the northern hemisphere), the neutral sheet is raised above equatorial plane at local midnight but is depressed below that plane near the flanks of the tail (see Figure 11.6.7). Of course, the opposite effect occurs for $\psi < 0$ (winter in the northern hemisphere).

Three-dimensional model calculations (Voigt, 1984) are in qualitative agreement with Fairfield's original findings. Shown in Figure 11.6.8 are magnetotail cross sections with computed tail current stream lines that form the tail plasma sheet. Tail radius is 20 Re, and dipole tilt angle is $\psi = 35°$ (summer in the northern hemisphere). From top to bottom, tail cross sections are taken at $x_{GSM} = -15$ Re, -30 Re, and -60 Re. The plasma sheet thickness increases in $|y_{GSM}|$ direction toward the flanks of the tail. In fact, Fairfield (1979) found that more magnetic flux crosses the equatorial plane near the dawn and dusk flanks ($B_z = 3.5$ nT) than near local midnight ($B_z = 1.8$ nT).

The middle current stream line in each panel represents the neutral sheet which is associated with highest plasma and currents densities. The plasma sheet flattens with increasing distance down the tail. This effect is due to the reduced influence of the tilted geomagnetic dipole far away from the Earth. Voigt's (1984) X_{GSM}-dependent theoretical plasma sheet model has been largely confirmed by Dandouras' (1988) statistical data analysis.

A wavy structure of the neutral sheet has been observed by Nakagawa and Nishida (1989). By analyzing IMP 6 magnetic field data from about 20 Re tailward from the Earth, Nakagawa and Nishida found that the neutral sheet experiences a wavy motion in the Y_{GSM}-Z_{GSM} plane that propagates in the dawn-dusk direction in the presence of a dawn-dusk B_y component of the magnetic field. The authors conjecture that duskward plasma flows in the neutral sheet might cause the neutral sheet to become Klevin-Helmholtz unstable which might explain the observed wavy motion.

6.4. THE DISTANT TAIL

Earth's magnetotail has been observed by the ISEE 3 spacecraft up to a distance of about 220 Re from the Earth (for a summary of the ISEE 3 far tail mission, see Tsurutani and von Rosenvinge, 1984). At 220 Re from the Earth the geotail still has a surprisingly well-ordered structure: stretched magnetic field lines of opposite polarity in the two tail lobes are separated by a cross-tail current sheet. Tail flaring ceases at about $|X_{GSM}| = 120 \pm 10$ Re, and the magnetic field strength in the tail lobes is observed as 9.2 nT at that distance. The plasma sheet thickness is 6.8

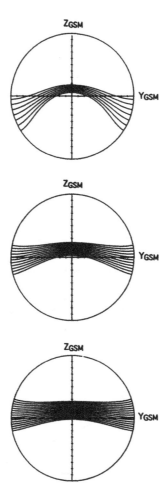

Figure 11.6.8 Magnetotail cross sections with computed tail current stream lines that form the tail plasma sheet (Voigt, 1984). Tail radius is 20 Re, and dipole tilt angle is $\psi = 35°$ (summer in the northern hemisphere). From top to bottom, tail cross sections are taken at $X_{GSM} = -15$ R_E, -30 R_E, and -60 R_E. The plasma sheet thickness increases in $|Y_{GSM}|$ direction toward the flanks of the tail. The middle current stream line in each panel represents the neutral sheet which is associated with highest plasma and currents densities. The plasma sheet flattens with increasing distance down the tail. (From Voight, G.-H. (1984). *J. Geophys. Res.*, 89, 2169. 1984 Copyright by the American Geophysical Union. With permission.)

Re at about $|X_{GSM}| = 100$ Re (Fairfield, 1992). The magnetic B_z component at the center of the plasma sheet is found to be 0.6 nT during magnetically quiet periods and almost zero during disturbed periods (Fairfield, 1992).

The average tail radius is observed to be 30 ± 2.5 Re at $|X_{GSM}| = 130$ to 225 Re (Slavin et al., 1985). On the other hand, Fairfield (1992a) argues that the ISEE 3 observations match expectations if the average radius of the distant magnetotail is about 24 Re. Sibeck et al. (1985) demonstrated that the distant magnetotail has a flattened, ellipsoidal shape with the east-west dimension exceeding the north-south dimension. In addition, the entire tail is twisted. Figure 11.6.9 shows a cross section of the distant magnetotail viewed from the Earth (Sibeck et al., 1985). There are two windows, one on either side of the geotail, through which magnetosheath plasma may enter the tail region. Magnetosheath and magnetotail field lines are interconnected through the windows. Without a window, the magnetopause would lie along the dashed line.

The twist of the entire tail under the presence of a magnetosheath B_y component has been illustrated by Cowley (1981). He argues that, in response to the magnetosheath B_y torque, the twist increases with increasing distance from the Earth. Cowley's concept is sketched in Figure

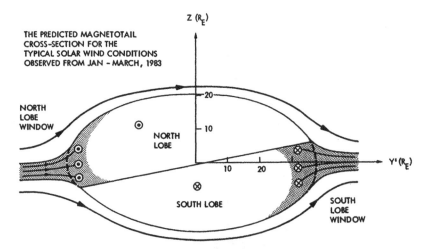

Figure 11.6.9 Cross section of the distant magnetotail viewed from the Earth (Sibeck et al., 1985). The magnetotail has a flattened, ellipsoidal shape with the east-west dimension exceeding the north-south dimension. The entire tail is twisted owing to a torque exerted by the IMF. There are two windows, one on either side of the geotail, through which magnetosheath plasma may enter the tail region. Magnetosheath and magnetotail field lines are interconnected through the windows. Without a window, the magnetopause would lie along the dashed line. (From Sibeck, D.G. et al. (1985). *J. Geophys. Res.,* 90, 9561. 1985 Copyright by the American Geophysical Union. With permission.)

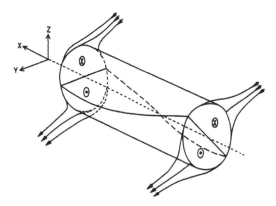

Figure 11.6.10 Sketch of the twist in the geomagnetic tail lobes and central current sheet (Cowley, 1981). The twist of the geotail increases with increasing distance from the Earth in response to the IMF torque. The twist shown here is due to an IMF $B_y > 0$. An opposite twist may occur when $B_y < 0$. In the extreme limit of an infinitely long geotail, one would expect the twisting angle of the tail plasma sheet to approach 90° with respect to the IMF orientation. (From Cowley, S.W.H. (1981). *Planet. Space Sci.,* 29, 79. With permission.)

11.6.10. In the extreme limit of an infinitely long geotail, one would expect the twisting angle of the tail plasma sheet to approach 90° with respect to the magnetosheath field orientation. Macwan (1992) analyzed magnetic field and plasma data from ISEE 3 and found that, with respect to the untwisted plasma sheet, a twist of more than 90° has occurred frequently on both closed and open magnetotail field lines. A comprehensive theoretical description of the twisted magnetotail does not exist yet.

The distant neutral line referred to in Section 1.2 (see Figure 11.1.5) has been observed to be V-shaped in the X_{GSM}-Y_{GSM} plane such that the bottom of the V points toward the Earth at local midnight (Slavin et al., 1985). Moreover, the distant neutral line moves between $X_{GSM} = -115$ and -220 Re (Macwan, 1992).

Modern magnetospheric **B**-field models must take into account the existence of the *distant X-line* (neutral line). Toffoletto and Hill (1993) have developed such a model for the open magnetosphere. That model includes a representation of the slow-mode expansion fan inside the

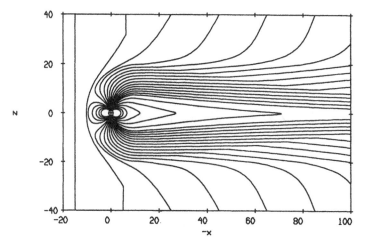

Figure 11.6.11 Three-dimensional magnetospheric **B**-field model developed by Toffoletto and Hill (1993). Plotted are magnetic field lines in the X_{GSM}-Z_{GSM} plane. The IMF is southward in this example. Plasma sheet field lines converge toward the distant X-line which is located at the center of the tail plasma sheet. The two tail lobes show features of the expansion fan. Both the position of the V-shaped X-line in the X_{GSM}-Y_{GSM} plane, as well as the form of the V, are free model parameters. (From Toffoletto, F.R., and Hill, T.W. (1984). *J. Geophys. Res.,* 1989 Copyright by the American Geophysical Union. With permission.)

high-latitude tail magnetopause, according to a concept introduced by Siscoe and Sanchez (1987). The **B**-field model also includes a magnetotail field which represents the magnetosphere's response to the tailward stress exerted by the solar wind on open field lines. Figure 11.6.11 displays the magnetic field line configuration of that model in the X_{GSM}-Z_{GSM} plane. The IMF is southward in this example. Note that plasma sheet field lines converge toward the distant X-line which is located at the center of the tail plasma sheet. The two tail lobes show features of the expansion fan. Both the position of the V-shaped X-line, as well as the form of the V, are free model parameters.

7. POLAR CAP AND AURORAL OVAL

Earth's magnetotail plasma sheet, mapped along magnetic field lines to the ionosphere, occupies a region from 68 to about 70° geomagnetic latitude on the nightside of the Earth. The polar cap region consists of open field lines that extend far out into the magnetotail. Viewed in terms of the magnetic field topology, the low-latitude termination of the polar cap is defined by the boundary between open and closed magnetic field lines. That boundary is oval-shaped. The appearance of northern lights and the ionospheric footprints of field-aligned Birkeland currents coincide qualitatively with the location of the boundary between open and closed magnetic field lines.

7.1. CLOSED AND OPEN FIELD LINES
In the following, a magnetic field line is defined as closed when it originates from a point in one hemisphere and ends at its conjugate point in the other hemisphere. In other words, a closed field line has both ends rooted in Earth's ionosphere. In contrast, an open field line has only one end connected to the Earth; the other end extends far out into the geotail where it ultimately connects to the IMF. The two domains of fundamentally different magnetic field topologies are separated by a boundary between open and closed field lines.

Figure 11.7.1 shows a geomagnetic polar diagram of the boundary between open and closed magnetic field lines in the northern ionosphere. The curves plotted here result from a theoretical three-dimensional magnetospheric **B**-field model (Voigt, 1976). The boundary is oval-shaped

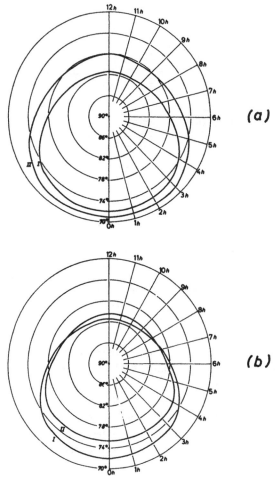

Figure 11.7.1 Geomagnetic polar diagram of the boundary between open and closed field lines in the northern hemisphere. The boundaries result from a theoretical **B**-field model (Voigt, 1976). The curves in curve I are identical in both panels and indicate the boundary for an average quiet magnetosphere and zero dipole tilt angle ($\psi = 0°$). Panel (a) is for an untilted geomagnetic dipole $\psi = 0°$. Here curve II is computed for a compressed magnetosphere with enhanced magnetic activity. Panel (b) is for an average quiet magnetosphere. Here curve II indicates the boundary for the dipole tilt $\psi = -30°$ (winter in the northern hemisphere). (From Voigt, G.-H. (1976). *The Scientific Sattelite Programme During the International Magnetosphere Study,* Knott and Battrick, Eds., Reidel, Dordrect, Holland, p. 38. With permission.)

which explains the term "polar oval". Both in the real magnetosphere and in the theoretical model, the polar oval changes in size, shape, and location, depending on the magnetic activity level of the entire magnetosphere.

From a theoretical point of view, the polar oval is determined by (1) the magnetic boundary conditions as seen in Equation 4.2 specified at the magnetopause, (2) the state of compression of the magnetosphere characterized by the subsolar stand-off distance in Equation 5.10, and (3) the magnetotail current system, the details of which follow from quasistatic MHD equilibrium and convection theories.

The top panel in Figure 11.7.1 displays the last closed field line boundary for the untilted geomagnetic dipole ($\psi = 0°$). Here curve I indicates the boundary for an average quiet magnetosphere and curve II for a compressed magnetosphere with enhanced magnetic activity. In this particular example, the latitudinal shift from curve I to curve II is larger on the *dayside than on* the nightside owing to the reduction of the stand-off distance from 11 to 8 Re during a magnetic storm.

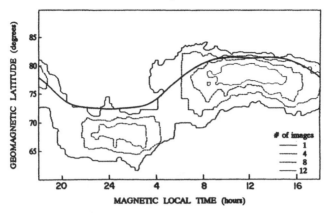

Figure 11.7.2 Number densities of 30 quiet-time auroral images in corrected geomagnetic coordinates (Oznovich et al., 1993). The bold solid line is the last closed field line boundary for the untilted Earth's dipole ($\psi = 0°$) adopted from the theoretical curve I in Figure 11.7.1. (From Oznovich I., et al. (1993). *J. Geophys. Res.*, 98, 3789. 1993 Copyright by the American Geophysical Union. With permission.)

The bottom panel in Figure 11.7.1 shows the boundary for an average quiet magnetosphere. Here curve I indicates the boundary for the dipole tilt $\psi = 0°$, and curve II for $\psi = -30°$ (winter in the northern hemisphere). In the transition from curve I to curve II, the nightside shift exceeds the dayside shift when the geomagnetic dipole is tilted toward winter. The opposite result appears in the summer hemisphere. Curve I is identical in both panels, and the polar coordinate system is centered at the geomagnetic dipole axis.

The appearance of northern lights seems to coincide with the location of the last closed field line boundary. This is demonstrated in Figure 11.7.2 where plotted are number densities of 30 quiet time auroral images in corrected geomagnetic coordinates (Oznovich et al., 1993). The bold solid line is the last closed field line boundary for the untilted Earth's dipole ($\psi = 0°$) adopted from the theoretical curve I in Figure 11.7.1. Based on that result, one has to conclude that under average quiet geomagnetic conditions northern lights appear most frequently close to the high-latitude border of the closed field line region.

If the ionosphere and the magnetosphere are linked via magnetic field lines, one finds that the dayside part of the last closed field line boundary maps to the subsolar region at the dayside magnetopause. The nightside part of the boundary maps to the earthward (inner) edge of the tail plasma sheet which is sketched in the bottom part of Figure 11.7.3. In the real magnetosphere the boundary between closed and open field lines has not been observed directly. Instead, the location of that boundary is inferred indirectly from plasma distribution functions measured by spacecraft crossing the polar ionosphere at low altitudes.

7.2. FIELD-ALIGNED BIRKELAND CURRENTS

Magnetic field lines and field-aligned currents parallel to the magnetic field, $\mathbf{j}_\| = (\mathbf{j} \cdot \mathbf{B})\, \mathbf{B}/B^2$, provide the coupling between the ionosphere and the magnetosphere. The electric currents in the ionosphere-magnetosphere system, $\mu_0\, \mathbf{j} = \nabla \times \mathbf{B}$, are most conveniently expressed in terms of their perpendicular and parallel components with respect to the magnetic field, namely,

$$\mathbf{j}_\perp = \frac{\mathbf{B} \times \nabla P}{B^2} \tag{7.1}$$

$$|\mathbf{j}_\||= -\frac{\mathbf{B}}{|B|}\left(\nabla P \times \nabla \int_1^2 \frac{ds}{B(s)}\right) \tag{7.2}$$

where the integral, $V = \int ds/B(s)$, denotes the volume of a closed magnetic flux tube of unit

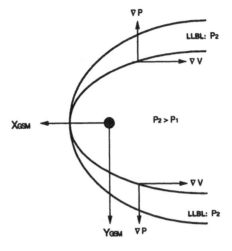

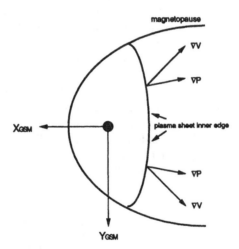

Figure 11.7.3 Characteristic boundaries in the magnetospheric equatorial plane where field-aligned Birkeland currents from Equation 7.2 are generated. Indicated are the vectors ∇V of the flux tube volume and ∇P of the thermal plasma pressure. Upper panel: Source region for region 1 field-aligned currents. Strong plasma pressure gradients are expected at the inner demarcation line of the LLBL near the equatorial flanks of the tail. Bottom panel: Source region for region 2 field-aligned currents. The flux tube volume increases radially outward from the Earth. Strong plasma pressure gradients are expected perpendicular to the inferred earthward edge of the tail plasma sheet. Birkeland currents are strongest near dawn and dusk, where the two vectors ∇P and ∇V contribute most to the cross product in Equation 7.2.

magnetic flux. The integration is performed along a magnetotail field line from the ionosphere in one hemisphere to the conjugate point in the opposite hemisphere.

Expression 7.1 follows directly from the equilibrium condition in Equation 6.1, and Equation 7.2 follows from the condition

$$\nabla \cdot (j_{\perp} + j_{\parallel}) = 0 \qquad (7.3)$$

and an integration along a closed magnetotail field line (Vasyliunas, 1970). A derivation of Equation 7.2 has been given, for example, by Schindler and Birn (1978) and by Wolf (1983).

Because most magnetospheric **B**-field models do not explicitly include the thermal plasma pressure, it is useful to combine Equations 6.1 and 7.2 to obtain the equivalent expression

$$\mathbf{j}_{\parallel} = -\mathbf{B}\left(\mathbf{j}_{\perp} \cdot \nabla \int_{1}^{2} \frac{ds}{B(s)}\right) \tag{7.4}$$

where the integral $\int ds/B(s)$ is a geometrical information inherent in the **B**-field model at hand. Figure 11.7.3 shows characteristic boundaries in the magnetospheric equatorial plane where pressure gradients are large, and thus field-aligned Birkeland currents in Equation 7.2 are generated most effectively. Sketched are locations where the vectors ∇V and ∇P are most prominent. In a typical magnetotail configuration, flux tube volumes, $V = \int ds/B(s)$, increase radially outward from the Earth, simply because magnetic field lines become longer with increasing distance from the Earth.

The upper panel in Figure 11.7.3 shows the source region for so-called region 1 field-aligned currents. Sketched is the low-latitude boundary layer (LLBL) near the equatorial flanks of the tail. The LLBL is a plasma-dominated layer adjacent to the equatorial magnetopause; the strongest plasma pressure gradients are expected near the inner demarcation line of the LLBL.

The relevance of pressure gradients for generating region 1 Birkeland currents in the LLBL is not without controversy, though. Sonnerup (1980) and others, for example, claim that pressure gradients (∇P) are less important than plasma flows ($\rho\, \partial v/\partial t + \rho\, (v \cdot \nabla)\, v$). In their view, the LLBL is dominated by plasma flows that are caused by a viscous or diffusive momentum transfer between the magnetosheath and the adjacent LLBL. Fortunately, the different source mechanisms do not affect the general direction of region 1 currents generated in the LLBL.

The bottom panel in Figure 11.7.3 sketches the source region for so-called region 2 field-aligned currents. Shown is the earthward (inner) edge of the tail plasma sheet where the plasma-dominated tail region ($\beta > 1$) merges with the innermost part of the magnetosphere which is dominated by the geomagnetic vacuum dipole field ($\beta < 1$). Strong plasma pressure gradients are expected perpendicular to the earthward edge of the tail plasma sheet, but Birkeland currents are strongest near dawn and dusk, where the vectors ∇P and ∇V contribute most to the cross product in Equation 7.2.

If the source regions for Birkeland currents, sketched in Figure 11.7.3, are mapped along magnetic field lines from the magnetotail region into the ionosphere, one finds that the region 1 currents map to higher geomagnetic latitudes than the region 2 currents. Note that, at a given longitude, the signs of the vector products $\nabla P \times \nabla V$ in Figure 11.7.3 are opposite for region 1 and region 2 currents. The corresponding changes in the current flow directions are confirmed by Birkeland currents observed in the ionosphere (Iijima and Potemra, 1978).

Figure 11.7.4 shows footprints of region 1 and region 2 field-aligned Birkeland currents in the northern ionosphere (Iijima and Potemra, 1978). Black areas indicate currents flowing from the magnetosphere into the ionosphere, and dotted areas indicate currents flowing in the opposite direction. Distribution and flow directions of Birkeland currents represent statistical averages of hundreds of Triad spacecraft measurements during quiet and weakly disturbed geomagnetic conditions (left panel) and during magnetically active periods (right panel).

Viewed now from the ionosphere, one concludes that high-latitude (region 1) currents map along magnetic field lines into the LLBL in the magnetotail equatorial plane. In contrast, the low-latitude (region 2) currents map into the earthward (inner) edge of the tail plasma sheet. The orientations of the two vectors, ∇P and ∇V, in Figure 11.7.3 are consistent with the current flow directions in Figure 11.7.4 which can be verified by inspecting Expression 7.2.

Iijima and Potemra (1978) demonstrated that the large-scale characteristics of the Birkeland current system persists during all phases of geomagnetic activity, including magnetospheric substorms. Region 1 currents are located near the poleward boundary of the field-aligned current region, and region 2 currents are located near the equatorward boundary. During magnetically

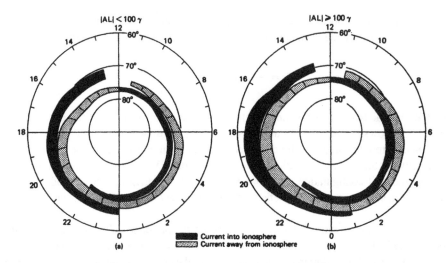

Figure 11.7.4 Patterns of region 1 (high-latitude) and region 2 (low-latitude) field-aligned Birkeland currents in the northern ionosphere (Iijima and Potemra, 1978). Black areas indicate currents flowing from the magnetosphere into the ionosphere, and dotted areas indicate currents flowing in the opposite direction. The left panel summarizes satellite observations during quiet and weakly disturbed geomagnetic conditions, the right panel during magnetically active periods. (From Iijima, T., and Potemra, T.A. (1978). *J. Geophys. Res.*, 83, 599. 1978 Copyright by the American Geophysical Union. With permission.)

active periods the average latitudinal width of regions 1 and 2 increases by 20 to 30%, and the centers of these regions shift equatorward by 2 to 3° with respect to the quiet-time values. The equatorward shift is consistent with the corresponding shift of the boundary between open and closed magnetic field lines (see top panel in Figure 11.7.1) and thus reflects a reconfiguration of the entire magnetospheric **B**-field during magnetically active periods.

A rapid generation of strong field-aligned currents is expected during a magnetic substorm in the course of which part of the near Earth tail current sheet collapses. The corresponding current disruption is sketched in Figure 11.7.5. When the tail currents become disrupted, owing to the onset of a tearing mode instability, for example, then strong gradients ∇P and ∇V occur at the site of the disruption, and the disrupted currents close through the ionosphere via field-aligned Birkeland currents. That mechanism sketched in Figure 11.7.5 was originally suggested by McPherron et al. (1973).

8. MAGNETOSPHERIC CONVECTION

Magnetospheric convection, or plasma circulation, is a slowly time-dependent process, whereby closed magnetotail flux tubes near local midnight move adiabatically toward the Earth. Theoretical arguments suggest that the thermal plasma pressure P in earthward convecting flux tubes increases beyond critical values that cannot be balanced by the magnetic field pressure in the magnetotail lobes. This phenomenon is called the "pressure-balance crisis" and might be responsible for periodic onsets of magnetospheric substorms. Thus the convecting magnetosphere presumably never reaches a steady state.

8.1. IONOSPHERE–MAGNETOSPHERE COUPLING

The first computational scheme of magnetospheric convection in the inner magnetosphere, and its coupling to the low-latitude ionosphere, is due to Vasyliunas (1970). His convection scenario is pictured in Figure 11.8.1; it simply states that convective plasma flow patterns in Earth's magnetotail correspond to equivalent flow patterns in the ionosphere and vice versa.

A three-dimensional computer code of the above scheme is the Rice Convection Model (RCM) created and maintained at Rice University in Houston, Texas (Spiro and Wolf 1983). The code

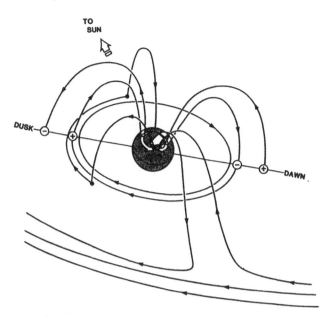

Figure 11.7.5 Disruption of the near Earth part of the tail current sheet during a magnetospheric substorm. When the tail currents become disrupted, the currents close through the ionosphere via field-aligned Birkeland currents. (From McPherron. R. L. et al. (1973). *J. Geophys. Res.,* 78, 3131. Copyright 1973 by the American Geophysical Union. With permission.)

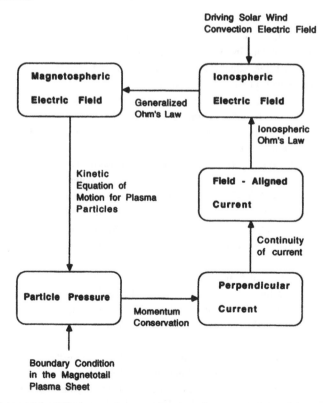

Figure 11.8.1 Mathematical scheme for a self-consistent calculation of magnetospheric convection (after Vasyliunas, 1970). The physical quantities in boxes are calculated at each time step. Each line joining two boxes is labeled with the physical principle or MHD equation that provides the link between the two quantities. Quantities entering the loop from the outside are boundary conditions that must be specified.

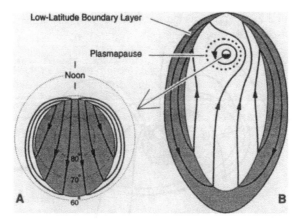

Figure 11.8.2   Sketch of average plasma convection patterns in the high-latitude ionosphere (panel A) and in the magnetospheric equatorial plane (panel B). The two flow patterns correspond to each other through field-aligned currents in Equation 7.4 flowing along magnetic field lines from the ionosphere into the LLBL and vice versa. (From Hill, T. W., and Dessler, A. J. (1991). *Science,* 252, 345. With permission.)

for computing convection time sequences goes around the loop in Figure 11.8.1 every time step Δt. The physical quantities in boxes are computed at each time step. Each line joining two boxes is labeled with the physical principle or MHD equation that provides the link between the two quantities. Quantities entering the loop from the outside are boundary conditions that must be specified.

1. A given magnetospheric **B**-field model is used together with an initial plasma distribution at the outermost boundary in the magnetospheric equatorial plane to compute the perpendicular current density in Equation 7.1. Ideally, the **B**-field model satisfies the MHD equilibrium conditions in Equations 6.1 through 6.3.
2. Current continuity in Equation 7.3 allows one to calculate the field-aligned Birkeland currents in Equation 7.4 into the ionosphere. Note that Expression 7.4 contains the flux tube volume, $V = \int ds/B(s)$, as a geometrical information inherent in the **B**-field model.
3. The ionospheric Ohm's law can now be solved by using the calculated field-aligned currents, a model for the ionospheric height-integrated conductivity and boundary conditions for the polar-cap potential. This leads to the ionospheric potential distribution $V(\mathbf{x})$.
4. The computed potential distribution $V(\mathbf{x})$ is mapped along the magnetic field lines into the magnetospheric equatorial plane to give the electric field distribution there.
5. Then the plasma distribution in the equatorial plane is recalculated by following the $\mathbf{E} \times \mathbf{B}$ plus gradient and curvature drift velocities of the magnetospheric plasma for the time step Δt.
6. If during the time interval Δt the magnetic field has changed, a new magnetic field configuration will be used together with the newly calculated plasma distribution to satisfy the equilibrium conditions in Equations 6.1 through 6.3 as closely as possible. This step concludes the computational loop, and one can start again at Step 1.

Let us now look at the convection process in the high-latitude ionosphere. If the magnetosphere is open and thus magnetospheric and IMF lines interconnect, then the solar wind electric field, $\mathbf{E} = -\mathbf{v} \times \mathbf{B}$, maps down to the ionospheric polar cap where it causes the ionospheric plasma to flow in closed loops according to panel A in Figure 11.8.2. The corresponding flow pattern in the magnetospheric equatorial plane is shown in panel B. The connection between the two regions is maintained through field-aligned currents in Equation 7.4 flowing along magnetic field lines from the ionosphere into the equatorial plane (Hill and Dessler, 1991). Using a closed tear-drop model (e.g., Figure 11.1.2) the gray-shaded ionospheric polar cap maps into the gray-shaded LLBL in the magnetospheric equatorial plane.

In the equatorial plane, magnetotail flux tubes move toward the Earth around local midnight but move anti-sunward near the flanks of the tail. The earthward motion follows the $\mathbf{E} \times \mathbf{B}$ and gradient-curvature drifts, until the gradient-curvature drift velocities force magnetic flux tubes to drift around the Earth. Near the inner edge of the tail plasma sheet (see Figure 11.7.3), the combined effects of both gradient and curvature drifts become comparable to, or even dominate, the earthward $\mathbf{E} \times \mathbf{B}$ drift. Because gradient and curvature drifts depend on the particle energy, low-energetic particles can drift closer to the Earth than high-energetic particles can. Consequently, each plasma energy has its own so-called "Alfvén layer" which is defined as a boundary beyond which particles with that energy cannot convect closer toward the Earth.

8.2. CONVECTION AND THE PRESSURE BALANCE INCONSISTENCY

Erickson and Wolf (1980) showed that standard magnetospheric B-field models of the type shown in Figure 11.4.1 or in Figure 11.6.2 are all inconsistent with the thermodynamic (adiabatic) condition in Equation 10.7. They found that the thermal plasma pressure P in earthward convecting flux tubes increases beyond critical values that cannot be balanced by the magnetic field pressure in the tail lobes. The discrepancy between plasma sheet and lobe field pressures becomes larger and larger for any given flux tube approaching the Earth. Erickson and Wolf (1980) suggested that this "pressure-balance crisis" might be responsible for the onset of magnetospheric substorms.

To test the above hypothesis, Erickson (1984) computed time series of earthward convecting magnetotail flux tubes near local midnight. He used a self-consistent, two-dimensional MHD equilibrium model in the quasistatic approximation. That model is based on solutions to the Grad-Shafranov equation 6.4 with the additional thermodynamic condition in Equation 10.7 which in two dimensions assumes the equivalent form

$$P(\alpha, t) \cdot (V(\alpha, t))^{\gamma} = C(\alpha) \tag{8.1}$$

where $\gamma = 5/3$ is the adiabatic exponent. The quantity $C(\alpha)$ is different for each flux tube but remains constant for any given flux tube throughout the convection process. The quantity $V(\alpha, t)$ is the above-mentioned flux tube volume

$$V = \int_1^2 \frac{ds}{B(s)} \tag{8.2}$$

The integration ds is performed along a closed magnetic field line from one ionospheric end to the conjugate ionospheric end in the other hemisphere. In two dimensions (see Section 6.2), each B-field line corresponds to an equivalent value of the magnetic vector potential $\alpha(x, z)$.

Erickson's (1984) two-dimensional convection time sequence is shown in Figure 11.8.3. The three magnetospheric field configurations represent snapshots taken at three different times after convection had started. The time labels refer to three different values of the magnetopause flux function, namely, $A_{mp} = 0$ (top), -12 (middle), and -22 (bottom). The flux function A in Figure 11.8.3 corresponds to the function $\alpha(x, z)$ introduced in Section 6.2. There are equal amounts of magnetic flux between adjacent tail field lines plotted. The $A = 0$ field line is labeled, and the dashed field line is $A = +30$ for reference, in order to show how field lines actually move earthward during the convection process.

There is an interesting consequence of the convection experiment shown in Figure 11.8.3: the adiabatic condition in Equation 8.1 forces the system to develop a distinct minimum in the equatorial field strength B_z near the earthward edge of the magnetotail plasma sheet. By lowering

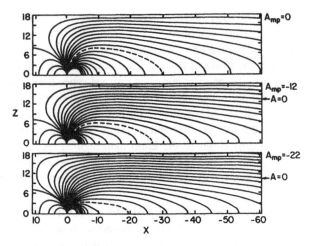

Figure 11.8.3 Time sequence of global magnetospheric convection. Tail field lines convect under the influence of the driving solar wind electric field specified at the magnetopause. The magnetic field configuration must change in time in order to satisfy the adiabatic condition in Equation 8.1. Note that the tail field lines become more and more stretched, and that the dashed field line (A = 30) moves toward the Earth, as convection proceeds. (From Erickson, G. M. (1984). *Magnetic Reconnection, Geophys. Monogr. Ser.,* 30, E. W. Hones, Ed., pp. 296–302, AGU, Washington, D.C. 1984 Copyright by the American Geophysical Union. With permission.)

the magnetic field strength there, the magnetosphere increases the flux tube volume seen in Equation 8.2 in order to satisfy the adiabatic condition in Equation 8.1 in a region where otherwise the pressure balance crisis would occur.

It is important to understand that the pressure balance crisis is not an artifact of two-dimensional MHD equilibrium models. Even in a three-dimensional model, in which flux tubes could easily drift around the Earth in a realistic fashion, the pressure crisis would occur. By drifting around the Earth, the flux tube volume does not increase sufficiently to prevent the crisis from occurring.

The conjecture that magnetospheric convection leads to a magnetic field minimum in the equatorial plane near $X_{GSM} = -10$ R_e is of crucial importance for understanding the substorm mechanism (see Section 9.1). The "near earth neutral line model" of substorms suggests that the magnetotail becomes unstable to the tearing mode in the region of the **B**-field minimum (Hau and Wolf, 1987). This then leads to the conclusion that magnetospheric convection and substorms are two facets of the same global physical process.

9. MAGNETOSPHERIC SUBSTORMS

The following key features of a typical substorm have been summarized by Lui (1992). One identifies three substorm phases, namely, growth, expansion, and recovery phase.

The growth phase usually begins with the start of southward turning of the solar wind magnetic field (IMF). In the ionosphere, the polar cap size increases as the ionospheric electrojets in the polar region intensify. In the magnetosphere, the cross section of the magnetotail enlarges. Plasma sheet thinning develops in the near Earth magnetotail between $X_{GSM} = -5$ and -15 Re downstream.

At substorm expansion onset one of the nightside auroral arcs, typically the most equatorward one in the midnight sector, brightens suddenly, breaks up, and expands poleward. The auroral electrojets are simultaneously enhanced with the auroral activity. In the near Earth magnetosphere, the stretched tail-like configuration developed during the growth phase relaxes abruptly toward a more dipolar field geometry. Simultaneously the plasma sheet thickens, and energetic particles are injected earthward. The immediate cause for the onset of the substorm expansion phase is still controversial, but the suddenness of the onset suggests a global MHD instability that disrupts the cross-tail current (see Figure 11.7.5).

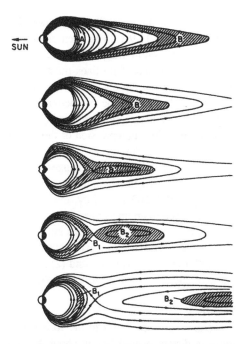

Figure 11.9.1 Sketch of the convection process in Earth's geomagnetic tail. The time sequence starts in the top diagram with the field line **B.** That field line moves earthward (second diagram) and forms a neutral line (third diagram) which leads to an onset of the tearing mode instability, i.e., a magnetospheric substorm. As a consequence of the substorm-type instability a plasmoid forms and moves rapidly tailward. (From Wolf, R. A. and Spiro, R. W. (1993). *High-Latitude Space Plasma Physics,* V. Hultqvist and T. Hagfors, Eds., pp. 19–38, Plenum Press, New York. With permission.)

The recovery phase begins when the poleward motion of the aurora ceases. In the mid-tail region, between $x_{GSM} = -15$ and -80 Re downstream, the plasma sheet suddenly thickens with fast plasma flows, predominantly field aligned, at the plasma sheet boundary.

A magnetospheric substorm is a transient process, initiated on the night side of the Earth, in which a significant amount of energy derived from the solar wind-magnetosphere interaction is deposited in the auroral ionosphere and magnetosphere.

That qualitative definition of a substorm has been formulated by Rostoker et al. (1980). Most probably more than one physical mechanism accounts for the triggering of a magnetic substorm in Earth's magnetotail. Contending substorm theories are among the most debated issues in magnetospheric physics today; this section briefly summarizes some more recent ones.

9.1. NEAR EARTH NEUTRAL LINE MODEL

The convection and substorm scenario outlined in Section 8 constitutes the essence of what is called the "near earth neutral line" hypothesis: the development of a minimum in the equatorial magnetic field strength leads to a magnetic field configuration that eventually becomes tearing-mode unstable. This leads to the formation of a magnetic neutral line at the center of the tail plasma sheet.

The sequence of events is sketched in Figure 11.9.1; it starts in the top panel where the field line **B** denotes an ordinary stretched tail field line. Subsequently this field line convects toward the Earth, whereby the magnetosphere goes through the configurational changes shown quantitatively in Figure 11.8.3. Then, in the third panel, a magnetic neutral X-line has formed owing to the preceding development of the magnetic field minimum discussed in Section 8.2. The instability associated with the formation of the neutral line can, of course, not be described by the quasistatic convection theory. Instead, it must be calculated on the level of the MHD theory (see Section 10.3). As a consequence of the instability a plasmoid forms which finally moves rapidly tailward (panels 4 and 5 in Figure 11.9.1). This scenario implies that the formation of a neutral line is an essential element of the entire magnetospheric convection cycle.

For a tearing instability to occur, magnetic field lines must be stretched such that the field component perpendicular to the central current sheet will become minimized. The following mechanisms may stretch magnetic field lines in the magnetotail: (1) an increase in the overall plasma sheet pressure (see Figure 11.6.2); (2) a southward turning of the IMF; and (3) an increase in the solar wind ram pressure resulting in a compression of the magnetopause shape. All three effects increase the tail current density accordingly.

Therefore, substorms can be externally triggered via changes of magnetopause boundary conditions. Substorms can also be internally triggered via the process of quasistatic convection and the occurrence of a subsequent global MHD instability, even if the magnetopause boundary conditions (and hence, the solar wind) remain unchanged.

Although the near earth neutral line model of substorms intrigues by its theoretical elegance, it has difficulties in explaining (1) the occurrence of consistently earthward plasma flow inside of $X_{GSM} = -15$ Re (1 Re = 1 earth radius = 6371 km) and (2) the onset of magnetic turbulence at $X_{GSM} = -15$ Re at the time of substorm onset (Fairfield, 1990, 1992b).

9.2. BOUNDARY LAYER MODEL

The boundary layer model ascribes substorm onset to the Kelvin-Helmholtz instability arising from enhanced reconnection at a distant neutral line at about $X_{GSM} = -100$ Re downstream (Rostoker and Eastman, 1987). The model is based on the existence of the plasma sheet boundary layer as the primary region of plasma transport in the magnetotail.

Earthward flow of magnetotail plasma creates a velocity shear with the opposing flow of the LLBL. A Kelvin-Helmholtz unstable boundary results which generates substorm-associated field-aligned currents and auroral forms. A current wedge is formed along the plasma sheet boundary, and multiple surges are emphasized. The estimated growth time is about 2 to 3 min, comparable with the onset time scale of substorm expansion.

9.3. THERMAL CATASTROPHE MODEL

The thermal catastrophe model addresses the thermodynamic state of the plasma sheet (Goertz and Smith, 1989). Plasma sheet heating is due to resonant absorption of Alfvén waves incident on the plasma sheet boundary in the tail. The plasma sheet is suggested to evolve through a succession of equilibrium states parameterized by a quantity dependent on the incident power flux of Alfvén waves, the local plasma density, and the convective velocity. Expansion onset occurs when this parameter reaches a critical value, rendering the plasma sheet opaque to wave transmission which then leads to sudden plasma heating. In the mathematical model, that sudden heating is called the "the thermal catastrophe".

The model is based on the observation that the temperature of the central plasma sheet is well correlated with geomagnetic activity (Huang and Frank, 1986). It is among the most quantitative of substorm theories. Its strengths include the ability to explain both plasma sheet heating and the sudden onset of substorms (Fairfield, 1990). A primary question about the theory is whether adequate wave energy is incident on the plasma sheet boundary layer to drive a substorm.

9.4. CURRENT DISRUPTION MODEL

In the current disruption model (Chao et al., 1977; Lui et al., 1990, 1991) a "cross-field current instability" impedes the cross-tail current flow in the tail without necessarily forming an X-type neutral line. Particle injection comes from collapse of a stretched field line as a result of current disruption. The deflation of the plasma sheet is communicated downstream by the generation of a rarefraction wave, causing plasma sheet thinning farther downstream of the current disruption region.

If ions are streaming across the tail and are carrying significant cross-tail current late in the substorm growth phase, they will be susceptible to the cross-field current instability; it will produce electromagnetic waves that will disrupt the currents. The instability is found to have a growth rate on the order of the substorm time scale and is most likely to occur in the region near, or just beyond, geosynchronous orbit (6.6 Re).

9.5. PLASMA ANISOTROPIES

Mitchell et al. (1990) argue that during the early growth phase of a substorm a significant fraction of the cross-tail current is due to a pressure anisotropy $(P_\parallel \gg P_\perp)$ in convecting magnetotail flux tubes. The anisotropy is dominated by electrons of energies near 1 keV. For times of slow adiabatic convection, prior to substorm expansion phase onset, Tsyganenko (1989) assumed conservation of the first two adiabatic invariants and showed that, as the magnetotail configuration becomes more stretched in the convection process, a pressure anisotropy with $P_\parallel > P_\perp$ can develop in the region near $X_{GSM} = -8$ Re. This is the region where intense cross-tail currents develop prior to substorm onset.

Hill and Voigt (1992) showed that the condition of MHD equilibrium implies tight constraints on the degree of anisotropy that is permissible in a magnetotail field geometry. The domain of stable magnetic field configurations with respect to the mirror and firehose instabilities is highly dependent on the plasma β_0 value in the equatorial plasma sheet. Mirror and firehose stable magnetic field configurations have to satisfy the two conditions

$$\text{Firehose stable: } \frac{P_\parallel - P_\perp}{B^2} \leq \frac{1}{\mu_0} \tag{9.1}$$

$$\text{Mirror stable: } \frac{P_\parallel - P_\perp}{B^2} \geq -\frac{1}{2\,\mu_0}\frac{P_\parallel}{P_\perp} \tag{9.2}$$

The scalar pressure P_0 defined by the trace of the pressure tensor $\mathbf{P}$, namely,

$$P_0 = \frac{1}{3}\,(2\,P_{\perp 0} + P_{\parallel 0}) \tag{9.3}$$

is used to define the plasma β-parameter in Equation 1.1 at the center of the plasma sheet, namely,

$$\beta_0 = \frac{2\,\mu_0\,P_0}{B_{z0}^2} \tag{9.4}$$

The combination of Equations 9.1 through 9.4 leads to expressions for permissible values of plasma anisotropies depending on the plasma-loading β_0 in the magnetotail plasma sheet. Here, the anisotropy factor is

$$\tau_0 = \frac{P_{\perp 0}}{P_{\parallel 0}} \tag{9.5}$$

in the equatorial plane of symmetry. Then Equations 9.1 and 9.2 read

$$\text{Firehose stable: } \tau_0 \geq \frac{3\,\beta_0 - 2}{3\,\beta_0 + 4} \tag{9.6}$$

$$\text{Mirror stable: } \tau_0 \leq \frac{(3\,\beta_0 + 2)}{6\,\beta_0} + \sqrt{\left(\frac{(3\,\beta_0 + 2)}{6\,\beta_0}\right)^2 + \frac{1}{3\,\beta_0}} \tag{9.7}$$

It is obvious that for very high values of β_0 in Equations 9.6 and 9.7 the anisotropy in Equation 9.5 ceases to exist because τ_0 approaches unity (Hill and Voigt, 1992). This might explain the fact that one commonly observes isotropic plasma down in the magnetotail, but substantial anisotropies in the region of the equatorial ring current where plasma β values are much smaller, namely, of the order $\beta_0 \approx 1 - 5$. If for one or the other reason the magnetic field configuration near geosynchronous orbit becomes suddenly stretched (i.e., β_0 increases) then that configuration no longer satisfies, say, the condition in Equation 9.6. A violation of Equation 9.6 constitutes an MHD instability and thus perhaps the onset of a substorm expansion.

From a theoretical point of view, it is not confirmed that a current sheet with $P_\parallel > P_\perp$ slightly beyond the marginal firehose limit becomes unstable. At the marginal firehose limit, $P_\parallel = P_\perp$, the current density reaches extremely high values, but that high cross-tail current is confined to a very narrow region at the center of the plasma sheet. The current "singularity" is embedded in an otherwise stable thick geotail plasma sheet.

Each of the above substorm models is incomplete in providing an entire substorm development because it is derived from observations taken in different regions in space. A major attempt toward a synthesis of compatible features among those models is due to Lui (1991, 1992).

10. MAGNETOHYDRODYNAMICS

The MHD theory combines fluid mechanics and electromagnetism. The MHD description of the interaction between plasmas and magnetic fields does not provide us with any insight into microscopic, kinetic plasma effects of that interaction. In particular, physical processes at transition regions (such as the discontinuities discussed in Section 3) remain unexplained. Yet the MHD description provides us with a large-scale, macroscopic picture of the entire solar wind-magnetosphere-ionosphere system. MHD models promise to become some of the most powerful research tools in large-scale magnetospheric physics.

10.1. BASIC EQUATIONS IN NONCONSERVATIVE FORM

In nonconservative form, the equations of one-fluid, ideal MHD for isotropic scalar pressure read as follows:

$$\rho \frac{\partial \mathbf{v}}{\partial t} + \rho (\mathbf{v} \cdot \nabla) \mathbf{v} = -\nabla P + (\mathbf{j} \times \mathbf{B}) \tag{10.1}$$

$$\nabla \times \mathbf{B} = \mu_0 \mathbf{j} \tag{10.2}$$

$$\nabla \cdot \mathbf{B} = 0 \tag{10.3}$$

$$\frac{\partial \mathbf{B}}{\partial t} = \nabla \times (\mathbf{v} \times \mathbf{B}) \tag{10.4}$$

$$\frac{\partial \rho}{\partial t} + \mathbf{v} \cdot \nabla \rho + \rho (\nabla \cdot \mathbf{v}) = 0 \tag{10.5}$$

$$\frac{\partial P}{\partial t} + \mathbf{v} \cdot \nabla P + \gamma P (\nabla \cdot \mathbf{v}) = 0 \tag{10.6}$$

The combination of Equations 10.5 and 10.6 recovers the thermodynamic equation of state, namely,

$$\frac{d}{dt} \left(\frac{P}{\rho^\gamma} \right) = 0 \tag{10.7}$$

The adiabatic exponent in Equations 10.6 and 10.7 is $\gamma = 5/3$ for a monatomic plasma with three degrees of freedom. The thermodynamic condition does not follow directly from MHD; it is instead an additional physical assumption needed to complete the system of the one-fluid MHD equations.

The above set of equations is based on several assumptions. First of all, the plasma is seen as a continuous fluid. This is a valid approximation if characteristic scale lengths L for changes of physical quantities are much greater than the ion gyroradius. Secondly, the plasma must be close to the state of local thermodynamic equilibrium, that is, the plasma distribution function must (locally) be Maxwellian or nearly Maxwellian. Finally, the plasma is seen as a single fluid. This approximation is valid when the bulk part of the plasma consists of only one ion species (e.g., protons), and when electrons and protons have roughly the same temperature, such that the thermal pressure can be written as $P = (n_p + n_e)kT$, and the mass density can be written as $\rho = n_p\, m_p + n_e\, m_e \approx n_p\, m_p$. The validity of the induction equation in the form of Equation 10.4 deserves particular attention. Ohm's law

$$E = -(v \times B) + \frac{1}{\sigma}\, j \tag{10.8}$$

combined with Equations 10.2 through 10.4 leads to the diffusion equation

$$\frac{\partial B}{\partial t} = \nabla \times (v \times B) + \frac{1}{\sigma}\, \nabla^2 B \tag{10.9}$$

The first term in Equation 10.9 describes plasma transport, the second term plasma diffusion. If L denotes the characteristic scale length for variations, then ideal MHD applies when

$$\frac{1}{L}|v|\,|B| \gg \frac{1}{\sigma}\frac{1}{L^2}\,|B| \tag{10.10}$$

or when, in other words,

$$R = \sigma\, L\, |v| \gg 1 \tag{10.11}$$

The quantity R is called the magnetic Reynolds number. In the solar wind-magnetosphere system, characteristic scale lengths L are usually quite large, so that ideal MHD applies even when the electric conductivity σ is finite. It is also evident that ideal MHD breaks down at discontinuities, such as the magnetopause, because there the scale length L is sometimes no larger than a couple of ion gyroradii.

10.2. BASIC EQUATIONS IN CONSERVATIVE FORM

It is mentioned in Section 3 that at discontinuities the Rankine-Hugoniot jump conditions must hold. Therefore, most modern numerical MHD codes use the MHD equations in conservative form, namely,

$$\frac{\partial \rho}{\partial t} = -\nabla \cdot (\rho v) \tag{10.12}$$

$$\frac{\partial}{\partial t}(\rho v) = -\nabla \cdot \left[\rho vv + I\left(P + \frac{B^2}{2\,\mu_0}\right) - \frac{BB}{\mu_0}\right] \tag{10.13}$$

$$\frac{\partial U}{\partial t} = -\nabla \cdot \left[\left(U + P + \frac{B^2}{2\,\mu_0}\right)v - \frac{(v \cdot B)}{\mu_0}\,B\right] \tag{10.14}$$

$$\frac{\partial \mathbf{B}}{\partial t} = \nabla \times (\mathbf{v} \times \mathbf{B}) \qquad (10.15)$$

$$\nabla \cdot \mathbf{B} = 0 \qquad (10.16)$$

$$U = \frac{1}{2} \rho v^2 + \frac{P}{\gamma - 1} + \frac{B^2}{2 \mu_0} \qquad (10.17)$$

Equations 10.12 through 10.14 denote the conservation of mass, momentum, and energy, respectively. Expression 10.17 denotes the total energy U of the system where the three terms are the kinetic, internal, and magnetic energy densities, respectively.

10.3. MHD SIMULATIONS

Global MHD simulations have been carried out by several groups (Lyon et al., 1981; Wu et al., 1981; Brecht et al., 1982; Wu, 1983, 1984; Ogino, 1986; Fedder and Lyon, 1987; Usadi et al., 1993). For reasons of physical consistency and numerical convergence, some of the referenced MHD codes use the MHD Equations 10.12 through 10.17 in conservative form.

As an example, Figure 11.10.1 shows equatorial and meridian cross sections of a three-dimensional magnetosphere generated by a large-scale MHD simulation (Usadi et al., 1993). Shown in Figure 11.10.1 are plasma flow lines and magnetic field lines over a 1.5-h period for the case when a southward IMF is present in the solar wind. A southward IMF supports magnetospheric convection (plasma circulation). The convection process (see Section 8) leads to periodic substorms which are caused by large-scale instabilities in the magnetotail plasma sheet.

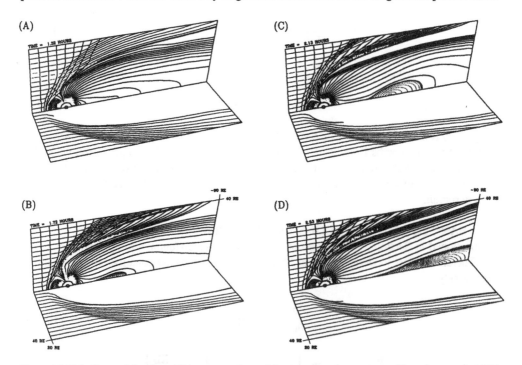

Figure 11.10.1 Equatorial and meridian cross sections of the magnetosphere generated by a large-scale MHD simulation (Usadi et al., 1993). Shown are plasma flow lines and magnetic field lines over a 1.5-h period for the case when a southward IMF is present in the solar wind. Because of reconnection at the center of the plasma sheet, a plasmoid develops and is accelerated toward the tail boundary of the simulation box. It has not been determined under what physical conditions the MHD simulation could reach a steady state. (From Usadi, A. et al. (1993). *J. Geophys. Res.*, 98, 7503. 1993 Copyright by the American Geophysical Union. With permission.)

One particular manifestation of such an instability is the occurrence of magnetic reconnection at the center of the plasma sheet. In that process, a plasmoid develops and is accelerated tailward (in the MHD simulation the plasmoid eventually reaches the end of the simulation box).

Time-dependent MHD simulations of the entire solar wind-magnetosphere-ionosphere system generate, in principle, self-consistent, large-scale magnetic field configurations of Earth's magnetosphere. Thus today's MHD codes will soon come close to answering the ultimate question posed in the introduction of this chapter: "What will happen, when Earth's vacuum dipole field is exposed to the steady stream of solar wind plasma originating from the Sun?"

ACKNOWLEDGMENTS

This work was supported by NSF through grant ATM-91-14269 (Magnetospheric Physics Program) and by NASA through grant NAGW-2824 (SR&T Program).

REFERENCES

Ashour-Abdalla, M., Berchem, J. P., Büchner, J., and Zelenyi, L. M. (1993). Shaping of the magnetotail from the mantle: global and local structuring, *J. Geophys. Res.*, 98, 5651.

Barraclough, D. R. (1987). International geomagnetic reference field: the fourth generation, *Phys. Earth Planet Int.*, 48, 279.

Basinska, E. M., Burke, W. J., Maynard, N. C., Hughes, W. J., Winningham, J. D., and Hanson, W. B. (1992). Small-scale electrodynamics of the cusp with northward interplanetary magnetic field, *J. Geophys. Res.*, 97, 6369.

Beard, D. B. (1973). The interactions of the solar wind with planetary magnetic fields: basic principles and observations, *Planet. Space Sci.*, 21, 1475.

Behannon, K. W. (1968). Mapping of the Earth's bow shock and magnetic tail by Explorer 33, *J. Geophys. Res.*, 73, 907.

Behannon, K. W. (1970). Geometry of the geomagnetic tail, *J. Geophys. Res.*, 75, 743.

Birn, J., Yur, G., Rahman, H. U., and Minami, S. (1992). On the termination of the closed field line region of the magnetotail, *J. Geophys. Res.*, 97, 14833.

Brecht, S. H., Lyon, J. G., Fedder, J. A., and Hain, K. (1982). A time dependent three-dimensional simulation of the earth's magnetosphere: Reconnection events, *J. Geophys. Res.*, 87, 6098.

Chao, J. K., Kan, J. R., Lui, A. T. Y., and Akasofu, S.-I. (1977). A model for thinning of the plasma sheet, *Planet. Space Sci.*, 25, 703.

Cap, F. F., and Leubner, M. P. (1974). A model of the magnetopause using an angular distribution function for the incident particles, *J. Geophys. Res.*, 79, 5304.

Carovillano, R. L., and Forbes, J. M. (Eds.) (1983). *Solar-Terrestrial Physics*, D. Reidel, Dordrecht, The Netherlands.

Chapman, S., and Ferraro, V. C. A. (1931). A new theory of magnetic storms, *Terr. Magn. Atmos. Electr.*, 36, 171.

Choe, J. Y., Beard, D. B., and Sullivan, E. C. (1973). Precise calculation of the magnetosphere surface for the tilted dipole, *Planet. Space Sci.*, 21, 485.

Cowley, S. W. H. (1981). Magnetospheric asymmetries associated with the y-component of the IMF, *Planet. Space Sci.*, 29, 79.

Crooker, N. U., Siscoe, G. L., and Toffoletto, F. R. (1990). A tangent subsolar merging line, *J. Geophys. Res.*, 95, 3787.

Dandouras, J. (1988). On the average shape and position of the geomagnetic neutral sheet and its influence on plasma sheet statistical studies, *J. Geophys. Res.*, 93, 7345.

Dungey, J. W. (1961). Interplanetary magnetic field and the auroral zone, *Phys. Rev. Lett.* 6, 47.

Erickson, G. M. (1984). On the cause of X-line formation in the near-earth plasma sheet: results of adiabatic convection of plasma sheet plasma, in *Magnetic Reconnection, Geophys. Monogr. Ser.*, 30, E. W. Hones, Ed., pp. 296–302, AGU, Washington, D.C.

Erickson, G. M., and Wolf, R. A. (1980). Is steady-state convection possible in the Earth's magnetosphere?, *Geophys. Res. Lett.*, 7, 897.

Fairfield, D. H. (1971). Average and unusual locations of the earth's magnetopause and bow shock, *J. Geophys. Res.*, 76, 6700.

Fairfield, D. H. (1979). On the average configuration of the geomagnetic tail, *J. Geophys. Res.*, 84, 1950.

Fairfield, D. H. (1980). A statistical determination of the shape and position of the geomagnetic neutral sheet, *J. Geophys. Res.*, 85, 775.

Fairfield, D. H. (1990). Recent advances in magnetospheric substorm research, *J. Atmos. Terr. Phys.*, 52, 1155.

Fairfield, D. H. (1991). Solar wind control of the size and shape of the magnetosphere, *J. Geomagn. Geoelectr.*, 43, Suppl., 117.

Fairfield, D. H. (1992a). On the structure of the distant magnetotail: ISEE-3, *J. Geophys. Res.*, 97, 1403.

Fairfield, D. H. (1992b). Advances in magnetospheric storm and substorm research: 1989–1991, *J. Geophys. Res.*, 97, 10865.

Fedder, J. A., and Lyon, J. G. (1987). The solar wind-magnetosphere-ionosphere current-voltage relationship, *Geophys. Res. Lett.*, 14, 880.

Formisano, V., Domingo, V., and Wenzel, K.-P. (1979). The three-dimensional shape of the magnetopause, *Planet. Space Sci.*, 27, 1137.

Frank, L. A. (1967). On the extraterrestrial ring current during geomagnetic storms, *J. Geophys. Res.*, 72, 3753.

Fuchs, K., and Voigt, G.-H. (1979). Self-consistent theory of a magnetospheric B-field model, in *Quantitative Modeling of Magnetospheric Processes, Geophys. Monogr. Ser.*, Vol. 21, W. P. Olson, Ed., pp. 86–95, AGU, Washington, D.C.

Goertz, C. K., and Smith, R. A. (1989). Thermal catastrophe model of substorms, *J. Geophys. Res.*, 94, 6581.

Grad, H. (1964). Some new variational properties of hydromagnetic equilibria, *Phys. Fluids*, 7, 1283.

Hasegawa, A., and Sato, T. (1989). *Space Plasma Physics*, Springer-Verlag, Tokyo.

Hau, L.-N., and Wolf, R. A. (1987). Effects of a localized minimum in the equatorial field strength on resistive tearing instability in the geomagnetotail, *J. Geophys. Res.*, 92, 4745.

Hill, T. W., and Dessler, A. J. (1991). Plasma motions in planetary magnetospheres, *Science*, 252, 345.

Hill, T. W., and Voigt, G.-H. (1992). Limits on plasma anisotropy in a tail-like magnetic field, *Geophys. Res. Lett.*, 19, 2441.

Huang, C. Y., and Frank, L. A. (1986). A statistical study of the central plasma sheet: implications for substorm models, *Geophys. Res. Lett.*, 13, 652.

Hudson, P. D. (1971). Rotational discontinuities in anisotropic plasma, *Planet. Space Sci.*, 19, 1693.

Hudson, P. D. (1973). Rotational discontinuities in anisotropic plasma II, *Planet. Space Sci.*, 21, 475.

Iijima, T., and Potemra, T. A. (1978). Large-scale characteristics of field aligned currents associated with substorms, *J. Geophys. Res.*, 83, 599.

Johnson, F. S. (1960). The gross character of the geomagnetic field in the solar wind, *J. Geophys. Res.*, 65, 3049.

Kamide, Y., and Slavin, J. A. (1986). *Solar Wind-Magnetosphere Coupling*, Terra Scientific, Tokyo.

Kertz, W. (1971). *Einführung in die Geophysik II*, Bibliographisches Institut, Mannheim, Germany.

Kosik, J. C. (1977). An analytical approach to the Choe-Beard magnetosphere, *Planet. Space Sci.*, 25, 457.

Landau, L. D., and Lifshitz, E. M. (1984). *Electrodynamics of Continuous Media*, 2nd ed., Pergamon Press, London.

Lennartsson, W., and Shelley, E. G. (1986). Survey of 01.- to 16-keV/e plasma sheet ion composition, *J. Geophys. Res.*, 91, 3061.

Levi, R. H., Petschek, H. E., and Siscoe, G. L. (1964). Aerodynamic aspects of the magnetospheric flow, *AIAA J.*, 2, 2065.

Lui, A. T. Y. (1991). A synthesis of magnetospheric substorm models, *J. Geophys. Res.*, 96, 1849.

Lui, A. T. Y. (1992). Magnetospheric substorms, *Phys. Fluids B*, 4, 2257.

Lui, A. T. Y., Mankofsky, A., Chang, C.-L., Papadopoulos, K., and Wu, C. S. (1990). A current disruption mechanism the neutral sheet: a possible trigger for substorm expansion, *Geophys. Res. Lett.*, 17, 745.

Lui, A. T. Y., Chang, C.-L., Mankofsky, A., Wong, H.-K., and Winske, D. (1991). A cross-field current instability for substorm expansions, *J. Geophys. Res.*, 96, 11389.

Lyon, J. G., Brecht, S. H., Huba, J. D., Fedder, J. A., and Palmadesso, P. J. (1981). Computer simulation of a geomagnetic substorm, *Phys. Rev. Lett.*, 46, 1038.

Macwan, S. E. (1992). A determination of the twisting of the Earth's magnetotail at distances 115–220 Re: ISEE-3, *J. Geophys. Res.*, 97, 19239.

McPherron, R. L., Russell, C. T., and Aubry, M. P. (1973). Satellite studies of magnetospheric substorms on August 15, 1968; 9. Phenomenological model for substorms, *J. Geophys. Res.*, 78, 3131.

Mead, G. D., and Beard, D. B. (1964). Shape of the geomagnetic field solar wind boundary, *J. Geophys. Res.*, 69, 1169.

Mihalov, J. D., and Sonett, C. P. (1968). The cislunar geomagnetic tail gradients in 1967, *J. Geophys. Res.*, 73, 6837.

Mitchell, D. G., Williams, D. J., Huang, C. Y., Frank, L. A., and Russell, C. T. (1990). Current carriers in the near-Earth cross-tail current sheet during substorm growth phase, *Geophys. Res. Lett.*, 17, 583.

Nakagawa, T., and Nishida, A. (1989). Southward magnetic field in the neutral sheet produced by wavy motions propagating in the dawn-dusk direction, *Geophys. Res. Lett.*, 16, 1265.

Ness, N. F., Scearce, C. S., and Seek, J. B. (1964). Initial results of the IMP 1 magnetic field experiments, *J. Geophys. Res.*, 69, 3531.

Nishida, A., Ed. (1982). *Magnetospheric Plasma Physics*, D. Reidel, Dordrecht, The Netherlands.

Ogino, T. (1986). A three-dimensional MHD simulation of the interaction of the solar wind with earth's magnetosphere: the generation of field aligned currents, *J. Geophys. Res.*, 91, 6791.

Olson, W. P. (1969). The shape of the tilted magnetopause, *J. Geophys. Res.*, 74, 5642.

Oznovich, I., Eastes, R. W., Huffman, R. E., Tur, M., and Glaser, I. (1993). The aurora at quiet magnetospheric conditions: repeatability and dipole tilt dependence, *J. Geophys. Res.*, 98, 3789.

Parks, G. K. (1991). *Physics of Space Plasmas,* Addison-Wesley, Reading, PA.

Peddie, N. W. (1982). International geomagnetic reference field: the third generation, *J. Geomagn. Geoelectr.,* 34, 309.

Pilipp, W. G., and Morfill, G. (1978). The formation of the plasma sheet resulting from plasma mantle dynamics, *J. Geophys. Res.,* 83, 5670.

Rostoker, G., Akasofu, S.-I., Foster, J., Greenwald, R. A., Kamide, Y., Kawasaki, K., Lui, A. T. Y., McPherron, R. L., Russell, C. T. (1980). Magnetospheric substorms: definition and signatures, *J. Geophys. Res.,* 85, 1663.

Rostoker, G., and Eastman, T. E. (1987). A boundary layer model for magnetospheric substorms, *J. Geophys. Res.,* 92, 12187.

Russell, C. T. (1971). Geophysical coordinate transformations, *Cosmic Electrodyn.,* 2, 184.

Sanchez, E., Summers, D., and Siscoe, G. L. (1990). Downstream evolution of an open MHD magnetotail boundary, *J. Geophys. Res.,* 95, 20743.

Schindler, K., and Birn, J. (1978). Magnetospheric physics, *Phys. Rep.,* 47, 109.

Sckopke, N. (1972). A study of self-consistent ring-current models, *Cosmic Electrodyn.,* 3, 330.

Sibeck, D. G., Siscoe, G. L., Slavin, J. A., Smith, E. J., Tsurutani, B. T., and Bame, S. J. (1985). Magnetic field properties of the distant magnetotail magnetopause and boundary layers, *J. Geophys. Res.,* 90, 9561.

Sibeck, D. G., Lopez, R. E., and Roelof, E. C. (1991). Solar wind control of the magnetopause shape, location, and motion, *J. Geophys. Res.,* 96, 5489.

Siscoe, G. L., and Sanchez, E. (1987). An MHD model for the complete open magnetotail boundary, *J. Geophys. Res.,* 92, 7405.

Slavin, J. A., Smith, E. J., Sibeck, D. G., Baker, D. N., Zwickl, R. D., and Akasofu, S.-I. (1985). An ISEE 3 study of average and substorm conditions in the distant magnetotail, *J. Geophys. Res.,* 90, 10875.

Sonnerup, B. U. Ö. (1980). Theory of the low-latitude boundary layer, *J. Geophys. Res.,* 85, 2017.

Spiro, R. W., and Wolf, R. A. (1983). Electrodynamics of convection in the inner magnetosphere, in *Magnetospheric Currents, Geophys. Monogr. Ser.,* 28, 247, American Geophysics Union, Washington, D.C.

Spreiter, J. R., and Jones, W. P. (1963). On the effect of a weak interplanetary magnetic field on the interaction between the solar wind and the geomagnetic field, *J. Geophys. Res.,* 68, 3555.

Spreiter, J. R., Summers, A. L., and Alksne, A. Y. (1966). Hydromagnetic flow around the magnetosphere, *Planet. Space Sci.,* 14, 223.

Spreiter, J. R., and Alksne, A. Y. (1969). Plasma flow around the magnetosphere, *Rev. Geophys.,* 7, 11.

Stern, D. P. (1985). Parabolic harmonics in magnetospheric modeling: the main dipole and ring current, *J. Geophys. Res.,* 90, 10851.

Toffoletto, F. R., and Hill, T. W. (1989). Mapping of the solar wind electric field to the Earth's polar caps, *J. Geophys. Res.,* 94, 329.

Toffoletto, F. R., and Hill, T. W. (1993). A nonsingular model of the open magnetosphere, *J. Geophys. Res.,* 98, 1339.

Toffoletto, F. R., and Voigt, G.-H. (1993). The stand-off distance of Earth's magnetosphere (abstract), *Eos Trans. AGU,* 74, Spring Meeting Suppl., 279.

Tsurutani, B. T., and von Rosenvinge, T. T. (1984). ISEE-3 distant geotail results, *Geophys. Res. Lett.,* 11, 1027.

Tsyganenko, N. A. (1989). On the re-distribution of the magnetic field and plasma in the near nightside magnetosphere during a substorm growth phase, *Planet. Space Sci.,* 37, 183.

Usadi, A., Kageyama, A., Watanabe, K., and Sato, T. (1993). A global simulation of the magnetosphere with a long tail: southward and northward interplanetary magnetic field, *J. Geophys. Res.,* 98, 7503.

Vasyliunas, V. M. (1970). Mathematical models of magnetospheric convection and its coupling to the ionosphere, in *Particles and Fields in the Magnetosphere,* B. M. McCormac, Ed., p. 60, D. Reidel, Dordrecht, The Netherlands.

Voigt, G.-H. (1976). Influence of magnetospheric parameters on geosynchronous field characteristics, last closed field lines and dayside neutral points, in *The Scientific Satellite Programme During the International Magnetospheric Study,* Knott and Battrick, Eds., p. 38, D. Reidel, Dordrecht, The Netherlands.

Voigt, G.-H. (1978). A static-state field line reconnection model for the Earth's magnetosphere, *J. Atmos. Terr. Phys.,* 40, 355.

Voigt, G.-H. (1981). A mathematical magnetospheric field model with independent physical parameters, *Planet. Space Sci.,* 29, 1.

Voigt, G.-H. (1984). The shape and position of the plasma sheet in Earth's magnetosphere, *J. Geophys. Res.,* 89, 2169.

Voigt, G.-H. (1986). Magnetospheric equilibrium configurations and slow adiabatic convection, in *Solar Wind-Magneto-sphere Coupling,* Y. Kamide and J. A. Slavin, Eds., pp. 233–273, Terra Scientific, Tokyo.

Voigt, G.-H., and Fuchs, K. (1979). A macroscopic model for field line interconnection between the magnetosphere and the interplanetary space, in *Quantitative Modeling of Magnetospheric Processes,* W. P. Olson, Ed., *Geophys. Monogr. Ser.,* 21, 448, American Geophysics Union, Washington, D.C.

Voigt, G.-H., and Wolf, R. A. (1988). Quasi-static magnetospheric MHD processes and the "ground state" of the magnetosphere, *Rev. Geophys.,* 26, 823.

Voigt, G.-H., and Toffoletto, F. R. (1993). The magnetopause shape and magnetic normal component: their effects on the magnetotail plasma sheet (abstract), *Eos Trans. AGU,* 74, Spring Meeting Suppl., 279.

Wolf, R. A. (1983). The quasi-static (slow-flow) region of the magnetosphere, in *Solar-Terrestrial Physics*, Carovillano and Forbes, Eds., p. 303, D. Reidel, Dordrecht, The Netherlands.

Wolf, R. A., and Spiro, R. W. (1983). The role of the auroral ionosphere in magnetospheric substorms, in *High-Latitude Space Plasma Physics*, V. Hultqvist and T. Hagfors, Eds., pp. 19–38, Plenum Press, New York.

Wu, C. C. (1983). Shape of the magnetosphere, *Geophys. Res. Lett.*, 10, 545.

Wu, C. C. (1984). The effects of dipole tilt on the structure of the magnetosphere, *J. Geophys. Res.*, 89, 11048.

Wu, C. C., Walker, R. J., and Dawson, J. M. (1981). A three-dimensional MHD model of the earth's magnetosphere, *Geophys. Res. Lett.*, 8, 523.

Chapter 12

Magnetospheric Electric Fields and Currents

Wolfgang Baumjohann

CONTENTS

1. PLASMA CONVECTION

Plasma convection in the magnetosphere is one of the major macroscopic dynamic processes in the magnetosphere. The large-scale pattern of plasma motion and electric fields in the magnetosphere and its consequences were first described using purely theoretical arguments by Axford and Hines (1961) and Dungey (1961). After 3 decades of magnetospheric measurements the essential features of magnetospheric convection suggested by them are still valid.

1.1. SOLAR WIND-MAGNETOSPHERE COUPLING

The Earth's magnetosphere is a cavity filled with hot dilute plasma embedded in the fast-flowing denser but colder solar wind plasma. Due to the intrinsic terrestrial magnetic field, the solar wind cannot directly penetrate the outer boundary of the magnetosphere, the magnetopause, but is deflected around it after having been slowed down to subsonic velocities at the Earth's bow shock.

The energy coupling processes at the magnetopause can be divided primarily into two different categories: (1) magnetic reconnection or field line merging, which implies interaction between the interplanetary magnetic field (IMF) and the terrestrial field at the dayside magnetopause (Dungey, 1961); (2) viscous-like interaction which implies tangential momentum transfer from the magnetosheath plasma through the magnetopause via some kind of viscosity generated by micro- or macro-instabilities (Axford and Hines, 1961). Figure 12.1.1 sketches these two processes and the resulting plasma flow in the magnetosphere. Magnetic reconnection will drive a tailward plasma flow on open field lines across the polar caps and the magnetospheric lobes, while viscous-like processes will drive a tailward plasma flow in the low-latitude boundary layers which are threaded by closed field lines. For both processes the convection cycle will be completed by sunward convection in the inner magnetosphere.

Satellite missions have provided *in situ* evidence for quasisteady reconnection (Paschmann et al., 1979) and of patchy, impulsive reconnection, so-called flux transfer events (Russell and Elphic,

0-8493-2520-X/95/$0.00+$.50
© 1995 by CRC Press

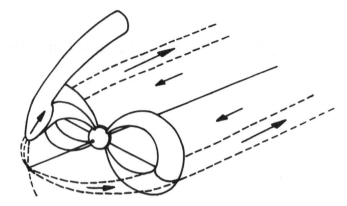

Figure 12.1.1 Sketch illustrating the two basic processes contributing to solar wind-driven convection in the magnetosphere: reconnection and viscous-like interaction. (From Cowley, S. W. H., *Rev. Geophys. Space Phys.*, 20, 531–565, 1982. Copyright by the American Geophysical Union. With permission.)

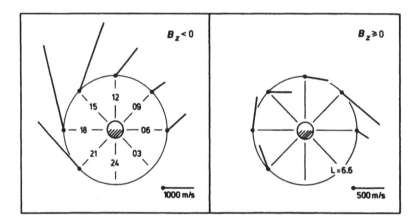

Figure 12.1.2 Plasma drift vectors (in the corotating frame of reference) as measured by the GEOS-2 electron gun experiment for positive and negative IMF B_z components. (From Baumjohann, W., and G. Haerendel, *J. Geophys. Res.*, 90, 6370–6378, 1985. Copyright by the American Geophysical Union. With permission.)

1979). Reconnection is a process that has a strong dependence on the angle between the IMF and the terrestrial field: it operates most efficiently for antiparallel field orientation and ceases for purely parallel fields. This way its effects on magnetospheric convection can easily be distinguished from those generated by viscous processes which are independent from the IMF.

Figure 12.1.2 shows the magnetospheric convection pattern at a radial distance of 6.6 R_E for periods when the IMF is southward directed ($B_z < 0$) and for periods of northward B_z (Baumjohann and Haerendel, 1985). If viscous-like interactions would dominate magnetospheric convection the plasma flow at synchronous orbit should be sunward directed regardless of the IMF direction. One sees, however, sunward flow for negative B_z (when conditions for reconnection are favorable) and more or less clockwise plasma circulation during periods of northward B_z. During the latter intervals the solar wind dynamo obviously works rather inefficiently and the convection in the Earth's magnetosphere is strongly influenced by ionospheric dynamo action (see Chapter II/9).

Results along the same line were obtained earlier for the correlation between the cross-polar cap potential Φ_{pc} measured by low-altitude satellites and the east-west component of the interplanetary electric field, i.e., the north-south component of the IMF (Reiff et al., 1981). These studies show a significant dependence of Φ_{pc} on the IMF B_z component, indicating that the

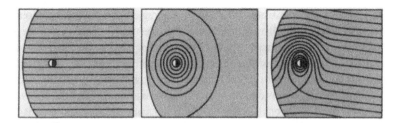

Figure 12.1.3 Equipotential contours of the magnetospheric electric field. From left to right: homogeneous sunward convection field, corotation field, and total field.

observed changes in Φ_{pc} are reconnection associated. Viscous interaction plays, at most, a minor role.

We will leave the discussion of the solar wind-magnetosphere-coupling at this stage. The reader interested in a more extensive overview on this topic is referred to reviews by Haerendel and Paschmann (1982), Baumjohann and Paschmann (1987), and Siscoe (1988).

1.2. MAGNETOSPHERIC CONVECTION

As mentioned in the last section, magnetic reconnection at the Earth's magnetopause generates an electric convection field inside the magnetosphere. This field is directed from dawn to dusk and, at synchronous orbit, has typical amplitudes of some 0.1 mV/m (Baumjohann and Haerendel, 1985). The total potential difference between the dawn and dusk magnetopause (or, equivalently, across the polar cap) amounts to about 50 to 100 kV.

As sketched on the left-hand side of Figure 12.1.3, this field can be assumed as homogeneous in a zero-order approximation. Thus, the convection potential can be written as:

$$\Phi_{con} = -E_0 r \sin\psi \tag{1.1}$$

Here, E_0 is the electric field strength, r is the radial distance, and ψ denotes azimuth. A more realistic form, which also takes into account the shielding of the inner magnetosphere from the convection field, is (Volland, 1973):

$$\Phi_{con} = -E_0 r^\gamma \sin\psi \tag{1.2}$$

where the so-called shielding factor γ has a value in the range 2 to 3 (Baumjohann et al., 1985).

To get the full magnetospheric plasma convection, one has to add the corotation field to the convection field. The Earth's rotation forces the ionospheric plasma via viscous coupling into corotation and thus creates electric polarization fields which are directed opposite to the electromotive force $(\Omega \times r) \times B$, where Ω is the angular velocity of the Earth's rotation. Because the magnetic field lines are equipotential contours, this field is mapped into the magnetosphere, too. Its potential distribution is radially symmetric and sketched in the middle panel of Figure 12.1.3. The corotation potential can be described as

$$\Phi_{cor} = -\frac{kR_E}{r} \tag{1.3}$$

where $k = 92$ kV, $R_E = 6.4 \cdot 10^6$ m, and r is measured in m.

The total potential $\Phi_{con} + \Phi_{cor}$ is depicted on the right-hand side of Figure 12.1.3, where one clearly sees two topologically different regions. Close to the Earth a region of closed equipotential contours exists. Here, within the so-called plasmasphere, corotation dominates. The plasma content of a flux tube is nearly constant and, accordingly, the plasma density is rather high (several thousand particles per cubic centimeter). Outside of this region, the potential contours are open and a magnetic flux tube will, at some time, encounter the magnetopause and lose its plasma. This is the reason for the sharp density gradient between the plasmasphere and the outer region, where typical plasma densities are of the order of several particles per cubic centimeter only.

An interesting point is the stagnation point in the left-hand diagram of Figure 12.1.3, i.e., the point on the evening side where an equipotential contour crosses itself. Here the eastward directed corotation and the westward directed convection have the same velocity and the plasma is stagnant. Since $\Phi_{con} = \Phi_{cor}$ and $\sin\psi = -1$ at this point, its distance from the Earth's surface can easily be calculated as

$$r = \frac{3.81 \, R_E}{\sqrt{E_0}} \tag{1.4}$$

for E_0 given in mV/m. Therefore, the radial distance of the stagnation point depends on the strength of the convection electric field: whenever the convection electric field increases, the radius of the plasmasphere decreases and vice versa.

1.3. MAGNETOSPHERE-IONOSPHERE COUPLING

The magnetosphere is dominated by a collision-free plasma, while the ionosphere is the region where the effects of collisions of charged particles with neutral particles cannot be neglected and electrical conductivities transverse to the geomagnetic field maximize. The magnetic field connects electrically the ionosphere and the magnetosphere, causing an exchange or coupling of energy and momentum between the two regions. In a sense, the magnetosphere-ionosphere coupling is the interaction of different physical processes taking place in either of these two regions. The strong coupling occurs because the two regions are connected by the same magnetic field lines.

The self-consistent logic governing the entire magnetosphere-ionosphere coupling system, along with the elements constituting the system are shown in Figure 12.1.4 (Vasyliunas, 1970). One can understand the physical mechanisms fully only if one examines the coupled system in its entirety. However, because the entire coupled system is very complex, one often looks at one aspect only. The impost important aspects are the field-aligned currents $j_\parallel$. In the ionosphere these field-aligned currents must be balanced by the divergence of ionospheric conduction currents **J**. These three-dimensional currents require a specific distribution of the space charge and ionospheric electric field **E**, which can be computed from the given field-aligned currents.

Namely,

$$\nabla \cdot \mathbf{J} = \nabla \cdot \Sigma \, \mathbf{E} = j_\parallel \tag{1.5}$$

where Σ is the ionospheric conductivity tensor. To close the logical loop of Figure 12.1.4, the ionospheric electric field must in turn be consistent with the driving magnetospheric electric field through the generalized Ohm's law.

Solving Equation 1.5 is already a challenging task (Mishin et al., 1979; Kamide et al., 1981) and trying to simulate the complete logic diagram of Figure 12.1.4 is even more complicated (Harel et al., 1981). Thus we will leave the subject at this stage and refer the interested reader to a recent monograph by Kamide and Baumjohann (1993).

2. LARGE-SCALE CURRENT SYSTEMS

Following Maxwell's law, the distortion of the terrestrial dipole field into the typical magnetospheric shape is accompanied by electrical currents. As schematically shown in Figure 12.2.1, the compression of the terrestrial magnetic field on the dayside is caused by currents flowing

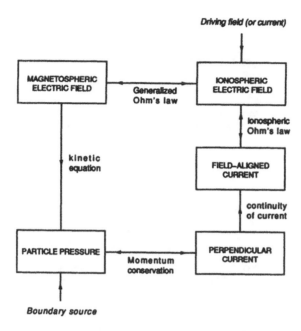

Figure 12.1.4 Logic diagram for a self-consistent treatment of electrodynamic processes in the coupled magnetosphere-ionosphere system. The boxes indicate quantities to be determined and each line joining two boxes is labeled with the physical principle that governs the relation between the two quantities. (From Vasyliunas, V. M., *Particles and Fields in the Magnetosphere,* B. M. McCormack, Ed., pp. 60–71, Reidel, Dordrecht, 1970. With permission.)

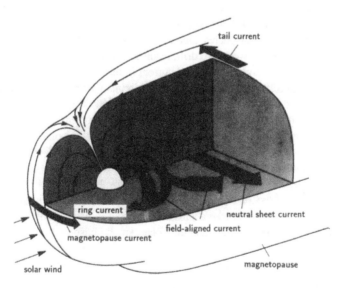

Figure 12.2.1 Overview on current circuits flowing in the magnetosphere and on the magnetopause. (From Baumjohann, W., and Haerendel, G. J., *Naturwissenschaften,* 74, 181–187, 1987. With permission.)

perpendicular to the dipole field lines across the magnetopause surface. These currents are often called Chapman-Ferraro currents, because Chapman and Ferraro were the first to postulate such a current system some 60 years ago. The tail-like field of the nightside magnetosphere is accompanied by the tail current flowing on the tail surface and the neutral sheet current in the central plasma sheet, both of which are connected and form a Θ-like current system, if seen from along the Earth-sun line.

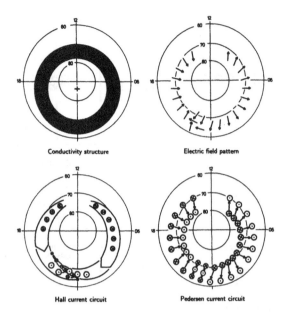

Figure 12.2.2 Schematic diagram of conductivity structure, electric field pattern (direction only), and current flow within the convection auroral electrojet system. Coordinates are invariant latitude and magnetic local time.

Another large-scale current system which influences the configuration of the inner magnetosphere is the ring current. This current flows encircling the Earth in a westward direction at radial distances of several R_E and is mainly carried by protons. The protons are trapped particles which have energies of some tens of keV, bouncing back and forth between their magnetic mirror points in the northern and southern hemispheres. Due to gradients in the plasma pressure and magnetic field, these protons experience a westward drift, leading to a westward electric current. The outer portions of the ring current merge with the tail current in the plasma sheet.

In addition to these purely magnetospheric current circuits which all flow perpendicular to the ambient magnetic field, there is a another set of currents which flows along magnetic field lines. These field-aligned currents, which often are called Birkeland currents, connect the current systems in the magnetosphere and its boundaries to those flowing in the polar ionosphere. The field-aligned currents are essential for the exchange of energy and momentum between these regions.

There are quite a number of high-latitude current systems which can coexist at auroral latitudes and in the polar cap, depending on the state of the magnetosphere, quiet or disturbed, and of the solar wind, especially the direction of the IMF. Some of these current circuits are global in nature, involving the whole polar ionosphere or a substantial part of it. Most notable are the convection electrojets, the substorm current wedge, and the polar cap and cusp currents.

2.1. CONVECTION ELECTROJETS

The most prominent current system at auroral latitudes is the global convection electrojet system, carrying a total current of the order of a million amperes. Our present knowledge about conductivity structure, electric fields, and current flow associated with the convection auroral electrojet system is summarized in Figure 12.2.2. The auroral electrojets flow throughout the whole auroral oval, which is, in fact, an off-center ring shifted by an average 4° from the invariant magnetic pole toward magnetic midnight (Meng et al., 1977).

Inside the auroral oval the ionospheric conductivity is enhanced above the solar UV-induced level due to the ionization of neutral atoms and molecules by precipitating electrons and, to a lesser extent, ions. The energetic particles drift toward and around the Earth and precipitate, depending on their energy and pitch angle, in different local time sectors (McDiarmid et al.,

1975). The precipitation pattern is reflected in the conductivity structure, depending on the energy of the precipitating particles. The weakest conductivities are found near the noon sector while the conductivity maximum lies in the midnight sector where typical values of 7 to 10 S and 10 to 20 S for Pedersen and Hall conductances (height-integrated conductivities), respectively, were found by, for example, Wallis and Budzinski (1981).

The electric field pattern in the auroral oval reflects the large-scale pattern of magnetospheric plasma convection associated with the solar wind-magnetosphere coupling (c.f. Section 1). The transport of open and closed flux tubes depicted in Figure 12.1.1 results in a convection pattern with two cells. The electric field associated with this two-cell system of plasma transport has typical values of between 20 mV/m during quiet times and 50 mV/m for active geomagnetic conditions at auroral latitudes. It is poleward directed in the afternoon and early evening sector, points equatorward in the postmidnight and morning sector, and rotates from north over west to south in the premidnight sector (Foster et al., 1981; Zi and Nielsen, 1982). This region of field rotation is called the Harang discontinuity region (c.f. Heppner, 1972; Maynard, 1974).

In contrast to the conductivities and electric fields, it is difficult to directly measure the ionospheric and field-aligned currents. Thus the intensity and the distribution of these currents have to be estimated from ground-based and satellite measurements of magnetic fields. Numerous studies of this kind have been made in the past (Iijima and Potemra, 1976; Hughes and Rostoker, 1979; Baumjohann et al., 1980; Kamide et al., 1981; Kunkel et al., 1986) and yielded fairly consistent results, as far as the average picture is concerned. They are summarized in the two lower panels of Figure 12.2.2.

Both eastward and westward electrojets are primarily Hall currents which originate around noon and are fed by downward field-aligned currents. Their sheet current densities range between 0.5 and 1 A/m and increase toward midnight due to the increasing Hall conductance. The eastward electrojet terminates in the region of the Harang discontinuity where it partially flows up magnetic field lines and partially rotates northward, joining the westward electrojet. The westward current originates in the noon sector, flows through the morning and midnight sector, and typically extends into the evening sector along the poleward border of the auroral oval where it also diverges as upward field-aligned currents.

The Pedersen current with typical densities of 0.3 to 0.5 A/m flows northward in the eastward electrojet region and is connected to sheets of downward and upward field-aligned currents in the southern and northern half of the afternoon-evening northern hemisphere auroral oval, respectively. In the midnight-to-noon sector the Pedersen current flows equatorward and field-aligned currents provide continuity by flowing upward in the southern and downward in the poleward half of the auroral oval. Iijima and Potemra (1976) named the sheet of field-aligned currents in the poleward half of the auroral oval Region 1 currents and those in the equatorward half Region 2 currents. In the Harang discontinuity, the evening and morning side Pedersen current circuits overlap, leading to three sheets of field-aligned currents (Iijima and Potemra, 1978; Robinson et al., 1985; Kunkel et al., 1986).

2.2. SUBSTORM CURRENT WEDGE

During magnetospheric substorms the ionospheric current flow is affected in two ways (Baumjohann, 1983; Rostoker et al., 1987; see also Chapter II/1) to accommodate for the enhanced energy input from the solar wind via increased dissipation. On the one hand, the Hall current flow in the convection auroral electrojets described in Section 2.1 (see the left panel of Figure 12.2.3) increases in direct relation to the energy input from the solar wind along with the driven process. Additionally, sporadic unloading of energy previously stored in the magnetotail leads to the formation of a substorm current wedge (see right panel of Figure 12.2.3) with strongly enhanced westward current flow in the midnight sector.

During a substorm, the solar wind energy which is not directly dissipated via enhanced convection in the auroral electrojets is stored in the tail magnetosphere, mainly in the form of magnetic

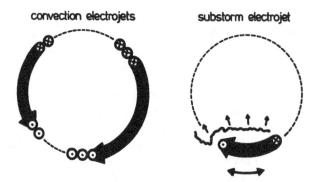

Figure 12.2.3 Schematic illustration describing location, flow direction and field-aligned current closure of the convection electrojets, and the substorm current wedge. (From Baumjohann, W., *Adv. Space Res.*, 2(10), 55–62, 1983. With permission.)

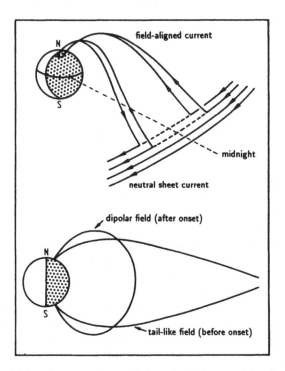

Figure 12.2.4 Upper panel: the substorm current wedge short-circuits the neutral sheet current in the magnetotail. Lower panel: reconfiguration of the nightside magnetospheric magnetic field associated with the disruption of the cross-tail current.

energy by enhancing the neutral sheet-tail current circuit shown in Figure 12.2.1 (McPherron, 1979). This is accompanied by a distortion of the nightside magnetospheric field lines into a strongly tail-like configuration even at smaller radial distances where the field usually is rather dipolar, for example, at geostationary distances of 6.6 R_E. As sketched in Figure 12.2.4, after a 30- to 60-min long phase of storage (the so-called growth phase) the magnetic tension is suddenly released, at expansion phase onset, by short-circuiting the neutral sheet current in a azimuthally limited region of the tail and diverting its current along magnetic field lines and through the midnight sector auroral ionosphere.

As indicated in Figure 12.2.3, the substorm current wedge expands, in general, poleward and westward during the course of the expansion phase. Its western edge is always collocated with

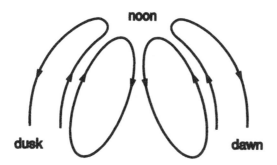

Figure 12.2.5 Sketch of the ionospheric Hall current pattern in the southern hemisphere for purely northward IMF. (From Zanetti, L. J. et al., *J. Geophys. Res.*, 89, 7453–7458, 1984. Copyright by the American Geophysical Union. With permission.)

the head of the westward traveling surge, a unique auroral form typically associated with the substorm expansion phase. The fine-scale structure of electric fields, ionospheric conductances, and ionospheric and field-aligned currents, especially near the head of the surge, is quite well-known today. It will be detailed later in Section 3.2.

2.3. POLAR CAP AND CUSP CURRENTS

While the nightside auroral currents discussed above are governed by processes taking place in the magnetospheric equatorial plane, the currents flowing in and near the polar cusp and in the dayside polar cap are connected along magnetic field lines to the vicinity of the magnetopause. They are thus governed primarily by the processes of solar wind-magnetosphere coupling described in Section 1.1. The solar wind and especially the IMF exert a controlling influence on these dayside high-latitude currents.

For nearly purely northward IMF, reconnection cannot occur on closed dayside magnetopause field lines, but rather has to operate on the open lobe field lines north of the polar cusp (Reiff and Burch, 1985). In this case also a two-cell convection and current pattern tends to evolve (Maezawa, 1976), but the pattern is confined to open polar cap field lines and its direction is reversed. Together with the S_q^p current system, which is a quiet time current system confined to latitudes north of the auroral oval independent of the IMF (Mishin et al., 1979), the total current system exhibits (for IMF $B_y = 0$) a four-cell pattern as shown in Figure 12.2.5, with antisunward current and sunward convection, respectively, across the polar cap (Zanetti et al., 1984).

In the polar cusp and cap region, not only the IMF B_z but also the B_y component exert a controlling influence. As can be seen in Figure 12.2.6 (Cowley, 1983), the sunward polar cap Hall current is displaced toward dawn (dusk) for positive (negative) B_y. Moreover, in the polar cusp region a zonal current, the DPY current (Svalgaard, 1973), appears for $B_y \neq 0$. This current is eastward (westward) directed for a positive (negative) B_y component.

The IMF B_y effects and thus the DPY currents can be explained by the dawn-dusk magnetic tension of the magnetosheath end of the merged field line. Figure 12.2.6 illustrates this situation. Beginning on the dayside, the tension on newly opened flux tubes or field lines has a net east-west component in the presence of a B_y field. The magnetic tension is oppositely directed in the two hemispheres. In response, oppositely directed azimuthal plasma flows and, hence, Hall currents appear in the polar cusp region.

2.4. CURRENT CLOSURE IN THE MAGNETOSPHERE

Soon after the experimental verification that the ionospheric current is not solely closed within the ionosphere, but that particularly the Pedersen current is closed via field-aligned currents, people started to speculate on where these field-aligned currents are closed in the magnetosphere.

Sugiura (1975) was one of the first to come up with a qualitative model for current closure, displayed in Figure 12.2.7. In his model the field-aligned current of the equatorward belt of Figure

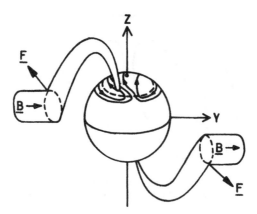

Figure 12.2.6   Effects of the IMF B_y component on the plasma convection in the dayside polar cap. Shown is a schematic view of newly opened flux tubes on the dayside, illustrating the field line tension, F, which results in oppositely directed azimuthal plasma flows and Hall currents in the northern and southern cusp regions. (From Cowley, S. W. H., *High-Latitude Space Plasma Physics*, B. Hultqvist and T. Hagfors, Eds., pp. 225–249, Plenum Press, New York, 1983. With permission.)

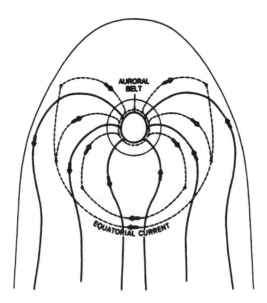

Figure 12.2.7   A suggestion by Sugiura (1975) on how two systems of field-aligned currents are closed in the magnetosphere. (From Sugiura, M., *J. Geophys. Res.*, 80, 2057–2068, 1975. Copyright by the American Geophysical Union. With permission.)

12.2.2 is closed by a westward ring current in the near Earth equatorial plane. On the other hand, the field-aligned current of the poleward belt extends deep into the magnetotail and flows along the high-latitude boundary of the plasma sheet or the low-latitude boundary layer adjacent to the magnetosheath.

However, it took more than another decade before an experimental verification of this model could be achieved, since it is very difficult to extract the spatial magnetic field gradients caused by field-aligned currents from temporal gradients in the highly variable magnetotail plasma. A statistical study of field-aligned currents in and near the plasma sheet boundary layer using ISEE 1 and 2 data (Ohtani et al., 1988) lead to the result that at tailward distances beyond 15 R_E field-aligned currents with typical sheet current densities of 5 to 10 mA/m exist irrespective of the level of magnetic activity. The polarity of these currents is shown in Figure 12.2.8. Nearly all morning sector currents flow toward the Earth, while most of the field-aligned currents in the

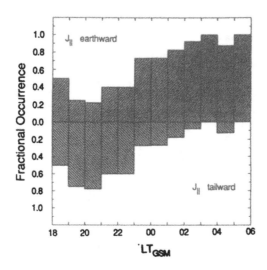

Figure 12.2.8   Fractional occurrence of field-aligned current sheet crossings with earthward or tailward polarity as a function of magnetic local time. (From Ohtani, S. et al., *J. Geophys. Res.,* 93, 9709–9720, 1988. Copyright by the American Geophysical Union. With permission.)

evening sector flow into the tail, consistent with the direction of the field-aligned currents in the poleward half of the belt in Figure 12.2.2. It is thus very likely that the currents identified by Ohtani et al. (1988) constitute the magnetotail extension of the region 1 field-aligned current belt. Where this current loop is finally closed by cross-field current flow remains to be determined.

Iijima et al. (1990) were able to close the current loop arising from the equatorward region 2 belt of field-aligned current. Figure 12.2.9 shows their statistical results of radial (field-aligned) and azimuthal (cross-field) current flow in the near Earth magnetotail, at radial distances of 4 to 9 R_E. The upper panel shows current away from the Earth throughout the postmidnight and morning sectors (00 to 12 MLT). The afternoon and evening sectors (12 to 24 MLT) are dominated by current flowing toward the Earth. The directions of these currents are consistent with the field-aligned current flow in the equatorward half of the auroral oval. The lower panel of Figure 12.2.9 indicates that the field-aligned current can be closed via westward azimuthal current flow across the field lines and that most of this closure current is concentrated in the nightside magnetosphere, at radial distances of about 7 to 9 R_E.

3.   AURORAL FIELDS AND CURRENTS

The precipitation of auroral particles into the polar ionosphere and thus the structure of conductivity enhancements are in reality much more complicated than described in Section 2. Very often the conductivity is highly enhanced in relatively small regions due to localized particle precipitation associated with discrete aurora, making the conductivity distribution highly inhomogeneous and distorting the electric field structure.

There are three main types of auroral structures, appearing in different local time sectors. Discrete auroral arcs are seen throughout the nightside oval. Break-up aurora and a westward traveling surge prevail around midnight and in the late evening sector of the auroral oval, respectively. In the morning sector, extending up to the dawn meridian, omega bands can often be found.

Associated with such auroral forms are spatially confined high-conductivity regions, with typical scale sizes ranging from several tens of kilometers in auroral arcs and patches to several hundreds of kilometers for westward traveling surges and omega bands. These high-conducting patches severely alter and distort the electric field pattern associated with the large-scale eastward and westward auroral electrojets.

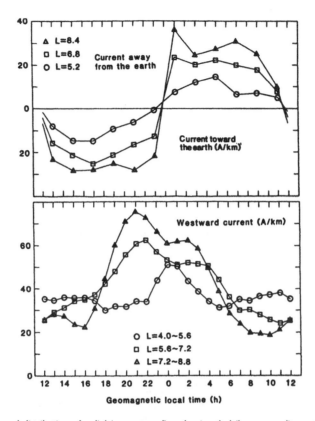

Figure 12.2.9 Diurnal distribution of radial (upper panel) and azimuthal (lower panel) current flow in the magnetosphere during disturbed geomagnetic conditions. (From Iijima, T. et al., *J. Geophys. Res.*, 95, 991–999, 1990. Copyright by the American Geophysical Union. With permission.)

3.1. AURORAL ARCS

Because auroral arcs are very thin in their north-south extent, one needs incoherent scatter radars or, even better, rocket-borne measurements to study their electrodynamics in detail (Marklund et al., 1982; Ziesolleck et al., 1983). In all auroral arcs the Hall and Pedersen conductances are significantly enhanced inside the arc, reaching peak values of 50 to 100 S. However, electric field and ionospheric current patterns are quite different in pre- and postmidnight arcs (see Figure 12.3.1).

In the evening sector the meridional component of the electric field is directed northward. It reaches peak values of the order of 100 mV/m just south of the arc, drops below typical convection electric fields (see Section 2.1) inside the arc, and recovers north of it. The zonal electric field component is typically eastward directed and nearly constant through the arc. In the morning sector, the meridional electric field has a southward direction. It is again reduced below convection level inside the arc, but in this local time sector it reaches its peak values north of the arc. The zonal component is directed toward west in this case and is again much less variable than the meridional electric field.

The zonal current is mainly a Hall current. In both local time sectors the zonal arc current flows in the same direction as the auroral electrojet in that particular local time sector. It is notable, however, that the current density is strongly enhanced inside the arc, reaching peak values of 1 to 2 A/m. The northward (in the evening) or southward (in the morning) directed Pedersen current typically does not show any systematic variations with respect to the arc.

Figure 12.3.1 (Baumjohann, 1983) shows how the measured field may be decomposed *into* an ambient convection electric field and an arc-associated field. In the evening sector the ambient

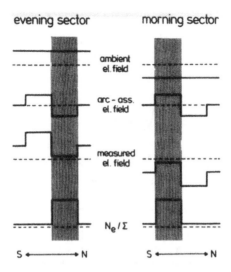

Figure 12.3.1 Schematic diagram of conductances and meridional electric field components (positive northward) measured around auroral arcs in the evening and morning sector and a possible decomposition of the measured field into an ambient and an arc-associated electric field. The dashed lines denote the approximate zero levels; the shading indicates the location of the arc. (From Baumjohann, W., *Adv. Space Res.,* 2(10), 55–62, 1983. With permission.)

electric field is northward directed and thus the arc-associated meridional field must be antisymmetric with respect to the arc's southern edge: it is northward directed within a few kilometers south of the arc and southward inside the arc. Auroral arcs in the morning sector have the same antisymmetric field pattern, but the line of symmetry, where the arc associated electric field changes its direction from north to south, is now located at the northern border of the arc.

The superposition of the ambient and the arc-associated electric fields sketched in Figure 12.3.1 may also explain the observations made for midnight sector auroral arcs located just south of the Harang discontinuity (Behm et al., 1979; Stiles et al., 1980). Here the northward electric field does not reach convection levels north of the arc, because the background northward convection field decreases with latitude toward the Harang discontinuity.

As sketched in Figure 12.3.2, the peculiar structure of the arc-associated electric field may be the combined effect of the two current continuity mechanisms, polarization electric fields, and enhanced field-aligned currents (Marklund, 1984). The decrease of the ambient field inside the arc is most likely the result of a polarization electric field which serves for current continuity across the arc by reducing the ambient meridional field and thus the meridional current inside the arc (see Section 3.2). The strong increase in the meridional field just north or south of the arc may be the low-altitude signature of an asymmetric inverted-V electric potential structure above the arc (Marklund, 1984), which accelerates particles and serves for current continuity of an enhanced meridional current inside the arc via enhanced field-aligned currents at its borders. Most often both mechanisms operate concurrently, but there are also cases where either polarization effects or enhanced field-aligned currents prevail.

3.2. WESTWARD TRAVELING SURGES

The westward traveling surge is one of the most prominent manifestations of auroral substorms. It characterizes the leading edge of the intense and dynamic westward electrojet appearing during the substorm expansion phase, i.e., the ionospheric part of the substorm current wedge described in Section 2.2. An intense upward field-aligned current is flowing at the head of the surge and is carried primarily by keV electrons precipitating along field lines (Meng et al., 1978). The ionospheric conductivities, electric fields, and currents associated with the surge are highly variable

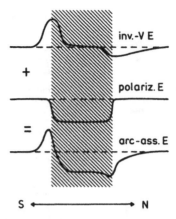

Figure 12.3.2 Possible interpretation of the arc-associated meridional electric field pattern for the evening auroral arc. The upper panel gives the low-altitude signature of an asymmetric inverted-V potential, the middle panel the polarization electric field, and the lower panel shows their superposition (all positive northward). The dashed lines denote the approximate zero levels; the shading indicates the arc location. (From Marklund, G., *Planet. Space Sci.*, 30, 179–197, 1982. With permission.)

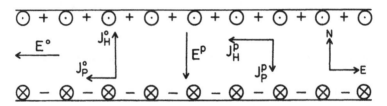

Figure 12.3.3 Generation of a polarization electric field and an incomplete Cowling channel in a region of enhanced ionization.

in time and, in contrast to the auroral electrojet or auroral arc current system, have strong gradients along the azimuthal axis.

Baumjohann et al. (1981) were the first to combine two-dimensional observations of ionospheric electric fields and ground magnetic fields and to model the three-dimensional current systems associated with break-up aurora around midnight. Their observations and models match in detail the features to be expected during the generation of a (incomplete) Cowling channel in the highly conducting region described first by Boström (1975) and summarized in Figure 12.3.3. Due to the conductivity enhancement, the westward component of the primary electric field, **E**, drives an enhanced northward Hall current, J_H. The excess Hall current deposits positive charges at the northern border of the highly conductive channel while negative charges build up at its southern boundary. These charges are removed, to some extent, by field-aligned current sheets, but the remainder gives rise to a southward polarization electric field, E^p. E^p drives a southward Pedersen current, J_P^p, which balances that part of J_H which is not continued via field-aligned currents. The westward currents due to the primary convection and secondary polarization electric field add up to an intense westward Cowling current (see also Yasuhara et al., 1985). Conductivity gradients at the western end eastern boundary of the channel lead to intense localized upward field-aligned currents at the western edge and less intense and more wide-spread downward field-aligned currents on the eastern side.

Inhester et al. (1981) and Opgenoorth et al. (1983a) showed that this type of current circuit was also valid for westward traveling surges observed in the early and late evening sector. Typical numbers taken from the work of Baumjohann et al. (1981), Inhester et al. (1981), Opgenoorth et al. (1983a), and Kirkwood et al. (1988) yield the following values for conductances, fields, and currents. Within the active region the Hall and conductances reach peak values of more than

100 S. The pre-breakup electric field pattern is distorted by the superposition of a southward polarization electric field with a strength of up to 50 mV/m in the area covered by break-up aurora. The ionospheric current has sheet current densities of 500 to 1000 mA/m, comparable with typical westward electrojet values. The westward component of the ionospheric current is connected to very localized and intense (about 5 to 10 μA/m^2) upward field-aligned currents at the western border and more wide-spread downward field-aligned current of lower density (1 to 2 μA/m^2) in the eastern half of the break-up region. The northward current is connected to field-aligned current sheets of 1 to 2 μA/m^2 at the southern and northern boundaries of the active region.

The westward traveling surge current circuit is the auroral form current system, which is best understood. As described in Section 2.2, the upward and downward field-aligned currents at its western and eastern edge seem to be naturally associated with near Earth reconnection in a limited local time sector (Birn and Hones, 1981; Sato et al., 1983). Numerical simulations by, for example, Rothwell et al. (1984) and Kan and Sun (1985) show that bouncing Alfvén waves near a conductivity gradient can reproduce the westward expansion of the surge quite well. These simulations have the additional advantage that they also explain the Pi2 pulsations, which are intimately related to the sudden development of the break-up current circuit (Baumjohann and Glassmeier, 1984; see also Chapter II/14).

3.3. OMEGA BANDS

While auroral breakup and the westward traveling surge prevail in the premidnight sector during the substorm expansion phase, eastward propagating sequences of omega bands are the dominant feature in the postmidnight sector during substorm recovery. The auroral omega bands are formed like the greek letter Ω and are associated with undulations in the westward electrojet. Their magnetic signature has been denoted as Ps 6 pulsations because of the periodic nature of the magnetogram recorded when a sequence of omega bands passes overhead.

Earlier work (Kawasaki and Rostoker, 1979; Gustafsson et al., 1981) had indicated that the current system associated with these particular auroral structures is three-dimensional. André and Baumjohann (1982) and Opgenoorth et al. (1983b) were the first to analyze simultaneous two-dimensional measurements of electric and magnetic fields of eastward drifting omega bands. They found that the electric and magnetic field pattern was actually stationary but drifting with the omega band. The undulated electrojet pattern is caused by a longitudinally alternating azimuthal electric field component of about 10 mV/m which is superimposed on the southward convection electric field.

This became evident by a decomposition of the measured electric field pattern into a uniform, steady background field and a second, highly structured electric field pattern which drifts eastward with the velocity of the omega band. Moreover, in both cases and also a later study by Buchert et al. (1988) the drift velocities of the omega band and the associated electric and magnetic field structures (typically 700 m/s) equals the $\mathbf{E} \times \mathbf{B}$ velocity in the uniform background electric field. Hence, the eastward motion of the omega bands is most likely caused by an $\mathbf{E} \times \mathbf{B}$ drift of the precipitating particles in the southward convection electric field.

Models of the three-dimensional current system associated with a series of omega bands were proposed by Opgenoorth et al. (1983b) and by Buchert et al. (1990). In both models, the ionospheric conductance is strongly enhanced within the auroral tongues and the electric field has an alternating azimuthal component which is directed toward the bright aurora (and high conductance) inside the tongue. The most recent current system model is displayed in Figure 12.3.4. The meandering of the westward electrojet is clearly visible. Inside the bright tongue region, the current flow is even purely southward, with an intensity of 1000 to 2000 mA/m. It is also in the tongue, where strong upward field-aligned currents reach densities of about 2 μA/m^2.

As shown above, the distribution of conductances, electric fields, and currents in and around omega bands is reasonably well known today. The physical processes leading to the generation of these active auroral forms are far less understood. At present, there is only the idea that these

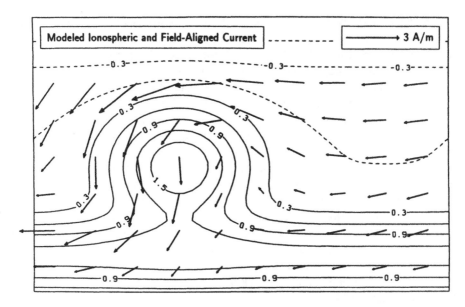

Figure 12.3.4 Modeled ionospheric (arrows; in A/m) and field-aligned current distribution in and around an omega band (solid isocontours denote upward, dashed isocontours zero or downward field-aligned current; isocontours are labeled in μA/m². (From Buchert, S. et al., *J. Geophys. Res.*, 95, 3733–3743, 1990. Copyright by the American Geophysical Union. With permission.)

forms are generated by the Kelvin-Helmholtz instability. Lyons and Walterscheid (1985) suggested that this instability develops at the edge of an intense neutral wind jet stream in the highly conducting region. Buchert et al. (1988) found, however, that the neutral wind velocities for their event are far below the threshold required for this mechanism to operate. Rostoker and Samson (1984) proposed that the instability develops at the boundaries of the plasma sheet, but again, it is unclear whether the velocity shear is sufficient to excite the instability and, furthermore, if omega bands map at all to this interface.

REFERENCES

André, D., and W. Baumjohann, Joint two-dimensional observations of ground magnetic and ionospheric electric fields associated with auroral zone currents. 5. Current systems associated with eastward drifting omega bands, *J. Geophys.*, 50, 194–201, 1982.

Axford, W. I., and C. O. Hines. A unifying theory of high-latitude geophysical phenomena and geomagnetic storms, *Can. J. Phys.*, 39, 1433–1463, 1961.

Baumjohann, W., Ionospheric and field-aligned current systems in the auroral zone: a concise review, *Adv. Space Res.*, 2(10), 55–62, 1983.

Baumjohann, W., and K.-H. Glassmeier, The transient response mechanism and Pi2 pulsations at substorm onset: review and outlook, *Planet. Space Sci.*, 32, 1361–1370, 1984.

Baumjohann, W., and G. Haerendel, Magnetospheric convection observed between 0600 and 2100 LT: solar wind and IMF dependence, *J. Geophys. Res.*, 90, 6370–6378, 1985.

Baumjohann, W., and G. Haerendel, Erdmagnetismus und extraterrestrische Vorgänge, *Naturwissenschaften*, 74, 181–187, 1987.

Baumjohann, W., and G. Paschmann, Solar wind-magnetosphere coupling: processes and observations, *Phys. Scr.*, T18, 61–72, 1987.

Baumjohann, W., J. Untiedt, and R. A. Greenwald, Joint two-dimensional observations of ground magnetic and ionospheric electric fields associated with auroral zone currents. 1. Three-dimensional current flows associated with a substorm-intensified eastward electrojet, *J. Geophys. Res.*, 85, 1963–1978, 1980.

Baumjohann, W., R. J. Pellinen, H. J. Opgenoorth, and E. Nielsen, Joint two-dimensional observations of ground magnetic and ionospheric electric fields associated with auroral zone currents: current systems associated with local auroral break-ups, *Planet. Space Sci.*, 29, 431–447, 1981.

Baumjohann, W., G. Haerendel, and F. Melzner, Magnetospheric convection observed between 0600 and 2100 LT: variations with Kp, *J. Geophys. Res.*, 90, 393–398, 1985.

Behm, D. A., F. Primdahl, L. J. Zanetti, R. L. Arnoldy, and L. J. Cahill, Jr., Ionospheric electric currents in the late evening plasma flow reversal, *J. Geophys. Res.*, 84, 5339–5343, 1979.

Birn, J., and E. W. Hones, Jr., Three-dimensional computer modeling of dynamic reconnection in the geomagnetic tail, *J. Geophys. Res.*, 86, 6802–6808, 1981.

Boström, R., Mechanisms for driving Birkeland currents, in *Physics of the Hot Plasma in the Magnetosphere*, B. Hultqvist and L. Stenflo, Eds., pp. 431–447, Plenum Press, New York, 1975.

Buchert, S., W. Baumjohann, G. Haerendel, C. LaHoz, and H. Lühr, Magnetometer and incoherent scatter observations of an intense Ps 6 pulsation event, *J. Atmos. Terr. Phys.*, 49, 357–367, 1988.

Buchert, S., G. Haerendel, and W. Baumjohann, A model for the electric fields, currents and conductances during a Ps 6 pulsation event, *J. Geophys. Res.*, 95, 3733–3743, 1990.

Cowley, S. W. H., The causes of convection in the Earth's magnetosphere: a review of developments during the IMS, *Rev. Geophys. Space Phys.*, 20, 531–565, 1982.

Cowley, S. W. H., Interpretation of observed relations between solar wind characteristics and effects at ionospheric altitudes, in *High-Latitude Space Plasma Physics*, B. Hultqvist and T. Hagfors, Eds., pp. 225–249, Plenum Press, New York, 1983.

Dungey, J. W., Interplanetary magnetic fields and the auroral zones, *Phys. Rev. Lett.*, 6, 47–48, 1961.

Foster, J. C., J. R. Doupnik, and G. S. Stiles, large scale patterns of auroral ionospheric convection observed with the Chatanika radar, *J. Geophys. Res.*, 86, 11,357–11,371, 1981.

Gustafsson, W. Baumjohann, and I. Iversen, Multi-method observations and modelling of the three-dimensional currents associated with a very strong Ps6 event, *J. Geophys.*, 49, 138–145, 1981.

Haerendel, G., and G. Paschmann, Interaction of the solar wind with the dayside magnetosphere, in *Magnetospheric Plasma Physics*, A. Nishida, Ed., pp. 49–142, D. Reidel, Dordrecht, 1982.

Harel, M., R. A. Wolf, P. H. Reiff, R. W. Spiro, W. J. Burke, F. J. Rich, and M. Smiddy, Quantitative simulation of a magnetospheric substorm, 1. Model logic and overview, *J. Geophys. Res.*, 86, 2217–2241, 1981.

Heppner, J. P., The Harang discontinuity in auroral belt ionospheric currents, *Geophys. Publ.*, 29, 105–120, 1972.

Hughes, T. J., and G. Rostoker, A comprehensive model current system for high-latitude magnetic activity. I. The steady state system, *Geophys. J. R. Astron. Soc.*, 58, 525–569, 1979.

Iijima, T. and T. A. Potemra, The amplitude distribution of field-aligned currents at northern high latitudes observed by Triad, *J. Geophys. Res.*, 81, 2165–2174, 1976.

Iijima, T. and T. A. Potemra, Large-scale characteristics of field-aligned currents associated with substorms, *J. Geophys. Res.*, 83, 599–615, 1978.

Iijima, T., T. A. Potemra, and L. J. Zanetti, Large-scale characteristics of magnetospheric equatorial currents, *J. Geophys. Res.*, 95, 991–999, 1990.

Inhester, B., W. Baumjohann, R. A. Greenwald, and E. Nielsen, Joint two-dimensional observations of ground magnetic and ionospheric electric fields associated with auroral zone currents. 3. Auroral zone currents during the passage of a westward travelling surge, *J. Geophys.*, 49, 155–162, 1981.

Kamide, Y., and W. Baumjohann, *Magnetosphere-Ionosphere Coupling*, Springer-Verlag, Heidelberg, 1993.

Kamide, Y., A. D. Richmond, and S. Matsushita, Estimation of ionospheric electric fields, ionospheric currents, and field-aligned currents from ground magnetic records, *J. Geophys. Res.*, 86, 801–813, 1981.

Kan, J. R., and W. Sun, Simulation of the westward travelling surge and Pi2 pulsations during substorms, *J. Geophys. Res.*, 90, 10,911–10,922, 1985.

Kawasaki, K., and G. Rostoker, Perturbation magnetic fields and current systems associated with eastward drifting auroral structures, *J. Geophys. Res.*, 84, 1464–1480, 1979.

Kirkwood, S., H. J. Opgenoorth, and J. S. Murphree, Ionospheric conductivities, electric fields and currents associated with auroral substorms measured by the Eiscat radar, *Planet. Space Sci.*, 36, 1359–1380, 1988.

Kunkel, T., W. Baumjohann, J. Untiedt, and R. A. Greenwald, Electric fields and currents at the Harang discontinuity: A case study, *J. Geophys.*, 59, 73–86, 1986.

Lyons, L. R., and R. L. Walterscheid, Generation of auroral omega bands by shear instability of the neutral winds, *J. Geophys. Res.*, 90, 12,321–12,329, 1985.

Maezawa, K., Magnetospheric convection induced by the positive and negative Z components of the interplanetary magnetic field: quantitative analysis using polar cap magnetic records, *J. Geophys. Res.*, 81, 2289–2303, 1976.

Marklund, G., Auroral arc classification scheme based on the observed arc-associated electric field pattern, *Planet. Space Sci.*, 32, 193–211, 1984.

Marklund, G., I. Sandahl, and H. Opgenoorth, A study of the dynamics of a discrete auroral arc, *Planet. Space Sci.*, 30, 179–197, 1982.

Maynard, N. C., Electric field measurements across the Harang discontinuity, *J. Geophys. Res.*, 79, 4620–4631, 1974.

McDiarmid, I. B., J. R. Burrows, and E. E. Budzinski, Average characteristics of magnetospheric electrons (150 eV to 200 keV) at 1400 km, *J. Geophys. Res.*, 80, 73–79, 1975.

McPherron, R. L., Magnetospheric substorms, *Rev. Geophys. Space Phys.*, 17, 657–681, 1979.

Meng, C.-I., R. H. Holzworth, and S. I. Akasofu, Auroral circle: delineating the poleward boundary of the quiet auroral belt, *J. Geophys. Res.*, 82, 164–172, 1977.

Meng, C.-I., A. L. Snyder, and H. W. Kroehl, Observations of auroral westward traveling surges and electron precipitations, *J. Geophys. Res.*, 83, 575–585, 1978.

Mishin, V. M., A. D. Bazarzhapov, and G. B. Shpynev, Electric fields and currents in the Earth's magnetosphere, in *Dynamics of the Magnetosphere*, S.-I. Akasofu, Ed., pp. 249–286, D. Reidel, Dordrecht, 1979.

Ohtani, S., S. Kokubun, R. C. Elphic, and C. T. Russell, Field-aligned current signatures in the near-tail region. 1. ISEE observations in the plasma sheet boundary layer, *J. Geophys. Res.*, 93, 9709–9720, 1988.

Opgenoorth, H. J., R. J. Pellinen, W. Baumjohann, E. Nielsen, G. Marklund, and L. Eliasson, Three-dimensional current flow and particle precipitation in a westward travelling surge (observed during the Barium-Geos rocket experiment), *J. Geophys. Res.*, 88, 3138–3152, 1983a.

Opgenoorth, J. Oksman, K. U. Kaila, E. Nielsen, and W. Baumjohann, On the characteristics of eastward drifting omega bands in the morning sector of the auroral oval, *J. Geophys. Res.*, 88, 9171–9185, 1983b.

Paschmann, G., B. U. Ö. Sonnerup, I. Papamastorakis, N. Sckopke, G. Haerendel, S. J. Bame, J. R. Asbridge, J. T. Gosling, C. T. Russell, and R. C. Elphic, Plasma acceleration at the Earth's magnetopause: evidence for reconnection, *Nature*, 282, 243–246, 1979.

Reiff, P. H., and J. L. Burch, IMF B_y-dependent plasma flow and Birkeland currents in the dayside magnetosphere. 2. A global model for northward and southward IMF, *J. Geophys. Res.*, 90, 1595–1609, 1985.

Reiff, P. H., R. W. Spiro, and T. W. Hill, Dependence of polar cap potential drop on interplanetary parameters, *J. Geophys. Res.*, 86, 7639–7648, 1981.

Robinson, R. M., F. Rich, and R. R. Vondrak, Chatanika radar and S3-2 measurements of auroral zone electrodynamics in the midnight sector, *J. Geophys. Res.*, 90, 8487–8499, 1985.

Rostoker, G., and J. C. Samson, Can substorm expansive phase effects and low frequency Pc magnetic pulsations be attributed to the same source mechanism?, *Geophys. Res. Lett.*, 11, 271–274, 1984.

Rostoker, G., S.-I. Akasofu, W. Baumjohann, Y. Kamide, and R. L. McPherron, The roles of direct input of energy from the solar wind and unloading of stored magnetotail energy in driving magnetospheric substorms, *Space Sci. Rev.*, 46, 93–111, 1987.

Rothwell, P. L., M. B. Silevitch, and L. P. Block, A model for the propagation of the westward travelling surge, *J. Geophys. Res.*, 89, 8941–8948, 1984.

Russell, C. T., and R. C. Elphic, ISEE observations of flux transfer events at the dayside magnetopause, *Geophys. Res. Lett.*, 6, 33–36, 1979.

Sato, T., T. Hayashi, R. J. Walker, and M. Ashour-Abdalla, Neutral sheet current interruption and field-aligned current generation by three-dimensional driven reconnection, *Geophys. Res. Lett.*, 10, 221–224, 1983.

Siscoe, G. L., The magnetospheric boundary, in *Physics of Space Plasmas (1987)*, T. Chang, G. B. Crew, and J. R. Jasperse, Eds., pp. 3–78, Scientific Publishers Inc., Cambridge, MA, 1988.

Stiles, G. S., J. C. Foster, and J. R. Doupnik, Prolonged radar observations of an auroral arc, *J. Geophys. Res.*, 85, 1223–1234, 1980.

Sugiura, M., Identifications of the polar cap boundary and the auroral belt in the high-altitude magnetosphere: a model for field-aligned currents, *J. Geophys. Res.*, 80, 2057–2068, 1975.

Svalgaard, L., Polar cap magnetic variations and their relationship with the interplanetary sector structure, *J. Geophys. Res.*, 78, 2064–2078, 1973.

Vasyliunas, V. M., Mathematical models of magnetospheric convection and its coupling to the ionosphere, in *Particles and Fields in the Magnetosphere*, B. M. McCormack, Ed., pp. 60–71, D. Reidel, Dordrecht, 1970.

Volland, H., A semi-empirical model of large-scale magnetospheric electric fields, *J. Geophys. Res.*, 78, 171–180, 1973.

Wallis, D. D., and E. E. Budzinski, Empirical models of height integrated conductivities, *J. Geophys. Res.*, 86, 125–137, 1981.

Yasuhara, F., Y. Kamide, and J. F. Vickrey, On the efficiency of the Cowling mechanism in the auroral electrojet, *Geophys. Res. Lett.*, 12, 389–392, 1985.

Zanetti, L. J., T. A. Potemra, T. Iijima, W. Baumjohann, and P. F. Bythrow, Ionospheric and Birkeland current distributions for northward interplanetary magnetic field: inferred polar convection, *J. Geophys. Res.*, 89, 7453–7458, 1984.

Zi, M., and E. Nielsen, Spatial variation of electric fields in the high-latitude ionosphere, *J. Geophys. Res.*, 87, 5202–5206, 1982.

Ziesolleck, C., W. Baumjohann, K. Brüning, C. W. Carlson, R. I. Bush, Comparison of height-integrated current densities derived from ground-based magnetometer and rocket-borne observations during the Porcupine F3 and F4 flights, *J. Geophys. Res.*, 88, 8063–8070, 1983.

Chapter 13

Magnetospheric LF-, VLF-, and ELF-Waves

Vikas S. Sonwalkar

CONTENTS

0-8493-2520-X/95/$0.00+$.50
© 1995 by CRC Press

1. INTRODUCTION

The Earth's magnetosphere supports a wide variety of wave phenomena. These waves are important partly because they influence the behavior of the magnetosphere and partly because we use them as experimental tools in our investigations of the upper atmosphere. These waves may be electromagnetic, electrostatic, or magnetosonic and most are generated by the conversion of free energy within the plasma into wave energy through a variety of plasma-wave processes. Often the term plasma wave is used to denote a wave that is generated within the magnetosphere or which has its characteristics significantly modified by the magnetospheric plasma. In this chapter we treat plasma waves in the frequency range extending from a few hertz to a few megahertz and which have their sources within the magnetosphere. Thus, the frequencies of the majority of the plasma waves discussed here lie in ELF (3 to 3000 kHz), VLF (3 to 30 kHz), and LF (30 to 300 kHz) ranges. We include in this chapter results from VLF wave injection experiments at Siple Station, Antarctica, in which the Siple transmitter signals are found to trigger emissions within the magnetosphere. However, waves of extra-magnetospheric origin such as whistlers generated by lightning discharges or man-made signals from ground transmitters and power lines, which are also observed in the magnetosphere as propagating whistler-mode signals, are treated in other chapters (II/7 and II/10) of this book. Similarly, geomagnetic micropulsation in the frequency range below ~1 Hz is also treated elsewhere in the book (Chapter II/14).

To understand the origin and propagation of plasma waves, it is important to have a basic understanding of the medium — the Earth's magnetosphere — in which these waves are generated and propagate. The Earth's magnetosphere (and ionosphere) is described in Chapters II/8, II/9, II/11, and II/12 of this handbook. Figure 13.1.1 shows a sketch of the various regions and the boundaries of the Earth's magnetosphere in a noon-midnight meridian cross section. The figure also indicates where different wave phenomena are observed in the magnetosphere. Though no standard nomenclature exists for the plasma waves observed in the magnetosphere, the names given to these diverse wave phenomena are generally indicative of one or more properties or features that each kind exhibits: frequency range, spectral characteristics, region and local time of occurrence, and plasma wave modes.

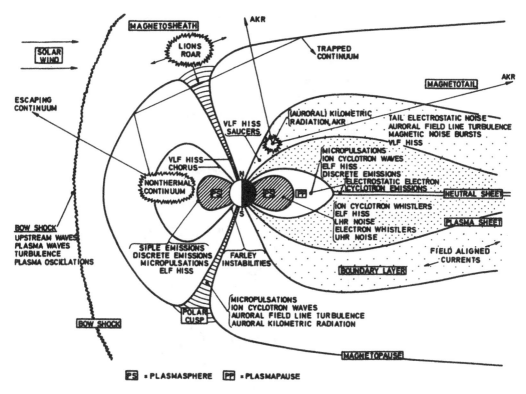

Figure 13.1.1   Regions of plasma wave occurrence located in a noon-midnight meridian cross section of the Earth's magnetosphere. (From Shawhan, S. D., *Solar System Plasma Physics,* 1979, Chap. III.1.6. With permission.)

This chapter is organized as follows: Section 2 briefly describes the experimental methods; Section 3 provides a theoretical background on the propagation and generation mechanisms of plasma waves; Section 4 presents the ground and spacecraft observations of plasma waves, followed by Section 5 on concluding remarks.

2. EXPERIMENTAL METHODS

Experimental methods in plasma wave research include both the ground-based and satellite-borne experiments, which provide complementary sets of observations. Rockets and balloons are also used to perform *in situ* measurements at ionospheric or stratospheric heights where satellite measurements are not possible. However, because of their relatively short time of flight, rocket and balloon experiments are generally used only during certain campaigns aimed at achieving a more detailed understanding of a specific phenomenon. While the bulk of past plasma wave observations has come from *passive experiments,* ground-based or space-borne *active experiments,* in which a radio transmitter or an electron gun is used to perturb the medium, have in recent years come to play an important part in the studies of various processes in the ionosphere and magnetosphere.

2.1. GROUND-BASED EXPERIMENTS

Because the majority of the signals of magnetospheric origin that reach the ground propagate roughly along the geomagnetic field lines (Section 3.2), the latitude of a ground station is generally determined by the region of the magnetosphere to be explored. Often observations from multiple ground stations are required — stations at magnetically conjugate points have been used to conduct active wave injection experiments (Helliwell and Katsufrakis, 1974), and multiple ground

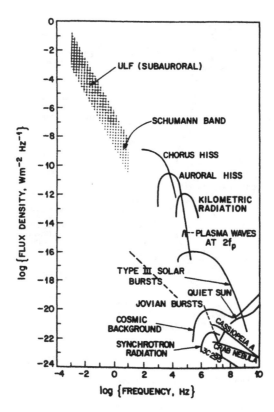

Figure 13.2.1 Power flux levels for various frequency ranges of naturally occurring waves in the Earth's environment and in astrophysical sources as observed at Earth. (From Lanzerotti, L. J., *Upper Atmosphere Research in Antarctica, Antarctic Research Series, Vol.* 29, 130, 1978. Copyright by the American Geophysical Union. With permission.)

stations separated by a few hundred kilometers have been used to study the location and the size of duct exit points (Tsuruda et al., 1982). The choice of instrumentation depends on the nature of the scientific questions posed. For example, low data rate multichannel spectrum analyzers are used in a global noise survey conducted at many stations around the Earth (Fraser-Smith and Helliwell, 1985), whereas wideband VLF receiving systems are used to study the detailed properties of VLF emissions and wave-particle interactions (Helliwell 1988).

Figure 13.2.1 illustrates the frequency ranges and power levels of various natural ELF/VLF/LF waves produced in the magnetosphere as well as geomagnetic pulsations (ULF) measured at geomagnetic mid-latitudes. Substantially higher (by factors of $\sim 10^1 - 10^2$) power levels are frequently observed in the auroral zones. The power levels over the entire frequency range vary roughly with the inverse frequency squared. The figure aids in choosing appropriate design parameters (sensitivity and dynamic range) for the measuring instruments described below.

ELF/VLF measurements are performed using crossed loop antennas and a vertical whip antenna, which are sensitive to the two horizontal magnetic field and the vertical electric field components of the incident wave. The electric field information is useful in radio direction finding at the station. Typical linear dimensions of the antennas are of the order of ~ 10 m. Broadband ELF/VLF receivers typically have a flat frequency response from a few hundred hertz to almost 100 kHz. The sensitivity of the VLF system (loop and preamp) used during IGY is plotted as a dashed line in Figure 13.2.1. It is necessary to have accurate timing information for use in the correlation of VLF events received at different stations and on satellites. An amplitude-modulated pilot tone around 9 to 10 kHz is used to provide time marks on the second and station identification and time of day information in Morse code every minute. Broadband analog signals from VLF receivers have traditionally been recorded on standard 1/4-in. magnetic tapes. The availability of inexpensive digital recording media and technology has made it possible to record and store wideband data digitally on various digital media.

Beside the basic broadband equipment listed above, ground stations have additional equipment for measuring special types of VLF signals. These include narrowband (200 Hz BW) filters for monitoring the strength of VLF transmitters for propagation studies and wideband (approximately octave bandwidth) hiss filters for measuring the strength of naturally occurring VLF signals. Data from these instruments are recorded on 8-channel chart recorders. The details of the design and performance of a typical ground-receiving system as well as that of the Siple wave-injection facility (Section 4.3) can be found in the literature (Helliwell and Katsufrakis, 1978; Helliwell, 1988).

2.2. SATELLITE-BORNE EXPERIMENTS

Principal factors that go into designing a space-borne experiment to investigate plasma waves are: (1) orbit characteristics and (2) instrumentation (antennas and receivers).

Important orbit characteristics are: apogee and perigee altitude, inclination, orbital period, local time and latitude of initial perigee, precession of perigee, and life time. For any given satellite mission, the orbit characteristics of the spacecraft are determined primarily by the region of the magnetosphere to be explored. Thus, a low inclination orbit is preferred to investigate equatorial phenomena, whereas a high inclination ($\sim 90°$) is used to investigate polar regions and interactions along magnetic field lines.

It has been recognized that in order to identify various wave modes and to locate the sources of plasma waves, it is necessary to measure both frequency (f) and wave vector ($\mathbf{k}$) spectra (Shawhan, 1970, 1983). In general, measurements from three orthogonal electric and three orthogonal magnetic wave sensors are needed to obtain complete information on the local electromagnetic field. In practice, however, the most essential information — frequency spectra, wave polarization, wave normal direction with respect to the geomagnetic field, and energy flow (Poynting vector) direction — can be obtained with fewer than six measured wave components.

We describe here briefly plasma wave instrumentation employed on the DE 1 satellite (Shawhan et al., 1981). This package included many features of the instruments used on past satellites as well as several innovative new developments. Five antennas were used to measure plasma waves: (1) a 200-m tip-to-tip electric dipole (E_x) perpendicular to the spacecraft spin axis; (2) a 9-m tip-to-tip electric dipole (E_z) parallel to the spin axis; (3) a 0.6-m short electric antenna (E_s); (4) a search coil magnetometer (B_z) parallel to the spin axis; and (5) a 1-m^2 loop antenna (B_x) perpendicular to spin axis. The E_x and E_z antennas were sensitive from DC to 2 MHz and were used to detect nearly static magnetic fields. All five antennas were used to measure AC electric and magnetic fields in the LEF/VLF frequency range, with the E_x and E_z antennas used to measure the electric field up to 400 kHz.

The Plasma Wave Instrument (PWI) on the DE 1 satellite consists of four receivers: step frequency correlator (SFC), low frequency correlator (LFC), wideband analog receiver (WBR), and linear wave receiver (LWR). The step frequency correlator provided 128 logarithmically spaced frequency channels from 100 Hz to 400 kHz and low frequency correlator provides eight logarithmically spaced frequency channels from 1.78 Hz to 100 Hz. These receivers performed amplitude and correlation measurements of signals from selected pair of sensors. By correlating magnetic and electric sensors it was possible to discriminate between electrostatic and electromagnetic emissions, a measurement important for determining the source mechanisms and propagation characteristics of plasma waves. Correlation of the crossed electric antennas allowed determination of the wave polarization for all electric wave emissions. For electrostatic waves the comparison of amplitudes between the three electric antennas of 200-, 9-, and 0.6-m length provided an estimate of wave length and phase velocity, since these waves have wavelengths comparable to the antenna lengths. Measurement of the correlation change with spacecraft spin

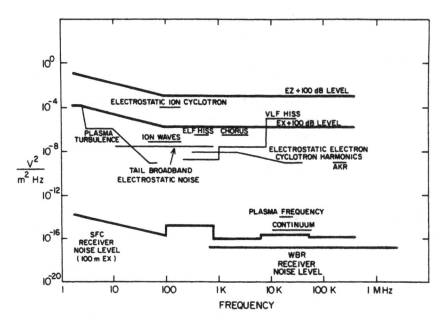

Figure 13.2.2   Electric field spectral density response of the PWI electric field sensors and receivers. Also indicated are the maximum spectral densities of a variety of plasma wave phenomena. (From Shawhan, S. D., Gurnett, D. A., Odem, D. L., Helliwell, R. A., and Park, C. G., *Space Sci. Instrum.,* 5, 535, 1981. With permission.)

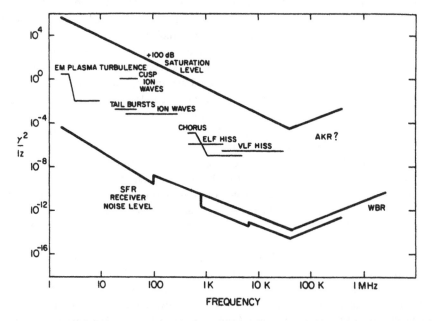

Figure 13.2.3   Magnetic field spectral density response of the PWI magnetic field sensors and receivers. Also indicated are the maximum spectral densities of a variety of plasma wave phenomena. (From Shawhan, S. D., Gurnett, D. A., Odem, D. L., Helliwell, R. A., and Park, C. G., *Space. Sci. Instrum.,* 5, 535, 1981. With permission.)

provided estimates of wave normal vector directions and of the Poynting vector. To acquire high time and frequency resolution for the identification and analysis of specific types of plasma wave phenomena, the wideband receivers (WBR and LWR) were used.

Figures 13.2.2 and 13.2.3 show the overall dynamic range and sensitivity with frequency for the sensor-receiver system described above and the maximum spectral densities for a variety of

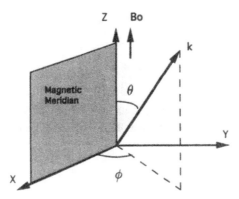

Figure 13.3.1 Typical geometry used in describing wave propagation in the Earth's magnetosphere.

plasma wave phenomena observed in the magnetosphere. Together with Figure 13.2.1 these figures provide a comparison of typical ground and spacecraft receiver characteristics as well as spectral densities of various plasma waves as observed on the ground and on a satellite.

3. PROPAGATION, GENERATION, AND DAMPING OF PLASMA WAVES

In this section we provide the theoretical background necessary to understand the experimental observations and their interpretations presented in the next section. Wave propagation in a magnetoplasma, discussed in the first two subsections, is a mature subject and is extensively treated in many texts (Budden, 1985; Stix, 1992; Swanson, 1989; Walker, 1993). On the other hand, the generation mechanisms and origins of the many plasma wave phenomena discussed in the third subsection are still not well understood.

3.1. PROPAGATION OF PLASMA WAVES: CHARACTERISTICS MODES AND FREQUENCIES

Magnetospheric plasma consisting of electrons and ions of finite temperature and permeated by a magnetic field can support a large variety of electromagnetic, electrostatic, and magnetosonic wave modes that cannot exist in free space. In particular, all types of waves and oscillations predicted by linear theory for both cold and warm magnetoplasma have been observed in the magnetosphere. The range of frequencies for which the various modes are observed to exist can be understood in terms of the characteristic frequencies of the plasma.

3.1.1. Waves in a Cold Uniform Magnetoplasma

In linear theory, solutions to Maxwell's equations are obtained for plane waves of the form

$$\mathbf{E} = \mathbf{E}_0 \exp[i(\mathbf{k} \cdot \mathbf{r} - \omega t)] \tag{3.1}$$

where, $\mathbf{E}_0$, k, and ω are the wave polarization, wave normal vector, and angular wave frequency ($\omega = 2\pi f$), respectively. As a first approximation, the magnetospheric plasma is considered to be made up of cold plasma consisting of electrons and ions with zero temperature. It is also assumed that the plasma is homogeneous and of infinite extent. The plasma parameters that determine the nature of the solutions in this case are: electron (or plasma) density (N_e), electron mass (m_e), electron charge (e), ion masses (m_i), charges (e_i), ion composition (relative abundance of various ion species), and the ambient geomagnetic field, $\mathbf{B}_0$. For propagation in a warm or thermal plasma, the temperatures of the electrons and ions also have to be taken into account. Figure 13.3.1 shows the geometry. The angle between the static magnetic field $\mathbf{B}_0$ and the wave normal vector $\mathbf{k}$ is called the wave normal angle, θ.

It is convenient to define wave refractive index μ as follows:

$$\mu = \frac{c|\mathbf{k}|}{\omega} \tag{3.2}$$

where c is the velocity of light. Comparing Equations 3.1 and 3.2 we see that phase velocity, υ_p $= (\omega/|\mathbf{k}|)$, of a mode is given by c/μ and that the wave normal vector $\mathbf{k}$ can be written in terms of the refractive index μ as follows:

$$\mathbf{k} = \frac{\omega\mu}{c}\,\mathbf{u}(\theta, \phi) \tag{3.3}$$

where $\mathbf{u}(\theta, \phi)$ is the unit vector in the direction of the wave normal vector. The relation 3.3 between $\mathbf{k}$ and ω is called the *dispersion relation.*

The solution to Maxwell's equation for a plane wave in a cold magnetoplasma is obtained in terms of a general expression for the wave refractive index (Stix, 1992).

$$A\mu^4 - B\mu^2 + C = 0 \tag{3.4}$$

where

$$A = S \sin^2 \theta + P \cos^2 \theta \tag{3.5}$$
$$B = RL \sin^2 \theta + PS(1 + \cos^2 \theta) \tag{3.6}$$
$$C = PRL \tag{3.7}$$

The quantities R (for right), L (for left), S (for sum), D (for difference), and P (for plasma) are defined in terms of the characteristic plasma frequencies: the plasma frequency, ω_p, and the gyrofrequency, ω_h. These frequencies are determined by the electron density (N_e), ion species (mass m_i and density N_i) and the ambient magnetic field strength (B). Table 13.3.1 gives the names and formulas of these frequencies for a two-component (electron and proton) plasma.

$$R = 1 - \sum_k \frac{\omega_{pk}^2}{\omega^2}\left(\frac{\omega}{\omega + \varepsilon_k \omega_{hk}}\right) \tag{3.8}$$

$$L = 1 - \sum_k \frac{\omega_{pk}^2}{\omega^2}\left(\frac{\omega}{\omega - \varepsilon_k \omega_{hk}}\right) \tag{3.9}$$

$$S = \frac{1}{2}\,(R + L) \tag{3.10}$$

$$D = \frac{1}{2}\,(R - L) \tag{3.11}$$

$$P = 1 - \sum_k \frac{\omega_{pk}^2}{\omega^2} \tag{3.12}$$

where ω_{pk} and ω_{hk} are the plasma and gyrofrequencies of species k. ε_k is $+1$ for positive ions and -1 for electrons and negative ions.

An alternative expression for the refractive index is given by:

$$\tan^2 \theta = \frac{-P(\mu^2 - R)(\mu^2 - L)}{(S\mu^2 - RL)(\mu^2 - P)} \tag{3.13}$$

Table 13.3.1 **Characteristic Plasma Frequencies**

Name	Formula (MKS)	Handy formula
Electron plasma frequency, f_{pe}	$\left(\dfrac{N_e^2}{\varepsilon_0 m_e}\right)$	$9000\sqrt{(N,\ electrons/cm^3)}$ Hz
Ion plasma frequency, f_{pi}	$\left(\dfrac{N_e^2}{\varepsilon_0 m_i}\right)$	$f_{pe}(m_e/m_i)^{1/2}$ Hz
Electron gyrofrequency, f_{he} (electron cyclotron frequency, f_{ce})	$\dfrac{eB}{m_e}$	$2.8 \times 10^6 (B,\ G)$ Hz
Ion gyrofrequency, f_{hi} (ion cyclotron frequency, f_{ci})	$\dfrac{eB}{m_i}$	$f_{he}(m_e/m_i)$ Hz
$P = 0$ cut-off frequency, $f_{P=0}$	f_{pe}	
$R = 0$ cut-off frequency, $f_{R=0}$	$\dfrac{f_{ce} - f_{ci}}{2} + \left[\left(\dfrac{f_{ce} + f_{ci}}{2}\right)^2 + f_{pe}^2\right]^{\frac{1}{2}}$	
$L = 0$ cut-off frequency, $f_{L=0}$	$\dfrac{f_{ci} - f_{ce}}{2} + \left[\left(\dfrac{f_{ci} + f_{ce}}{2}\right)^2 + f_{pe}^2\right]^{\frac{1}{2}}$	
$R \to \infty$ resonance (electron cyclotron resonance)	f_{ce}	
$L \to \infty$ resonance (ion cyclotron resonance)	f_{ci}	
Lower hybrid resonance frequency, f_{LHR}	$\left[\dfrac{m_i}{m_e}\left(\dfrac{1}{f_{pe}^2} + \dfrac{1}{f_{he}^2}\right)\right]^{-1/2}$	
Upper hybrid resonance frequency, f_{UHR}	$[f_{pe}^2 + f_{he}^2]^{1/2}$	

From Equation 3.4 we note that for a given set of plasma parameters, wavefrequency and wave normal direction, four solutions exist for the value of the wave refractive index. Each of the four solutions represents a unique wave mode. Because of the ambient magnetic field, the magnetospheric plasma is anisotropic, and therefore the refractive index for each mode is a function of wavefrequency as well as wave normal vector direction (Figure 13.3.1). Depending on the values of the plasma parameters, however, some of the modes may represent propagating modes (real value of μ) and the others will be evanescent (imaginary value of μ). For example, from Equation 3.13 we note that when $\theta = 0$, i.e., for propagation parallel to $\mathbf{B}_0$, $\mu^2 = R$ and $\mu^2 = L$ are obtained as the two propagating modes ($P = 0$ gives nonpropagating plasma oscillations); for $\theta = \pi/2$, $\mu^2 = RL/S$ and $\mu^2 = P$ are obtained as the two propagating modes. For a collisional plasma, the refractive index assumes complex values, with imaginary part representing the wave attenuation (Budden, 1985).

A characteristic polarization (right- or left-hand elliptic) is associated with each mode:

$$\frac{E_y}{E_x} = \frac{iD}{\mu^2 - S} \tag{3.14}$$

$$\frac{E_z}{E_x} = \frac{\mu^2\cos\theta\sin\theta}{\mu^2\sin^2\theta - P} \tag{3.15}$$

It can be shown that, generally, waves below the plasma frequency with wave normal angle near $\sim0°$ have their energy primarily in the wave magnetic field, and the waves are circularly polarized. Waves with wave normal near $\sim90°$ tend to have their energy primarily in the electric field component and are linearly polarized. For wave normal angles between 0 and 90°, the waves are right- or left-hand elliptically polarized.

For a cold plasma different wave modes can be labeled in various ways. The most commonly used labeling scheme (Stix, 1992) is described below.

1. Fast (*F*) or slow (*S*) according to the magnitude of the phase velocity at wave normal angles between 0 and 90°.
2. Right (*R*) or left (*L*), according to the circular polarization for propagation parallel to the static magnetic field.
3. Ordinary (*O*) or extraordinary (*X* or *E*), according to the dispersion relation for propagation at right angles ($\theta = 90°$) to the static magnetic field. The dispersion relation for ordinary wave is the same as that for waves in an unmagnetized plasma.

We note here that in magnetospheric plasma wave research, the above scheme is not always used, and various modes detected have been labeled based on various other associations (Section 3.1.2).

As the wave propagates it may encounter regions where values of the plasma parameters are such that μ^2 goes to zero or infinity; the former is called a cutoff and the latter a resonance. In going through a cutoff μ^2 goes through zero, and the transition is made from a region of possible propagation to a region of evanescence. Generally, a reflection occurs in this circumstance.

According to Equations 3.4 and 3.7 cutoff occurs when

$$P = 0 \quad \text{or} \quad R = 0 \quad \text{or} \quad L = 0 \tag{3.16}$$

Resonance occurs for propagation at certain frequencies and at certain wave normal directions. In the transition region between propagation and evanescence which occurs when μ^2 goes through ∞, absorption and/or reflection may occur. From Equation 3.13 we note that resonance occurs at the angle θ which satisfies the criterion

$$\tan^2 \theta = \frac{-P}{S} \tag{3.17}$$

For $\theta = 0$, Equation 3.17 we obtain resonances at $R \to \infty$ (electron cyclotron resonance at ω_{he}) and at $L \to \infty$ (ion cyclotron resonance at ω_{hi}). For $\theta = \pi/2$, we obtain resonances at $S \to 0$, called lower (at f_{LHR}) and upper hybrid (at f_{UHR}) resonances.

At a given point in space, cutoffs and resonances define ranges of frequency and wave normal direction over which different modes may exist. The frequencies of various resonances and cut offs, which can be expressed in terms of the characteristic plasma frequencies, are given in Table 13.3.1. for a two-component (electrons and protons) plasma.

3.1.2. Magnetospheric Plasma Wave Modes

We describe here some of the more important cold and warm plasma wave modes observed in the magnetosphere. Figures 13.3.2 and 13.3.3 show the frequency ranges of six commonly observed wave modes in the magnetosphere near the equatorial and polar regions, respectively. As is evident in these figures, and also from Table 13.3.1, an important feature to note is that the electron plasma and gyrofrequencies are higher, by a factor of $\sim 10^2 - 10^3$, than ion plasma and gyrofrequencies. This difference is a result of the large ratio of ion-to-electronic masses. Thus there is a range of "high" frequencies where ion effects may be neglected and a range of "low" frequencies where electron effects may be neglected. When thermal effects are included, many more modes including electrostatic modes appear. Therefore, it is convenient to discuss these modes according to the following: (1) cold plasma wave modes near the electron plasma and gyrofrequencies, when ion effects can be neglected; (2) cold plasma wave modes near the ion

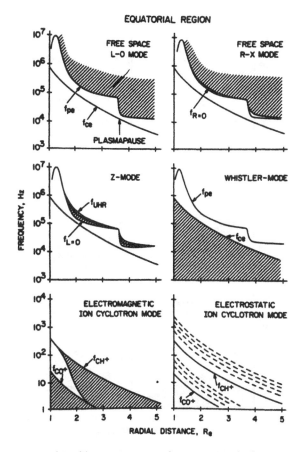

Figure 13.3.2 Frequency range of six of the most important plasma wave modes for a representative plasma density profile near the equatorial plane. The plasma density is usually larger near the equator than over the polar region. (From Gurnett, D. A. and Inan, U. S., *Rev. Geophys.*, 26, 285, 1988. Copyright by the American Geophysical Union. With permission.)

plasma and gyrofrequencies, when electron effects can be neglected; and (3) warm plasma wave modes that appear when the plasma is allowed to have a finite temperature. Cold and warm plasma wave modes for frequencies much below ion cyclotron frequencies $f \ll f_{ci}$, when the ambient magnetic field plays a dominant role, are known as hydromagnetic and magnetosonic modes. These modes cover the frequency range ULF and below and are treated in Chapter II/14.

3.1.2.1. Cold Plasma Wave Modes Near Characteristic Electron Frequencies

Linear theory predicts four plasma wave modes for a cold plasma at frequencies near the electron gyro and plasma frequencies. These modes are the (1) free space $L - O$ mode, (2) free space $R - X$ mode, (3) Z mode, and (4) whistler mode. Plasma waves belonging to each of these modes are commonly observed in the magnetosphere (Gurnett and Inan, 1988).

The term free space is used to indicate that these modes connect smoothly to free-space modes and their velocity becomes the light velocity in free space. $L - O$ and $R - X$ are used to indicate the polarization and the nature of the dispersion relations as discussed in Section 3.1.1. As shown in the top panel of Figure 13.3.2 and 13.3.3 and in Table 13.3.1, the $L - O$ and $R - X$ free-space modes have lower cut offs at the electron plasma frequency f_{pe} and the $R = 0$ cutoff, $f_{R=0}$, respectively. The highest frequency waves found in both the polar and the equatorial regions of the Earth's magnetosphere are found to propagate in these modes covering a frequency range ~100 kHz to 2 MHz. HF wave propagation in the ionosphere is also in this category.

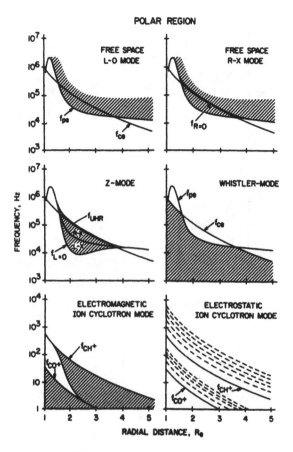

Figure 13.3.3   Frequency range of six of the most important plasma wave modes for a representative plasma density profile over the polar region. (From Gurnett, D. A. and Inan, U. S., *Rev. Geophys.,* 26, 285, 1988. Copyright by the American Geophysical Union. With permission.)

The Z mode is named after the Z trace observed on ground ionograms (Ratcliffe, 1959). The Z mode is bounded by the upper hybrid resonance, f_{UHR} and the $L = 0$ cutoff, $f_{L=0}$. Its index of refraction for parallel propagation ($\theta = 0$) undergoes a complicated change near the plasma frequency, and the labeling of the mode changes from $R - X$ above to $L - O$ below plasma frequency. By comparing the expressions (Table 13.3.1) for the lower and upper cutoffs, we see that the bandwidth of the Z mode is a very sensitive function of the plasma frequency to gyro-frequency ratio. When $f_{pe}/f_{ce} \gg 1$ the bandwidth is very narrow, whereas when $f_{pe}/f_{ce} \ll 1$ the bandwidth is very broad. Near equatorial regions of relatively high plasma density ($f_{pe} \geq f_{ce}$), Z-mode emissions appear in a narrowband near f_{UHR} and have often been called UHR emissions. Because of their narrow bandwidth they cannot propagate large distances without reaching regions where the mode no longer can be sustained by the medium, and therefore these emissions have been frequently considered "local" radiation, even though they represent a propagating mode. In contrast to this situation, the Z mode has a broad frequency range in the auroral regions due to relatively low plasma densities, permitting propagation over large distances both horizontally and vertically.

The whistler mode is named after lightning generated "whistlers" which propagate in this mode (Helliwell, 1965). The whistler mode propagates at frequencies below either f_{pe} or f_{ce}, whichever is lower. It has a lower cutoff at f_{LHR} when propagating perpendicular to the ambient magnetic field ($\theta = 90°$), and the waves undergo a reflection when the wave frequency is equal to the local f_{LHR}. It has a right-hand circular (R) polarization when propagating parallel to the magnetic field. Waves propagating in this mode are found throughout the magnetosphere, including the equatorial and polar regions, polar cusp, near the magnetopause, and in the magnetosheath and the bow shock.

3.1.2.2.   Cold Plasma Wave Modes Near Characteristic Ion Frequencies

At frequencies well below f_{pe} and f_{ce} and near $f_{LHR} \simeq (f_{ce}f_{ci})^{1/2}$, ion effects on propagation become evident. As mentioned above, whistler mode signals propagating with wave normal angles near $\theta = 90°$ reflect when $f = f_{LHR}$. A quasi-electrostatic mode (generally whistler mode signals propagating with large wave normal angles) with lower cutoff at f_{LHR} is frequently observed in the ionosphere and inner magnetosphere ($R_e \leq 6$). At even lower frequencies, below the ion cyclotron frequencies, each species of ion is associated with an electromagnetic mode called an ion cyclotron mode. This mode has a resonance at its ion cyclotron frequency, and thus exists only below f_{ci}. The bottom left panels of Figures 13.3.2 and 13.3.3 show the ion cyclotron modes introduced by protons (H^+) and singly charged oxygen ions (O^+), which are the dominant species in most parts of the ionosphere and the magnetosphere. Although not shown, an ion cyclotron branch also occurs in association with helium ions (He^+), which usually are minor constituents. Ion cyclotron waves are labeled as $L - O$ waves and are also known as slow (S) mode, or pure Alfven mode. Ion cyclotron waves are found in various regions of the magnetosphere including equatorial and auroral regions and polar cusps.

There is another mode, $R - X$ mode associated with ions, which is also known as the modified Alfven mode or fast (F) mode. Unlike the ion cyclotron mode, this mode is not affected by ion cyclotron resonance and exists at frequencies well above the ion cyclotron frequencies, undergoes a reshaping transition at the lower hybrid frequency, and disappears only at the electron cyclotron resonance.

3.1.2.3.   Finite Temperature Plasma Wave Modes

The magnetospheric background plasma has a finite temperature which may be different for electrons and ions and may range from ~500 to 10^7 K. It is found that finite temperature conditions introduce only minor modifications to the characteristics of the cold plasma modes described above (Stix, 1992). However, finite temperature plasmas can sustain wave modes, particularly electrostatic modes, that have no counterparts in a cold plasma. Finite temperature effects become important when the thermal velocities of ions and electrons become comparable to the wave phase velocity c/n.

At higher frequencies, most important modes associated with thermal effects are electrostatic electron cyclotron modes which occur near half integral $(n + 1/2)f_{ce}$, harmonics of the electron cyclotron frequency (Kennel et al., 1970). Electrostatic wave modes are also found near the electron plasma frequency (Rodriguez and Gurnett, 1975). At lower frequencies, important hot plasma modes are called electrostatic ion cyclotron (EIC) modes which occur between the harmonics of ion cyclotron frequencies. The approximate frequencies of these waves is shown in the right bottom panels of figures 13.3.2 and 13.3.3. Another type of electrostatic wave, called ion acoustic waves, occur at frequencies below the ion plasma frequency (f_{pi}). Hot plasma effects modify electromagnetic waves propagating near the harmonics of ion cyclotron frequencies, and the modified dispersion relations are often referred to as electromagnetic ion Bernstein modes (Fredricks, 1968). Both EIC and electromagnetic ion Bernstein modes usually involve waves propagating perpendicular to $\mathbf{B_0}$.

3.2.   PROPAGATION OF PLASMA WAVES IN THE MAGNETOSPHERE

Wave propagation in a magnetoplasma is both anisotropic and dispersive. In such a medium, the direction of propagation (called ray direction) of a wavepacket of finite temporal duration and

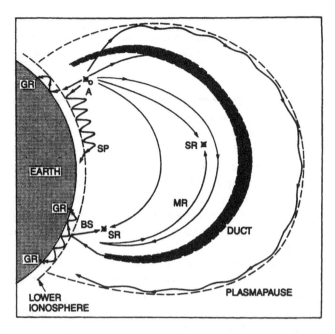

Figure 13.3.4 Schematic depicting various propagation paths from the VLF transmitter aboard the Activny satellite (A) to a satellite receiver (SR) aboard the DE 1 satellite and to several ground receiver sites (GR). A VLF signal transmitted from Activny can reach DE 1 either by a direct nonducted propagation path, by a magnetospherically reflected path (MR), or by back scattering (BS) of a ducted signal. The ground reception of the signals can take place by a ducted path, by a subprotonospheric (SP) propagation mode, or by guiding along the plasmapause. (From Sonwalkar, V. S., Inan, U. S., Bell, T. F., Helliwell, R. A., Molchanov, O. A., and Green, J. L., *J. Geophys. Res.,* 99, 6173, 1994. Copyright by the American Geophysical Union. With permission.)

spatial extent is different from the wave normal direction. The expressions for the group velocity $\mathbf{v}_g$ and the ray direction α with respect to wave normal direction are given by

$$\mathbf{v}_g = \frac{\partial \omega}{\partial \mathbf{k}} = \mathbf{k}' \frac{\partial \omega}{\partial |k|} + \theta' \frac{1}{|k|} \frac{\partial \omega}{\partial \theta} \tag{3.18}$$

$$\tan\alpha = \frac{-1}{\mu} \frac{\partial \mu}{\partial \theta} \tag{3.19}$$

where $\mathbf{k}'$ and θ' are unit vectors along and perpendicular to wave normal vector $\mathbf{k}$. For a given mode, the group velocity and ray direction can be determined using relations 3.18 and 3.19 in conjunction with Equations 3.3 and 3.4.

The magnetosphere is an inhomogeneous medium with its plasma density, composition, and the magnitude and direction of the geomagnetic field varying from point to point. In most instances, however, the medium properties change slowly over many wavelengths of a plasma wave mode, and thus locally one can still consider propagation of plasma waves in terms of the propagation modes described for an infinite homogeneous plasma. This approximation, called the ray approximation, enables one to derive ray equations which are used to describe long distance propagation in a slowly varying magnetoplasma (Budden, 1985). In carrying out ray-tracing simulations on modern computers, a realistic magnetosphere incorporating various plasma density gradients such as the plasmapause, ducts, and the vertical and horizontal density gradients in the ionosphere are employed.

Figure 13.3.4 schematically illustrates the propagation of whistler-mode signals from a source location A within the magnetosphere (in this case the source is assumed to be on the Activny satellite) to various observation points in the magnetosphere and on the ground. Most of the rays from a magnetospheric source cannot reach a ground receiver because they either undergo total internal reflection at the Earth-ionosphere boundary, where the refractive index changes suddenly from a large value (~10 to 100) to unity, or they undergo LHR reflections in the magnetosphere. However, in the presence of field-aligned columns of enhanced or depressed ionization called ducts, some of the rays can be guided to the low altitude ionosphere with their wave normal at small angles with the local vertical, and these rays can be observed at a ground receiver after propagating within the Earth-ionosphere waveguide (Helliwell, 1965). The waves can also reach the ground directly from the source if the source is at low altitudes and radiates such that some of the wave energy reaches the Earth-ionosphere boundary with sufficiently small wave normal angles with respect to the local vertical. Other modes of guiding energy from a magnetospheric source to the ground are guiding along the plasmapause (Inan and Bell, 1977) and a subprotonospheric mode (SP); (Smith, 1964; Raghuram, 1975). Experience indicates that most of the whistler-mode waves observed on the ground, however, propagate in one or more ducts and in general a signal observed at the ground is a sum of the signals that have exited from various ducts located at different latitudes and longitudes.

Rays starting from A can reach a high altitude satellite (SR) via a direct nonducted path or via a nonducted magnetospherically reflected path (MR) after an LHR reflection (Kimura, 1966; Smith and Angerami, 1968). In nonducted propagation, a plane wave tends to propagate across the geomagnetic field lines with its wave normal making a large angle with respect to the geomagnetic field. A low altitude satellite (SR) can receive signals via direct or magnetospherically reflected nonducted propagation; however, it can also receive ducted signals that have backscattered from the ionosphere-waveguide boundary (BS); (Rastani et al., 1985).

Various aspects of the ionospheric plasma can profoundly affect the propagation of whistler-mode energy from the magnetosphere to the ground and vice versa. Thus, 1 to 10 km scale density gradients in the ionosphere lead to closely spaced multiple nonducted paths between a ground source and a satellite receiver (Sonwalkar et al., 1984). 10 to 100 m scale irregularities in the ionosphere lead to conversion of part of the incident whistler-mode energy to quasi-electrostatic lower hybrid waves (Bell and Ngo, 1990). Highly collisional D-region plasma can lead to a 10- to 20-dB reduction in wave energy that passes through this region (upward or downward). A consequence of D-region absorption is that many magnetospheric wave phenomena are better observed at nighttime when the D-region absorption is minimum (Helliwell, 1965).

Here we have reviewed several aspects of whistler-mode propagation in the Earth's magnetosphere. These are treated in depth in Chapter II/7 of this handbook. Kimura (1989) has reviewed 2- and 3-D ray-tracing studies in the Earth's magnetosphere for a large variety of wave modes including the whistler mode, the ion cyclotron mode, and the free-space and Z modes.

3.3. GENERATION AND DAMPING OF PLASMA WAVES

In this section we discuss mechanisms that have been proposed to explain the generation of a variety of plasma waves observed in the magnetosphere. We limit ourselves, however, to providing a qualitative description, because it is not possible to review the vast literature that exists on this subject. Moreover, excepting a few cases (such as lightning-generated whistlers), the generation mechanisms of most of the observed plasma waves are still the subject of intense research and at times controversy. For more detailed discussions of this topic we refer readers to several excellent textbooks and review articles (Stix, 1992; Fredricks, 1975; Gendrin, 1975; Hasegawa, 1975; Shawhan, 1979a,b; Al'pert, 1983; Lyons and Williams, 1984; Melrose, 1986; Gurnett and Inan, 1988; Anderson and Kurth, 1989, Swanson, 1989; Kurth, 1991).

3.3.1. General Considerations

The ultimate sources of energy for naturally occurring plasma waves are: (1) the solar wind; (2) the Earth's rotational energy; and (3) the Earth's magnetic field. Conversion of energy from these primary sources manifests itself in the form of electric fields and particle kinetic energy. A number of processes operating in the magnetosphere disturb the particle distribution function from its equilibrium state (generally a Maxwellian distribution), leading to new particle distributions which are unstable to the growth of plasma waves. An unstable distribution function of particles can result from any one of several processes such as plasma diffusion and convection across the magnetic field, pitch angle scattering, plasma drifts resulting from gradients of plasma density and electric field, particle acceleration in electric fields, and disturbances arising in a mixture of cold (ionosphere) and hot (solar wind) plasma.

Resonant conversion of kinetic energy of particles to wave energy or vice versa can take place by two different mechanisms, depending on whether the particle motion along the geomagnetic field (longitudinal motion) or the particle motion transverse to the magnetic field is the controlling factor. The former mechanism leads to flow or beam instabilities (or damping) and the latter to gyroresonance (or cyclotron resonance) instabilities (or damping). A general resonance condition for a plasma consisting of ions and electrons is given by:

$$\omega - k_\parallel \upsilon_\parallel + n\omega_{Hi} = 0, \, n = 0, \pm 1, \pm 2 \ldots \qquad (3.20)$$

where, $k_\parallel$ and $\upsilon_\parallel$ are, respectively, the components of the wave normal vector $\mathbf{k}$ and particle velocity $\mathbf{v}$ along the geomagnetic field, and $\omega_{Hi} = |(q_i B)/m_i|$ is the gyrofrequency of a plasma particle with charge q_i and mass m_i. The wavefrequency ω is not an independent variable but is a unique function of $\mathbf{k}$. For any given mode this relation is given by the dispersion relation 3.3. For $n = 0$ in Equation 3.20 we obtain the Landau resonance condition for beam instabilities where the parallel wave phase velocity equals the particle parallel velocity. All other values of n represent cyclotron resonances (or nth order gyroresonances), in which the wavefrequency in the frame of reference of a particle equals some harmonic of the particle's gyrofrequency. The principle cyclotron harmonic resonance (gyroresonance), given by $n = -1$, occurs for a wavefrequency in the counter-streaming particle's frame of reference equal to its gyrofrequency. It can be shown that when the wave growth rate is much smaller than the wavefrequency, only particles of species i satisfying resonance condition 3.20 interact with the waves (Kennel and Englemann, 1966). Thus under gyroresonance, electrons and ions produce waves near their respective gyrofrequencies.

In the following section we discuss important instabilities and damping mechanisms responsible for the generation and decay of plasma waves.

3.3.2. Gyroresonance Instabilities

An anisotropic particle distribution in which the temperature $T_\perp$ in the direction transverse to the geomagnetic field is higher than the longitudinal temperature $T_\parallel$ (i.e., particle distribution is anisotropic in pitch angle, the angle between the particle velocity vector and the geomagnetic field), represents an unstable particle distribution with free energy which can be liberated if the conditions of gyroresonance are satisfied. Energetic particles trapped in the Earth's radiation belts represent an example of anisotropic particle distribution. Ion cyclotron waves found near the plasmapause are believed to be generated by a gyroresonance mechanism. In this case, the free energy sources are hot ions injected from the region of the geomagnetic tail. Whistler-mode emissions generated near the magnetic equator including mid-latitude hiss, discrete, quasiperiodic, periodic emissions, and chorus are also believed to be generated by gyroresonance instabilities (these are sometimes called whistler-mode instabilities). The energy source in most of these cases is the energetic electron (~1 to 100 keV) trapped in the magnetosphere. In the case of chorus, the energy sources are 5 to 150 keV electrons injected in the dayside region beyond the plasmapause during the period of heightened magnetic activity. A Doppler-shifted cyclotron mechanism has been proposed for the auroral kilometric radiation. The energetic electrons precipitating in the auroral regions leading to optical emissions are believed to be the energy source for the generation of AKR. Electron cyclotron harmonic waves found near the plasmapause are generated when a particle distribution with anisotropic pitch angle distribution or a beam of energetic particles exists.

3.3.3. Streaming Instabilities

The sources of free energy in streaming instabilities are one or more components of plasma streaming through the other thermal and energetic components. Depending on the characteristics of the flowing component, these instabilities are classed as current driven, beam, two stream, and drift wave (Shawhan, 1979a). The fundamental condition for the excitation of streaming instability is the presence of a "bump" on the tail type of particle distribution function, i.e., compared to a Maxwellian distribution, this distribution function has an enhanced population centered at some velocity v_{max} different from the average velocity. The energy of the plasma waves generated is derived from the particles belonging to that part of the distribution function where the phase-space density increases with increase in the particle velocity. Plasma waves are created with phase velocities near v_{min}, where v_{min} is near the lower end of the "bump" where the phase space density is minimum. A necessary condition for the generation of waves is that Landau resonance condition ($n = 0$ in Equation 3.20) is satisfied. For waves propagating parallel to the geomagnetic field, this condition reduces to $v_{min} = c/\mu$, where c is the velocity of light and μ is the wave refractive index. For frequencies near one of the characteristic resonances of cold plasma modes (the electron and ion gyrofrequencies and plasma frequencies, and the upper and lower hybrid frequencies) refractive index tends to infinity, giving a low phase velocity and thus making it possible to stimulate steaming instabilities with relatively low energy particles.

In the bow shock, magnetosheath, and in the plasma sheet of the distant geomagnetic tail, directed ion fluxes and *currents* (as measured by magnetometers) have been detected. The current is probably carried by the ions. The observed bow shock turbulence, tail broadband electrostatic noise, and auroral field line turbulence may be attributed to ion cyclotron or ion-acoustic instabilities deriving energy from ion fluxes. In the auroral regions precipitating electrons (<15 keV) often constitute a field-aligned *beam* which can generate plasma waves at lower and upper hybrid frequencies. Auroral hiss may be produced by precipitating beams of electrons with energies greater than 10 keV. It has been proposed that VLF saucers which have been observed to propagate upward may be produced by beams of suprathermal electrons (<5 eV) moving upward. It has been suggested that part of the auroral kilometric radiation with left-hand polarization may result from the auroral electron beam and current system. At the bow shock, solar wind ions and electrons are reflected to produce a counterstreaming, *two-stream*, distribution which may be responsible for the generation of whistler-mode noise below the ion gyrofrequency, and oscillations near electron gyro and plasma frequencies. The *drift instability* can also be classified as a kind of streaming instability. It is a result of the bulk particle motion caused by large scale gradients of plasma density and temperature. The free energy of a drifting plasma may be converted into vibrational energy of the particles and plasma waves. The ion cyclotron waves in the polar cap and micropulsations of the magnetic field are presumably associated with this instability mechanism.

3.3.4. Coherent Wave Instability

Most of the theoretical analyses of plasma instabilities have used linear or quasilinear mechanisms. However, some of the observed wave phenomena in the magnetosphere are distinctly nonlinear. These generally involve quasimonochromatic waves of natural or artificial origin present in the magnetosphere. Interactions of these waves with energetic particles are highly nonlinear and distributed (i.e., inhomogeneity is important) and lead to a wide variety of wave phenomena. Most of the whistler-mode discrete emissions including chorus and artificially stimulated emissions belong to this category. While it is generally agreed that cyclotron resonance is responsible for the generation of this class of emissions, the underlying nonlinear mechanism is largely unknown. Helliwell (1988) has called the underlying instability coherent wave instability (CWI). This instability is further discussed in Section 4.3.5.

3.3.5. Kelvin-Helmholtz Instability

Nonuniformity in a plasma often provides free energy for plasma instabilities which work to smear out original nonuniformity. These instabilities are called macroscopic or macroinstabilities, the most important of which for the generation of space plasma waves are (1) drift instability mentioned above and (2) the Kelvin-Helmholtz instability.

The Kelvin-Helmholtz instability is caused by velocity shear and its commonplace manifestation is the waves on the surface of water caused by winds. Two kinds of Kelvin-Helmholtz instabilities are relevant in the magnetospheric dynamics. One is the hydromagnetic instability which may be excited at the boundary of the magnetosphere by the solar wind flow and is considered to be a major cause of Pc3 to 5 magnetic pulsations (see Chapter II/14). The other is the electrostatic Kelvin-Helmholtz instability excited by $\mathbf{E} \times \mathbf{B}$ flow in a nonuniform electric field in the auroral sheet. This instability has been evoked to explain certain features of auroral field-line turbulence.

3.3.6. Cerenkov Radiation

An electron moving with uniform velocity v_z in the direction of geomagnetic field $\mathbf{B}_0$ emits Cerenkov radiation if the following coherence condition is satisfied.

$$v_z \cos\theta = v_{ph} = \frac{\omega}{k} \tag{3.21}$$

where ω is the angle between the wave normal direction and the direction of the motion of the electrons, and v_{ph} is the phase velocity of the wave in the medium. Emission of Cerenkov radiation essentially transverse to the geomagnetic field is possible in the magnetospheric plasma, because for certain modes and for certain ranges of plasma parameters, v_{ph} is much less than the speed of light, and the Equation 3.21 can easily be satisfied. Incoherent or coherent Cerenkov instabilities have been proposed as generation mechanisms for the observed Z-mode radiation in the auroral regions.

3.3.7. Damping of Plasma Waves: The Landau and Cyclotron Damping

The magnetospheric plasma is essentially collisionless and therefore the most important damping mechanisms for space plasma waves are the Landau and cyclotron damping which occur when the resonance condition given by Equation 3.20 is satisfied for some region of the particle distribution. Qualitatively, in both cases a beam of particles traveling with a velocity in the neighborhood of the critical velocity given by Equation 3.20 gives to or absorbs energy from the electromagnetic field, depending on whether the beam drift velocity is greater or smaller than the wave phase velocity. In general, equilibrium particle distributions lead to wave damping, and the particle distribution away from thermal equilibrium (such as those described in Sections 3.3.2 and 3.3.3 above) lead to wave growth.

4. OBSERVATIONS

We classify observations of plasma waves of magnetospheric origin into three categories: (1) the ground observations of naturally occurring plasma waves; (2) spacecraft observations of naturally occurring plasma waves; and (3) observations of artificially stimulated emissions.

The reason for dividing the ground and spacecraft observations into separate categories is that somewhat distinct and apparently uncorrelated wave activity is detected at ground stations and on satellites. As noted below, this situation arises from certain fundamental differences in the nature of observations made from space and those from the ground and in the nature of propagation of plasma waves from the magnetosphere to a ground station.

1. Ground observations are made at one or more fixed locations, often permitting studies of temporal changes in plasma wave activity at longitudes close to a ground station over time scales of hours. Satellite observations, on the other hand, allow the detection of plasma waves over a much larger region of the magnetosphere covered by the satellite orbit. However, because of the motion of the satellite, it is difficult to differentiate between spatial and temporal changes in wave activity. Satellite experiments also permit *in situ* measurements of the plasma environment which are necessary to identify the wave modes and generation mechanisms of plasma waves.

2. Only a small fraction of magnetospheric plasma waves can be observed on the ground. Electrostatic waves that travel short distances due to heavy Landau damping and the free-space mode radiations such as AKR and TMR which are radiated away from the Earth cannot be seen on the Earth. Most of the nonducted whistler-mode waves do not reach the ground because they undergo reflections within the magnetosphere. Thus, most of the plasma waves observed on the ground are ducted whistler-mode waves that have propagated inside ducts (see Section 3.2).

Satellites, on the other hand, can detect all kinds of waves generated within the magnetosphere including free-space mode radiation, whistler-mode waves, electrostatic waves, and others. However, the ducted whistler mode waves that are commonly observed on the ground are rarely observed on satellites because ducts occupy a relatively small volume of the magnetosphere ($\sim$0.01%); (Burgess and Inan, 1993), and thus the probability of a satellite being inside a duct is extremely small. This explains why the satellite and ground observations of plasma waves are often so distinct. We note, however, that a low altitude satellite can at times receive ducted signals that have backscattered from the Earth-ionosphere boundary.

4.1. GROUND OBSERVATIONS OF NATURALLY OCCURRING PLASMA WAVES

Historically the plasma waves (ELF/VLF emissions) observed on the ground have been classified according to their characteristic dynamic spectra. Figure 13.4.1. depicts the classification scheme used by Helliwell (1965). Some of these wave phenomena, such as auroral and mid-latitude hiss and chorus are frequently observed on satellites and are also discussed in the next subsection. Others, such as periodic and quasiperiodic emissions and discrete emissions, are rarely detected on a spacecraft.

4.1.1. Auroral Hiss (VLF Hiss)

Observed both on the ground and on satellites, auroral hiss is a high latitude VLF phenomenon occurring in the frequency range from a few kilohertz to several hundred kilohertz in the evening and night hours near the auroral oval region (Helliwell, 1965; Sahzin and Hayakawa, 1993). Its spectrum resembles that of a band-limited thermal or fluctuation noise and can be identified aurally by a hissing sound. The bottom panel of Figure 13.4.2 shows a spectrogram of auroral hiss observed at the South Pole station, Antarctica. Auroral hiss often shows no substantial changes in its spectrum for periods of minutes or even hours, and is then called "steady" or "continuous" hiss; occasionally it exhibits marked variations in amplitudes over periods of the order of a second and is then called impulsive hiss. The impulsive hiss is observed in the frequency range extending to several hundred kilohertz, whereas the continuous hiss is limited to frequencies below $\sim$30 kHz. In both types of hiss, the maximum power spectral density is achieved around $\sim$10 kHz. The average value of power spectral density of VLF hiss observed at auroral zone ground stations is of the order of 10^{-16} Wm^{-2}Hz^{-1} (Jorgenson, 1968). Auroral hiss is found to be confined to

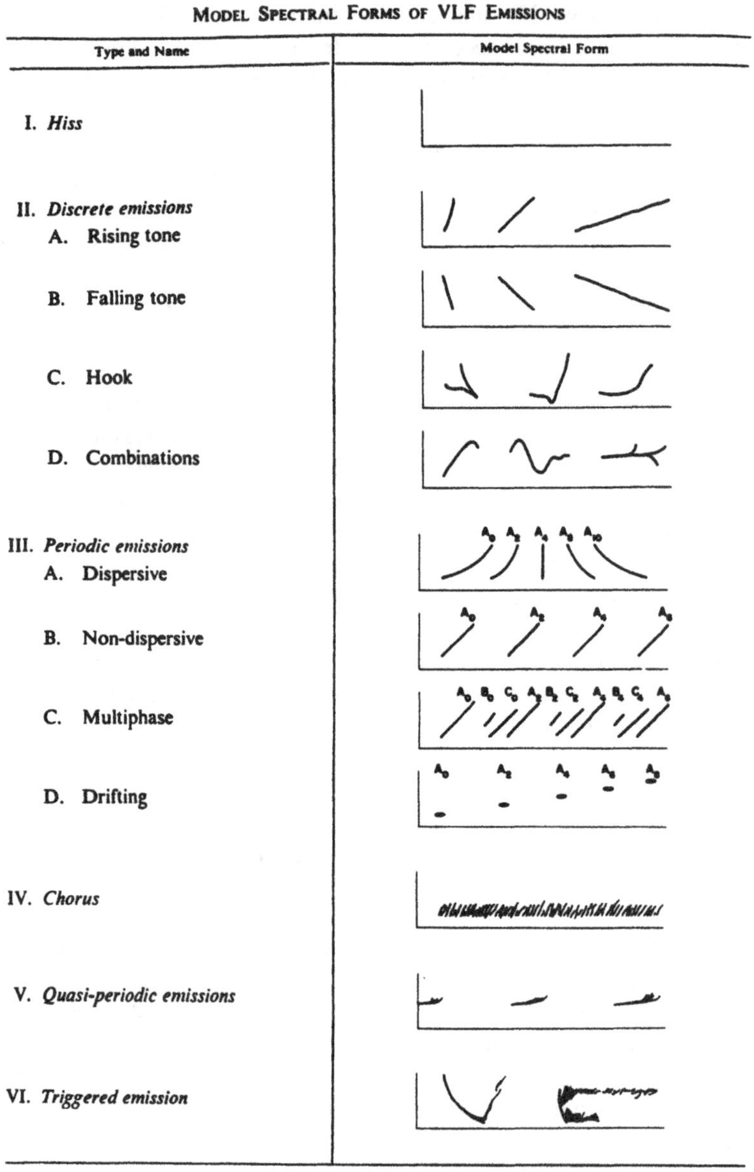

MODEL SPECTRAL FORMS OF VLF EMISSIONS

Type and Name	Model Spectral Form
I. *Hiss*	
II. *Discrete emissions* A. Rising tone	
B. Falling tone	
C. Hook	
D. Combinations	
III. *Periodic emissions* A. Dispersive	
B. Non-dispersive	
C. Multiphase	
D. Drifting	
IV. *Chorus*	
V. *Quasi-periodic emissions*	
VI. *Triggered emission*	

Figure 13.4.1 Classification of VLF emissions observed on the ground. (From Helliwell, R. A., *Whistlers and Related Ionospheric Phenomena*, Stanford University Press, 1965. With permission.)

very narrow latitudinal extent. There is very little or no correlation between auroral hiss emissions recorded at two stations separated by ~5° in latitude (Swift and Kan, 1975).

There is a close association between auroral hiss and visible aurora (Helliwell, 1965; Rosenberg, 1968; Swift and Kan, 1975; Nishino et al., 1982). Impulsive hiss is found to originate from localized regions of bright electron auroras at the ionospheric level where one finds rapid changes in luminosities and motion, whereas an increase in the continuous (steady) hiss is observed when a steady auroral arc is located within a few hundred kilometers of the ground station. Satellite observations and generation mechanisms of auroral hiss are described in Section 4.2.2.3.

4.1.2. Mid-Latitude Hiss

Mid-latitude hiss is observed at ground stations at invariant latitudes of about 20 to 70° and peaks near latitudes 55 to 65° corresponding to the plasmapause location (Helliwell 1965; Hayakawa

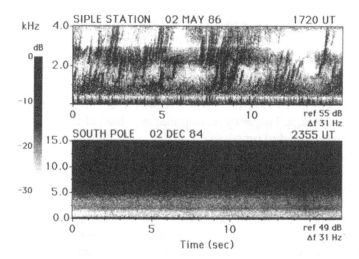

Figure 13.4.2 Top panel: double-banded ELF/VLF chorus observed at Siple, Antarctica (λ_g = 76°S, ϕ_g = 84°W, λ_m = 60°S). In this case chorus bands are accompanied by weak bands of mid-latitude hiss. Bottom panel: continuous auroral hiss observed at South Pole, Antarctica (λ_g = 90°S, λ_m = 79°S).

and Sazhin, 1992). It usually appears as a band-limited white noise in a 3 to 8 kHz band lasting for an hour or so with a relatively smooth rise and fall. During geomagnetically active periods, mid-latitude hiss can last for several hours showing amplitude fluctuations on a time scale of tens of minutes. The peak amplitude of mid-latitude hiss is lower than that of auroral hiss, about 1 μV/mHz$^{-1/2}$ at 5 kHz and rarely varies more than a factor of two or so. Steady mid-latitude hiss shows a local time variation, peaking around 0500 and 2000 local time.

Mid-latitude hiss is frequently accompanied by whistlers echoing in the same path (duct) and amplified in the same frequency band. Dowden (1971) suggested that at least part of the mid-latitude hiss may be generated by the superposition of highly dispersed, unresolved, overlapping whistlers continuously echoing from hemisphere to hemisphere along a duct. However, to overcome geometric loss of 10 to 20 dB corresponding to the small fraction of reflected energy being reduced, a magnetospheric amplification of 10 to 20 dB is required to maintain steady hiss intensity over time scales of tens of minutes (Thompson and Dowden, 1978). On the other hand, ground-based, direction-finding measurements suggest that mid-latitude hiss is generated just inside the plasmapause (Hayakawa et al., 1986). In both cases electron cyclotron instability of whistler-mode waves is believed to play an important role in the generation or amplification of mid-latitude hiss. Satellite observations of mid-latitude hiss are presented in Section 4.2.1.2.

4.1.3. Discrete Emissions and Chorus

One of the most interesting kinds of magnetospheric plasma waves are discrete emissions, lasting a few seconds or less and covering a bandwidth of a few hundred hertz to a few kilohertz in a 0 to 10 kHz band (Helliwell, 1965; Andreson and Kurth, 1989; Sazhin and Hayakawa, 1992). Depending on their dynamic spectral form (Figure 13.4.1), principal elementary types of discrete emissions have been called rising tones (risers), falling tones (fallers), and hooks. Elements with these three basic forms may be joined in various ways to form combinations of discrete emissions. While isolated discrete emissions are occasionally observed on the ground, they are rarely observed on satellites. The most common discrete emission observed both on the ground and on satellites is the whistler-mode chorus whose spectral form looks like a closely spaced superposition of the three elementary types of discrete emissions, though in most cases it consists of superposition of risers. The top panel of Figure 13.4.2 shows a spectrogram of chorus recorded at Siple, Antarctica.

The earliest observations of a chorus were made at mid-latitude stations where it was found to peak in the early morning hours (dawn chorus; Isted and Millington, 1957). Later observations showed that both the frequency range and local time maximum of chorus intensity were dependent on the latitude of the ground station (Helliwell, 1965; Francis et al., 1983). The frequency range of chorus emissions generally decreases with increasing latitude; chorus observed at mid- or subauroral latitude (mid-latitude chorus) has a typical frequency range of 2 to 4 kHz, whereas that observed at polar latitude (polar latitude) is found between 500 to 1500 Hz. The peak intensity of chorus varies between 10^{-16} to 10^{-14} W/m²/Hz, with polar chorus being the strongest and auroral chorus the weakest. Frequently accompanied by a band of hiss, mid-latitude chorus is more common during disturbed periods.

Many studies show a one-to-one correlation between electron precipitation and individual chorus elements (risers; Rosenberg et al., 1971, 1990; Roeder et al., 1985). In these studies VLF chorus elements observed at Siple Station, Antarctica or at Roberval, Canada were correlated with X-ray bursts caused by electrons in a 30- to 200-keV range. Park et al. (1981) reported simultaneous observations of VLF chorus at Siple and electrons with energies below 50 keV observed on the ATS satellite ($L = 6.6$). A one-to-one correlation was also observed at Siple Station between chorus elements in the 2- to 4-kHz range and optical emissions at $\lambda = 427.8$ nM (Helliwell et al., 1980b). These data in general were found to be consistent with scattering of electrons into the loss cone over Siple by emissions that were triggered by waves propagating away from the equator after reflection in the ionosphere over Siple Station.

Satellite observations and generation mechanisms of chorus are discussed in Section 4.2.1.2.

4.1.4. Periodic Emissions

A sequence of discrete emissions or clusters of discrete emissions showing regular spacing is called periodic emissions (Helliwell, 1965; Sazhin and Hayakawa, 1992). Usually their period is constant, but on occasion it changes slowly. The top panel of Figure 13.4.3 shows a spectrogram of periodic emissions recorded at Eights, Antarctica. They generally occur at frequencies below 5 to 10 kHz with a few kilohertz bandwidth and a period in the range from 2 to 6 s. Periodic emissions are observed at latitudes that correspond to closed magnetic field lines and are rarely observed on satellites. Observations of periodic emissions at conjugate stations Byrd-Hudson Bay showed that the emissions appeared at the two stations with a delay of about 0.8 s (Lokken et al., 1961). As shown in Figure 13.4.1, depending on their spectral forms, periodic emissions are classified into various categories. If the period varies with frequency, the periodic emissions are called dispersive; if there is little or no variation in period with frequency, they are called non-dispersive. Two or more interleaved sets with the same period are called multiphase periodic emissions.

The periodic signals, like other VLF emissions observed on the ground, are assumed to have propagated in field-aligned ducts. By comparing the emission period with the two-hop whistler transit time, it was established that the period of the emissions is identical to the two-hop whistler transit time (Dowden and Helliwell, 1962). For the dispersive type, the period between bursts varied systematically with frequency; these variations were often exactly the same as that found in the associated echoing whistlers. Thus, each burst was simply the result of whistler-mode echoing in a duct. As for nondispersive periodic emissions, Helliwell (1965) suggested that each burst was triggered by the previous one, and that the whistler-mode cyclotron instability may be the trigger mechanism.

4.1.5. Quasiperiodic Emissions

Quasiperiodic (QP) emissions are a sequence of repeated noise bursts of a relatively long period, in which each burst may consist of a number of discrete events such as periodic emissions or chorus or may consist of diffuse hiss (Helliwell, 1965). The bottom panel of Figure 13.4.3 shows a spectrogram of QP emissions observed at Eights, Antarctica. The period between bursts is measured in tens of seconds, usually between 10 and 200 s, and is irregular compared with that of periodic emissions. Although observed frequently on ground, there have been relatively few reports of QP emissions on satellites (see Sazhin and Hayakawa, 1994 and references therein).

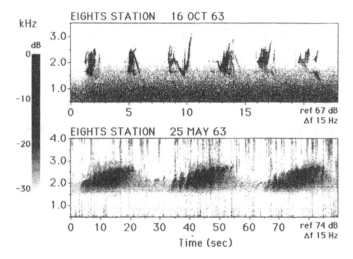

Figure 13.4.3 Periodic (top panel) and quasiperiodic (bottom panel) emissions observed at Eights, Antarctica ($\lambda_g = 79°S$, $\phi_g = 77°W$, $\lambda_m = 64°S$). In the example shown, individual bursts of quasiperiodic emissions consist of periodic emissions.

QP emissions have periods comparable to those of micropulsations, and these two phenomena are often closely associated in time of occurrence. Therefore, QP emissions have been classified based on whether (type 1 or QP1) or not (type 2 or QP2) magnetic pulsations were associated with them. Based on ground observations, Kitamura et al. (1969) and Sato (1980) have described the general features of QP emissions. QP1 emissions occur in the frequency range 0.3 to 1.5 kHz whereas QP2 emissions occur at somewhat higher frequencies ranging from 1.5 to 6 kHz. Both types show a diurnal maximum of occurrence near local noon and QP2 shows an additional maximum in the local afternoon-evening sector. The period of both kinds of emissions was found to increase from morning to evening and the period was found to increase with increasing latitude of the observing station. QP2 emissions are observed more frequently under quiet magnetic conditions ($K_p \sim 1$) compared to QP1 emissions which are observed more frequently for K_p of 2 to 3. Both types of QP emissions have been observed at magnetically conjugate stations.

Various theoretical models to explain the generation of QP1 emissions assume that (1) discrete elements that make up QP emissions are generated in the magnetosphere by the gyroresonance whistler-mode instability and (2) magnetic pulsations modulate this generation process leading to the observed period of QP emissions. The intensity of VLF emissions generated by whistler-mode instability depends on the energetic electron distribution function, the cold plasma density, and the geomagnetic field at the interaction region, generally assumed to be near the equator. A comprehensive model that considers a variety of ways of modulating the plasma parameters by magnetic pulsations has been considered by Sazhin (1987). The magnetic pulsations that modulate whistler-mode waves could be standing compressional hydromagnetic waves, as pointed out by Sato and Fukinishi (1981). As far as QP2 emissions are concerned, they could be generated by the same mechanism as QP1 emissions except that the associated magnetic pulsations are not detectable at ground stations. This is possible if magnetic pulsations do not form standing waves in the magnetosphere or if they have short wavelengths so that they are shielded by the ionosphere. Studies of QP emissions on GEOS 1 satellite support this picture, because the distinction between QP1 and QP2 is not so clear when observed in the magnetosphere (Tixier and Cornilleau-Wehrlin, 1986).

4.1.6. Triggered Emissions
Any ELF/VLF emission that appears to have been initiated or triggered by another event is called a triggered emission (Helliwell, 1965). Triggering sources include whistlers, discrete emissions, and signals from ground transmitters. Most triggered emissions follow the apparent source with

one exception known as precursor, which precedes the associated whistler (Park and Helliwell, 1977; Paschal, 1990; Chmyrev et al., 1990). In all cases of precursor observation, it was found that the precursor was associated with a two-hop whistler. Whistler-mode instability is believed to be responsible for the generation of triggered emissions. Triggered emissions have been investigated most extensively in the context of ground-based active experiments which are described in Section 4.3.

4.2. SPACECRAFT OBSERVATIONS OF NATURALLY OCCURRING PLASMA WAVES

We have categorized the plasma waves observed on spacecraft according to the distinctive plasma regions where they are generally observed; the reason being that the generation and propagation of the waves is closely related to the properties of the cold and hot plasma distributions found in the respective regions. The magnetosphere is divided in four broad but overlapping regions: (1) equatorial magnetosphere up to $\sim 10~R_e$ including low-to-subauroral latitudes ($<60°$); (2) auroral and polar region including auroral field lines and extending from a few hundred kilometers to several thousand kilometers; (3) distant tail region including the plasma and neutral sheets and the magnetosheath; and (4) the bow shock region. Though distinctive wave activity is associated with each of these regions, some of the plasma waves are found in more than one region; for example, half integral electron gyrofrequency emissions are found in the equatorial region near the plasmapause as well as in the plasma sheet in the tail region. In such cases we have placed them in the region where they are either observed most frequently or are found most intense, with references to other regions where those waves are also found. Within this grouping, plasma waves associated with electron characteristic frequencies are described first, followed by those associated with LHR frequency and ion characteristic frequencies.

Figure 13.1.1 shows the regions of plasma wave occurrence located in a noon-midnight meridian cross section of the Earth's magnetosphere. Although not shown in Figure 13.1.1, the plasma waves observed are similar in the north and south half of the magnetosphere.

Figure 13.4.4 shows plasma waves observed on the CRRES satellite (inclination = 18°, apogee = 33,582 km, perigee = 384 km) close to the magnetic equator (Anderson, 1994). Electron gyrofrequency along the orbit is also plotted to help identify various wave modes described in Section 3 (Figure 13.3.2). The plasmapause location is easily identified as the lower cut off of the UHR band. Most prominent wave activity in the equatorial region is whistler mode waves at frequencies below f_{ce}, primarily consisting of plasmaspheric hiss (ELF hiss), whistlers and chorus emissions (Carpenter, 1994). At higher frequencies electrostatic $(n + 1/2)f_{ce}$ emissions (labeled as electron cyclotron harmonics), UHR emissions (Z mode), terrestrial myriametric radiation ($L - O$ mode, labeled as escaping continuum radiation), and trapped continuum radiation are observed (Gurnett and Inan, 1988). At lower frequencies, the noise labeled equatorial VLF noise is actually composed of two parts: electromagnetic emissions from about 50 to 100 Hz and electrostatic emissions from 100 Hz to about 1 kHz.

Figure 13.4.5 shows a spectrogram for a high altitude DE 1 pass over the auroral zone in which three types of auroral plasma wave emissions can be seen: auroral kilometric radiation ($R - X$ mode), Z-mode radiation, and auroral hiss (VLF hiss; Gurnett et al., 1983). The latter is identified separately from Z-mode radiation by the sharp upper cutoff at f_{pe}. This figure should be compared with Figure 13.3.3 to identify the different wave modes. At lower altitudes in the auroral regions, broadband electric and magnetic noise is observed over the evening and morning auroral zones at frequencies close to ion cyclotron frequencies.

Near the bow shock, three different kinds of wave phenomena are found: electron plasma oscillations, electromagnetic whistler-mode waves, and electrostatic turbulence (Rodriguez and Gurnett, 1976).

In the distant tail region, broadband electrostatic noise is found in the plasma sheet boundary layer; whistler-mode lion's roar is found in the magnetosheath, and electrostatic $(n + 1/2)f_{ce}$ emissions are found in a region near the neutral sheet (Anderson and Kurth, 1989; Kurth, 1991).

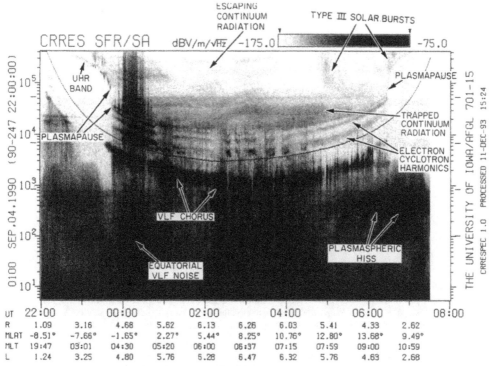

Figure 13.4.4 A 10-h CRRES PWE electric field spectrogram for orbit 100 showing various plasma waves observed in the equatorial region. The geomagnetic field was becoming disturbed following 2 to 3 very quiet days. The dark dashed line plotted on the spectrogram shows the electron cyclotron frequency (f_{ce}) calculated from the CRRES fluxgate magnetometer experiment. (From Anderson, R. R. *Dusty Plasma, Noise, and Chaos in Space and Laboratory,* Ed., H. Kicuchi, Plenum Publishing, New York, 1994. With permission.)

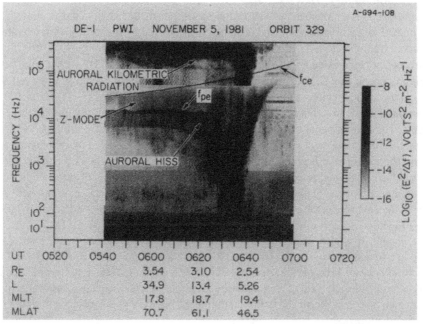

Figure 13.4.5 A spectrogram for high-altitude DE 1 pass over the auroral zone in which three types of auroral plasma wave emissions can be seen: auroral kilometric radiation, *Z*-mode radiation, and auroral hiss. (From Gurnett, D. A. and Inan, U. S., *Rev. Geophys.,* 26, 285, 1988. Copyright by the American Geophysical Union. With permission.)

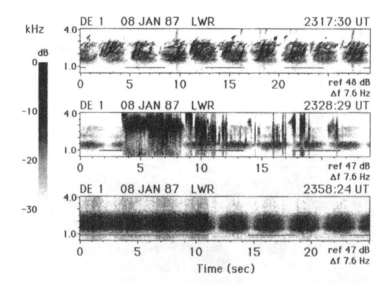

Figure 13.4.6 DE 1 Observations of plasma wave near the equatorial region close to the plasmapause: (top panel) during the early part of the trajectory, when the satellite was outside the plasmasphere ($L \sim 5.0$ to 5.3, $\lambda_m \sim 16$ to 20°) hiss and chorus were observed; (middle panel) as the satellite moved close to the plasmapause ($L \sim 4.6$ to 5.0, $\lambda_m \sim 10$ to 15°), impulsive electromagnetic emissions and hiss were observed; (bottom panel) inside the plasmasphere ($L < 4.5$) only plasmaspheric hiss was observed. The top two panels show the electric field spectra and the bottom panel shows the magnetic field spectra for the first 11 s and the electric field spectra for the last 14 s. The spin fading at $\sim$3 s of the signals has been used to determine wave normal directions of hiss. From Sonwalkar, V. S. and Inan, U. S., *J. Geophys. Res.*, 93, 7493, 1988. Copyright by the American Geophysical Union. With permission.)

4.2.1.   Waves Found Near the Magnetic Equator up to $\sim$10 to 12 R_e

4.2.1.1.   Plasmaspheric Hiss (ELF Hiss)

Plasmaspheric hiss is a broadband and structureless whistler-mode radiation that is almost always present in the Earth's plasmasphere and is commonly observed by magnetospheric satellites, but not observed from the ground (Thorne et al., 1973; Hayakawa and Sazhin, 1992). It is observed in the frequency range extending from a few hundred hertz to 2 to 3 kHz with a peak below 1 kHz. Plasmaspheric hiss should be distinguished from auroral hiss, which is observed at low altitudes in the auroral regions and covers a much wider frequency band ($\sim$100 Hz to 100 kHz) and from mid-latitude hiss that is observed from ground stations and on satellites in the 2 to 10 kHz range. The bottom panel of Figure 13.4.6 shows a spectrogram of plasmaspheric hiss observed on the DE 1 satellite at $\sim$4 R_e near the magnetic equator. Detailed morphological studies of plasmaspheric hiss have been performed using data from the high altitude satellites (see Sonwalkar and Inan, 1988 and references therein). The general conclusions of these studies can be summarized as follows: (1) plasmaspheric hiss is found throughout the plasmasphere and is stronger in the daytime sector compared to the midnight-to-dawn sector, and generally peaks at high ($>$40°) latitudes; (2) it often shows a sharp cutoff at the plasmapause for frequencies below $\sim$1 kHz, though it is also observed outside the plasmasphere at higher frequencies; (3) it is observed in a frequency band $\sim$100 Hz to $\sim$3 kHz with a peak intensity (10^{-7} to 10^{-5} (nT)^2/Hz) at frequencies below 1 kHz; (4) the upper cutoff of hiss is controlled equatorially while the lower cutoff is controlled locally; and (5) hiss intensity is correlated with magnetic substorm activity. It decreases soon after a substorm and builds up during the recovery phase; detectable waves persist even during quiet periods.

Recently many studies have been devoted to determining the wave normal directions of plasmaspheric hiss (see Storey et al., 1991 and references therein). The general conclusion from these studies is that most often hiss propagates with oblique wave normal angles, even close to the geomagnetic equator, a region believed to be favorable for the generation of hiss.

The generation mechanism of plasmaspheric hiss remains unknown (Storey et al., 1991). Using quasilinear theory of cyclotron instability (Kennel and Petschek, 1966) for growth rate calculations, Church and Thorne (1983) and Huang et al. (1983) found that the estimated gains are not sufficient to explain the observed intensities of hiss if it is assumed to be generated from ambient thermal noise. Moreover, these studies assumed generation of hiss near equator with wave normal aligned with the geomagnetic field, an assumption in contradiction with the wave normal measurements mentioned above. Recently, Sonwalkar and Inan (1989) have provided evidence that lightning-generated whistlers can generate hiss-like signals and that lightning could be an embryonic source of plasmaspheric hiss. Draganov et al. (1992), using ray-tracing simulations, have argued that passive propagation of whistler-mode energy injected into the magnetosphere from the lightning discharges may explain the plasmaspheric hiss intensities observed in the magnetosphere.

4.2.1.2. Mid-Latitude Hiss

A band of mid-latitude hiss between $\sim$2 to 10 kHz is often observed from the equator to subauroral latitudes, between $\sim$2 to 3 R_e to $\sim$8 to 9 R_e, and at all local times except for $L > 5$ in the midnight-to-dawn sector in the $\pm50°$ geomagnetic latitude range (Taylor and Gurnett, 1968; Dunckel and Helliwell, 1969; Hayakawa and Sazhin, 1992). On low altitude satellites, such as ISIS 1, 2, and Ariel, mid-latitude hiss has been observed at latitudes ranging from $\sim$50° to about 65°. On the OV1-14 satellite, hiss was observed in the 3.9 to 10.4 kHz band in the slot region ($L \sim$ 2.5 to 3.6) within the plasmasphere. The maximum power spectral densities of 1.1 pT/Hz$^{1/2}$ and 2.8 μV/Hz$^{1/2}$ on the magnetic and electric antennas occurred in the 5.6-kHz channel. Mid-latitude hiss is often accompanied by chorus and the upper cut off of the combined band of hiss and chorus is found to be proportional to the equatorial electron gyrofrequency (Dunckel and Helliwell, 1969). Possible generation mechanisms of mid-latitude hiss have been discussed in Section 4.1.2.

4.2.1.3. Chorus

Chorus has been observed from ground stations, low altitude satellites in the topside ionosphere, and high altitude equatorial satellites. The top panel of Figure 13.4.6 shows an example of whistler-mode chorus observed on the DE 1 satellite just outside the plasmasphere at $L = 4.3$. In this example, chorus is accompanied by a weak band of hiss at frequencies below $\sim$3 kHz.

Low altitude ($\leq$0.5 R_e) satellite observations of chorus have been made on many satellites. (See Sazhin and Hayakawa, 1992, and references therein.) These observations showed that in general chorus occurred primarily at mid- to high latitudes, consistent with ground observations at mid- and polar latitudes. Detailed studies using data from high altitude OGO 1, 3, and 5 satellites were carried out by Burtis and Helliwell (1975, 1976) and by Tsurutani and Smith (1974, 1977). Principal conclusions of these studies are: (1) chorus occurs in the frequency range from a few hundred hertz to 5 to 6 kHz for $4 < L < 10$ mainly from 0300 to 1500 local time, at higher L at noon than at dawn, moving to lower L during geomagnetic substorms, all in accordance with ground observations of VLF chorus; (2) most intense chorus is found within $\pm15°$ of the geomagnetic equator; (3) chorus is bandlimited and the center frequency varies as L^{-3}; (4) peak chorus intensities range from 1 to 50 pT and that chorus amplitude is inversely related to frequency, implying lower intensities at lower L values; (5) chorus often occurs in two bands with the center frequency of each band closely related to the equatorial gyrofrequency (see Figure 13.4.4); and (6) the majority of the individual chorus emissions rise in frequency at a rate between 0.2 and 2.0 kHz/s; this rate increases with K_p and decreases with L.

A new type of discrete emission at $L < 4$ has recently been observed in the magnetosphere by Poulsen and Inan (1988). The emission elements are confined to a bandwidth of 1 to 5 kHz, and the lower cut-off frequency varies with L and equals $\sim$0.2 to 0.5 equatorial gyrofrequency. The dispersion of these emissions is often different from that of the chorus elements at higher L, and these emissions are found both inside and outside the plasmapause.

Burton and Holzer (1974) using OGO 5 search coil data found that close to the geomagnetic equator chorus wave normal angles were within 20° of the geomagnetic field. Away from the equator, the wave normal fell into two groups, one within 40° of the geomagnetic field, which they interpreted as ducted chorus, and the other at all wave normal angle from 0 to 90° of the geomagnetic field, which they interpreted as nonducted chorus. Goldstein and Tsurutani (1984) and Hayakawa et al. (1984) found that for $f/f_{he} < 0.3$ chorus wave normals are nearly parallel to the geomagnetic field near the magnetic equator.

A close association between mid-latitude hiss and chorus was noted by Dunckel and Helliwell (1969) who, using OGO 1 data, found that during magnetically quiet periods often mid-latitude hiss and chorus were simultaneously present within $L = 10$ for all local times except for $L > 5$ in the midnight-to-dawn sector. Koons (1981) and Hattori et al. (1991) reported that chorus emissions were often observed to start at frequencies that lie within a hiss band (see Figure 13.4.6).

Chorus is found to be correlated with energetic particles in the keV range. Oliven and Gurnett (1968), using Injun 3 data, found that microbursts of $E > 40$ keV electrons were always accompanied by a group of chorus emissions, though chorus was not always accompanied by microbursts. Tsurutani and Smith (1974) noted that the distribution of chorus as a function of local time and L was strikingly similar to the distribution of enhanced, trapped, and precipitated substorm electrons with energies $E \geq 10 - 40$ keV.

Various theories presented to explain the existence of chorus generally agree that energy for chorus generation comes from energetic electrons in the ~1 to 100 keV range and that a mechanism based on cyclotron resonance is probably the most likely one (Anderson and Kurth, 1989). However, as noted by Sazhin and Hayakawa (1992) in a recent review, none of the theories presented thus far explain all or even most of the observed features of chorus.

4.2.1.4. Terrestrial Myriametric Radiation (TMR) and UHR Emissions

Terrestrial myriametric radiation, also called escaping continuum radiation or terrestrial non-thermal radiation, was first detected in the frequency range 30 to 110 kHz by Brown (1973) and in the range 5 to 20 kHz by Gurnett and Shaw (1973). TMR is a weak broadband electromagnetic radiation generated by intense upper hybrid emissions near the plasmapause in the morning and early afternoon. Examples of TMR and of UHR emissions observed on the CRRES satellite are shown in Figure 13.4.4. When the radiation is first generated as a free-space mode in the magnetosphere, it is predominantly in the $L - O$ mode, although some $R - X$ mode components are also present which are possibly produced after multiple reflections at the magnetospheric cavity walls (Jones, 1982; Shaw and Gurnett, 1980).

TMR is generated by waves at the upper hybrid frequency (Z-mode radiation; Kurth et al. 1979). The free energy source for the upper hybrid waves is believed to be a loss cone anisotropy in the trapped electron distribution. Path-integrated growth rate calculations show that the favored region for the growth of these waves is near the equatorial plane. Both linear and nonlinear theories have been advanced to explain the conversion of local UHR emissions to escaping myriametric radiation. A linear mechanism, proposed by Jones (1976), involves conversion of the Z-mode upper hybrid radiation to $L - O$ mode myriametric radiation via a tunneling process. This linear tunneling process has been well known in ionospheric radio propagation as the "radio window" (Budden, 1985). The radio window is located at the electron plasma frequency and occurs whenever the plasma density gradient is sufficiently steep to make the gap between Z mode and $L - O$ mode comparable or smaller than one wavelength. The equatorial beaming and the left-hand polarization of the myriametric radiation predicted by Jones' theory was confirmed by the DE 1 data (Jones et al., 1987; Gurnett et al., 1988). A more detailed recent analysis of the DE 1 data, however, shows several features that are not predicted by Jones' model (Morgan and Gurnett, 1991). Various linear and nonlinear theories of generation of AKR and TMR are reviewed by Melrose (1981) and Lee (1989).

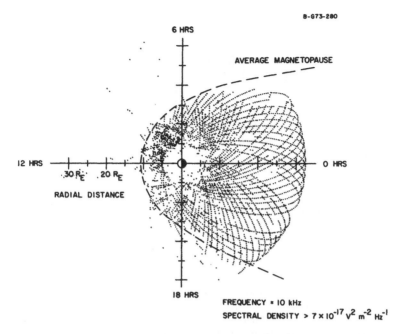

Figure 13.4.7   Region of occurrence of the trapped $f > f_p$ electromagnetic noise as determined from the 10-kHz electric field spectrum analyzer channel aboard the IMP 6 satellite. The coordinates are geocentric radial distance and local time. In order to prevent whistlers and other impulsive noise phenomena from contributing to this survey, only field strengths with a peak to average ratio less than 1.2 are included. (From Gurnett, D. A. and Shaw, R. R., *J. Geophys. Res.*, 78, 8136, 1973. Copyright by the American Geophysical Union. With permission.)

4.2.1.5.   Trapped Nonthermal Continuum Radiation

The IMP 6 satellite detected a weak electromagnetic background radiation trapped between the plasmapause and magnetopause, called trapped nonthermal continuum radiation (TNCR; Gurnett and Shaw, 1973). Figures 13.4.4 and 13.4.7 show, respectively, the typical noise-like spectra and the region of the magnetosphere where TNCR was detected. The radiation has a low frequency cut off near the electron plasma frequency f_{pe} and an upper frequency cutoff close to harmonics nf_{He} of the electron gyrofrequency. Near the plasmapause these emissions were sometimes associated with UHR emissions. The radiation is abruptly cutoff near the Earth's dayside magnetopause ($\sim 10\ R_e$), and on the nightside the radiation has been observed out to $R \sim 32\ R_e$. This radiation almost always exists and has a stable broadband intensity of 5 to 10 μV/m.

The radiation is trapped in the plasmatrough between the plasmapause and magnetopause, because outside this region (that is inside the plasmasphere and in the solar wind) the electron density and therefore the electron plasma frequency is higher than inside; thus the waves generated in the trough and in the tail cannot escape from these regions.

4.2.1.6.   Half-Integral Gyroresonance Emissions

Electrostatic emissions near $(n + 1/2)f_{ce}$, also known as electrostatic electron cyclotron harmonic waves, have been observed in several different regions of the magnetosphere. Figure 13.4.4 shows the CRRES satellite observations of these emissions (Paranicas et al., 1992; Anderson, 1994). These emissions were first detected by Kennel et al. (1970) on the OGO 5 satellite, most frequently at $3/2 f_{ce}$ where their intensity is the highest. Subsequent studies (Fredricks and Scarf, 1973; Scarf et al., 1973b; Coroniti et al., 1971) using OGO 5 data showed: (1) these emissions exist within 40 to 50° latitude on either side of the magnetic equator between L-values of 5 to 10 and between 1900 to 0700 local time; (2) they were most intense near the equator at midnight and their field strength varied over 1 to 10 mV/m; (3) the intensity of these emissions increases by several orders

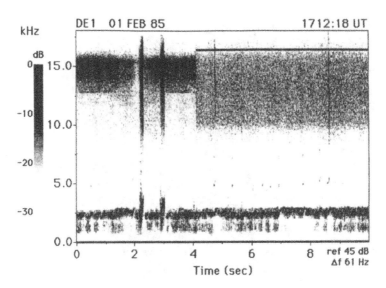

Figure 13.4.8 Examples of wideband impulsive emissions observed on the DE 1 satellite close to the magnetic equator ($L \sim 4.6$, $\lambda_m \sim 12°$). Association of electromagnetic impulsive emissions on the electric (first 4 s) and magnetic (last 6 s) antennas with individual chorus elements (~ 2.5 kHz) is clearly seen. The emission frequency range extends from well below to well above the local gyrofrequency ($f_{he} \sim 10.5$ kHz).

of magnitude during magnetic substorms and can reach up to 100 mV/m; and (4) they consist of several discrete emissions in a narrow frequency band $\Delta f \sim 200$ Hz.

Electrostatic waves near half-integral gyroresonances have also been detected by IMP 7 and 8 in the geomagnetic tail and in the region of transition from the plasma sheet to the neutral sheet (Scarf et al., 1974b; Gurnett et al., 1976).

Several theories have been advanced to explain the generation of these emissions (Ashour-Abdalla and Kennel, 1978; Rönmark et al., 1978; Kurth et al., 1980). In these studies it was assumed that the electron distribution function consists of a cold plasma portion with Maxwellian distribution and a hot plasma portion with a specified density and a temperature, and $\partial f / \partial v_\perp > 0$ between 0 and some maximum value. A general result of these linear theories which have made use of general dispersion relation for electrostatic waves (Harris, 1959) is that an instability is possible for frequencies in the harmonic bands between nf_{ce} and $(n + 1)f_{ce}$, in agreement with wave observations. The highest unstable frequency is expected to be near or somewhat above the local upper hybrid frequency.

Kennel et al. (1970) suggested that these waves may be an important cause of pitch angle diffusion for 0.1 to 10 keV electrons on auroral field lines, and Scarf et al. (1973) presented evidence that the waves can cause significant pitch angle diffusion of electrons up to 80 keV in energy during geomagnetic disturbances.

4.2.1.7. Wideband Electrostatic/Electromagnetic Impulsive Emissions

Plasma wave data from the ISEE 1 2 spacecrafts revealed that whistler-mode chorus emissions ($f < 1$ kHz) in the dayside magnetosphere between 9 to 12 R_e and $\pm30°$ magnetic latitude were often accompanied by bursts of electrostatic waves in the 3 to 10 kHz frequency band with an upper frequency cut off slightly below the local plasma frequency f_{pe} (Reinleitner et al., 1982, 1983). Typical broadband electric fields of the bursts were of the order of 50 μV/m and the electric and magnetic fields of the associated chorus emissions were about 300 μV/m and 40 pT.

Chorus-related bursts have been also observed on the DE 1 satellite near and outside the plasmapause in $\pm40°$ magnetic latitude range at all local times (Ondoh et al., 1989; Sonwalkar et al., 1990). Unlike ISEE 1 observations, impulsive emissions observed on the DE 1 satellite were electromagnetic, with a measurable magnetic field component. Figure 13.4.8 shows a spectrogram of impulsive electromagnetic burst and the associated chorus elements observed on the DE 1 satellite. The burst frequency extended from well below to well above the local gyrofrequency and beyond the upper cut off of the DE 1 receiver (16 kHz). The electric field in the 1-kHz band centered at 12 kHz was 1 to 10 μV/m, and $cB/E \sim 5$. The bursts were detected in association with chorus or hiss (Figure 13.4.6, middle panel) or at times without any associations.

The resistive medium instability (Briggs, 1964) mechanism suggested by Reinleitner et al. (1983) explains the electrostatic burst observed on the ISEE 1 satellite, but fails to account for the observed magnetic field component on the DE 1 satellite. An alternate mechanism that involves spacecraft discharge triggered by a beam of electrons has been offered to explain the generation of wideband electromagnetic bursts (Sonwalkar et al., 1990).

4.2.1.8. Ion Cyclotron Waves

Electromagnetic ion cyclotron waves were detected on the Explorer 45 and Hawkeye satellites close to the plasmapause in the vicinity of the geomagnetic equator (Taylor et al., 1975; Taylor and Lyons, 1976; Kintner and Gurnett, 1977). Most occurrences of ion cyclotron waves were associated with the presence of a ring current of ions, formed in the magnetosphere during a magnetic substorm. Their frequencies generally lie near the ion cyclotron frequencies and are left-handed polarized. These early observations detected waves near the proton cyclotron frequency, which is ~ 10 Hz near the plasmapause. Subsequently, waves were detected that were associated with cyclotron frequencies of heavier He^+ and O^+ ions around ≤ 1 Hz (Mauk et al., 1981; Roux et al., 1982; Quinn et al., 1986). Figure 13.4.9 shows an example of ion cyclotron waves detected on the DE 1 satellite inside the plasmasphere.

Ring current ions are believed to be the free energy source for the generation of electromagnetic ion cyclotron waves. During magnetic storms, ions are injected into the inner regions of the magnetosphere producing an east-west ring current that typically decays over a period of several days (Joselyn and Lyons, 1976). Cornwall et al. (1970) first suggested that the ring current ion distribution is unstable to the growth of ion cyclotron waves in the high density region inside the plasmasphere and that these waves scatter ions into the atmosphere causing low latitude auroral displays known as stable auroral red (SAR). This mechanism explains the generation of waves as well as the decay of the ring current. Several studies have shown that ion cyclotron waves heat ions and play an important role in the energy transfer between various ion species in the ring current (Gendrin and Roux, 1980; Mauk et al., 1981; Roux et al., 1982; Quinn et al., 1986).

4.2.1.9. Electromagnetic Ion Bernstein Waves

Using data from the OGO 3 satellite, Russell et al. (1970) identified an intense band of electromagnetic noise near the magnetic equator at frequencies from 10 to 100 Hz. It was later discovered that this noise consisted of many narrowband emissions near harmonics of the H^+, He^+, and O^+ ion cyclotron frequencies (Gurnett, 1976). These emissions are most intense in the frequency range around a few hundred hertz, between f_{ci} and f_{LHR}. The existence of ion cyclotron harmonic effects provided strong evidence that these waves belong to electromagnetic ion Bernstein modes (Section 3.1.2.3). These waves, which occur a few times a year, are found during magnetically quiet as well as disturbed periods, lasting for a few days to a week and then gradually disappear.

Olsen (1981) showed that equatorial ion Bernstein modes are closely related to a highly anisotropic low energy (20 to 50 eV) ion population that is present near the equatorial plane. Using wave and particle data from the DE 1 satellite, Olsen et al. (1987) have demonstrated a close correspondence between plasma heating and the occurrence of ion Bernstein waves. Specific mechanisms by which ion Bernstein mode waves can heat the ions are discussed by Curtis (1985).

4.2.2. Waves Generated in Auroral and Polar Regions

4.2.2.1. Auroral Kilometric Radiation (AKR)

Auroral kilometric radiation (AKR) is an intense radio emission generated along the auroral field lines in association with discrete auroral arcs with highest intensities between 100 and 400 kHz,

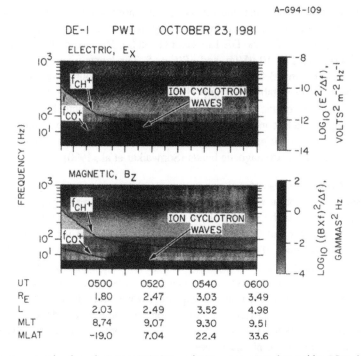

Figure 13.4.9 An example of an electromagnetic ion cyclotron wave event detected by DE 1 deep inside the plasmasphere. Ion cyclotron waves in this case occur from 0505 to 0520 UT as the spacecraft crossed magnetic equator at about 2.2 R_e, deep inside the plasmasphere. The emission frequency is about 3 to 6 Hz, slightly above the O^+ cyclotron frequency. This event is associated with strong heating of O^+ ions in the vicinity of the magnetic equator. (From Gurnett, D. A. and Inan, U. S. *Rev. Geophys.*, 26, 285, 1988. Copyright by the American Geophysical Union. With permission.)

but also observed at frequencies as low as 20 kHz and as high as 2 MHz (Dunckel et al., 1970; Brown, 1973; Gurnett 1974; Kurth et al., 1975). AKR is primarily an $R - X$ mode radiation that escapes from the Earth into interplanetary space. Gurnett (1974) found the total power to be as high as 10^9 W, concluding that the Earth is an intense planetary radio source with AKR intensities comparable to decametric radiation from Jupiter. Figure 13.4.5 shows an example of AKR observed on a typical auroral pass of the DE 1 satellite.

Based on the observed lower cut off of the AKR radiation, Gurnett and Green (1978) concluded that AKR was generated in the $R - X$ mode. Mellott et al. (1984) showed, that although the most intense radiation is usually right-hand polarized, a weak left-hand polarized ($L - O$ mode) is also present with intensities less than $R - X$ mode radiation by a factor of ~50. Benson et al. (1988c) suggested that $L - O$ mode radiation may be the result of a weak (2%) conversion of $R - X$ mode radiation to $L - O$ mode radiation in the source region, possibly due to scattering by density irregularities or some other relatively inefficient mode conversion process. Wave data from the ISIS 1 and DE 1 satellites indicated the presence of a harmonic structure, particulary the existence of a second harmonic in the AKR radiation (Benson, 1985; Mellott et al., 1986).

Direct measurements of AKR within the auroral source region were obtained from the low altitude (apogee ~3500 km) ISIS 1 satellite (Benson and Calvert, 1979; Benson et al., 1980; Calvert, 1981a). Several new features of AKR were discovered: (1) AKR is generated within regions of local depletion of the electron density, where the ratio $f_{pe}/f_{ce} < 0.2$; (2) radiation is generated just above the electron gyrofrequency at a wave normal angle perpendicular to the geomagnetic field; (3) AKR is generated in association with keV electron precipitation of the inverted-V form associated with discrete aurora; and (4) the source region is located along the auroral field lines at a geocentric altitude of 1.5 to 3.0 R_e. The association with discrete aurora

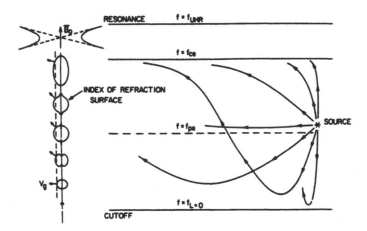

Figure 13.4.10 Geometric constructions used to analyze the ray path of *Z*-mode radiation. The *Z*-mode radiation can propagate large distances horizontally and has a tendency to be refracted asymptotically to the level slightly below $f = f_{ce}$. (From Gurnett, D. A., Shawhan, S. D., and Shaw, R. R., *J. Geophys. Res.,* 88, 329, 1983. Copyright by the American Geophysical Union. With permission.)

has recently been confirmed by Huff et al. (1988) who successfully traced the direction-finding results on AKR back to discrete auroral arcs observed by the DE 1 auroral imaging instrument (Frank et al., 1981).

Wu and Lee (1979) proposed a cyclotron laser mechanism to explain the generation of AKR. They demonstrated that a direct amplification of the $R - X$ and $L - O$ mode radiation is possible when a population of energetic electrons (1 to 10 keV) possesses a loss-cone distribution. A series of papers by Benson and co-workers (Benson and Wong, 1987; Benson et al., 1988a–c) provides a number of observations from low altitude satellites and from the ground which support a number of variations of AKR, which are predicted by this mechanism including the detection of cyclotron laser-generated whistler-mode emissions at ground stations in the 100 kHz to 1 MHz frequency range.

Another new observation critical to the understanding of AKR radiation was provided by interferometry observations made by ISEE 1 and 2, suggesting that the minimum bandwidth of individual AKR emissions could be 5 Hz or less. Such an observation implies a source thickness of less than one wavelength and supports the notion of radio lasing in emission process (Baumback and Calvert, 1987). AKR continues to be investigated intensively, both observationally and theoretically (Kurth, 1991; Roux et al., 1993).

4.2.2.2. Z-Mode Radiation

DE 1 observations of high altitude auroral phenomena in radial distances between 2 and 5 R_e have revealed that broadband Z mode radiation is observed on essentially every pass over the auroral zone and polar cap (Gurnett et al., 1983). As shown in Figure 13.4.5, the Z-mode radiation has a sharp upper cutoff near the electron gyrofrequency and extends downward in frequency to the $L = 0$ cutoff frequency (Section 3.1.2.1). The frequency ranges of whistler mode auroral hiss and Z-mode radiation often overlap, making it difficult to distinguish between these two radiations. In many cases, however, auroral hiss has a sharp upper cutoff at f_{pe}, and in such cases these two can be easily distinguished (Figure 13.4.5).

Z-mode radiation is believed to be generated in the auroral region (Gurnett et al., 1983). Simple ray path considerations illustrated in Figure 13.4.10 account for most of the observed characteristics of the Z-mode radiation. The ability of the radiation to propagate horizontally with relatively little refraction, except near regions where the cut-off frequencies f_{ce} and $f_{L=0}$ are close to the radiation frequency, explains why the radiation can be detected over a large region of the polar cap far away from the source in the auroral regions. The intensities of both Z-mode radiation and auroral hiss are within 10 to 20 dB of each other and are relatively steady, with little change on consecutive passes over the auroral region, suggesting that Z mode radiation is generated in the same general region as the auroral hiss.

Because Z-mode radiation becomes quasi-electrostatic and has very low phase velocities for frequencies between f_{ce} and f_{UHR}, most theories have used incoherent or coherent Cerenkov instabilities as the generation mechanism for Z-mode radiation (Taylor and Shawhan, 1974; Jones, 1976; Kaufman et al., 1978; Maggs and Lotko, 1981). In general, waves generated at frequencies above f_{ce} cannot reach regions below cfe_{ce} where Z-mode radiation is observed. However, if electron density gradients perpendicular to the magnetic field exist, as is expected in the field-aligned structures associated with the auroral zone, the Z mode can propagate smoothly across the level where $f = f_{ce}$ with essentially no attenuation (Gurnett et al., 1983). Direction-finding studies by Hashimoto et al. (1990) indicate that Z-mode radiation is generated in the auroral zone and may originate from auroral hiss via a coupling window that occurs near the electron gyrofrequency. On the other hand, Hui and Seyler (1993) have recently proposed that Z-mode radiation as well as electrostatic whistler-mode radiation may be excited by AKR via a parametric process.

4.2.2.3. Auroral Hiss (VLF Hiss)

Auroral hiss is a whistler-mode radiation found in auroral regions ($\lambda_m \sim 60$ to $85°$) in a few hundred hertz to ~100 kHz frequency range at all local times, but more often and stronger between 1200 to 2400 local time (Gurnett et al., 1983; Hayakawa and Sazhin, 1992). Figure 13.4.5 shows typical auroral hiss spectra observed on the DE 1 satellite as it passed through auroral regions.

Consistent with ground observations, low altitude satellite observations revealed that the auroral hiss spectrum falls into two categories, continuous (or structureless) hiss and impulsive hiss. Auroral hiss often shows a sharp cutoff at the lower hybrid resonance (LHR) frequency. Maximum power spectral densities recorded by Injun 5 and Alouette 2 satellites were 1.5 to 2.0 $\times 10^{-11}$ W/m^2 (Gurnett and Frank, 1972a; Barrington et al., 1971). The auroral hiss emissions recorded by satellites have a characteristic V-shaped lower boundary on their spectra (Smith, 1969; Gurnett et al., 1969). The duration of V-shaped hiss events was ~100 s. Low altitude satellite measurements have shown that auroral hiss emissions are correlated with low energy (100 eV to 10 keV) inverted-V electron precipitation (Hoffman and Laaspere, 1972). Using Injun 5 data, Gurnett and Frank (1972b) found that the auroral hiss power flux abruptly decreased below the receiver noise level when the electron flux decreased to $10^4 = 10^5$ el/cm²-sec-ster-ev.

The most detailed information on the source region and propagation of auroral hiss has come from polar orbiting DE 1 satellite which provided measurements over 1.1 to 4.65 R_e (Gurnett et al., 1983). Typical broadband electric fields of auroral hiss measured on DE 1 were about 30 to 300 µV/m. DE 1 observations showed that the auroral hiss spectrum had a characteristic "funnel shape" as shown in Figure 13.4.5 with an upper frequency cut off at f_{He} or f_{pe}, whichever was lower. As shown in Figure 13.4.11, the funnel-shaped spectrum can be explained in terms of propagation of auroral hiss at wave normal angles close to the resonance cone. In fact, using ray tracing, Gurnett et al. (1983) were able to place a lower limit of 1.7 to 1.9 R_e on the generation region of auroral hiss.

Because auroral hiss propagates at large wave normal angles, it is believed that the whistler-mode radiation is produced by a plasma instability associated with the Landau resonance which provides the largest growth rates for whistler-mode radiation propagating at large wave normal angles (Maggs, 1976; Lotko and Maggs, 1981). Maggs' (1976) analysis based on different beam models for energetic electrons showed that continuous hiss will be generated at higher altitudes than impulsive hiss, consistent with observations. Maggs also predicted a limiting value of electron flux 10^5 el/cm²-sec-ster-ev below which auroral hiss would not be observed. According to the Landau mechanism, auroral hiss observed on the DE 1 satellite above the source region must be

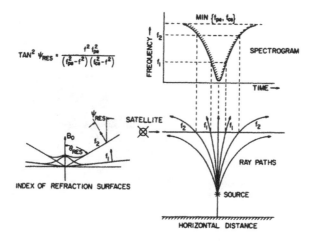

$$TAN^2 \ \psi_{RES} = \frac{f^2 \ f_{pe}^2}{(f_{pe}^2 - f^2)(f_{ce}^2 - f^2)}$$

Figure 13.4.11   Geometric constructions used to analyze the ray path of auroral hiss. For wave normal angles near the resonance cone, whistler mode emissions from a point source produce a funnel-shaped spectrum qualitatively similar to the funnel-shaped auroral hiss events observed by DE 1. (From Gurnett, D. A., Shawhan, S. D., and Shaw, R. R., *J. Geophys. Res.*, 88, 329, 1983. Copyright by the American Geophysical Union. With permission.)

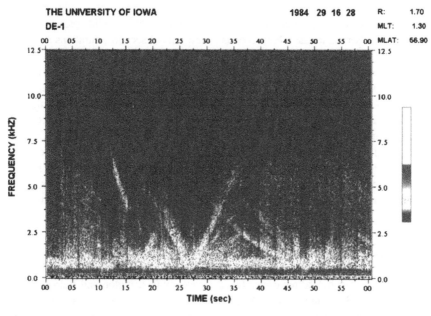

Figure 13.4.12   An example of saucer-shaped VLF emission observed on the DE 1 satellite. (Courtesy of W. S. Kurth.)

accompanied by upward moving flux of low energy electrons. A search of such upward-directed moving electrons was carried out by Lin et al. (1984) using DE 1 data, and several events were found in which upward moving electron beams were associated with funnel shaped auroral hiss emissions.

4.2.2.4.   VLF Saucers

VLF saucers are whistler-mode emissions observed in the auroral regions in the frequency range ~100 Hz to 20 kHz and in the altitude range from 500 to 4000 km (Smith, 1969; Mosier and Gurnett, 1969; James, 1976). As shown in Figure 13.4.12, the lower boundary of their spectra

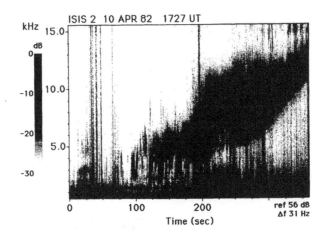

Figure 13.4.13 Spectrogram showing an LHR noise band observed on the ISIS 2 satellite (1400 km, circular orbit). The lower cutoff is at the local f_{LHR}. The sudden jump in the lower cutoff near 200 s mark indicates the plasmapause crossing.

have a characteristic saucer or V shape. Using data from ISIS 1 and 2 satellites, Yoshino et al. (1981) observed saucer emissions in a latitude range 55° to the geomagnetic pole on the nightside and 65° to the geomagnetic pole on the dayside. On a low altitude satellite pass through auroral regions they are observed to last a few seconds to few tens of seconds, durations much less than typical durations of V-shaped auroral hiss.

Whistler-mode propagation at large wave normal angles from a source region to spacecraft can explain the saucer shape of the lower boundary of the spectrum (the same mechanism that explains funnel- or V-shaped auroral hiss discussed above). James (1976), using data from ISIS 2 and Alouette 2, determined the source location of saucers at 1000 km (compared to 5000 to 10,000 km for auroral hiss generation region). A typical source region had 0.5 km extent in the horizontal and 10 km in the vertical direction. The total power radiated from the source was estimated to be ~10 mW. It is believed that VLF saucers are generated by beams of suprathermal electrons with energies not exceeding 5 eV via a coherent Cerenkov process (James, 1976). This mechanism is consistent with quasi-electrostatic (large wave normal angles) propagation of saucer emissions.

4.2.2.5. Lower Hybrid Waves

Satellite observations have often shown quasi-electrostatic plasma wavebands with a sharp cut off near the local lower hybrid resonance (f_{LHR}) frequency (Barrington and Belrose, 1963; Barrington et al., 1963; Brice and Smith, 1965; Laaspere and Taylor, 1970; Laaspere and Johnson, 1973). These bands are observed at mid- to high latitudes in the altitude range of ~1000 to a few thousand kilometers. These waves are often excited by whistlers, or they could simply be hiss emissions with a cutoff at f_{LHR}. These emissions are recorded mainly on electric antennas reflecting their electrostatic character. Figure 13.4.13 is a spectrogram showing an LHR noise band observed on the ISIS 2 satellite. The cold plasma theory predicts a resonance at the lower hybrid frequency (Section 3.1.1) for whistler-mode waves propagating perpendicular to the static magnetic field. At frequencies higher than the local LHR frequency, the resonance takes place at wave normals smaller than 90°. This explains the sharp lower cutoff at f_{LHR} as well as the quasi-electrostatic nature of the band (Brice and Smith, 1965).

Several mechanisms have been proposed to explain the generation of lower hybrid waves. In one mechanism, lower hybrid waves are assumed to be excited by energetic electrons beams. It is known that energetic electrons, injected into the supra-auroral region from the plasma sheet, are accelerated by field-aligned DC electric fields toward the ionosphere. The energies of these electron beams can reach the keV range. A number of electrostatic modes, including those with frequencies near the lower hybrid frequency, can be excited by wave-particle resonances involving the positive slope region of the electron distribution functions (Maggs, 1976; Maggs and Lotko, 1981). Mechanisms involving parametric instabilities have also been proposed (see Lee and Kuo, 1984 and references therein). In these mechanisms, an electromagnetic whistler-mode wave decays into an ion acoustic type wave and an electrostatic lower hybrid wave with wave normal on the resonance cone. Recently Bell and Hgo (1990) have proposed a passive liner mode conversion mechanism in which lower hybrid waves are excited by a source whistler mode waves by scattering from 10 to 100 m scale plasma density irregularities (Gross and Muldrew, 1984). It is possible that for a given situation one or more of these mechanisms may operate to generate LHR waves.

4.2.2.6. Auroral Field-Line Turbulence

Intense quasistatic electric field turbulence is commonly detected along auroral field lines in the altitude range of ~450 to 23,000 km (Heppner, 1969; Gurnett and Inan, 1988; Temerin and Kintner, 1989). The wave power spectra associated with this noise are generally characterized by a power law such that the power as a function of frequency is given by $P(f) \sim f^m$ where m, called spectral index, is some negative number of the order of -1 to -4 or so. The similarity of this kind of power dependence to Navier-Stokes turbulence for fluids with $m = -5/3$ (Kolmogorov, 1941) explains why this noise is called turbulence. Broadband low frequency electric noise in a frequency range from a few hertz to several hundred hertz commonly exists outside the plasmapause, throughout the auroral zone and polar cap, with the peak amplitudes as high as ~100 mV/m within the auroral zone. The noise is most intense around ~1 Hz and decreases monotonically with increasing frequency, with measurable intensities up to frequencies of several hundred hertz. Electron density fluctuations and the field-aligned currents are also found to coexist with the electric field noise. In the regions of largest amplitudes a magnetic field component is also detected with a Poynting vector suggestive of Alfven waves.

Figure 13.4.14 shows an examples of broadband electric and magnetic field noise observed over the evening and morning auroral zones during a low altitude DE 1 pass over the southern polar region (Gurnett et al., 1984). The noise is most intense below the O^+ ion cyclotron frequency. The magnetic spectrum is usually steeper than the electric field spectrum. At low altitudes below ~2 R_e, both the electric and magnetic field fluctuations are perpendicular to the static magnetic field and the Poynting flux is always directed downward. Integrated over the entire auroral zone, the electromagnetic energy flow associated with these fluctuations is substantial, of the order of ~10^8 W.

Temerin (1978) showed that much of the low-frequency electric field spectrum is due to spatial structure and consists of short wavelength quasistatic turbulence that is Doppler shifted upward in frequency by the spacecraft motion relative to the plasma. While several theories have been presented to explain the origins of this noise, none explains all the observed features of the turbulence (Temerin and Kintner, 1989). In some cases the spectral index is of the order of $-5/3$ implying a Kolmogorov process associated with velocity shears (Kelvin-Helmholtz instability). In yet another case the electric field fluctuations are associated with magnetic field fluctuations implying the existence of kinetic Alfven waves (Gurnett et al., 1984). In some other cases the spectral index is much smaller, of the order of -0.5, which does not agree with any of the local theories.

4.2.3. Waves Generated in the Distant Magnetosphere: Tail Region, Magnetosheath, Plasma Sheet

4.2.3.1. Broadband Electrostatic Noise

Broadband electrostatic noise (BEN), first discovered by Scarf et al. (1974) and named by Gurnett et al. (1976), is characterized by its wide range in frequency, from 10 Hz to several kilohertz, very impulsive with large peak-to-average field ratios, and high field intensity of the order of

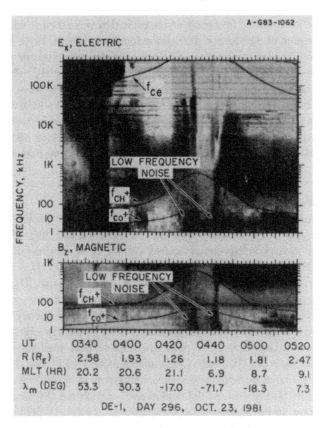

Figure 13.4.14 Typical DE 1 electric and magnetic field spectrograms showing the low-frequency electric and magnetic field noise observed at low altitudes over the auroral zones. Note that the low-frequency noise occurs over both the evening and morning auroral regions, but is almost completely absent over the polar cap. (From Gurnett, D. A., Huff, R. L., Menietti, J. D., Burch, J. L., Winningham, J. D., and Shawhan, S. D., *J. Geophys. Res.*, 89, 8971, 1984. Copyright by the American Geophysical Union. With permission.)

1 mV/m (Gurnett et al., 1976; Gurnett and Frank, 1977; Parks et al., 1984; Cattell and Mozer, 1986). On a spectrogram, this noise can easily be distinguished from other plasma waves by sharp protuberance out of the lower intensity surrounding noise, such as auroral hiss. This noise is confined to discrete auroral field lines or to the edges of the plasma sheet in the magnetotail (plasma sheet boundary layer), regions where substantial field-aligned currents are frequently observed. Figures 13.4.15 and 13.4.16 show, respectively, a typical spectrogram and spectral density of BEN observed on the IMP 8 satellite. BEN occurs during both quiet and active magnetic times and appears to be a permanent feature of the boundary layer. The wavepower is highest at low frequencies near the lower hybrid frequency f_{LHR}, with a peak intensity at about 10 to 50 Hz and fall off in intensity as the frequency increases. The upper cutoff is usually near or above the local electron plasma frequency f_{pe}, although the high frequency waves are confined to a narrow region adjacent to the edge of the plasma sheet boundary layer. BEN is almost always accompanied by a cospatial yet distinct emission with a conspicuous magnetic signature, referred to as whistler-mode magnetic noise bursts or lion's roar.

Grabbe and Eastman (1984), using simultaneous observations of BEN and ion beams from the ISEE 1 satellite, proposed that ion beams might serve as a free energy source for various instabilities. Using linear theory, they showed that ion beams streaming through hot plasma sheet background electrons and ions can excite waves over a wide range of frequencies. With the inclusion of cold electrons the upper cutoff extends to f_{pe}, and waves remain strongly unstable over a broad frequency range (Grabbe, 1987; Schriver and Ashour-Abdalla, 1987). In recent years it has been proposed that BEN served to heat the ionospheric ions as a source of the hot central

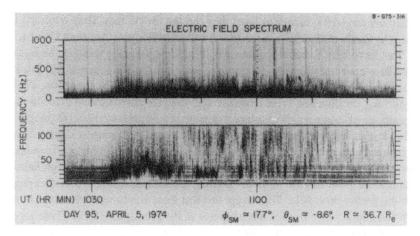

Figure 13.4.15 Frequency-time spectrogram of the broadband electrostatic noise (BEN). Note the low-frequency cutoff at about 10 Hz and the marked decrease in the intensity above about 400 Hz. The electron gyrofrequency during this period varies from about 400 to 500 Hz. (From Gurnett, D. A., Frank, L. A., and Lepping, R. P., *J. Geophys. Res.*, 81, 6059, 1976. Copyright by the American Geophysical Union. With permission.)

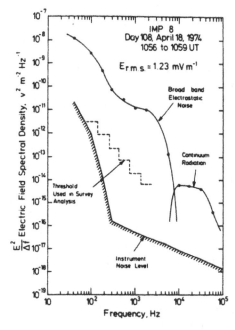

Figure 13.4.16 A typical spectrum of the broadband electrostatic noise during a period of relatively high intensity, from 1056 to 1059 UT on day 108. The electron gyrofrequency during this period is about 300 Hz. (From Gurnett, D. A., Frank, L. A., and Lepping, R. P., *J. Geophys. Res.*, 81, 6059, 1976. Copyright by the American Geophysical Union. With permission.)

plasma sheet and the boundary layer and causes isotropic auroral ion precipitation on the nightside (Dusenbery and Lyons, 1989; Schriver and Ashour-Abdalla, 1990). Recent work on BEN has been reviewed by Kurth (1991).

4.2.3.2. Lion's Roar

Lion's roar (that is how it is aurally perceived) is an intense sporadic whistler-mode wave burst observed in the Earth's magnetosheath. They are found in the 100 to 200 Hz frequency range and typically last for ~ 10 s, though at times they can last for ≥ 5 min. They were first observed (and named) by Smith et al. (1969) in the search coil magnetometer data from OGO 1, 3, and 5 satellites. They have been found at a distance ~ 10 to 15 R_e from the Earth in the Earth's magnetosheath (Rodriguez and Gurnett, 1971), in the polar cusp region (Gurnett and Frank, 1972), and in the distant magnetotail near the neutral sheet (Gurnett et al., 1976; Anderson, 1984; Cattell et al., 1986). Long duration lion's roar emissions (≥ 5 min) are often detected immediately behind or in the nearby downstream region of subsolar quasiperpendicular bow shocks (Rodriguez, 1985). Observed more frequently during disturbed periods, they are the strongest signals found in the Earth's magnetosheath with intensities of the order of 10 to 100 pT (Smith and Tsurutani, 1976).

Gurnett and Frank (1972), using Injun 5 data, detected lion's roar in the high latitude magnetosheath near the polar cusp at 70 to 80° geomagnetic latitude and at ≤ 3000 km propagating parallel to the geomagnetic field. They suggested that these waves were generated in the magnetosheath and trapped there in open tubes of force of the geomagnetic field. These waves propagated along $\mathbf{B}_0$, being guided by field-aligned irregularities, and reach polar zone at ~ 3000-km altitude. Simultaneous measurement of electric and magnetic fields obtained on Injun 5 were $E \sim 10$ to 30 mV/m and $H \sim 10$ to 30 pT, consistent with the whistler-mode propagation. Tsurutani et al. (1982) used plasma, wave, and magnetic field data from the ISEE 1 satellite to conclude that lion's roar near the magnetopause is generated by cyclotron instability of anisotropic thermal electrons when the local plasma critical energy falls to values close to or below the electron thermal energy as a result of the decrease in the ambient geomagnetic field magnitude. Anderson and Kurth (1989) have reviewed lion's roar phenomenon.

4.2.4. Plasma Waves Associated with the Bow Shock

Plasma wave detectors aboard the OGO 5 and Imp 6 spacecraft detected three different kinds of wave phenomena near the Earth's bow shock: electron plasma oscillations, electromagnetic whistler-mode waves, and electrostatic turbulence (Fredricks et al., 1968; Rodriguez and Gurnett, 1975, 1976; Onsager et al., 1990).

Intense narrowband plasma oscillations upstream of the bow shock occur in the 10 to 30 kHz frequency range with electric field strengths varying between 0.1 to 10 mV/m and with the electric field vector oriented parallel to the local magnetic field direction. This electric field direction is consistent with the direction expected if the plasma oscillations are produced by a two-stream instability from upstreaming solar wind electrons (~ 1 keV) reflected from the bow shock. Broadband electrostatic noise typically extending from the local plasma frequency (20 to 30 kHz) down to ~ 1 kHz is also frequently observed in the upstream of the bow shock. The electric field vector of these waves is also oriented parallel to the local magnetic field. Because these broadband waves are closely associated with narrowband plasma oscillations, and because they have the same upper cut off and polarization, it is believed that these waves are simply shorter wavelength electron plasma oscillations that are strongly Doppler shifted downward in frequency from the local plasma frequency (Rodriguez and Gurnett, 1975).

In the shock transition region, both electric and magnetic fields were observed. Figure 13.4.17 shows electric and magnetic field spectral densities of the waves observed in the transition region. The electric field spectrum showed two distinct components: one component with a broad peak centered between 200 and 800 Hz, and the other component (≤ 200 Hz $\sim f_{he}$) increasing monotonically with decreasing frequency. The magnetic field spectrum also showed a distinct steepening of the spectrum, indicative of an upper cut-off frequency, at about 100 to 200 Hz ($\sim f_{he}$).

The low frequency (< 200 Hz) component of the electric field and the magnetic field have been interpreted as whistler-mode waves. The typical intensity of these whistler-mode waves is 3.0×10^{-9} V²m^{-2}Hz^{-1}. The electric to magnetic field energy density ratio $\varepsilon_E/\varepsilon_B \simeq 10^{-3}$ to 10^{-4} of the monotonic component is consistent with the electric to magnetic field ratio expected for whistler-mode waves. Electrostatic waves associated with the bow shock transition region extend from 200 Hz to 4 kHz with a broad peak in the electric field spectrum between 200 and 800 Hz.

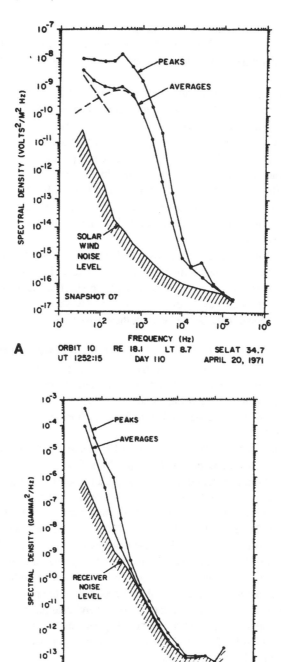

Figure 13.4.17 (A) Shock electric field spectrum $E^2(\omega)_{sh}$ selected on the basis of the rms electric field amplitude. The peak spectrum obtained during the averaging time of the average measurements is also shown. The two distinct components in the shock spectrum are shown by dashed lines. (B) The shock magnetic field spectrum $B^2(\omega)_{sh}$ for the shock crossing of 1252 UT. The characteristic frequency dependence f^{-4} is present in both the peak and the average spectrum. The spectrum steepens at about 200 Hz, below the local electron gyrofrequency at about 350 Hz. (From Rodriguez, P. and Gurnett, D. A., *J. Geophys. Res.,* 80, 19, 1975. Copyright by the American Geophysical Union. With permission.)

These are almost purely electrostatic waves with $\varepsilon_E/\varepsilon_B \simeq 10^2$ to 10^4. The average spectral density at the peak is about 10^{-9} V^2m^{-2} Hz^{-1}. IMP 6 spin modulation measurements show that the electric field directions of the electrostatic waves tend to be oriented parallel to the static magnetic field.

The rms field strengths of electrostatic turbulence in the Earth's bow shock are found to correlate with specific solar wind parameters measured upstream the bow shock. The source mechanism for the generation of electrostatic waves is believed to be ion two-stream instability producing ion acoustic waves associated with proton heating (Greenstadt and Fredricks, 1974). Low frequency (20 to 200 Hz) electric field turbulence corresponding to the whistler mode also shows a strong correlation with solar wind parameters in a manner similar to that shown by the electrostatic waves. This indicates that the entire electric field turbulence in the bow shock is primarily composed of electrostatic turbulence, and the lower frequency part below electron gyrofrequency is transformed into electromagnetic whistler mode in the region of high temperature gradients (Rodriguez and Gurnett, 1976).

In the magnetosheath downstream of the bow shock electrostatic waves with similar properties were observed on IMP 6. However, their intensities were two to four orders of magnitude less than those found in the shock region. A similarity of the spectrum and the orientation of the electric field direction of electrostatic turbulence in the bow shock and in the magnetosheath suggest that same mechanism may be operative in these two regions.

4.3. ACTIVE EXPERIMENTS: ARTIFICIALLY STIMULATED EMISSIONS (ASE)

Active experiments involving injection of man-made waves, particle beams, and chemical releases in the magnetosphere have come to play an important role in studies of various processes in the Earth's magnetosphere (Hultqvist, and Falthammar, 1990). We describe here the results from VLF wave injection experiments conducted from Siple Station, Antarctica (Helliwell and Katsufrakis, 1978; Helliwell, 1988, 1994).

As discussed in Section 4.1, a large variety of VLF emissions observed on the ground are whistler-mode signals in a limited bandwidth, typically a few kilohertz. Their high intensity, narrow bandwidths and discrete nature indicate the presence of a previously unknown kind of wave-particle interaction that converts the kinetic energy of charged particles to coherent (narrow bandwidth) electromagnetic radiation. Helliwell (1988) has therefore called this process coherent wave instability (CWI). Energetic charged particles of the radiation belts are precipitated into the ionosphere through resonant scattering by these same waves, causing enhanced thermal ionization, X-rays, light, and heat. Finally, because whistler-mode waves generated within the magnetosphere as a result of wave-particle interactions can propagate large distances within the magnetosphere and can be observed on the ground, they reveal the properties of the plasma through which they travel and thus can be used as remote sensing tools for hot and cold plasma diagnostics. Helliwell (1974) stated that the applications of wave injection experiments included (1) study of emission generation mechanisms (CWI), (2) control of energetic particle population, (3) diagnostics of hot and cold plasma, and (4) VLF communications.

4.3.1. Description of the Experiment

Figure 13.4.18 describes the essential ingredients of the active experiments performed from Siple Station, Antarctica. The experiments cover the frequency range 1 to ~6 kHz and were carried out between Siple Station, Antarctica (transmitter location) and its conjugate area (receiver location) including Roberval and Lake Mistissini, Canada, as shown in Figure 13.4.18a. Since the ground-based experiments depend on the presence of suitable field-aligned ducts, it is necessary that transmitting and receiving sites be located near the ends of the same field line.

A 21.2-km horizontal dipole supported above the snow surface and aligned with the local magnetic east-west direction was used. The antenna-transmitter system, tunable in ~1.5 to 20 kHz provided about 2% efficiency at 6 kHz. The transmitter (called Zeus) output power ranged between 100 to 150 kW from 5 to 20 kHz. Thus, with an effective input power of 50 kW it

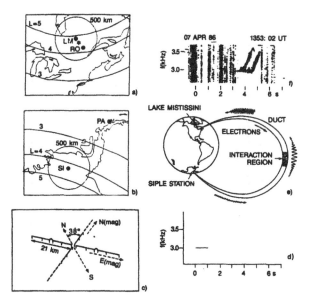

Figure 13.4.18 Elements of field-aligned whistler-mode, wave-injection experiments between Siple Station, Antarctica (SI), and Lake Mistissini, Quebec, Canada (LM). A 1-s pulse (lower right) from SI is received at LM (upper right) after propagation along two ducts in which wave growth and triggering of rising emissions occurs. A VLF receiver located at Palmer (PA) provides data on low L-shell whistler-mode paths and on subionospheric propagation from Siple Station. (From Helliwell, R. A., *Rev. Geophys.*, 26, 551, 1988. Copyright by the American Geophysical Union. With permission.)

would be possible to radiate ~1 kW upward into the magnetosphere. Operation of Zeus began at Siple Station ($L = 4.3$) in 1973. A higher power, more versatile transmitter (called Jupiter) was installed in 1978 and operated until February 1988. A further improvement in the Siple Station facility was made in 1986, with the addition of crossed dipoles (each 42-km long) for the generation of variable polarization, depicted in Figure 13.4.18e.

Because the results of the active wave-particle injection experiments depend on the frequency spectrum of the transmitted signal, various modulation formats have been used: single frequency pulses of varying duration, frequency ramps and parabolas of varying df/dt and d^2f/dt^2, frequency steps, and multiple frequencies within a certain bandwidth consisting of a set of discrete or randomly varying frequencies. In addition to frequency, the amplitude and the polarization of the transmitted signal are also varied. These various kinds of injected signals permit a detailed investigation of different aspects of CWI such as wave growth, saturation, triggering, threshold, and wave suppression.

In these experiments, while the frequency, power, polarization, and modulation are specified by the experimenter, the parameters of the medium are variable and not controllable. In particular for successful active experiments, both the presence of ducts near the L-shell and longitude of the Siple Station is required, as well presence of enough energetic particles to produce growth and triggering. As a general rule the 2 or 3 days of quieting following a significant magnetic disturbance provides good conditions for growth and triggering (Carpenter and Bao, 1983).

4.3.2. Experimental Observations from Conjugate-Stations Experiments

We describe here a select number of experiments and their interpretations followed by a summary of the main results obtained over the 15 years of the Siple Station operation. For more details see Helliwell (1988, 1994).

The most basic features of CWI can be illustrated by considering a step function single frequency pulse input to the magnetosphere. Single frequency excitation of the CWI by a three-second long amplitude-modulated pulse is illustrated in Figure 13.4.19. Initially the received pulse

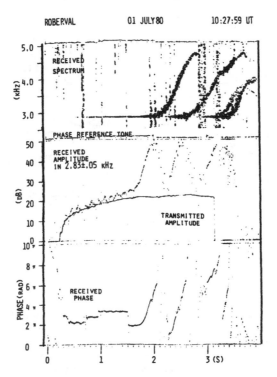

Figure 13.4.19 Threshold effect. (A) Dynamic spectrum of a 3-s pulse transmitted from Siple Station and observed at Lake Mistissini. (B) Amplitudes of transmitted pulse (solid line) and received pulse in a 100-Hz bandwidth centered on carrier at 2830 Hz. (C) Phase of received signal; phase jumps are caused by sferics (vertical lines in Figure 13.4.19A) that perturb the local phase reference signal. (From Paschal, E. W. and Helliwell, R. A., *J. Geophys. Res.,* 89, 1667, 1984. Copyright by the American Geophysical Union. With permission.)

tends to follow the increasing amplitude of the transmitted pulse, with the addition of an 8-Hz fluctuation that has been attributed to beating between the received Siple pulse and a harmonic of the induction field from the local power line. As the Siple input pulse approaches its peak value, there is an abrupt onset (at ~1.7 s) of exponential growth having a nearly constant rate of ~110 dB/s. This is the threshold for growth of the CWI, an effect not yet understood but which has been ascribed to the growth suppressing effect of ambient nonducted noise in the duct (Mielke and Helliwell, 1992). Saturation occurs after a total temporal growth of 23 dB, followed by a rising emission which carries away all of the stimulated wave energy, leaving behind the un-amplified carrier wave which begins the same cycle of growth and triggering. This process is repeated again, giving three sequences of exponential growth, saturation, and triggering. Although it is somewhat obscured by other signals, a weak termination faller is triggered at the end of the input signal. Such "end triggering", as it is called, is often used as a marker of the propagation time of its parent pulse. A key feature of growth and triggering in single-frequency experiments is the phase behavior shown in Figure 13.4.19C. (Each of the sharp jumps in the phase seen up to ~1.5 s is the result of a sferic, seen on the top panel as a vertical line, that resets the phase reference.) As the threshold is crossed at ~1.7 s, the phase begins to advance parabolically with time, corresponding to a linear increase in frequency of ~35 Hz/s. Just after saturation is reached, the phase (and frequency) rises more rapidly, and the emission is said to have begun.

Frequency ramps have been used in many experiments to distinguish between multiple paths, i.e., signals propagating within different ducts. Frequency ramps, both rising and falling, shown in Figure 13.4.20, exhibit growth and triggering on multiple paths. The transmitted format is shown in the bottom panel. Entrainment and cutoff of emissions often occurs at harmonics of the Canadian power grid, as shown in Figure 13.4.21, and has been attributed to weak whistler-mode radiation from the power grid.

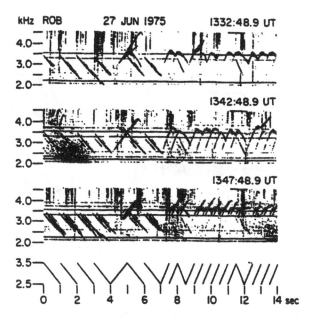

Figure 13.4.20 Rising and falling ramps (transmitter format, lowest panel) show growth and triggering on multiple paths; selective entrainment of emissions (inverted hooks) by rising ramp propagating in the same duct. (From Helliwell, R. A., *Conf. Proc., XXIV General Assembly of the Int. Union of Radio Sci. (URSI)*, Kyoto, Japan, 1994, in press. Copyright by the American Geophysical Union. With permission.)

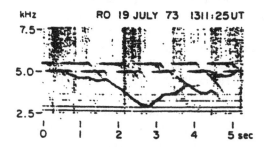

Figure 13.4.21 Triggered emissions from Siple pulses on 5000 Hz are entrained or cut off at harmonics of the Canadian power grid. (From Helliwell, R. A., *Conf. Proc., XXIV General Assembly of the Int. Union of Radio Sci. (URSI)*, Kyoto, Japan, 1994, in press. Copyright by the American Geophysical Union. With permission.)

Multifrequency experiments in which two or more carriers, closely spaced in frequency, are added together have been used to understand wave-wave interactions in the magnetosphere. In general, it has been found that the addition of any signal, CW or noise, to another CW signal will reduce or eliminate the growth of the latter if its frequency differs by no more than ~100 Hz. Suppression of CWI growth was discovered accidentally on a 30-s long pulse during a period of strong echoing (Raghuram et al., 1977a). It has been found that even natural background hiss can be suppressed by a transmitter signal (Raghuram et al., 1977b). One of the most interesting results of multifrequency experiments is that signal growth increases as the signal bandwidth is reduced (Helliwell et al., 1986b). For example, simulated random noise with a bandwidth more than ~100 Hz shows little or no growth, but if the bandwidth is decreased, growth and triggering increase as shown in Figure 13.4.22.

Many of the multifrequency experiments show evidence of subharmonics between the harmonics of the frequency spacing between the carriers. The subharmonic structure varies with the frequency spacing and relative amplitudes of two or more input carriers (Sa and Helliwell, 1988). Analysis has shown that half-harmonics and one third harmonics are common even when the amplitude ratio of the two injected carriers is 40 dB. If only one Siple carrier is transmitted a

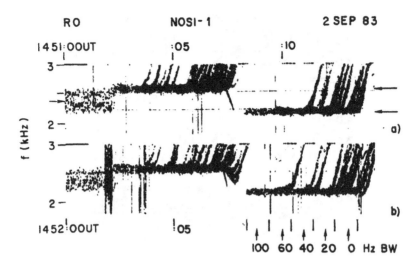

Figure 13.4.22 Two samples of a 400-Hz band of simulated hiss, each followed by two sequences of five contiguous 1-s pulses of hiss with 100-, 60-, 40-, 20-, and 0-Hz bandwidth. The average intensity increases about 10 dB as the bandwidth is reduced to zero. Arrows mark centers of hiss bands. (From Helliwell, R. A., *Rev. Geophys.,* 26, 551, 1988. Copyright by the American Geophysical Union. With permission.)

similar sideband structure is still observed, with line spacings that can be explained in terms of the carrier and the ambient harmonics of 60 Hz. In an extension of these sideband results, it has been proposed that VLF emission triggering itself can be attributed to the sidebands created by the Siple signal and background power line radiation (Sa, 1990).

The most significant results from wave-injection experiments from Siple Station over the 15-year period (1973 to 1988) can be summarized as follows: (1) exponential temporal growth, saturation, and emission triggering (Helliwell and Katsufrakis, 1974; Helliwell, 1974; Stiles and Helliwell, 1977; Paschal and Helliwell, 1984); (2) entrainment (Helliwell and Katsufrakis, 1974); (3) suppression of signals by adjacent signals (Raghuram et al., 1977a,b; Helliwell et al., 1986b); (4) threshold for the onset of temporal growth (Helliwell et al., 1980a; Mielke and Helliwell, 1992); (5) parabolic phase advance for emission triggering from single-frequency transmission (Paschal and Helliwell, 1984); (6) generation of sidebands and subharmonics (Park et al., 1981; Helliwell et al., 1986a; Sa and Helliwell, 1988); and (7) generation of chorus-like emissions from artificially generated hiss band (Helliwell et al., 1986).

4.3.3. Spacecraft Observations of Siple Transmitter Signals
Various satellites such as Explorer 45, IMP 6, EXOS B, ISIS 1 and 2, ISEE 1, and DE 1 have been used in conjunction with the Siple Station transmitter to study nonducted propagation and wave-particle interaction in the Earth's magnetosphere (Inan et al., 1977; Kimura et al., 1983; Bell et al., 1981, 1983a,b, 1985; Bell and Ngo, 1990; Bell et al., 1991a,b; Inan and Helliwell, 1982; Inan and Bell, 1985; Sonwalkar et al., 1984; Sonwalkar and Inan, 1986; Rastani et al., 1985). The primary results from these experiments can be summarized as follows:

1. Siple signals seldom trigger emissions in the nonducted mode. In the few cases when emissions triggering is observed, emissions show much larger df/dt than that observed for the ducted case (Bell et al., 1981). However, it has been observed that amplification and triggering that occurs inside of ducts as shown in Figure 13.4.19 produces signals that reflect back from the lower edge of the ionosphere into the magnetosphere in the nonducted mode where they can be readily observed by satellites (Rastani et al., 1985).

2. At lower altitudes spectral broadening of Siple signals with a bandwidth of <1 kHz was observed on the ISIS satellite (Bell et al., 1983). Figure 13.4.23 shows an example of spectral broadening of a Siple Station ramp signal. These emissions are excited only when

the transmitter frequency is above the local LHR frequency. Subsequent observations from many satellites and theoretical analysis show that the excited waves are short wavelength (5 m < λ < 100 m) electrostatic lower hybrid (LH) waves excited by electromagnetic whistler-mode waves through scattering from magnetic field-aligned irregularities located in the topside ionosphere and magnetosphere. These LH waves exhibit very large Doppler shifts which cause the observed spectral broadening (~300 to 1000 Hz) of the signal as received at the moving satellite (Inan and Bell, 1985; Bell and Ngo, 1990; Bell et al., 1991a, 1993). Figure 13.4.24 shows the region throughout which whistler-mode waves have been observed to excite LHR waves in the magnetosphere.

3. Multipath propagation is not limited to ducts. From ISEE 1 observations of Siple Station VLF pulses, evidence was found for closely spaced multiple nonducted paths of propagation between the ground transmitter and ISEE 1 (Sonwalkar et al., 1984). The amplitude fading patterns of the composite received signal were interpreted in terms of ionospheric field-aligned irregularities of 1 to 10 km horizontal scale suggesting that such measurements could provide a new method of remote sensing of such irregularities. An important consequence of multiple path propagation is that coherence length for nonducted wave-particle interaction is reduced to a few hundred kilometers (compared to a few thousand kilometers in the ducted case).

4.3.4. Theoretical Modeling of Coherent Wave Instability (CWI)

The similarities between many features of naturally occurring discrete emissions (spectral forms such as risers, fallers, hooks, etc.) and those of ASE suggest that similar mechanisms may be

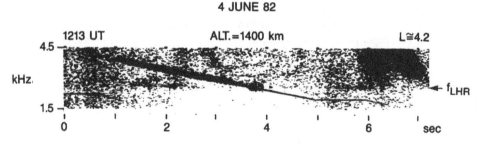

Figure 13.4.23 Low hybrid waves excited by a transmitter pulse of descending frequency. Amplitude and apparent bandwidth of the lower hybrid waves increase as $f \rightarrow f_{LHR}$ (From Bell, T. F., Helliwell, R. A., and Hudson, M. K., *J. Geophys. Res.*, 96, 11379, 1991. Copyright by the American Geophysical Union. With permission.)

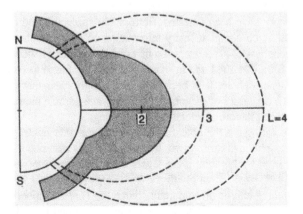

Figure 13.4.24 Regions of the magnetosphere and topside ionosphere where lower hybrid waves have been excited by VLF transmitters pulses (shaded area). (Courtesy of T. F. Bell.)

operating in both cases and thus many theories of CWI may also apply to the generation of naturally occurring discrete emissions. From the experimental results described above, it is clear that the wave-particle and wave-wave interactions involved in these phenomena are highly non-linear and take place over a large region, i.e., the interactions are distributed (Helliwell, 1967; Dungey, 1972; Gendrin, 1972; Matsumoto, 1979; Carlson, 1987; Anderson and Kurth, 1989).

Most of the theoretical treatments of CWI assume cyclotron resonance as a mechanism for wave growth, as first suggested by Brice (1963) and Dungey (1963). Helliwell (1967) proposed a phenomenological theory of discrete emissions which introduced many new features that subsequent theories adopted. In particular this theory included a nonlinear saturation mechanism and linked the frequency evolution of a triggered emission to the inhomogeneity of the medium. This theory also showed that the equator is the most favorable region for maximizing the wave growth. Subsequently, this theory was improved to include feedback and other realistic features (Helliwell and Crystal, 1973; Helliwell and Inan, 1982; Carlson et al., 1985; Carlson, 1987).

Several theories assume that electron trapping in the wave potential wave is necessary for wave growth (Nunn, 1971, 1974, 1984; Das and Kulkarni, 1975; Dowden et al., 1978; Molvig et al., 1986). These theories, however, fail to provide a saturation mechanism and require a threshold intensity (a few pT at $L = 4$) for the applied wave. While no direct measurements of the wave intensities inside a duct are available, the nonducted wave intensities measured on the DE 1 satellite at $L = 4$ are one to two orders of magnitude smaller than this value (Rastani et al., 1985; Mielke and Helliwell, 1992).

Complexity of modeling CWI has led to the use of computer simulations based essentially on Maxwell's equations and the laws of motion applied to an inhomogeneous medium, often with time-varying parameters (see Carlson, 1987 and references therein). Progress in computer simulations has been limited by the computer capacity needed to accommodate a realistic energetic electron distribution function in such a model. One of the recent simulations of CWI (using the LTS model) employs a limited $v_{\parallel}$ distribution in a fully inhomogeneous medium at $L = 4$ (Carlson et al., 1990). This paper predicts observed growth rates, phase advance and saturation levels prior to triggering, and includes many references to related work. However, the multifrequency effects such as triggering, entrainment, sideband generation, suppression (including quiet bands), and wavelet fusion are not fully explained, and therefore constitute a significant challenge to theoreticians and modelers (Helliwell, 1994).

5. CONCLUDING REMARKS

This chapter shows that a great advance has been made in the last 35 years since the launch of the first scientific satellite in measuring the characteristics of a large variety of plasma waves in all parts of the magnetosphere. Almost all kinds of wave modes and resonances supported by cold plasma wave theory have been detected. Both *in situ* spacecraft and remote ground observations have provided much complementary information on magnetospheric wave activity as a function of various geophysical parameters. To first order, free energy sources for the generation of these waves have been identified and in most cases they are found to be an energetic particle population of the magnetosphere. It is becoming increasingly clear that plasma waves play a fundamental role in the dynamics of magnetosphere via wave-particle interactions and contribute to particle diffusion, precipitation, acceleration, and heating.

We conclude by pointing out the potentially important role that lightning may be playing in the physics of magnetospheric plasma waves. It has been widely assumed that most of the plasma waves are generated from background noise levels within the magnetosphere via wave-particle interactions with energetic particles supplying the free energy (Lyons and Williams, 1984). Recent results show that lightning can be an important source of (1) plasmaspheric hiss believed to be responsible for the slot region in the radiation belts (Sonwalkar and Inan, 1989; Draganov et al.,

1993), (2) lower hybrid waves that can heat and accelerate protons to suprathermal temperatures (Bell and Ngo, 1990; Bell et al., 1991b), and (3) ULF magnetic fields that can influence the generation and amplification of geomagnetic pulsations (Fraser-Smith, 1993). In addition, lightning-induced electron precipitation events (LEP) regularly occur throughout the plasmasphere and are important on a global scale as a loss process for the radiation belt electrons (Inan et al., 1990). Approximately 2000 thunderstorms are active near the Earth's surface at any given time, and on the average, lighting strikes the Earth ~100 times per second (Volland, 1984). The average lightning discharge radiates an intense pulse of ~20 GW peak power which propagates through the lower atmosphere and into the ionospheric and magnetospheric plasmas, generating new waves, heating, accelerating, and precipitating components of the charged particles comprising these plasmas. Thus, future investigations should consider electromagnetic energy released in the thunderstorms as a potential free energy source for the generation of magnetospheric plasma waves.

REFERENCES

Al' pert, I. A. L. (1983). The near-earth and interplanetary plasma, Cambridge University Press, New York.

Anderson, R. R. (1984). Plasma waves at and near the neutral sheet, Proceedings of the Conference on Achievements of the IMS, *Eur. Space Agency Spec. Publ., ESA SP-217,* 199.

Anderson, R. R. (1994). CRRES plasma wave observations during quiet times, during geomagnetic disturbances, and during chemical release, *Dusty Plasma, Noise, and Chaos in Space and Laboratory,* Ed., H. Kicuchi, Plenum Press, New York.

Anderson, R. R., and Kurth, W. S. (1989). Discrete electromagnetic emissions in planetary magnetospheres, *Plasma Waves and Instabilities at Comets and in Magnetospheres, Geophys. Monograph 53,* Eds., B. T. Tsurutani and H. Oya, American Geophysical Union, Washington, D.C., 81.

Ashour-Abdalla, M., and Kennel, C. F. (1978). Nonconvective and convective electron cyclotron harmonic instabilities, *J. Geophys. Res.,* 83, 1531.

Ashour-Abdalla, M., Chanteur, M. G., and Pellat, R. (1975). A contribution to the theory of the electrostatic half-harmonic electron gyrofrequency waves in the magnetosphere, *J. Geophys. Res.,* 80, 2775.

Barrington, R. E., and Belrose, J. S. (1963). Preliminary results from the very-low-frequency receiver aboard Canada's Alouette satellite, *Nature,* 198, 651.

Barrington, R. E., Hartz, T. R., and Harvey, R. W. (1971). Diurnal distribution of ELF, VLF and LF noise at high latitudes as observed by Alouette-2, *J. Geophys. Res.,* 76, 5278.

Barrington, R. E., Belrose, J. S., and Keeley, D. A. (1963). Very low frequency noise bands observed by the Alouette 1 satellite, *J. Geophys. Res.,* 68, 6539.

Baumback, M. M., and Calvert, W. (1987). The minimum bandwidths of auroral kilometric radiation, *Geophys. Res. Lett.,* 14, 119.

Bell, T. F. (1984). The nonlinear gyroresonance interaction between energetic electrons and coherent wave propagating at an arbitrary angle with respect to the earth's magnetic field, *J. Geophys. Res.,* 89, 905.

Bell, T. F. and Ngo, H. D. (1990). Electrostatic lower hybrid waves excited by electromagnetic whistler mode waves scattering from planar magnetic-field-aligned plasma density irregularities, *J. Geophys. Res.,* 95, 149.

Bell, T. F., Inan, U. S., Sonwalkar, V. S., and Helliwell, R. A. (1991a). DE-1 Observations of lower hybrid waves excited by VLF whistler mode waves, *Geophys. Res. Lett.,* 18, 393.

Bell, T. F., Helliwell, R. A., and Hudson, M. K. (1991b). Lower hybrid waves excited through linear mode coupling and the heating of ions in the auroral and subauroral magnetosphere, *J. Geophys. Res.,* 96, 11379.

Bell, T. F., Katsufrakis, J. P., and James, H. G. (1985). A new type of VLF emission triggered at low altitude in the subauroral region by Siple Station VLF transmitter signals, *J. Geophys. Res.,* 90, 12, 183.

Bell, T. F., Inan, U. S., Kimura, I., Matsumoto, H., and Hashimoto, K. (1983a). EXOS-B/Siple VLF wave-particle interaction experiments. 2. Transmitter signals and associated emission, *J. Geophys. Res.,* 88, 295.

Bell, T. F., James, H. G., Inan, U. S., and Katsufrakis, J. P. (1983b). The apparent spectral broadening of VLF transmitter signals during transionospheric propagation, *J. Geophys. Res.,* 88, 4813.

Bell, T. F., Inan, U.S., and Helliwell, R. A. (1981). Conducted coherent VLF waves and associated triggered emissions observed on the ISEE-1 satellite, *J. Geophys. Res.,* 86, 4649.

Benson, R. F. (1985). Auroral kilometric radiation: wave modes, harmonics, and source region density structures, *J. Geophys. Res.,* 90, 2753.

Benson, R. F., and Wong, H. K. (1897). Low-altitude ISIS 1 observations of auroral radio emissions and their significance to the cyclotron maser instability, *J. Geophys. Res.*, 92, 1218.

Benson, R. F., and Calvert, W. (1979). ISIS 1 observations at the source of auroral kilometric radiation, *Geophys. Res. Lett.*, 6, 479.

Benson, R. F., Deach, M. D., Hunsucker, R. D., and Romick, G. J. (1988a). Ground-level detection of low- and medium-frequency auroral radio emissions, *J. Geophys. Res.*, 93, 277.

Benson, R. F., Deach, M. D., Hunsucker, R.D., and Romick, G. J. (1988b). Correction to "Ground-level detection of low-and medium-frequency auroral radio emissions," *J. Geophys. Res.*, 93, 2767.

Benson, R. F., Mellot, M. M., Huff, R. L., and Gurnett, D. A. (1988c). Ordinary mode auroral kilometric radiation fine structure observed by DE 1, *J. Geophys. Res.*, 93, 7515.

Benson, R. F., Calvert, W., and Klumpar, D. M. (1980). Simultaneous wave and particle observations in the auroral kilometric radiation source region, *Geophys. Res. Lett.*, 7, 959.

Brice, N. M. (1964). Fundamentals of VLF emission generation mechanisms, *J. Geophys. Res.*, 69(21), 1515.

Brice, N. M., and Smith, R. L. (1965). Lower hybrid resonance emissions, *J. Geophys. Res.*, 70, 71.

Briggs, R. J. (1964). *Electron-Stream Interaction with Plasmas*, M.I.T. Press, Cambridge, MA.

Brown, L. W. (1973). The galactic radio spectrum between 130 kHz and 2600 kHz, *Astrophys. J.*, 180, 359.

Budden, K. G. (1985). *The Propagation of Radio Waves*, Cambridge University Press, Cambridge, U.K.

Burgess, W. C., and Inan, U. S. (1993). The role of ducted whistlers in the precipitation loss and equilibrium flux of radiation belt electrons, *J. Geophys. Res.*, 98, No. A9, 15,643.

Burtis, W. J., and Helliwell, R. A. (1976). Magnetospheric chorus: occurrence patterns and normalized frequency, *Planet. Space. Sci.*, 24, 1007.

Burtis, W. J., and Helliwell, R. A. (1975). Magnetospheric chorus: amplitude and growth rate, *J. Geophys. Res.*, 80, 3265.

Burton, R. K., and Holzer, R. E. (1974). The origin and propagation of chorus in the outer magnetosphere, *J. Geophys. Res.*, 79, 1014.

Calvert, W. (1981a). The signature of auroral kilometric radiation on ISIS 1 ionograms, *J. Geophys. Res.*, 86, 76.

Carlson, C. R. (1987). Simulation and modeling of whistler mode wave growth through cyclotron resonance with energetic electrons in the magnetosphere, Ph.D. thesis, Stanford University, Stanford, CA.

Carlson, C. R., Helliwell, R. A., and Inan, U. S. (1990). Space-time evolution of whistler mode wave growth in the magnetosphere, *J. Geophys. Res.*, 95, 15,073.

Carlson, C. R., Helliwell, R. A., and Carpenter, D. L. (1985). Variable frequency VLF signals in the magnetosphere: associated phenomena and plasma diagnostics, *J. Geophys. Res.*, 90, 1507.

Carpenter, D. L. (1994). Plasma wave observations and the plasmasphere, in Chapter 2 of *Plasmasphere*, Ed., J. F. Lemaire and C. K. Gringanz, Cambridge University Press, Cambridge, U.K.

Carpenter, D. L., Foster, J. C., Rosenberg, T. J., and Lanzerotti, L. J. (1975). A subauroral and mid-latitude view of substorm activity, *J. Geophys. Res.*, 80(31), 4279.

Carpenter, D. L., and Bao, Z. T. (1983). Occurrence properties of ducted whistler-mode signals from the new VLF transmitter at Siple Station, Antarctica, *J. Geophys. Res.*, 88, 7051.

Cattell, C. A., and Mozer, F. S. (1986). Experimental determination of the dominant wave mode in the active near-earth magnetotail, *Geophys. Res. Lett.*, 13, 221.

Cattell, C. A., Mozer, F. S., Anderson, R. R., Hones, E. W., Jr., and Sharp, R. D. (1986a). ISEE observations of the plasma sheet boundary, plasma sheet and neutral sheet. 1. Electric field, magnetic field, plasma and ion composition, *J. Geophys. Res.*, 91, 5663.

Cattell, C. A., Mozer, F. S., Anderson, R. R., Hones, E. W., Jr., and Sharp, R. D. (1986b). ISEE observations of the plasma sheet boundary, plasma sheet and neutral sheet. 2. Waves, *J. Geophys. Res.*, 91, 5681.

Chmyrev, V. M., Draganov, A. B., Taranenko, Yu. N. (1990). On the theory of whistler precursors on a VLF transmitter signal, *J. Atmos. Terr. Phys.*, 52, 9, 797.

Church, S. R., and Thorne, R. M (1983). On the origin of plasmaspheric hiss: ray path integrated amplification, *J. Geophys. Res.*, 88, 7941.

Cornwall, J. M., Coroniti, F. V., and Thorne, R. M. (1970). Turbulent loss of ring current protons, *J. Geophys. Res.*, 75, 4699.

Coroniti, F. V., Fredricks, R. W., Kennel, C. F., and Scarf, F. L. (1971). Fast time-resolved spectral analysis of VLF banded emissions, *J. Geophys. Res.*, 76, 2366.

Curtis, S. A. (1985). Equatorial trapped plasmasphere ion distributions and transverse stochastic acceleration, *J. Geophys. Res.*, 90, 1765.

Das, A. C. and Kulkarni, V. H. (1975). Frequency-time structure of VLF emissions, *Planet. Space Sci.*, 23, 41.

Dowden, R. L. (1971). Distinction between mid latitude VLF hiss and discrete emissions, *Planet. Space Sci.*, 19, 374.

Dowden, R. L., and Helliwell, R. A. (1962). Very-low-frequency discrete emissions received at conjugate points, *Nature*, 195, 64.

Dowden, R. L., McKey, A. C., Amon, L. E. S., Koons, H. C., and Dazey, M. H. (1978). Linear and nonlinear amplification in the magnetosphere during a 6.6-kHz transmission, *J. Geophys. Res.*, 83, 169.

Draganov, A. B., Inan, U. S., Sonwalkar, V. S., and Bell, T. F. (1992). Magnetospherically reflected whistlers as a source of plasmaspheric hiss, *Geophys. Res. Lett.*, 19, 233–236.

Draganov, A. B., Inan, U. S., Sonwalkar, V. S., and Bell, T. F. (1993). Whistlers and plasmaspheric hiss: wave directions and three-dimensional propagation, *J. Geophys. Res.*, 98, 11,401.

Dungey, J. W. (1972). Instabilities in the magnetosphere, in *Solar-Terrestrial Physics/1970*, Ed., E. R. Dyer, Proc. International Symposium on Solar-Terrestrial Physics, Leningrad, May 1970, pp. 219, D. Reidel, Dordrecht, Holland.

Dungey, J. W. (1963). Loss of Van Allen electrons due to whistlers, *Planet. Space Sci.*, 11, 591.

Dunckel, N., and Helliwell, R. A. (1969). Whistler mode emissions on the OGO 1 satellite, *J. Geophys. Res.*, 74, 6731.

Dunckel, N., Ficklin, B., Rorden, L., and Helliwell, R. A. (1970). Low-frequency noise observed in the distant magnetosphere with OGO 1, *J. Geophys. Res.*, 75, 1854.

Dusenbery, P. B., and Lyons, L. R. (1989). Ion diffusion coefficients from resonant interactions with broadband turbulence in the magnetotail, *J. Geophys. Res.*, 94, 2484.

Francis, C. R., Strangeways, H. J., and Bullough, K. (1983). Discrete VLF emissions (7–9 kHz) displaying unusual banded and periodic structure, *Planet. Space Sci.*, 31, 537.

Frank, L. A., Craven, J. D., Ackerson, K. L., English, M. R., Eather, R. H., and Carovillano, R. L. (1981). Global auroral imaging instrumentation for the Dynamics Explorer Mission, *Space Sci. Instrum.*, 5, 369.

Fraser-Smith, A. C. (1993). ULF magnetic fields generated by electrical storms and their significance to geomagnetic pulsation generation, *Geophys. Res. Lett.*, 20, 467.

Fraser-Smith, A. C., and Helliwell, R. A. (1985). The Stanford University ELF/VLF radiometer project: measurement of the global distribution of ELF/VLF electromagnetic noise, *Proc. IEEE Int. Symp. Electromag. Compatability*, Wakefield, MA, August.

Fredricks, R. W. (1968). Structure of generalized ion Bernstein modes from the full electromagnetic dispersion relation, *J. Plasma Phys.*, 2, 365.

Fredricks, R. W. (1975). Wave-particle interactions in the outer magnetosphere: a review, *Space Sci. Rev.*, 17, 741.

Fredricks, R. W., and Scarf, F. L. (1973). Recent studies of magnetospheric electric field emissions above the electron gyrofrequency, *J. Geophys. Res.*, 78, 310.

Fredricks, R. W., Kennel, C. F., Scarf, F. L., Crook, G. M., and Green, I. M. (1968). Detection of electric field turbulence in the Earth's bow shock, *Phys. Rev. Lett.*, 21, 1761.

Gendrin, R. (1975). Waves and wave-particle interactions in the magnetosphere: a review, *Space Sci. Rev.*, 18, 145.

Gendrin, R., and Roux, A. (1980). Energization of helium ions by proton induced hydromagnetic waves, *J. Geophys. Res.*, 85, 4577.

Goldstein, B. E., and Tsurutani, B. T. (1984). Wave normal directions of chorus near the equatorial source region, *J. Geophys. Res.*, 89, 2798.

Grabbe, C. L. (1987). Numerical study of the spectrum of broad-band electrostatic noise in the magnetotail, *J. Geophys. Res.*, 92, 1185.

Grabbe, C. L., and Eastman, T. E. (1984). Generation of broadband electrostatic noise by ion beam instabilities in the magnetotail, *J. Geophys. Res.*, 89, 3865.

Greenstadt, E. W., and Fredricks, R. W. (1974). Plasma instability modes related to the earth's bow shock, *Magnetospheric Physics*, Ed., B. M. McCormac, D. Reidel, Dordrecht, Holland.

Gross, S. H., and Muldrew, D. B. (1984). Uniformly spaced field-aligned ionization ducts, *J. Geophys. Res.*, 89, 8986.

Gurnett, D. A. (1976). Plasma wave interactions with energetic ions near the magnetic equator, *J. Geophys. Res.*, 81, 2765.

Gurnett, D. A. (1975). The Earth as a radio source: the nonthermal continuum, *J. Geophys. Res.*, 80, 2751.

Gurnett, D. A. (1974). The earth as a radio source: terrestrial kilometric radiation, *J. Geophys. Res.*, 79, 4227.

Gurnett, D. A., and Inan, U. S. (1988). Plasma wave observations with the Dynamics Explorer 1 spacecraft, *Rev. Geophys.*, 26, 285.

Gurnett, D. A., and Green, J. L. (1978). On the polarization and origin of auroral kilometric radiation, *J. Geophys. Res.*, 83, 689.

Gurnett, D. A., and Frank, L. A. (1977). A region of intense plasma wave turbulence on auroral field lines, *J. Geophys. Res.*, 82, 1031.

Gurnett, D. A., and Shaw, R. R. (1973). Electromagnetic radiation trapped in the magnetosphere above the plasma frequency, *J. Geophys. Res.*, 78, 8136.

Gurnett, D. A., and Frank, L. A. (1972a). VLF hiss and related plasma observations in the polar magnetosphere, *J. Geophys. Res.*, 77, 172.

Gurnett, D. A., and Frank, L. A. (1972b). ELF noise bands associated with auroral electron precipitation, *J. Geophys. Res.*, 77, 3411.

Gurnett, D. A., Calvert, W., Huff, R. L., Jones, D., and Sugiura, M. (1988). The polarization of escaping terrestrial continuum radiation, *J. Geophys. Res.*, 93, 12,817.

Gurnett, D. A., Huff, R. L., Menietti, J. D., Burch, J. L., Winningham, J. D., and Shawhan, S. D. (1984). Correlated low-frequency electric and magnetic noise along the auroral field lines, *J. Geophys. Res.*, 89, 8971.

Gurnett, D. A., Shawhan, S. D., and Shaw, R. R. (1983). Auroral hiss, Z mode radiation, and auroral kilometric radiation in the polar magnetosphere: DE1 observations, *J. Geophys. Res.*, 88, 329.

Gurnett, D. A., Frank, L. A., and Lepping, R. P. (1976). Plasma waves in the distant magnetotail, *J. Geophys. Res.*, 81, 6059.

Gurnett, D. A., Pfeiffer, G. W., Anderson, R. R., Mosier, S. R., and Cauffman, D. P. (1969). Initial observations of VLF electric and magnetic fields with the Injun-5 satellite, *J. Geophys. Res.*, 74, 4631.

Harris, E. G. (1959). Unstable plasma oscillations in a magnetic field, *Phys. Rev. Lett.*, 2, 34.

Hasegawa, A. (1975). *Plasma Instabilities and Nonlinear Effects*, Springer-Verlag, Berlin.

Hashimoto, K., and Calvert, W. (1990). Observation of the Z mode with DE 1 and its analysis by three-dimensional ray tracing, *J. Geophys. Res.*, 95, 3933.

Hattori, K., Hayakawa, M., Lagoutte, D., Parrot, M., and Lefeuvre, F. (1991b). An experimental evidence of the role of hiss in the generation of chorus in the outer magnetosphere, as based on spectral analysis and direction finding, *Proc. Natl. Inst. Polar Res. Symp. Upper Atmos. Phys.*, 4, 20.

Hayakawa, M. (1994). Whistler, in *Handbook of Atmospheric Electrodynamics, Vol. II*, Ed., H. Volland, CRC Press, Boca Raton, FL.

Hayakawa, M., and Sazhin, S. S. (1992). Mid-latitude and plasmaspheric hiss: a review, *Planet. Space Sci.*, 40, 1325.

Hayakawa, M., Parrot, M., and Lefeuvre, F. (1986). The wave normals of ELF hiss emissions observed on board GEOS 1 at the equatorial and off-equatorial regions of the plasmasphere, *J. Geophys. Res.*, 91, 7989.

Hayakawa, M., Yamanaka, Y., Parrot, M., and Lefeuvre, F. (1984). The wave normals of magnetospheric chorus emissions observed on board GEOS 2, *J. Geophys. Res.*, 89, 2811.

Helliwell, R. A. (1994). 40 years of whistlers, Conference Proceedings, XXIV *General Assembly of the Int. Union of Radio Sci.* (URSI), Kyoto, Japan, 1993, in press.

Helliwell, R. A. (1988). VLF wave stimulation experiments in the magnetosphere from Siple Station, Antarctica, *Rev. Geophys.*, 26, 551.

Helliwell, R. A., and Katsufrakis, J. P. (1978). Controlled wave-particle interaction experiments, Paper 5 in *Upper Atmosphere Research in Antarctica, Antarctic Research Series Vol. 29*, Eds., L. J. Lanzerotti and C. G. Park, American Geophysical Union, Washington, D.C.

Helliwell, R. A. (1974). Controlled VLF wave injection experiments in the magnetosphere, *Space Sci. Rev.*, 15, 781.

Helliwell, R. A. (1967). A theory of discrete VLF emissions from the magnetosphere, *J. Geophys. Res.*, 72(19), 4773.

Helliwell, R. A. (1965). *Whistlers and Related Ionospheric Phenomena*, Stanford University Press, Stanford, CA.

Helliwell, R. A., and Inan, U. S. (1982). VLF wave growth and discrete emission triggering in the magnetopshere: a feedback model, *J. Geophys. Res.*, 87, 3537.

Helliwell, R. A., and Katsufrakis, J. P. (1974). VLF wave injection into the magnetosphere from Siple Station, Antarctica, *J. Geophys. Res.*, 79, 2511.

Helliwell, R. A., and Crystal, T. L. (1973). A feedback model of cyclotron interaction between whistler-mode waves and energetic electrons in the magnetosphere, *J. Geophys. Res.*, 78(31), 7357.

Helliwell, R. A., Inan, U. S., Katsufrakis, J. P., and Carpenter, D. L. (1986a). Beat excitation of whistler mode sidebands using the Siple VLF transmitter, *J. Geophys. Res.*, 91, 143.

Helliwell, R. A., Carpenter, D. L., Inan, U. S., and Katsufrakis, J. P. (1986b). Generation of band-limited noise using the Siple transmitter: a model for magnetospheric hiss, *J. Geophys. Res.*, 91, 4381.

Helliwell, R. A., Carpenter, D. L., and Miller, T. R. (1980a). Power threshold for growth of coherent VLF signals in the magnetosphere, *J. Geophys. Res.*, 85, 3360.

Helliwell, R. A., Mende, S. B., Dolittle, J. H., Armstrong, W. C., and Carpenter, D. L. (1980b). Correlations between λ4278 optical emissions and VLF wave events observed at $L \simeq 4$ in the Antarctic, *J. Geophys. Res.*, 85, 3376.

Heppner, J. P. (1969). Magnetospheric convection patterns inferred from high latitude activity, in *Atmospheric Emissions*, Eds., B. M. McCormac and A. Omholt, p. 251, Reinhold, New York.

Hoffman, R. A., and Laaspere, T. (1972). Comparison of very-low-frequency auroral hiss with precipitating low-energy electrons by the use of simultaneous data from two OGO-4 experiments, *J. Geophys. Res.*, 77, 640.

Huang, C. Y., Goertz, C. K., and Anderson, R. R. (1983). A theoretical study of plasmaspheric hiss generation, *J. Geophys. Res.*, 88, 7927.

Huff, R. L., Calvert, W., Craven, J. D., Frank, L. A., and Gurnett, D. A. (1988). Mapping of auroral kilometric radiation sources to the aurora, *J. Geophys. Res.*, 93, 11,445.

Hui, C.-H. and Seyler, C. E. (1993). Parametric excitation of the electrostatic whistler and Z mode by auroral kilometric radiation, *J. Geophys. Res.*, 98, A7, 11701.

Hultqvist, B., and Falthammar, C.-G., Eds. (1990). *Magnetospheric Physics. Achievements and Prospects. Proceedings of the Craford Symposium*, Plenum, New York.

Inan, U. S., and Bell, T. F. (1985). Spectral broadening of VLF transmitter signals observed on DE1: a quasi-electrostatic phenomenon?, *J. Geophys. Res.*, 90, 1771.

Inan, U. S., and Helliwell, R. A. (1982). DE-1 observations of VLF transmitter signals and wave-particle interactions in the magnetosphere, *Geophys. Res. Lett.*, 9, 917.

Inan, U. S., and Bell, T. F. (1977). The plasmapause as a VLF wave guide, *J. Geophys. Res.*, 82(19), 2819.

Inan, U. S., Knifsend, F. A., and Oh, J. (1990). Subionospheric VFL "Imaging" of lightning-induced electron precipitation from the magnetosphere, *J. Geophys. Res.*, 95, 17217.

Inan, U. S., Bell, T. F., and Rodriguez, J. V. (1991). Heating and ionization of the lower ionosphere by lightning, *Geophys. Res. Lett.*, 18, 705.

Inan, U. S., Bell, T. F., Carpenter, D. L., and Anderson, R. R. (1977). Explorer 45 and Imp 6 observations in the magnetosphere of injected waves from the Siple Station VLF transmitter, *J. Geophys. Res.*, 82(7), 1177.

Isted, G. A., and Millington, G. (1957). The "dawn chorus" in radio observation, *Nature*, 180, 716.

James, H. G. (1976). VLF saucers, *J. Geophys. Res.*, 81, 501.

Jones, D. (1982). Terrestrial myriametric radiation from the Earth's plasma pause, *Planet. Space Sci.*, 30, 399.

Jones, D. (1976). Source of terrestrial non-thermal radiation, *Nature*, 260, 686.

Jones, D., Calvert, W., Gurnett, D. A., and Huff, R. L. (1987). Observed beaming of terrestrial myriametric radiation, *Nature*, 328, 391.

Jorgensen, T. S. (1968). Interpretation of auroral hiss measured on OGO-2 and at Byrd Station in terms of incoherent Cerenkov radiation, *J. Geophys. Res.*, 73, 1055.

Joselyn, J. A., and Lyons, L. R. (1976). Ion-cyclotron wave growth calculated from satellite observations of the proton ring current during storm recovery, *J. Geophys. Res.*, 81, 2275.

Kaufman, R. L., Dusenbery, P. B., and Thomas, B. J. (1978). Stability of the auroral plasma: parallel and perpendicular propagation of electrostatic waves, *J. Geophys. Res.*, 83, 5663.

Kennel, C. F., and Englemann, F. (1966). Velocity space diffusion from weak plasma turbulence in a magnetic field, *Phys. Fluids*, 9, 2377.

Kennel, C. F., and Petschek, H. E. (1966). Limit on the stably trapped particle fluxes, *J. Geophys. Res.*, 71, 1.

Kennel, C. F., Scarf, F. L., Fredricks, R. W., McGehee, J. H., and Coroniti, F. V. (1970). VLF electric-field observations in the magnetosphere, *J. Geophys. Res.*, 75, 6136.

Kimura, I. (1989). Ray paths of electromagnetic and electrostatic waves in the earth and planetary magnetospheres, *Plasma Waves and Instabilities at Comets and in Magnetospheres, Geophys. Monogr.*, 53, Eds., B. T. Tsurutani and H. Oya, 161.

Kimura, I. (1966). Effects of ions on whistler-mode ray tracing, *Radio Sci.*, 1, 269.

Kimura, I., Matsumoto, H., Mokai, T., Hashimoto, K., Bell, T. F., Inan, U. S., Helliwell, R. A., and Katsufrakis, J. P. (1983). EXOS B/Siple station VLF wave-particle interaction experiments. 1. General description and wave-particle correlations, *JGR*, 88(A1), 282.

Kintner, P. M., Jr., and Gurnett, D. A. (1977). Observation of ion cyclotron waves within the plasmasphere, *J. Geophys. Res.*, 82, 2314.

Kitamura, T., Jacobs, J. A., and Watanabe, T. (1969). An investigation of quasiperiodic emissions, *J. Geophys. Res.*, 74, 5652.

Kolmogorov, A. N. (1941). The local structure of turbulence in incompressible viscous fluids for very high Reynolds numbers, *Co. Re. Acad. Sci. U.S.S.R.*, 30, 301.

Koons, H. C. (1981). The role of hiss in magnetospheric chorus emissions, *J. Geophys. Res.*, 86, 6745.

Kurth, W. S. (1991). Magnetospheric radio and plasma wave research: 1987–90, *Rev. Geophys.*, 29, Suppl., 1,075.

Kurth, W. S., Frank, L. A., Ashour-Abdalla, M., Gurnett, D. A., and Burek, B. G. (1980). Observations of a free-energy source for intense electrostatic waves, *Geophys. Res. Lett.*, 7, 293.

Kurth, W. S., Craven, J. D., Frank, L. A., and Gurnett, D. A. (1979). Intense electrostatic waves near the upper hybrid resonance frequency, *J. Geophys. Res.*, 84, 4145.

Kurth, W. S., Baumback, M. M., and Gurnett, D. A. (1975). Direction-finding measurements of auroral kilometric radiation, *J. Geophys. Res.*, 80, 2764.

Laaspere, T., and Johnson, W. C. (1973). Additional results from an OGO-6 experiment concerning ionospheric electric and electromagnetic fields in the range 20 Hz to 540 kHz, *J. Geophys. Res.*, 78, 2926.

Laaspere, T., and Taylor, H. A. (1970). Comparison of certain VLF noise phenomena with the lower hybrid frequency calculated from simultaneous ion composition measurements, *J. Geophys. Res.*, 75, 97.

Lanzerotti, L. J. (1978). Studies of geomagnetic pulsations, in *Upper Atmosphere Research in Antarctica, Antarctic Research Series, Vol. 29*, 130.

Lee, L. C. (1989). Theories of non-thermal radiations from planets, *Plasma Waves and Instabilities at Comets and in Magnetospheres, Geophys. Monogr. 53*, Eds., B. T. Tsurutani and H. Oya, 239.

Lee, Me. C., and Kuo, S. P. (1984). Production of lower hybrid waves and field-aligned plasma density striations by whistlers, *J. Geophys. Res.*, 89, 10,873.

Lin, C. S., Burch, J. L., Shawhan, S. D., and Gurnett, D. A. (1984). Correlation of auroral hiss and upward electron beams near the polar cusp, *J. Geophys. Res.*, 89, 925.

Lokken, J. E., Shand, J. A., Sir, C. S., Wright, K. C. B., Martin, L. H., Brice, N. M., and Helliwell, R. A. (1961). Stanford-Pacific Naval Laboratory conjugate point experiment. *Nature*, 192, 319.

Lotko, W., and Maggs, J. E. (1981). Amplification of electrostatic noise in cyclotron resonance with an adiabatic auroral beam, *J. Geophys. Res.*, 86, 3449.

Lyons, L. R., and Williams, D. J. (1984). Quantitative aspects of magnetospheric physics, D. Reidel, Boston.

Lyons, L. R., Thorne, R. M., and Kennel, C. F. (1972). Pitch angle diffusion of radiation belt electrons within the plasmasphere, *J. Geophys. Res.*, 77, 3455.

Maggs, J. E. (1976). Coherent generation of VLF hiss, *J. Geophys. Res.*, 81, 1707.

Maggs, J. E., and Lotko, W. (1981). Altitude dependent model of the auroral beam and beam-generated electrostatic noise, *J. Geophys. Res.*, 86, 3439.

Matsumoto, H. (1979). Nonlinear whistler-mode interaction and triggered emissions in the magnetosphere: a review, in *Wave Instabilities in Space Plasmas*, Eds. P. J. Palmadesso and K. Papadopoulos, pp. 163, D. Reidel, Hingham, MA.

Mauk, B. H., McIlwain, C. E., and McPherron, R. L. (1981). Helium cyclotron resonance within the Earth's magnetosphere, *Geophys. Res. Lett.*, 8, 103.

Mellott, M. M., Huff, R. L., and Gurnett, D. A. (1986). DE-1 observations of harmonic auroral kilometric radiation, *J. Geophys. Res.*, 91, 13,732.

Mellott, M. M., Calvert, W., Huff, R. L., Gurnett, D. A., and Shawhan, S. D. (1984). DE-1 observation of ordinary mode and extraordinary mode auroral kilometric radiation, *Geophys. Res. Lett.*, 11, 1188.

Melrose, D. B. (1986). *Instabilities in Space and Laboratory Plasmas*, Cambridge University Press, Cambridge, U.K.

Melrose, D. B. (1981). A theory for the nonthermal radio continuum in the terrestrial and Jovian magnetospheres, *J. Geophys. Res.*, 86, 30.

Mielke, T. A., and Helliwell, R. A. (1992). An experiment on the threshold effect in the coherent wave instability, *Geophys. Res. Lett.*, 19, No. 20, 2075–2078.

Molvig, K., Hilfer, G., and Myczkowski, J. (1986). Self-consistent theory of triggered whistler emissions, *EOS Trans.*, *Am. Geophys. Union*, 67, 44, 1166.

Morgan, D. D., and Gurnett, D. A. (1991). The source location and beaming of terrestrial continuum radiation, *J. Geophys. Res.*, 96, 9595.

Mosier, S. R., and Gurnett, D. A. (1972). Observed correlation between auroral and VLF emissions, *J. Geophys. Res.*, 77, 1137.

Nishino, M., Tanaka, Y., Iwai, A., Kamada, T., and Hirasawa, T. (1982). Comparison between the arrival direction of auroral hiss and the location of aurora observed at Syowa station, *Mem. Natl. Inst. Polar Res. Tokyo*, Special Issue No. 22, 35.

Nunn, D. (1984). The quasi static theory of triggered VLF emissions. *Planet. Space Sci.*, 32, 325.

Nunn, D. (1974). A self-consistent theory of triggered VLF emissions, *Planet. Space Sci.*, 22, 349.

Nunn, D. (1971). A theory of VLF emissions. *Planet. Space Sci.*, 19, 1141.

Oliven, M. N., and Gurnett, D. A. (1968). Microburst phenomena. 3. An association between microbursts and VLF chorus, *J. Geophys. Res.*, 73, 2355.

Olsen, R. C. (1981). Equatorially trapped plasma populations, *J. Geophys. Res.*, 86, 11,235.

Olsen, R. C., Shawhan, S. D., Gallagher, D. L., Green, J. L., Chappell, C. R., and Anderson, R. R. (1987). Plasma observation at the Earth's magnetic equator, *J. Geophys. Res.*, 92, 2385.

Ondoh, T., Nakamura, Y., Watanabe, S., Aikyo, K. (1989). Impulsive plasma waves observed by DE 1 in nightside magnetosphere, *J. Geophys. Res.*, 94, A4, 3779.

Onsager, T. G., Holzworth, R. H., Koons, H. C., Bauer, O. H., Gurnett, D. A., Anderson, R. R., Luhr, H., Carlson, C. W. (1990). *J. Geophys. Res.*, 94, A10, 13397.

Paranicas, C., Hughes, W. J., Singer, H. J., Anderson, R. R. (1992). *J. Geophys. Res.*, 97, A9, 13889.

Park, C. G., and Helliwell, R. A. (1977). Whistler precursors: a possible catalytic role of power line radiation, *J. Geophys. Res.*, 82(25), 3634.

Park, C. G., Lin, C. S., and Parks, G. K. (1981). A ground-satellite study of wave-particle correlations, *J. Geophys. Res.*, 86, 37.

Parks, G. K., McCarthy, M., Fitzenreiter, R. J., Etcheto, J., Anderson, K. A., Anderson, R. R., Eastman, T. E., Frank, L. A., Gurnett, D. A., Huang, C., Lin, R. P., Lui, A. T. Y., Ogilvie, K. W., Pedersen, A., Reme, H., and Williams, D. J. (1984). Particle and field characteristics of the high-latitude plasma sheet boundary layer, *J. Geophys. Res.*, 89, 8885.

Paschal, E. W., and Helliwell, R. A. (1984). Phase measurements of whistler mode signals from the Siple VLF transmitter, *J. Geophys. Res.*, 89, 1667.

Paschal, E. W. (1990). Whistler precursors on a VLF transmitter signal, *J. Geophys. Res.*, 95, A1, 225.

Poulsen, W. L., and Inan, U. S. (1988). Satellite observations of a new type of discrete VLF emission at $L < 4$, *J. Geophys. Res.*, 93, 1817.

Quinn, J. M., Shelley, E. G., Collin, H. L., and Gurnett, D. A. (1986). The generation of an O^+ conic distribution by equatorially confined waves as observed by DE-1, *EOS Trans. AGU*, 67, 1138.

Raghuram, R. (1975). A new interpretation of subprotonospheric whistler characteristics, *J. Geophys. Res.*, 80, 4729.

Raghuram, R., Bell, T. F., Helliwell, R. A., and Katsufrakis, J. P. (1977a). Quiet band produced by VLF transmitter signals in the magnetosphere, *Geophys. Res. Lett.*, 4(5), 199.

Raghuram, R., Bell, T. F., Helliwell, R. A., and Katsufrakis, J. P. (1977b). Echo-induced suppressions of coherent VLF transmitter signals in the magnetosphere, *J. Geophys. Res.*, 82(19), 2787.

Rastani, K., Inan, U. S., and Helliwell, R. A. (1985). DE-1 observations of Siple transmitter signals and associated sidebands, *J. Geophys. Res.*, 90, 4128.

Ratcliffe, J. A. (1959). *The Magneto-Ionic Theory and Its Applications to the Ionosphere*, Cambridge University Press, Cambridge, MA.

Reinleitner, L. A., Gurnett, D. A., and Gallgher, D. L. (1982). Chorus related electrostatic bursts in the earth's outer magnetosphere, *Nature*, 295, 46.

Reinleitner, L. A., Gurnett, D. A., and Eastman, T. E. (1983). Electrostatic bursts generated by electrons in Landau resonance with whistler-mode chorus, *J. Geophys. Res.*, 88, 3079.

Rodriguez, P. (1985). Long duration lion roars associated with quasi-perpendicular bow shocks, *J. Geophys. Res.*, 90, 241.

Rodriguez, P., and Gurnett, D. A. (1976). Correlation of bow shock plasma wave turbulence with solar wind parameters, *J. Geophys. Res.*, 81, 2871.

Rodriguez, P., and Gurnett, D. A. (1975). Electrostatic and electromagnetic turbulence associated with the Earth's bow shock, *J. Geophys. Res.*, 80, 19.

Rodriguez, P., and Gurnett, D. A. (1971). An experimental study of very-low-frequency mode coupling and polarization reversal, *J. Geophys. Res.*, 76, 960.

Roeder, J. L., Benbrook, J. R., Bering, E. A., and Sheldon, W. R. (1985). X ray microbursts and VLF chorus, *J. Geophys. Res.*, 90, 10,975.

Rönnmark, K., Borg, H., Christiansen, P. J., Gough, M. P., and Jones, D. (1978). Banded electron cyclotron harmonic instability. A first comparison of theory and experiment, *Space Sci. Rev.*, 22, 401.

Rosenberg, T. J. (1968). Correlated bursts of VLF hiss, auroral height and X-rays, *Planet. Space Sci.*, 16, 1419.

Rosenberg, T. J., Wei, R., Detrick, D. L., and Inan, U. S. (1990). Observations and modeling of wave-induced microburst electron precipitation, *J. Geophys. Res.*, 95, 6467.

Rosenberg, T. J., Helliwell, R. A., and Katsufrakis, J. P. (1971). Electron precipitation associated with discrete very-low-frequency emissions, *J. Geophys. Res.*, 76, 8445.

Roux, A., Perrout, S., Rauch J. L., deVilledary, C., Kresmser, G., Knorth, A., and Young, D. T. (1982). Wave particle interactions near Ω_{He}^+ onboard GEOS 1 and 2: generation of ion cyclotron waves and heating of He$^+$ ions, *J. Geophys. Res.*, 87, 8174.

Roux, A., Hilgers, A., DeFeraudy, H., LeQueau, D., Louarn, P., Perraut, S., Bahnsen, A., Jespersen, M., and Ungstrup, E. (1993). Auroral kilometric radiation sources: in situ and remote observations from Viking, *J. Geophys. Res.*, 98, A7, 11657.

Russell, C. T., Holzer, R. E., and Smith, E. J. (1970). OGO 3 observations of ELF noise in the magnetosphere. 2. The nature of the equatorial noise, *J. Geophys. Res.*, 75, 755.

Sá, L. A. D. (1990). A wave-particle-wave interaction mechanism as a cause of VLF triggered emissions, *J. Geophys. Res.*, 95, 12277.

Sá, L. A. D., and Helliwell, R. A. (1988). Structure of VLF whistler mode sideband waves in the magnetosphere, *J. Geophys. Res.*, 93.

Sato, N. (1980). Quasi-periodic (QP) ELF-VLF emissions observed in high latitudes, *Mem. Natl. Inst. Polar Res.*, A, 17.

Sato, N. and Fukinishi, H. (1981). Interaction between ELF-VLF emissions and magnetic pulsations: classification of quasi-periodic ELF-VLF emissions based on frequency-time spectra, *J. Geophys. Res.*, 86, 19.

Sazhin, S. S. (1987). An analytical model of quasiperiodic ELF-VLF emissions, *Planet. Space Sci.*, 35, 1267.

Sazhin, S. S., and Hayakawa, M. (1992). Magnetospheric chorus emissions: a review, *Planet. Space Sci.*, 40, 681.

Sazhin, S. S., and Hayakawa, M. (1993). ''Auroral hiss: a review.'' *Planet. Space Sci.*, 41, 2, 153.

Sazhin, S. S., and Hayakawa, M. (1994). Periodic and quasiperiodic VLF emissions, *J. Atmos. Terr. Phys.*, 56, 735.

Scarf, F. L., Frank, L. A., Ackerson, K., and Lepping, K. (1974b). Plasma wave turbulence at distant crossings of the plasma sheet boundaries and neutral sheet, *Geophys. Res. Lett.*, 1, 189.

Scarf, E. L., Fredricks, R. W., Kennel, C. F., and Coroniti, F. V. (1973a). Satellite studies of magnetospheric substorms on August 15, 1968: OGO 5 plasma wave observations, *J. Geophys. Res.*, 78, 3119.

Scarf, E. L., Fredricks, R. W., Russell, C. T., Kivelson, M., Neugebauer, M., and Chappell, C. R. (1973b). Observation of a current-driven plasma instability at the outer zone-plasma sheet boundary, *J. Geophys. Res.*, 78, 2150.

Schriver, D., and Shour-Abdalla, M. (1990). Cold plasma beating in the plasma sheet boundary layer: theory and simulations, *J. Geophys. Res.*, 95, 3987.

Schriver, D., and Ashour-Abdalla, M. (1987). Generation of high-frequency broadband electrostatic noise: the role of cold electrons, *J. Geophys. Res.*, 92, 5807.

Shaw, R. R., and Gurnett, D. A. (1980). A test of two theories for the low-frequency cutoffs of nonthermal continuum radiation, *J. Geophys. Res.*, 85, 4571.

Shawhan, S. D. (1979a). *Magnetospheric Plasma Waves, Solar System Plasma Physics*, Eds., E. N. Parker, C. F. Kennel, and L. J. Lanzerotti, North-Holland Publishing, Amsterdam.

Shawhan, S. D. (1979b). Magnetospheric plasma wave research 1975–1978, *Rev. Geophys. Space Phys.*, 17, 4, 705.

Shawhan, S. D. (1970). The use of multiple receivers to measure the wave characteristics of VLF noise in space, *Space Sci. Rev.*, 10, 689.

Shawhan, S. D. (1983). Estimation of wave vector characteristics, *Adv. Space Res.*, 2, 7, 31.

Shawhan, S. D., Gurnett, D. A., Odem, D. L., Helliwell, R. A., and Park, C. G. (1981). The plasma wave and quasi-static electric field instrument (PWI) for Dynamics Explorer-A, *Space Sci. Instrum.*, 5, 535.

Smith, E. J., and Tsurutani, B. T. (1976). Magnetosheath lion roars, *J. Geophys. Res.*, 81, 2261.

Smith, E. J., Holzer, R. E., and Russell, C. T. (1969). Magnetic emissions in the magnetosheath at frequencies near 100 Hz, *J. Geophys. Res.,* 74, 3027.

Smith, R. L. (1964). An explanation of subprotonospheric whistlers, *J. Geophys. Res.,* 69, 5019.

Smith, R. L. (1969). VLF observations of auroral beams as sources of a class of emissions, *Nature,* 224, 351.

Smith, R. L., and Angerami, J. J. (1968). Magnetospheric properties deduced from OGO 1 observations of ducted and non-ducted whistlers, *J. Geophys. Res.,* 73(1), 1.

Sonwalkar, V. S., and Inan, U. S. (1989). Lightning as an embryonic source of VLF hiss, *J. Geophys. Res.,* 94, 6986.

Sonwalkar, V. S., and Inan, U. S. (1988). Wave normal direction and spectral properties of whistler mode hiss observed on the DE 1 satellite, *J. Geophys. Res.,* 93, 7493.

Sonwalkar, V. S., and Inan, U. S. (1986). Measurements of Siple transmitter signals on the DE 1 satellite: wave normal direction and antenna effective length, *J. Geophys. Res.,* 91, 154.

Sonwalkar, V. S., Helliwell, R. A., and Inan, U. S. (1990). Wideband VLF electromagnetic bursts observed on the DE 1 satellite, *Geophys. Res. Lett.,* 17, No. 11, 1861.

Sonwalkar, V. S., Inan, U. S., Bell, T. F., Helliwell, R. A., Molchanov, O. A., Green, J. L. (1994). DE 1 VLF Observations during Activity Wave Injection Experiments, *J. Geophys. Res.,* 99, 6173.

Sonwalkar, V. S., Bell, T. F., Helliwell, R. A., and Inan, U. S. (1984). Direct multiple path magnetospheric propagation: a fundamental property of nonducted VLF waves, *J. Geophys. Res.,* 89, 2823.

Stiles, G. S., and Helliwell, R. A. (1977). Stimulated growth of coherent VLF waves in the magnetosphere, *J. Geophys. Res.,* 82(4), 523.

Stix, T. H. (1992). Waves in plasmas, *Am. Inst. Phys.,* New York.

Storey, L. R. O., and Lefeuvre, F., Parrot, M., Cairo, L., and Anderson, R. R. (1991). Initial survey of the wave distribution functions for plasmaspheric hiss observed by ISEE 1, *J. Geophys. Res.,* 96, 19,469.

Swanson, D. G. (1989). *Plasma Waves,* Academic Press, Boston.

Swift, D. W., and Kan, J. R. (1975). A theory of auroral hiss and implications on the origin of auroral electrons, *J. Geophys. Res.,* 80, 985.

Taylor, W. W. L., Parady, B. K., and Cahill, L. J., Jr. (1975). Explorer 45 observations of 1- to 30-Hz magnetic fields during magnetic storms, *J. Geophys. Res.,* 80, 1271.

Taylor, W. W. L., and Lyons, L. R. (1976). Simultaneous equatorial observations of 1- to 30 Hz waves and pitch angle distributions of ring current ions, *J. Geophys. Res.,* 81, 6177.

Taylor, W. W. L., and Shawhan, S. D. (1974). A test of incoherent Cerenkiov radiation for VLF hiss and other magnetospheric emissions, *J. Geophys. Res.,* 79, 105.

Taylor, W. W. L., and Gurnett, D. A. (1968). The morphology of VLF emissions observed with the INJUN 3 satellite, *J. Geophys. Res.,* 73, 5615.

Temerin, M. (1978). The polarization, frequency, and wavelengths of high latitude turbulence, *J. Geophys. Res.,* 83, 2609.

Termerin, M., and Kintner, P. M. (1989). Review of ionospheric turbulence, *Plasma Waves and Instabilities at Comets and in Magnetospheres, Geophys. Monogr.* 53, Eds., B. T. Tsurutani and H. Oya, 65.

Thompson, R. J., and Dowden, R. L. (1978). Ionospheric whistler propagation, *J. Atmos. Terr. Phys.,* 40, 215, 1978.

Thorne, R. M., Smith, E. J., Burton, R. K., and Holzer, R. E. (1973). Plasmaspheric hiss, *J. Geophys. Res.,* 78, 1581.

Tixier, M., and Cornilleau-Wehrlin, N. (1986). How are the VLF quasi-periodic emissions controlled by harmonics of field line oscillations? The result of a comparison between ground and GEOS satellite measurements, *J. Geophys. Res.,* 91, 6899.

Tsuruda, K., Machida, S., Terasawa, T., Nishida, A., and Maezawa, K. (1982). High spatial attenuation of the Siple transmitter signal and natural VLF chorus observed at ground-based chain stations near Roberval, Quebec, *J. Geophys. Res.,* 87, 742.

Tsurutani, B. T., and Smith, E. J. (1977). Two types of magnetospheric ELF chorus and their substorm dependence, *J. Geophys. Res.,* 82, 5112.

Tsurutani, B. T., and Smith, E. J. (1974). Postmidnight chorus: a substorm phenomenon, *J. Geophys. Res.,* 79, 118.

Tsurutani, B. T., Smith, E. J., Anderson, R. R., Ogilvie, K. W., Scudder, J. D., Baker, D. N., and Bame, S. J. (1982a). Lion roars and non-oscillatory drift mirror waves in the magnetosphere, *J. Geophys. Res.,* 87, 6060.

Volland, H. (1984). *Atmospheric Electrodynamics,* Springer-Verlag, New York.

Walker, A. D. M. (1993). *Plasma Waves in the Magnetosphere,* Springer-Verlag, Berlin.

Wu, C. S., and Lee, L. C. (1979). A theory of terrestrial kilometric radiation. *Astrophys. J.,* 230, 621.

Yoshino, T., Ozaki, T., and Fukinishi, H. (1981). Occurrence distribution of VLF hiss and saucer emissions over the southern polar region, *J. Geophys. Res.,* 86, 864.

Chapter 14

ULF Pulsations

Karl-Heinz Glassmeier

CONTENTS

1. OBSERVATIONS OF ULF PULSATIONS

1.1. INTRODUCTION

On September 1, 1859, when Balfour Stewart recorded the great magnetic storm which occurred the same day, he was certainly aware of the potential importance of the observation made. Indeed, at around the same time when the Kew observatory recorded this major variation of the Earth's magnetic field Carrington observed intense sunspot activity, an observation indicating strong solar-terrestrial relations as far as geomagnetism is concerned.

Whether Balfour Stewart was aware of the importance of his observations with respect to our present understanding of ultra low-frequency (ULF) waves in space plasmas is not known. However, his observations, part of which are reproduced in Figure 14.1.1, are the first published records of so-called geomagnetic pulsations, a phenomenon today interpreted as excitation and propagation of magnetohydrodynamic waves in the Earth's magnetosphere. Since Balfour Stewart, many researchers have dealt with the study of geomagnetic pulsations, among them Rolf Harang and Gustav Angenheister in the early days of geomagnetic research.

It took a long time and the scientific capabilities of colleagues such as Hannes Alfvén (1942) and Jim Dungey (1954) to provide us with a clue to a definite understanding of geomagnetic

0-8493-2520-X/95/$0.00+$.50
© 1995 by CRC Press

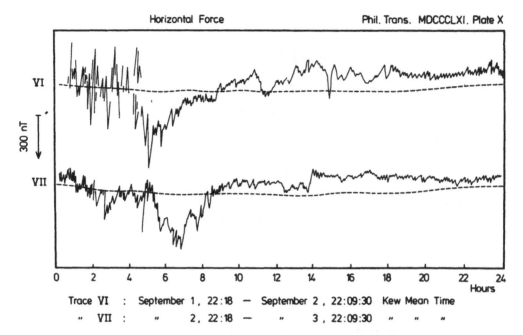

Figure 14.1.1 Reproduction of Balfour Stewart's historical observation of the great magnetic storm in 1859 and associated ULF waves. Shown is the horizontal perturbation during the time interval September 1, 22:18 until September 3, 22:09:30 Kew Mean Time.

pulsations. It was Alfvén who first proposed the existence of magnetohydrodynamic waves, or electromagnetic hydrodynamic waves, as he called them first. Dungey suggested the possible connection between geomagnetic pulsations and magnetohydrodynamic waves and already provided an elegant theory to describe many of their characteristics. Understanding these characteristics is certainly necessary as ULF waves transport energy, momentum, and, most important, information about magnetospheric dynamics. In this respect magnetohydrodynamic waves are as important as elastic waves when studying the structure of the solid earth.

This chapter intends to be an up-to-date overview on ULF pulsations as well as a tutorial for a wider community. It is a biased presentation as this review is based on my own approach to the subject as well as some of my own contributions to the field. Among other earlier presentations of the subject the reader is referred to the monographs by Jacobs (1970), Lanzerotti and Southwood (1979), Southwood and Hughes (1983), or the more recent reviews by Samson (1991) and Allan and Poulter (1992).

The structure of this review is as follows. First, we will summarize important observational findings on ULF pulsations. Second, an approach to the theory and propagation of magnetohydrodynamic waves is given, with special attention paid to those observations described. Third, possible excitation mechanisms are introduced.

1.2. OBSERVATIONAL EXAMPLES AND CLASSIFICATION

As outlined above ULF pulsations have first been identified in ground magnetic field observations. Today, these pulsations of the Earth magnetic field manifest themselves in ionospheric electric fields, magnetospheric magnetic and electric field measurements as well as particle observations on board spacecraft. Examples of such observations are given in Figures 14.1.2 through 14.1.6. Figure 14.1.2 shows the D component of a so-called geomagnetic giant pulsation. These pulsations occur preferentially in the morningside magnetosphere, that is between about 2:00 to 8:00 magnetic local time (MLT). They are striking because of their nearly monochromatic waveform with

November 19, 1976

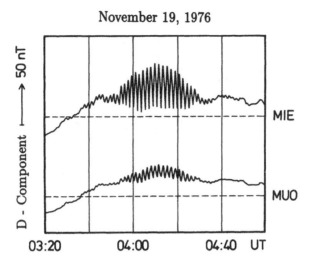

Figure 14.1.2 A giant pulsation event, recorded on November 19, 1976, at two stations of the Scandinavian Magnetometer Array in the Finmark.

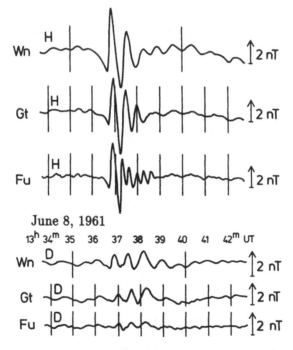

Figure 14.1.3 An example of a pulsation event with latitude-dependent period, recorded at the geomagnetic observatories Wingst (Wn). Göttingen (Gt), and Fürstenfeldbruck (Fu). Wingst is the northern-most observatory. (Modified from Voelker, H., *Naturwissenschaften,* 49, 8–9, 1962.)

periods of about 100 s (Annexstad and Wilson, 1968; Green, 1979) and amplitudes up to several tens of nT. The example shown in Figure 14.1.2 has been recorded at two stations of the Scandinavian Magnetometer Array (Küppers et al., 1979) only about 100 km apart. Between both stations the wave amplitude has significantly changed, which indicates a rather localized wavefield of extension 100 to 200 km in north-south and about 500 to 1000 km in east-west direction (Glassmeier, 1980).

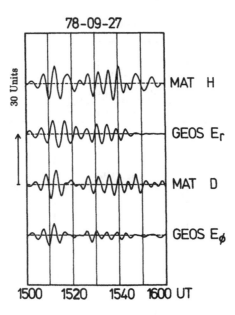

Figure 14.1.4   A Pc5 pulsation event, observed at the ground at a station of the Scandinavian Magnetometer Array in the Finmark as a geomagnetic pulsation and in the magnetosphere by the geostationary satellite GEOS 2 as a variation of the electric field. Units of the magnetic and electric field are nT and 0.1 mV/m, respectively.

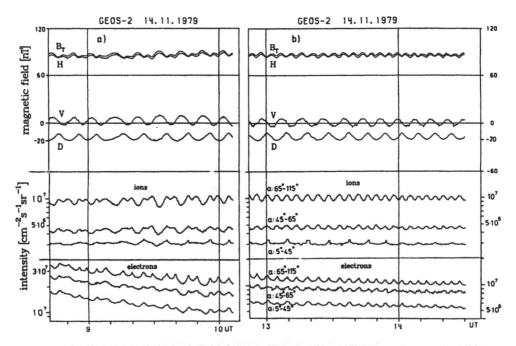

Figure 14.1.5   Observations of the magnetospheric magnetic field as well as electron and ion flux variations as observed on board the geostationary satellite GEOS 2 during a Pc5 pulsation. The components of the magnetic field vector (H, D, V) denote variations along the ambient field direction, at right angle to this in azimuthal direction, and along the radial direction. BT denotes the total magnetic field magnitude variation. The integral ion flux density is given for energies larger than 27 keV and various pitch angles, that of the electrons for energies larger than 22 keV. During the interval 8:42 to 10:06 UT (left panel) the satellite was 1.5° south of the geomagnetic equator, while between 12:52 to 14:32 UT (right panel) it was at the equator. (Modified from Woch, J. et al., *Planet. Space Sci.*, 36, 383–393, 1988.)

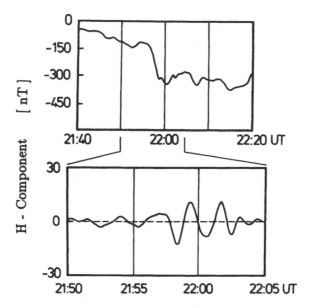

Figure 14.1.6 Typical variations of the H-component of the geomagnetic field during a magnetospheric substorm (upper panel) and a high-pass filtered time series of the same record, showing a Pi2 pulsation event (bottom panel). The substorm onset is characterized by the sudden decrease of the H-component at about 21:57 UT.

The example exhibited in Figure 14.1.3 is a pulsation event of more global extension. It is a damped-type oscillation with a period changing with geomagnetic latitude in at least the H component recorded at the geomagnetic observatories Wingst (northern-most station), Göttingen, and Fürstenfeldbruck (southern-most station; Voelker, 1962).

In the early days of ULF pulsation research these fluctuations of the Earth's magnetic field were usually called "geomagnetic micropulsations". This term is somewhat misleading as Figure 14.1.4 shows, where ground-magnetic and electric field observations made on board the spacecraft GEOS 2 during a Pc5 event (concerning the ULF pulsation nomenclature see below) are displayed. The pulsation is characterized in this case by about 0.1 mV/m electric field oscillations in the magnetospheric equatorial plane too, and Figure 14.1.4 allows one to deduce the magnetospheric (rather than purely ionospheric) and electromagnetic nature of ULF pulsations.

The magnetohydrodynamic character of ULF pulsations may be inferred from Figure 14.1.5 where satellite magnetic field and proton as well as electron flux intensity measurements during a pulsation event are shown (Woch et al., 1988). Components H, V, and D denote magnetic field components in a mean-magnetospheric-magnetic-field-aligned coordinate system, with H denoting field-aligned component, D the azimuthal component, counted positive eastward, and V the radial component. The particle observations are given for different pitch angles and cover the energy range $E > 27$ keV (protons) and $E > 22$ keV (electrons). The first time interval in Figure 14.1.5 covers observations made somewhat south of the geomagnetic equator, while the second interval shows those made at the geomagnetic equator. Important features of the magnetic field measurements are the approximately 90° phase shift between the two transverse components V and D, seen in both intervals. In contrast, V and H are in-phase south and about 70° out-of-phase at the geomagnetic equator. Furthermore, taking the particle measurements at pitch-angles 65 to 115° as indicative of the transverse plasma pressure, allows one to deduce an anti-phase relationship between plasma pressure and magnetic field magnitude (BT component) variations. These phase relationships are well in accordance with the curvature-driven drift mirror instability suggested by Woch et al. (1988) and as discussed in more detail in Section 3.2.

Another type of ULF pulsation, a so-called Pi2 pulsation, is exhibited in Figure 14.1.6. The upper part of the figure shows the H component of an isolated magnetospheric substorm (see

Table 14.1.1 **Classification of ULF Pulsations**

Group	Period range (s)	Amplitude[a] (nT)
Pc1	0.2–5	0.001–1
Pc2	5–10	0.1–1
Pc3	10–45	≤ 10
Pc4	45–150	5–20
Pc5	150–600	≤ 10–50
Pi1	1–40	0.01–1
Pi2	40–150	1–50

[a] Amplitude values are only approximate values; they depend much on the location where the observation has been made, etc.

Baumjohann, Chapter II/12), while the bottom part gives one the high-pass filtered magnetogram around substorm onset, i.e., the Pi2 pulsation with period about 150 s and amplitude about 10 nT. Pi2 pulsations are thus damped-type oscillations, reminiscent to those shown in Figure 14.1.3 (but of a much different physical nature), and are most clearly identified in geomagnetic mid-latitude records, where the substorm signature is not as prominent as in high latitudes.

The ULF pulsation examples shown here already elucidate interesting features of these indicators of magnetospheric activity. Their amplitudes may easily reach about 1% of the Earth magnetic field, their periods vary significantly with the waveforms exhibiting quasimonochromatic as well as damped-typed character, and the activity is testified in magnetic and electric field observations as well as particle measurements.

In the past 3 decades a vast phenomenology of ULF pulsations has been developed (see Saito, 1969; Jacobs, 1970), which was very useful in order to classify and structure all the numerous observations made. Table 14.1.1 gives the most important (because of its widespread use) classification mainly based on grouping pulsations according to their period and whether they accompany substorms (Pi) or not (Pc). More up-to-date classifications, especially with respect to satellite observations may be found in Anderson (1993).

1.3. SPATIAL STRUCTURE AND OCCURRENCE PROBABILITY

Knowledge of the spatial structure of ULF pulsation wavefields is of paramount importance when discussing possible source mechanisms of these waves. Of particular interest are spatial variations along the ambient magnetospheric magnetic field $\vec{B}$ as well as transverse to it, that is in north-south direction at the ground (radial direction in space) and east-west direction (azimuthal in space). Much emphasis has been put into studies of the harmonic structure of pulsation wavefields, i.e., the variation along $\vec{B}$. Lanzerotti and Fukunishi (1974) reviewed early work on this subject and found that ground magnetic observations demonstrate overwhelming occurrence of odd mode waves. According to their analysis, magnetic fieldlines oscillate as guitar strings fixed in the ionosphere and exhibit a maximum of the magnetic field (nodal point of the electric field) at the equator.

Due to occasional two-satellite (ATS 1 and OGO 5) encounters, Hughes et al. (1977) identified a second harmonic standing field-line oscillation much as did Singer et al. (1982) and Hughes and Grad (1984). Singer and Kivelson (1979), using magnetic field and ion flux measurements made on board OGO 5, found evidence that their observed Pc5 pulsations are fundamental mode field-line oscillations. Also Junginger et al. (1984) concluded from their statistical study of electric field measurements made on board the satellite GEOS 2 that Pc5 pulsations are fundamental mode eigenoscillations.

Furthermore, simultaneous existence of a fundamental and second harmonic mode of the same field line have been reported by Singer et al. (1979) and Junginger and Baumjohann (1984). A particular instructive example of the simultaneous existence of fundamental and higher harmonic

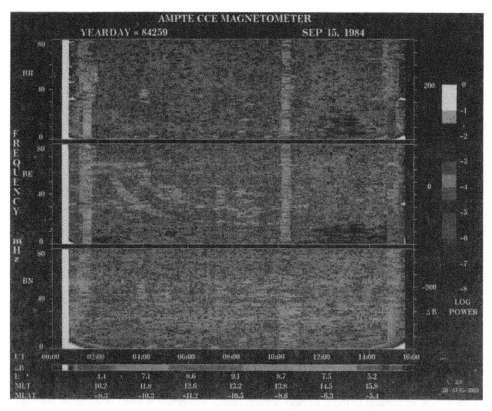

Figure 14.1.7 Three-component dynamic power spectra of magnetic field data from the AMPTE/CCE satellite for a full orbit from 0:25 to 16:04 UT on September 15, 1984. The bottom panel represents B, the difference in field magnitude between the observed total field and the value determined from the IGRF-80 model, the international geomagnetic reference field. Apogee is at the center of the figure. The components (BR, BN, BE) denote radial, compressional (i.e., along the ambient field), and azimuthal components. (The figure is courtesy of M. Engebretson.)

eigenoscillations on the same field-line shell is given in Figure 14.1.7, where a spectrogram of ULF waves in the 0- to 80-mHz frequency range is shown, which has been prepared from magnetic field data obtained by the elliptically orbiting AMPTE/CCE satellite (Engebretson et al., 1986). A geomagnetic coordinate system has been used where BR and BN denote the radial and compressional components, respectively, and BE the azimuthal component. While there is no evidence of harmonic structures in BR and BN, the bright falling and rising tones seen in the BE spectrogram are evidence of fundamental and higher harmonic field-line oscillations. The observed eigenperiods change with the McIlwain shell parameter L, indicating larger periods for larger L values and vice versa. This behavior of the oscillation period with L value, that is geomagnetic latitude, is reminiscent to that one displayed in Figure 14.1.3 for the case of a damped-type pulsation. However, in this latter case the period of the H component (radial component) changes, while the harmonic field-line oscillations shown in Figure 14.1.7 are purely azimuthal perturbations. As observations in Figure 14.1.3 are made at the ground below the ionosphere and those in Figure 14.1.7 above the ionosphere in the magnetosphere any theoretical model on ULF pulsations has to explain not only the observed frequency variation but also the influence of the ionosphere.

Studies of ULF pulsation wavefields across the ambient magnetic field have been made on the ground (Green, 1978), in the ionosphere (Walker et al., 1979), and in space (Singer et al., 1979). A landmark in ground-based observations which led to the modern development of ULF pulsation theory is the work by Samson et al. (1971), whose essential results are shown in

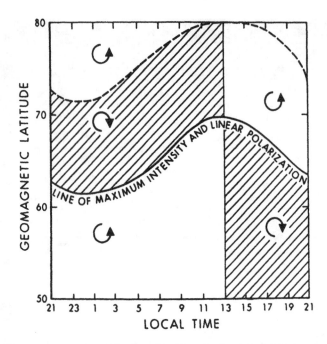

Figure 14.1.8 Schematic representation of the diurnal and latitudinal variation of the location of maximum amplitude and sense of rotation of the horizontal polarization of Pc4 to 5 pulsations. (Modified from Samson, J.C. et al., *J. Geophys. Res.*, 76, 3675–3683, 1971. Copyright by the American Geophysical Union. With permission.)

Figure 14.1.8. According to these authors several clearly distinctive regions of different pulsation characteristics, in particular the horizontal polarization, can be found, depending on geomagnetic latitude and local time.

According to Samson et al. (1971) the Pc5 pulsation amplitude maximizes at auroral zone geomagnetic latitudes, with the latitude changing with local time. Across this line of maximum amplitude the sense of polarization changes from counterclockwise (clockwise) to clockwise (counterclockwise) in the prenoon (postnoon) dayside magnetosphere going from south to north. The change of polarization is equivalent to a change of the phase of one of the magnetic field perturbation components by 180°. Thus, amplitude maximum and this phase change are very much reminiscent of a resonance phenomenon.

Much attention has been paid to this resonance feature. In particular Walker et al. (1979) have been successful in verifying Samson et al.'s (1971) results using measurements of the ionospheric electric fields during ULF pulsation events. Figure 14.1.9 gives one of their results, and shows the north-south variation of spectral power and phase of the north-south component of a Pc5 pulsation recorded on February 2, 1977, around 10 MLT. The power maximizes at about 69.4° geographic latitude and exhibits a very narrow, only about 100-km wide distribution, while the phase changes by nearly 180° over the power maximum, much as expected for a resonance.

First analyses of the EW phase variations have been performed by Herron (1966). In the Pc3 to 4 period range Green (1976) reports about small EW phase differences, i.e., azimuthal wavenumbers $m \leq 10$. Olson and Rostoker (1978) analyzed Pc4 to 5 pulsations and derived the following empirical relation between m and pulsation frequency

$$m = (1.4 \pm 0.4) \cdot f + 0.26 \qquad (1.1)$$

with f given in millihertz. Both Green (1976) and Olson and Rostoker (1978), furthermore, report phase propagation toward the tail of the magnetosphere, i.e., to the west in the morning and to the east in the afternoon sector of the dayside magnetosphere. Such a change of phase propagation

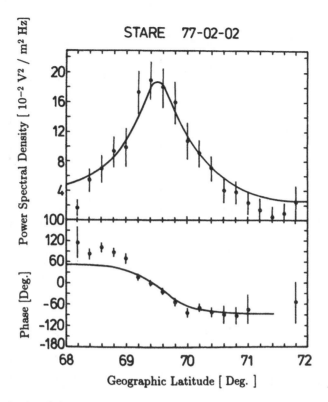

Figure 14.1.9 Amplitude and phase variation vs. geographic latitude (e.g., approximately geomagnetic latitude) of the EW component of the ionospheric electric field as seen by the Scandinavian Twin Auroral Radar system in the Pc5 pulsation period band in the late morning hours of February 2, 1977, in northern Scandinavia. (Modified from Walker, A.D.M. et al., *J. Geophys. Res.*, 84, 3373–3388, 1979. Copyright by the American Geophysical Union. With permission.)

direction across local magnetic noon corresponds to polarization changes as found by Samson et al. (1971; see Figure 8).

The Olson-Rostoker relation gives azimuthal wavenumbers of m ≈ 5.5 for a typical Pc5 pulsation with period T = 250 s, as shown in Figure 14.1.4. In contrast to these small m values, giant pulsations (Figure 14.1.2) or compressional Pc5 pulsations as displayed in Figure 14.1.5 usually are associated with rather large m values, m ≈ 20 to 30 (Glassmeier, 1980; Allan and Poulter, 1983). Such pulsations are therefore probably due to a different source mechanism than typical Pc5, as discussed in more detail later.

Not many studies of the two-dimensional amplitude structure of ULF pulsations have been performed up to now as this requires observations from an extended two-dimensional magnetometer array. Glassmeier et al. (1984c) and Glassmeier (1987) report about EW extensions of the pulsation wavefield for typical Pc5 pulsations of the order 5000 km, with the NS scale given by the already discussed resonance features (see Figure 14.1.9). Further detailed studies of the amplitude and phase variation of Pc3 to 5 pulsations have been conducted by, e.g., Ziesolleck et al. (1993) and Ziesolleck and Chamalaun (1994) using observations from the two-dimensional Australia Wide Array of Geomagnetism Stations (Chamalaun and Barton, 1990) at low latitudes. Results by Ziesolleck and Chamalaun (1994) indicate a peculiar spatial variation of the Pc3 to 4 amplitude, polarization, and phase structure, indicative of a spatially localized resonance structure deep within the plasmasphere much as observed at high latitudes.

In the Pc5 period range Ziesolleck et al. (1993) found that the Pc5 frequency is virtually independent of latitude and longitude with the amplitude considerably decreasing toward lower latitudes. Determination of phase differences between the observational sites yields small phase

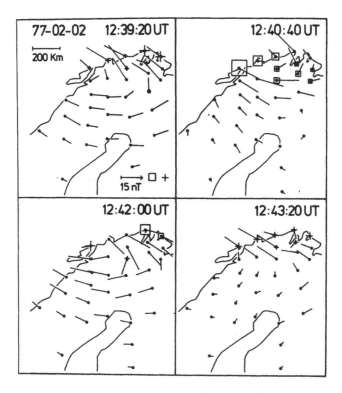

Figure 14.1.10 Ground-equivalent current vectors as observed by the Scandinavian Magnetometer Array during the February 2, 1977, Pc5 pulsation event (see also Figure 14.1.9). The pulsation period was about 250 s. The current vectors have their origin at the place of measurement. Squares and crosses denote negative and positive Z component.

variations across the entire magnetometer array. This is quite in contrast to the resonance structure observed for Pc3 to 4 pulsations at these latitudes and for Pc5 pulsations at high latitudes. Azimuthal wavenumbers are rather low with $1 < m < 3$. As much observed at higher latitudes, phase propagation is usually toward the magnetotail. These observations indicate a more global oscillation of the entire plasmasphere for Pc5 oscillations.

A very convenient presentation of both north-south and east-west amplitude as well as phase variation is by using so-called ground-equivalent current vectors, i.e., horizontal magnetic perturbation vectors rotated clockwise by 90°. The example exhibited in Figure 14.1.10 is for the same pulsation event whose ionospheric electric field is shown in Figure 14.1.9. The current vectors are given at four successive time intervals, separated by about a quarter wave period. The north-south as well as east-west amplitude and phase variation is apparent. Figure 14.1.10 also indicates that ground-equivalent current patterns have a striking vorticity structure. Observing a ULF pulsation over a much more extended array than used in Figure 14.1.10 would result into a vortex-street pattern as discussed by Obertz and Raspopov (1968). Glassmeier (1980), Jede (1985), and McHenry et al. (1990). This vorticity feature of ULF pulsations will be discussed in more detail later when discussing the influence of the ionosphere.

Besides the spatial variation of ULF pulsation wavefields along the ambient magnetospheric magnetic field and in north-west and east-west direction, their variation with local time is a very important parameter which should be explainable by any theoretical model. Detailed discussions may be found in, e.g., Anderson (1993). As an example, Figure 14.1.11 displays the diurnal variation of ULF pulsation activity in the Pc5 period band observed at a ground-magnetometer station (Alta in Northern Scandinavia, see Lühr et al., 1984) in the northern auroral zone. A ULF

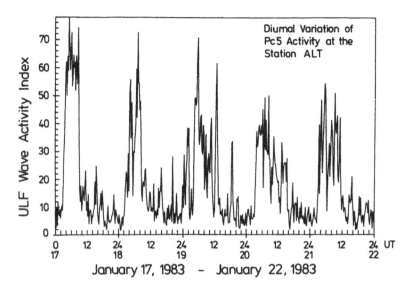

Figure 14.1.11   Diurnal variation of Pc5 pulsation activity as observed at a station in northern Norway during the time interval January 17, 1983 to January 22, 1983.

pulsation activity index has been defined as the percentage ratio of power of the H component in the 1- to 7-mHz to the total power in the 0.2- to 100-mHz frequency range (Stein and Glassmeier, 1991). The figure clearly shows that at successive days Pc5 pulsation activity is most prominent at around 6 MLT, that is these pulsations are obviously generated at the flanks of the magnetosphere. The clear repetition of the activity every 24 h, furthermore, indicates a stationary source region in the magnetosphere with the rotating Earth and the observing station moving across this region. Activity groups as displayed in Figure 14.1.11 show a 27-day recurrence probability, too, and Stein und Glassmeier (1991) found evidence of such groups being associated with the Earth passing solar wind high-speed flow regimes (Hundhausen, 1972), an observation allowing the understanding of the 27-day reoccurrence tendency.

Other types of ULF pulsations are characterized by different MLT dependencies. For example, compressional Pc5, i.e., those oscillations associated with significant variations of the magnetospheric magnetic field magnitude (see Figure 14.1.5), at geosynchronous orbit occur primarily in the afternoon and early evening hours and at dusk are correlated with storm-time development of partial ring currents (see Anderson, 1993 and references therein), which indicates a relation of such waves to the dynamics of the high β ring current plasma.

2.   MAGNETOHYDRODYNAMIC WAVES

2.1.   MHD WAVES IN A COLD HOMOGENEOUS PLASMA

A detailed review of MHD waves is outside the scope of this review, and the interested reader is referred to the excellent tutorial review by Siscoe (1983) or the already cited works. Here I will use a description aiming to demonstrate certain relations between elastic waves and hydromagnetic waves. In a cold, homogeneous, and magnetized nonrotating plasma small-amplitude, linear perturbations of the hydrodynamic as well as electromagnetic field will be treated. The basic equation to be tackled is the equation of motion

$$\rho_0 \frac{d\vec{\upsilon}}{dt} = \vec{j} \times \vec{B}_0 \tag{2.1}$$

where $\vec{v}$, $\vec{j}$ and $\vec{B}_0$ are the perturbation of the velocity field, the current associated with the wave, and the ambient magnetic induction field, respectively; ρ_0 is the unperturbed plasma density. The equation of motion has to be complemented by Maxwell's equations

$$\nabla \times \vec{E} = -\frac{\partial \vec{b}}{\partial t} \tag{2.2}$$

and

$$\nabla \times \vec{b} = \mu_0 \vec{j} \tag{2.3}$$

as well as Ohm's law in a plasma with infinite conductivity, that is the frozen-in theorem

$$\vec{E} = -\vec{v} \times \vec{B}_0 \tag{2.4}$$

where $\vec{E}$ and $\vec{b}$ are the perturbed electric and magnetic field, respectively. In the above equations it has been assumed that the wavefrequency is much less than the gyrofrequency of the plasma ions. This essentially justifies the dropping of the displacement current term in Equation 2.3. Only in very large magnetic fields, as they are in the Jovian magnetosphere close to the ionosphere, does the displacement current need to be included. Eliminating $\vec{b}$, $\vec{v}$ and $\vec{j}$ from the above Equations 2.1 through 2.4 gives a wave equation for the electric field of the form

$$\frac{\partial^2 \vec{E}}{\partial t^2} = \vec{v}_A \times \vec{v}_A \times \nabla \times \nabla \times \vec{E} \tag{2.5}$$

where the cross products are to be read from right to left. The vector $\vec{v}_A$ is the Alfvén velocity defined by

$$\vec{v}_A = \frac{\vec{B}_0}{\sqrt{\mu_0 \rho_0}} \tag{2.6}$$

An excursion into the field of elastic waves allows one to learn about an elegant way to find the normal modes of Equation 2.5 in case of a uniform magnetic field $\vec{B}_0 = (0, 0, B_0)$. Let $\vec{s}$ denote the displacement field of an elastic deformation. The corresponding wave equation reads

$$\frac{\partial^2 \vec{s}}{\partial t^2} = v_p^2 \Delta \vec{s} - v_t^2 \nabla \times \nabla \vec{s} \tag{2.7}$$

where v_p and v_t are the phase velocity of the compressional and shear mode, respectively. According to the Helmholtz theorem (Morse and Feshbach, p. 53, 1953) each vector field may be described by its sources and vortices:

$$\vec{s} = \vec{s}_{irr} + \vec{s}_{sf} \tag{2.8}$$

$$\nabla \cdot \vec{s}_{sf} = 0 \text{ and } \nabla \times \vec{s}_{irr} = 0$$

It can easily be proven that $\vec{s}_{irr}$ and $\vec{s}_{sf}$ describe the compressional and shear mode, respectively. This means that the two fundamental modes may be found using the Helmholtz theorem.

Does this apply for the MHD wave Equation 2.5 as well? From Equation 2.4 it follows that $\vec{E} \perp \vec{B}_0$, that is the electric field has only components transverse to $\vec{B}_0$, or $\vec{E} = \vec{E}_\perp$, where the subscript denotes the transverse part. With this and a little manipulation Equation 2.5 can be written as

$$\frac{\partial^2 \vec{E}_\perp}{\partial t^2} = -v_A^2 \nabla \times \nabla \times \vec{E}_\perp = -v_A^2 (\nabla(\nabla \cdot \vec{E}_\perp) - \Delta \vec{E}_\perp) \tag{2.9}$$

an equation exhibiting certain analogies with Equation 2.7. By splitting

$$\vec{E}_\perp = \vec{E}_{\perp,irr} + \vec{E}_{\perp,sf} \tag{2.10}$$

and having $\nabla_\perp \cdot \vec{E}_{\perp,sf} = 0$ and $(\nabla \times \vec{E}_{\perp,irr})_\parallel = 0$, Equation 2.9 may be transformed into two decoupled equations for $\vec{E}_{\perp,sf}$ and $\vec{E}_{\perp,irr}$ alone:

$$\frac{\partial^2 \vec{E}_{\perp,sf}}{\partial t^2} = \upsilon_A^2 \Delta \vec{E}_{\perp,sf} \tag{2.11}$$

and

$$\frac{\partial^2 \vec{E}_{\perp,irr}}{\partial t^2} = \upsilon_A^2 \frac{\partial^2 \vec{E}_{\perp,irr}}{\partial z^2} \tag{2.12}$$

where the z-direction is parallel to $\vec{B}_0$. As the Laplace operator appears in Equation 2.11 the mode described by $\vec{E}_{\perp,sf}$ is an isotropic mode, that is the fast or compressional mode. The mode described by $\vec{E}_{\perp,irr}$ and Equation 2.12 is an anisotropic mode as only the derivative along $\vec{B}_0$ appears. This is the well-known Alfvén mode. For a plane wave and the ansatz $\exp(i\vec{k}\vec{x} - i\omega t)$ the corresponding dispersion relations are

$$\omega^2 = \upsilon_A^2 k^2 \tag{2.13a}$$

for the fast mode, and

$$\omega^2 = \upsilon_A^2 k^2 \cos^2\theta \tag{2.13b}$$

for the Alfvén mode, where θ is the angle between $\vec{k}$ and $\vec{B}_0$. The above derivation shows that our excursion to elastic waves was very helpful. However, it should be noted that the split of the vector fields has been done for the displacement field in case of the elastic wave and for the electric field in the magnetohydrodynamic case. Thus the analogy should be used with care. A further difference appears. In case of the elastic waves the Helmholtz theorem is applied to a three-dimensional vector field, while in the MHD case it is applied to a two-dimensional field only, due to the electric field perturbation always being transverse to $\vec{B}_0$. This causes the use of the transverse part of $\nabla \cdot$ and the z component of the $\nabla \times$ operator in Equation 2.10. Further similarities between elastic waves and MHD waves are discussed by Krummheuer (1991).

Further important characteristics of MHD waves may be derived from the above equations concerning the Poynting flux $\vec{S}$, the magnetic induction perturbation, and the relation between $\vec{\upsilon}$ and $\vec{b}$. For the energy flux one finds

$$\vec{S}_{fast} \parallel \vec{k} \text{ and } \vec{S}_{Alfvén} \parallel \vec{B}_0 \tag{2.14}$$

As for the Alfvén mode $(\nabla \times \vec{E}_\perp)_\parallel = 0$ we have

$$\frac{\partial b_\parallel}{\partial t} = 0 \tag{2.15}$$

where the subscript $\parallel$ denotes the field-aligned component. This shows that the Alfvén mode is a purely transverse mode with respect to the magnetic field perturbation, while the fast mode may well carry a field-aligned perturbation, that is, it describes a magnetically compressive mode.

Of interest for the analysis of spacecraft observations is the Alfvén relation

$$\vec{\upsilon} = \pm \frac{\vec{b}}{\sqrt{\mu_0 \rho}} \qquad (2.16)$$

according to which velocity and magnetic induction field perturbations are related in the same way as the phase velocity and the ambient magnetic field. For waves propagating parallel to $\vec{B}_0$ the negative sign in Equation 2.16 holds, that is $\vec{\upsilon}$ and $\vec{b}$ are anti-correlated, while for waves propagating anti-parallel to $\vec{B}_0$ $\vec{\upsilon}$ and $\vec{b}$ are correlated. Thus the Alfvén relation allows the determination of the Alfvén wave energy propagation direction, using a single-point measurement only. This technique has widely been used in the past by, for example, Belcher and Davis (1971) or Glassmeier et al. (1989b) in the solar wind.

A further interesting distinction between the Alfvén mode and the fast mode exists with respect to the electrical current perturbation carried by the wave. Multiplying Equation 2.1 from the right by $\times B_0$ allows one to deduce the transverse current component $j_\perp$

$$\vec{j}_\perp = \vec{j}_p = \frac{\rho_0}{B_0^2} \frac{d\vec{E}}{dt} \qquad (2.17)$$

This transverse current is also known as polarization or inertial current. It is formally a displacement current and arises due to the temporal variations of $\vec{E}$. With Equations 2.10 a couple of interesting results can be derived. For the fast mode with

$$\nabla_\perp \cdot \vec{E}_{\perp,sf} = 0$$

a consequence is the relation

$$\nabla_\perp \cdot \vec{j}_{p,fast} = 0 \qquad (2.18)$$

that is, the fast mode polarization current is entirely closed in planes perpendicular to B_0. In other words, the fast mode does not carry field-aligned currents. For the Alfvén mode we find

$$\nabla_\perp \cdot \vec{j}_{p,Alfvén} = -\frac{\partial j_\parallel}{\partial z} \neq 0 \qquad (2.19)$$

if $\vec{k}$ is not parallel to $\vec{B}_0$. Thus, the Alfvén mode carries field-aligned currents. This is schematically displayed in Figure 14.2.1 and is of eminent importance for the understanding of magnetospheric dynamics and magnetosphere-ionosphere coupling. As discussed in detail by, for example, Southwood and Hughes (1983) field-aligned currents act to couple different plasma regimes such as the distant magnetosphere and the ionosphere. Any change in this coupling associated with a change of the existing field-aligned current system must be communicated between the two plasma regimes via Alfvén wave propagation. This represents an interesting source of ULF pulsations as discussed in Section 3.4.

Equations 2.17 and 2.19 do not allow insight into the current carriers of both the polarization as well as the field-aligned currents. To get this information the plasma model must be expanded. Consideration of a two-fluid model, that is ions and electrons, gives one the equations of motions

$$m_e \frac{d\vec{\upsilon}_e}{dt} = -e(\vec{E} + \vec{\upsilon}_e \times \vec{B}_0) \qquad (2.20a)$$

and

$$m_i \frac{d\vec{\upsilon}_i}{dt} = e(\vec{E} + \vec{\upsilon}_i \times \vec{B}_0) \qquad (2.20b)$$

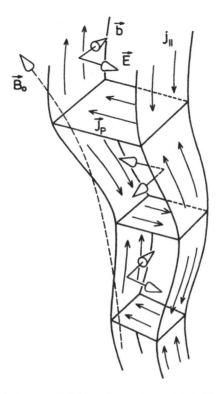

Figure 14.2.1 Electromagnetic fields and currents associated with an Alfvén wave.

where e is the electron charge. For ULF waves as discussed in this work $\omega \ll \Omega_i$ and Equations 2.20 transform into

$$\vec{v}_e = -\frac{1}{\Omega_e B_0} \frac{d\vec{E}_\perp}{dt} + \frac{\vec{E} \times \vec{B}_0}{B_0^2} + \frac{\Omega_e}{\omega^2 B_0} \frac{d\vec{E}_\parallel}{dt} \qquad (2.21a)$$

and

$$\vec{v}_i = \frac{1}{\Omega_i B_0} \frac{d\vec{E}_\perp}{dt} + \frac{\vec{E} \times \vec{B}_0}{B_0^2} - \frac{\Omega_i}{\omega^2 B_0} \frac{d\vec{E}_\parallel}{dt} \qquad (2.21b)$$

where indices $\perp$ and $\parallel$ denote components transverse and parallel to the ambient magnetic field, and $\Omega_{i,e}$ are the ion and electron gyrofrequencies, respectively. The first terms on the right-hand side of Equation 2.21a,b denote the polarization or inertial drift, the middle terms the charge independent $\vec{E} \times \vec{B}_0$ drift, and the last terms the field-parallel motion of the particles. Both the polarization drift as well as the field-aligned motion give rise to an electric current in the plasma. As the polarization drift depends on $1/\Omega_{i,e}$ the associated transverse polarization current is mainly carried by the ions, while the field-aligned current due to its dependence on $\Omega_{i,e}$ is mainly carried by electrons. The two-fluid approach thus allows identification of the carriers of currents associated with Alfvén and fast mode.

Alfvén and fast mode waves as discussed above occur in a cold plasma only. The cold plasma approximation is suitable for most regions of the Earth's magnetosphere. However, in regions of the ring current, that is in high plasma β regimes, the thermal pressure might not be neglible in the equations of motion, Equation 2.1. Pressure gradient terms $-\nabla p$ must be considered, too. Due to

$$(\nabla \times \vec{E}_{\perp,irr})_\parallel = 0$$

and because of Equation 2.4 one has $\nabla_\perp \cdot \vec{v} = 0$. Therefore, the Alfvén mode is unaffected by the pressure term. The fast mode, however, will couple with the acoustic mode. A more detailed analysis shows that two magneto-acoustic modes occur, the fast mode of the warm plasma and the so-called slow mode with the dispersion relation (Siscoe, 1983)

$$\frac{\omega^2}{k^2} = \frac{1}{2}(c_s^2 + v_A^2) \pm \frac{1}{2}\sqrt{(c_s^2 + v_A^2)^2 - 4c_s^2 v_A^2 \cos^2\theta} \qquad (2.22)$$

Here c_s is the acoustic wave speed, and the positive sign corresponds to the fast mode and the negative sign to the slow mode.

The fast mode propagates almost isotropically, while the slow mode is a strongly anisotropic mode. Both modes are associated with magnetic field magnitude as well as plasma density fluctuations. In case of the fast mode magnetic field and density oscillate in-phase, while for the slow mode they are anti-correlated. Concerning the slow mode it should be noted that its phase speed is rather small. Thus this mode is heavily Landau damped and is usually not observed in space plasmas. Wave events exhibiting an anti-correlation between pressure or density and magnetic field magnitude are usually caused by high β (see Figure 14.1.5) plasma instabilities as discussed below.

2.2.   MHD WAVES IN INHOMOGENEOUS PLASMAS

MHD waves as discussed in the previous section are those as observable in homogeneous plasmas, i.e., under conditions of uniform background magnetic field and plasma density. However, planetary magnetospheres are regions of both strongly inhomogeneous magnetic field and plasma density. In the following sections several aspects of the influence of such inhomogeneities on wave propagation are discussed.

2.2.1.   The Density Step

The most simple inhomogeneity encountered in space plasmas is that of a step in plasma density. The terrestrial plasmapause is the most prominent example of such a step, the magnetopause another one. To analyze the reflection and transmission of MHD waves at such a discontinuity we treat the special case of an Alfvén wave obliquely incident on a plane separating two half spaces with different plasma densities. The uniform magnetic field is perpendicular to the plane, i.e., along its normal. At the boundary three conditions must be fulfilled. First, we have to consider the continuity of the tangential component of the electric field. This results in

$$\vec{E}_T = \vec{E}_I + \vec{E}_R \qquad (2.23)$$

where subscripts T, I, and R denote the transmitted, incident, and reflected wave, respectively. Second, we have to consider current continuity at the boundary, which gives

$$\nabla_\perp \cdot \vec{j}_{p,1} = \nabla_\perp \cdot \vec{j}_{p,2} \qquad (2.24)$$

For a plane harmonic wave and with Equation 2.17 condition 2.24 may be written as

$$\frac{\rho_1}{\rho_2}\vec{k}_T \cdot \vec{E}_T = \vec{k}_I \cdot \vec{E}_I - \vec{k}_R \cdot \vec{E}_R$$

The third condition at the boundary is the continuity of the tangential component of the propagation vector

$$k_{\perp,T} = k_{\perp,I} = k_{\perp,R} = k_\perp \qquad (2.25)$$

Combining Equations 2.23 through 2.25 allows us to deduce relations between the electric field of the reflected and transmitted wave with respect to that of the incident wave

$$\vec{E}_R = -\frac{\rho_1 - \rho_2}{\rho_1 + \rho_2} \vec{E}_I \qquad (2.26)$$

and

$$\vec{E}_T = \frac{2\rho_2}{\rho_1 + \rho_2} \vec{E}_I \qquad (2.27)$$

At the plasmapause typical density ratios of $\rho_1/\rho_2 \approx 10000$ are observed. Thus, $E_R \approx 0.998\ E_I$, i.e., almost perfect reflection occurs. This description of wave reflection and transmission at a density step applies to the Alfvén wave incidence only. More complete discussions may be found in the work by Stein (1971).

2.2.2. The Weakly Inhomogeneous Plasma

As compared to the density step a weakly inhomogeneous plasma is the other extreme occurring in space plasma systems. We regard the plasma as weakly inhomogeneous if the spatial scale L of plasma density and magnetic field variations is much larger than the wavelength λ, i.e., we require $\lambda \ll L$. In such cases WKB approximations are applicable and internal reflections can be neglected. Considering wave propagation along a flux tube with varying cross section A, magnetic field magnitude, and plasma density as well as neglecting internal reflections gives one

$$\vec{S}_P \cdot A = const. \qquad (2.28)$$

where $\vec{S}_P$ is the Poynting flux. Magnetic flux conservation along the tube yields

$$B_0 \cdot A = const. \qquad (2.29)$$

Combining Equations 2.28 and 2.29, and taking into account

$$S_P = \frac{v_A b^2}{\mu_0}$$

where b is the wave magnetic field perturbation, the following relation between b and the flux tube plasma density results

$$b \propto \rho^{1/4} \qquad (2.30)$$

that is the wave magnetic field varies as the fourth root of the density. With some care the above discussion is applicable to the terrestrial cusp regions, where spacecraft and ground-based observations exhibit very pronounced ULF wave activity (see Glassmeier, 1989a for a review).

2.2.3. The Ionosphere as a Boundary

Ground-based observations of ULF pulsations are severely influenced by the terrestrial ionosphere. Planetary ionospheres are thus other important examples of inhomogeneous situations causing drastic changes in wave propagation characteristics. The dominant effect is that of the electrical conductivity, mainly that in the E region. Many theoretical studies concerning the influence of the conducting ionosphere have been published in the past years (Nishida, 1964; Huges, 1974; Glassmeier, 1984a).

In the following section we will discuss the effect of the high-latitude ionosphere on incident Alfvén waves. A Cartesian system is adopted with the x-axis directed to the north, the y-direction pointing toward east, and the z-direction positively downward. In this coordinate system the ionosphere will be regarded as a highly conducting sheet lying in the x-y plane at a height $z = -h$, where $h = 110$ km is the height of the current carrying E region. The height-integrated Hall and Pedersen conductances are denoted by Σ_H and Σ_P, respectively. A sheet ionosphere approach is justified because the extension of the terrestrial E region has a scale of about 20 km, while typical ULF pulsation wavelengths are of the order of tens of thousand of kilometers.

Above this sheet ionosphere we assume a uniform semi-infinite hydromagnetic region as a model magnetosphere while the region below the ionosphere is regarded as a semi-infinite, non-conducting atmosphere. The magnetospheric magnetic field $\vec{B}_0$ is assumed uniform and parallel to the z-axis. In the ionosphere the curl-free electric field $\vec{E}$ of an Alfvén wave gives rise to a sheet current system

$$\vec{I}(x,y) = \Sigma_P(x,y)\vec{E} - \Sigma_H(x,y)\vec{E} \times \vec{e}_B \tag{2.31}$$

where $\vec{e}_B$ is the unit vector along $\vec{B}_0$. Displacement currents are neglected due to the large periods of ULF pulsations. In the magnetosphere the Alfvén wave carries transverse polarization and field-aligned currents as described by Equations 2.17 and 2.19. Assuming a plane harmonic wave, i.e., $\vec{E} \propto exp(i\vec{k} \cdot \vec{x} - i\omega t)$ the field-aligned current $j_\parallel = jz$ may be written as

$$j_\parallel = \nabla_\perp \cdot \left(-\frac{i}{\mu_0\omega}\frac{\partial\vec{E}}{\partial z}\right) \tag{2.32}$$

At the interface between ionosphere and magnetosphere the boundary conditions for the density step discussed above hold. In particular we have to consider current continuity

$$\nabla \cdot \vec{I} = j_\parallel \tag{2.33}$$

The electric field in the ionosphere $\vec{E}$ is given as the sum of those of the incident and reflected wave, $\vec{E} = \vec{E}_I + \vec{E}_R$. This and Equations 2.31 through 2.33 allow one to deduce the following relation between $\vec{E}_I$ and $\vec{E}_R$

$$(\Sigma_A + \Sigma_P)\nabla_\perp \cdot \vec{E}_R + \nabla_\perp\Sigma_P \cdot \vec{E}_R - (\nabla_\perp\Sigma_H \times \vec{E}_R)z$$
$$= (\Sigma_A - \Sigma_P)\nabla_\perp \cdot \vec{E}_I - \nabla_\perp\Sigma_P \cdot \vec{E}_I + (\nabla_\perp\Sigma_H \times \vec{E}_I)z \tag{2.34}$$

where $\Sigma_A = 1/(\mu_0\upsilon_A)$ is the Alfvén wave conductance. This equation allows one to determine $\vec{E}_R$ for any distribution of $\Sigma_H(x,y)$ and $\Sigma_P(x,y)$ once $\vec{E}_I$ is known. Numerical solutions are discussed in Glassmeier (1984) and Glassmeier (1987).

If Σ_P and Σ_H are uniform, Equation 2.34 reduces to

$$(\Sigma_A + \Sigma_P)\nabla_\perp \cdot \vec{E}_R = (\Sigma_A - \Sigma_P)\nabla_\perp \cdot \vec{E}_I$$

or

$$\vec{E}_R = \frac{\Sigma_A - \Sigma_P}{\Sigma_A + \Sigma_P}\vec{E}_I = R \cdot \vec{E}_I \tag{2.35}$$

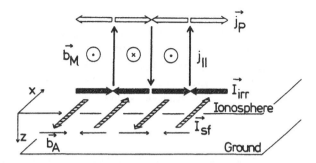

Figure 14.2.2 Electric currents associated with an Alfvén wave reflected at the ionosphere and the resulting magnetic field above and below the ionosphere.

an equation first derived by Scholer (1970). With typical values for the terrestrial magnetosphere and ionosphere, $\Sigma_A = 0.05S$ and $\Sigma_P = 4S$, a reflection coefficient $R = -0.975$ results. The terrestrial ionosphere is thus an almost perfect reflector, just like the density step at the plasmapause. The similarity of Equations 2.26 and 2.35 is noteworthy.

A reflection coefficient close to -1 implies not only a 180° phase shift between the electric fields of the incident and the reflected waves, but also that the total ionospheric electric field is rather small. As the plasma motion is related to the electric field via $\vec{v} = \vec{E} \times \vec{B}_0/B_0^2$ a small value of $\vec{E}$ implies small $\vec{v}$ too. Thus, the oscillating field lines are almost anchored in the ionosphere.

The transmission characteristics of the ionosphere can best be studied by analyzing the total ionospheric sheet current system $\vec{I}$. This current system can be determined using Equation 2.31 once $\vec{E}_R$ has been computed using Equation 2.34. According to the Helmholtz theorem $\vec{I}$ can be separated into a curl-free and a source-free contribution, a separation already used above when deriving the Alfvén and fast modes. In general we have

$$\vec{I} = \vec{I}_{irr} + \vec{I}_{sf} \tag{2.36}$$

and a rather tedious method is required to determine both contributions (Glassmeier, 1984). In case of a uniform conductance distribution, however, both parts are given by

$$\vec{I}_{irr} = \Sigma_P \vec{E} \tag{2.37a}$$

and

$$\vec{I}_{sf} = \Sigma_H \vec{E} \times \vec{e}_B \tag{2.37b}$$

As schematically shown in Figure 14.2.2 the current system build up by the magnetospheric polarization currents of the wave, its field-aligned currents, and irrotational part $\vec{I}_{irr}$ of the ionospheric sheet current system is a poloidal current system whose magnetic field effect is not detectable below the ionosphere. A more formal proof of this can be found in McHenry and Clauer (1987). Here only the source-free part contributes to the magnetic field (Figure 13). Again, for a uniform conductance distribution this results in an interesting effect. The Pedersen and Hall current vectors, i.e., the irrotational and source-free contributions $\vec{I}_{irr}$ and $\vec{I}_{sf}$ to $\vec{I}$ are perpendicular to each other. Thus, the magnetospheric magnetic field $\vec{b}_M$ and the magnetic field below the ionosphere, $\vec{b}_A$, are perpendicular to each other, too. The ionosphere therefore acts as a 90° rotator of the wave polarization vector, an effect most clearly worked out by Hughes (1974). However, in general, the ionosphere may cause a rotation of the polarization vector by any angle between 0 to 90°, depending on the conductance distributions (Glassmeier, 1987). An observational hint toward the ionospheric rotation effect is the latitude variation of the resonant ULF wave eigen-period of the azimuthal component in space and the H component at the ground, that is the latitudinal variation is seen in components rotated by 90°.

Below the ionosphere the wave magnetic field is curl-free due to only the source-free part of $\vec{I}$ contributing to the magnetic field there. Thus, the atmospheric magnetic field of an incident Alfvén wave can be described by a scalar magnetic potential W with

$$\Delta W = 0 \tag{2.38}$$

As the horizontal variation of W is prescribed by the incident wave the vertical variation of W and therefore $\vec{b}_A$ is that of a surface wave. In particular, Equation 2.38 indicates that the ionosphere acts as a spatial low-pass filter. Below the ionosphere the wave magnetic field varies as $\vec{b}_A \propto \exp(-k_\perp z)$, that is wavefields with large horizontal variations k or large m values are spatially damped by this filter effect.

2.2.4. Magnetospheric Eigenoscillations

As the terrestrial ionosphere and plasmapause act as almost perfect reflectors of MHD waves it is tempting to study standing wave modes in the magnetosphere, i.e., to look for eigenoscillations of magnetospheric systems. Dungey (1954) was the first to tackle this problem in detail. Searching for such eigenoscillations requires the solution of the general wave Equation 2.5, using appropriate boundary conditions. Dungey (1954) studied the special case of a dipole magnetosphere with a constant plasma density and an ambient magnetic field

$$\vec{B}_0(\vec{r}) = \frac{\mu_0 M}{4r^3} \cdot (\sin\theta \; \vec{e}_\theta + 2\cos\theta \; \vec{e}_r) \tag{2.39}$$

where M is the magnetic dipole moment. In spherical coordinates and using Equation 2.39 the general wave Equation 2.5 is

$$\left(\frac{\mu_0\rho_0}{B_0^2}\frac{\partial^2}{\partial t^2} - \frac{\partial^2}{\partial r^2} - \frac{\sin\theta}{r^2}\frac{\partial}{\partial\theta}\frac{1}{\sin\theta}\frac{\partial}{\partial\theta}\right)r\sin\theta \; E_\phi = \sin\theta\left(B_{0,r}\frac{\partial}{\partial\theta} - B_{0,\theta}r\frac{\partial}{\partial r}\right)\frac{1}{r\sin\theta}\frac{\partial v_\phi}{\partial\phi} \tag{2.40}$$

$$\left(\mu_0\rho_0\frac{\partial^2}{\partial t^2} - \frac{1}{r^2\sin^2\theta}\left((\vec{B}_0\cdot\nabla)(r^2\sin^2\theta)(\vec{B}_0\cdot\nabla) + B_0^2\frac{\partial^2}{\partial\phi^2}\right)\right)\frac{v_\phi}{r\sin\theta}$$
$$= \frac{1}{r^3\sin^3\theta}\left(\frac{B_{0,r}}{r}\frac{\partial}{\partial\theta} - B_{0,\theta}\frac{\partial}{\partial r}\right)r\sin\theta\frac{\partial E_\phi}{\partial\phi} \tag{2.41}$$

where $v_\phi = -E_\theta/B_{0,r}$ has been used. Equations 2.40 and 2.41 are coupled partial differential equations describing the azimuthal component of the wave electric field E_ϕ and the plasma velocity perturbation v_ϕ. The coupling between E_ϕ and v_ϕ is given by the right-hand sides of the equations and depends on the structure of the ambient magnetic field as well as the perturbation wave field.

A general solution of Equations 2.40 and 2.41 has not yet been achieved. Numerical solutions have been published by Lee and Lysak (1989, 1990). Analytical solutions are known only for several special cases. An important example is that of an axisymmetric perturbation, i.e., for the case $\partial/\partial\phi = 0$. For this case the coupling terms on the right-hand sides of Equations 2.40 and 2.41 vanish and one gets the following set of decoupled equations

$$\left(\frac{\mu_0\rho_0}{B_0^2}\frac{\partial^2}{\partial t^2} - \frac{\partial^2}{\partial r^2} - \frac{\sin\theta}{r^2}\frac{\partial}{\partial\theta}\frac{1}{\sin\theta}\frac{\partial}{\partial\theta}\right)r\sin\theta \; E_\phi = 0 \tag{2.42}$$

$$\left(\mu_0\rho_0\frac{\partial^2}{\partial t^2} - \frac{1}{r^2\sin^2\theta}((\vec{B}_0\cdot\nabla)(r^2\sin^2\theta)(\vec{B}_0\cdot\nabla))\right)\frac{v_\phi}{r\sin\theta} = 0 \tag{2.43}$$

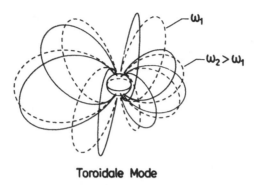

Toroidale Mode

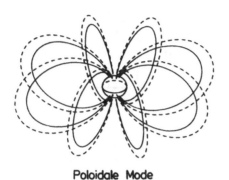

Poloidale Mode

Figure 14.2.3 Toroidal and poloidal mode eigenoscillations of the dipol magnetosphere. Dashed lines are undisturbed field lines, solid ones describe the perturbed situation. Toroidal eigenoscillations are shown for two different field line tori, an inner and an outer one.

The above equations were first derived by Dungey (1954) and describe toroidal (υ_ϕ) and poloidal (E_ϕ) oscillations. As $\upsilon_\phi \propto b_\phi$ the toroidal mode is associated with axisymmetric azimuthal transverse magnetic field perturbations, while the poloidal one is associated with variations of $\vec{b}$ in the radial and field-aligned directions, i.e., it is associated with variations of the magnetic field magnitude as well. Both modes are schematically shown in Figure 14.2.3, and they can be described as a twisting (toroidal) and a breathing (poloidal) eigenmode of the dipole magnetosphere.

A more detailed discussion of the toroidal mode shows that in Equation 2.43 only differential operations $\vec{B} \cdot \nabla$ occur, that is merely variations along magnetic field lines exist. Each field line shell is thus oscillating independently from those lying further inside or outside. The oscillation period is therefore a function of the radial distance of the field-line shell. Under nominal magnetospheric conditions one expects an increase of the oscillation period with radial distance or increasing geomagnetic latitude, at least at auroral latitudes. This theoretical finding fits very nicely with observations displayed in Figure 14.1.3 and taking into account the previously described 90° rotation of the polarization vector of magnetospheric ULF waves when passing the ionosphere.

Siebert (1993) has recently questioned the validity of Dungey's (1954) approach. Equation 2.43 contains differential operations $\vec{B}_0 \cdot \nabla$ along field lines only while, for example, the wave-frequency may vary across field lines due to changes of the plasma density. Furthermore, the decoupled toroidal motion of the field lines is associated with a radial electric field $\vec{E}_r$ as well as a radial polarization current $\vec{j}_r$. Both quantities are noncontinuous functions of the radial distance. Current continuity, however, requires a field-aligned current $j_\| = \partial j_r / \partial_r$. As jr is a noncontinuous function of r, the operation $\partial / \partial r$ is not defined. This mathematical inconsistency requires an extension of the physical model used to describe ULF waves. Siebert (1993) in particular emphasizes the importance of including field-aligned electric fields and is able to show that this immediately relaxes the mathematical difficulties mentioned above.

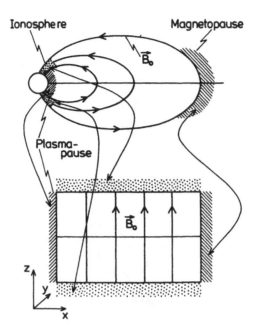

Figure 14.2.4 The box model magnetosphere. (Modified from Southwood, D.J., and M. G. Kivelson, *J. Geophys. Res.*, 87, 1707–1710, 1982. Copyright by the American Geophysical Union. With permission.)

2.2.5. The Field-Line Resonance Phenomenon

The condition of axisymmetry imposed on Dungey's equations to obtain decoupled toroidal and poloidal eigenoscillation is a severe one. Most of the observed ULF pulsations exhibit a clear variation of the wavefield in azimuthal direction. Theoretical work by Tamao (1965), Southwood (1974), and Chen and Hasegawa (1974) on wave propagation in the magnetosphere including in particular non-axisymmetric conditions has led to the development of the field line resonance model of ULF pulsations.

A particularly elucidating approach is that given in Southwood (1974). The magnetosphere is reduced to a box magnetosphere as shown in Figure 14.2.4. Plasma density varies along the x-axis only, while the background magnetic field is uniform and along the z-axis, i.e., $B_0 = (0, 0, B_0)$. With the ansatz

$$\vec{E}_\perp = (E_x(x), E_y(x), 0) \exp(ily + ikz - i\omega t) \tag{2.44}$$

and defining

$$R^2(x) = \frac{\mu_0 \rho(x) \omega^2}{B_0^2} \tag{2.45}$$

the poloidal component E_y of the general wave Equation 2.5 is transformed into (Southwood, 1974)

$$\frac{\partial^2}{\partial x^2} E_y - l^2 \frac{\partial}{\partial x} R^2(x) \frac{1}{(R^2 - k^2)(R^2 - k^2 - l^2)} \frac{\partial E_y}{\partial x} + (R^2 - k^2 - l^2) E_y = 0 \tag{2.46}$$

This equation exhibits strong singularities found in the denominator of its second term. The following solutions are possible. Let us first assume the case $R^2 - k^2 - l^2 > 0$, from which

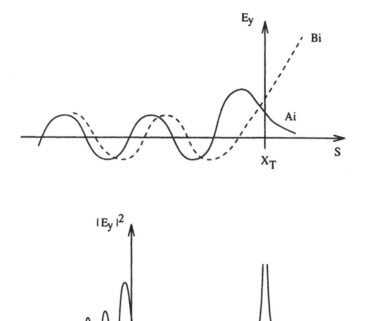

Figure 14.2.5 Airy function and field-line resonance phenomenon.

$R^2 - k^2 > 0$ follows. Thus, no singularities appear. Assuming further $l \approx 0$, Equation 2.46 reduces into

$$\frac{\partial^2 E_y}{\partial x^2} + (R^2 - k^2)E_y = 0 \tag{2.47}$$

which gives one for a linear density profile, i.e., $R^2(x) = \alpha_0^2 - \alpha^2 \cdot x$, with the definition of the turning point $R_T^2 = \alpha_0^2 - \alpha^2 x_T = k^2$, and the transformation $s = \alpha^{2/3}(x - x_T)$

$$\frac{\partial^2 E_y}{\partial s^2} - sE_y = 0 \tag{2.48}$$

This is an Airy or Stokes differential equation with the two principle solutions displayed in Figure 14.2.6. The solution Bi(s) is unphysical as it implies unlimited growth of E_y behind the wave turning point s = 0. Thus $E_y(s) = Ai(s)$ is the required solution. The turning point actually is the point of total reflection of the wavefield. Its appearance can be understood on the following grounds. The fast mode dispersion relation is given by

$$\frac{\omega^2}{l^2 + k^2 + r^2} = \upsilon_A^2 \tag{2.49}$$

where r is a wavenumber in x-direction. Now if υ_A^2 increases with x, that is with s in Figure 14.2.5, r^2 has to decrease as ω, l, and k stay constant. Eventually r^2 may become negative, which implies an imaginary wavenumber r. At this turning point the wave will be reflected.

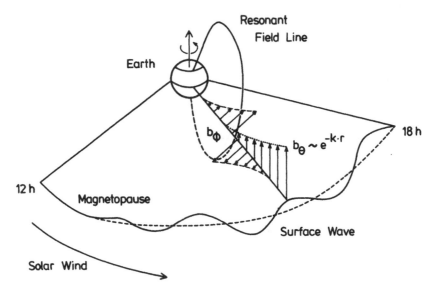

Figure 14.2.6 Kelvin-Helmholtz unstable surface waves and field-line resonances.

Considering the case $R^2 - k^2 - l^2 < 0$, $R^2 - k^2 = 0$ may occur. Assuming again a linear density profile, defining a resonance point $R_\omega^2 = \alpha_0^2 - \alpha^2 \cdot x_\omega = k^2$, and $s = l(x - x_\omega)$ Equation 2.46 transforms close to the resonance point $x - x_\omega = 0$ into

$$\frac{\partial^2 E_y}{\partial s^2} + \frac{1}{s}\frac{\partial E_y}{\partial s} - E_y = 0 \tag{2.50}$$

which is a modified Bessel equation. At the resonance point its solution exhibits a clear singularity, that is unlimited growth of E_y. To discuss the general solution of Equation 2.46 for the case $R^2 - k^2 - l^2 < 0$ is beyond the scope of this review. Schematically the solution is given in Figure 14.2.5. It can be seen that in front of the turning point the solution is reminiscent to an Airy function, while behind it singular behavior is observed at the resonance point.

This singular behavior, i.e., a peaked electric field distribution is very much reminiscent of the ionospheric electric field observations displayed in Figure 14.1.9. The following physical interpretation is tempting. The MHD wave propagating into the magnetosphere is a fast mode wave generated by, say, plasma instabilities at the magnetopause. Eventually the wave reaches the turning point where reflection occurs. If conditions are favorable, that is if a resonance point occurs, part of the fast mode wave energy can tunnel into the resonance, where coupling from the fast mode disturbance into an Alfvén mode perturbation takes place. This scenario is schematically shown in Figure 14.2.6, where the reflection or turning point is assumed to coincide at the magnetopause. At resonance a phase shift of 180° between the toroidal field components (b_ϕ in the figure, E_x in the box model) on both sides of the resonance is apparent. This phase shift can be understood using Maxwell's Equations 2.2 and 2.3 with Equation 2.17, and deriving the equation

$$(R^2 - k^2)E_x = il\left(\frac{\partial E_y}{\partial x} - ilE_x\right) = l\omega b_z \tag{2.51}$$

At resonance $R^2 = k^2$, thus $b_z = 0$, and for the electric field polarization we have

$$\frac{E_x}{E_y} = -\frac{i}{lE_y}\frac{\partial E_y}{\partial x} \tag{2.52}$$

The polarization therefore depends on l as well as $\partial E_y/\partial x$. As l changes sign across local noon for a Kelvin-Helmholtz instability generated fast mode wave at the magnetopause (see Section 3.1) and $\partial E_y/\partial x$ changes sign across the resonance point, a four-quadrant pattern emerges as a result of the field-line resonance phenomenon, much as observed by Samson et al. (1971; see Figure 8).

2.2.6. Recent Developments on the Mode Coupling Problem

The field-line resonance mechanism described above has developed into the paradigm of ULF pulsation theory. Following the work by Tamao (1965), Southwood (1974), and Chen and Hasegawa (1974) much further work has been done to understand the coupling processes between fast and Alfvén mode perturbations in the magnetosphere. Points of particular interest have been questions on which magnetospheric conditions support the evolution of resonances, which kind of magnetic topologies support decoupled oscillations, whether localized transverse field-line oscillations are propagated without coupling to compressional perturbations, etc.

For example, Inhester (1985) has provided a very detailed discussion of the problem, taking into account the effects of finite ionospheric conductivity as well as the effects of magnetospheric asymmetries. As discussed in Section 1.3 toroidal fundamental Pc5 pulsations occur primarily in the morningside magnetosphere, an asymmetry, probably associated with magnetospheric asymmetries caused by the existence of the magnetotail and the plasmaspheric bulge in the afternoon sector of the magnetosphere. As shown by Inhester (1985), such asymmetries of the background magnetospheric plasma are as important for the mode coupling problem as the asymmetry of the ULF pulsation sources. A lack of symmetry causes the polarization of the resonant oscillations to be in conflict with the requirement of magnetic incompressible perturbations.

Wright and Evans (1991) extensively studied the necessary conditions for magnetic field topologies to allow decoupling of transverse or compressional perturbations. For example, in a dipole field transverse azimuthal oscillations are decoupled from transverse radial or field-aligned oscillations if the perturbation is axisymmetric. However, this does not apply for the radial transverse component in the corresponding situation as it is always coupled to compressional oscillations due to the condition $\nabla \cdot \vec{b} = 0$. As an example, only for a spheroidal magnetic field topology can decoupled poloidal transverse oscillations exist.

Of particular interest for purposes of analyzing actual observations of ULF pulsations is the question of whether localized transverse MHD waves or torsional waves remain as transverse oscillations along magnetic field lines. Such waves are characterized by the condition $(\nabla \times \vec{E})_\parallel \neq 0$ or $\nabla_\perp \cdot \vec{b} = 0$. As discussed by Wright (1990) this condition can only be satisfied along field lines if the eigenperiods of the toroidal and poloidal field components can be matched, i.e., if the phase velocities in planes containing the toroidal and poloidal components are identical. Results from Wright (1990) can be applied, for example, when studying the propagation of localized ULF wavefields in magnetospheres. Examples of particular interest are torsional waves excited by pressure pulses at the magnetopause (see Section 3.4) or the relative motion of a conductive body in a magnetized plasma such as the moon Io in the Jovian magnetosphere (Neubauer, 1980).

The theory of field-line resonances and other works on the mode coupling demonstrate that MHD wave propagation in a realistic magnetospheric environment is governed by a plethora of types of couplings between fast and transverse oscillations. In fact, usage of the terms "fast mode" and "Alfvén mode" is often misleading as in most cases these pure eigenmodes do not exist. However, these two terms can be used as synonyms for the source-free and irrotational parts of the polarization current system associated with any MHD wave. A question of interest from the theoretical and experimental standpoint is whether the coupling processes between both parts of the polarization current system are reversible, e.g., one has to study whether in the asymptotic time limit one of the modes predominates. Radoski (1974) has demonstrated that in this case the energy of any compressional perturbation finally couples irreversible into transverse oscillations. Physically this result can be understood in terms of the number of degrees of freedom that the electric current associated with either perturbations has. As discussed in Section 2.1 the Alfvén mode is supported by transverse as well as field-aligned currents while the fast mode is driven by transverse currents only. Thus the Alfvén mode has one more degree of freedom in arranging current flow than does the fast mode.

An important application of this result has been discussed by Kivelson and Southwood (1985) and Allan and Poulter (1989). Interpretation of fundamental toroidal ULF pulsations as field-line resonances is widely accepted. As discussed in Section 3.1, a possible source mechanism for these waves is the Kelvin-Helmholtz instability at the Earth's magnetopause. This instability, however, generates a rather broad spectrum of unstable waves. Therefore, one would expect different field-line resonances to be excited at any time by this instability, that is a continuous spectrum of resonant field lines should emerge. However, usually only one is present, e.g., a discrete spectrum is observed, which raises serious questions on the applicability of either the field-line resonance or the Kelvin-Helmholtz hypothesis.

Kivelson and Southwood (1985) pointed out a very appealing way out of this dilemma. They suggested that excitation of field-line resonances does not necessarily come directly via coupling of Kelvin-Helmholtz instability generated surface waves to resonant transverse azimuthal oscillations, but via a coupling to so-called cavity modes. These global poloidal eigenoscillations are characterized by a global eigenfrequency (compare with Section 2.2.4), that is the Kelvin-Helmholtz instability gives first rise to fundamental and higher harmonic cavity modes. The excitation of these modes then acts as a frequency filter for the broad band wave sources. Finally, the cavity modes with a discrete spectrum coupled to field-line resonances, a coupling well in accordance with the above-mentioned work on the behavior of compressional waves in the asymptotic time limit.

Since Kivelson and Southwood's (1985) suggestion, much emphasis has been put into the identification of the required cavity modes. Only a few observations may have indicated the existence of such waves, among them observations by Kivelson et al. (1984) and Higbie et al. (1982) or the observations by Ziesolleck et al. (1993) for plasmaspheric cavity modes. A more indirect hint toward the existence of cavity modes was provided by Crowley et al. (1987) using measurements of the ionospheric Pedersen conductivity and damping rates of ULF pulsations. As discussed later, ionospheric joule heating is a major energy sink for ULF pulsations, and determination of the damping rates allows one to estimate the Pedersen conductivity (Glassmeier et al., 1984). If the estimated and the measured conductivity values do not agree this may indicate that the resonant waves are not only damped but also excited at the same time. The anticipated excitation is due to coupling of cavity wave mode energy into the resonant waves. It is this mechanism which led Crowley et al. (1987) to postulate the existence of a cavity mode during their observed resonant Pc5 pulsation event. As another hint toward the existence of cavity mode observations by Ziesolleck et al. (1993) on plasmaspheric field lines may be regarded too (see Section 1.3).

3. EXCITATION OF ULF PULSATIONS

Known excitation mechanisms for ULF pulsations can be grouped into four different classes: (1) plasma instabilities at the boundaries of the magnetosphere; (2) instabilities of the internal magnetospheric plasma distributions; (3) wave generation outside the magnetosphere and subsequent propagation into the magnetosphere; and (4) macroscopic changes of the magnetospheric configuration. Each of these excitation mechanisms as well as wave damping will be discussed below.

3.1. THE KELVIN-HELMHOLTZ INSTABILITY

Surface waves at the magnetopause, generated by the Kelvin-Helmholtz-instability, have already been suggested by Dungey (1954) as a possible excitation mechanism. A couple of observational

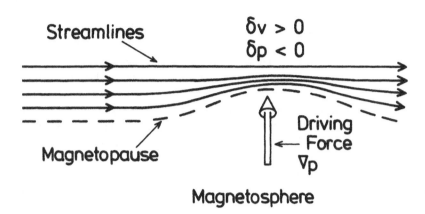

Figure 14.3.1 The Kelvin-Helmholtz instability mechanism.

facts support this solar-wind-related mechanism. First, analysis of east-west phase differences shows wave phase propagation toward the tail of the magnetosphere, both in the morning and afternoon sectors (see Olson and Rostoker, 1978). Second, ULF pulsations as shown in Figure 14.1.11 exhibit a tendency to occur at the flanks of the magnetosphere, especially in the morning sector. Third, ULF pulsation activity is clearly modulated by the solar wind flow velocity. Fourth, ULF wave polarization characteristics are controlled by solar wind flow as well.

The Kelvin-Helmholtz-instability is capable of explaining most of these observations. Therefore much work has been devoted to understanding this instability mechanism. For details, the reader is referred to work by Chandrasekhar (1961), Sen (1965), Walker (1981), Pu and Kivelson (1983), and Miura (1984). In this paper, I will only describe the physical mechanism. As displayed in Figure 14.3.1 the instability mechanism is based on the relative streaming of two plasma regimes separated by a tangential discontinuity. A small perturbation of the equilibrium position of the interface, in our case the magnetopause, results in diminishing the flow cross section of region 1 (see Figure 14.3.1). Flux conservation therefore requires an increase of the streaming velocity in this perturbation region, i.e. $\delta\upsilon > O$. Applying to first order Bernoulli's law

$$\frac{1}{2}\rho\upsilon^2 + p = const. \tag{3.1}$$

a flow velocity perturbation $\delta\upsilon > O$ is accompanied by a negative pressure perturbation $\delta v < O$. This pressure perturbation in turn supports the initial perturbation of the interface: instability results and waves are generated and fed by kinetic energy from the differential streaming.

A more detailed treatment (see above-referenced work) allows one to deduce the following equation as a necessary condition for instability

$$\frac{1}{\mu_0}(B_1^2 + B_2^2) < \frac{\rho_1\rho_2}{\rho_1 + \rho_2}\upsilon_0^2 \tag{3.2}$$

where B_1, B_2, ρ_1, and ρ_2 are magnetic field magnitude and plasma density on both sides of the interface, and $\upsilon_0 = \upsilon_1 - \upsilon_2$ denotes the relative flow velocity. Approximating $\upsilon_2 = 0$, $\rho_2 \approx \rho_1$, and $B_2 \approx B_1$ gives one from Equation 3.2

$$4\upsilon_A < \upsilon_1 \tag{3.3}$$

This relation implies that the flow velocity in region 1, or the magnetosheath, always has to be super-Alfvénic in order to allow the Kelvin-Helmholtz instability to develop. This has important implications. Close to the stagnation point, that is near magnetic local noon, the magnetosheath flow is sub-Alfvénic. Only toward the flanks of the magnetosphere does the flow speed become trans-Alfvénic again (Spreiter et al., 1966). Thus, the Kelvin-Helmholtz instability will occur at the flanks of the magnetosphere, i.e., around dawn and dusk, much as has been observed, at least for the dawnside. The observed Pc5 pulsation occurrence asymmetry (see Figure 14.1.11) is not yet fully understood. Here the structure of the magnetosheath magnetic field is of importance, as its direction determines the growth rates of the instability too. Also, as pointed out by Inhester (1985), field-line resonances may not occur under asymmetric background plasma condition as occurring in the afternoon magnetosphere due to the plasmaspheric bulge (see also Section 2.2.6).

As discussed by Southwood (1968) the unstable waves always propagate with the magneto-sheath flow, i.e., tailward. The stagnation point near local magnetic noon is therefore a natural point across which characteristics of toroidal fundamental mode Pc5 pulsations should change. The polarization change across noon (Figure 14.1.8) as first shown by Samson et al. (1971) is a result of the changing wave propagation direction.

How Kelvin-Helmholtz unstable waves are related to field-line resonances is schematically displayed in Figure 14.2.6. The unstable waves are surface waves of the fast mode type (b_ϕ-component). Energy may thus propagate across field lines up to the resonance point where toroidal oscillations (b_μ-component) are generated due to the above described linear mode coupling process.

3.2.   INTERNAL MAGNETOSPHERIC PLASMA INSTABILITIES

The Kelvin-Helmholtz instability discussed above is a specific mechanism that extracts energy from the shear flow to generate plasma wave energy in the low-frequency regime. Other very important mechanisms to transfer energy toward ULF waves are several different instabilities directly tapping the kinetic energy reservoir of the plasma particle motion. A favorable magnetospheric region for such energy transfer mechanisms is the ring current region with its trapped particle population (Lanzerotti and Schulz, 1974). Common mechanisms of wave-particle inter-action are Landau damping or the ion-cyclotron instability discussed in Chapter II/13. A similar but more general mechanism is that of the bounce-resonance instability first discussed by South-wood et al. (1969) as a possible mechanism to generate ULF pulsations.

In a dipole magnetic field configuration charged particles are described not only by their gyromotion around the local magnetic field direction but also by their bounce motion between the mirror points above the northern and southern ionospheres as well as a drift motion in azi-muthal direction (Lanzerotti and Schulz, 1974). The rate of change of particle energy, $\dot{W}$, of a magnetospheric particle moving adiabatically through a small amplitude ULF wavefield is given by

$$\frac{dW}{dt} = \mu \frac{\partial b_\parallel}{\partial t} + q\vec{E}_\perp \cdot \vec{v}_D + qE_\parallel v_\parallel \qquad (3.4)$$

where $b_\parallel$ and $E_\parallel$ are magnetic and electric field perturbations along the ambient magnetic field, and $\vec{v}_D$ is the particle drift velocity. Thus, in a uniform field wave-particle interaction is only possible if the wave is of the fast mode type ($b_\parallel \neq 0$) or a kinetic Alfvén wave with a large perpendicular wavenumber in which case $E_\parallel \neq 0$. This latter case would correspond to the clas-sical collisionless Landau damping mechanism.

Interaction with transverse waves, i.e., Alfvén modes, is only possible in nonuniform field geometries which support drift motions. The energy exchange is due to particle deceleration by the transverse electric field resonating with the drift motion. The resonance condition is given by

$$\omega - m\Omega_D = N\Omega_B \qquad (3.5)$$

where Ω_D and Ω_B are the drift and bounce frequencies, respectively; m is the azimuthal wavenumber of the ULF wavefield and N any integer number. The above resonance condition is equivalent to the corresponding ion-cyclotron resonance in that the drift motion replaces the field-aligned motion and provides the required Doppler shift to put particles and wavefield into bounce resonance.

Equation 3.5 is only a necessary condition for instability. In order to secure wave growth the resonant particle distribution, f_{res}, that is the part of the particle distribution approximately fulfilling condition 3.5 must be inverted with

$$\frac{\partial f_{res}}{\partial W} > 0 \qquad (3.6)$$

Hughes et al. (1978) provided the first convincing observational evidence for this type of instability to actually occur in the Earth's magnetosphere. They studied particle and field observations made on board the ATS-6 satellite and found that the observed quasimonochromatic transverse ULF waves are closely related to a bump-on-the-tail particle velocity distribution with Equation 3.6 fulfilled for resonant protons of energy about 100 keV. Subsequent analysis of the resonance condition 3.5 yields an azimuthal wavenumber $m \approx 100$ for the ULF wave generated. As discussed in Section 2.2.3 the ionosphere acts as a spatial low-pass filter with a cut-off wavelength of the order $1/110$ km^{-1}, that is the inverse of the E-region height. Very large m-value ULF pulsations as observed by Hughes et al. (1978) are thus barely detectable at the ground.

However, Glassmeier (1980) suggests that giant pulsations with m values of the order of 20 to 30 as displayed in Figure 14.1.2 are a possible candidate for ULF waves generated by the bounce resonance instability. Further observational support for this hypothesis is given by Glassmeier et al. (1992), who reported that bump-on-the-tail plasma distributions of ≈ 60 keV protons were associated with giant pulsations. The inverted distribution itself is thought to be the result of proton velocity dispersion with the protons injected into the Earth ring current by nighttime substorm activity. Evidence against the bounce resonance instability as a source mechanism for giant pulsations has been discussed by Takahashi et al. (1992). These authors point out that bounce resonance unstable waves should be even harmonic standing waves in the magnetosphere, while their observed giant pulsation event seems to indicate odd-harmonic mode structure.

The dependence of the bounce resonance mechanism on the mode structure is visualized in Figure 14.3.2. Whether a bouncing proton acquires or loses energy in a fundamental mode electric field structure (see left panel) very much depends on the phase between the bounce motion and the azimuthal phase of the wave. However, the symmetry of the system requires that net acceleration over a bounce period is zero whatever the relative phasing between the wave and the proton is (Southwood and Kivelson, 1982). If the ULF wave generated exhibits an even mode structure (e.g., second harmonic as in the right panel of Figure 14.3.2) particles in bounce resonance may gain or lose energy over many consecutive bounces. Thus a net wave-particle interaction results.

This dependence of the bounce resonance instability on the symmetry of the excited wave with respect to the magnetospheric equatorial plane has led several authors to discuss the mode structure of giant pulsations (Annexstad and Wilson, 1968; Green, 1979; Glassmeier, 1980; Hillebrand et al., 1982; Chisham and Orr, 1991; Takahashi et al., 1992). However, up to now no conclusive results have been derived, and future work is required to identify the excitation mechanism of giant pulsations and the importance of the bounce resonance instability.

While the above discussed bounce resonance instability is occurring due to deviations of the particle distribution function from local thermodynamic equilibrium, other instabilities are driven

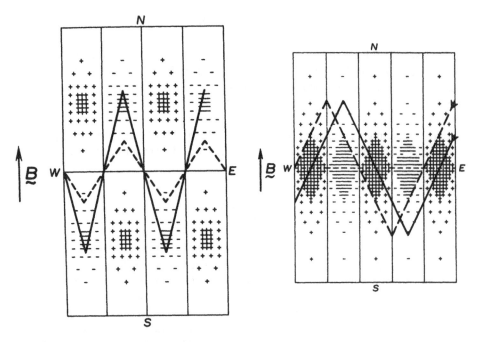

Figure 14.3.2 Schematic of the electric field intensity of a fundamental (left panel) and a second harmonic (right panel) standing mode ULF wave in the magnetosphere. The magnetic field lines are shown as vertical lines between the northern (N) and southern (S) ionospheres. The electric field intensity and direction is indicated by plus or minus symbols, corresponding to eastward or westward orientation. Heavy and dashed lines denote the guiding centers of bouncing particles with different phases relative to the wave (left panel) and different equatorial pitch angles (right panel). (Modified from Southwood, D.J., and M.G. Kivelson, *J. Geophys. Res.,* 87, 1707–1710, 1982. Copyright by the American Geophysical Union. With permission.)

by anisotropies of the plasma pressure. In particular the firehose and mirror instabilities are discussed, first in the seminal work of Chew et al. (1956). A common feature of space plasmas is that they very often exhibit differences in the plasma pressure transverse and along the ambient magnetic field. Excess transverse pressure may occur in mirror magnetic field topologies with particles lost along the field lines at the mirror points. Excess longitudinal pressure may result due to backstreaming particles reflected off the Earth bow shock, for example.

Firehose instability then occurs as a result of excess longitudinal pressure, in which case the centrifugal forces of field-aligned streaming particles in a transversely perturbed field are not completely balanced by magnetic tension forces (see Siscoe, 1983), thus giving rise to instability of Alfvén wave-like perturbations. Mirror instability occurs as a result of excess transverse pressure. Consider the situation depicted in Figure 14.3.3, which shows a mirror magnetic field situation associated with an initial compressional magnetic field and particle pressure perturbation almost perpendicular to the ambient field (Hasegawa, 1969). As the particle and magnetic pressure balance each other the perturbation is made up by a succession of high magnetic field magnitude and low plasma density regions followed by low magnetic field and high plasma density ones. Under isotropic pressure conditions this perturbed situation would rapidly relax. However, the mirror geometry, once set up, increases the pressure anisotropy due to particles with large field-aligned velocity components leaving the mirror geometry regions (Hasegawa, 1969). The initial perturbation is thereby amplified, and mirror instability results, if

$$\frac{\beta_\perp}{\beta_\parallel} > 1 + \beta_\perp^{-1} \tag{3.7}$$

is fulfilled, where $\beta_\perp$ and $\beta_\parallel$ are the transverse and longitudinal ratios of plasma to magnetic field pressure, respectively.

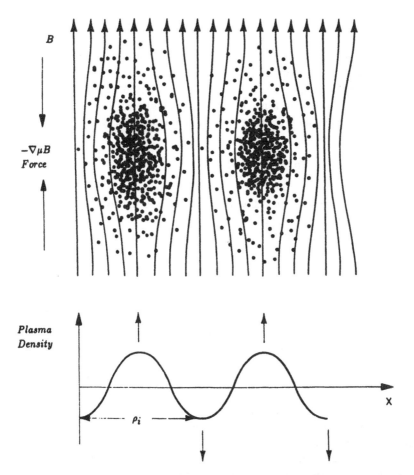

Figure 14.3.3 Schematic of the mirror instability mechanism. The upper panel displays the magnetic field topology and the associated particle distribution, the bottom panel shows the plasma density perturbation. (Modified from Hasegawa, A., *Plasma Instabilities and Nonlinear Effects,* Springer-Verlag, Berlin, 1975.)

As compared to the firehose instability, mirror modes are purely growing perturbations, that is $\omega = 0$, and are standing structures perpendicular to the ambient magnetic field, not traveling waves. The wavelength $\lambda_\perp$ is the order of several thermal ion gyroradii, i.e., mirror modes should be large m-value, nonoscillatory structures. This and their compressional structure make them attractive to understand compressional perturbations such as described in Figure 4.1.5 and as very often observed in the ring current (Barfield and McPherron, 1978; Allan et al., 1982; Woch et al., 1988).

However, the classical mirror mode gives rise to purely growing modes. Thus, a mechanism is required to couple these modes to propagating waves. As the ring current is expected to exhibit large spatial pressure gradients (Williams, 1987) a natural way to get propagating mirror modes is coupling to drift waves, a mechanism first proposed and studied by Hasegawa (1969), who called the resulting instability a drift mirror instability. This instability, which describes the growth of large m-value compressional waves with plasma density and magnetic field magnitude in antiphase (see Figure 14.3.3), has been successfully used to explain many features of storm-time ULF pulsations in the ring current (Kremser et al., 1981). Still, as the drift mirror instability predicts the growth of predominantly compressional waves with only a minor transverse component as compared to the field-aligned, compressional perturbation, observations of large transverse components cannot be explained. This needs a modification of the classical drift mirror instability by allowing coupling of the compressional modes to shear Alfvén perturbations. The most advanced studies of this are presented by Pokhotelov et al. (1986) and Woch et al. (1988), who also include the effect of magnetic field curvature, which results in a lower threshold of the instability as compared to Equation 3.7.

Coupled compressional Alfvén waves and drift mirror modes as described by the curvature-driven drift mirror instability differ significantly from the classical Alfvén, fast, and slow modes in that forces different from the polarization current (see Equation 2.10) driven Lorentz force $\vec{j}_p \times \vec{B}_0$ act as restoring forces. In a magnetized warm plasma with pressure gradients, pressure anisotropies, and field curvature the transverse drift current is given by

$$\vec{j}_D = \frac{\vec{B}}{B^2} \times \left(\nabla p_\perp + \frac{p_\| - p_\perp}{B^2}(\vec{B} \cdot \nabla)\vec{B} \right) \tag{3.8}$$

where the first term describes the diamagnetic current, and the second term denotes the field curvature and pressure anisotropy effect. It is the interplay of these different restoring forces which constitutes the complexity of ULF wave modes and instability in the ring current region of the Earth's and other planets' magnetospheres.

3.3. THE SOLAR WIND SOURCE

Studies of solar wind properties reveal the existence of many different kinds of plasma waves, among them species within the ULF pulsation frequency range (Hundhausen, 1972). Thus, it seems natural to assume propagation of such waves into the magnetosphere and plasmasphere as another possible source mechanism. Theoretical studies of the propagation of ULF waves through the Earth's bow shock, magnetosheath, and magnetopause have been done by, e.g., Kwok and Lee (1984). The idea of direct penetration of solar-wind-generated waves into the magnetosphere is supported by numerous observational studies. For example, Takahashi et al. (1984) report about a modest correlation between toroidal Pc3 power and the cone angle of the interplanetary magnetic field (IMF), while Engebretson et al. (1987) found a good correlation between the radial IMF component and multiharmonic toroidal dayside resonances. An association of toroidal multiharmonic and radial IMF was demonstrated in a superposed epoch analysis of IMF and harmonic Pc3 events, too (Anderson et al., 1991). As Yumoto et al. (1984) found that compressional Pc3 pulsations are well correlated with the IMF magnitude, it seems reasonable to attribute the toroidal oscillations studied by Takahashi et al. (1984), Engebretson et al. (1987), or Anderson et al. (1991) to the coupling between compressional, that is fast mode waves and toroidal field-line resonances as theoretically described above.

Another similarly indirect way of energy transfer from the solar wind to the inner magnetosphere and ionosphere has been suggested by Engebretson et al. (1991). These authors propose an ionospheric link whereby fluctuating field-aligned currents associated with magnetosheath wave activity drive multiharmonic ULF pulsations through a coupling of the so-called region 1 and region 2 currents systems (see Baumjohann, Chapter II/12 for a description of these current systems). A major ingredient of Engebretson's et al. idea is the generation of fluctuating particle precipitation and subsequent periodic modification of the ionospheric conductivity (compare Heacock and Hunsucker, 1977).

The above-mentioned correlations between solar wind parameters and ULF pulsations do not directly prove the transfer of wave energy into the magnetosphere. However, they imply a very close relationship between conditions required for the generation of waves upstream of the Earth bow shock and ULF waves within the dayside magnetosphere. In particular, the correlation with the IMF cone angle has been assumed as a clear hint of waves generated upstream of the bow shock by electromagnetic ion/ion instabilities in the quasiparallel regions of the shock (see Gary, 1991 for a review) representing an important source for ULF pulsations.

The study of the relation between solar wind wave activity therefore elucidates that energy transfer is not only caused by propagation of waves from the interplanetary medium to the magnetosphere, but points toward different transformations of energy from, for example, the fast mode into the Alfvén mode or from wave into particle kinetic energy and back to wave energy via ionospheric processes. Much future work should be devoted to further study such energy transfer processes as they may play an important part for the overall understanding of the energetics of the magnetosphere-ionosphere-atmosphere system.

3.4. MACROSCOPIC CHANGES OF THE MAGNETOSPHERIC CONFIGURATION

The excitation mechanisms discussed hitherto are always based on plasma instabilities, either generated outside the magnetosphere in the solar wind, at the magnetospheric boundary, or within the magnetosphere. A common feature of all these mechanisms is that they cause more or less continuous ULF pulsations. Usually more impulsive pulsations are indicative of major changes in the overall configuration of the magnetosphere. Prominent examples are storm sudden commencement associated Pc5 pulsations and related transients as well as impulsive nightside Pi2 pulsations.

The position of the dayside magnetopause is determined by the dynamic pressure of the solar wind. Sudden changes of this pressure as occurring during storm sudden commencements (McPherron, 1991) cause sudden changes of the position of the magnetopause, as a result of which transient ULF pulsations are generated. An example is displayed in Figure 14.1.3. As the SSC is a large-scale phenomenon, changing the position of the entire magnetopause, it is expected that the generated waves are almost axisymmetric, i.e., they are characterized by very low azimuthal wavenumbers m. This allows the generation of almost decoupled toroidal and poloidal magnetospheric eigenoscillations. At the ground below the ionosphere the toroidal oscillations appear as H-component perturbations due to the ionospheric rotation effect. The observed change of the period of the H component may then be explained as the result of decoupled toroidal magnetic field-line shell oscillations much as described by Equation 2.42.

ULF wave features of a similar origin as SSC associated Pc5 pulsations are traveling magnetospheric convection twin-vorticies (Friis-Christensen et al., 1988; Glassmeier et al., 1989c; Glassmeier, 1992). In ground-magnetic observations these twin-vortices appear as transient magnetic field variations. Two-dimensional magnetometer array observations exhibit a pair of clockwise and counterclockwise rotating ground-equivalent current patterns moving overhead an observer in tailward direction with velocities comparable to magnetosheath flow velocities (when mapped from the ground to the magnetopause region).

Observations by Friis-Christensen et al. (1988) suggest that these convection twin-vortices are the result of solar wind pressure perturbations causing partial (in contrast to the more global SSC) compressions and expansions of the dayside magnetopause. Glassmeier and Heppner (1992) and Glassmeier (1992) have developed a theoretical model to explain important features of these transient waves. According to them pressure pulses cause a perturbation of the dayside Chapman-Ferraro magnetopause current system. Neglecting inertial effects the divergence of this perturbed current system is given by (Glassmeier, 1992)

$$\nabla_\perp \cdot \vec{\delta j}_{MP} = -\frac{\mu_0}{B_0^2}(\nabla \delta p) \cdot \vec{j}_{MP} \qquad (3.9)$$

where δp is the external dynamic pressure perturbation, and $\vec{j}_{MP}$ the magnetopause current. This equation has a straight-forward interpretation: any divergence $\nabla \cdot \vec{\delta j}_{MP}$ is associated with field-aligned currents generated by the pressure perturbation. The first term on the right-hand side of the above equation may be coined a source term, the second one the structure term. Field-aligned current flow requires both the source as well as the structure term to be non-zero. This is only accomplished if, $\vec{j}_{MP} \neq 0$, i.e., if there is a magnetopause, and if $\nabla \delta p \neq 0$ along the magnetopause. In terms of the Dungey problem discussed in Section 2.2.4 this means that coupling between a fast mode (the pressure perturbation) and an Alfvén mode (carrying field-aligned current) requires

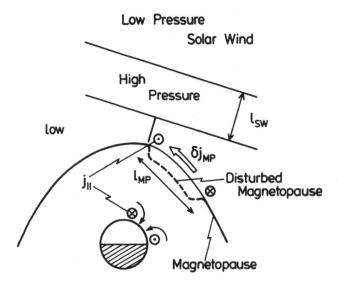

Figure 14.3.4 Schematic representation of the interaction of a spatially localized pressure pulse, i.e., a partial compression of the dayside magnetopause and an Alfvén mode pulse with its system of upward and downward field-aligned currents, generated due to the magnetopause current perturbation δj_{MP}.

a nonaxisymmetric perturbation in a nonuniform medium. In our case, the nonuniformity is guaranteed by the existence of the Chapman-Ferraro currents. Thus, coupling from a fast-mode perturbation into the Alfvén mode or generation of field-aligned currents can merely occur in the plane of the magnetopause.

Field-aligned currents generated in this way will be closed in the ionosphere by Pedersen currents whose magnetic-field effect is not detectable at the ground. There, only the magnetic effect of the Hall currents can be observed (see Section 2.2.3). As a single field-aligned current gives rise to a Hall-current vortex the observed twin-vortex structures must be interpreted as a pair of moving field-aligned currents of opposite flow direction much as is shown in Figure 14.3.4. A pair of field-aligned currents is generated if a localized pressure perturbation region in the solar wind, i.e., a succession of low-high-low pressure regions as observed by Friis-Christensen et al. (1988) hits the dayside magnetopause, causing first a compression, then an expansion of the magnetosphere, i.e., giving rise to $\nabla \delta p < 0$ and $\nabla \delta p > 0$. More complicated structures may be observed if inertial effects cannot be neglected as compared to the pressure effects (Kivelson and Southwood, 1991; Glassmeier, 1992).

The transient magnetic field variations discussed here, furthermore, show how wide an application the theoretical tools described in Section 2 have. The discussed source mechanism, external pressure perturbation, and coupling to field-aligned currents should give rise to an Alfvén-pulse propagating back and forth between the ionosphere and the perturbation region at the magnetopause. As a consequence of this a damp-typ oscillation is expected, different from the observed transient variation. The reason for this not to be observed is the moving nature of the source; the pressure perturbation is propagating along the magnetopause with the magnetosheath flow tailward. The effects of a moving ULF pulsation source have not yet been studied in detail and are subject to future work.

However, a pulse propagating back and forth between the ionosphere and the magnetosphere is most probably responsible for the generation of Pi2 pulsations (see Figure 14.1.6). These pulsations are related to magnetospheric substorms in the magnetotail (Baumjohann and Glassmeier, 1984). The substorm process (McPherron, 1991 for a review), on the other hand, is associated with current diversion of the neutral sheet current along geomagnetic field lines (McPherron, 1979). The substorm current wedge thus forming (see Figure 12.2.4 of Baumjohann,

Chapter II/12) is connected with drastic changes of the field-aligned current system in the nightside magnetosphere, caused by magnetic reconnection processes in the magnetotail. To build up the substorm current wedge, i.e., to accommodate the neutral sheet current diversion a large-amplitude Alfvén wave carrying the required field-aligned currents is necessary. This Alfvén pulse is propagating back and forth between the magnetotail and the ionosphere until the new current system in the magnetotail has been established, and the substorm current wedge has been built up.

The Pi2 generation process as outlined above, of course, gives only a brief description of the actual physical process occurring. Further details are described in works by Lester et al. (1984), Samson and Harold (1985), or Yumoto et al. (1990). It is important to note here the similarities and differences between the night- and dayside magnetosphere. In both cases a current wedge is forming due to current diversion of the neutral sheet and Chapman-Ferrano currents, respectively. The major difference, however, is a more or less stationary source region in the Pi2 case, while for the transient magnetic field variation one must assume a moving source.

3.5. ULF WAVE DAMPING

There are at least three major damping mechanisms for magnetospheric ULF pulsations: damping due to Joule heating in the ionosphere, particle acceleration due to wave-particle interactions, and mode coupling as a damping mechanism for a particular mode. Joule heating as a sink for ULF wave energy has theoretically been studied by Newton et al. (1978). The damping is due to energy losses to the wave caused by Pedersen currents $\vec{I}_P$ in the ionosphere

$$P_J = \vec{I}_P \cdot \vec{E} \tag{3.10}$$

where the ionospheric wave electric field $\vec{E}$ is approximately given by

$$\vec{E} \approx \frac{2\Sigma_P}{\Sigma_A}\vec{E}_I \tag{3.11}$$

as $\Sigma_A \ll \Sigma_P$ in most cases (see also Section 2.2.3). Therefore, one has

$$P_J \approx 4\frac{\Sigma_A^2}{\Sigma_P}E_I^2 \tag{3.12}$$

Thus, damping due to ionospheric Joule heating increases with decreasing ionospheric Pedersen conductance.

Observational evidence for Joule heating as the major damping mechanism of ULF pulsations has been given by Glassmeier et al. (1984b), who also demonstrated the usage of damping rate measurements as a viable tool to diagnose the ionospheric conductivity. Further evidence for the effectiveness of Joule heating has been provided by Lathuilliere et al. (1986) who studied variations of the ionospheric temperature during ULF pulsation events using incoherent backscatter radar observations. They found clear evidence of periodic temperature changes associated with the waves, that is they were able to directly trace the heat production mechanism.

4. SUMMARY AND OUTLOOK

An attempt has been made to review our current understanding of ULF pulsations as one of the major dynamic features of the magnetosphere. Both with respect to observations as well as theory the field of ULF pulsations may be regarded as one of the most developed fields in extraterrestrial physics. Open questions remain with respect to at least four different future areas of research: the characteristics of ULF waves in very high geomagnetic latitudes, that is in the polar cusp and cap regions, ULF wave coupling between high and low geomagnetic latitudes, features of such waves in other than the Earth's magnetosphere, and the propagation of localized MHD wave pulses in magnetospheric plasma regimes as well as nonlinear effects.

The study of polar cusp and cap pulsations is of importance for any future understanding of energy transfer processes from the dayside and magnetotail magnetosheath into the magnetosphere and ionosphere (Glassmeier, 1989a; Olson et al., 1986). It is also in these regions where the understanding of the propagation of localized wave pulses will largely contribute to analysis of dynamic features such as flux transfer events (Scholer, 1991) or dayside auroral transient events (Sandholt and Egeland, 1988). Coupling between low and high latitudes needs further study as wave energy is transported via the Earth-ionosphere waveguide. The Alfvén resonator theory will receive more attention in the future (Pilipenko, 1990). With respect to a comparative theory of planetary magnetospheres the study of ULF pulsations at other planets is very important (Khurana, 1994), and will provide one with new theoretical problems such as propagation in very large background magnetic fields as in the case of the Jovian magnetosphere (Glassmeier et al., 1989c) or in dusty plasmas as in the Kronian magnetosphere or interstellar clouds (Pilipp et al., 1987). Wider astrophysical applications are apparent, too. Finally, the nonlinear effects of ULF pulsations on, e.g., mass transport in the magnetosphere deserve further attention (Allan, 1992).

ACKNOWLEDGMENT

This work was supported through grants from the Deutsche Forschungsgemeinschaft and the German Space Agency DARA. Thanks are due to M. Stellmacher, B. Tsurutani, and H. Volland for carefully reading the manuscript.

REFERENCES

Alfvén, H., On the existence of electromagnetic-hydromagnetic waves, *Ark. Mat. Astron. Fys.*, 29 B, No. 2, 1942.

Allan, W., E.M. Poulter, and E. Nielsen, STARE observations of a Pc5 pulsation with large azimuthal wave number, *J. Geophys. Res.*, 87, 6163, 1982.

Allan, W., and E.M. Poulter, Pc5 pulsations associated with ring current proton drifts: STARE radar observations, *Planet. Space Sci.*, 31, 1279–1289, 1983.

Allan, W., E.M. Poulter, and S.P. White, Impulse excited hydromagnetic cavity and field-line resonances in the magnetosphere, *Planet. Space Sci.*, 35, 1193, 1987.

Allan, W., and E.M. Poulter, Damping of magnetospheric cavity modes: a discussion, *J. Geophys. Res.*, 94, 11843–11853, 1989.

Allan, W., Ponderomotive mass transport in the magnetosphere, *J. Geophys. Res.*, 97, 8483–8493, 1992.

Allan, W., and E.M. Poulter, ULF waves — their relationship to the structure of the Earth's magnetosphere, *Rep. Prog. Phys.*, 55, 553–598, 1992.

Anderson, B.J., T.A. Potemra, L.J. Zanetti, and M.J. Engebretson, Statistical correlations between Pc3-5 pulsations and solar wind/IMF parameters and geomagnetic indices, in *Physics and Space Plasmas, SPI Conference Proceedings and Reprint Series*, Vol. 10, T. Chang, G.B. Crew, and J.R. Jasperse, Eds., Scientific Publishers, Cambridge, MA, pp. 419–429, 1991.

Anderson, B.J., Statistical studies of Pc3-5 pulsations and their relevance for possible source mechanisms of ULF waves, *Ann. Geophys.*, in press.

Annexstad, J.O., and C.R. Wilson, Characteristics of pg micropulsations at conjugate points, *J. Geophys. Res.*, 73, 1805–1818, 1968.

Barfield, J.N., and R.L. McPherron, Stormtime Pc5 magnetic pulsations observed at synchronous orbit and their correlation with the partial ring current, *J. Geophys. Res.*, 83, 739, 1978.

Baumjohann, W., and K.H. Glassmeier, The transient response mechanism and Pi2 pulsations at substorm onset: review and outlook, *Planet. Space Sci.*, 32, 1361–1370, 1984.

Belcher, J.W., and L. Davis, Large-amplitude Alfvén waves in the interplanetary medium, *J. Geophys. Res.*, 76, 3534–3563, 1971.

Chamalaun, F.H., and C.E. Barton, Comprehensive mapping of Australia's geomagnetic variations, *EOS Trans. AGU*, 71, 1867, 1990.

Chandrasekhar, S., *Hydrodynamic and Hydromagnetic Stability*, Dover, New York, 1961.

Chen, L., and A. Hasegawa, A theory of longperiod magnetic pulsations. 1. Steady state excitation of field line resonance, *J. Geophys. Res.*, 79, 1024–1032, 1974.

Chew, G.F., M.L. Goldberger, and F.E. Low, The Boltzmann equation and the one-fluid hydromagnetic equations in the absence of particle collisions, *Proc. R. Soc. London, Ser. A*, 112–118, 1956.

Chisham, G., and D. Orr, Statistical studies of giant pulsations (Pgs): harmonic mode, *Planet. Space Sci.*, 39, 999, 1991.

Crowley, G., N. Wade, J.A. Waldock, T.R. Robinson, and T.B. Jones, High-time resolution observations of periodic frictional heating associated with a Pc5 micropulsation, *Nature*, 316, 528–530, 1985.

Crowley, G., W.J. Hughes, and T.B. Jones, Observational evidence of cavity modes in the Earth's magnetosphere, *J. Geophys. Res.*, 92, 12233, 1987.

Dungey, J.W., Electrodynamics of the outer atmosphere, *Penn. State Univ. Ionos. Res. Lab. Sci. Rep.*, No. 69, 1954.

Engebretson, M.J., L.J. Zanetti, and T.A. Potemra, Harmonically structured ULF pulsations observed by the AMPTE/CCE magnetic field experiment, *Geophys. Res. Lett.*, 13, 905–908, 1986.

Engebretson, M.J., L.J. Zanetti, T.A. Potemra, W. Baumjohann, H. Lühr, and M.H. Acua, Simultaneous observation of Pc3-4 pulsations in the solar wind and in the Earth's magnetosphere, *J. Geophys. Res.*, 92, 10053–10062, 1987.

Engebretson, M.J., L.J. Cahill, R.L. Arnoldy, B.J. Anderson, T.J. Rosenberg, D.L. Carpenter, U.S. Inan, and R.H. Eather, The role of the ionosphere in coupling upstream ULF wave power into the dayside magnetosphere, *J. Geophys. Res.*, 96, 1527–1542, 1991.

Fejer, J.A., Hydromagnetic stability at a fluid velocity discontinuity between compressible fluids, *Phys. Fluids*, 7, 499–503, 1964.

Friis-Christensen, E., M.A. McHenry, C.R. Clauer, and S. Vennerstrom, Ionospheric traveling convection vortices observed near the polar cleft: a triggered response to sudden changes in the solar wind, *Geophys. Res. Lett.*, 15, 253–256, 1988.

Fujita, S., and V.L. Patel, Eigenmode analysis of coupled magnetohydrodynamic oscillations in the magnetosphere, *J. Geophys. Res.*, 97, 13777–13788, 1992.

Gary, S.P., Electromagnetic ion/ion instabilities and their consequences in space plasmas: a review, *Space Sci. Res.*, 56, 373–415, 1991.

Glassmeier, K.H., Magnetometer array observations of a giant pulsation event, *J. Geophys.*, 48, 127–138, 1980.

Glassmeier, K.H., On the influence of ionospheres with non-uniform conductivity distribution on hydromagnetic waves, *J. Geophys.*, 54, 125–137, 1984a.

Glassmeier, K.H., H. Volpers, and W. Baumjohann, Ionospheric joule dissipation as a damping mechanism for high-latitude ULF pulsations: observational evidence, *Planet. Space Sci.*, 32, 1463–1468, 1984b.

Glassmeier, K.H., Reconstruction of the ionospheric influence on ground-based observations of a short-duration ULF pulsation event, *Planet. Space Sci.*, 36, 801–817, 1987.

Glassmeier, K.H., ULF pulsations in the polar cusp and cleft, in *Electromagnetic Coupling in the Polar Clefts and Caps*, A. Egeland and P.E. Sandholt, Eds., Kluwer Academic Publishers, Dordrecht, 1989a, pp. 167–187.

Glassmeier, K.H., A.J. Coates, M.H. Acuna, M.L. Goldstein, A.D. Johnstone, F.M. Neubauer, and H. Reme, Spectral characteristics of low-frequency plasma turbulence upstream of comet p/Halley, *J. Geophys. Res.*, 94, 37–48, 1989b.

Glassmeier, K.H., M. Hönisch, and J. Untiedt, Ground-based and satellite observations of traveling magnetospheric convection twin-vortices, *J. Geophys. Res.*, 94, 2520–2528, 1989c.

Glassmeier, K.H., F.M. Neubauer, M.H. Acuna, and N.F. Ness, Standing hydromagnetic waves in the Io plasma torus: Voyager 1 observations, *J. Geophys. Res.*, 94, 15064–15076, 1989d.

Glassmeier, K.H., Traveling magnetospheric convection twin-vortices: observations and theory, *Ann. Geophys.*, 10, 547–565, 1992.

Glassmeier, K.H., and C. Heppner, Traveling magnetospheric convection twin-vortices: Another case study, global characteristics, and a model, *J. Geophys. Res.*, 97, 3977–3992, 1992.

Glassmeier, K.H., M.v. Seggern, R.v. Stein, and A. Korth, Die Protonen-Sprung-Resonanz-Instabilität als möglicher Anregungsmechanismus von Riesenpulsationen, *Kleinheubacher Berichte*, 35, 217–224, 1992.

Green, C.A., The longitudal phase variation of mid-latitude pc3-4 micropulsations, *Planet. Space Sci.*, 24, 79–85, 1976.

Green, C.A., Meridional characteristics of a Pc4 pulsation event in the plasmasphere, *Planet. Space Sci.*, 26, 955–967, 1978.

Green, C.A., Observations of pg pulsations in the northern auroral zone and at lower latitude conjugate regions, *Planet. Space Sci.*, 27, 63–77, 1979.

Hasegawa, A., Drift mirror instability in the magnetosphere, *Phys. Fluids*, 12, 2642, 1969.

Hasegawa, A., *Plasma Instabilities and Nonlinear Effects*, Springer-Verlag, Berlin, 1975.

Heacock, R.P., and R.D. Hunsucker, Pi1-2 magnetic field pulsations on dayside cleft fieldlines, *Nature*, 269, 313–314, 1977.

Herron, T.J., Phase characteristics of geomagnetic micropulsations, *J. Geophys. Res.*, 71, 871–889, 1966.

Higbie, P.R., D.N. Baker, R.D. Zwickl, R.D. Belian, J.R. Asbridge, J.F. Fennell, B. Wilken, and C.W. Arthur, The global Pc5 event of November 14–15, *J. Geophys. Res.*, 87, 2337, 1982.

Hillebrand, O., J. Münch, and R.L. McPherron, Ground-satellite correlative study of a giant pulsation event, *J. Geophys.*, 51, 129, 1982.

Hughes, W.J., The effect of the atmosphere and ionosphere on long period magnetospheric micropulsations, *Planet. Space Sci.*, 22, 1157–1172, 1974.

Hughes, W.J., R.L. McPherron, and C.T. Russell, Multiple satellite observations of pulsation resonance structure in the magnetosphere, *J. Geophys. Res.*, 82, 492–498, 1977.

Hughes, W.J., D.J. Southwood, and B. Mauk, Alfvén waves generated by an inverted plasma energy distribution, *Nature*, 275, 43–44, 1978.

Hughes, W.J., and R.J.L. Grad, A second harmonic geomagnetic field line resonance at the inner edge of plasma sheet: Geos 1, ISEE 1, and ISEE 2 observations, *J. Geophys. Res.*, 89, 2755–2764, 1984.

Hundhausen, A.J., *Solar Wind and Coronal Expansion*, Springer-Verlag, Heidelberg, 1972.

Inhester, B., Resonance absorption of Alfvén oscillations in a nonaxisymmetric magnetosphere, *J. Geophys. Res.*, 91, 1509–1518, 1985.

Jacobs, J.A., *Geomagnetic Micropulsations*, Springer-Verlag, Berlin, 1970.

Jede, B., Koordinierte Untersuchung eines Pc5 Pulsationsereignisse mit Bodenmagnetometern, Riometern und Beobachtungen eines geostationären Satelliten, Diploma thesis, University Münster, 1985.

Junginger, H. et al., A statistical study of dayside magnetospheric field fluctuations with periods between 150 and 600 s, *J. Geophys. Res.*, 89, 10757, 1984.

Junginger, H., and W. Baumjohann, Resonant harmonic Alfvén waves in the magnetosphere: a case study, *J. Geophys. Res.*, 89, 2765, 1984.

Khurana, K.A., ULF waves in other magnetospheres — observations and possible source mechanisms, *Ann. Geophys.*, in press.

Kivelson, M.G., J. Etcheto, and J.G. Trotignon, Global compressional oscillations of the terrestrial magnetosphere: the evidence and a model, *J. Geophys. Res.*, 89, 9851–9856, 1984.

Kivelson, M.G., and D.J. Southwood, Resonant ULF waves: a new interpretation, *Geophys. Res. Lett.*, 12, 49–52, 1985.

Kivelson, M.G., and D.J. Southwood, Ionospheric travelling vortex generation by solar wind buffeting of the magnetosphere, *J. Geophys. Res.*, 96, 1661–1667, 1991.

Kremser, G., A. Korth, J.A. Fejer, B. Wilken, A.V. Gurevich, and E. Amata, Observations of quasi-periodic flux variations of energetic ion and electrons associated with Pc5 geomagnetic pulsations, *J. Geophys. Res.*, 86, 3345, 1981.

Krummheuer, F., Ausbreitung von MHD-Wellen unter inhomogenen Bedingungen, in *Plasmaphysik im Sonnensystem*, Glassmeier, K.H., and M. Scholer, Eds., B.I. Wissenschaftsverlag, Mannheim, 1991.

Küppers, F., J. Untiedt, W. Baumjohann, K. Lange, and A.G. Jones, A two-dimensional magnetometer array for ground-based observations of auroral zone electric currents during the International Magnetometer Study (IMS), *J. Geophys.*, 46, 429–540, 1979.

Kwok, Y.C., and L.C. Lee, Transmission of magnetohydrodynamic waves through the rotational discontinuity of Earth s magnetopause, *J. Geophys. Res.*, 889, 10697–10708, 1984.

Lanzerotti, L.J., and H. Fukunishi, Modes of MHD waves in the magnetosphere, *Rev. Geophys. Space Phys.*, 12, 724–729, 1974.

Lanzerotti, L.J., and D.J. Southwood, Hydromagnetic waves, in *Solar System Plasma Physics*, Vol. III, Lanzerotti, L.J., C.F. Kennel, and E.N. Parker, Eds., North Holland, Dordrecht, 1979.

Lathuillere, C., F. Glangeaud, and Z.Y. Zhao, Ionospheric ion heating by ULF Pc5 magnetic pulsations, *J. Geophys. Res.*, 91, 1619–1626, 1986.

Lee, D.-L., and R.L. Lysak, Magnetospheric ULF wave coupling in the dipole model: the impulsive excitation, *J. Geophys. Res.*, 94, 17097–17103, 1989.

Lee, D.-L., and R.L. Lysak, Effects of azimuthal asymmetry on ULF waves in the dipole magnetosphere, *Geophys. Res. Lett.*, 17, 53–56, 1990.

Lester, M., W.J. Hughes, and H.J. Singer, Longitudinal structure in Pi2 pulsations and the substorm current wedge, *J. Geophys. Res.*, 89, 5489–5494, 1984.

Lühr, H., S. Thürey, and N. Klöcker, The EISCAT magnetometer cross: operational aspects — first results, *Geophys. Surv.*, 6, 305, 1984.

McHenry, M.A., and C.R. Clauer, Modeled ground magnetic signatures of flux transfer events, *J. Geophys., Res.*, 92, 11231–11240, 1987.

McHenry, M.A., C.R. Clauer, E. Friis-Christensen, P.T. Newell, and J.D. Kelly, Ground observations of magnetospheric boundary layer phenomena, *J. Geophys. Res.*, 95, 14995–15005, 1990.

McPherron, R.L., Magnetospheric substorms, *Rev. Geophys. Space Phys.*, 17, 657, 1979.

McPherron, R.L., Magnetospheric substorm and magnetic storms, in *Geomagnetism*, Vol. IV, Jacobs, J.A. (Ed.), Academic, London, 1991.

Miura, A., Anomalous transport by magnetohydrodynamic Kelvin-Helmholtz instabilities in the solar wind-magnetosphere interaction, *J. Geophys. Res.*, 89, 801, 1984.

Morse, P.M., and H. Feshbach, *Methods of Theoretical Physics*, Part I, McGraw-Hill, New York, 1953.

Neubauer, F.M., Nonlinear standing Alfvén wave current system at Io: theory, *J. Geophys. Res.*, 85, 1171–1178, 1980.

Newton, R.S., D.J. Southwood, and W.J. Hughes, Damping of geomagnetic pulsations by the ionosphere, *Planet. Space Sci.*, 26, 201, 1978.

Nishida, A., Ionospheric screening effect and storm sudden commencement, *J. Geophys. Res.*, 69, 1861–1874, 1964.

Nishida, A., Possible origin of transient dusk-to-dawn electric field in the nightside magnetosphere, *J. Geophys. Res.*, 84, 3409, 1979.

Obertz, P., and O.M. Raspopov, Study of the spatial characteristics of type Pc5 geomagnetic pulsations, *Geomagn. Aeron.*, 8, 424–427, 1968.

Olson, J.V., and G. Rostoker, Longitudal phase variations of Pc4-5 pulsations, *J. Geophys. Res.*, 83, 2481–2488, 1978.

Olson, J.V., ULF signatures of the polar cusp, *J. Geophys. Res.*, 91, 10055, 1986.

Pilipenko, V.A., ULF waves on the ground and in space, *J. Atmos. Terr. Phys.*, 52, 1193–1209, 1990.

Pilipp, W., T.W. Hartquist, O. Havnes, and G.E. Morfill, The effects of dust on the propagation and dissipation of Alfvén waves in interstellar clouds, *Astrophys. J.*, 314, 341–351, 1987.

Pokhotelov, O.A., V.A. Pilipenko, Yu.M. Nezlina, J. Woch, G. Kremser, A. Koth, and E. Amata, Excitation of high-β plasma instabilities at the geostationary orbit. Theory and observations, *Planet. Space Sci.*, 34, 695, 1986.

Pu, Z., and M.G. Kivelson, Kelvin-Helmholtz instability at the magnetopause, *J. Geophys. Res.*, 88, 853, 1983.

Radoski, H., A theory of latitude dependent geomagnetic pulsations: the asymptotic fields, *J. Geophys. Res.*, 79, 595, 1974.

Saito, T., Geomagnetic pulsations, *Space Sci. Rev.*, 10, 319–412, 1969.

Samson, J.C., J.A. Jacobs, and G. Rostoker, Latitude-dependent characteristics of long-period geomagnetic micropulsations, *J. Geophys. Res.*, 76, 3675–3683, 1971.

Samson, J.C., and B.G. Harold, Characteristic time constants and velocities of high-latitude Pi2 s, *J. Geophys. Res.*, 90, 12173–12181, 1985.

Samson, J.C., Geomagnetic pulsations and plasma waves in the Earth's magnetosphere, in *Geomagnetism* 4, J.A. Jacobs, Ed., Academic Press, London, 1991, pp. 481.

Sandholt, P.E., and A. Egeland, Auroral and magnetic variations in the polar cusp and cleft-signatures of magnetopause boundary-layer dynamics, *Astrophys. Space Sci.*, 144, 171–199, 1988.

Scholer, M., On the motion of artificial ion clouds in the magnetosphere, *Planet. Space Sci.*, 18, 977–1007, 1970.

Scholer, M., Numerical models of magnetic reconnection, *Geophys. Astrophys. Fluid Dynam.*, 62, 51–68, 1991.

Schulz, M., and L.J. Lanzerotti, *Particle Diffusion in the Radiation Belts*, Springer-Verlag, Heidelberg, 1974.

Sen, A.K., Stability of the magnetospheric boundary, *Planet. Space Sci.*, 13, 131–141, 1965.

Siebert, M., Zur Theorie der inhomogenen MHD-Wellenausbreitung im kalten Plasma, *Kleinheubacher Berichte*, 36, 149–162, 1993.

Singer, H.J., and M.G. Kivelson, The latitudinal structure of Pc5 waves in space: magnetic and electric field observations, *J. Geophys. Res.*, 84, 7213–7222, 1979.

Singer, H.J. et al., Satellite observations of the spatial extent and structure of Pc3,4,5 pulsation near the magnetospheric equator, *Geophys. Res. Lett.*, 6, 889–892, 1979.

Singer, H.J. et. al., Standing hydromagnetic waves observed by ISEE 1 and 2: Radial extent and harmonic, *J. Geophys. Res.*, 87, 3519–3527, 1982.

Siscoe, G., Magnetohydrodynamics, in *Solar-Terrestrial Physics*, Carovillano, R.L., and T. Forbes, Eds., Reidel, Dordrecht, 1983.

Southwood, D.J., The hydromagnetic stability of the magnetic spheric boundary, *Planet. Space Sci.*, 16, 587–605, 1968.

Southwood, D.J., J.W. Dungey, and R.J. Etherington, Bounce resonant interaction between pulsations and trapped particles, *Planet. Space Sci.*, 17, 349–361, 1969.

Southwood, D.J., Some features of field line resonances in the magnetosphere, *Planet. Space Sci.*, 22, 483–491, 1974.

Southwood, D.J., and W.F. Stuart, Pulsations at the substorm onset, in *Dynamics of the Magnetosphere*, Akasofu, S.-I., Ed., D. Reidel, Dordrecht, 1979.

Southwood, D.J., and M.G. Kivelson, Charged particle behaviour in low-frequency geomagnetic pulsations. 2. Graphical approach, *J. Geophys. Res.*, 87, 1707–1710, 1982.

Southwood, D.J., and W.J. Hughes, Theory of hydromagnetic waves in the magnetosphere, *Space Sci. Rev.*, 35, 301–366, 1983.

Spreiter, J.R., A.L. Summers, and A.Y. Alksne, Hydromagnetic flow around the magnetosphere, *Plant. Space Sci.*, 14, 223, 1966.

Stein, R.F., Reflection, refraction, and coupling of MHD waves at a density step, *Astrophys. J.*, Suppl. Ser., No. 192, 22, 419–444, 1971.

Stein, R.v., and K.H. Glassmeier, Concerning the correlations between Pc5 pulsations and high speed streams in the solar wind (Abstract), IUGG XX. General Assembly, GAM 3.1, Vienna, 1991.

Stewart, B., On the great magnetic disturbance which extended from August 28 to September 7, 1859, *Philos. Trans. London*, 423–430, 1861.

Takashi, K.T., R.L. McPherron, and T. Terasawa, Dependence on the spectrum of Pc3-4 pulsations on the interplanetary magnetic field, *J. Geophys. Res.*, 89, 2770–2780, 1984.

Takahashi, K.T., N. Sato, J. Warnecke, H. Lühr, H.E. Spence, and Y. Tonegawa, On the standing wave mode of giant pulsations, *J. Geophys. Res.*, 97, 10717–10732, 1992.

Tamao, T., Transmission and coupling resonance of hydromagnetic disturbances in the non-uniform Earth's magnetosphere, Sci. Rep. Tohoku Univ., Series 5, *Geophysics,* Vol. 17, No. 2, 43–72, 1965.

Voelker, H., Zur Breitenabhängigkeit der Perioden erdmagnetischer Pulsationen, *Naturwissenschaften,* 49, 8–9, 1962.

Walker, A.D.M., R.A. Greenwald, C.A. Green, and W. Stuart, STARE radar auroral observation of Pc5 geomagnetic pulsations, *J. Geophys. Res.*, 84, 3373–3388, 1979.

Walker, A.D.M., The Kelvin-Helmholtz instability in the low-latitude boundary layer, *Planet. Space Sci.,* 29, 1119, 1981.

Williams, D., The Earth's ring current: present situation and future thrusts, *Phys. Scr.,* T 18, 140–151, 1987.

Woch, J., G. Kremser, A. Korth, O.A. Pokhotelov, V.A. Pilipenko, Yu.M. Nezlina, and E. Amaty, Curvature-driven drift mirror instability in the magnetosphere, *Planet. Space Sci.,* 36, 383–393, 1988.

Wright, A.N., On the existence of transverse MHD oscillations in an inhomogeneous magnetoplasma, *J. Plasma Phys.,* 43, 83–99, 1990.

Wright, A.N., and N.W. Evans, Magnetic geometries that carry decoupled transverse or compressional magnetic field oscillations, *J. Geophys. Res.*, 96, 209–216, 1991.

Yumoto, K., Long-period magnetic pulsations generated in the magnespheric boundary layer, *Planet. Space Sci.,* 1205–1218, 1984.

Yumoto, K., K. Takahashi, T. Sakurai, P.R. Sutcliff, S. Kokubun, H. Lühr, T. Saito, M. Kuwashima, and N. Sato, Multiple ground-based and satellite observations of global Pi2 magnetic pulsations, *J. Geophys. Res.*, 95, 15175–15184, 1990.

Ziesolleck, C.W.S., and F.H. Chamalaun, A two-dimensional array study of low latitude Pc5 geomagnetic pulsations, *J. Geophys. Res.,* in press.

Ziesolleck, C.W.S., F.W. Menk, B.J. Fraser, and P.W. Webb, Spatial characteristics of low latitude Pc3-4 geomagnetic pulsations, *J. Geophys. Res.*, 98, 197–207, 1993.

INDEX

Printed and bound by CPI Group (UK) Ltd, Croydon, CR0 4YY

22/10/2024

01777638-0013